B

ORDINARY DIFFERENTIAL EQUATIONS

Second Edition

PHILIP HARTMAN

1982
Birkhäuser
Boston • Basel • Stuttgart

Professor Philip Hartman*
Department of Mathematics
The Johns Hopkins University
Baltimore, Maryland 21218

This work is a reprint of the second edition.

CIP - Kurztitelaufnahme der Deutschen Bibliothek

Hartman, Philip:
Ordinary differential equations / Philip Hartman.
- Reprint of the 2. ed. -
Boston ; Basel ; Stuttgart : Birkhäuser, 1982.
    ISBN 3-7643-3068-6

Library of Congress Cataloging in Publication Data

Hartman, Philip, 1915-
    Ordinary differential equations.
    Reprint. Originally published: Baltimore, Md., 1973.
    Bibliography: p.
    Includes index.
    1. Differential equations.   I. Title.
QA372.H33   1982   515.3′52   81-21738
ISBN 3-7643-3068-6   AACR2

©Birkhäuser Boston, 1982

ISBN 3-7643-3068-6

Printed in USA

To the memory
of my parents

To the patience
of Sylvia, Judith, and Marilyn

Preface to the First Edition

This book is based on lecture notes of courses on ordinary differential equations which I have given from time to time for advanced undergraduates and graduate students in mathematics, physics, and engineering. It assumes a knowledge of matrix theory and, if not a thorough knowledge of, at least a certain maturity in the handling of functions of real variables.

I was never tempted to scatter asterisks liberally throughout this book and claim that it could serve as a sophomore–junior–senior textbook, for I believe that a course of this type should give way to basic courses in analysis, algebra, and topology.

This book contains more material than I ever covered in one year but not all of the topics which I treated in the many courses. The contents of these courses always included the subject matter basic to the theory of differential equations and its many applications to other disciplines (as, for example, differential geometry). A "basic course" is covered in Chapter I; §§ 1–3 of Chapter II; §§ 1–6 and 8 of Chapter III; Chapter IV except for the "Application" in § 3 and part (ix) in § 8; §§ 1–4 of Chapter V; §§ 1–7 of Chapter VII; §§ 1–3 of Chapter VIII; §§ 1–12 of Chapter X; §§ 1–4 of Chapter XI; and §§ 1–4 of Chapter XII.

Many topics are developed in depth beyond that found in standard textbooks. The subject matter in a chapter is arranged so that more difficult, less basic, material is usually put at the end of the chapter (and/or in an appendix). In general, the content of any chapter depends only on the material in that chapter and the portion of the "basic course" preceding it. For example, after completing the basic course, an instructor can discuss Chapter IX, or the remainder of the contents of Chapter XII, or Chapter XIV, etc. There are two exceptions: Chapter VI, Part I, as written, depends on Chapter V, §§ 5–12; Part III of Chapter XII is not essential but is a good introduction to Chapter XIII.

Exercises have been roughly graded into three types according to difficulty. Many of the exercises are of a routine nature to give the student an opportunity to review or test his understanding of the techniques just explained. For more difficult exercises, there are hints in the back of the book (in some cases, these hints simplify available proofs). Finally, references are given for the most difficult exercises; these serve to show extensions and further developments, and to introduce the student to the literature.

The theory of differential equations depends heavily on the "integration of differential inequalities" and this has been emphasized by collecting some of the main results on this topic in Chapter III and § 4 of Chapter IV. Much of the material treated in this book was selected to illustrate important techniques as well as results: the reduction of problems on differential equations to problems on "maps" (cf. Chapter VII, Appendix, and Chapter IX); the use of simple topological arguments (cf. Chapters VIII, § 1; X, §§ 2–7; and XIV, § 6); and the use of fixed point theorems and other basic facts in functional analysis (cf. Chapters XII and XIII).

I should like to acknowledge my deep indebtedness to the late Professor Aurel Wintner from whom and with whom I learned about differential equations, first as a student and later as a collaborator. My debt to him is at once personal, in view of my close collaboration with him, and impersonal, in view of his contributions to the resurgence of the theory of ordinary differential equations since the Second World War.

I wish to thank several students at Hopkins, in particular, N. Max, C. C. Pugh, and J. Wavrik, for checking parts of the manuscript. I also wish to express my appreciation to Miss Anna Lea Russell for the excellent typescript created from nearly illegible copy, numerous revisions, and changes in the revisions.

My work on this book was partially supported by the Air Force Office of Scientific Research.

PHILIP HARTMAN

Baltimore, Maryland
August, 1964

Preface to the Second Edition

This edition is essentially that of 1964 (John Wiley and Sons, Inc.) with some corrections and additions. The only major changes are in Chapter IX. A few pertinent additions have been made to the bibliography as a Supplement, but no effort has been made to bring it up to date.

I should like to thank Professors T. Butler, K.T. Chen, W.A. Coppel, L. Lorch, M.E. Muldoon, L. Nicolson and C. Olech for corrections and/or suggestions, and Mrs. Margaret A. Einstein for able secretarial assistance.

P.H. (10/73)

Contents

It is unfortunate that, with the decline of graduate education in the last decade, publishers have permitted many excellent books to go out of print. I am very grateful to Birkhäuser for their offer to republish my book. I have used this opportunity to correct a few more minor errors for this edition.

P.H. (12/81)

Chapter I

Preliminaries

1. Preliminaries

Consider a system of d first order differential equations and an initial condition

$$(1.1) \qquad y' = f(t, y), \qquad y(t_0) = y_0,$$

where $y' = dy/dt$, $y = (y^1, \ldots, y^d)$ and $f = (f^1, \ldots, f^d)$ are d-dimensional vectors, and $f(t, y)$ is defined on a $(d + 1)$-dimensional (t, y)-set E. For the most part, it will be assumed that f is continuous. In this case, $y = y(t)$ defined on a t-interval J containing $t = t_0$ is called a solution of the initial value problem (1.1) if $y(t_0) = y_0$, $(t, y(t)) \in E$, $y(t)$ is differentiable, and $y'(t) = f(t, y(t))$ for $t \in J$. It is clear that $y(t)$ then has a continuous derivative. These requirements on y are equivalent to the following: $y(t_0) = y_0$, $(t, y(t)) \in E$, $y(t)$ is continuous and

$$(1.2) \qquad y(t) = y_0 + \int_{t_0}^{t} f(s, y(s)) \, ds$$

for $t \in J$.

An initial value problem involving a system of equations of mth order,

$$(1.3) \quad z^{(m)} = F(t, z, z^{(1)}, \ldots, z^{(m-1)}), \, z^{(j)}(x_0) = z_0^{(j)} \text{ for } j = 0, \ldots, m - 1,$$

where $z^{(j)} = d^j z/dt^j$, z and F are e-dimensional vectors, and F is defined on an $(me + 1)$-dimensional set E, can be considered as a special case of (1.1), where y is a $d = me$-dimensional vector, symbolically, $y = (z, z^{(1)}, \ldots, z^{(m-1)})$ (or more exactly, $y = (z^1, \ldots, z^e, z^{1\prime}, \ldots, z^{e\prime}, \ldots, z^{e(m-1)})$); correspondingly, $f(t, y) = (z^{(1)}, \ldots, z^{(m-1)}, F(t, y))$ and $y_0 = (z_0, z_0^{(1)}, \ldots, z_0^{(m-1)})$. For example, if $e = 1$ so that z is a scalar, (1.3) becomes

$$y^{1\prime} = y^2, \ldots, y^{m-1\prime} = y^m, \quad y^{m\prime} = F(t, y^1, \ldots, y^m),$$

$$y^j(t_0) = z_0^{(j-1)} \quad \text{for } j = 1, \ldots, m,$$

where $y^1 = z$, $y^2 = z'$, $\ldots$, $y^m = z^{(m-1)}$.

1

The first set of questions to be considered will be (1) local existence (does (1.1) have a solution $y(t)$ defined for t near t_0?); (2) existence in the large (on what t-ranges does a solution of (1.1) exist?); and (3) uniqueness of solutions.

The significance of question (2) is clear from the following situation: Let y, f be scalars, $f(t, y)$ defined for $0 \leq t \leq 1, |y| \leq 1$. A solution $y = y(t)$ of (1.1), with $(t_0, y_0) = (0, 0)$, may exist for $0 \leq t \leq \frac{1}{2}$ and increase from 0 to 1 as t goes from 0 to $\frac{1}{2}$, then one cannot expect to have an extension of $y(t)$ for any $t > \frac{1}{2}$. Or consider the following scalar case where $f(t, y)$ is defined for all (t, y):

$$(1.4) \qquad\qquad y' = y^2, \qquad y(0) = c \ (> 0).$$

It is easy to see that $y = c/(1 - ct)$ is a solution of (1.4), but this solution exists only on the range $-\infty < t < 1/c$, which depends on the initial condition.

In order to illustrate the significance of the question of uniqueness, let y be a scalar and consider the intial value problem

$$(1.5) \qquad\qquad y' = |y|^{\frac{1}{2}}, \qquad y(0) = 0.$$

This has more than one solution, in fact, it has, e.g., the solution $y(t) \equiv 0$

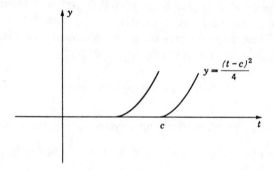

Figure 1.

and the 1-parameter family of solutions defined by $y(t) = 0$ for $t \leq c$, $y(t) = (t - c)^2/4$ for $t \geq c$, where $c \geq 0$; see Figure 1. This situation is typical in that if (1.1) has more than one solution, then it has a "continuum" of solutions; cf. Theorem II 4.1.

2. Basic Theorems

This section introduces some conventions, notions and theorems to be used later. The proofs of most of the theorems will be omitted.

The symbols O, o will be used from time to time where, e.g., $f(t) = O(g(t))$ as $t \to \infty$ means that there exists a constant C such that $|f(t)| \leq C|g(t)|$ for large t, while $f(t) = o(g(t))$ as $t \to \infty$ means that $C > 0$ can be chosen arbitrarily small (so that if $g(t) \neq 0$, $f(t)/g(t) \to 0$ as $t \to \infty$).

"Function" below generally means a map from some specified set of a vector space R^e into a space R^d, not always of the same dimension. R^d denotes a normed, real d-dimensional vector space of elements $y = (y^1, \ldots, y^d)$ with norm $|y|$. Unless otherwise specified, $|y|$ will be the norm

$$(2.1) \qquad |y| = \max(|y^1|, \ldots, |y^d|),$$

and $\|y\|$ the Euclidean norm.

If y_0 is a point and E a subset of R^d, then dist (y_0, E), the distance from y_0 to E, is defined to be inf $|y_0 - y|$ for $y \in E$. If E_1, E_2 are two subsets of R^d, then dist (E_1, E_2) is defined to be inf $|y_1 - y_2|$ for $y_1 \in E_1$, $y_2 \in E_2$, and is called the distance between E_1 and E_2. If E_1 (or E_2) is compact and E_1, E_2 are closed and disjoint, then dist $(E_1, E_2) > 0$.

If E is an open set or a closed parallelepiped in R^d, $f \in C^n(E)$, $0 \leq n < \infty$, means that $f(y)$ is continuous on E and that the components of f have continuous partial derivatives of all orders $k \leq n$ with respect to $y^1, \ldots, y^d$.

A function $f(y, z) = f(y^1, \ldots, y^d, z^1, \ldots, z^e)$ defined on a (y, z)-set E, where $y \in R^d$, is said to be *uniformly Lipschitz continuous on E with respect to y* if there exists a constant K satisfying

$$(2.2) \qquad |f(y_1, z) - f(y_2, z)| \leq K|y_1 - y_2| \qquad \text{for all} \quad (y_j, z) \in E$$

with $j = 1, 2$. Any constant K satisfying (2.1) is called a *Lipschitz constant* (for f on E). (The admissible values of K depend, of course, on the norms in the f- and y-spaces.)

A family F of functions $f(y)$ defined on some y-set $E \subset R^d$ is said to be *equicontinuous* if, for every $\epsilon > 0$, there exists a $\delta = \delta_\epsilon > 0$ such that $|f(y_1) - f(y_2)| \leq \epsilon$ whenever $y_1, y_2 \in E$, $|y_1 - y_2| \leq \delta$ and $f \in F$. The point of this definition is that δ_ϵ does not depend on f but is admissible for all $f \in F$. The most frequently encountered equicontinuous families F below will occur when all $f \in F$ are uniformly Lipschitz continuous on E and there exists a $K > 0$ which is a Lipschitz constant for all $f \in F$; in which case, δ can be chosen to be $\delta = \epsilon/K$.

Lemma 2.1. *If a sequence of continuous functions on a compact set E is uniformly convergent on E, then it is uniformly bounded and equicontinuous.*

Cantor Selection Theorem 2.1. *Let $f_1(y), f_2(y), \ldots$ be a uniformly bounded sequence of functions on a y-set E. Then for any countable set $D \subset E$, there exists a subsequence $f_{n(1)}(y), f_{n(2)}(y), \ldots$ convergent on D.*

In order to prove Cantor's theorem, let D consist of the points $y_1, y_2, \ldots$. Also assume that $f_n(y)$ is real-valued; the proof for the case that $f_n(y) = (f_n^1(y), \ldots, f_n^d(y))$ is a d-dimensional vector is similar. The sequence of numbers $f_1(y), f_2(y), \ldots$ is bounded, thus, by the theorem of Bolzano-Weierstrass, there is a sequence of integers $n_1(1) < n_1(2) < \ldots$ such that $\lim f_n(y_1)$ exists as $k \to \infty$, where $n = n_1(k)$. Similarly there is a subsequence $n_2(1) < n_2(2) < \ldots$ of $n_1(1), n_1(2), \ldots$ such that $\lim f_n(y_2)$ exists as $k \to \infty$ for $n = n_2(k)$. Continuing in this fashion one obtains successive subsequences of positive integers, such that if $n_j(1) < n_j(2) < \ldots$ is the jth one, then $\lim f_n(y_i)$ exists on $k \to \infty$, where $n = n_j(k)$ and $i = 1, \ldots, j$. The desired subsequence is the "diagonal sequence" $n_1(1) < n_2(2) < n_3(3) < \ldots$. Variants of this proof will be referred to as the "standard diagonal process."

The next two assertions usually have the names Ascoli or Arzela attached to them.

Propagation Theorem 2.2.   *On a compact y-set E, let $f_1(y), f_2(y), \ldots$ be a sequence of functions which is equicontinuous and convergent on a dense subset of E. Then $f_1(y), f_2(y), \ldots$ converges uniformly on E.*

Selection Theorem 2.3.   *On a compact y-set $E \subset R^d$, let $f_1(y), f_2(y), \ldots$ be a sequence of functions which is uniformly bounded and equicontinuous. Then there exists a subsequence $f_{n(1)}(y), f_{n(2)}(y), \ldots$ which is uniformly convergent on E.*

This last theorem can be obtained as a consequence of the preceding two. By applying Theorem 2.3 to a suitable subsequence, we obtain the following:

Remark 1.   If, in the last theorem, $y_0 \in E$ and f_0 is a cluster point of the sequence $f_1(y_0), f_2(y_0), \ldots$, then the subsequence $f_{n(1)}(y), f_{n(2)}(y), \ldots$ in the assertion can be chosen so that the limit function $f(y)$ satisfies $f(y_0) = f_0$.

Remark 2.   If, in Theorem 2.3, it is known that all (uniformly) convergent subsequences of $f_1(y), f_2(y), \ldots$ have the same limit, say $f(y)$, then a selection is unnecessary and $f(y)$ is the uniform limit of $f_1(y), f_2(y), \ldots$. This follows from Remark 1.

Theorem 2.3 and the following consequences of it will be used repeatedly.

Theorem 2.4.   *Let $y, f \in R^d$ and $f_0(t, y), f_1(t, y), f_2(t, y), \ldots$ be a sequence of continuous functions on the parallelepiped $R : t_0 \leqq t \leqq t_0 + a, |y - y_0| \leqq b$ such that*

$$(2.3) \qquad f_0(t, y) = \lim_{n \to \infty} f_n(t, y) \qquad \text{uniformly on } R.$$

Let $y_n(t)$ be a solution of

$$(2.4_n) \qquad\qquad y' = f_n(t, y), \qquad y(t_n) = y_n,$$

on $[t_0, t_0 + a]$, *where* $n = 1, 2, \ldots,$ *and*

(2.5) $\qquad\qquad t_n \to t_0, \qquad y_n \to y_0 \qquad$ as $\qquad n \to \infty.$

Then there exists a subsequence $y_{n(1)}(t), y_{n(2)}(t), \ldots$ *which is uniformly convergent on* $[t_0, t_0 + a]$. *For any such subsequence, the limit*

(2.6) $\qquad\qquad\qquad\qquad y_0(t) = \lim_{k \to \infty} y_{n(k)}(t)$

is a solution of (2.4_0) *on* $[t_0, t_0 + a]$. *In particular, if* (2.4_0) *possesses a unique solution* $y = y_0(t)$ *on* $[t_0, t_0 + a]$, *then*

(2.7) $\qquad y_0(t) = \lim_{n \to \infty} y_n(t) \qquad$ *uniformly on* $[t_0, t_0 + a]$.

Proof. Since $f_1, f_2, \ldots$ are continuous and (2.3) holds uniformly on R, there is a constant K such that $|f_n(t, y)| \leq K$ for $n = 0, 1, \ldots$ and $(t, y) \in R$; Lemma 2.1. Since $|y_n{}'(t)| \leq K$, it is clear that K is a Lipschitz constant for $y_1, y_2, \ldots$, so that this sequence is equicontinuous. It is also uniformly bounded since $|y_n(t) - y_0| \leq b$. Thus the existence of uniformly convergent subsequences follows from Theorem 2.3. By (2.3), Lemma 2.1, and the uniformity of (2.6), it is easy to see that

$$f_{n(k)}(t, y_{n(k)}(t)) \to f_0(t, y(t))$$

uniformly on $[t_0, t_0 + a]$ as $k \to \infty$. Thus term-by-term integration is applicable to

$$y_n(t) = y_n + \int_{t_n}^{t} f_n(s, y_n(s))\, ds$$

where $n = n(k)$ and $k \to \infty$. It follows that the limit (2.6) is a solution of (2.4_0).

As to the last assertion, note that the assumed uniqueness of the solution $y_0(t)$ of (2.4_0) shows that the limit of every (uniformly) convergent subsequence of $y_1(t), y_2(t), \ldots$ is the solution $y_0(t)$. Hence a selection is unnecessary and (2.7) holds by Remark 2 above.

Implicit Function Theorem 2.5. *Let* x, y, f, g *be d-dimensional vectors and* z *an e-dimensional vector. Let* $f(y, z)$ *be continuous for* (y, z) *near a point* (y_0, z_0) *and have continuous partial derivatives with respect to the components of* y. *Let the Jacobian* $\det (\partial f^j / \partial y^k) \neq 0$ *at* $(y, z) = (y_0, z_0)$. *Let* $x_0 = f(y_0, z_0)$. *Then there exist positive numbers,* ϵ *and* δ, *such that if* x *and* z *are fixed,* $|x - x_0| < \delta$ *and* $|z - z_0| < \delta$, *then the equation* $x = f(y, z)$ *has a unique solution* $y = g(x, z)$ *satisfying* $|y - y_0| < \epsilon$. *Furthermore,* $g(x, z)$ *is continuous for* $|x - x_0| < \delta, |z - z_0| < \delta$ *and has continuous partial derivatives with respect to the components of* x.

For a sharper form of this theorem, see Exercise II 2.3.

3. Smooth Approximations

In some situations, it will be convenient to extend the definition of a function f, say, given continuous on a closed parallelepiped, or to approximate it uniformly by functions which are smooth (C^1 or C^∞) with respect to certain variables. The following devices can be used to obtain such extensions or approximations (which have the same bounds as f).

Let $f(t, y)$ be defined on $R: t_0 \leq t \leq t_1$, $|y| \leq b$ and let $|f(t, y)| \leq M$. Let $f^*(t, y)$ be defined for $t_0 \leq t \leq t_1$ and all y by placing $f^*(t, y) = f(t, y)$ if $|y| \leq b$ and $f^*(t, y) = f(t, by/|y|)$ if $|y| > b$. It is clear that $f^*(t, y)$ is continuous for $t_0 \leq t \leq t_1$, y arbitrary, and that $|f^*(t, y)| \leq M$. In some cases, it is more convenient to replace f^* by an extension of f which is 0 for large $|y|$. Such an extension is given by $f^0(t, y) = f^*(t, y)\varphi^0(|y|)$, where $\varphi^0(s)$ is a continuous function for $t \geq 0$ satisfying $0 \leq \varphi^0(s) \leq 1$ for $s \geq 0$, $\varphi^0(s) = 1$ for $0 \leq s \leq b$, and $\varphi^0(s) = 0$ for $s \geq b + 1$.

In order to approximate $f(t, y)$ uniformly on R by functions $f^\epsilon(t, y)$ which are, say, smooth with respect to the components of y, let $\varphi(s)$ be a function of class C^∞ for $s \geq 0$ satisfying $\varphi(s) > 0$ for $0 \leq s < 1$ and $\varphi(s) = 0$ for $s \geq 1$. Then there is a constant $c > 0$ depending only on $\varphi(s)$ and the dimension d, such that for every $\epsilon > 0$,

$$(3.1) \qquad c\epsilon^{-d} \int_{-\infty}^{+\infty} \cdots \int_{-\infty}^{+\infty} \varphi(\epsilon^{-2} \|y\|^2)\, dy^1 \ldots dy^d = 1,$$

where $\|y\| = (\Sigma |y^k|^2)^{1/2}$ is the Euclidean length of y. Put

$$(3.2) \quad f^\epsilon(t, y) = c\epsilon^{-d} \int_{-\infty}^{+\infty} \cdots \int_{-\infty}^{+\infty} f^0(t, \eta)\varphi(\epsilon^{-2} \|y - \eta\|^2)\, d\eta^1 \ldots d\eta^d,$$

where $\eta = (\eta^1, \ldots, \eta^d)$, so that

$$(3.3) \quad f^\epsilon(t, y) = c\epsilon^{-d} \int_{-\infty}^{+\infty} \cdots \int_{-\infty}^{+\infty} f^0(t, y - \eta)\varphi(\epsilon^{-2} \|\eta\|^2)\, d\eta^1 \ldots d\eta^d.$$

Since $f^\epsilon(t, y)$ is an "average" of the values of f^0 in a sphere $\|\eta - y\| \leq \epsilon$ for a fixed t, it is clear that $f^\epsilon \to f^0$ as $\epsilon \to 0$ uniformly on $t_0 \leq t \leq t_1$, y arbitrary. Note that $|f^\epsilon| \leq M$ for all $\epsilon > 0$ and that $f^\epsilon(t, y) = 0$ for $|y| \geq b + 1 + \epsilon$. Furthermore, $f^\epsilon(t, y)$ has continuous partial derivatives of all orders with respect to $y^1, \ldots, y^d$.

The last formula can be used to show that if $f^0(t, y)$ has continuous partial derivatives of order k with respect to $y^1, \ldots, y^d$, then the corresponding partial derivatives of $f^\epsilon(t, y)$ tend uniformly to those of $f^0(t, y)$ as $\epsilon \to 0$.

4. Change of Integration Variables

In order to avoid an interruption to some arguments later, it is convenient to mention the following:

Lemma 4.1. *Let* t, u, U *be scalars;* $U(u)$ *a continuous function on* $A \leqq u \leqq B$; $u = u(t)$ *a continuous function of bounded variation on* $a \leqq t \leqq b$ *such that* $A \leqq u(t) \leqq B$. *Then, for* $a \leqq t \leqq b$,

$$(4.1) \qquad \int_a^t U(u(s)) \, du(s) = \int_{u(a)}^{u(t)} U(u) \, du,$$

where the integral on the left is a Riemann-Stieltjes integral and that on the right is a Riemann integral.

The point of the lemma is the fact that the change of variables $t \to u$ given by $u = u(t)$ is permitted even when $u(t)$ is not monotone (and not absolutely continuous).

Proof. It is clear that the relation (4.1) holds for $a \leqq t \leqq b$ if $u(t)$ has a continuous derivative. For in this case, both integrals in (4.1) vanish at $t = a$ and have the same derivative $U(u(t))u'(t)$. If $u(t)$ does not have a continuous derivative, let $u_1(t)$, $u_2(t)$, ... be a sequence of continuously differentiable functions on $a \leqq t \leqq b$ satisfying $A \leqq u_n(t) \leqq B$, $u_n(t) \to u(t)$ as $n \to \infty$ uniformly on $[a, b]$, and such that the sequence of total variations of $u_1(t)$, $u_2(t)$, ... over $[a, b]$ is bounded. (The existence of functions $u_1(t)$, $u_2(t)$, ... follows from the last section.) Then (4.1) holds if $u(t)$ is replaced by $u_n(t)$. Term-by-term integration theorems applied to both sides of the resulting equation lead to (4.1).

Notes

SECTION 2. Theorems 2.2 and 2.3 go back to Ascoli [1] and Arzela [1], [2].

Chapter II

Existence

1. The Picard-Lindelöf Theorem

Various types of existence proofs will be given. One of the most simple and useful is the following.

Theorem 1.1. *Let* $y, f \in \mathbf{R}^d$; $f(t, y)$ *continuous on a parallelepiped* $R: t_0 \leqq t \leqq t_0 + a, |y - y_0| \leqq b$ *and uniformly Lipschitz continuous with respect to* y. *Let* M *be a bound for* $|f(t, y)|$ *on* R; $\alpha = \min(a, b/M)$. *Then*

$$(1.1) \qquad y' = f(t, y), \qquad y(t_0) = y_0$$

has a unique solution $y = y(t)$ *on* $[t_0, t_0 + \alpha]$.

It is clear that there is a corresponding existence and uniqueness theorem if R is replaced by $t_0 - a \leqq t \leqq t_0, |y - y_0| \leqq b$. It is also clear from these "right" and "left" existence theorems that if R is replaced by $|t - t_0| \leqq a, |y - y_0| \leqq b$, then (1.1) has a unique solution on $|t - t_0| \leqq \alpha$, since the solutions on the right and left fit together.

The choice of $\alpha = \min(a, b/M)$ in Theorem 1.1 is natural. On the one hand, the requirement $\alpha \leqq a$ is necessary. On the other hand, the requirement $\alpha \leqq b/M$ is dictated by the fact that if $y = y(t)$ is a solution of (1.1) on $[t_0, t_0 + \alpha]$, then $|y'(t)| \leqq M$ implies $|y(t) - y_0| \leqq M(t - t_0)$, which does not exceed b if $t - t_0 \leqq b/M$.

Remark 1. Note that in Theorem 1.1, $|y|$ can be any norm on $\mathbf{R}^d$, not necessarily the norm (2.1) or the Euclidean norm.

For another proof of "uniqueness," see Exercise III 1.1.

Proof by Successive Approximations. Let $y_0(t) \equiv y_0$. Suppose that $y_k(t)$ has been defined on $[t_0, t_0 + \alpha]$, is continuous, and satisfies $|y_k(t) - y_0| \leqq b$ for $k = 0, \ldots, n$. Put

$$(1.2) \qquad y_{n+1}(t) = y_0 + \int_{t_0}^{t} f(s, y_n(s))\, ds.$$

Then, since $f(t, y_n(t))$ is defined and continuous on $[t_0, t_0 + \alpha]$, the same holds for $y_{n+1}(t)$. It is also clear that

$$|y_{n+1}(t) - y_0| \leqq \int_{t_0}^{t} |f(s, y_n(s))|\, ds \leqq M\alpha \leqq b.$$

8

Hence $y_0(t), y_1(t), \ldots$ are defined and continuous on $[t_0, t_0 + \alpha]$, and $|y_n(t) - y_0| \leqq b$.

It will now be verified by induction that

$$(1.3_n) \quad |y_{n+1}(t) - y_n(t)| \leqq \frac{MK^n(t - t_0)^{n+1}}{(n + 1)!} \quad \text{for } t_0 \leqq t \leqq t_0 + \alpha,$$

$n = 0, 1, \ldots$, where K is a Lipschitz constant for f. Clearly, (1.3_0) holds. Assume $(1.3_0), \ldots, (1.3_{n-1})$. By (1.2),

$$y_{n+1}(t) - y_n(t) = \int_{t_0}^t [f(s, y_n(s)) - f(s, y_{n-1}(s))] \, ds$$

for $n \geqq 1$. Thus, the definition of K implies that

$$|y_{n+1}(t) - y_n(t)| \leqq K \int_{t_0}^t |y_n(s) - y_{n-1}(s)| \, ds$$

and so, by (1.3_{n-1}),

$$|y_{n+1}(t) - y_n(t)| \leqq \frac{MK^n}{n!} \int_{t_0}^t (s - t_0)^n \, ds = \frac{MK^n(t - t_0)^{n+1}}{(n + 1)!}$$

This proves (1.3_n).

In view of (1.3_n), it follows that

$$y_0 + \sum_{n=0}^{\infty} [y_{n+1}(t) - y_n(t)] = y(t)$$

is uniformly convergent on $[t_0, t_0 + \alpha]$; that is,

$$(1.4) \qquad y(t) = \lim_{n \to \infty} y_n(t) \qquad \text{exists uniformly.}$$

Since $f(t, y)$ is uniformly continuous on R, $f(t, y_n(t)) \to f(t, y(t))$ as $n \to \infty$ uniformly on $[t_0, t_0 + \alpha]$. Thus term-by-term integration is applicable to the integrals in (1.2) and gives

$$(1.5) \qquad y(t) = y_0 + \int_{t_0}^t f(s, y(s)) \, ds.$$

Hence (1.4) is a solution of (1.1).

In order to prove uniqueness, let $y = z(t)$ be any solution of (1.1) on $[t_0, t_0 + \alpha]$. Then

$$(1.6) \qquad z(t) = y_0 + \int_{t_0}^t f(s, z(s)) \, ds.$$

An obvious induction using (1.2) gives

$$(1.7) \quad |y_n(t) - z(t)| \leqq \frac{MK^n(t - t_0)^{n+1}}{(n + 1)!} \quad \text{for } t_0 \leqq t \leqq t_0 + \alpha$$

and $n = 0, 1, \ldots$. If $n \to \infty$ in (1.7), it follows from (1.4) that $|y(t) - z(t)| \leqq 0$; i.e., $y(t) \equiv z(t)$. This proves the theorem.

Remark 2. Since $z \equiv y$, (1.7) gives an estimate of the error of approximation

$$(1.8) \qquad |y_n(t) - y(t)| \leqq \frac{MK^n(t - t_0)^{n+1}}{(n + 1)!} \qquad \text{on } [t_0, t_0 + \alpha].$$

Exercise 1.1.   Show that if, in addition to the conditions of the theorem, $f(t, y)$ is analytic on R (i.e., in a neighborhood of every point $(t^0, y^0) \in R$, $f(t, y)$ is representable as a convergent power series in $t - t^0, y^1 - y^{01}, \ldots, y^d - y^{0d}$), then the solution $y = y(t)$ of (1.1) is analytic on $[t_0, t_0 + \alpha]$. The analogous theorem in which t and the components of y, f are allowed to be complex-valued is also valid.

Exercise 1.2.   If, in Theorem 1.1, v is near to y_0, then the initial value problem $y' = f(t, y), y(t_0) = v$ has a unique solution $y = y(t, v)$ on some interval $[t_0, t_0 + \beta]$ independent of v. Show that $y(t, v)$ is uniformly Lipschitz continuous with respect to (t, v) for $t_0 \leqq t \leqq t_0 + \beta$, v near y_0.

2. Peano's Existence Theorem

The next theorem to be proved drops the assumption of Lipschitz continuity and the assertion of uniqueness.

Theorem 2.1.   *Let $y, f \in R^d$; $f(t, y)$ continuous on $R: t_0 \leqq t \leqq t_0 + a$, $|y - y_0| \leqq b$; M a bound for $|f(t, y)|$ on R; $\alpha = \min (a, b/M)$. Then (1.1) possesses at least one solution $y = y(t)$ on $[t_0, t_0 + \alpha]$.*

In this theorem, $|y|$ can be any convenient norm on R^d.

Proof.   Let $\delta > 0$ and $y_0(t)$ a C^1 d-dimensional vector-valued function on $[t_0 - \delta, t_0]$ satisfying $y_0(t_0) = y_0, |y_0(t) - y_0| \leqq b$, and $|y_0'(t)| \leqq M$.

For $0 < \epsilon \leqq \delta$, define a function $y_\epsilon(t)$ on $[t_0 - \delta, t_0 + \alpha]$ by putting $y_\epsilon(t) = y_0(t)$ on $[t_0 - \delta, t_0]$ and

$$(2.1) \qquad y_\epsilon(t) = y_0 + \int_{t_0}^t f(s, y_\epsilon(s - \epsilon)) \, ds \qquad \text{on } [t_0, t_0 + \alpha].$$

Note first that this formula is meaningful and defines $y_\epsilon(t)$ for $t_0 \leqq t \leqq t_0 + \alpha_1$, $\alpha_1 = \min (\alpha, \epsilon)$, and, on this interval,

$$(2.2) \qquad |y_\epsilon(t) - y_0| \leqq b, \qquad |y_\epsilon(t) - y_\epsilon(s)| \leqq M |t - s|.$$

It then follows that (2.1) can be used to extend $y_\epsilon(t)$ as a C^0 function over $[t_0 - \delta, t_0 + \alpha_2]$, $\alpha_2 = \min (\alpha, 2\epsilon)$, satisfying (2.2). Continuing in this fashion, (2.1) serves to define $y_\epsilon(t)$ over $[t_0, t_0 + \alpha]$ so that $y_\epsilon(t)$ is a C^0 function on $[t_0 - \delta, t_0 + \alpha]$, satisfying (2.2).

It follows that the family of functions, $y_\epsilon(t), 0 <$ $\epsilon \leqq \delta$, is equicontinuous. Thus, by Theorem I 2.3, there is a sequence $\epsilon(1) > \epsilon(2) > \ldots$, such that $\epsilon(n) \to 0$ as $n \to \infty$ and

$$y(t) = \lim_{n \to \infty} y_{\epsilon(n)}(t) \quad \text{exists uniformly}$$

on $[t_0 - \delta, t_0 + \alpha]$. The uniform continuity of f implies that $f(t, y_{\epsilon(n)}(t - \epsilon(n))$ tends uniformly to $f(t, y(t))$ as $n \to \infty$; thus term-by-term integration of (2.1) where $\epsilon = \epsilon(n)$ gives (1.5). Hence $y(t)$ is a solution of (1.1). This proves the theorem.

An important consequence of Peano's existence theorem will often be used:

Corollary 2.1. *Let $f(t, y)$ be continuous on an open (t, y)-set E and satisfy $|f(t, y)| \leqq M$. Let E_0 be a compact subset of E. Then there exists an $\alpha > 0$, depending on E, E_0 and M, with the property that if $(t_0, y_0) \in E_0$, then (1.1) has a solution on $|t - t_0| \leqq \alpha$.*

In fact, if $a = \text{dist} (E_0, \partial E) > 0$, where ∂E is the boundary of E, then $\alpha = \min (a, a/M)$. In applications, when f is not bounded on E, the set E in this corollary is replaced by an open subset E^0 having compact closure in E and containing E_0.

Exercise 2.1 (*Polygonal Approximation*). Under the conditions of Theorem 2.1, define a set of functions $y_\Sigma(t)$ as follows: Let $\Sigma : t_0 < t_1 < \cdots < t_m = t_0 + \alpha$ be a mesh on $[t_0, t_0 + \alpha]$ with a degree of fineness $\delta(\Sigma) = \max (t_{k+1} - t_k)$ for $k = 0, \ldots, m - 1$. On $[t_0, t_1]$, put $y_\Sigma(t) = y_0 + (t - t_0)f(t_0, y_0)$. If $y_\Sigma(t)$ has been defined on $[t_0, t_k], k < m$, and $|y_\Sigma(t) - y_0| \leqq b$, put $y_\Sigma(t) = y_\Sigma(t_k) + (t - t_k)f(t_k, y_\Sigma(t_k))$ on $[t_k, t_{k+1}]$. This serves to define $y_\Sigma(t)$ on $[t_0, t_0 + \alpha]$ as a continuous piecewise linear function. Prove Theorem 2.1 by obtaining a solution of (1.1) as a limit of a suitable sequence $y_{\Sigma(1)}(t), y_{\Sigma(2)}(t), \ldots$, where $\delta(\Sigma(n)) \to 0$ as $n \to \infty$. [Note that if the solution of (1.1) is not unique, then not all solutions can be obtained by this procedure; cf. the scalar problem $y' = |y|^{1/2}, y(0) = 0$.]

Exercise 2.2 (*Another Proof*). There exists a sequence of continuous functions $f_1(t, y), f_2(t, y), \ldots$ on R which tend uniformly to $f(t, y)$ on R, $|f_n(t, y)| \leqq M$ for $(t, y) \in R$ and $n = 1, 2, \ldots$, and $f_n(t, y)$ is uniformly Lipschitz continuous with respect to y; cf. § I 3. Consider $y' = f_n(t, y)$, $y(t_0) = y_0$ and apply Theorems 1.1 and I 2.4. [In contrast to Exercise 2.1, all solutions of (1.1) can be obtained by the method of this exercise; cf. the proof of Theorem 4.1.]

Exercise 2.3. [This exercise gives a sharpened form of the Implicit Function Theorem I 2.5 (with no parameters z). In the statement of this theorem, the norm $\|A\|$ of a $d \times d$ matrix occurs. Let R^d be the (real) d-dimensional vector space with any convenient norm $|y|$. Then $\|A\|$ is defined to be $\|A\| = \max |Ay|$ for $|y| = 1$. This norm $\|A\|$ depends on the choice of the norm $|y|$ in $\mathbb{R}^d$.] Let $x = f(y)$ be a function of class C^1 on $D: |y| \leqq b$, and let $f(0) = 0$. Let the Jacobian matrix, $f_y(y) = (\partial f^j/\partial y^k)$ for $j, k = 1, \ldots, d$, be nonsingular on D and put $M = \max \|f_y^{-1}(y)\|$, $M_1 = \max \|f_y(y)\|$ for $y \in D$, where f_y^{-1} is the inverse of the matrix f_y and $\|f_y\|$, $\|f_y^{-1}\|$ denote the norms of the respective matrices. Let $D_1: |y| \leqq b/MM_1$. (Note that $MM_1 \geqq 1$. Why?) Then there exists a domain D_0 such that $D_1 \subset D_0 \subset D$ and $x = f(y)$ is a one-to-one map of [the closure of] D_0 onto [the closure of] the ball $\tilde{D}^0: |x| < b/M$; see Figure 1. Assuming the Implicit Function Theorem I 2.5, deduce the result just stated from

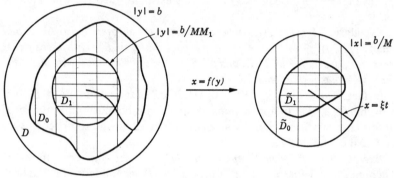

Figure 1.

Peano's Existence Theorem 2.1 by writing the equation $x = f(y)$ for y in the form $\xi t = f(y)$, where $\xi \neq 0$ is a constant vector, differentiating with respect to t to obtain the differential equation $y' = f_y^{-1}(y)\xi$, and considering the solution satisfying the initial condition $y(0) = 0$. (It is possible to avoid the use of the Implicit Function Theorem by using the results of § V 6.)

3. Extension Theorem

Let $f(t, y)$ be continuous on a (t, y)-set E and let $y = y(t)$ be a solution of

$$(3.1) \qquad y' = f(t, y)$$

on an interval J. The interval J is called a *right maximal interval* of existence for y if there does not exist an extension of $y(t)$ over an interval J_1 so that $y = y(t)$ remains a solution of (3.1); J is a proper subset of J_1; J, J_1 have different right endpoints. A *left maximal interval of existence* for y is defined similarly. A *maximal interval of existence* is an interval which is both a left and right maximal interval.

Theorem 3.1. *Let $f(t, y)$ be continuous on an open (t, y)-set E and let $y(t)$ be a solution of (3.1) on some interval. Then $y(t)$ can be extended (as a* *solution) over a maximal interval of existence* (ω_-, ω_+). *Also, if* (ω_-, ω_+) *is a maximal interval of existence, then* $y(t)$ *tends to the boundary* ∂E *of* E *as* $t \to \omega_-$ *and* $t \to \omega_+$.

The extension of $y(t)$ need not be unique and, correspondingly, $\omega_{\pm}$ depends on the extension. To say, e.g., that $y(t)$ tends to ∂E as $t \to \omega_+$ is interpreted to mean that if E^0 is any compact subset of E, then $(t, y(t)) \notin E^0$ when t is near ω_+.

Proof. Let $E_1, E_2, \ldots$ be open subsets of E such that $E = \bigcup E_n$; the closures $\bar{E}_1, \bar{E}_2, \ldots$ are compact, and $\bar{E}_n \subset E_{n+1}$ (e.g., let $E_n = \{(t, y):$ $(t, y) \in E, |t| < n, |y| < n$ and dist $((t, y), \partial E) > 1/n\})$. Corollary 2.1 implies that there exists an $\epsilon_n > 0$ such that if (t_0, y_0) is any point of $\bar{E}_n$, then all solutions of (3.1) through (t_0, y_0) exist on $|t - t_0| \leq \epsilon_n$.

Consider a given solution $y = y(t)$ of (3.1) on an interval J. If J is not a right maximal interval of existence, then $y(t)$ can be extended to an interval containing the right endpoint of J. Thus, in proving the existence of a right maximal interval of existence, it can be supposed that $y(t)$ is defined on a closed interval $a \leq t \leq b_0$ and that $y(t)$ does not have an extension over $a \leq t < \infty$.

Let $n(1)$ be so large that $(b_0, y(b_0)) \in \bar{E}_{n(1)}$. Then $y(t)$ can be extended over an interval $[b_0, b_0 + \epsilon_{n(1)}]$. If $(b_0 + \epsilon_{n(1)}, y(b_0 + \epsilon_{n(1)})) \in \bar{E}_{n(1)}$, then $y(t)$ can be extended over another interval $[b_0 + \epsilon_{n(1)}, b_0 + 2\epsilon_{n(1)}]$ of length $\epsilon_{n(1)}$. Continuing this argument, it is seen that there is an integer $j(1) \geq 1$ such that $y(t)$ can be extended over $a \leq t \leq b_1$, where $b_1 = b_0 + j(1)\epsilon_{n(1)}$ and $(b_1, y(b_1)) \notin \bar{E}_{n(1)}$.

Let $n(2)$ be so large that $(b_1, y(b_1)) \in \bar{E}_{n(2)}$. Then there exists an integer $j(2) \geq 1$ such that $y(t)$ can be extended over $a \leq t \leq b_2$, where $b_2 = b_1 + j(2)\epsilon_{n(2)}$ and $(b_2, y(b_2)) \notin \bar{E}_{n(2)}$.

Repetitions of this argument lead to sequences of integers $n(1) < n(2) < \ldots$ and numbers $b_0 < b_1 < \ldots$ such that $y(t)$ has an extension over $[a, \omega_+)$, where $\omega_+ = \lim b_k$ as $k \to \infty$, and that $(b_k, y(b_k)) \notin \bar{E}_{n(k)}$. Thus $(b_1, y(b_1)), (b_2, y(b_2)), \ldots$ is either unbounded or has a cluster point on the boundary ∂E of E.

To see that $y(t)$ tends to ∂E as $t \to \omega_+$ on a right maximal interval $[a, \omega_+)$, it must be shown that no limit point of a sequence $(t_1, y(t_1))$, $(t_2, y(t_2)), \ldots$, where $t_n \to \omega_+$, can be an interior point of E. This is a consequence of the following:

Lemma 3.1. *Let* $f(t, y)$ *be continuous on a* (t, y)-*set* E. *Let* $y = y(t)$ *be a solution of* (3.1) *on an interval* $[a, \delta)$, $\delta < \infty$, *for which there exists a sequence* $t_1, t_2, \ldots$ *such that* $a \leq t_n \to \delta$ *as* $n \to \infty$ *and* $y_0 = \lim y(t_n)$ *exists. If* $f(t, y)$ *is bounded on the intersection of* E *and a vicinity of the point* (δ, y_0), *then*

$$(3.2) \qquad\qquad y_0 = \lim y(t) \quad \text{as} \quad t \to \delta.$$

If, in addition, $f(\delta, y_0)$ is or can be defined so that $f(t, y)$ is continuous at (δ, y_0), then $y(t) \in C^1[a, \delta]$ and is a solution of (3.1) on $[a, \delta]$.

Proof. Let $\epsilon > 0$ be so small and $M_\epsilon > 1$ so large that $|f(t, y)| \leq M_\epsilon$ for (t, y) on the intersection of E and the parallelepiped $0 \leq \delta - t \leq \epsilon$, $|y - y_0| \leq \epsilon$. If n is so large that $0 < \delta - t_n \leq \epsilon/2M_\epsilon$ and $|y(t_n) - y_0| \leq \epsilon/2$, then

$$(3.3) \qquad |y(t) - y(t_n)| < M_\epsilon (\delta - t_n) \leq \tfrac{1}{2}\epsilon \qquad \text{for } t_n \leq t < \delta.$$

Otherwise, there is smallest t^1 such that $t_n < t^1 < \delta$, $|y(t^1) - y(t_n)| = M(\delta - t_n) \leq \tfrac{1}{2}\epsilon$. Hence $|y(t) - y_0| \leq \tfrac{1}{2}\epsilon + |y(t_n) - y_0| \leq \epsilon$ for $t_n \leq t < t^1$; thus $|y'(t)| \leq M_\epsilon$ for $t_n \leq t \leq t^1$. Consequently, $|y(t^1) - y(t_n)| \leq M_\epsilon(t^1 - t_n) < M_\epsilon(\delta - t_n)$. This proves (3.3), hence (3.2). The last part of the lemma follows from $y'(t) = f(t, y(t)) \to f(\delta, y_0)$ as $t \to \delta$.

Corollary 3.1. *Let $f(t, y)$ be continuous on a strip $t_0 \leq t \leq t_0 + a$ $(< \infty)$, $y \in \mathbf{R}^d$ arbitrary. Let $y = y(t)$ be a solution of (1.1) on a right maximal interval J. Then either $J = [t_0, t_0 + a]$ or $J = [t_0, \delta)$, $\delta \leq t_0 + a$, and $|y(t)| \to \infty$ as $t \to \delta$.*

More generally,

Corollary 3.2. *Let $f(t, y)$ be continuous on the closure $\bar{E}$ of an open (t, y)-set E and let (1.1) possess a solution $y = y(t)$ on a maximal right interval J. Then either $J = [t_0, \infty)$, or $J = [t_0, \delta]$ with $\delta < \infty$ and $(\delta, y(\delta)) \in \partial E$, or $J = [t_0, \delta)$ with $\delta < \infty$ and $|y(t)| \to \infty$ as $t \to \delta$.*

A somewhat different, but very useful, result is given by the following theorem.

Theorem 3.2. *Let $f(t, y)$ and $f_1(t, y)$, $f_2(t, y)$, ... be a sequence of continuous functions defined on an open (t, y)-set E such that*

$$(3.4) \qquad f_n(t, y) \to f(t, y) \qquad \text{as } n \to \infty$$

holds uniformly on every compact subset of E. Let $y_n(t)$ be a solution of

$$(3.5) \qquad y' = f_n(t, y), \qquad y(t_n) = y_{n0},$$

$(t_n, y_{n0}) \in E$, and let $(\omega_{n-}, \omega_{n+})$ be its maximal interval of existence. Let

$$(3.6) \qquad (t_n, y_{n0}) \to (t_0, y_0) \in E \qquad \text{as } n \to \infty.$$

Then there exist a solution $y(t)$ of

$$(3.7) \qquad y' = f(t, y), \qquad y(t_0) = y_0,$$

having a maximal interval of existence (ω_-, ω_+) and a sequence of positive integers $n(1) < n(2) < \ldots$ with the property that if $\omega_- < t^1 < t^2 < \omega_+$, then $\omega_{n-} < t^1 < t^2 < \omega_{n+}$ for $n = n(k)$ and k large, and

$$(3.8) \qquad y_{n(k)}(t) \to y(t) \qquad \text{as } k \to \infty$$

uniformly for $t^1 \leqq t \leqq t^2$. In particular,

(3.9) $\lim \sup \omega_{n-} \leqq \omega_- < \omega_+ \leqq \lim \inf \omega_{n+}$ as $n = n(k) \to \infty$.

Proof. Let $E_1, E_2, \ldots$ be open subsets of E such that $E = \bigcup E_n$, the closures $\bar{E}_1, \bar{E}_2, \ldots$ are compact and $\bar{E}_n \subset E_{n+1}$. Suppose that $(t_0, y_0) \in E_1$ and hence that $(t_n, y_{n0}) \in E_1$ for large n.

In the proof, $y(t)$ will be constructed only on a right maximal interval of existence $[t_0, \omega_+)$. The construction for a left maximal interval is similar.

By Corollary 2.1, there exists an ϵ_1, independent of n for large n, such that any solution of (3.5) [or (3.7)] for any point $(t_n, y_{n0}) \in E_1$ [or $(t_0, y_0) \in E_1$] exists on an interval of length $3\epsilon_1$ centered at $t = t_n$ [or $t = t_0$]. By Arzela's theorem, it follows that if $n(1) < n(2) < \ldots$ are suitably chosen, then the limit (3.8) exists uniformly for $t_0 \leqq t \leqq t_0 + \epsilon_1$ and is a solution of (3.7). If the point $(t_0 + \epsilon_1, y(t_0 + \epsilon_1)) \in E_1$ the sequence $n(1) < n(2) < \ldots$ can be replaced by a subsequence, again called $n(1) < n(2) < \ldots$, such that the limit (3.8) exists uniformly for $t_0 + \epsilon_1 \leqq t \leqq t_0 + 2\epsilon_1$ and is a solution of (3.1). This process can be repeated j times, where $(t_0 + m\epsilon_1, y(t_0 + m\epsilon_1)) \in E_1$ for $m = 0, \ldots, j - 1$ but not for $m = j$.

In this case, let $t_1 = t_0 + j\epsilon_1$ and choose the integer $r > 1$ so that $(t_1, y(t_1)) \in E_r$. Repeat the procedure above using a suitable $\epsilon_r > 0$ (depending on r but independent of n for large n) to obtain $y(t)$ on an interval $t_1 \leqq t \leqq t_1 + j_1\epsilon_r$, where $(t_1 + m\epsilon_r, y(t_1 + m\epsilon_r)) \in E_r$ for $m = 0, \ldots j_1 - 1$ but not for $m = j_1$. Put $t_2 = t_1 + j_1\epsilon_r$.

Repetitions of these arguments lead to a sequence of t-values $t_0 < t_1 < \ldots$ and a sequence of successive subsequences of integers:

$$n_1(1) < n_1(2) < \ldots$$
$$n_2(1) < n_2(2) < \ldots$$
$$\cdots\cdots\cdots\cdots\cdots$$

such that (3.8) holds uniformly for $t_0 \leqq t \leqq t_m$ if $n(k) = n_m(k)$. Put $\omega_+ = \lim t_m (\leqq \infty)$. Since $(t_{m+1}, y(t_{m+1})) \notin E_m$, for $m = 1, 2, \ldots$, $[t_0, \omega_+)$ is the right maximal interval of existence for $y(t)$. The usual diagonal process supplies the desired sequence $n(1) < n(2) < \ldots$. This proves the theorem.

4. H. Kneser's Theorem

The following theorem concerning the case of nonunique solutions of initial value problems will be proved in this section.

Theorem 4.1. *Let $f(t, y)$ be continuous on $R : t_0 \leqq t \leqq t_0 + a, |y - y_0| \leqq b$. Let $|f(t, y)| \leqq M, \alpha = \min(a, b/M)$ and $t_0 < c \leqq t_0 + \alpha$. Finally, let*

S_c be the set of points y_c for which there is a solution $y = y(t)$ of

(4.1) $y' = f(t, y), \qquad y(t_0) = y_0$

on $[t_0, c]$ such that $y(c) = y_c$; i.e., $y_c \in S_c$ means that y_c is a point reached at $t = c$ by some solution of (4.1). Then S_c is a continuum, i.e., a closed connected set.

Exercise 4.1. If y is a scalar, Theorem 4.1 has a very simple proof even without the assumption $t_0 \leq c \leq t_0 + \alpha$. The conclusion is that S_c is either empty, a point, or a closed y-interval. Prove Theorem 4.1 in this case by first showing that if $y_1, y_2 \in S_c$, so that (4.1) has solutions $y_j(t)$ on $[t_0, c]$ such that $y_j(c) = y_j$ for $j = 1, 2$, and if $y_1 < y^0 < y_2$, then $y^0 \in S_c$.

Proof. Let Σ denote the set of solutions of (4.1). These exist on $[t_0, c]$ and S_c is the set of points $y(c)$, where $y(t) \in \Sigma$. To see that the set S_c is closed, let $y_{nc} \to y_c$, $n \to \infty$, and $y_{nc} \in S_c$. Then $y_{nc} = y_n(c)$ for some $y_n(t) \in \Sigma$. By Theorem I 2.4, $y_1(t), y_2(t), \dots$ has a subsequence which is uniformly convergent to some $y(t) \in \Sigma$ on $[t_0, c]$. Clearly, $y_c = y(c) \in \Sigma$.

Suppose that the assertion is false, then S_c is not connected and is therefore the union of two nonempty, disjoint closed sets S^0, S^1. Since S_c is bounded, $\delta \equiv \text{dist}(S^0, S^1) > 0$, where $\text{dist}(S^0, S^1) = \inf |y^0 - y^1|$ for $y^0 \in S^0, y^1 \in S^1$. For any y, put $e(y) = \text{dist}(y, S^0) - \text{dist}(y, S^1)$, so that $e(y) \geq \delta > 0$ if $y \in S^1$ and $e(y) \leq -\delta < 0$ if $y \in S^0$. The function $e(y)$ is continuous and $e(y) \neq 0$ for $y \in S_c$.

Let $\epsilon > 0$ and $y(t) \in \Sigma$. There exists a continuous function $g(t, y)$ depending on ϵ and (the fixed) $y(t)$, defined for $t_0 \leq t \leq c$ and all y such that (i) $|g(t, y)| \leq M + \epsilon$; that (ii)

(4.2) $|f(t, y) - g(t, y)| \leq \epsilon \qquad$ on R;

that (iii) $g(t, y)$ is uniformly Lipschitz continuous with respect to y; and that (iv) $y = y(t)$ is a solution of

(4.3) $y' = g(t, y), \qquad y(t_0) = y_0.$

In order to see this, let $g^*(t, y)$ be a function with the properties (i)–(iii), but with $M + \epsilon$ replaced by M in (i) and ϵ by $\frac{1}{2}\epsilon$ in (4.2); cf. § I 3. Let $g(t, y) = g^*(t, y) + f(t, y(t)) - g^*(t, y(t))$. Then $|g(t, y) - g^*(t, y)| \leq |f(t, y(t)) - g^*(t, y(t))| \leq \frac{1}{2}\epsilon$, so that conditions (i), (ii) follow. Condition (iii) is clear and (iv) follows from $g(t, y(t)) = f(t, y(t)) = y'(t)$.

Let $y = y_0(t), y_1(t) \in \Sigma$, and $y_0(c) \in S^0, y_1(c) \in S^1$. For a given $\epsilon > 0$, let $g_0(t, y), g_1(t, y)$ be functions with the properties (i)–(iv), when $y(t) = y_0(t), y_1(t)$, respectively. Consider the 1-parameter family of initial value problems:

(4.4) $y' = g_\theta(t, y), \qquad y(t_0) = y_0,$

where $0 \leq \theta \leq 1$ and

(4.5) $$g_\theta(t, y) = \theta g_1(t, y) + (1 - \theta)g_0(t, y).$$

Since $g_\theta(t, y)$ is uniformly Lipschitz continuous with respect to y, (4.4) has a unique solution $y = y(t, \theta)$; see Theorem 1.1. As a consequence of Theorem 1.1 (where $b > 0$ is arbitrary), this solution exists on $[t_0, c]$ since g_θ is bounded, $|g_\theta(t, y)| \leq M + \epsilon$, on the strip $t_0 \leq t \leq c$ and all y.

Note that $|g_\theta(t, y)| \leq M + \epsilon, c \leq t_0 + \alpha$ imply that $|y(t, \theta) - y_0| \leq (M + \epsilon)\alpha$ for $t_0 \leq t \leq c$. Theorem I 2.4 implies that $y(t, \theta) \to y(t, \theta_0)$, $\theta \to \theta_0$, uniformly on $[t_0, c]$. In particular, $y(c, \theta)$, hence $e(y(c, \theta))$, is a continuous function of θ. Since $y(c, 0) = y_0(c) \in S^0, y(c, 1) = y_1(c) \in S^1$, so that $e(y(c, 0)) \leq -\delta < 0, e(y(c, 1)) \geq \delta > 0$, there exists a θ-value, $\theta = \eta, 0 < \eta < 1$ such that $e(y(c, \eta)) = 0$.

If $y_0(t), y_1(t)$ are fixed, a choice of an η depends only on ϵ, say $\eta = \eta(\epsilon)$. Let $\epsilon = 1/n, n > 1$, and let $g^n(t, y) = g_\theta(t, y)$, where $\theta = \eta(1/n)$. Thus (4.2) and (4.5) show that

(4.6) $$|f(t, y) - g^n(t, y)| \leq 1/n \qquad \text{on } R$$

and, by the choice of $\theta = \eta$,

(4.7) $$y' = g^n(t, y), \qquad y(t_0) = y_0$$

has a (unique) solution $y = y^{(n)}(t)$ on $[t_0, c]$ such that $e(y^{(n)}(c)) = 0$. The sequence $y^{(1)}(t), y^{(2)}(t), \ldots$ has a subsequence which is uniformly convergent, say to $y = y(t)$, on $[t_0, c]$. Since $|y^{(n)}(t) - y_0| \leq b$ for $t_0 \leq t \leq \min(c, t_0 + b/(M + 1/n))$ and $\min(c, t_0 + b/(M + 1/n)) \to c$ as $n \to \infty$, Theorem I 2.4 implies that $y(t)$ is a solution of (4.1) on $[t_0, c]$. Also $e(y(c)) = \lim e(y^{(n)}(c)) = 0$ as $n \to \infty$. But then $y(c) \in S_c$ and $e(y(c)) = 0$. This contradiction proves the theorem.

Exercise 4.2. Show by an example that S_c need not be a convex set if $d > 1$, where y is a d-dimensional vector; e.g., if $d = 2, S_c$ can be the boundary of a circle.

Exercise 4.3. (a) Let $f(t, y)$ be continuous for $t_0 \leq t \leq t_0 + a$ and all y. Let $t_0 < c \leq t_0 + a$ and assume that all solutions $y(t)$ of (4.1) exist on $t_0 \leq t \leq c$. Then S_c is a continuum. (b) Show, by an example, that S_c need not be connected if $d = 2$ and not all solutions of (4.1) exist for $t_0 \leq t \leq t_0 + c$.

Exercise 4.4. Let $f(t, y), c$ be as in Theorem 4.1 or in part (a) of Exercise 4.3 and $T_c = \{(t, y): t_0 \leq t \leq t_0 + c, y = y(t)$ for some solution of (4.1)$\}$. In particular, $S_c = \{y: (c, y) \in T_c\}$. (a) Let y^* be a boundary point of S_c. Show that (4.1) has a solution $y = y^*(t)$ such that $(t, y^*(t))$ is on the boundary of T_c for $t_0 \leq t \leq t_0 + c$ and that $y^*(c) = y^*$. This is a theorem of Fukuhara; see Kamke [2]. (b) Let (t_1, y_1) be a point of the boundary of T_c, where $t_0 < t_1 < c$. Show that there need not exist a solution $y(t)$ of (4.1) such that $y(t_1) = y_1$ and $(t, y(t))$ is on the boundary of T_c on any interval $t_0 \leqq t \leqq t_1 + \epsilon, \epsilon > 0$. This is a result of Fukuhara and Nagumo; see Digel [1].

5. Example of Nonuniqueness

In order to illustrate how bad the situation as to uniqueness can become, it will be shown that there exists a (scalar) function $U(t, u)$ continuous on the (t, u)-plane such that for *every* choice of initial point (t_0, u_0), the initial value problem

$$(5.1) \qquad\qquad u' = U(t, u), \qquad u(t_0) = u_0$$

has more than one solution on every interval $[t_0, t_0 + \epsilon]$ and $[t_0 - \epsilon, t_0]$ for arbitrary $\epsilon > 0$.

Let S_0 be the set of arcs,

$$(5.2) \quad u = 4i + \cos \pi t \quad \text{and} \quad u = 4i + 2 - \cos \pi t \quad \text{for} \quad -\infty < t < \infty,$$

$i = 0, \pm 1, \ldots$, considered to be made up of subarcs defined on the intervals of length $1, k \leqq t \leqq k + 1$, and $k = 0, \pm 1, \ldots$.

For every $n = 0, 1, 2, \ldots$, there will be constructed a set S_n of twice continuously differentiable arcs

$$(5.3) \quad u = u_{jk}(t), \qquad \frac{k}{2^n} \leqq t \leqq \frac{k+1}{2^n}, \qquad \text{and} \qquad j, k = 0, \pm 1, \ldots.$$

The symbol S_n will denote either the set of arcs (5.3) or the set of points on these arcs. The set S_n of arcs (5.3) will have the properties that (i)

$$(5.4) \qquad u_{jk}(t) < u_{j+1,k}(t) \qquad \text{for} \quad \frac{k}{2^n} < t < \frac{k+1}{2^n} \, ;$$

(ii) the arcs $u = u_{jk}(t)$ and $u = u_{j+1,k}(t)$ have exactly one endpoint in common; (iii) for any pair j, k, there is at least one index h such that $u_{h,k-1} = u_{h+1,k-1} = u_{jk}$ at $t = k/2^n$ and an index i such that $u_{i,k+1} = u_{i+1,k+1} = u_{jk}$ at $t = (k + 1)/2^n$; (iv) any two arcs of S_n which have a point in common have the same tangent at that point; hence (v) any continuous arc $u = u(t)$, say, on $a \leqq t \leqq b$, which is made up of arcs of S_n can be continued over $-\infty < t < \infty$, not uniquely, so as to have the same property and any such continuation is of class C^1 (and piecewise of class C^2); also, (vi) if $U_n(t, u)$ is defined on the point set S_n to be the slope of the tangent at the point $(t, u) \in S_n$, then $U_n(t, u)$ is uniformly continuous on S_n and arcs of (v) constitute the set of solutions of

$$(5.5) \qquad\qquad u' = U_n(t, u);$$

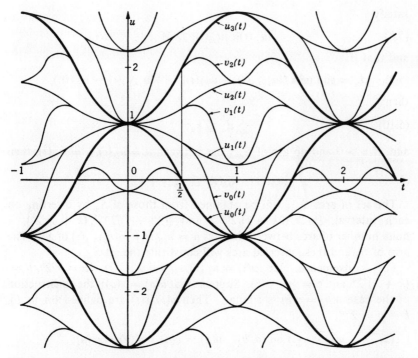

Figure 2. The heavy curved lines represent arcs of S_0. Heavy and light curved lines represent arcs of S_1 if $m_0 = 3$. The construction of the arcs of S_1, not in S_0, is indicated above: the arcs $u(t) = u_0(t)$, $u_1(t)$, $u_2(t)$, $u_3(t) = v(t)$ are defined on $[a, b] = [0, 1]$, the arcs $v_0(t)$, $v_1(t)$, $v_2(t)$ on $[c, b] = [\frac{1}{2}, 1]$. The sketch makes it clear how S_0 or S_1 divides the plane into sets G.

(vii) the sets S_0, S_1, ... satisfy $S_n \subset S_{n+1}$, so that $U_{n+1}(t, u)$ is an extension of $U_n(t, u)$; (viii) $S = \bigcup S_n$ and, in fact, the set of end points $(k/2^n, u_{jk}(k/2^n))$, for $j, k = 0, \pm 1, \dots$ and $n = 0, 1, \dots$, is dense in the plane; finally, (ix)

$$(5.6) \qquad\qquad U(t, u) = \lim_{n \to \infty} U_n(t, u)$$

which is defined on $S = \bigcup S_n$ has a (unique) continuous extension over the plane. Condition (ix) is the only nontrivial condition. (The construction of S is indicated in Figure 2.)

Let $\pi^2 = \epsilon_0 > \epsilon_1 > \dots$,

$$(5.7) \qquad\qquad M_n = \sum_{k=0}^{n} \epsilon_k \quad \text{and} \quad M = \sum_{k=0}^{\infty} \epsilon_k < \infty.$$

Suppose that S_n has already been constructed so that the functions (5.3) satisfy

(5.8) $$|u'_{jk}(t)|, \ |u''_{jk}(t)| \leqq M_n;$$

and that if

(5.9) $$d_n = \sup_{j,k,t} \max \ (|u_{j+1,k}(t) - u_{j,k}(t)|, \quad |u'_{j+1,k}(t) - u'_{jk}(t)|),$$

then

(5.10) $$d_n \leqq \epsilon_n;$$

and, if $n > 0$ and no arc of S_{n-1} lies between $u = u_{ik}(t)$, $u = u_{hk}(t)$, then

(5.11) $$|u'_{ik}(t) - u'_{hk}(t)| \leqq d_{n-1} + \epsilon_n.$$

The set of arcs S_{n+1} will be obtained from those of S_n by inserting on each interval, $[k/2^n, (2k + 1)/2^{n+1}]$ and $[(2k + 1)/2^{n+1}, (k + 1)/2^n]$, a finite number of arcs between the arcs $u = u_{jk}(t)$, $u = u_{j+1,k}(t)$ of S_n. The arcs of S_n and these inserted arcs will constitute the set S_{n+1}.

For convenience, let $u(t) = u_{jk}(t), v(t) = u_{j+1,k}(t)$, $a = k/2^n$, $b = (k + 1)/2^n$, and $c = \frac{1}{2}(a + b)$. Suppose that $u(a) = v(a)$; the construction in the case $u(b) = v(b)$ is similar. Then $u(t), v(t)$ are defined on $[a, b]$, $b - a = 2^{-n}$;

(5.12) $$u(t) < v(t) \text{ on } (a, b], \quad u(a) = v(a), \quad u'(a) = v'(a);$$

(5.13) $$|u'(t)|, |u''(t)|, |v'(t)|, |v''(t)| \leqq M_n;$$

(5.14) $$|u(t) - v(t)|, \quad |u'(t) - v'(t)| \leqq d_n \leqq \epsilon_n.$$

Let $m = m_n > 0$ be an integer to be specified below. For $i = 0, 1, \ldots,$ m, put

(5.15) $$u_i(t) = \frac{(m - i)u(t) + iv(t)}{m} = u(t) + [v(t) - u(t)]\frac{i}{m}$$

on $[a, b]$. Then $u_0(t) = u(t)$, $u_m(t) = v(t)$, and

(5.16) $$u(t) \leqq u_i(t) < u_{i+1}(t) \leqq v(t) \text{ on } (a, b],$$

(5.17) $$u_i(a) = u(a), u_i{}'(a) = u'(a) \text{ for } i = 0, 1, \ldots, m.$$

It is clear from (5.14) that

(5.18) $$|u_h - u_i| \leqq |h - i|\frac{d_n}{m} \leqq d_n,$$

(5.19) $$|u_i{}'| \leqq M_n, \ |u_h{}' - u_i{}'| \leqq |h - i|\frac{d_n}{m} \leqq d_n,$$

(5.20) $$|u_i{}''| \leqq M_n.$$

For $i = 0, 1, \ldots, m - 1$ and $c = \frac{1}{2}(a + b)$, put

(5.21) $v_i(t) = u_i(t) \sin^2 2^{n+1}\pi(t - c) + u_{i+1}(t) \cos^2 2^{n+1}\pi(t - c)$

on $[c, b]$, so that $c - a = b - c = 1/2^{n+1}$ implies that

(5.22) $u_i(t) < v_i(t) < u_{i+1}(t)$ on (c, b),

(5.23) $v_i = u_{i+1}, v_i' = u_{i+1}'$ at $t = c$,   and $v_i = u_i, v_i' = u_i'$ at $t = b$.

The relations in (5.23) involving derivatives follow from

(5.24) $v_i' = u_i' \sin^2 2^{n+1}\pi(t - c) + u_{i+1}' \cos^2 2^{n+1}\pi(t - c)$
$\qquad\qquad + 2^{n+1}\pi(u_i - u_{i+1}) \sin 2^{n+2}\pi(t - c)$.

From (5.24) and (5.18)–(5.20),

(5.25) $|v_i'| \leqq M_n + 2^{n+1}\pi \dfrac{d_n}{m}$, $|v_i''| \leqq M_n + (2^{n+2} + 2^{2n+3}\pi)\pi \dfrac{d_n}{m}$.

Also, by (5.21),
$$v_i - u_i = (u_{i+1} - u_i) \cos^2 2^{n+1}\pi(t - c),$$
$$v_i - u_{i+1} = (u_i - u_{i+1}) \sin^2 2^{n+1}\pi(t - c),$$

so that (5.18)–(5.19) give

(5.26) $|v_i - u_h| \leqq \dfrac{d_n}{m}$, $|v_i' - u_h'| \leqq (1 + 2^{n+1}\pi) \dfrac{d_n}{m}$ for $h = i, i + 1$.

Finally, let $m = m_n$ be chosen so large that

(5.27) $$(2^{n+2} + 2^{2n+3}\pi)\pi \dfrac{d_n}{m} < \dfrac{\epsilon_{n+1}}{3}.$$

In order to obtain S_{n+1} from S_n, let the arcs $u = u_i(t)$, $i = 0, \ldots, m$, on $[a, c]$ and the arcs $u = u_i(t)$, $i = 0, \ldots, m$, and $u = v_h(t)$, $h = 0, \ldots, m - 1$, on $[c, b]$ be inserted between $u = u(t)$, $u = v(t)$. It is clear from (5.19)–(5.20), (5.25)–(5.26), and (5.27) that the analogues of (5.8) and (5.10) hold if n is replaced by $n + 1$. Also the analogue of (5.11) follows from (5.19), (5.26), (5.27).

This completes the construction of the sequence $S_0 \subset S_1 \subset \ldots$. It is clear that $S = \bigcup S_n$ is dense in the (t, u)-plane.

The continuity of $U(t, u)$, given by (5.6), will now be considered. Let $p \geqq n \geqq 0$. The set of arcs S_n divide the plane into closed sets G of the form $G = \{(t, u) : \gamma \leqq t \leqq \delta, u^n(t) \leqq u \leqq v^n(t)\}$, where no point of S_n is interior to G; $u = u^n(t)$ and $u = v^n(t)$ on $[\gamma, \delta]$, $\delta - \gamma = 2/2^n$, are arcs each made up of two arcs of S_n; $u^n = v^n$ at $t = \gamma, \delta$; and $u^n < v^n$ on (γ, δ).

Let $(t_0, u_p) \in G \cap S_p$ and let (t^1, u^1) be any point of the boundary of G. The difference

$$\Delta_p = |U_p(t_0, u_p) - U_p(t^1, u^1)|$$

will be estimated. Consider first the case that $p = n$. Then (t_0, u_p) is on the boundary of G, say $u_p = u^n(t_0)$. Since $U_n(t, u^n(t)) = u^{n\prime}(t)$, it is seen by (5.8) that

$$|U_n(t_0, u_n) - U_n(\gamma, u(\gamma))| \leq M_n |t_0 - \gamma| \leq \frac{2M}{2^n} .$$

Thus, in the case that $p = n$, $\Delta_n \leq 4M/2^n$.

Let $p > n$. It can be supposed that $(t_0, u_p) \in S_p - S_{p-1}$. Let $u_n = u^n(t_0)$, and $u_n \leq u_{n+1} \leq \cdots \leq u_p$, where (t_0, u_j) is the highest point of the segment $t = t_0$, $u_{j-1} \leq u \leq u_p$ which in S_j, $j = n+1, \ldots, p$. Then, by (5.11),

$$|U_p(t_0, u_{j+1}) - U_p(t_0, u_j)| \leq d_j + \epsilon_{j+1} \leq 2\epsilon_j.$$

Hence

$$|U_p(t_0, u_p) - U_p(t_0, u_n)| \leq 2 \sum_{j=n}^{\infty} \epsilon_j.$$

If this is combined with $\Delta_n \leq 4M/2^n$, it follows that

$$\Delta_p \leq \eta_n \text{ where } \eta_n = \frac{4M}{2^n} + 2 \sum_{j=n}^{\infty} \epsilon_j.$$

Consider now two points (t_i, u_i), $i = 0, 1$, in $S_p, p \geq n$. Each of these points (t_i, u_i) is contained in a region $G = G_i$ of the type just considered. There exist points (t^i, u^i) on the boundary of G_i such that

$$|t^0 - t^1|^2 + |u^0 - u^1|^2 \leq |t_0 - t_1|^2 + |u_0 - u_1|^2$$

(where, e.g., $(t^0, u^0) = (t^1, u^1)$ if $G_0 = G_1$). Thus the above estimate for Δ_p implies that

$$|U(t_0, u_0) - U(t_1, u_1)| \leq 2\eta_n + |U_n(t^0, u^0) - U_n(t^1, u^1)|.$$

Since $U_n(t, u)$ is uniformly continuous on S_n, it follows from the last three formula lines that $U(t, u)$ is uniformly continuous on S. Hence $U(t, u)$ has a continuous extension, denoted also by $U(t, u)$, on the (t, u)-plane.

It will now be verified that (5.1) has the asserted property. It is clear that any continuous arc $u = u(t)$ on an interval $[c, d]$ made up of subarcs of S is a solution of (5.1). The Extension Theorem 3.1 or the case $d = 1$ of Theorem 4.1 shows that if (t_0, u_0) is any point of a set G of the type just considered, then (5.1) has a solution $u = u(t)$ over $[\gamma, \delta]$ satisfying $u = u^n = v^n$ at $t = \gamma, \delta$. Such a solution can be continued to the left of $t = \gamma$ [right of $t = \delta$] in a nonunique manner by using arcs of S_n. If n is sufficiently large, the interval $[\gamma, \delta]$ containing t_0 can be made arbitrarily small. This completes the verification.

Notes

For references and discussion of the history of existence theorems, see Painlevé [1], Vessiot [1], Müller [3], and Kamke [4, I, pp. 2, 33].

SECTION 1. Theorem 1.1 goes back to Cauchy (see Moigno [1]) and Lipschitz [1]. Its proof by successive approximations is due to Picard [1] and Lindelöf [1], although this method had been used earlier in special cases by Liouville and by Cauchy. The existence theorem in Exercise 1.1 is often associated with the names of Cauchy and Poincaré who used the method of majorants. A proof by successive approximations was given by von Escherich [2].

SECTION 2. Theorem 2.1 is due to Peano [2]. Simplifications in the proof have been made by Mie, de la Vallée Poussin, Arzela, Montel, and Perron. The proof in the text utilizes a device of Tonelli [1]. The polygonal approximations in Exercise 2.1 go back to Cauchy (and the method of Cauchy-Lipschitz).

The result of Exercise 2.3 is due to Ważewski [4] but the proof in the Hints may be new; cf. Nevanlinna [1] and Heinz [3]. Results of this type often involve a "degree of continuity" for $f_y(y)$ at $y = 0$; cf. Yamabe [1], Bartle [1], and Sternberg and Wintner [1].

SECTION 4. Theorem 4.1* is a theorem of H. Kneser [1]; the proof in the text is that of Müller [2]; for related results, see Pugh [1]. The suggestion given in Hints for Exercise 4.2 is due to C. C. Pugh (unpublished); cf. also Fukuhara and Nagumo [1]. Exercise 4.3(a) is a result of Kamke [2]. Exercise 4.4 is due to Fukuhara [1]; see also Kamke [2], Fukuhara [2], and Fukuhara and Nagumo [1]. A related example is given in Digel [1].

SECTION 5. The first example of this type was given by Lavrentieff [1]; the example in the text is that of Hartman [27].

---

*For a generalization and another proof, see Stampacchia [S1].

Chapter III

Differential Inequalities and Uniqueness

The most important techniques in the theory of differential equations involve the "integration" of differential inequalities. The first part of this chapter deals with basic results of this type which will be used throughout the book. In the second part of this chapter immediate applications are given, including the derivation of some uniqueness theorems.

In this chapter u, v, U, V are scalars; y, z, f, g are d-dimensional vectors.

1. Gronwall's Inequality

One of the simplest and most useful results involving an integral inequality is the following.

Theorem 1.1. *Let $u(t)$, $v(t)$ be non-negative, continuous functions on $[a, b]$; $C \geqq 0$ a constant; and*

(1.1)
$$v(t) \leqq C + \int_a^t v(s)u(s)\, ds \qquad \text{for } a \leqq t \leqq b.$$

Then

(1.2)
$$v(t) \leqq C \exp \int_a^t u(s)\, ds \qquad \text{for } a \leqq t \leqq b;$$

in particular, if $C = 0$, then $v(t) \equiv 0$.

For a generalization, see Corollary 4.4.

Proof. *Case* (i), $C > 0$. Let $V(t)$ denote the right side of (1.1), so that $v(t) \leqq V(t)$, $V(t) \geqq C > 0$ on $[a, b]$. Also, $V'(t) = u(t)v(t) \leqq u(t)V(t)$. Since $V > 0$, $V'/V \leqq u$, and $V(a) = C$, an integration over $[a, t]$ gives $V(t) \leqq C \exp \int_a^t u(s)\, ds$. Thus (1.2) follows from $v(t) \leqq V(t)$.

Case (ii), $C = 0$. If (1.1) holds with $C = 0$, then Case (i) implies (1.2) for every $C > 0$. The desired result follows by letting C tend to 0.

Exercise 1.1. Show that Theorem 1.1 implies the uniqueness assertion of Theorem II 1.1.

2. Maximal and Minimal Solutions

Let $U(t, u)$ be a continuous function on a plane (t, u)-set E. By a *maximal* solution $u = u^0(t)$ of

$$(2.1) \qquad u' = U(t, u), \qquad u(t_0) = u_0$$

is meant a solution of (2.1) on a maximal interval of existence such that if $u(t)$ is any solution of (2.1), then

$$(2.2) \qquad u(t) \leqq u^0(t)$$

holds on the common interval of existence of u, u^0. A minimal solution is similarly defined.

Lemma 2.1. *Let $U(t, u)$ be continuous on a rectangle $R: t_0 \leqq t \leqq t_0 + a$, $|u - u_0| \leqq b$; let $|U(t, u)| \leqq M$ and $\alpha = \min(a, b/M)$. Then (2.1) has a solution $u = u^0(t)$ on $[t_0, t_0 + \alpha]$ with the property that every solution $u = u(t)$ of $u' = U(t, u)$, $u(t_0) \leqq u_0$ satisfies (2.2) on $[t_0, t_0 + \alpha]$.*

In view of the proof of the Extension Theorem II 3.1, this lemma implies existence theorems for maximal and minimal solutions (which will be stated only for an open set E):

Theorem 2.1. *Let $U(t, u)$ be continuous on an open set E and $(t_0, u_0) \in E$. Then (2.1) has a maximal and a minimal solution.*

Proof of Lemma 2.1. Let $0 < \alpha' < \alpha$. Then, by Theorem II 2.1,

$$(2.3) \qquad u' = U(t, u) + 1/n, \qquad u(t_0) = u_0$$

has a solution $u = u_n(t)$ on an interval $[t_0, t_0 + \alpha']$ if n is sufficiently large. By Theorem I 2.4, there is a sequence $n(1) < n(2) < \cdots$ such that

$$(2.4) \qquad u^0(t) = \lim_{k \to \infty} u_{n(k)}(t)$$

exists uniformly on $[t_0, t_0 + \alpha']$ and is a solution of (2.1).

It will be verified that (2.2) holds on $[t_0, t_0 + \alpha']$. To this end, it is sufficient to verify

$$(2.5) \qquad u(t) \leqq u_n(t) \qquad \text{on } [t_0, t_0 + \alpha']$$

for all large fixed n. If (2.5) does not hold, there is a $t = t_1$, $t_0 < t_1 < t_0 + \alpha'$ such that $u(t_1) > u_n(t_1)$. Hence there is a largest t_2 on $[t_0, t_1)$, where $u(t_2) = u_n(t_2)$, so that $u(t) > u_n(t)$ on $(t_2, t_1]$. But (2.3) implies that $u_n'(t_2) = u'(t_2) + 1/n$, so that $u_n(t) > u(t)$ for $t(> t_2)$ near t_2. This contradiction proves (2.5). Since $\alpha' < \alpha$ is arbitrary, the lemma follows.

Remark. The uniqueness of the solution $u = u^0(t)$ shows that $u_n(t) \to u^0(t)$ uniformly on $[t_0, t_0 + \alpha']$ as $n \to \infty$ continuously.

3. Right Derivatives

The following simple lemmas will be needed subsequently.

Lemma 3.1. *Let* $u(t) \in C^1[a, b]$. *Then* $|u(t)|$ *has a right derivative* $D_R |u(t)|$ *for* $a \leqq t < b$, *where*

$$(3.1) \qquad D_R |u(t)| = \lim h^{-1}(|u(t + h)| - |u(t)|) \qquad \text{as } 0 < h \to 0,$$

and $D_R |u(t)| = u'(t) \operatorname{sgn} u(t)$ *if* $u(t) \neq 0$ *and* $D_R |u(t)| = |u'(t)|$ *if* $u(t) = 0$. *In particular,* $|D_R |u(t)|| = |u'(t)|$.

The assertion concerning $D_R |u(t)|$ is clear if $u(t) \neq 0$. The case when $u(t) = 0$ follows from $u(t + h) = h(u'(t) + o(1))$ as $h \to 0$, so that $|u(t + h)| = h(|u'(t)| + o(1))$ as $0 < h \to 0$.

Lemma 3.2. *Let* $y = y(t) \in C^1[a, b]$. *Then* $|y(t)|$ *has a right derivative* $D_R |y(t)|$ *and* $|D_R |y(t)|| \leqq |y'(t)|$ *for* $a \leqq t < b$.

Since $|y(t)| = \max (|y^1(t)|, \dots, |y^d(t)|)$, there are indices k such that $|y^k(t)| = |y(t)|$. In the following, k denotes any such index. By the last lemma, $|y^k(t)|$ has a right derivative, so that

$$|y^k(t + h)| = |y(t)| + h(D_R |y^k(t)| + o(1)) \qquad \text{as } 0 < h \to 0.$$

For small $h > 0$, $|y(t + h)| = \max_k |y^k(t + h)|$, so that by taking the $\max_k$ in the last formula line,

$$|y(t + h)| = |y(t)| + h(\max_k D_R |y^k(t)| + o(1)) \qquad \text{as } 0 < h \to 0.$$

Thus $D_R |y(t)|$ exists and is $\max_k D_R |y^k(t)|$. Since $|D_R |y^k(t)|| = |y^{k'}(t)| \leqq |y'(t)|$, Lemma 3.2 follows.

Exercise 3.1. Show that Lemma 3.2 is correct if $|y|$ is replaced by the Euclidean length of y.

4. Differential Inequalities

The next theorem concerns the integration of a differential inequality. It is one of the results which is used most often in the theory of differential equations.

Theorem 4.1. *Let* $U(t, u)$ *be continuous on an open* (t, u)-*set* E *and* $u = u^0(t)$ *the maximal solution of* (2.1). *Let* $v(t)$ *be a continuous function on* $[t_0, t_0 + a]$ *satisfying the conditions* $v(t_0) \leqq u_0$, $(t, v(t)) \in E$, *and* $v(t)$ *has a right derivative* $D_R v(t)$ *on* $t_0 \leqq t < t_0 + a$ *such that*

$$(4.1) \qquad \qquad D_R v(t) \leqq U(t, v(t)).$$

Then, on a common interval of existence of $u^0(t)$ *and* $v(t)$,

$$(4.2) \qquad \qquad v(t) \leqq u^0(t).$$

Remark 1. If the inequality (4.1) is reversed and $v(t_0) \geqq u_0$, then the conclusion (4.2) must be replaced by $v(t) \geqq u_0(t)$, where $u = u_0(t)$ is the minimal solution of (2.1). Correspondingly, if in Theorem 4.1 the function $v(t)$ is continuous on an interval $t_0 - \alpha \leqq t \leqq t_0$ with a left derivative $D_L v(t)$ on $(t_0 - \alpha, t_0]$ satisfying $D_L v(t) \leqq U(t, v(t))$ and $v(t_0) \geqq u_0$, then again (4.2) must be replaced by $v(t) \geqq u_0(t)$.

Remark 2. It will be clear from the proof that Theorem 4.1 holds if the "right derivative" D_R is replaced by the "upper right derivative" where the latter is defined by replacing "lim" by "lim sup" in (3.1).

Proof of Theorem 4.1. It is sufficient to show that there exists a $\delta > 0$ such that (4.2) holds for $[t_0, t_0 + \delta]$. For if this is the case and $u^0(t), v(t)$ are defined on $[t_0, t_0 + \beta]$, it follows that the set of t-values where (4.2) holds cannot have an upper bound different from β.

Let $n > 0$ be large and let $\delta > 0$ be chosen independent of n such that (2.3) has a solution $u = u_n(t)$ on $[t_0, t_0 + \delta]$. In view of the proof of Lemma 2.1, it is sufficient to verify that $v(t) \leqq u_n(t)$ on $[t_0, t_0 + \delta]$, but the proof of this is identical to the proof of (2.5) in § 2.

Corollary 4.1. *Let $v(t)$ be continuous on $[a, b]$ and possess a right derivative $D_R v(t) \leqq 0$ on $[a, b]$. Then $v(t) \leqq v(a)$.*

Corollary 4.2. *Let $U(t, u), u^0(t)$ be as in Theorem 4.1. Let $V(t, u)$ be continuous on E and satisfy*

$$(4.3) \qquad V(t, u) \leqq U(t, u).$$

Let $v = v(t)$ be a solution of

$$(4.4) \qquad v' = V(t, v), \qquad v(t_0) = v_0(\leqq u_0)$$

on an interval $[t_0, t_0 + a]$. Then (4.2) holds on any common interval of existence of $v(t)$ and $u^0(t)$ to the right of $t = t_0$.

It is clear from Remark 1 that if $v(t)$ is extended to an interval to the left of $t = t_0$, then, on such an interval, (4.2) must be replaced by $v(t) \geqq u_0(t)$ where $u_0(t)$ is a minimal solution of (2.1) with $u_0 \geqq v(t_0)$.

Corollary 4.3. *Let $u^0(t)$ be the maximal solution of $u' = U(t, u), u(t_0) = u^0;$ $u = u_0(t)$ the minimal solution of*

$$(4.5) \qquad u' = -U(t, u), \qquad u(t_0) = u_0(\geqq 0).$$

Let $y = y(t)$ be a C^1 vector-valued function on $[t_0, t_0 + \alpha]$ such that $u_0 \leqq |y(t_0)| \leqq u^0, (t, |y(t)|) \in E$ and

$$(4.6) \qquad |y'(t)| \leqq U(t, |y(t)|)$$

on $[t_0, t_0 + \alpha]$. Then the first [second] of the two inequalities

$$(4.7) \qquad u_0(t) \leqq |y(t)| \leqq u^0(t)$$

holds on any common interval of existence of $u_0(t)$ and y $[u^0(t)$ and $y]$.

This is an immediate consequence of Theorem 4.1 and Remark 1 following it, since $|y(t)|$ has a right derivative satisfying $-|y'(t)| \leq D_R |y(t)| \leq |y'(t)|$ by Lemma 3.2. (In view of Exercise 3.1, this corollary remains valid if $|y|$ denotes the Euclidean norm.)

Exercise 4.1. (*a*) Let $f(t, y)$ be continuous on the strip $S : a \leq t \leq b$, y arbitrary, and let $f^k(t, y^1, \dots, y^d)$ be nondecreasing with respect to each of the components y^i, $i \neq k$, of y. Assume that the solution of the initial value problem $y' = f(t, y)$, $y(a) = y_0$ is unique for a fixed y_0, and that this solution $y = y(t)$ exists on $[a, b]$. Let $z(t) = (z^1(t), \dots, z^d(t))$ be continuous on $[a, b]$ such that $z^k(t)$ has a right derivative for $k = 1, \dots, d$, $z^k(a) \leq y_0^k$ and $D_R z^k(t) \leq f^k(t, z(t))$ for $a \leq t \leq b$ [or $z^k(a) \geq y_0^k$ and $D_R z^k(t) \geq f^k(t, z(t))$ for $a \leq t \leq b$]. Then $z^k(t) \leq y^k(t)$ [or $z^k(t) \geq y^k(t)$] for $a \leq t \leq b$. (This is applicable if $g(t, y)$ is continuous on S, $z(t)$ is a solution of $z' = g(t, z)$ and $z^k(a) \leq y_0^k$, $g^k(t, y) \leq f^k(t, y)$ on S [or $z^k(a) \geq y_0^k$, $g^k(t, y) \geq f^k(t, y)$ on S].) See Remark in Exercise 4.3.

(*b*) If, in part (*a*), all initial value problems associated with $y' = f(t, y)$ have unique solutions, $f^k(t, y)$ is increasing with respect to y^i, $i \neq k$ and $k = 1, \dots, d$, and $z^j(a) < y_0^j$ [or $z^j(a) > y_0^j$] for at least one index j, then $z^k(t) < y_0^k(t)$ [or $z^k(t) > y_0^k(t)$] for $a < t \leq b$, $k = 1, \dots, d$.

(*c*) If, in addition to the assumptions of (*a*), there is an index h such that $f^h(t, y)$ is nondecreasing with respect to y^h, then $y_0^h(t) - z^h(t)$ is nondecreasing [or nonincreasing] on $a \leq t \leq b$.

(*d*) If the assumptions of (*b*) and (*c*) hold, then $y_0^h(t) - z^h(t)$ is increasing [or decreasing] on $a \leq t \leq b$.

(*e*) Let u, U denote real-valued scalars and $y = (y^1, \dots, y^d)$ a real d-dimensional vector. Let $U(t, y)$ be continuous for $a \leq t \leq b$ and arbitrary y such that solutions of $u^{(d)} = U(t, u, u', \dots, u^{(d-1)})$ are uniquely determined by initial conditions and that $U(t, y^1, \dots, y^d)$ is nondecreasing with respect to each of the first $d - 1$ components y^j, $j = 1, \dots, d - 1$, of y. Let $u_1(t)$, $u_2(t)$ be two solutions of $u^{(d)} = U$ on $[a, b]$ satisfying $u_1^{(j)}(a) \leq u_2^{(j)}(a)$ for $j = 0, \dots, d - 1$. Then $u_1^{(j)}(t) \leq u_2^{(j)}(t)$ for $j = 0, \dots, d - 1$ and $a \leq t \leq b$; furthermore, $u_2^{(j)}(t) - u_1^{(j)}(t)$ is nondecreasing for $j = 0, \dots, d - 2$ and $a \leq t \leq b$.

Exercise 4.2. Let $f(t, y)$, $g(t, y)$ be continuous on a strip, $a \leq t \leq b$, and y arbitrary, such that $f^k(t, y) < g^k(t, y)$ for $k = 1, \dots, d$ and that, for each $k = 1, \dots, d$, either $f^k(t, y^1, \dots, y^d)$ or $g^k(t, y^1, \dots, y^d)$ is nondecreasing with respect to y^i, $i \neq k$. On $a \leq t \leq b$, let $y = y(t)$ be a solution of $y' = f(t, y)$, $y(a) = y_0$ and $z = z(t)$ a solution of $z' = g(t, z)$, $z(a) = z_0$, where $y_0^k \leq z_0^k$ for $k = 1, \dots, d$. Then $y^k(t) \leq z^k(t)$ for $a \leq t \leq b$.

Exercise 4.3. Let $f(t, y)$ be continuous for $t_0 \leq t \leq t_0 + a$, $|y - y_0| \leq b$ such that $f^k(t, y^1, \dots, y^d)$ is nondecreasing with respect to each y^i,

$i \neq k$. Show that $y' = f(t, y), y(t_0) = y_0$ has a maximal [minimal] solution $y_0(t)$ with the property that if $y = y(t)$ is any other solution, then $y^k(t) \leq y_0^k(t)$ $[y^k(t) \geq y_0^k(t)]$ holds on the common interval of existence. Remark: The assumption in Exercise 4.1(a) that the solution of the initial value problem $y' = f(t, y), y(a) = y_0$, is unique can be dropped if $y(t)$ is replaced by the maximal solution [or minimal solution] $y_0(t)$.

Exercise 4.4.   Let $f(t, y), g(t, y)$ be linear in y, say $f^k(t, y) = \Sigma a_{kj}(t)y^j + f^k(t)$ and $g^k(t, y) = \Sigma b_{kj}(t)y^j + g^k(t)$, where $a_{kj}(t), b_{kj}(t), f^k(t), g^k(t)$ are continuous for $a \leq t \leq b$. Let $y(t), z(t)$ be solutions of $y' = f(t, y)$, $y(a) = y_0$ and $z' = g(t, z), z(a) = z_0$, respectively. (These solutions exist on $[a, b]$; cf. Corollary 5.1.)   What conditions on $a_{jk}(t), b_{jk}(t), f^k(t), g^k(t)$, y_0, z_0 imply that $|z^k(t)| \leq y^k(t)$ on $[a, b]$ for $k = 1, \ldots, d$?

Theorem 4.1 has an "integrated" analogue which, however, requires the monotony of U with respect to u. This theorem is a generalization of Theorem 1.1:

Corollary 4.4.   *Let $U(t, u)$ be continuous and nondecreasing with respect to u for $t_0 \leq t \leq t_0 + a$, u arbitrary. Let the maximal solution $u = u^0(t)$ of (2.1) exist on $[t_0, t_0 + a]$. On $[t_0, t_0 + a]$, let $v(t)$ be a continuous function satisfying*

$$(4.8) \qquad v(t) \leq v_0 + \int_{t_0}^t U(s, v(s)) \, ds,$$

where $v_0 \leq u_0$. Then $v(t) \leq u^0(t)$ holds on $[t_0, t_0 + a]$.

Proof.   Let $V(t)$ be the right side of (4.8), so that $v(t) \leq V(t)$, and $V'(t) = U(t, v(t))$. By the monotony of U, $V'(t) \leq U(t, V(t))$. Hence Theorem 4.1 implies that $V(t) \leq u^0(t)$ on $[t_0, t_0 + a]$; thus $v(t) \leq u^0(t)$ holds.

Exercise 4.5.   Corollary 4.4 is false if we omit: $U(t, u)$ is nondecreasing in u.

Exercise 4.6.   State the analogue of Corollary 4.4 for the case that the constant v_0 in (4.8) is replaced by a continuous function $v_0(t)$.

Exercise 4.7.   Let y, f, z be d-dimensional vectors; $f(t, y)$ continuous for $t_0 \leq t \leq t_0 + a$ and y arbitrary such that $f^k(t, y^1, \ldots, y^d)$ is nondecreasing with respect to each $y^j, j = 1, \ldots, d$. Let the maximal solution $y_0(t)$ of $y' = f(t, y), y(t_0) = y_0$ exist on $[t_0, t_0 + a]$; cf. Exercise 4.3. Let $z(t)$ be a continuous vector-valued function such that $z^k(t) \leq y_0^k + \int_{t_0}^t f^k(s, z(s)) \, ds$ for $t_0 \leq t \leq t_0 + a$. Then $z^k(t) \leq y_0^k(t)$ on $[t_0, t_0 + a]$.

5. A Theorem of Wintner

Theorem 4.1 and its corollaries can be used to help find intervals of existence of solutions of some differential equations.

Theorem 5.1.   *Let $U(t, u)$ be continuous for $t_0 \leq t \leq t_0 + a, u \geq 0$, and let the maximal solution of (2.1), where $u_0 \geq 0$, exist on $[t_0, t_0 + a]$, e.g.,*

*let $U(t, u) = \psi(u)$, where $\psi(u)$ is a positive, continuous function on $u \geq 0$
such that*

(5.1)
$$\int^{\infty} du/\psi(u) = \infty.$$

*Let $f(t, y)$ be continuous on the strip $t_0 \leq t \leq t_0 + a$, y arbitrary, and
satisfy*

(5.2)
$$|f(t, y)| \leq U(t, |y|).$$

Then the maximal interval of existence of solutions of

(5.3)
$$y' = f(t, y), \qquad y(t_0) = y_0,$$

where $|y_0| \leq u_0$, is $[t_0, t_0 + a]$.

Remark 1. It is clear that (5.2) is only required for large $|y|$. Admissible choices of $\psi(u)$ are, for example, $\psi(u) = Cu$, $Cu \log u$, ... for large u and a constant C.

Proof. (5.2) implies the inequality (4.6) on any interval on which $y(t)$ exists. Hence, by Corollary 4.3, the second inequality in (4.7) holds on such an interval and so the main assertion follows from Corollary II 3.1.

In order to complete the proof, it has to be shown that the function $U(t, u) = \psi(u)$ satisfies the condition that the maximal solution of

(5.4)
$$u' = \psi(u), \qquad u(t_0) = u_0 (\geq 0)$$

exists on $[t_0, t_0 + a]$ by virtue of (5.1). Since $\psi > 0$, (5.4) implies that for ·any solution $u = u(t)$,

(5.5)
$$t - t_0 = \int_{t_0}^{t} u'(s)\,ds/\psi(u(s)) = \int_{u_0}^{u(t)} du/\psi(u).$$

Note that $\psi > 0$ implies that $u'(t) > 0$ and $u(t) > 0$ for $t > t_0$. By Corollary II 3.1, the solution $u(t)$ can fail to exist on $[t_0, t_0 + a]$ only if it exists on some interval $[t_0, \delta)$ and satisfies $u(t) \to \infty$ as $t \to \delta$ ($\leq a$). If this is the case, however, $t \to \delta$ in (5.5) gives a contradiction for the left side tends to $\delta - t_0$ and the right side to ∞ by (5.1). This completes the proof.

Remark 2. The type of argument in the proof of Theorem 5.1 supplies a priori estimates for solutions $y(t)$ of (5.3). For example, if $\psi(u)$ is the same as in the last part of Theorem 5.1, let

$$\Psi(u) = \int_{u_0}^{u} ds/\psi(s) \qquad \text{for } u \geq u_0$$

and let $u = \Phi(v)$ be the function inverse to $v = \Psi(u)$. Then $|f(t, y)| \leq \psi(|y|)$ implies that a solution $y(t)$ of (5.3) satisfies

$$|y(t)| \leq \Phi(t - t_0) \qquad \text{for } t_0 \leq t \leq t_0 + a;$$

cf. (5.5).

Exercise 5.1.   Let $f(t, y)$ be continuous on the strip $t_0 \leqq t \leqq t_0 + a$, y arbitrary.   Let $|f(t, y)| \leqq \varphi(t)\psi(|y|)$, where $\varphi(t) \geqq 0$ is integrable on $[t_0, t_0 + a]$ and $\psi(u)$ is a positive continuous function on $u \geqq 0$ satisfying (5.1).   Show that the assertion of Theorem 5.1 and an analogue of Remark 2 are valid.

Corollary 5.1.   *If $A(t)$ is a continuous $d \times d$ matrix function and $g(t)$ a continuous vector function for $t_0 \leqq t \leqq t_0 + a$, then the (linear) initial value problem*

$$(5.6) \qquad y' = A(t)y + g(t), \qquad y(t_0) = y_0$$

has a unique solution $y = y(t)$, and $y(t)$ exists on $t_0 \leqq t \leqq t_0 + a$.

This is a consequence of Theorem II 1.1 and Theorem 5.1 with the choice of $\psi(u) = C(1 + u)$ for some large C.

In a scalar case, Theorem 5.1 can be "read backwards":

Corollary 5.2.   *Let $U(t, u)$, $V(t, u)$ be continuous functions satisfying (4.3) on $t_0 \leqq t \leqq t_0 + a$, u arbitrary.   Let some solution $v = v(t)$ of (4.4) on $[t_0, \delta)$, $\delta \leqq t_0 + a$, satisfy $v(t) \to \infty$ as $t \to \delta$.   Then the maximal solution $u = u^0(t)$ of (2.1) has a maximal interval of existence $[a, \omega_+)$, where $\omega_+ \leqq \delta$, and $u^0(t) \to \infty$ as $t \to \omega_+$.*

6. Uniqueness Theorems

One of the principal uses of Theorem 4.1 and its corollaries is to obtain uniqueness theorems. The following result is often called Kamke's general uniqueness theorem.

Theorem 6.1.   *Let $f(t, y)$ be continuous on the parallelepiped $R: t_0 \leqq t \leqq t_0 + a$, $|y - y_0| \leqq b$.   Let $\omega(t, u)$ be a continuous (scalar) function on $R_0: t_0 < t \leqq t_0 + a$, $0 \leqq u \leqq 2b$, with the properties that $\omega(t, 0) = 0$ and that the only solution $u = u(t)$ of the differential equation*

$$(6.1) \qquad u' = \omega(t, u)$$

on any interval $(t_0, t_0 + \epsilon]$ satisfying

$$(6.2) \qquad u(t) \to 0 \quad \text{and} \quad \frac{u(t)}{t - t_0} \to 0 \qquad \text{as } t \to t_0 + 0$$

is $u(t) \equiv 0$.   For (t, y_1), $(t, y_2) \in R$ with $t > t_0$, let

$$(6.3) \qquad |f(t, y_1) - f(t, y_2)| \leqq \omega(t, |y_1 - y_2|).$$

Then the initial value problem

$$(6.4) \qquad y' = f(t, y), \qquad y(t_0) = y_0$$

has at most one solution on any interval $[t_0, t_0 + \epsilon]$.

In Theorem 6.1, we can also conclude uniqueness for initial value problems $y' = f(t, y)$, $y(t_1) = y_1$ for $t_1 \neq t_0$. Theorem 6.1 remains valid if Euclidean norms are employed.

Exercise 6.1. Show that Theorem 6.1 is false if (6.2) is replaced by $u(t)$, $u'(t) \to 0$ as $t \to t_0 + 0$.

Proof. The fact that

(6.5) $$\omega(t, 0) = 0 \quad \text{for } t_0 < t \leqq t_0 + a$$

implies of course that $u(t) \equiv 0$ is a solution of (6.1).

Suppose that, for some $\epsilon > 0$, (6.4) has two distinct solutions $y = y_1(t)$, $y_2(t)$ on $t_0 \leqq t \leqq t_0 + \epsilon$. Let $y(t) = y_1(t) - y_2(t)$. By decreasing ϵ, if necessary, it can be supposed that $y(t_0 + \epsilon) \neq 0$ and $|y(t_0 + \epsilon)| < 2b$. Also $y(t_0) = y'(t_0) = 0$. By (6.3), $|y'(t)| \leqq \omega(t, |y(t)|)$ on $(t_0, t_0 + \epsilon]$. It follows from Corollary 4.3 (and the Remark 1 following Theorem 4.1) that if $u = u_0(t)$ is the minimal solution of the initial value problem $u' = \omega(t, u)$, $u(t_0 + \epsilon) = |y(t_0 + \epsilon)|$, where $0 < |y(t_0 + \epsilon)| < 2b$, then

(6.6) $$|y(t)| \geqq u_0(t)$$

on any subinterval of $(t_0, t_0 + \epsilon]$ on which $u_0(t)$ exists; see Figure 1.

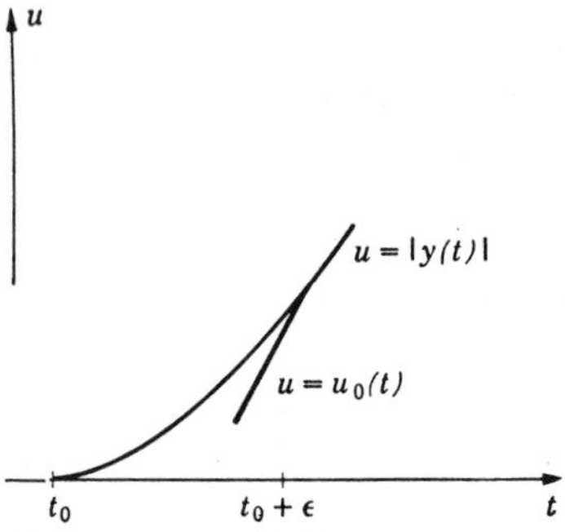

Figure 1.

By the proofs of the Extension Theorem II 3.1 and Lemma 2.1, $u_0(t)$ can be extended, as the minimal solution, to the left until $(t, u_0(t))$ approaches arbitrarily close to a point of ∂R_0 for some t-values. During the extension (6.6) holds, so that $(t, u_0(t))$ comes arbitrarily close to some point $(\delta, 0) \in \partial R_0$ for certain t-values, where $\delta \geqq t_0$. If $\delta > t_0$, then (6.5) shows that $u_0(t)$ has an extension over $(t_0, t_0 + \epsilon]$ with $u_0(t) = 0$ for $(t_0, \delta]$. Thus, in any case, the left maximum interval of existence of $u_0(t)$ is $(t_0, t_0 + \epsilon]$. It follows from (6.6) that $u_0(t) \to 0$ and $u_0(t)/(t - t_0) \to 0$ as $t \to t_0 + 0$. By the assumption concerning (6.1), $u_0(t) \equiv 0$. Since this contradicts $u_0(t_0 + \epsilon) = |y(t_0 + \epsilon)| \neq 0$, the theorem follows.

Corollary 6.1 (Nagumo's Criterion). *If* $t_0 = 0$, *then* $\omega(t, u) = u/t$ *is admissible in Theorem 6.1 (i.e., the conclusion of Theorem 6.1 holds if* (6.3) *is replaced by*

(6.7) $$|f(t, y_1) - f(t, y_2)| \leqq \frac{|y_1 - y_2|}{t - t_0}$$

for (t, y_1), $(t, y_2) \in R$ *with* $t > t_0$).

Exercise 6.2. The function $\omega(t, u) = u/t$ in Corollary 6.1 cannot be replaced by $\omega(t, u) = Cu/t$ for any constant $C > 1$. Show that if $C > 1$, then there exist continuous real-valued functions $f(t, y)$ on $0 \leqq t \leqq 1$, $|y| \leqq 1$ with the properties that

$$|f(t, y_1) - f(t, y_2)| \leqq \frac{C\,|y_1 - y_2|}{t} \qquad \text{for } t > 0,$$

but that $y' = f(t, y)$, $y(0) = 0$ has more than one solution.

Corollary 6.2 (Osgood's Criterion). *If $t_0 = 0$, then $\omega(t, u) = \varphi(t)\psi(u)$ is admissible in Theorem 6.1 if $\varphi(t) \geqq 0$ is continuous for $0 < t \leqq a$; $\psi(u)$ is continuous for $u \geqq 0$ and $\psi(0) = 0$, $\psi(u) > 0$ if $u > 0$; and $\int_{+0} \varphi(t)\, dt < \infty, \int_{+0} du/\psi(u) = \infty$.*

Actually, the continuity condition on $\varphi(t)$ in this corollary can be weakened. The analogous uniqueness theorem can be proved directly if $\varphi(t)$ is only assumed to be integrable over $0 < t \leqq a$.

Exercise 6.3 [Generalization of Corollaries 6.1 and (6.2)]. Let $t_0 = 0$.
(*a*) If $\varphi(t) \geqq 0$ is continuous for $0 < t \leqq a$, show that $\omega(t, u) = \varphi(t)u$ is admissible in Theorem 6.1 if and only if $\lim \inf \left[\int_t^a \varphi(s)\, ds + \log t \right] < \infty$
as $t \to +0$. (*b*) Let $\varphi(t) \geqq 0$ be continuous for $0 < t \leqq a$; $\psi(u)$ continuous for $0 \leqq u \leqq 2b$, $\psi(0) = 0$, $\psi(u) > 0$ for $0 < u \leqq b$, and $\int_{+0} du/\psi(u) = \infty$. Show that $\omega(t, u) = \varphi(t)\psi(u)$ is admissible in Theorem 6.1 if, for every $C > 0$, $\lim \sup t^{-1}\Phi\left(C + \int_t^a \varphi(s)\, ds \right) > 0$ as $t \to 0$, where $u = \Phi(v)$ is the function inverse to $\Psi(u) = \int_u^{2b} ds/\psi(s)$.

Exercise 6.4. Let $\psi(u)$ be continuous for $|u| \leqq 1$, $\psi(0) = 0$. Show that the initial value problem $u' = \psi(u)$, $u(0) = 0$ has a unique solution $u(t) \equiv 0$ unless there exists an ϵ, $0 < \epsilon \leqq 1$, such that either $\psi(u) \geqq 0$ for $0 \leqq u \leqq \epsilon$ and $1/\psi(u)$ is (Lebesgue) integrable over $[0, \epsilon]$ or $\psi(u) \leqq 0$ for $-\epsilon \leqq u \leqq 0$ and $1/\psi(u)$ is (Lebesgue) integrable over $[-\epsilon, 0]$.

Exercise 6.5. Let f, ω be as in Theorem 6.1. Show that there exists a function $\omega_0(t, u)$ which is continuous on the *closure* of R_0[is nondecreasing with respect to u if ω is], and satisfies the conditions on $\omega(t, u)$; thus $\omega_0(t, 0) \equiv 0$; the only solution of $u' = \omega_0(t, u)$ and $u(t_0) = 0$ on any interval $[t_0, t_0 + \epsilon]$ is $u(t) \equiv 0$; and $|f(t, y_1) - f(t, y_2)| \leqq \omega_0(t, |y_1 - y_2|)$. (Note that, since ω_0 is continuous on the closure of R_0, any solution of $u' = \omega_0(t, u)$ on $(t_0, t_0 + \epsilon]$ satisfying (6.2) is necessarily continuously differentiable and is the usual type of solution on $[t_0, t_0 + \epsilon]$.)

Exercise 6.6. (*a*) Let $\epsilon_0, \ldots, \epsilon_{d-1}$ be non-negative constants such that $\epsilon_0 + \cdots + \epsilon_{d-1} = 1$. Let $U(t, y) = U(t, y^1, \ldots, y^d)$ be a real-valued continuous function on $R: 0 \leqq t \leqq a$ and $|y^k| \leqq b$ for $k = 1, \ldots, d$

such that $|U(t, y_1) - U(t, y_2)| \leqq \sum_{k=1}^{d} \epsilon_{k-1}(d - k + 1)! t^{-(d-k+1)} |y_1{}^k - y_2{}^k|$ if $t > 0$. Show that the dth order (scalar) equation $u^{(d)} = U(t, u, u', \ldots, u^{(d-1)})$ has at most one solution (on any interval $0 \leqq t \leqq \epsilon \leqq a$) satisfying given initial conditions $u(0) = u_0$, $u' = u_0', \ldots, u^{(d-1)}(0) = u_0^{(d-1)}$, where $u_0, u_0', \ldots, u^{(d-1)}$ are d given numbers on the range $|u| \leqq b$. (b) Note that part (a) remains correct if the constants $\epsilon_0, \ldots, \epsilon_{d-1}$ are replaced by continuous non-negative functions $\epsilon_0(t), \ldots, \epsilon_{d-1}(t)$ such that $\epsilon_0(t) + \cdots + \epsilon_{d-1}(t) \leqq 1$.

Exercise 6.7.   (a) Let $f(t, y)$ be continuous for $R: 0 \leqq t \leqq a$, $|y| \leqq b$. On $R_0: 0 < t \leqq a$, $|u| \leqq 2b$, let $\omega_1(t, u)$, $\omega_2(t, u)$ be continuous non-negative functions which are nondecreasing in u for fixed t, satisfy $\omega_j(t, 0) = 0$, and

$$|f(t, y_1) - f(t, y_2)| \leqq \omega_j(t, |y_1 - y_2|) \qquad \text{for } j = 1, 2.$$

Let there exist continuous non-negative functions $\alpha(t)$, $\beta(t)$ for $0 \leqq t \leqq a$ satisfying $\alpha(0) = \beta(0) = 0$, $\beta(t) > 0$ for $0 < t < a$, and $\alpha(t)/\beta(t) \to 0$ as $t \to 0$. Suppose that each solution $u(t)$ of $u' = \omega_1(t, u)$ for small $t > 0$ with the property that $u(t) \to 0$ as $t \to 0$ satisfies $u(t) \leqq \alpha(t)$ on its interval of existence. Finally, suppose that the only solution of $v' = \omega_2(t, v)$ for small $t > 0$ satisfying $v(t)/\beta(t) \to 0$ as $t \to 0$ is $v(t) \equiv 0$. Then the initial value problem $y' = f(t, y)$, $y(0) = 0$ has exactly one solution. (b) Prove that $\omega_1(t, u) = Cu^\lambda$, $\omega_2(t, u) = ku/t$ are admissible if $k > 0$, $0 < \lambda < 1$, $k(1 - \lambda) < 1$ with $\alpha(t) = C(1 - \lambda)t^{1/(1-\lambda)}$, $\beta(t) = t^k$.

The following involves a "one-sided inequality" and gives "one-sided uniqueness."

Theorem 6.2   *Let $f(t, y)$ be continuous for $t_0 \leqq t \leqq t_0 + a$, $|y - y_0| \leqq b$. Considering y, f to be Euclidean vectors, suppose that*

$$(6.8) \qquad [f(t, y_2) - f(t, y_1)] \cdot (y_2 - y_1) \leqq 0$$

for $t_0 \leqq t \leqq t_0 + a$ and $|y_i - y_0| \leqq b$, $i = 1, 2$, where the dot denotes scalar multiplication. Then (6.4) has at most one solution on any interval $[t_0, t_0 + \epsilon]$, $\epsilon > 0$.

When it is desired to obtain uniqueness theorems for intervals $[t_0 - \epsilon, t_0]$. it is necessary to assume the reverse inequality in (6.8).

Corollary 6.3.   *Let $U(t, u)$ be a continuous real-valued function for $t_0 \leqq t \leqq t_0 + a$, $|u - u_0| \leqq b$ which is nonincreasing with respect to u (for fixed t). Then the initial value problem $u' = U(t, u)$, $u(t_0) = u_0$ has at most one solution on any interval $[t_0, t_0 + \epsilon]$, $\epsilon > 0$.*

Proof of Theorem 6.2.   Let $y = y_1(t)$, $y_2(t)$ be solutions of (6.4) on $[t_0, t_0 + \epsilon]$. Let $\delta(t) = \|y_2(t) - y_1(t)\|^2 = (y_2 - y_1) \cdot (y_2 - y_1)$ be the square of the Euclidean length of $y_2(t) - y_1(t)$, so that $\delta(t_0) = 0$, $\delta(t) \geqq 0$.

But $\delta'(t) = 2(y_2' - y_1') \cdot (y_2 - y_1) \leqq 0$ by (6.8). Hence $\delta(t) = 0$ on $[t_0, t_0 + \epsilon]$ as was to be proved.

Exercise 6.8 (One-sided Generalization of Nagumo's Criterion and of Theorem 6.2). Theorem 6.2 remains valid if condition (6.8) is relaxed to

$$[f(t, y_2) - f(t, y_1)] \cdot (y_2 - y_1) \leqq \frac{\|y_2 - y_1\|^2}{t - t_0}$$

for $t_0 < t \leqq t_0 + a$ [or if the right side is replaced by $\frac{1}{2}\omega(t, \|y_2 - y_1\|^2)$ or $\|y_1 - y_2\| \omega(t, \|y_1 - y_2\|)$ with ω as in Theorem 6.1].

7. van Kampen's Uniqueness Theorem

In the following uniqueness theorem, conditions are imposed on a family of solutions rather than on $f(t, y)$ in

$$(7.1) \qquad\qquad y' = f(t, y), \qquad y(t_0) = y_0.$$

Theorem 7.1. *Let $f(t, y)$ be continuous on a parallelepiped $R: t_0 \leqq t \leqq t_0 + a, |y - y_0| \leqq b$. Let there exist a function $\eta(t, t_1, y_1)$ on $t_0 \leqq t, t_1 \leqq t_0 + a, |y_1 - y_0| \leqq \beta(< b)$ with the properties (i) that, for a fixed (t_1, y_1), $y = \eta(t, t_1, y_1)$ is a solution of*

$$(7.2) \qquad\qquad y' = f(t, y), \qquad y(t_1) = y_1;$$

(ii) that $\eta(t, t_1, y_1)$ is uniformly Lipschitz continuous with respect to y_1; finally, (iii) that no two solution arcs $y = \eta(t, t_1, y_1)$, $y = \eta(t, t_2, y_2)$ pass through the same point (t, y) unless $\eta(t, t_1, y_1) \equiv \eta(t, t_2, y_2)$ for $t_0 \leqq t \leqq t_0 + a$. Then $y = \eta(t, t_0, y_0)$ is the only solution of (7.1) for $t_0 \leqq t_1 \leqq t_0 + a, |y_1 - y_0| \leqq \beta$.

Exercise 7.1. Show that the existence of a continuous $\eta(t, t_1, y_1)$ satisfying (i) and (iii) [but not (ii)] does not imply the uniqueness of the solution of (7.1).

Exercise 7.2. When $f(t, y)$ is uniformly Lipschitz continuous with respect to y, it can be shown that a function $y = \eta(t, t_1, y_1)$ satisfying the conditions of the theorem exists (for small $\beta > 0$); e.g., cf. Exercise II 1.2. Show that the converse is not correct, i.e., the existence of $\eta(t, t_1, y_1)$ satisfying (i)–(iii) does not imply that $f(t, y)$ is uniformly Lipschitz continuous with respect to y (for y near y_0).

Proof. Let $y(t)$ be any solution of (7.1). It will be shown that $y(t) = \eta(t, t_0, y_0)$ for small $t - t_0 \geqq 0$.

Condition (ii) means that there exists a constant K such that

$$(7.3) \qquad\qquad |\eta(t, t_1, y_1) - \eta(t, t_1, y_2)| \leqq K |y_1 - y_2|$$

for $t_0 \leqq t, t_1 \leqq t_0 + a$ and $|y_1 - y_0| \leqq \beta, |y_2 - y_0| \leqq \beta$.

Let $|f(t, y)| \leqq M$ on R. Then any solution $y = y(t)$ of (7.1) satisfies $|y(t) - y_0| \leqq M(t - t_0) \leqq \frac{1}{2}\beta$ if $t_0 \leqq t \leqq t_0 + \beta/2M$. Thus $\eta(t, s, y(s))$ is defined and $|\eta(t, s, y(s)) - y(s)| \leq M |t - s| \leq \tfrac{1}{2}\beta$ if $t_0 \leq t,\ s \leq t_0 + \beta/2M$. Hence

(7.4) $|\eta(t, s, y(s)) - y_0| \leq \beta$ if $t_0 \leq t,\ s \leq t_0 + \gamma$,

where $\gamma = \min(a, \beta/2M)$. Condition (iii) means that any point on any of the arcs $y = \eta(t, t_1, y_1)$ can be used to determine this arc. Thus (7.3),

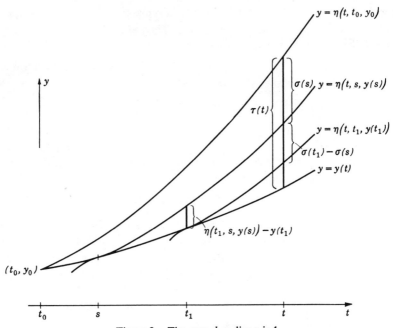

Figure 2.   The case $d = \dim y$ is 1.

with $y_1 = y(t_1)$ and $y_2 = \eta(t_1, s, y(s))$, implies that

(7.5) $|\eta(t, t_1, y(t_1)) - \eta(t, s, y(s))| \leq K |y(t_1) - \eta(t_1, s, y(s))|$

if $t_0 \leq t,\ t_1,\ s \leq t_0 + \gamma$; cf. Figure 2.

Let t be fixed on $t_0 \leq t \leq t_0 + \gamma$. It will be shown that

(7.6) $\tau(t) \equiv \eta(t, t_0, y_0) - y(t) = 0.$

To this end, put

(7.7) $\sigma(s) = \eta(t, t_0, y_0) - \eta(t, s, y(s))$ for $t_0 \leq s \leq t\ (\leq t_0 + \gamma)$,

so that $\sigma(t_0) = 0$ and $\sigma(t) = \tau(t)$. Then (7.5) and (7.7) imply that

(7.8) $|\sigma(t_1) - \sigma(s)| \leq K |y(t_1) - \eta(t_1, s, y(s))|.$

Since $y = \eta(t, s, y(s))$ is a solution of $y' = f$ through the point $(s, y(s))$, it is seen that $\eta(t_1, s, y(s)) = y(s) + (t_1 - s) [f(s, y(s)) + o(1)]$ as $t_1 \to s$. Also, $y(t_1) = y(s) + (t_1 - s) [f(s, y(s)) + o(1)]$ as $t_1 \to s$. Hence (7.8) gives $\sigma(t_1) - \sigma(s) = Ko(1) |t_1 - s|$ as $t_1 \to s$; i.e., $d\sigma/ds$ exists and is 0. Thus $\sigma(s)$ is the constant $\sigma(t_0) = 0$ for $t_0 \leqq s \leqq t$. In particular, $\tau(t) = \sigma(t)$ satisfies (7.6), as was to be proved.

Exercise 7.3 (*One-sided Analogue of Theorem* 7.1). Let $f(t, y)$ be continuous on $R: t_0 \leqq t \leqq t_0 + a$, $|y - y_0| \leqq b$. Let there exist a function $\eta(t, t_1, y_1)$ on $t_0 \leqq t_1 \leqq t \leqq t_0 + a$, $|y_1 - y_0| \leqq \beta(< b)$ with the properties (i) that, for fixed (t_1, y_1), $y = \eta(t, t_1, y_1)$ is a solution of (7.2) and (ii) that there exists a constant K such that for max $(t_1, t_2) \leqq t^* \leqq t \leqq t_0 + a$.

$$|\eta(t, t_1, y_1) - \eta(t, t_2, y_2)| \leqq K |\eta(t^*, t_1, y_1) - \eta(t^*, t_2, y_2)|.$$

Then $y = \eta(t, t_1, y_1)$ is the only solution of (7.2) for sufficiently small intervals $[t_1, t_1 + \epsilon]$, $\epsilon > 0$, to the right of t_1 (but not necessarily to the left of t_1).

8. Egress Points and Lyapunov Functions

Let $f(t, y)$ be continuous on an open (t, y)-set Ω and let Ω_0 be an open subset of Ω. Let $\partial\Omega_0$ and $\bar{\Omega}_0$ denote the boundary and closure of Ω_0, respectively. A point $(t_0, y_0) \in \partial\Omega_0 \cap \Omega$ is called an *egress point* [or an *ingress point*] of Ω_0 with respect to the system

$$(8.1) \qquad\qquad y' = f(t, y)$$

if, for some solution $y = y(t)$ of (8.1) satisfying $y(t_0) = y_0$, there exists an $\epsilon > 0$ such that $(t, y(t)) \in \Omega_0$ for $t_0 - \epsilon < t < t_0$ [or for $t_0 < t < t_0 + \epsilon$]. If, in addition, $(t, y(t)) \notin \bar{\Omega}_0$ for $t_0 < t < t_0 + \epsilon$ [or for $t_0 - \epsilon < t < t_0$] for a small $\epsilon > 0$ for *every* such $y(t)$, then (t_0, y_0) is called a *strict egress point* [or *strict ingress point*]. A point $(t_0, y_0) \in \partial\Omega_0 \cap \Omega$ will be referred to as a *nonegress point* if it is not an egress point.

Lemma 8.1. *Let $f(t, y)$ be continuous on an open set Ω and Ω_0 an open subset of Ω such that $\partial\Omega_0 \cap \Omega$ is either empty or consists of nonegress points. Let $y(t)$ be a solution of (8.1) satisfying $(t^0, y(t^0)) \in \Omega_0$ for some t^0. Then $(t, y(t)) \in \Omega_0$ on a right maximal interval of existence $[t^0, \omega_+)$.*

If the conclusion is false, there is a least value $t_0(> t^0)$ of t, where $(t_0, y(t_0)) \in \partial\Omega_0 \cap \Omega$. But then $(t_0, y(t_0))$ is an egress point, which contradicts the assumption and proves the lemma.

Let $u(t, y)$ be a real-valued function defined in a vicinity of a point $(t_1, y_1) \in \Omega$. Let $y(t)$ be a solution of (8.1) satisfying $y(t_1) = y_1$. If $u(t, y(t))$ is differentiable at $t = t_1$, this derivative is called the *trajectory derivative* of u at (t_1, y_1) along $y = y(t)$ and is denoted by $\dot{u}(t_1, y_1)$. When $u(t, y)$ has continuous partial derivatives, its trajectory derivative exists and can be calculated without finding solutions of (8.1). In fact,

$$(8.2) \qquad \dot{u}(t, y) = \partial u/\partial t + (\text{grad } u) \cdot f(t, y),$$

where the dot denotes scalar multiplication and grad $u = (\partial u/\partial y^1, \ldots, \partial u/\partial y^d)$ is the gradient of u with respect to y.

Let $(t_0, y_0) \in \partial\Omega_0 \cap \Omega$ and let $u(t, y)$ be a function of class C^1 on a neighborhood N of (t_0, y_0) in Ω such that $(t, y) \in \Omega_0 \cap N$ if and only if $u(t, y) < 0$. Then a necessary condition for (t_0, y_0) to be an egress point is that $\dot{u}(t_0, y_0) \geq 0$ and a sufficient condition for (t_0, y_0) to be a strict egress point is that $\dot{u}(t_0, y_0) > 0$. Further, a sufficient condition for (t_0, y_0) to be a nonegress point is that $\dot{u}(t, y) \leq 0$ for $(t, y) \in \Omega_0$.

When the system under consideration

$$(8.3) \qquad\qquad y' = f(y),$$

is autonomous (i.e., when the right side does not depend on t), definitions are similar. For example, let $f(y)$ be continuous on an open y-set Ω, Ω_0 an open subset of Ω, and $y_0 \in \partial\Omega_0 \cap \Omega$. The point y_0 is called an egress point of Ω_0 with respect to (8.3) if, for some solution $y(t)$ of (8.3) satisfying $y(0) = y_0$, there exists an $\epsilon > 0$ such that $y(t) \in \Omega_0$ for $-\epsilon < t < 0$. If, in addition, $y(t) \notin \bar{\Omega}_0$ for $0 < t < \epsilon$ for some $\epsilon > 0$, then y_0 is called a strict egress point. A lemma analogous to Lemma 8.1 is clearly valid here.

For an application of these notions, consider a function $f(y)$ defined on an open set containing $y = 0$. A function $V(y)$ defined on a neighborhood of $y = 0$ is called a *Lyapunov function* if (i) it has continuous partial derivatives; (ii) $V(y) \gtreqless 0$ according as $|y| \gtreqless 0$; and (iii) the trajectory derivative of V satisfies $\dot{V}(y) \leq 0$.

Theorem 8.1. *Let $f(y)$ be continuous on an open set containing $y = 0$, $f(0) = 0$, and let there exist a Lyapunov function $V(y)$. Then the solution $y \equiv 0$ of (8.3) is stable (in the sense of Lyapunov).*

Lyapunov stability of the solution $y \equiv 0$ means that if $\epsilon > 0$ is arbitrary, then there exists a $\delta_\epsilon > 0$ such that if $|y_0| < \delta_\epsilon$, then a solution $y(t)$ of (8.3) satisfying the initial condition $y(0) = y_0$ exists and satisfies $|y(t)| < \epsilon$ for $t \geq 0$. If in addition, $y(t) \to 0$ as $t \to \infty$, then the solution $y \equiv 0$ of (8.3) is called *asymptotically stable* (in the sense of Lyapunov). Roughly speaking, Lyapunov stability of $y \equiv 0$ means that if a solution $y(t)$ starts near $y = 0$ it remains near $y = 0$ in the future ($t \geq 0$); and Lyapunov asymptotic stability of $y \equiv 0$ means that, in addition, $y(t) \to 0$ as $t \to \infty$.

Proof. Let $\epsilon > 0$ be any number such that the set $|y| \leq \epsilon$ is in the open set on which f and V are defined. For any $\eta > 0$, let $\delta(\eta)$ be chosen so that $0 < \delta(\eta) < \epsilon$ and $V(y) < \eta$ if $|y| < \delta(\eta)$.

Reference to Figure 3 will clarify the following arguments. Since $V(y)$ is continuous and positive on $|y| = \epsilon$, there is an $\eta = \eta_\epsilon > 0$ such that

$V(y) > \eta$ for $|y| = \epsilon$. Let Ω_0 be the open set $\{y : |y| < \epsilon,\ V(y) < \eta\}$. The boundary $\partial\Omega_0$ is contained in the set $\{y : |y| < \epsilon,\ V(y) = \eta\}$. The function $u(y) = V(y) - \eta$ satisfies $u(y) < 0$ at a point y, $|y| < \epsilon$, if and only if $y \in \Omega_0$. Clearly $\dot{u} = \dot{V} \leq 0$. Hence, no point of $\partial\Omega_0$ is an egress point. Consequently, by the analogue of Lemma 8.1, a solution $y(t)$ of (8.3) satisfying $y(0) \in \Omega_0$ remains in Ω_0 on its right maximal interval of existence $[0, \omega_+)$. Since Ω_0 is contained in the sphere $|y| \leq \epsilon$ in Ω, it follows that $\omega_+ = \infty$; Corollary II 3.2.

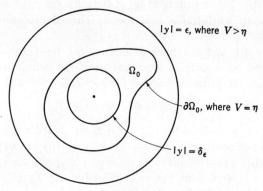

Figure 3.

Finally, put $\delta_\epsilon = \delta(\eta_\epsilon) > 0$, so that $V(y) < \eta$ if $|y| < \delta_\epsilon < \epsilon$. Thus $|y(0)| < \delta_\epsilon$ implies that $y(0) \in \Omega_0$, hence $y(t)$ exists and $y(t) \in \Omega_0$ for $t \geq 0$. In particular, $|y(t)| < \epsilon$ for $t \geq 0$. This proves the theorem.

Exercise 8.1. Let $f(y)$ be continuous on an open set containing $y = 0$ and let $f(0) = 0$. Let (8.3) possess a continuous first integral $V(y)$ [i.e., a function which is constant along solutions $y = y(t)$ of (8.3)] such that $V(y)$ has a strict extremum (maximum or minimum) at $y = 0$. Then the solution $y \equiv 0$ of (8.3) is stable.

Theorem 8.2. *If, in Theorem 8.1, $\dot{V}(y) \lessgtr 0$ according as $|y| \gtrless 0$, then the solution $y \equiv 0$ of (8.3) is asymptotically stable (in the sense of Lyapunov).*

Proof. Use the notation of the last proof. Let $y(t)$ be a solution of (8.3) with $|y(0)| < \delta_\epsilon$. Since $\dot{V} \leq 0$, it follows that $V(y(t))$ is nonincreasing and tends monotonically to a limit, say $c \geq 0$, as $t \to \infty$.

Suppose first that $c = 0$. Then $y(t) \to 0$ as $t \to \infty$. For otherwise, there is an $\epsilon_0 > 0$ such that $\epsilon_0 \leq |y(t)| \leq \epsilon$ for certain large t-values. But there exists a constant $m_0 > 0$ such that $V(y) > m_0$ for $\epsilon_0 \leq |y| \leq \epsilon$; thus $V(y(t)) > m_0 > 0$ for certain large t-values. This is impossible; hence, $y(t) \to 0$ as $t \to \infty$.

Suppose, if possible, that $c > 0$, so that $0 < c \leq \eta$ and $V(y) < \frac{1}{2}c$ if $|y| < \delta(\frac{1}{2}c) < \epsilon$. Hence $|y(t)| \geq \delta(\frac{1}{2}c)$ for large t. But the assumption on

$\dot{V}$ implies that there exists an $m > 0$ such that $\dot{V}(y) \leqq -m < 0$ if $\delta(\tfrac{1}{2}c) \leqq$ $|y| \leqq \epsilon$. In particular, $\dot{V}(y(t)) \leqq -m < 0$, for all large t. This is impossible. Hence $c = 0$ and $y(t) \to 0$ as $t \to \infty$. This proves the theorem.

A result analogous to Theorem 8.1 in which the conclusion is that the solution $y = 0$ is not stable is given by the following:

Exercise 8.2. Let $f(y)$ be continuous on an open set E containing $y = 0$ and let $f(0) = 0$. Let there exist a function $V(y)$ on E satisfying $V(0) = 0$, having continuous partial derivatives and a trajectory derivative such that $\dot{V}(y) \leqq 0$ according as $|y| \geqq 0$ on E. Let $V(y)$ assume negative values for some y arbitrarily near $y = 0$. Then the solution $y \equiv 0$ is not (Lyapunov) stable.

Theorems 8.1 and 8.2 have analogues for nonautonomous systems which depend on a suitable modification of the definition of Lyapunov function: Let $f(t, y)$ be continuous for $t \geqq T$, $|y| \leqq b$ and satisfy

$$(8.4) \qquad f(t, 0) = 0 \qquad \text{for } t \geqq T.$$

A function $V(t, y)$ defined for $t \geqq T$, $|y| \leqq b$ is called a *Lyapunov function* if (i) $V(t, y)$ has continuous partial derivatives; (ii) $V(t, 0) = 0$ for $t \geqq T$ and there exists a continuous function $W(y)$ on $|y| \leqq b$ such that $W(y) \geqq 0$ according as $|y| \geqq 0$, and $V(t, y) \geqq W(y)$ for $t \geqq T$; (iii) the trajectory derivative of V satisfies $\dot{V}(t, y) \leqq 0$.

Theorem 8.3. *Let* $f(t, y)$ *be continuous for* $t \geqq T$, $|y| \leqq b$ *and satisfy* (8.4). *Let there exist a Lyapunov function* $V(t, y)$. *Then the solution* $y \equiv 0$ *of* (8.1) *is uniformly stable* (*in the sense of Lyapunov*).

Here, *Lyapunov stability* means that if $\epsilon > 0$ is arbitrary, then there exists a $\delta_\epsilon > 0$ and a $t_\epsilon \geqq T$ such that if $y(t)$ is a solution of (8.1) satisfying $|y(t^0)| < \delta_\epsilon$ for some $t^0 \geqq t_\epsilon$, then $y(t)$ exists and $|y(t)| < \epsilon$ for all $t \geqq t^0$. If, in addition, $y(t) \to 0$ as $t \to \infty$, then the solution $y \equiv 0$ is called *Lyapunov asymptotically stable*. The modifier "uniform" for "stability" or "asymptotic stability" means that t_ϵ can be chosen to be T for all $\epsilon > 0$.

Theorem 8.4. *Let* $f(t, y)$, $V(t, y)$ *be as in Theorem* 8.3. *In addition, assume that there exists a continuous* $W_1(y)$ *for* $|y| \leqq b$ *such that* $W_1(y) \geqq 0$ *according as* $|y| \geqq 0$ *and that* $\dot{V}(t, y) \leqq -W_1(y)$ *for* $t \geqq T$. *Then the solution* $y \equiv 0$ *of* (8.1) *is uniformly asymptotic stable* (*in the sense of Lyapunov*).

Exercise 8.3. (*a*) Prove Theorem 8.3. (*b*) Prove Theorem 8.4.

9. Successive Approximations

The proof of Theorem II 1.1 suggests the question as to whether or not a solution of

$$(9.1) \qquad y' = f(t, y), \qquad y(t_0) = y_0$$

can always be obtained as the limit of the sequence (or a subsequence) of the successive approximations defined in § II 1. That the answer is in the negative is shown by the following example for a scalar initial value problem

$$(9.2) \qquad u' = U(t, u), \qquad u(0) = 0,$$

where $U(t, u)$ will be defined for $t \geq 0$ and all u.

Consider the approximations $u_0(t) \equiv 0$ and

$$u_{n+1}(t) = \int_0^t U(s, u_n(s))\, ds \qquad \text{if } n \geq 0.$$

Let $U(t, 0) = -2t$, hence $u_1(t) = -t^2$; put $U(t, -t^2) = 0$, hence $u_2(t) = 0$. Then $u_{2n}(t) = 0$ and $u_{2n+1}(t) = -t^2$ for $n \geq 0$. It only remains to complete the definition of $U(t, u)$ as a continuous function to obtain the desired example.

One possible completion of this definition is to let $U(t, u) = -2t$ if $u \geq 0$, $U(t, u) = 0$ if $u \leq -t^2$, and to be a linear function of u when $-t^2 \leq u \leq 0, t > 0$ fixed. In this way, we obtain an example in which $U(t, u)$ is nonincreasing with respect to u (for fixed $t \geq 0$). In this case, the solution of (9.2) is unique (Corollary 6.3) although no subsequence of the successive approximations converge to a solution.

If the solutions of (9.1) are unique by virtue of Theorem 6.1 with a monotone ω, then successive approximations converge to a solution.

Theorem 9.1.[*] *Let R, R_0, f, ω be as in Theorem 6.1. Let $|f(t, y)| \leq M$ on R and $\alpha = \min (a, b/M)$. Let $\omega(t, u)$ be nondecreasing with respect to u. Then $y_0(t) = y_0$ and*

$$(9.3) \qquad y_n(t) = y_0 + \int_{t_0}^t f(s, y_{n-1}(s))\, ds \qquad \text{if } n \geq 1,$$

are defined and converge uniformly on $[t_0, t_0 + \alpha]$ to the solution $y = y(t)$ of (9.1).

Proof. By Exercise 6.5, it can be supposed that $\omega(t, u)$ is continuous on the closure of R_0 and is nondecreasing with respect to u for fixed t.

The sequence of approximations (9.3) are uniformly bounded and equicontinuous on $[t_0, t_0 + \alpha]$ and hence possesses uniformly convergent subsequences. If it is known that $y_n(t) - y_{n-1}(t) \to 0$ as $n \to \infty$, then (9.3) implies that the limit of any such subsequence is the unique solution $y(t)$ of (9.1). It then follows that the full sequence $y_0, y_1, \ldots$ converges uniformly to $y(t)$; cf. Remark 2 following Theorem I 2.3. Thus, in order to prove Theorem 9.1, it suffices to verify that $\lambda(t) \equiv 0$, where

$$(9.4) \qquad \lambda(t) = \lim \sup |y_n(t) - y_{n-1}(t)| \qquad \text{as } n \to \infty.$$

---

*See Notes, page 44.

Since $|f| \leq M$ on R,

$$|y_n(t_1) - y_{n-1}(t_1)| \leq |y_n(t_2) - y_{n-1}(t_2)| + 2M |t_1 - t_2|.$$

The right side is at most $\lambda(t_2) + \epsilon + 2M |t_1 - t_2|$ for large n if $\epsilon > 0$. Hence $\lambda(t_1) \leq \lambda(t_2) + \epsilon + 2M |t_1 - t_2|$. Since $\epsilon > 0$ is arbitrary and t_1, t_2 can be interchanged, $|\lambda(t_1) - \lambda(t_2)| \leq 2M |t_1 - t_2|$. In particular, $\lambda(t)$ is continuous for $t_0 \leq t \leq t_0 + \alpha$.

By the relation (9.3),

$$y_{n+1}(t) - y_n(t) = \int_{t_0}^{t} [f(s, y_n(s)) - f(s, y_{n-1}(s))] \, ds.$$

Hence, by (6.3),

$$|y_{n+1}(t) - y_n(t)| \leq \int_{t_0}^{t} \omega(s, |y_n(s) - y_{n-1}(s)|) \, ds.$$

For a fixed t on the range $t_0 < t \leq t_0 + \alpha$, there is a sequence of integers $n(1) < n(2) < \ldots$ such that $|y_{n+1}(t) - y_n(t)| \to \lambda(t)$ as $n = n(k) \to \infty$ and that $\lambda_1(s) = \lim |y_n(s) - y_{n-1}(s)|$ exists uniformly on $t_0 \leq s \leq t_0 + \alpha$ as $n = n(k) \to \infty$. Thus,

$$\lambda(t) \leq \int_{t_0}^{t} \omega(s, \lambda_1(s)) \, ds.$$

Since $\lambda_1(s) \leq \lim \sup |y_n(s) - y_{n-1}(s)| = \lambda(s)$ and $\omega(t, u)$ is monotone in u,

$$\lambda(t) \leq \int_{t_0}^{t} \omega(s, \lambda(s)) \, ds.$$

By Corollary 4.4, $\lambda(t) \leq u_0(t)$, where $u_0(t)$ is the maximal solution of

$$u' = \omega(t, u), \qquad u(t_0) = 0.$$

Since this initial value problem has the unique solution $u_0(t) \equiv 0$, it follows that $\lambda(t) \equiv 0$. This proves the theorem.

Exercise 9.1. Show that under the conditions of Exercise 6.7(a), the successive approximations $y_0(t) = 0$ and (9.3), where $t_0 = 0$ and $y_0 = 0$, converge uniformly on $0 \leq t \leq \min (a, b/M)$ to the solution of $y' = f(t, y)$, $y(0) = 0$.

Exercise 9.2. For two vectors, $y = (y^1, \ldots, y^d)$ and $z = (z^1, \ldots, z^d)$, use the notation $y \geq z$ if $y^k \geq z^k$ for $k = 1, \ldots, d$. Let $f = (f^1, \ldots, f^d)$ and $y = (y^1, \ldots, y^d)$. Assume that $f(t; y)$ is continuous on $R: 0 \leq t \leq a$, $|y| \leq b$ and that $f(t, y_1) \leq f(t, y_2)$ if $y_1 \leq y_2$. (a) Define two sequences of successive approximations $y_{0\pm}(t), y_{1\pm}(t), \ldots$ on $0 \leq t \leq \alpha = \min (a, b/M)$, where $y_{0\pm}(t) = \pm M(1, \ldots, 1)t$ and

$$y_{n\pm}(t) = \int_{0}^{t} f(s, y_{n-1\pm}(s)) \, ds \text{ for } n = 1, 2, \ldots .$$

Show that $y_{0+}(t) \geqq y_{1+}(t) \geqq \ldots$ and $y_{0-}(t) \leqq y_{1-}(t) \leqq \ldots$ and that both sequences converge uniformly to solutions of $y' = f(t, y)$, $y(0) = 0$. (b) Show that $y_{0\pm}(t)$ can be replaced by continuous functions $y_{0\pm}(t)$ on $0 \leqq t \leqq \alpha$ satisfying $|y_{0\pm}(t)| \leqq b$ and

$$y_{0+}(t) \geqq \int_0^t f(s, y_{0+}(s))\,ds, \qquad y_{0-}(t) \leqq \int_0^t f(s, y_{0-}(s))\,ds$$

(e.g., $y_{0-}(t) \equiv y_0$ is admissible if $f(t, y_0) \geqq 0$).

Exercise 9.3. (a) Using the notation $y \geqq z$ introduced in Exercise 9.2, let $f(t, y)$ be continuous for $t \geqq 0$ and all y and satisfy $f(t, y_1) \leqq f(t, y_2)$ if $y_1 \leqq y_2$. Let $y(t)$ be a solution of $y' = -f(t, y)$ satisfying $y(t) \leqq y(0)$ for $t \geqq 0$; cf., e.g., § XIV 2. Consider the successive approximations $y_0(t)$, $y_1(t), \ldots$ defined by $y_0(t) \equiv y(0)$, $y_n(t) = y(0) - \int_0^t f(s, y_{n-1}(s))\,ds$ for $n = 1, 2, \ldots$. Let $z_n(t)$ denote the "error" $z_n(t) = y_n(t) - y(t)$. Show that $(-1)^n z_n(t) \geqq 0$ for $n = 0, 1, \ldots$ and $(-1)^n z_n'(t) \geqq 0$ for $n = 1, 2, \ldots$ and $t \geqq 0$. (Convergence of the successive approximations is not asserted.) (b) Let $E_n(t) = \sum_{m=0}^{n} (-1)^m t^m/m!$ be the nth partial sum of the MacLaurin series for e^{-t}. Show that $(-1)^n(E_n(t) - e^{-t}) \geqq 0$ for $n = 0, 1, \ldots$ and $t \geqq 0$.

Exercise 9.4. Let $U(t, u)$ be real-valued and continuous for $t \geqq 0$ and arbitrary u and $U(t, u)$ nondecreasing with respect to u for fixed t. Let u_0, u_0' be fixed numbers and $u(t)$ a solution of $u'' = -U(t, u)$. Define successive approximations for $u(t)$ by putting $u_0(t) = u_0 + u_0' t$ and

$$u_n(t) = u_0(t) - \int_0^t (t - s)U(s, u_{n-1}(s))\,ds \qquad \text{for } n = 1, 2, \ldots .$$

Then $u_0(t)$, $u_1(t), \ldots$ are defined for $t \geqq 0$. (a) Suppose that $u(t)$ satisfies $u(t) \leqq u_0 + u_0' t$ on its right maximal interval of existence $[0, \omega_+)$. Show that $\omega_+ = \infty$ and that the "error" $v_n(t) = u_n(t) - u(t)$ satisfies $(-1)^n v_n(t) \geqq 0$, $(-1)^n v_n'(t) \geqq 0$ for $n = 1, 2, \ldots$ and $t \geqq 0$. (Convergence of the successive approximations is not asserted.) (b) Let $C_n(t) = \sum_{m=0}^{n} (-1)^m t^{2m}/(2m)!$ and $S_n(t) = \sum_{m=0}^{n} (-1)^m t^{2m+1}/(2m + 1)!$ be the nth partial sums of the Maclaurin series for $\cos t$ and $\sin t$, respectively. Show that $(-1)^n [C_n(t) - \cos t] \geqq 0$ and $(-1)^n [S_n(t) - \sin t] \geqq 0$ for $n = 0, 1, \ldots$ and $t \geqq 0$. (c) Let $U(t, u) = q(t)u$, where $q(t) \geqq 0$ is continuous and nondecreasing for $t \geqq 0$. Using Theorem XIV 3.1$_\infty$ and the remarks following it, show that (a) is applicable if $u_0 \geqq 0$ and $u_0' \geqq 0$ [i.e., show that $u(t) \leqq u_0 + u_0' t$ for $t \geqq 0$].

Notes

SECTION 1. Theorem 1.1 goes back essentially to Peano [1]. A special case was stated and proved by Gronwall [1]; a slightly more general form of the theorem (which is contained in Corollary 4.4) is given by Reid [1, p. 290]. The proof in the text is that of Titchmarsh [1, pp. 97–98].

SECTION 2. Maximal and minimal solutions were considered by Peano [1]; see Perron [4].

SECTION 4. Differential inequalities of the type (4.1) occur in the work of Peano [1] and of Perron [4]. Theorem 4.1 and its proof are taken from Kamke [1] and are essentially due to Peano. Exercises 4.2 and 4.3 are results of Kamke [2]; see Ważewski [7]. A special case of Corollary 4.4 is given by Bihari [1]. Exercise 4.6 is a result of Opial [1].

SECTION 5. Results of the type in Theorem 5.1 and Exercise 5.1 were first given by Wintner [1], [4].

SECTION 6. Theorem 6.1 is due to Kamke [1]. An earlier version, in which it is assumed that $\omega(t, u)$ is continuous also for $t = 0$, was given by Perron [6]. (Exercise 6.5, due to Olech [2], shows that, in a certain sense, Perron's theorem is not less general than Kamke's.) For the case $d = 1$, earlier results of the type of Perron's were given by Bompiani [1] and Iyanaga [1]. For Exercise 6.1, see Szarski [1]. For Corollary 6.1, see Nagumo [1]; a less sharp form was first proved by Rosenblatt [1] with $\omega(t, u) = Cu/t$ and $0 < C < 1$. An example of the type required in Exercise 6.2 was given by Perron [8]. For Corollary 6.2, see Osgood [1]. For Exercise 6.3(a), see Lévy [1, pp. 46–47]. For Exercise 6.4, see Wallach [1]. For Exercise 6.5, see Olech [2]. For a particular case of Exercise 6.6, see Wintner [22]. For Exercise 6.7, part (a), see F. Brauer [1], who generalized the result of part (b) due to Krasnosel'skiĭ and S. G. Krein [1].

For other uniqueness theorems related to those of this section, see F. Brauer and S. Sternberg [1]. These involve estimates for a function $V(t, |y_2(t) - y_1(t)|)$ instead of $|y_1(t) - y_2(t)|$. For earlier references on the subject of uniqueness theorems, see Müller [3] and Kamke [4, pp. 2 and 33].

SECTION 7. Theorem 7.1 is a result of van Kampen [2].

SECTION 8. The terminology "egress point" and "ingress point" is that of Wazewski [5]. Exercise 8.1 is due to Dirichlet [1]; it was first given by Lagrange [1, pp. 36–44] under the assumption that $V(y)$ is analytic and that the Hessian matrix $(\partial^2 V/\partial y^i \, \partial y^j)$ of V at $y = 0$ is definite. This result is the forerunner of Lyapunov's Theorem 8.1. Theorems 8.1 and 8.2, Exercise 8.2, and Theorems 8.3 and 8.4 are due to Lyapunov [2] (and constitute the basis for his "direct" or "second" method); cf. LaSalle and Lefschetz [1]. For references and recent developments on this subject, see W. Hahn [1], Antosiewicz [1], Massera [2], and Krasovskiĭ [4].

SECTION 9. The example of nonconvergent successive approximations is due to Müller [1]. Theorem 9.1,* as stated, is due to Viswanatham [1]. Earlier versions and special cases are to be found in Rosenblatt [1], van Kampen [3] (cf. also Haviland [1]), Dieudonné [1], Wintner [2], LaSalle [1], Coddington and Levinson [1], and Ważewski [8]. The reduction of the proof of Theorem 9.1 to the verification that $\lambda(t) \equiv 0$ is due to Dieudonné (and independently to Wintner) and is used by the authors following them. Exercise 9.1 is a result of F. Brauer [1] and generalizes Luxemburg [1]. Exercise 9.2(a) is due to Müller [1]; cf. LaSalle [1] for part (b). For Exercise 9.3(a), cf. Hartman and Wintner [16]. For Exercise 9.4, cf. Wintner [16].

---

*If we omit "ω is nondecreasing in u", Theorem 9.1 remains valid when d = dim $y > 1$, but not if d = 1; Evans and Feroe [S1].

Chapter IV

Linear Differential Equations

In this chapter, u, v, p are scalars; c, y, z, f, g are (column) d-dimensional vectors; and A, B, Y, Z are matrices. The scalars, components of the vectors, and elements of the matrices will be supposed to be complex-valued.

1. Linear Systems

This chapter will be concerned with some elementary facts about linear systems of differential equations in the homogeneous case,

$$(1.1) \qquad y' = A(t)y,$$

and in the inhomogeneous case,

$$(1.2) \qquad y' = A(t)y + f(t).$$

Throughout this chapter, $A(t)$ is a continuous $d \times d$ matrix and $f(t)$ a continuous vector on a t-interval $[a, b]$. Recall the following fundamental fact stated as Corollary III 5.1.

Lemma 1.1. *The initial value problem* (1.2) *and*

$$(1.3) \qquad y(t_0) = y_0,$$

$a \leqq t_0 \leqq b$, *has a unique solution* $y = y(t)$ *and* $y(t)$ *exists on* $a \leqq t \leqq b$.

The fact that the elements of $A(t)$ and components of y are complex-valued does not affect the applicability of Corollary III 5.1. For example, (1.2) is equivalent to a real linear system for a $2d$-dimensional vector made up of the real and imaginary parts of the components of y. Actually, the simplest proof of Lemma 1.1 is a direct one employing the standard successive approximations:

Exercise 1.1. Prove Lemma 1.1 by using successive approximations. (This proof also gives the majorization $|y(t)| \leqq e^{K|t-t_0|}|y_0|$ if K denotes a constant such that $|A(t)y| \leqq K|y|$ for all vectors y and $a \leqq t \leqq b$; cf. (4.2) below.)

Exercise 1.2. Let $A(t) = (a_{jk}(t))$ be (not necessarily continuous but) integrable over $[a, b]$; i.e., let the entries $a_{jk}(t)$ be Lebesgue integrable over $[a, b]$. Show by successive approximations that Lemma 1.1 remains correct. Here a solution $y(t)$ is interpreted as a continuous solution of the integral equation

$$y(t) = y_0 + \int_{t_0}^{t} A(s)y(s)\, ds$$

or, equivalently, $y(t)$ satisfies (1.3) and is absolutely continuous on $[a, b]$ with its derivative $y'(t)$ satisfying (1.1) except on a null set (i.e., a set of Lebesgue measure 0).

The uniqueness of solutions of (1.1), (1.3) implies that

Corollary 1.1. *If* $y = y(t)$ *is a solution of* (1.1) *and* $y(t_0) = 0$ *for some* t_0, $a \leqq t_0 \leqq b$, *then* $y(t) \equiv 0$.

For the solutions of (1.1) and (1.2), there is the obvious theorem:

Theorem 1.1 (Principles of Superposition)   (i) *Let* $y = y_1(t)$, $y_2(t)$ *be solutions of* (1.1), *then any linear combination* $y = c_1 y_1(t) + c_2 y_2(t)$ *with constant coefficients* c_1, c_2 *is a solution of* (1.1). (ii) *If* $y = y_1(t)$ *and* $y = y_0(t)$ *are solutions of* (1.1) *and* (1.2), *respectively, then* $y = y_0(t) + y_1(t)$ *is a solution of* (1.2); *conversely, if* $y = y_0(t)$, $y^0(t)$ *are solutions of* (1.2), *then* $y = y_0(t) - y^0(t)$ *is a solution of* (1.1).

The vector equation (1.1) can be replaced by a matrix differential equation,

$$(1.4) \qquad\qquad Y' = A(t)Y,$$

where Y is matrix with d rows and k (arbitrary) columns. It is clear that a matrix $Y = Y(t)$ is a solution of (1.4) if and only if each column of $Y(t)$, when considered as a column vector, is a solution of (1.1).

Corollary 1.1 and the principle of superposition imply that if $Y = Y(t)$ is a $d \times k$ matrix solution of (1.4), then rank $Y(t)$ does not depend on t. That is, if $y_1(t), \ldots, y_r(t)$ are r solutions of (1.1), then the constant vectors $y_1(t_0), \ldots, y_r(t_0)$ are linearly independent for *some* t_0 if and only if they are linearly independent for *every* t_0 on $a \leqq t_0 \leqq b$.

Below, unless otherwise specified only $d \times d$ matrix solutions $Y = Y(t)$ of (1.4) will be considered. In this case, either det $Y(t) \equiv 0$ or det $Y(t) \neq 0$ for all t. This fact can be strengthened as follows:

Theorem 1.2 (Liouville).   *Let* $Y = Y(t)$ *be a* $d \times d$ *matrix solution of* (1.4), $\Delta(t) = \det Y(t)$, *and* $a \leqq t_0 \leqq b$. *Then, on* $[a, b]$,

$$(1.5) \qquad\qquad \Delta(t) = \Delta(t_0) \exp \int_{t_0}^{t} \operatorname{tr} A(s)\, ds.$$

For a square matrix $A = (a_{jk})$, the trace of A is defined to be the sum of its diagonal elements, $\operatorname{tr} A = \Sigma\, a_{jj}$.

Proof. Let $A(t) = (a_{jk}(t))$, $j, k = 1, \ldots, d$. The usual expansion for the determinant $\Delta(t) = \det Y(t)$, where $Y(t) = (y_k^j(t))$, and the rule for differentiating the product of d scalar functions show that

$$\Delta'(t) = \sum \det Y_j(t),$$

where $Y_j(t)$ is the matrix obtained by replacing the jth row $(y_1^j(t), \ldots, y_d^j(t))$ of $Y(t)$ by its derivative $(y_1^{j\prime}(t), \ldots, y_d^{j\prime}(t))$. Since $y_k^{j\prime}(t) = \sum a_{ji}y_k^i$ by (1.1), it is seen that the jth row of $Y_j(t)$ is the sum of $a_{jj}(t)$ times the jth row of $Y(t)$ and a linear combination of the other rows of $Y_j(t)$. Hence $\det Y_j(t) = a_{jj}(t) \det Y(t)$ and so, $\Delta'(t) = (\text{tr } A(t)) \Delta(t)$. This gives (1.5).

By a *fundamental matrix* $Y(t)$ of (1.1) or (1.4) is meant a solution of (1.4) such that $\det Y(t) \neq 0$. In order to obtain a fundamental matrix $Y(t)$, let $Y(t)$ be a matrix with columns $y_1(t), \ldots, y_d(t)$, where $y = y_j(t)$ is a solution of (1.1) belonging to a given initial condition $y_j(t_0) = y_{j0}$, where $y_{10}, \ldots, y_{d0}$ are (constant) linearly independent vectors. It is clear that all fundamental matrices $Y(t)$ can be obtained in this fashion. Let $Y(t) = Y(t, t_0)$ denote the particular fundamental matrix satisfying

$$(1.6) \qquad\qquad Y(t_0, t_0) = I.$$

Exercise 1.3. Let $A(t)$ be a continuous $d \times d$ matrix for $t \geqq 0$ such that every solution $y(t)$ of (1.1) is bounded for $t \geqq 0$. Let $Y(t)$ be a fundamental matrix of (1.1). Show that $Y^{-1}(t)$ is bounded if and only if $\text{Re} \left[\int_0^t \text{tr } A(s) \, ds \right]$ is bounded from below.

If $Y(t)$ is a solution of (1.4) and c is a constant vector, the principle of superposition states that

$$(1.7) \qquad\qquad y(t) = Y(t)c$$

is a solution of (1.1). Furthermore, if $Y(t)$ is a fundamental solution of (1.4), then every solution of (1.1) is of the form (1.7) with $c = Y^{-1}(t_0)y(t_0)$; that is,

$$(1.8) \qquad\qquad y(t) = Y(t) Y^{-1}(t_0)y(t_0).$$

In particular, if $Y(t) = Y(t, t_0)$, then

$$(1.9) \qquad\qquad y(t) = Y(t, t_0)y(t_0).$$

More generally, if $Y = Y_0(t)$ is a matrix solution of (1.4) and C is a constant $d \times d$ matrix, then $Y(t) = Y_0(t)C$ is a solution of (1.4). When $Y_0(t)$ is a fundamental solution of (1.4), all $d \times d$ matrix solutions of (1.4) are obtained in this fashion and all fundamental solutions are obtained in this way with a choice of C, $\det C \neq 0$.

Lemma 1.2. *Let* $Y(t) = Y(t, t_0)$ *be the fundamental solution of* (1.4) *satisfying* (1.6). *Then, for* $t_0, t \in [a, b]$,

(1.10) $$Y(t, t_0) = Y(t, s) Y(s, t_0).$$

Proof. By the remarks just made, where $C = Y(s, t_0)$, the right side is a fundamental matrix and reduces to $Y(s, t_0)$ at $t = s$. Since the left side of (1.10) is a matrix with the same properties, the relation (1.10) is clear from uniqueness (Lemma 1.1).

2. Variation of Constants

A linear change of dependent variables in (1.1) or (1.2) will often be used.

Theorem 2.1. *Let* $Z(t)$ *be a continuously differentiable, nonsingular* $d \times d$ *matrix for* $a \leq t \leq b$. *Under the linear change of variables* $y \to z$, *where*

(2.1) $$y = Z(t)z,$$

(1.2) *is transformed into*

(2.2) $$z' = Z^{-1}(t)[A(t)Z(t) - Z'(t)]z + Z^{-1}(t)f(t).$$

In particular, if $Z(t)$ *is a fundamental matrix for*

(2.3) $$z' = B(t)z,$$

where $B(t) = Z'(t)Z^{-1}(t)$ *is continuous for* $a \leq t \leq b$, *then* (2.2) *becomes*

(2.4) $$z' = Z^{-1}(t)[A(t) - B(t)]Z(t)z + Z^{-1}(t)f(t).$$

The equation (2.2) is clear, for (2.1) implies that $y' = Z'z + Zz'$, so that $z' = Z^{-1}(y' - Z'z)$ and $y' = Ay + f$ from (1.2).

In the particular case where $A(t) = B(t), Z(t) = Y(t)$, and the latter is a fundamental matrix for (1.1), the change of variables (2.1) is called a "*variation of constants*," i.e., the replacement of the constant vector c in (1.7) by a variable vector z. In this case, (2.4) reduces to $z' = Y^{-1}(t)f(t)$, so that its solutions are given by a quadrature

$$z(t) = c + \int_{t_0}^{t} Y^{-1}(s)f(s) \, ds,$$

where $c = z(t_0)$ is a constant vector. In view of (2.1), this gives the first part of the following corollary. The last part follows from (1.10).

Corollary 2.1. *Let* $Y(t)$ *be a fundamental matrix of* (1.1). *Then the solutions of* (1.2) *are given by*

(2.5) $$y = Y(t)\left[c + \int_{t_0}^{t} Y^{-1}(s)f(s) \, ds\right].$$

In particular, if $Y(t) = Y(t, t_0)$, (2.5) *becomes*

(2.6) $$y = Y(t, t_0)c + \int_{t_0}^{t} Y(t, s)f(s)\, ds.$$

Formula (2.5) or (2.6) shows that the solutions of (1.2) are determined by a quadrature if the solutions of (1.1) are known. For arbitrary c, $Y(t)c$ in (2.5) is merely an arbitrary solution of (1.1).

Exercise 2.1. Let $A(t)$ be continuous on some t-interval (not necessarily closed or bounded), $Y(t)$ a fundamental matrix for (1.1), and $Z(t)$ a continuously differentiable, nonsingular matrix. Under the change of variables, $y = Z(t)z$, let (1.1) become $z' = C(t)z$; cf. (2.2) where $f(t) = 0$. For any matrix A, let A^* denote the complex conjugate transpose of A and $A^H = \frac{1}{2}(A + A^*)$, the Hermitian part of A. (*a*) Show that if $Z(t)$ is unitary [i.e., $Z^*(t) = Z^{-1}(t)$], then $C^H(t) = Z^*(t)A^H(t)Z(t)$ [since the derivative of $Z^*(t)Z(t) = I$ is 0]. (*b*) Let $Z(t) = Y(t)Q(t)$, so that $Q = Y^{-1}Z$ is continuously differentiable and nonsingular. Show that $C(t) = -Q^{-1}(t)Q'(t)$. In particular, $C(t)$ is triangular [or diagonal] if $Q(t)$ is triangular [or diagonal].

Exercise 2.2 (*Continuation*). (*a*) Show that there exists a unitary $Z(t)$ such that $C(t)$ is triangular; in this case, $C(t)$ is bounded if $A(t)$ is bounded. (*b*) Show that there exists a bounded $Z(t)$ such that $C(t)$ is diagonal. It is not claimed that $Z(t)$ can be chosen so that $Z^{-1}(t)$ is bounded.

3. Reductions to Smaller Systems

If a set of r linearly independent solutions of (1.1) is known, the determination of all solutions of (1.1) can be reduced essentially to the problem of determining the solutions of a linear homogeneous system of $d-r$ differential equations. The simplest formulae giving this reduction are, however, "local," i.e., applicable only on subintervals of $[a, b]$ and vary from subinterval to subinterval.

Let $Y = Y_r(t)$ be a $d \times r$ matrix solution of (1.4). Corresponding to a given point $t = t_0$ of $[a, b]$, renumber the components of y so that if $Y_r(t) = (y_k{}^j(t))$, $j = 1, \ldots, d$ and $k = 1, \ldots, r$, and

(3.1) $$Y_r(t) = \begin{pmatrix} Y_{r1}(t) \\ Y_{r2}(t) \end{pmatrix},$$

where $Y_{r1}(t)$ is an $r \times r$ matrix, then $\det Y_{r1}(t_0) \neq 0$. Let $[\gamma, \delta]$ be any subinterval of $[a, b]$ containing t_0 on which $\det Y_{r1}(t) \neq 0$. Define the $d \times d$ matrix

(3.2) $$Z(t) = \begin{pmatrix} Y_{r1}(t) & 0 \\ Y_{r2}(t) & I_{d-r} \end{pmatrix},$$

where I_{d-r} is the unit $(d - r) \times (d - r)$ matrix. Then $\det Z(t) = \det Y_{r1}(t) \neq 0$ on $[\gamma, \delta]$. A simple calculation shows that $Z^{-1}(t)$ is a matrix of the form

$$(3.3) \qquad Z^{-1}(t) = \begin{pmatrix} Y_{r1}^{-1}(t) & 0 \\ Z_{r2}(t) & I_{d-r} \end{pmatrix},$$

where $Z_{r2}(t)$ is the $(d - r) \times r$ matrix

$$(3.4) \qquad Z_{r2}(t) = -Y_{r2}(t) Y_{r1}^{-1}(t).$$

Note that $Z'(t) = (Y_r'(t), 0) = A(t)(Y_r(t), 0)$; i.e.,

$$(3.5) \qquad Z'(t) = A(t) \begin{pmatrix} Y_{r1}(t) & 0 \\ Y_{r2}(t) & 0 \end{pmatrix}.$$

Introduce the change of variables $y = Z(t)z$ into (1.1), then the case $f(t) = 0$ of (2.2) gives the resulting differential equation for z. Writing the right side of (2.2), with $f = 0$, as $Z^{-1}(t)[A(t)Z(t) - Z'(t)]z$ gives, by virtue of (3.2) and (3.5),

$$(3.6) \qquad z' = Z^{-1}(t)A(t) \begin{pmatrix} 0 & 0 \\ 0 & I_{d-r} \end{pmatrix} z.$$

Let $A_{11}(t)$, $A_{22}(t)$ be square $r \times r$, $(d - r) \times (d - r)$ matrices such that

$$(3.7) \qquad A(t) = \begin{pmatrix} A_{11}(t) & A_{12}(t) \\ A_{21}(t) & A_{22}(t) \end{pmatrix};$$

and let

$$(3.8) \qquad z = \begin{pmatrix} z_1 \\ z_2 \end{pmatrix},$$

where z_1 is an r-dimensional vector, z_2 a $(d - r)$-dimensional vector. Then (3.3) and (3.8) show that (3.6) splits into

$$(3.9) \qquad z_1' = Y_{r1}^{-1}(t)A_{12}(t)z_2,$$

$$(3.10) \qquad z_2' = [Z_{r2}(t)A_{12}(t) + A_{22}(t)]z_2.$$

Note that (3.10) is a linear homogeneous system for the $(d - r)$-dimensional vector z_2 and that z_1 is given by a quadrature when z_2 is known. In (3.10), $Z_{r2}(t)$ is given by (3.4). The reduction of (1.1) to (3.10) is of course only valid on an interval $[\gamma, \delta]$ where $Y_{r1}(t)$ is nonsingular. The result can be summarized as follows:

Lemma 3.1. *Let* $Y = Y_r(t)$ *be a* $d \times r$ *matrix solution of* (1.4) *on* $[a, b]$ *such that if* $Y_r(t)$ *is written as* (3.1), *then* $\det Y_{r1}(t) \equiv \det (y_k{}^j(t))$, *where* $k, j = 1, \ldots, r$, *does not vanish on* $[\gamma, \delta] \subset [a, b]$. *Then the change of* *variables* (2.1) *in terms of* (3.2) *reduces* (1.1) *to* (3.9)–(3.10) *on* $[\gamma, \delta]$ *in which* A_{12}, A_{22} *and* Z_{r2} *are defined by* (3.7) *and* (3.4), *respectively.*

An application. Consider a system (1.1) in which $A(t) = (a_{jk}(t))$ satisfies

$$(3.11) \qquad a_{j,j+1}(t) \neq 0 \text{ and } a_{jk}(t) \equiv 0 \text{ for } k \geq j + 2$$

on $[a, b]$. It is readily verified that a solution $y = y(t)$ of (1.1) is known as soon as its first component $u = y^1(t)$ is known.

Corollary 3.1. *Let* $A(t)$ *satisfy* (3.11) *on* $[a, b]$ *and let* (1.1) *possess* d *solutions* $y_1(t), \ldots, y_d(t)$ *with the property that, for* $k = 1, \ldots, d$ *and* $a \leq t \leq b$,

$$(3.12) \qquad W_k(t) \equiv \det (y_j{}^i(t)) \neq 0 \text{ for } i, j = 1, \ldots, k.$$

Then (1.1) *is equivalent to the single differential equation of* d*th order for* $u = y^1$ *of the form*

$$(3.13) \qquad (a_{d-1} \ldots \{a_2[a_1(a_0 u)']'\}' \ldots)' = 0,$$

where

$$(3.14) \qquad a_j = \frac{W_j^2}{a_{j,j+1} W_{j-1} W_{j+1}} \quad \text{for } j = 0, \ldots, d - 1$$

and $W_{-1} = W_0 = 1$, $a_{01} = 1$.

Proof. This will be seen to follow by $d - 1$ successive applications of the reduction process described in Lemma 3.1 with $r = 1$. Introduce the notation

$$(3.15) \qquad W_{i_1 \ldots i_k;\, j_1 \ldots j_k} = \det (y_{j_m}^{i_n}) \quad \text{for } m, n = 1, \ldots, k,$$

$$(3.16) \qquad W_{jk}^\alpha = W_{12 \ldots \alpha-1, j; 12 \ldots \alpha-1, k} \quad \text{for } j, k = \alpha, \ldots, d.$$

Then, by a standard formula for minors of the "adjoint determinant,"

$$(3.17) \qquad W_{\alpha\alpha}^\alpha W_{jk}^\alpha - W_{j\alpha}^\alpha W_{\alpha k}^\alpha = W_{jk}^{\alpha+1} W_{\alpha-1}.$$

In order to verify this, let a symbol (e.g., $W_{\alpha-1}$ or W_{jk}^α) denote either a matrix or its determinant. Let γ_{mn} denote the cofactor of the (n,m)th element in the $(\alpha + 1) \times (\alpha + 1)$ determinant $W_{jk}^{\alpha+1}$. In particular, $\gamma_{\alpha\alpha} = W_{jk}^\alpha$, $\gamma_{\alpha,\alpha+1} = -W_{\alpha k}^\alpha$, $\gamma_{\alpha+1,\alpha} = -W_{j\alpha}^\alpha$, and $\gamma_{\alpha+1,\alpha+1} = W_{\alpha\alpha}^\alpha$. Consider the product of the determinants $W_{jk}^{\alpha+1}\Gamma$, where

$$\Gamma = \det \begin{pmatrix} & & & \gamma_{1\alpha} & \gamma_{1\alpha+1} \\ & I_{\alpha-1} & & \cdot & \cdot \\ & & & \cdot & \cdot \\ & & & \cdot & \cdot \\ 0 & \ldots & 0 & \gamma_{\alpha\alpha} & \gamma_{\alpha,\alpha+1} \\ 0 & \ldots & 0 & \gamma_{\alpha+1,\alpha} & \gamma_{\alpha+1,\alpha+1} \end{pmatrix} = \begin{vmatrix} \gamma_{\alpha\alpha} & \gamma_{\alpha,\alpha+1} \\ \gamma_{\alpha+1,\alpha} & \gamma_{\alpha+1,\alpha+1} \end{vmatrix}$$

On the other hand, by matrix multiplication,

$$W_{jk}^{\alpha+1}\Gamma = \det\begin{pmatrix} & & 0 & 0 \\ & W_{\alpha-1} & \cdot & \cdot \\ & & \cdot & \cdot \\ & & 0 & 0 \\ y_1^{\ \alpha} & \cdots & y_{\alpha-1}^\alpha & W_{jk}^{\alpha+1} & 0 \\ y_1^{\ j} & \cdots & y_{\alpha-1}^j & 0 & W_{jk}^{\alpha+1} \end{pmatrix} = W_{\alpha-1}(W_{jk}^{\alpha+1})^2.$$

The last two displays give (3.17).

In order to systematize notation, write $y_{(1)}$, $A^1 = (a_{jk}^1)$ for y, A, respectively. Thus $W_1 = y_{(1)1}^1(t) \neq 0$. Introduce new variables $y_{(2)}$ by the variation of constants associated with (3.2), $r = 1$; namely,

(3.18) T_1: $y_{(1)}^1 = y_{(1)1}^1(t)y_{(2)}^1$, $y_{(1)}^{(k)} = y_{(1)1}^k(t)y_{(2)}^1 + y_{(2)}^k$ for $k = 2, \ldots, d$.

Consider $y_{(2)}$ to be a $(d-1)$-dimensional vector $y_{(2)} = (y_{(2)}^2, \ldots, y_{(2)}^d)$, so that $y_{(2)}^1$ is not considered a component of $y_{(2)}$. Then (1.1) reduces, by (3.9)–(3.10) and (3.4), to

(D$_2^0$) $$y_{(2)}^{1\prime} = \sum_{k=2}^d a_{1k}^1 \frac{y_{(2)}^k}{y_{(1)1}^1},$$

(D$_2$) $y_{(2)}' = A^2(t)y_{(2)}$, $\quad a_{jk}^2 = a_{jk}^1 - \dfrac{y_{(1)1}^j}{y_{(1)1}^1} a_{1k}^1 \quad$ for $j, k = 2, \ldots, d$;

so that solutions of (D$_2^0$) are determined by quadratures when solutions of (D$_2$) are known. Using (3.18) and the known solutions $y_{(1)1}, \ldots, y_{(1)d}$ of (1.1), we obtain $d-1$ solutions $y_{(2)j} = (y_{(2)j}^2(t), \ldots, y_{(2)j}^d(t))$ of (D$_2$) for $j = 2, \ldots, d$; namely,

(3.19) $$y_{(2)j}^k = \frac{y_{(1)1}^1 y_{(1)j}^k - y_{(1)1}^k y_{(1)j}^1}{y_{(1)1}^1} = \frac{W_{kj}^2}{W_1}.$$

In particular, $y_{(2)2}^{(2)}(t) = W_2/W_1 \neq 0$, and this procedure can be repeated to reduce the order of (D$_2$). Suppose that the successive changes of variables $T_1, \ldots, T_{\alpha-1}$ have been defined, each of the form

(3.20) T_α:
$$y_{(\alpha)}^\alpha = y_{(\alpha)\alpha}^\alpha(t)y_{(\alpha+1)}^\alpha,$$
$$y_{(\alpha)}^k = y_{(\alpha)\alpha}^k(t)y_{(\alpha+1)}^\alpha + y_{(\alpha+1)}^k, \qquad k = \alpha+1, \ldots, d,$$

where $y_{(\alpha+1)} = (y_{(\alpha+1)}^{\alpha+1}, \ldots, y_{(\alpha+1)}^d)$ is a $(d-\alpha)$-dimensional vector and $y_{(\alpha+1)}^\alpha$ is not considered a component of $y_{(\alpha+1)}$. Suppose that there results a system of differential equations for $y^\alpha_{(\alpha+1)}, y_{(\alpha+1)}$ of the form

$(D^0_{\alpha+1})$ $\qquad\qquad y^{\alpha'}_{(\alpha+1)} = \sum_{k=\alpha+1}^{d} a^\alpha_{\alpha k} \dfrac{y^k_{(\alpha)}}{y^\alpha_{(\alpha)\alpha}},$

$(D_{\alpha+1})$ $\qquad y'_{(\alpha+1)} = A^{\alpha+1}(t)y_{(\alpha+1)}, \quad a^{\alpha+1}_{jk} = a^\alpha_{jk} - \dfrac{y^j_{(\alpha)\alpha}}{y^\alpha_{(\alpha)\alpha}} a^\alpha_{\alpha k},$

where $j, k = \alpha + 1, \ldots, d$, and suppose that $d - \alpha + 1$ solutions $y_{(\alpha)r} = (y^\alpha_{(\alpha)r}(t), \ldots, y^d_{(\alpha)r}(t))$ of (D_α) are given by

(3.21_α) $\qquad\qquad y^k_{(\alpha)j} = \dfrac{W^\alpha_{kj}}{W_{\alpha-1}}$ $\quad$ for $j, k = \alpha, \ldots, d$.

In particular, since $W^\alpha_{\alpha\alpha} = W_\alpha$,

(3.22) $\qquad\qquad y^\alpha_{(\alpha)\alpha}(t) = \dfrac{W_\alpha}{W_{\alpha-1}}.$

It is readily verified from T_α and (3.17) that $(D_{\alpha+1})$ has $d - \alpha$ solutions $y_{(\alpha+1)j} = (y^{\alpha+1}_{(\alpha+1)j}(t), \ldots, y^d_{(\alpha+1)j}(t))$ for $j = \alpha + 1, \ldots, d$ given by $(3.21_{\alpha+1})$. For, by (3.21_α) and (3.22), the relations in (3.20) show that $y^\alpha_{(\alpha+1)} = y^\alpha_{(\alpha)} W_{\alpha-1}/W_\alpha$ and

$$y^k_{(\alpha+1)} = \frac{W_\alpha y^k_{(\alpha)} - W^\alpha_{\alpha k} y^\alpha_{(\alpha)}}{W_\alpha} \quad \text{for } k = \alpha + 1, \ldots, d.$$

The replacement of $y^k_{(\alpha)}, y^\alpha_{(\alpha)}$ by $y^k_{(\alpha)j}, y^\alpha_{(\alpha)j}$ given in (3.21_α) and the use of $W_\alpha = W^\alpha_{\alpha\alpha}$ and (3.17) give $(3.21_{\alpha+1})$.

Note that $y_{(d)}$ is a 1-dimensional vector $y_{(d)} = y^d_{(d)}$ and that (D_d) which is a linear homogeneous equation has, by (3.22), the general solution $y^d_{(d)}(t) = cW_d/W_{d-1}$, where c is an arbitrary constant. Thus (D_d) is equivalent to

(3.23) $\qquad\qquad \left(\dfrac{W_{d-1}}{W_d} y^d_{(d)}\right)' = 0.$

The assumption (3.11), which has not been used so far, and an induction on α show that $a^\alpha_{jk} = a_{jk}$ if $d \geq k > j = \alpha, \ldots, d - 1$. Hence

(3.24) $\qquad a^\alpha_{\alpha,\alpha+1} = a_{\alpha,\alpha+1}$ $\quad$ and $\quad$ $a^\alpha_{\alpha k} = 0$ $\qquad$ for $k \geqq \alpha + 2$,

so that, by (3.22), $(D^0_{\alpha+1})$ reduces to

(3.25) $\qquad\qquad y^{\alpha'}_{(\alpha+1)} = \dfrac{W_{\alpha-1} a_{\alpha,\alpha+1} y^{\alpha+1}_{(\alpha+1)}}{W_\alpha}.$

Note that, by virtue of (3.22), the first equation for T_α in (3.20) gives

(3.26) $\qquad\qquad y^\alpha_{(\alpha)} = \dfrac{W_\alpha y^\alpha_{(\alpha+1)}}{W_{\alpha-1}}$ $\quad$ for $\alpha = 1, \ldots, d - 1$.

Hence, by (3.25),

$$y_{(\alpha+1)}^{\alpha'} = \frac{W_{\alpha-1}W_{\alpha+1}a_{\alpha,\alpha+1}y_{(\alpha+2)}^{\alpha+1}}{W_\alpha^2} \quad \text{for } \alpha = 1, \ldots, d-2,$$

or, by (3.14),

$$(3.27) \qquad y_{(\alpha+1)}^{\alpha'} = \frac{y_{(\alpha+2)}^{\alpha+1}}{a_\alpha} \quad \text{for } \alpha = 1, \ldots, d-2.$$

By (3.18), $u = y_{(1)}^1$ determines $y_{(2)}^1 = W_0 u/W_1 = a_0 u$ which satisfies $(a_0 u)' = y_{(3)}^2/a_1$ by (3.27). Thus $y_{(3)}^2 = a_1(a_0 u)'$ satisfies $[a_1(a_0 u)']' = y_{(4)}^3/a_2$ by (3.27). Repetitions of this argument give

$$y_{(d)}^{d-1'} = (a_{d-2} \ldots \{a_2[a_1(a_0 u)']'\} \ldots)'.$$

Note that (3.25), with $\alpha = d-1$, shows that $y_{(d)}^{d-1'}$ is

$$W_{d-2}a_{d-1,d}^{\cdot}y_{(d)}^d / W_{d-1}.$$

Hence the desired result (3.13) follows from (3.23).

4. Basic Inequalities

Let the norm $\|A\|$ of the matrix A be defined by

$$(4.1) \qquad \|A\| = \sup |Ay| \quad \text{for } |y| = 1.$$

This norm of A depends on the norm $|y|$ of the vector y. For the choice of either the norm $|y| = \max(|y^1|, \ldots, |y^d|)$ or the Euclidean norm, there is the following estimate for solutions of (1.2):

Lemma 4.1. *Let $y = y(t)$ be a solution of* (1.2) *and $t, t_0 \in [a, b]$. Then*

$$(4.2) \qquad |y(t)| \leq \left\{ |y(t_0)| + \left| \int_{t_0}^t |f(s)| \, ds \right| \right\} \exp \left| \int_{t_0}^t \|A(s)\| \, ds \right|.$$

Proof. By (1.2), the inequality $|y'| \leq \|A(t)\| \cdot |y| + |f(t)|$ holds and is an analogue of (III 4.6). Thus, if $u^0(t)$ is the (unique) solution of

$$(4.3) \qquad u' = \|A(t)\| u + |f(t)|$$

satisfying $u(t_0) = u_0$ with $u_0 = |y(t_0)|$, i.e.,

$$u^0(t) = \left\{ u_0 + \int_{t_0}^t |f(s)| \left(\exp - \int_{t_0}^s \|A(r)\| \, dr \right) ds \right\} \exp \int_{t_0}^t \|A(s)\| \, ds,$$

Corollary III 4.3 and the remark following it imply that $|y(t)| \leq u^0(t)$ for $t_0 \leq t \leq b$. This gives (4.2) for $t_0 \leq t$. Similarly, if $u = u_0(t)$ is the solution of

$$u' = -\|A(t)\| u - |f(t)|$$

satisfying $u(t_0) = u_0(= |y(t_0)|)$, i.e.,

$$u_0(t) = \left\{ u_0 - \int_{t_0}^t |f(s)| \left(\exp \int_{t_0}^s \|A(r)\| \, dr \right) ds \right\} \exp \left(- \int_{t_0}^t \|A(s)\| \, ds \right),$$

then $|y(t)| \geq u_0(t)$ for $t_0 \leq t$. Interchanging t and t_0 in the inequality so obtained gives (4.2) for $t \leq t_0$.

Corollary 4.1. *Let $A_0(t)$; $A_1(t), A_2(t), \ldots$ be a sequence of continuous $d \times d$ matrices and $f_0(t)$; $f_1(t), f_2(t), \ldots$ a sequence of continuous vectors on $[a, b]$ such that $A_n(t) \to A_0(t)$ and $f_n(t) \to f_0(t)$ as $n \to \infty$ uniformly on $[a, b]$. Let $y = y_n(t)$ be the solution of*

$$(4.4_n) \qquad\qquad y' = A_n(t)y + f_n(t), \qquad y(t_n) = y_n,$$

where $a \leq t_n \leq b$ and $(t_n, y_n) \to (t_0, y_0)$ as $n \to \infty$. Then $y_n(t) \to y_0(t)$ uniformly on $[a, b]$ as $n \to \infty$.

Proof. It is clear that $\|A_1(t)\|$, $\|A_2(t)\|$, $\ldots$ and $|f_1(t)|, |f_2(t)|, \ldots$ are uniformly bounded on $[a, b]$. Thus Lemma 4.1 implies that $|y_1(t)|$, $|y_2(t)|, \ldots$ is uniformly bounded, say $|y_n(t)| \leq c$ for $n = 1, 2, \ldots$, and $a \leq t \leq b$. The right side $f_n(t, y) = A_n(t)y + f_n(t)$ of the differential equation in (4.4_n) tends uniformly to $f(t, y) = A_0(t)y + f_0(t)$ as $n \to \infty$ for $a \leq t \leq b$, $|y| \leq c$. Thus Corollary 4.1 follows from Theorem I 2.4.

For a matrix A, let A^* denote the complex conjugate transpose of A and $A^H = \frac{1}{2}(A + A^*)$, the Hermitian part of A. Let

$$y \cdot z = \sum_{k=1}^d y^k \bar{z}^k$$

denote the scalar product of the pair of vectors y, z (so that in the case of vectors with complex-valued components, $y \cdot z$ is the complex conjugate of $z \cdot y$). In particular,

$$(4.5) \qquad\qquad Ay \cdot z = y \cdot A^*z.$$

Finally, let μ_0, μ^0 be the least and greatest eigenvalue of A^H, i.e., the least and greatest zero of the polynomial $\det (A^H - \lambda I)$ in λ. (The fact that A^H is Hermitian implies that its eigenvalues are real.) Equivalently, if $\|y\|$ denotes the Euclidean norm, then μ_0, μ^0 are given by

$$(4.6) \quad \mu_0 = \inf A^H y \cdot y \qquad \text{and} \qquad \mu^0 = \sup A^H y \cdot y \qquad \text{for } \|y\| = 1$$

or

$$(4.7) \quad \mu_0 = \inf \operatorname{Re}(Ay \cdot y) \quad \text{and} \quad \mu^0 = \sup \operatorname{Re}(Ay \cdot y) \quad \text{for } \|y\| = 1,$$

where "Re a" denotes the real part of the complex number a.

When $y(t), z(t)$ are differentiable vector-valued functions, then $(y(t) \cdot z(t))' = y'(t) \cdot z(t) + y(t) \cdot z'(t)$; in particular, $(\|y(t)\|^2)' = y'(t) \cdot y(t) + y(t) \cdot y'(t) = 2 \operatorname{Re}[y'(t) \cdot y(t)]$. This implies

Lemma 4.2. *Let* $\|y\|$ *denote the Euclidean norm of* y; $\mu_0(t)$, $\mu^0(t)$ *the least, greatest eigenvalue of the Hermitian part* $A^H(t)$ *of* $A(t)$; *and* $y(t)$ *a solution of* (1.2). *Then*

$$(4.8) \qquad \mu_0(t)\, \|y\| - \|f(t)\| \leqq \|y\|' \leqq \mu^0(t)\, \|y\| + \|f(t)\|$$

(*where* $\|y\|'$ *denotes either a left or right derivate of* $\|y\|$). *Hence, for* $a \leqq t_0 \leqq t \leqq b$,

$$(4.9) \qquad \|y(t)\| \leqq \|y(t_0)\| \exp \int_{t_0}^{t} \mu^0(s)\, ds + \int_{t_0}^{t} \|f(s)\| \left(\exp \int_{s}^{t} \mu^0(r)\, dr \right) ds,$$

$$(4.10) \qquad \|y(t)\| \geqq \|y(t_0)\| \exp \int_{t_0}^{t} \mu_0(s)\, ds - \int_{t_0}^{t} \|f(s)\| \left(\exp \int_{s}^{t} \mu_0(r)\, dr \right) ds.$$

Proof. The inequalities (4.8) follow from (1.2). For, on the one hand, $y(t) \neq 0$ implies that $\|y(t)\|' = \frac{1}{2}(\|y(t)\|^2)'/\|y(t)\| = \mathrm{Re}\,[y'(t) \cdot y(t)]/\|y(t)\|$; on the other hand, $y(t) = 0$ implies that $|D\,\|y(t)\|\,| = \|f(t)\|$ for $D = D_R$ or $D = D_L$. The continuity of $A(t)$ implies that of $\mu_0(t)$ and $\mu^0(t)$, so that (4.9) and (4.10) are consequences of (4.8) by Theorem III 4.1 and Remark 1 following it.

Consider a function $y(t)$ for $t \geqq 0$. A number τ is called a (Lyapunov) *order number* for $y(t)$ if, for every $\epsilon > 0$, there exist positive constants $C_0(\epsilon)$, $C(\epsilon)$ such that

$$(4.11) \qquad \|y(t)\| \leqq C(\epsilon) e^{(\tau + \epsilon)t} \qquad \text{for all large } t.$$

$$(4.12) \qquad \|y(t)\| \geqq C_0(\epsilon) e^{(\tau - \epsilon)t} \qquad \text{for some arbitrarily large } t.$$

When $y(t) \neq 0$ for large t, an equivalent formulation of (4.11) and (4.12) is

$$(4.13) \qquad \limsup_{t \to \infty} t^{-1} \log \|y(t)\| = \tau.$$

Lemmas 4.1 and 4.2 show that if $A(t)$ is continuous for $t \geqq 0$, then a sufficient condition for every solution $y(t) \not\equiv 0$ of the homogeneous system (1.1) to possess an order number τ is that

$$t^{-1} \int_0^t \|A(s)\|\, ds \qquad \text{be bounded,}$$

in which case

$$|\tau| \leqq \limsup_{t \to \infty} t^{-1} \int_0^t \|A(s)\|\, ds;$$

or, more generally, that

$$t^{-1} \int_0^t \mu^0(s)\, ds, \qquad -t^{-1} \int_0^t \mu_0(s)\, ds \qquad \text{be bounded from above,}$$

in which case

$$\liminf_{t\to\infty} t^{-1}\int_0^t \mu_0(s)\,ds \leqq \tau \leqq \limsup_{t\to\infty} t^{-1}\int_0^t \mu^0(s)\,ds.$$

5. Constant Coefficients

Let R be a constant $d \times d$ matrix and consider the system of differential equations

$$(5.1) \qquad\qquad y' = Ry.$$

Let $y_1 \neq 0$ be a constant vector, λ a complex number. By substituting

$$(5.2) \qquad\qquad y = y_1 e^{\lambda t}$$

into (5.1), it is seen that a necessary and sufficient condition, in order that (5.2) be a solution of (5.1), is that

$$(5.3) \qquad\qquad Ry_1 = \lambda y_1;$$

i.e., that λ be an eigenvalue and that $y_1 \neq 0$ be a corresponding eigenvector of R. Thus to each eigenvalue λ of R, there corresponds at least one solution of (5.1) of the form (5.2). If R has simple elementary divisors (i.e., if R has linearly independent eigenvectors $y_1, \ldots, y_d$ belonging to the respective eigenvalues $\lambda_1, \ldots, \lambda_d$), then

$$(5.4) \qquad\qquad Y = (y_1 e^{\lambda_1 t}, \ldots, y_d e^{\lambda_d t})$$

is a fundamental matrix for (5.1).

In the general case, a fundamental matrix can be found as follows: Successive approximations for a solution of the initial value problem (5.1) and $y(0) = c$ are

$$(5.5) \qquad y_0(t) = c, \; y_n(t) = c + \int_0^t Ry_{n-1}(s)\,ds \qquad \text{for } n \geqq 1,$$

so that an induction shows that

$$(5.6) \qquad\qquad y_n(t) = \left(I + Rt + \frac{R^2 t^2}{2!} + \cdots + \frac{R^n t^n}{n!}\right)c.$$

This suggests the following definition: For any $d \times d$ matrix B, let e^B denote the matrix defined by

$$(5.7) \qquad\qquad e^B = I + B + \frac{B^2}{2!} + \cdots + \frac{B^n}{n!} + \cdots,$$

where the matrix series on the right can be considered as d^2 (scalar) series, one for each of the elements of e^B. If $B = (b_{jk})$, then (4.1) shows that

$|b_{jk}| \leqq \|B\|$ for $j, k = 1, \ldots, d$. Since (4.1) clearly implies $\|B^n\| \leqq \|B\|^n$, it follows that if $B^n = (b_{jk}^{(n)})$, then $|b_{jk}^{(n)}| \leqq \|B\|^n$. Thus each of the d^2 series for the elements of e^B is convergent. The standard proof for the functional equation of the exponential function shows that

$$(5.8) \qquad\qquad e^{B+C} = e^B e^C \quad \text{if} \quad BC = CB.$$

This makes it clear that (5.6) converges uniformly on any bounded t-interval to $y = e^{Rt}c$, which is then a solution of (5.1) by (5.5). In other words,

$$(5.9) \qquad\qquad Y(t) = e^{Rt}$$

is a fundamental matrix for (5.1) and $Y(0) = I$.

Consider the inhomogeneous equation

$$(5.10) \qquad\qquad y' = Ry + f(t)$$

corresponding to (5.1). In this case formula (2.6) for the general solution of (1.2) reduces to

$$(5.11) \qquad\qquad y = e^{R(t-t_0)}y(t_0) + \int_{t_0}^{t} e^{R(t-s)}f(s)\, ds$$

for the general solution of (5.19).

Let Q be a constant, nonsingular matrix. The change of variables

$$y = Qz$$

transforms (5.1) into

$$z' = Jz, \quad \text{where} \quad J = Q^{-1}RQ.$$

This has the fundamental matrix $Z = e^{Jt}$. It is readily verified from $R = QJQ^{-1}$ and the definition (5.7) that

$$(5.12) \qquad e^{Rt} = Qe^{Jt}Q^{-1}, \quad \text{where} \quad J = Q^{-1}RQ.$$

Since a fundamental matrix, say e^{Rt}, can be multiplied on the right by a constant nonsingular matrix, say Q, to obtain another fundamental matrix, it follows that

$$(5.13) \qquad\qquad Y(t) = Qe^{Jt}$$

is also a fundamental matrix for (5.1).

Let Q be chosen so that J is in a Jordan normal form, i.e.,

$$(5.14) \qquad\qquad J = \operatorname{diag}[J(1), \ldots, J(g)],$$

where $J(j)$ is a square $h(j) \times h(j)$ matrix with all of its diagonal elements equal to a number $\lambda = \lambda(j)$ and, if $h(j) > 1$, its subdiagonal elements equal to 1 and its other elements equal to 0:

$$(5.15) \qquad J(j) = \begin{pmatrix} \lambda & 0 & 0 & \ldots & 0 & 0 & 0 \\ 1 & \lambda & 0 & \ldots & 0 & 0 & 0 \\ 0 & 1 & \lambda & \ldots & 0 & 0 & 0 \\ \cdot & \cdot & \cdot & \cdots & \cdot & \cdot & \cdot \\ 0 & 0 & 0 & \ldots & 0 & 1 & \lambda \end{pmatrix} = \lambda I_h + K_h,$$

where $\lambda = \lambda(j)$, $h = h(j)$, and K_h is the nilpotent square matrix

$$(5.16) \qquad K_h = \begin{pmatrix} 0 & 0 & 0 & \ldots & 0 & 0 & 0 \\ 1 & 0 & 0 & \ldots & 0 & 0 & 0 \\ 0 & 1 & 0 & \ldots & 0 & 0 & 0 \\ \cdot & \cdot & \cdot & \cdots & \cdot & \cdot & \cdot \\ 0 & 0 & 0 & \ldots & 0 & 1 & 0 \end{pmatrix};$$

and $h(1) + \cdots + h(g) = d$. $J(j)$ is just the scalar $\lambda = \lambda(j)$ and $K_h = 0$ if $h(j) = 1$.

From $J = \text{diag}\,[J(1), \ldots, J(g)]$ follows $J^n = \text{diag}\,[J^n(1), \ldots, J^n(g)]$; hence

$$(5.17) \qquad \exp Jt = \text{diag}\,[\exp J(1)t, \ldots, \exp J(g)t].$$

In view of (5.8) and (5.15), $\exp J(j)t = e^{\lambda t} \exp K_h t$ where $\lambda = \lambda(j)$, $h = h(j)$. Note that $K_h{}^2$ is obtained from K_h by moving the 1's from the subdiagonal to the diagonal below this; $K_h{}^3$ is obtained by moving the 1's to the next lower diagonal; etc. In particular, $K_h{}^h = 0$. Hence

$$(5.18)$$

$$\exp J(j)t = e^{\lambda t} \begin{pmatrix} 1 & 0 & \ldots & 0 \\ t & 1 & \ldots & 0 \\ \dfrac{t^2}{2!} & t & \ldots & 0 \\ \cdot & \cdot & \cdots & \cdot \\ \cdot & \cdot & \cdots & \cdot \\ \dfrac{t^{h-1}}{(h-1)!} & \dfrac{t^{h-2}}{(h-2)!} & \ldots & 1 \end{pmatrix}, \quad h = h(j) \quad \text{and} \quad \lambda = \lambda(j).$$

The formulae (5.12), (5.17) and (5.18) completely determine the fundamental matrix (5.9) of (5.1).

It follows from these formulae that if $y(t)$ is a solution of (5.1), then its components are linear combinations of the exponentials $e^{\lambda(1)t}, \ldots, e^{\lambda(g)t}$ with polynomials in t as coefficients. These polynomial coefficients cannot, of course, be chosen arbitrarily.

Thus the problem of determining the solutions of (5.1) is the algebraic one of determining the Jordan normal form J of R and a matrix Q such that $J = Q^{-1}RQ$. The simple case, at the beginning of this section, leading to (5.4) corresponds to a situation where $h(j) = 1$ for $j = 1, \ldots, g$ and $g = d$, so that $\exp J(j)t$ is the 1×1 matrix (scalar) $e^{\lambda t}$, $\lambda = \lambda(j)$.

Note that, in any case, if the eigenvalues of R are $\lambda_1, \lambda_2, \ldots, \lambda_d$, then (5.1) has d linearly independent solutions $y_1(t), \ldots, y_d(t)$ such that the order number of $y_j(t)$ is $\tau = \operatorname{Re} \lambda_j$ for $j = 1, \ldots, d$; cf. (4.13).

6. Floquet Theory

The case of variable, but periodic, coefficients can theoretically be reduced to the case of constant coefficients. This is the essence of the following.

Theorem 6.1.   *In the system*

$$(6.1) \qquad\qquad y' = P(t)y,$$

let $P(t)$ be a continuous $d \times d$ matrix for $-\infty < t < \infty$ which is periodic of period p,

$$(6.2) \qquad\qquad P(t + p) = P(t).$$

Then any fundamental matrix $Y(t)$ of (6.1) has a representation of the form

$$(6.3) \qquad Y(t) = Z(t)e^{Rt}, \qquad \text{where } Z(t + p) = Z(t)$$

and R is a constant matrix (and $Z(t)$, R are $d \times d$ matrices).

If y_0 is an eigenvector of R belonging to an eigenvalue λ, so that $e^{Rt}y_0 = y_0e^{\lambda t}$, then the solution $y = Y(t)y_0$ of (6.1) is of the form $z_1(t)e^{\lambda t}$ where the vector $z_1(t) = Z(t)y_0$ has the period p. More generally, it follows, from the structure of e^{Rt} discussed in the last section, that if $y(t)$ is any solution of (6.1), then the components of $y(t)$ are linear combinations of terms of the type $\alpha(t)t^k e^{\lambda t}$, where $\alpha(t + p) = \alpha(t)$, k is an integer $0 \leq k \leq d - 1$, and λ is an eigenvalue of R.

Neither the matrix R nor its eigenvalues are uniquely determined by the system (6.1). For example, the representation (6.3) can be replaced by $Y(t) = Z(t)e^{-2\pi it}e^{(R+2\pi iI)t}$, i.e., $Z(t)$ by $Z(t)e^{-2\pi it}$ and R by $R + 2\pi iI$. On the other hand, the eigenvalues of e^R are uniquely determined by the system (6.1). This will be clear from the argument below which shows that e^R is determined by the fundamental matrix $Y(t)$, while if $Y(t)$ is replaced by the arbitrary fundamental matrix $Y(t)C_0$, where C_0 is a nonsingular constant matrix, then e^R is replaced by $C_0^{-1}e^R C_0$. The eigenvalues $\sigma_1, \ldots, \sigma_d$ of $C = e^{Rp}$ are called the *characteristic roots* of the system (6.1). If $\lambda_1, \ldots, \lambda_d$ are eigenvalues of R, then $e^{\lambda_1}, \ldots, e^{\lambda_d}$ are the eigenvalues of e^R so that if the numeration is chosen correctly, $\sigma_1 = e^{\lambda_1 p}, \ldots, \sigma_d = e^{\lambda_d p}$. The numbers $\lambda_1, \ldots, \lambda_d$ which are determined by (6.1) only modulo $2\pi i/p$ are the *characteristic exponents* of (6.1). It is seen from (6.3) that (6.1) has d linearly independent solutions $y_1(t), \ldots, y_d(t)$ such that the order number of $y_j(t)$ is $\operatorname{Re} \lambda_j = p^{-1} \operatorname{Re} \log \sigma_j$ for $j = 1, \ldots, d$.

Proof of Theorem 6.1. Since $Y(t)$ is a fundamental matrix of (6.1), it follows from (6.2) that $Y(t + p)$ is also a fundamental matrix of (6.1). Hence, by the remark preceding Lemma 1.2, there exists a constant nonsingular matrix C such that

$$(6.4) \qquad\qquad Y(t + p) = Y(t)C.$$

It will be shown that $\det C \neq 0$ implies that there is a (nonunique) matrix R such that

$$(6.5) \qquad\qquad C = e^{Rp};$$

i.e., C has a logarithm Rp. If this is granted, (6.4) can be written as

$$(6.6) \qquad\qquad Y(t + p) = Y(t)e^{Rp}.$$

Define $Z(t)$ by (6.3), i.e., by $Z(t) = Y(t)e^{-Rt}$; cf. (5.8). Then $Z(t + p) = Y(t + p)e^{-R(t+p)} = [Y(t + p)e^{-Rp}]e^{-Rt} = Y(t)e^{-Rt}$ by (6.6). Thus $Z(t + p) = Z(t)$, as claimed.

In order to complete the proof, it is necessary to verify the existence of an R satisfying (6.5). Since (6.5) is equivalent to $QCQ^{-1} = \exp pQRQ^{-1}$, it is sufficient to suppose that C is in a Jordan normal form. In fact, the considerations of the last section show that it is sufficient to consider the case that C is a matrix of the form $J = \lambda I + K$, where the elements of K are 0 except for those on the subdiagonal which are 1; cf. (5.14) and (5.17). Also $\det J = \lambda^d \neq 0$ implies that $\lambda \neq 0$. Writing $J = \lambda(I + K/\lambda)$ and noting that $\log(1 + t) = t - t^2/2 + t^3/3 - \cdots$, we are led to expect that a logarithm of J is given by

$$(6.7) \qquad \log J = (\log \lambda)I + S, \qquad \text{where } S = -\sum_{j=1}^{\infty} \frac{(-K)^j}{\lambda^j j};$$

i.e., $J = \lambda e^S$ or equivalently, $I + K/\lambda = e^S$. This can be readily verified as follows:

Note that the series in (6.7) is, in fact, a finite sum since $K^d = 0$. The formal rearrangement of power series to obtain

$$\sum_{n=0}^{\infty} \left[-\sum_{j=1}^{\infty} \frac{(-t)^j}{j} \right]^n \frac{1}{n!} = 1 + t,$$

i.e., $\exp[\log(1 + t)] = 1 + t$, is valid for $|t| < 1$. It is clear that the same formal calculation gives $e^S = 1 + K/\lambda$. Furthermore, this formal calculation is permissible, since the powers of K commute and there is obviously no question of convergence to be considered.

7. Adjoint Systems

Consider again the system (1.1). If A^* is the complex conjugate transpose of A, the system

$$(7.1) \qquad\qquad z' = -A^*(t)z$$

is called the system adjoint to (1.1). The corresponding inhomogeneous system is

$$(7.2) \qquad\qquad z' = -A^*(t)z - g(t).$$

There are several results relating (1.1) and (7.1). The first of these is

Lemma 7.1. *A nonsingular, $d \times d$ matrix $Y(t)$ is a fundamental matrix for (1.1) if and only if $(Y^*(t))^{-1} \equiv (Y^{-1}(t))^*$ is a fundamental matrix for (7.1).*

This follows from the fact that if $Y(t)$ has a continuous derivative Y', then $(Y^{-1}(t))' = -Y^{-1}(t)Y'(t)Y^{-1}(t)$ as can be seen by differentiating $Y(t)Y^{-1}(t) \equiv I$. Thus, if $Y(t)$ is a fundamental matrix for (1.1), so that $Y' = AY$, then $(Y^{-1})' = -Y^{-1}A$ and taking the complex conjugate transpose of this relation gives $(Y^{*-1})' = -A^*(Y^{*-1})$. The converse is proved similarly.

Exercise 7.1. Show that (1.1) has a fundamental matrix $Y(t)$ which is unitary, $Y = Y^{*-1}$, if and only if (1.1) is self-adjoint; i.e., if and only if $A(t)$ is skew Hermitian, $A = -A^*$. In this case, if $y = y(t)$ is a solution of (1.1), then $\|y(t)\|$ is a constant.

Lemma 7.2 (Green's Formula). *Let $A(t)$, $f(t)$, $g(t)$ be continuous for $a \leqq t \leqq b$; $y(t)$ a solution of (1.2); $z(t)$ a solution of (7.2). Then, for $a \leqq t \leqq b$,*

$$(7.3) \qquad \int_a^t [f(s) \cdot z(s) - y(s) \cdot g(s)] \, ds = y(t) \cdot z(t) - y(a) \cdot z(a),$$

where the dot denotes scalar multiplication.

This relation is proved by showing that both sides have the same derivative, since $Ay \cdot z = y \cdot A^*z$.

8. Higher Order Linear Equations

In this section let $p_0(t), p_1(t), \ldots, p_{d-1}(t), h(t)$ be continuous, real- or complex-valued functions for $a \leq t \leq b$. The linear homogeneous differential equation,

$$(8.1) \qquad u^{(d)} + p_{d-1}(t)u^{(d-1)} + \cdots + p_1(t)u' + p_0(t)u = 0,$$

and the corresponding inhomogeneous equation,

$$(8.2) \qquad u^{(d)} + p_{d-1}(t)u^{(d-1)} + \cdots + p_0(t)u = h(t),$$

will be considered. The treatment of these equations reduces to that of (1.1) and (1.2) by letting $y = (u^{(0)}, u^{(1)}, \ldots, u^{(d-1)})$, where $u = u^{(0)}$,

$$(8.3) \qquad A(t) = \begin{pmatrix} 0 & 1 & 0 & 0 & \ldots & 0 \\ 0 & 0 & 1 & 0 & \ldots & 0 \\ . & . & . & . & \ldots & . \\ 0 & 0 & 0 & 0 & \ldots & 1 \\ -p_0 & -p_1 & -p_2 & -p_3 & \ldots & -p_{d-1} \end{pmatrix}$$

and $f(t) = (0, \ldots, 0, h(t))$. It seems worthwhile, however, to summarize the essential facts for this important special case:

(i) The initial value problem $u(t_0) = u_0, u'(t_0) = u_0', \ldots, u^{(d-1)}(t_0) = u_0^{(d-1)}$ belonging to (8.2), where $u_0, u_0', \ldots, u_0^{(d-1)}$ are d arbitrary numbers, has a unique solution $u = u(t)$ which exists on $a \leq t \leq b$. In particular, if (8.2) is replaced by (8.1) and $u_0 = u_0' = \cdots = u_0^{(d-1)} = 0$, then $u(t) \equiv 0$; hence no solution $u(t) \not\equiv 0$ of (8.1) has infinitely many zeros on $a \leq t \leq b$.

(ii) (Principle of superposition) (a) Let $u = u_1(t), u_2(t)$ be two solutions of (8.1), then any linear combination $u = c_1 u_1(t) + c_2 u(t)$ with constant coefficients c_1, c_2 is also a solution of (8.1); (b) if $u = u(t)$ and $u = u_1(t)$ are solutions of (8.1) and (8.2), respectively, then $u = u(t) + u_1(t)$ is a solution of (8.2); conversely, if $u = u_0(t), u_1(t)$ are solutions of (8.2), then $u = u_0(t) - u_1(t)$ is a solution of (8.1).

When the functions $u_1(t), \ldots, u_k(t)$ possess continuous derivatives of order $k - 1$, their Wronskian or Wronskian determinant, $W(t) = W(t, u_1, \ldots, u_k)$, is defined to be det $(u_i^{(j-1)}(t))$ for $i, j = 1, \ldots, k$,

$$W(t) = \det \begin{pmatrix} u_1 & u_2 & \cdots & u_k \\ u_1' & u_2' & \cdots & u_k' \\ . & . & \cdots & . \\ u_1^{(k-1)} & u_2^{(k-1)} & \cdots & u_k^{(k-1)} \end{pmatrix}$$

A set of k continuous functions $u_1(t), \ldots, u_k(t)$ on $a \leq t \leq b$ is said to be linearly dependent if there exists constants $c_1, \ldots, c_k$, *not all* 0, such that $c_1 u_1(t) + \cdots + c_k u_k(t) \equiv 0$ for $a \leq t \leq b$. Otherwise, the functions $u_1(t), \ldots, u_k(t)$ are called linearly independent. It is clear that if $u_1, \ldots, u_k$ have continuous derivatives of order $k - 1$, then a necessary condition for $u_1, \ldots, u_k$ to be linearly dependent is that $W(t; u_1, \ldots, u_k) \equiv 0$ for $a \leq t \leq b$. It is also clear that the converse is false (e.g., $u_1(t), u_2(t)$ can be linearly independent on $0 \leq t \leq 1$ with $u_1(t) = 0$ for $0 \leq t \leq \frac{1}{2}$ and $u_2(t) = 0$ for $\frac{1}{2} \leq t \leq 1$ so that $W(t; u_1, u_2) = 0$ for $0 \leq t \leq 1$). For $k = d$ solutions of (8.1), however, the following holds:

(iii) Let $u_1(t), \ldots, u_d(t)$ be solutions of (8.1) and $W(t) = W(t; u_1, \ldots, u_d)$. Then

$$(8.4) \qquad W(t) \exp \int_{t_0}^{t} p_{d-1}(s)\, ds = W(t_0) \qquad \text{for } a \leq t,\, t_0 \leq b,$$

and $u_1, \ldots, u_d$ are linearly dependent if and only if $W(t)$ vanishes at one point, in which case $W(t) \equiv 0$ for $a \leq t \leq b$.

Formula (8.4) is a particular case of (1.5) in view of (8.3). The last part of (iii) is a consequence of uniqueness (i) and the superposition principle (ii).

(iv) Let $u_1(t), \ldots, u_{d-1}(t)$ be $d - 1$ linearly independent solutions of (8.1). Then equation (8.1) is equivalent to the equation

$$(8.5) \qquad \left[W(t; u, u_1, \ldots, u_{d-1}) \exp \int_{a}^{t} p_{d-1}(s)\, ds \right]' = 0$$

for the unknown u. When $d = 2$, this equation is equivalent to

$$(u/u_1)' = u_1^{-2}(t) \exp\left(-\int^{t} p_1(s)\, ds \right) \qquad \text{where } u_1(t) \neq 0.$$

If d linearly independent solutions $u_1, \ldots, u_d$ of (8.1) are known, we can particularize (2.5) to find a formula giving solutions of (8.2) in terms of a quadrature. It is easier to verify the following directly:

(v) For a fixed $s, a \leq s \leq b$, let $u = u(t; s)$ be the solution of (8.1) determined by the initial conditions

$$(8.6) \qquad u = u' = \cdots = u^{(d-2)} = 0,\, u^{(d-1)} = 1 \qquad \text{at } t = s,$$

and let $u_0(t)$ be an arbitrary solution of (8.1). Then

$$(8.7) \qquad u(t) = u_0(t) + \int_{a}^{t} u(t; s) h(s)\, ds$$

is the solution of (8.2) satisfying $u^{(k)}(a) = u_0^{(k)}(a)$ for $k = 0, \ldots, d - 1$.

It is easy to deduce from (1.7) and Lemma 1.2 that $u(t; s)$ and its d derivatives $u', \ldots, u^{(d)}$ with respect to t are continuous functions of (t, s)

for $a \leq t, s \leq b$. In particular, the integral in (8.7) exists and is a function possessing d continuous derivatives with respect to t which can be calculated formally. A direct verification substituting (8.7) into (8.2), shows that (8.7) is the solution of (8.2) satisfying the specified initial conditions.

(vi) Consider a differential equation,

$$(8.8) \qquad u^{(d)} + a_{d-1}u^{(d-1)} + \cdots + a_1 u' + a_0 u = 0,$$

where $a_0, a_1, \ldots, a_{d-1}$ are constants. The associated *characteristic equation* is defined to be

$$(8.9) \qquad \lambda^d + a_{d-1}\lambda^{d-1} + \cdots + a_1 \lambda + a_0 = 0.$$

If y is the vector $y = (u^{(d-1)}, \ldots, u', u)$, then (8.8) is equivalent to the system (5.1) where R is the constant matrix

$$(8.10) \qquad R = \begin{pmatrix} -a_{d-1} & -a_{d-2} & -a_{d-3} & \cdots & -a_1 & -a_0 \\ 1 & 0 & 0 & \cdots & 0 & 0 \\ 0 & 1 & 0 & \cdots & 0 & 0 \\ \cdot & \cdot & \cdot & \cdots & \cdot & \cdot \\ 0 & 0 & 0 & \cdots & 1 & 0 \end{pmatrix}$$

Note that the components of y are written in the order reverse to that considered in connection with (8.1).

It is readily verified that $u = e^{\lambda t}$ is a solution of (8.8) if and only if (8.9) holds. Actually, (8.9) is identical with the characteristic equation $\det(\lambda I - R) = 0$ for R. In order to see this, consider the relation $Ry = \lambda y$ where $y \neq 0$. It is seen that this relation holds if and only if λ satisfies (8.9) and $y = c(\lambda^{d-1}, \ldots, \lambda^2, \lambda, 1)$ for some constant c. Thus λ is an eigenvalue of R if and only if λ satisfies (8.9). It follows that (8.9) is the characteristic equation of R when the roots of (8.9) are distinct. If the roots are not distinct, the coefficients $a_0, \ldots, a_{d-1}$ in (8.9) [and correspondingly in (8.10)] can be altered by arbitrarily small amounts so that the resulting polynomial has distinct roots and hence is the characteristic polynomial of the altered R. The desired conclusion follows by letting the arbitrarily small alterations tend to 0.

Exercise 8.1. By induction on d, give another proof that (8.9) is the same as the equation $\det(\lambda I - R) = 0$. (Still another proof follows from Exercise 8.2.)

If λ is a root of (8.9) of multiplicity m, $1 \leq m \leq d$, then $u = e^{\lambda t}$, $te^{\lambda t}, \ldots, t^{m-1}e^{\lambda t}$ are solutions of (8.8). In order to see this, let $L[u]$ denote the expression on the left of (8.8) if $u(t)$ is any function with d continuous derivatives. Thus (8.8) is equivalent to $L[u] = 0$. Let $F(\lambda)$ be the polynomial on the left of (8.9), so that $F = \partial F/\partial\lambda = \cdots = \partial^{m-1}F/\partial\lambda^{m-1} = 0$ at the given value of λ. Note that $L[e^{\lambda t}] = F(\lambda)e^{\lambda t}$ and, since the coefficients of (8.8) are constant, $L[t^k e^{\lambda t}] = L[\partial^k e^{\lambda t}/\partial\lambda^k] = \partial^k L[e^{\lambda t}]/\partial\lambda^k = \partial^k(F(\lambda)e^{\lambda t})/\partial\lambda^k = 0$ for $k = 0,\ldots, m-1$ at the given value of λ.

Thus if $\lambda(1),\ldots,\lambda(g)$ are the distinct roots of (8.9) and if $h(j)$ is the multiplicity of the root $\lambda(j)$, then d solutions of (8.8) are given by $u = t^k e^{\lambda(j)t}$ for $j = 1,\ldots,g$, and $k = 0,\ldots, h(j)-1$, where $h(1) + \cdots + h(g) = d$.

Exercise 8.2.   (a) Show that the functions $u = t^k e^{\lambda(j)t}$, where $j = 1,\ldots,$ g, and $k = 0,\ldots, h(j)-1$ are linearly independent. (b) Let $Y(t)$ be the fundamental matrix for $y' = Ry$ in which the successive columns are solution vectors $y = (u^{(d-1)},\ldots, u^{(0)})$ corresponding to u (or $u^{(0)}$) $= t^k e^{\lambda(j)t}/k!$ in the following order: first, $j = 1$ and $k = h(1)-1, h(1)-2,\ldots,0$; then $j = 2$ and $k = h(2)-1,\ldots,0$; etc. Let $J = \text{diag}\,[J(1),\ldots,J(g)]$ in the notation of § 5. Show that $Y^{-1}(t)RY(t) = J$; i.e., $Y(t)J = RY(t)$ or, equivalently, $Y(t)J = Y'(t)$. In particular, $J = Y^{-1}(0)RY(0)$ is a Jordan normal form for R. (c) For another proof of (a) and for use in the proof of Theorem X 17.5, show that

$$(8.11) \qquad \det Y(0) = \prod_{i<j}[\lambda(i) - \lambda(j)]^{h(i)h(j)}$$

(vii) When the coefficients $p_0(t),\ldots,p_{d-1}(t)$ of (8.1) are periodic of period p, the corresponding linear system (1.1) of first order has periodic coefficients of period p, by (8.3), and § 6 is applicable.

(viii) (Adjoint equations) Consider a dth order differential equation

$$(8.12) \qquad p_d(t)u^{(d)} + p_{d-1}(t)u^{(d-1)} + \cdots + p_1(t)u' + p_0(t)u = 0$$

with complex-valued coefficients on an interval $a \leq t \leq b$, where $p_k(t)$ has k continuous derivatives for $k = 0, 1,\ldots,d$. The adjoint equation of (8.12) is defined to be

$$(8.13) \quad (-1)^d(\bar{p}_d v)^{(d)} + (-1)^{d-1}(\bar{p}_{d-1}v)^{d-1} + \cdots + (-1)(\bar{p}_1 v)' + \bar{p}_0 v = 0.$$

Note that an integration by parts gives

$$\int_a^b p_k u^{(k)}\bar{v}\, dt = [p_k u^{(k-1)}\bar{v}]_a^b - \int_a^b u^{(k-1)}(p_k\bar{v})'\, dt,$$

and repetitions of this give

$$\int_a^b p_k u^{(k)}\bar{v}\, dt = \left[\sum_{j=0}^{k-1}(-1)^j u^{(k-j-1)}(p_k\bar{v})^{(j)}\right]_a^b + (-1)^k\int_a^b u(p_k\bar{v})^{(k)}\, dt.$$

Thus if $u(t), v(t)$ have continuous dth order derivatives and

$$(8.14) \qquad L[u] = \sum_{k=0}^d p_k(t)u^{(k)}(t), \qquad L^*[v] = \sum_{k=0}^d (-1)^k[\bar{p}_k(t)v(t)]^{(k)},$$

then

$$(8.15) \quad \int_a^b \{L[u]\bar{v} - u\bar{L}^*[v]\} \, dt = \left[\sum_{k=1}^d \sum_{j=0}^{k-1} (-1)^j u^{(k-j-1)} (p_k \bar{v})^{(j)} \right]_a^b.$$

This is called *Green's formula*. The differentiated form

$$(8.16) \qquad L[u]\bar{v} - u\bar{L}^*[v] = \left\{ \sum_{k=1}^d \sum_{j=0}^{k-1} (-1)^j u^{(k-j-1)} (p_k \bar{v})^{(j)} \right\}'$$

is called the *Lagrange identity*.

A corollary of (8.15) is the fact that if u, v are solutions of (8.12) and (8.13), respectively, then

$$(8.17) \qquad \sum_{k=0}^d \sum_{j=0}^{k-1} (-1)^j u^{(k-j-1)} (p_k \bar{v})^{(j)} \equiv \text{const.}$$

(ix) *(Frobenius factorization)* Suppose that (8.1) has d solutions $u_1(t), \ldots, u_d(t)$ such that

$$(8.18) \qquad W_k(t) \equiv W(t; u_1, \ldots, u_k) \neq 0 \qquad \text{for } k = 1, \ldots, d.$$

Then (8.2) can be written as

$$(8.19) \qquad a_d(a_{d-1} \ldots \{a_2[a_1(a_0 u)']'\}' \ldots)' = h(t),$$

where $a_j = W_j^2 / W_{j-1} W_{j+1}$ and $W_0 = W_{-1} = W_{d+1} = 1$. This is a consequence of Corollary 3.1.

Exercise 8.3 *(Pólya)*. Let $p_0(t), \ldots, p_{d-1}(t)$ in (8.1) be continuous and real-valued for $a \leq t \leq b$. Consider only real functions. Put

$$L[v](t) \equiv v^{(d)}(t) + p_{d-1}(t) v^{(d-1)}(t) + \cdots + p_0(t) v(t),$$

so that (8.2) takes the form $L[u] = h(t)$. If a function $v(t)$ has $k - 1$ (≥ 0) continuous derivatives, a point $t = t_0$ will be called a zero of $v(t)$ of a multiplicity at least k if $v(t_0) = v'(t_0) = \cdots = v^{(k-1)}(t_0) = 0$. Equation (8.1) will be said to have property (W) on (a, b) if (8.1) has d solutions $u_1(t), \ldots, u_d(t)$ satisfying (8.18) for $a < t < b$; actually this condition for $k = d$ is trivial by **(iii)**. (a) *Zeros and property* (W). Show that if no solution $u(t) \not\equiv 0$ of (8.1) has d zeros counting multiplicities on $[a, b)$ then (8.1) has property (W) on (a, b). In the remainder of this exercise, *assume that* (8.1) *has property* (W) *on* (a, b). (b) *Generalization of Rolle's theorem*. Let $v(t)$ have d continuous derivatives on (a, b) and at least $d + 1$ zeros counting multiplicities on (a, b). Then there exists at least one point $t = \theta$ of (a, b) where $L[v](\theta) = 0$. (c) *Partial converse for* (a). Show that no solution $u(t) \not\equiv 0$ of (8.1) has d zeros counting multiplicities on (a, b). (d) *Interpolation*. Let $k \geq 1$; $m_1 + \cdots + m_k = d$, where m_j (≥ 1) is an integer; $t_1 < \cdots < t_k$ points of (a, b) and $u_j^{(i)}$, $i = 1, \ldots, m_j$

and $j = 1, \ldots, k$, arbitrary numbers. Then (8.1) has a unique solution $u = u(t)$ satisfying $u^{(i)}(t_j) = u_j^{(i)}$ for $i = 0, \ldots, m_j - 1$ and $j = 1, \ldots, k$.

(e) *Equation* $L[u] = 1$. Same as (d) with (8.1) replaced by $L[u] = 1$.

(f) *Mean value theorem.* Let $k; m_1, \ldots, m_k$ and $t_1, \ldots, t_k$ be as in (d). Let $v(t)$ have d continuous derivatives on (a, b); $u(t)$ the unique solution of (8.1) satisfying $u^{(i)}(t_j) = v^{(i)}(t_j)$ for $i = 0, \ldots, m_j - 1$ and $j = 1, \ldots, k$; $u = u_0(t)$ the unique solution of $L[u] \equiv 1$ satisfying $u_0^{(i)}(t_j) = 0$ for $i = 0, \ldots, m_j - 1$ and $j = 1, \ldots, k$; and $a < t_0 < b$. If $[\gamma, \delta] \subset (a, b)$ contains t_1, t_k, and t_0, then there exists at least one point $t = \theta$ of (γ, δ) such that $v(t_0) = u(t_0) + u_0(t_0)L[v](\theta)$. (The assertions $(b) - (f)$ reduce to standard theorems if (8.1) reduces to the trivial equation $u^{(d)} = 0$ with the solutions $u = 1, t, \ldots, t^{d-1}$.)

Exercise 8.4. Let $p_0(t), \ldots, p_{d-1}(t)$ in (8.1) be continuous for $a < t < b$. Show that no solution $u(t) \not\equiv 0$ of (8.1) has d zeros counting multiplicities on (a, b) if and only if no solution $u(t) \not\equiv 0$ has d distinct zeros on (a, b). See Hartman [15].

9. Remarks on Changes of Variables

This section contains remarks which will be referred to in later chapters.

(i) If R is a constant $d \times d$ matrix, the usual Jordan normal form $J = Q^{-1}RQ$ for R under similarity transformations is described by (5.14)–(5.16). It will often be convenient to note that the 1's on the subdiagonal of K_h in (5.16) can be replaced by an arbitrary $\epsilon \neq 0$. This is a consequence of the following formula in which $J(j)$ is given in (5.15):

$$Q_\epsilon^{-1}J(j)Q_\epsilon = \lambda I_h + \epsilon K_h$$

if $Q_\epsilon = \mathrm{diag}\,(1, \epsilon^{-1}, \ldots, \epsilon^{1-h})$.

(ii) Consider a real nonlinear system of differential equations of the form

$$(9.1) \qquad y' = Ry + f(t, y),$$

where R is a constant matrix. Under a linear change of variables with constant coefficients,

$$(9.2) \qquad y = Qz, \qquad \det Q \neq 0,$$

(9.1) becomes

$$(9.3) \qquad z' = Jz + Q^{-1}f(t, Qz), \qquad J = Q^{-1}RQ.$$

Although R is a matrix with real entries, its eigenvalues need not be real. Correspondingly, there need not exist a real Q such that J is in a Jordan normal form. For a Q with complex entries, $f(t, Qz)$ may not be defined.

The point to be made, however, is that for many purposes the formal change of variables (9.2) using a Q with complex entries is permissible if (9.2) and (9.3) are suitably interpreted. Formal operations with (9.3) are then legitimate.

If Q is chosen so that $J = Q^{-1}RQ$ is in a Jordan normal form, then the columns of Q are eigenvectors of R or a power of R. Thus, if R has α real eigenvalues (counting multiplicities), $0 \leq \alpha \leq d$, then the other eigenvalues of R occur in pairs of complex conjugate numbers. Let $d - \alpha = 2\beta$. Correspondingly, it can be supposed that the first β columns of Q are, respectively, complex conjugates of the next β columns and that the last α columns are real. Thus, if Q_0 is the matrix

$$Q_0 = \begin{pmatrix} I_\beta & iI_\beta & 0 \\ I_\beta & -iI_\beta & 0 \\ 0 & 0 & I_\alpha \end{pmatrix}$$

where I_h is the unit $h \times h$ matrix, then QQ_0 is a matrix with real entries. The change of variables

(9.4) $$y = QQ_0 w$$

transforms (9.1) into

(9.5) $$w' = Q_0^{-1}JQ_0 w + Q_0^{-1}Q^{-1}f(t, QQ_0 w)$$

which is equivalent to

(9.6) $$(Q_0 w)' = J(Q_0 w) + Qf(t, QQ_0 w).$$

The differential equations (9.5) are real equations; the differential equations in (9.6) result by taking linear combinations of one or two equations in (9.5) with constant coefficients 1 or $\pm i$.

Below, equation (9.3) is to be interpreted as (9.6). This is equivalent to saying that, in (9.3), $z^{k+\beta} = \bar{z}^k$ for $k = 1, \ldots, \beta$ and $z^{k+2\beta}$ is real for $k = 1, \ldots, \alpha$. Thus we can consider the variables in (9.3) to be $w = (w^1, \ldots, w^d)$, where $w^k = \frac{1}{2}(z^k + \bar{z}^k)$ and $w^{k+\beta} = -\frac{1}{2}i(z^k - \bar{z}^k)$ for $k = 1, \ldots, \beta$, and $w^k = z^k$ for $2\beta \leq k \leq d$.

Exercise 9.1. Let R be a constant $d \times d$ matrix with eigenvalues $\lambda_1, \ldots, \lambda_d$ such that $\lambda_1, \ldots, \lambda_k$ are simple eigenvalues for some k, $1 \leq k \leq d$. Let $G(t)$ be a continuously differentiable $d \times d$ matrix for $t \geq 0$ such that $G(t) \to 0$ as $t \to \infty$ and $\int^\infty \|G'(t)\| \, dt < \infty$. (a) For large t, show that $R + G(t)$ has k simple eigenvalues $\lambda_j(t)$ such that $\lambda_j(t) \to \lambda_j$ as $t \to \infty$; $\lambda_j(t)$ is continuously differentiable and $\int^\infty |\lambda_j'(t)| \, dt < \infty$ for

$j = 1, \ldots, k$. (b) Let Q_0 be a constant, nonsingular matrix such that $Q_0^{-1}RQ_0 = \text{diag } [\lambda_1, \ldots, \lambda_k, E_0]$, where E_0 is a $(d - k) \times (d - k)$ matrix (e.g., suppose that $Q_0^{-1}RQ_0$ is a Jordan normal form for R). Let $e_r = (e_r{}^1, \ldots, e_r{}^d)$, where $e_r{}^j$ is 1 or 0 according as $j = r$ or $j \neq r$. Show that, for large t, $R + G(t)$ has an eigenvector $y_j(t)$, $[R + G(t)]y_j(t) = \lambda_j(t)y_j(t)$, such that $y_j(t) \rightarrow Q_0 e_j$ as $t \rightarrow \infty$, and $y_j(t)$ has a continuous derivative satisfying $\int^\infty \|y_j{}'(t)\| \, dt < \infty$ for $j = 1, \ldots, k$. (c) Show that, for large t, there exists a continuously differentiable, nonsingular matrix $Q(t)$ such that $Q(\infty) = \lim Q(t)$ as $t \rightarrow \infty$ exists and is nonsingular, $\int^\infty \|Q'(t)\| \, dt < \infty$, and $Q^{-1}(t)[R + G(t)]Q(t)$ has the form diag $[\lambda_1(t), \ldots, \lambda_k(t), E(t)]$, where $E(t)$ is a $(d - k) \times (d - k)$ matrix.

APPENDIX: ANALYTIC LINEAR EQUATIONS

10. Fundamental Matrices

This appendix deals with a linear system of differential equations

$$(10.1) \qquad\qquad y' = A(t)y$$

in which t is a complex variable and $A(t)$ is a matrix of single-valued, analytic functions on some open set E in the t-plane. "Analytic" is used here in the sense of "regular analytic." In a small neighborhood of a point $t_0 \in E$, (10.1) has a fundamental matrix $Y(t)$ which is an analytic function of t (i.e., which has elements that are analytic functions of t). This follows from a modification of the proof by successive approximations of the existence theorem, Lemma 1.1; cf. Exercise II 1.1. Hence, if E is simply connected, the monodromy theorem implies that $Y(t)$ exists and is a single-valued, analytic function of t on E.

Most of this appendix deals with the case when E is not simply connected but is a punctured disc $0 < |t| < a$.

Lemma 10.1. *Let $A(t)$ be a matrix of single-valued, analytic functions on the disc $0 < |t| < a$ and suppose that $A(t)$ [i.e., at least one of the elements of $A(t)$] is not analytic at $t = 0$. Then (10.1) cannot have a fundamental matrix $Y(t)$ which is single-valued, analytic on $0 < |t| < a$ and continuous at $t = 0$ with $\det Y(0) \neq 0$.*

Proof. If this is false, then $Y(t)$, $Y^{-1}(t)$ are analytic on $|t| < a$ and so is $A(t) = Y'(t)Y^{-1}(t)$ by Riemann's theorem on removable singularities.

Theorem 10.1. *Let $A(t)$ be a matrix of single-valued, analytic functions on $0 < |t| < a$. Then any fundamental matrix $Y(t)$ of (10.1) (which need* *not be single-valued) has a representation of the form*

(10.2) $$Y(t) = Z(t)t^R,$$

where $Z(t)$ is a matrix of single-valued, analytic functions on $0 < |t| < a$, R is a constant matrix, and

(10.3) $$t^R = e^{R \log t} = \sum_{n=0}^{\infty} \frac{(R \log t)^n}{n!}.$$

If T is a constant nonsingular matrix, then $Y(t)T$ is a fundamental matrix and

$$Y(t)T = Z(t)T(T^{-1}t^R T) = [Z(t)T]t^{T^{-1}RT}.$$

T can be chosen so that $T^{-1}RT$ is in a Jordan normal form. The form of the matrix $t^{T^{-1}RT} = \exp(T^{-1}RT \log t)$ can be seen by replacing t by $\log t$ in (5.18).

Proof. Let $Y(t)$ be a fundamental solution of (10.1) determined locally near a point $t = t_0$, $0 < |t_0| < a$, and continued analytically, possibly multiply valued. If the point t makes a circuit around $t = 0$ in $0 < |t| < a$, then the matrix returns with values, say $Y_0(t)$, for t near t_0. Since $A(t)$ is single-valued, $T_0(t)$ is also a fundamental matrix for (1.1), thus, there exists a nonsingular constant matrix C such that

(10.4) $$Y_0(t) = Y(t)C.$$

By analyticity, (10.4) holds for analytic continuations of $Y(t)$, $Y_0(t)$.

For a fixed r, $0 < r < a$, consider the matrix function $Y(\theta, r) = Y(re^{i\theta})$ of θ for $-\infty < \theta < \infty$. Then (10.4) means that $Y(\theta + 2\pi, r) = Y(\theta, r)C$ or that $Y(\theta + 2\pi, r) = Y(\theta, r)e^{2\pi iR}$ if $2\pi iR$ is a logarithm of C; cf. § 6. It is readily seen that $Z_0(\theta, r) = Y(\theta, r)e^{-iR\theta}$ is of period 2π in θ; see the argument following (6.6).

Note that $Z_0(\theta, r)r^{-R} = Y(\theta, r)e^{-iR\theta}r^{-R} = Y(t)t^{-R}$ is an analytic function of t, say $Z(t)$, and is single-valued since $Z_0(\theta, r)$ is of period 2π in θ. This proves Theorem 10.1.

If (10.2) is a fundamental matrix for (10.1), properties of $Z(t)$ and R will be investigated. To this end, a differential equation satisfied by Z will be calculated. Since R commutes with t^R and $(t^R)' = Rt^R/t$, it follows from (10.1) and (10.2) that

$$Z't^R + \frac{ZRt^R}{t} = AZt^R.$$

Hence

(10.5) $$Z = -\frac{ZR}{t} + A(t)Z.$$

The equation (10.5) is not the type of matrix differential equation considered above since Z occurs as a factor on the left of ZR and on the right of $A(t)Z$. In order to deal with (10.5), it is convenient to arrange the d^2 elements z_{jk} of the matrix $Z = (z_{jk})$ in some arbitrarily fixed order and to consider (10.5) as a linear, homogeneous system for a d^2-dimensional vector Z; say

$$(10.6) \qquad\qquad Z' = \tilde{A}(t)Z,$$

where $\tilde{A}(t)$ is a $d^2 \times d^2$ matrix. An element of $\tilde{A}(t)$ is a linear combination of elements of $A(t)$ and R/t with constant coefficients. Hence $\tilde{A}(t)$ is single-valued and analytic for $0 < |t| < a$. In particular, if the elements of $A(t)$ are analytic or have a simple pole at $t = 0$, then the same is true of $\tilde{A}(t)$.

In order to avoid an interruption to the arguments below, a simple algebraic lemma will be stated and proved here. Let B be a constant $d \times d$ matrix, X a variable $d \times d$ matrix and Y the commutator

$$(10.7) \qquad\qquad Y = BX - XB.$$

Consider the d^2 elements of X and Y arranged in a fixed (arbitrary) manner and (10.7) as a linear transformation from the d^2-dimensional X-space into itself. Thus (10.7) can be written as

$$(10.8) \qquad\qquad Y = \tilde{B}X,$$

where $\tilde{B}$ is a $d^2 \times d^2$ matrix and X, Y are d^2-dimensional vectors.

Lemma 10.2. *Let the d eigenvalues of B be $\lambda_1, \dots, \lambda_d$ counting multiplicities. Then the d^2 eigenvalues of $\tilde{B}$ are $\lambda_j - \lambda_k$ for $j, k = 1, \dots, d$.*

Proof. Let T be a nonsingular $d \times d$ matrix and let $C = T^{-1}BT$, so that B, C have the same eigenvalues. Let $\tilde{C}$ be related to C as $\tilde{B}$ is to B. The matrices $\tilde{C}$ and $\tilde{B}$ have the same eigenvalues. In order to see this, note that (10.8) is equivalent to $T^{-1}YT = \tilde{C}(T^{-1}XT)$, for (10.7) can be written as

$$(T^{-1}BT)(T^{-1}XT) - (T^{-1}XT)(T^{-1}BT) = T^{-1}YT.$$

Thus, e.g., $\tilde{B}X = \lambda X$ implies that $(\tilde{C}T^{-1}XT) = \lambda(T^{-1}XT)$.

Suppose first that the eigenvalues of B are distinct and choose T so that $C = \text{diag }[\lambda_1, \dots, \lambda_d]$. Then if $X = (x_{jk})$ and $Y = (y_{jk})$, it is seen that $Y = CX - XC$ is equivalent to $\dot{y}_{jk} = (\lambda_j - \lambda_k)x_{jk}$ for $j, k = 1, \dots, d$. Thus $\tilde{C}$ is a diagonal matrix with the d^2 diagonal elements $\lambda_j - \lambda_k$ for $j, k = 1, \dots, n$ and the lemma is proved in this case.

If the eigenvalues of B are not distinct, let $B_1, B_2, \dots$ be a sequence of matrices, each having distinct eigenvalues such that $B_n \to B, n \to \infty$. (The existence of $B_1, B_2, \dots$ is clear; it is sufficient to suppose that B is in a Jordan normal form and to change the diagonal elements by a small amount to obtain B_n.) Then the eigenvalues of B_n can be ordered $\lambda_{1n}, \ldots, \lambda_{dn}$ so that $\lambda_{jn} \to \lambda_j, n \to \infty$, for $j = 1, \ldots, d$. Correspondingly, the eigenvalues of $\tilde{B}_n$ tend to those of $\tilde{B}$. Hence the general case of the lemma follows from the special case treated above.

11. Simple Singularities

In (10.2), $Z(t)$ has a Laurent expansion about $t = 0$. The point $t = 0$ is called a *regular singular point* for (10.1) if (10.1) has a fundamental matrix (10.2) in which the elements of $Z(t)$ do not have an essential singularity at $t = 0$ (i.e., are analytic or have a pole at $t = 0$). In this case, we have

Corollary 11.1. *Let $A(t)$ be analytic and single-valued for $0 < |t| < a$ and let $t = 0$ be a regular singular point for (1.1). Then (1.1) has a fundamental matrix $Y(t)$ of the form*

$$Y(t) = Z_0(t)t^C,$$

where C is a constant matrix and $Z_0(t)$ is analytic for $|t| < a$.

In fact, if $Z(T)$ in (10.2) has at most a pole at $t = 0$, then (10.2) can be written as $Y(t) = Z(t)t^n t^{R-nI} = Z_0(t)t^C$, where $Z_0(t) = Z(t)t^n$ is analytic for some choice of the integer $n \geq 0$ and $C = R - nI$.

If the singularity of $A(t)$ [i.e., of each element of $A(t)$] at $t = 0$ is at most a pole of order one, then $t = 0$ will be called a *simple singularity* of (10.1). In this case (10.1) can be written as

$$(11.1) \qquad ty' = \left(B + \sum_{k=1}^{\infty} A_k t^k\right)y,$$

where $B, A_1, A_2, \ldots$ are constant matrices and

$$(11.2) \qquad B + \sum_{k=1}^{\infty} A_k t^k$$

is convergent for $|t| < a$.

If a new independent variable defined by $s = \log t$ is introduced, (11.1) becomes

$$(11.3) \qquad \frac{dy}{ds} = \left(B + \sum_{k=1}^{\infty} A_k e^{ks}\right)y, \qquad \operatorname{Re} s < a.$$

(For a treatment of this system for real s, see Chapter X.) If $A_1 = A_2 = \cdots = 0$, equation (11.3) is $dy/ds = By$ and has the solution $y = e^{Bs} = t^B$; cf. § 5 and Corollary 11.1.

Theorem 11.1 (Sauvage). *Let $t = 0$ be a simple singularity for (10.1), so that (10.1) is of the form (11.1) where (11.2) is convergent for $|t| < a$. Then $t = 0$ is a regular singular point for (11.1).*

This is an immediate consequence of the following lemma.

Lemma 11.1. *Let* $t = 0$ *be a simple singularity for* (10.1) *and let* $y(t)$ *be a single-valued, analytic solution of* (10.1) *for* $0 < |t| < a$. *Then* $y(t)$ *is analytic or has a pole at* $t = 0$.

For if this lemma is applied to the system (10.6), it follows that $Z(t)$ in (10.2) is analytic or has a pole at $t = 0$.

Proof of Lemma 11.1. Let θ be fixed, $0 \leq \theta < 2\pi$, and $t = re^{i\theta}$, so that $dt = e^{i\theta} dr$ and $y(re^{i\theta})$ is a solution of

$$\frac{dy}{dr} = e^{i\theta} A(re^{i\theta}) y$$

for $0 < r < a$. The assumption on $A(t)$ implies that there exists a constant c such that $\|A(re^{i\theta}) y\| \leq c \|y\|/r$ for $0 < r \leq \frac{1}{2}a$, where $\|y\|$ is the Euclidean norm of y. It follows that any solution $y \not\equiv 0$ satisfies

$$\frac{d\|y\|}{dr} \leq \frac{c \|y\|}{r}$$

for $0 < r \leq \frac{1}{2}a$. Hence $\|y(re^{i\theta})\| \leq C/r^c$ for $0 < r \leq \frac{1}{2}a$ and a suitable constant C. Thus if n is a positive integer, $n \geq c$, then $t^n y(t)$ is bounded for small $|t|$ and hence is analytic at $t = 0$. Thus $y(t)$ has at most a pole of order n. This proves the lemma.

The converse of Theorem 11.1 is false. For it is readily verified that the binary system

$$y^{1\prime} = y^2, \qquad y^{2\prime} = 2y^1/t^2$$

has the fundamental matrix

$$Y(t) = \begin{pmatrix} t^2 & t^{-1} \\ 2t & -t^{-2} \end{pmatrix},$$

which is of the form (10.2) with $Z(t) \equiv Y(t)$ and $R = 0$. Thus $t = 0$ is a regular singular point but is not a simple singularity. However, Theorem 11.1 has a partial converse.

Theorem 11.2. *Let* $Q(t)$ *be a* $d \times d$ *matrix of functions which are single-valued and analytic for* $0 < |t| < a$ *and such that* $t = 0$ *is a regular singular point for*

$$(11.4) \qquad\qquad w' = Q(t)w.$$

Then there exists a matrix $P(t)$, *which is a polynomial in* t *and satisfies* $\det P(t) \equiv 1$, *and a diagonal matrix* $D = \operatorname{diag} [\alpha(1), \ldots, \alpha(d)]$, *where* $\alpha(j) \geq 0$ *is an integer, such that the change of variables*

$$(11.5) \qquad\qquad w = T(t)y, \qquad T(t) = P(t)t^D,$$

transforms (11.4) *into the form* (11.1)–(11.2) *for which* $t = 0$ *is a simple singularity.*

Proof. Since $t = 0$ is a regular singularity of (11.4), there exists a fundamental matrix $W(t)$ of the form

(11.6) $$W(t) = X(t)t^R,$$

where R is a constant matrix and $X(t)$ is analytic for $|t| < a$.

Suppose first that $\det X(0) \neq 0$, then $X^{-1}(t)$ is analytic for $|t| < a$ and $Q(t) = W'(t)W^{-1}(t)$ is given by

(11.7) $$Q(t) = (X' + t^{-1}XR)X^{-1}.$$

Hence $Q(t)$ has at most a simple pole at $t = 0$, so that $t = 0$ is a simple singularity for (11.4) and the theorem follows with $P(t) = I$, $D = 0$.

Consider the case that $\det X(0) = 0$. Suppose, for a moment, that there exist matrices $P(t)$, D of the type specified, such that

(11.8) $$X(t) = P(t)t^D Z(t),$$

where $Z(t)$ is analytic for $|t| < a$ and $\det Z(0) \neq 0$. Thus

$$W(t) = T(t)Z(t)t^R \qquad \text{if } T(t) = P(t)t^D,$$

and so (11.5) transforms (11.4) into a system $y' = Q_0(t)y$ for which $Y(t) = Z(t)t^R$ is a fundamental matrix. By the analogue of (11.7),

$$Q_0(t) = (Z' + t^{-1}ZR)Z^{-1}.$$

Consequently, $t = 0$ is a simple singularity for the new system $y' = Q_0(t)y$, so that Theorem 11.2 will be proved if the following lemma is verified.

Lemma 11.2. *Let $X(t)$ be a matrix of functions analytic for $|t| < a$ such that $\det X(t) \not\equiv 0$. Then $X(t)$ has a representation of the form* (11.8), *where $P(t)$ is a matrix which is a polynomial in t and $\det P(t) \equiv 1$, $D = \operatorname{diag}[\alpha(1), \ldots, \alpha(d)]$ with $\alpha(j) \geq 0$ an integer, and the matrix $Z(t)$ is analytic for $|t| < a$ with $\det Z(0) \neq 0$.*

Remark 1. The relations (11.8), $\det P(t) = 1$, and $\det Z(0) \neq 0$ show that

(11.9) $$\alpha(1) + \cdots + \alpha(d) = \alpha,$$

where $\alpha \geq 0$ is the order of the zero of $\det X(t)$ at $t = 0$.

Proof. The equation (11.8) will be considered in the form

(11.10) $$P^{-1}(t)X(t) = t^D Z(t).$$

From the usual construction of the inverse of a matrix in terms of minors, divided by the determinant, it is clear that both $P(t)$ and $P^{-1}(t)$ have the properties just specified. Instead of constructing $P(t)$, it will be simpler to obtain $P^{-1}(t)$. Also, the proof will give a matrix $P^{-1}(t)$ such that $\det P^{-1}(t)$ is a constant ($\neq 0$), not necessarily 1. The normalization $\det P(t) \equiv 1$ is then obtained in a trivial way.

The matrix $P^{-1}(t)$ will be constructed as a product of a finite number of elementary matrices N of one of the following three types: (i) multiplication of a matrix on the left by N interchanges the jth and kth rows, e.g., multiplication by the matrix

$$N = \begin{pmatrix} 0 & 1 & 0 & . & . & . & 0 \\ 1 & 0 & 0 & . & . & . & 0 \\ 0 & 0 & 1 & . & . & . & 0 \\ . & . & . & . & . & . & . \\ 0 & 0 & 0 & . & . & . & 1 \end{pmatrix}$$

on the left interchanges the first and second rows; (ii) multiplication by N on the left multiplies the jth row by a number $\lambda \neq 0$, e.g., $N = \operatorname{diag}[1, \ldots, 1, \lambda, 1, \ldots, 1]$; and (iii) multiplication by N on the left replaces the jth row by the sum of the jth row and $p(t)$ times the kth row, where $p(t)$ is a polynomial and $k < j$; e.g., for $j = 2$ and $k = 1$,

$$N = \begin{pmatrix} 1 & 0 & 0 & . & . & . & 0 \\ p & 1 & 0 & . & . & . & 0 \\ 0 & 0 & 1 & . & . & . & 0 \\ . & . & . & . & . & . & . \\ 0 & 0 & 0 & . & . & . & 1 \end{pmatrix}$$

Each of these elementary matrices N satisfies $\det N = \text{const.} \neq 0$.

Let $X(t) = (x_{jk}(t))$ and let $\alpha \geqq 0$ be the order of the zero of $\det X(t)$ at $t = 0$. With the jth row of $X(t)$, associate an integer $a(j) \geqq 0$ such that $x_{jk}(t) = t^{a(j)}y_{jk}(t)$, where $y_{jk}(t)$ is analytic at $t = 0$ and at least one of the functions $y_{j1}(t), \ldots, y_{jd}(t)$ does not vanish at $t = 0$. In particular,

$$X(t) = t^E Y(t) \qquad \text{where } E = \operatorname{diag}[a(1), \ldots, a(d)];$$

so that $\det X(t) = t^n \det Y(t)$, where $n = a(1) + \cdots + a(d)$. Hence

$$(11.11) \qquad\qquad a(1) + \cdots + a(d) \leqq \alpha.$$

If equality holds in (11.11), then $\det Y(0) \neq 0$ and the lemma is trivial with $P(t) = I$ and $D = E$.

If inequality holds in (11.11), it will be shown that $X(t)$ can be multiplied on the left by a finite number of the matrices of the type N to obtain a matrix $X_0(t)$, such that if $\alpha(1), \ldots, \alpha(d)$ belong to $X_0(t)$ as $a(1), \ldots, a(d)$

belong to $X(t)$, then $\alpha(1) + \cdots + \alpha(d) = \alpha$. Thus $X_0(t) = P^{-1}(t)X(t)$ has the form $t^D Z(t)$, where $\det Z(0) \neq 0$.

After multiplying $X(t)$ on the left by matrices of type (i), it can be supposed that $a(1) \leq \cdots \leq a(d)$. Let $\epsilon_{jk} = y_{jk}(0)$, so that $x_{jk}(t) = t^{a(j)}(\epsilon_{jk} + \cdots)$. Then not all of the numbers $\epsilon_{11}, \ldots, \epsilon_{1d}$ are zero. Suppose that ϵ_{1m}, $1 \leq m \leq d$, is the first of these elements which is not zero. Then after multiplying $X(t)$ on the left by a matrix of type (ii), it can be supposed that $\epsilon_{1m} = 1$. Also, by replacing the jth row of $X(t)$ by the sum of the jth row and $p(t) = -\epsilon_{jm}t^{a(j)-a(1)}$ times the first row, it can be supposed that $\epsilon_{jm} = 0$ for $j = 2, \ldots, d$. Thus the matrix of the new elements $\epsilon_{jk} = y_{jk}(0)$ is of the form

$$
\begin{pmatrix}
0 & 1 & \epsilon_{13} & \cdots \\
\epsilon_{21} & 0 & \epsilon_{23} & \cdots \\
\epsilon_{31} & 0 & \epsilon_{33} & \cdots \\
\cdots & \cdot & \cdots & \cdots \\
\epsilon_{d1} & 0 & \epsilon_{d3} & \cdots
\end{pmatrix}
$$

if, e.g., $m = 2$. This procedure does not decrease the integers $a(1), \ldots, a(d)$.

If not all elements $\epsilon_{21}, \ldots, \epsilon_{2d}$ of the second row are zero, it can be supposed that if ϵ_{2n} is the first different from 0, then $\epsilon_{2n} = 1$ and $\epsilon_{jn} = 0$ for $j = 3, \ldots, d$.

This procedure can be applied to the second row, then the third row, . . . , unless all of the elements $\epsilon_{j1}, \ldots, \epsilon_{jd}$ in the jth row are zero. In this case, $a(j)$ can be replaced by a larger number, again called $a(j)$. If this occurs, the rows are again permuted to obtain the order $a(1) \leq \cdots \leq a(d)$ and the entire procedure repeated.

After a finite number of repetitions, the procedure of introducing a one in each of the d rows with zeros to the left and below it succeeds because, in view of the limitation $a(1) + \cdots + a(d) \leq \alpha$, it is only possible to increase an index $a(j)$ a finite number of times.

The first column $\epsilon_{11}, \epsilon_{21}, \ldots, \epsilon_{d1}$ contains at least one element different from zero. Otherwise, the ones occur in the last $d - 1$ columns, with only zeros below the ones. This implies that there is a row $\epsilon_{j1}, \ldots, \epsilon_{jd}$ with all zeros, contradicting the construction. If ϵ_{n1} is the first element of the first column such that $\epsilon_{n1} \neq 0$, then, by the construction, $\epsilon_{n1} = 1$ and $\epsilon_{m1} = 0$ for $n \neq m$. Move the nth row to the first position without disturbing the order of the other rows.

In the new matrix, the $d - 1$ elements $\epsilon_{22}, \epsilon_{32}, \ldots, \epsilon_{d2}$ of the second column are not all zero, in fact, exactly one (say, the mth) is one and the others are zero. This follows by the argument of the last paragraph.

Move the mth row to the second position. Continuing this procedure leads to a matrix $X_0(t)$ for which the corresponding (ϵ_{jk}) has ones on the principal diagonal and zeros below it, thus det $(\epsilon_{jk}) = 1$.

Thus the construction gives an $X_0(t) = P^{-1}(t)X(t)$ of the form $t^D Z(t)$, with $Z(0) = (\epsilon_{jk})$ having det $Z(0) = 1$ and $D = \text{diag}\,[\alpha(1), \ldots, \alpha(d)]$ where $\alpha(1), \ldots, \alpha(d)$ belong to $X_0(t)$ as $a(1), \ldots, a(d)$ belong to $X(t)$. This proves the lemma.

Exercise 11.1. Prove that if $Q(t)$ is a $d \times d$ matrix of single-valued, analytic functions on $0 < |t| < a$, then a necessary condition for $t = 0$ to be a regular singularity for (11.4) is that the elements of $Q(t)$ have at most a pole (not an essential singularity) at $t = 0$.

Exercise 11.2. Let $A_0(s)$ be a matrix of functions analytic for $a < |s| < \infty$. The point $s = \infty$ is called a simple singularity [or a regular singular point] for the system

$$(11.12) \qquad \frac{dy}{ds} = A_0(s)y$$

if $t = 0$, where $t = 1/s$, is a simple singularity [or a regular singular point] for $y' = A(t)y$, where $A(t) = -t^{-2}A_0(1/t)$. (*a*) Necessary and sufficient that $s = \infty$ be a simple singularity for (11.12) is that $A_0(s) \to 0$ as $|s| \to \infty$. (*b*) Let $A(t)$ be analytic for all $t \neq t_1, \ldots, t_n, \infty$. Necessary and sufficient conditions that $t = t_1, \ldots, t_n, \infty$ be simple singularities is that $A(t)$ be of the form $A(t) = (t - t_1)^{-1}R_1 + \cdots + (t - t_n)^{-1}R_n$, where R_1, $R_2, \ldots, R_n$ are constant matrices.

Theorem 11.3. *Let* (11.2) *be convergent for* $|t| < a$ *and let*

$$(11.13) \qquad y(t) = \sum_{n=0}^{\infty} y_n t^n$$

be a "formal" power series which satisfies (11.1) *in the sense that the formulae*

$$(11.14) \qquad By_0 = 0$$

$$(11.15_n) \qquad ny_n = By_n + \sum_{k=1}^{n} A_k y_{n-k},$$

$n = 1, 2, \ldots,$ *hold. Then* (11.13) *is convergent for* $|t| < a$.

Proof. Since (11.2) is convergent for $|t| < a$, there exist constants $c > 0$, $\rho > 0$ such that

$$(11.16) \qquad \|A_k\| \leqq c\rho^k, \qquad k = 1, 2, \ldots$$

in the sense that $|A_k y| \leqq c\rho^k |y|$ for all vectors y. (For example, ρ can be chosen to be any number such that $\rho > 1/a$.) Choose $r > 0$ so large that

$$(11.17) \qquad c \sum_{k=1}^{\infty} \left(\frac{\rho}{r}\right)^k \leqq 1.$$

Let m be an integer such that $|By| \leqq m|y|$ for all vectors y. Then $B - nI$ is nonsingular for $n > m$, in fact, $|By - ny| \geqq (n - m)|y| \neq 0$ for $y \neq 0$. Thus if $n > m$, the equation $By - ny = z$ has a solution y, for any given z, and

$$(11.18) \qquad |y| \leqq \frac{|z|}{(n - m)} \; ; \quad \text{in particular,} \; |y| \leqq |z| \quad \text{if } n > m.$$

Let $\gamma > 0$ be chosen so large that

$$(11.19_n) \qquad\qquad\qquad |y_n| \leqq \gamma r^n$$

holds for $n = 0, \dots, m$. It will be verified by induction that (11.19) holds for all n. Thus assume (11.19_n) for $n = 0, \dots, j - 1$ and $j - 1 \geqq m$. Then

$$By_j - jy_j = z_j, \qquad \text{where } z_j = -\sum_{k=1}^{j} A_k y_{j-k}.$$

By (11.16) and the induction hypotheses,

$$|z_j| \leqq \sum_{k=1}^{j} (c\rho^k)(\gamma r^{j-k}) \leqq \gamma r^j c \sum_{k=1}^{\infty} \left(\frac{\rho}{r}\right)^k,$$

so that $|z_j| \leqq \gamma r^j$ by (11.17). Hence (11.18) implies that $|y_j| \leqq |z_j| \leqq \gamma r^j$. This completes the induction.

Consequently (11.13) is convergent for $|t| < 1/r$ and is a solution of (11.1) for small $|t| > 0$. But then it is convergent for $|t| < a$ and is a solution of (11.1) for $0 < |t| < a$ as no solution of (11.1) has a singularity on $0 < |t| < a$.

Exercise 11.3. Show that Theorem 11.1 is false if (11.1) is replaced by (10.1) and it is not supposed that $t = 0$ is a simple singularity.

Corollary 11.2. *Let* (11.2) *be convergent for* $|t| < a$ *and suppose that if* $\lambda_1, \dots, \lambda_d$ *are the eigenvalues of* B, *then* $\lambda_j - \lambda_k = 0$ *or* $\lambda_j - \lambda_k$ *is not an integer for* $j, k = 1, \dots, d$. *Then* (11.1) *has a fundamental matrix of the form*

$$(11.20) \quad Y(t) = Z(t)t^B, \qquad \text{where} \quad Z(t) = I + Z_1 t + Z_2 t^2 + \cdots$$

is convergent for $|t| < a$.

Proof. The matrix (11.20) is a solution matrix for (11.1) if and only if Z satisfies the differential equation

$$(11.21) \qquad\qquad tZ' = (BZ - ZB) + \left(\sum_{k=1}^{\infty} A_k t^k\right)Z;$$

cf. (10.5). A formal series

$$(11.22) \qquad\qquad\qquad Z = \sum_{k=0}^{\infty} Z_k t^k$$

satisfies (11.21) if and only if the formulae

(11.23) $$BZ_0 - Z_0B = 0,$$

(11.24) $$nZ_n = BZ_n - Z_nB + \sum_{k=1}^{n} A_k Z_{n-k}$$

hold for $n = 1, 2, \ldots$.

Consider (11.21) and (11.23)–(11.24) as systems of equations for d^2-dimensional vectors, rather than for $d \times d$ matrices. For any constant matrix D,

$$\mu Z_n = BZ_n - Z_nB - D$$

has a unique solution if $\mu \neq \lambda_j - \lambda_k$ for $j, k = 1, \ldots, d$. For this set of equations can be viewed as $\tilde{B}Z_n - \mu Z_n = D$, where $\tilde{B}$ is a $d^2 \times d^2$ matrix with the d^2 eigenvalues $\lambda_j - \lambda_k$ for $j, k = 1, \ldots, d$ by Lemma 10.2.

Thus $Z_0 = I$ satisfies (11.23) and $Z_1, Z_2, \ldots$ can be obtained recursively from (11.24). The resulting series (11.22) is convergent for $|t| < a$ by Theorem 11.3. This proves Corollary 11.2.

When the eigenvalues $\lambda_1, \ldots, \lambda_d$ of B do not satisfy the conditions of Corollary 11.2, it is possible to change the dependent variable y,

(11.25) $$y = U(t)\eta, \qquad \text{where } \det U(t) \neq 0 \text{ for } t \neq 0,$$

so that (11.1) becomes a system

(11.26) $$t\eta' = \left(C + \sum_{k=1}^{\infty} D_k t^k\right)\eta$$

to which Corollary 11.2 is applicable. In this case, (11.1) has a fundamental matrix of the form

(11.27) $$Y(t) = U(t)Z(t)t^C,$$

where $U(t)$ is a polynomial in t and $Z(t)$ is of the same form as in (11.20). The precise result to be proved is the following:

Lemma 11.3. *Let* (11.2) *be convergent for* $|t| < a$. *Then there exists a matrix* $U(t)$ *with the properties that* $U(t)$ *is a polynomial in* t; *that* $\det U(t) \neq 0$ *for* $t \neq 0$; *and that* (11.25) *transforms* (11.1) *into a system* (11.26) *in which the factor of* η *is a convergent power series for* $|t| < a$ *and if* $\mu_1, \ldots, \mu_d$ *are the eigenvalues of* C, *then* $\mu_j - \mu_k = 0$ *or* $\mu_j - \mu_k$ *is not an integer for* $j, k = 1, \ldots, d$.

Remark 2. It will be clear from the proof that if $\lambda_1, \ldots, \lambda_d$ are the eigenvalues of B in (11.1), then $\mu_1, \ldots, \mu_d$ can be ordered so that $\lambda_j - \mu_j = n_j \geqq 0$ is an integer.

The proof below will not involve a knowledge of the solutions of (11.1) [and gives an algorism for the determination of a C in (11.26), i.e., of a C in Corollary 11.1]. Another proof of Lemma 11.3, but not of Remark 2 following it, can be obtained if one knows a fundamental matrix for (11.1) as in Corollary 11.1:

Exercise 11.4. Deduce Lemma 11.3 from Lemma 11.2.

Proof of Lemma 11.3. Suppose first that B is in a Jordan normal form $B = \text{diag}\,[J(1), \ldots, J(g)]$, where $J(j)$ is a Jordan block of the form (5.15) with $\lambda = \lambda(j)$, $h = h(j)$ for $j = 1, \ldots, g$. Let $B_2 = \text{diag}\,[J(2), \ldots, J(g)]$ so that B_2 is an $e \times e$ matrix, where $e = d - h(1)$.

Make the change of variables

$$(11.28) \qquad y = V(t)\eta, \qquad \text{where } V(t) = \text{diag}\,[tI_{h(1)}, I_e],$$

and $I_{h(1)}$, I_e are the unit matrices of the specified dimensions. Then (11.1) becomes

$$t\eta' = \left(V^{-1}BV - tV^{-1}V' + \sum_{k=1}^{\infty} V^{-1}A_k V t^k\right)\eta.$$

Rearranging the coefficient matrix according to the powers of t, this is of the form

$$(11.29) \qquad t\eta' = \left(C^1 + \sum_{k=1}^{\infty} A_k{}^1 t^k\right)\eta,$$

where C^1, $A_k{}^1$ are constant matrices and C^1 is given by

$$(11.30) \qquad C^1 = \begin{pmatrix} J(1) - I_{h(1)} & A_{12} \\ 0 & B_2 \end{pmatrix} \quad \text{if } A_1 = \begin{pmatrix} A_{11} & A_{12} \\ A_{21} & A_{22} \end{pmatrix}$$

and A_{11} is an $h(1) \times h(1)$ matrix and A_{22} is an $e \times e$ matrix. Thus if the eigenvalues of B are $\lambda(1), \ldots, \lambda(1)$ and $\lambda_{h(1)+1}, \ldots, \lambda_d$, then those of $\hat{C}^1$ are $\lambda(1) - 1, \ldots, \lambda(1) - 1$ and $\lambda_{h(1)+1}, \ldots, \lambda_d$.

If B is not in a Jordan normal form, the same result is achieved by a variation of constants $y = T_1 V_1(t)\eta$, where $T^{-1}BT_1$ is in a Jordan normal form. It is clear that the lemma follows with $U(t)$ of the form $U(t) = T_1 V_1(t) T_2 V_2(t) \ldots T_j V_j(t)$, where T_k is a constant nonsingular matrix and $V_k(t)$ is of the same type as $V(t)$ for $k = 1, \ldots, j$.

Theorem 11.4. *Let* (11.2) *be convergent for* $|t| < a$. *Let* λ *be a fixed complex constant and* $n_\lambda \,(\geqq 0)$ *the number of linearly independent vectors* y *satisfying* $By = \lambda y$. *Then the number* $N_\lambda \,(\geqq 0)$ *of linearly independent solutions of* (11.1) *of the form*

$$(11.31) \qquad y(t) = t^\lambda \sum_{n=0}^{\infty} y_n t^n$$

satisfies

$$(11.32) \qquad \max\,(n_\lambda, n_{\lambda+1}, \ldots) \leqq N_\lambda \leqq n_\lambda + n_{\lambda+1} + \cdots.$$

In particular, the number N_0 of linearly independent solutions $y(t)$ of (11.1) which are analytic at $t = 0$ satisfies

$$(11.33) \qquad n_0 \leqq \max(n_0, n_1, \ldots) \leqq N_0 \leqq n_0 + n_1 + \cdots.$$

It is not assumed that $y_0 \neq 0$ in (11.31). For a generalization of Theorem 11.4, see Theorem 13.1.

The proof of this theorem will depend on the proof of Lemma 11.3 and on the following lemma.

Lemma 11.4. *Let* (11.2) *be convergent for* $|t| < a$. *Then the number* $N_\lambda (\geqq 0)$ *of linearly independent solutions of* (11.1) *of the form* (11.31) *satisfies*

$$(11.34) \qquad N_\lambda \leqq n_\lambda + n_{\lambda+1} + \cdots.$$

If λ is an eigenvalue of B and $\lambda + 1, \lambda + 2, \ldots$ are not eigenvalues, then

$$(11.35) \qquad N_\lambda = n_\lambda$$

and $y_0 \neq 0$ in any solution (11.31) *of* (11.1).

Proof of Lemma 11.4. It can be supposed that $\lambda = 0$, otherwise the change of variables $y = t^\lambda \eta$ replaces (11.1) by

$$t\eta' = \left(B - \lambda I + \sum_{k=1}^{\infty} A_k t^k \right) \eta$$

and $\lambda + k$, $n_{\lambda+k}$ by k, n_k.

A function (11.13) is a solution of (11.1) if and only if (11.14) and (11.15$_n$), $n = 1, 2, \ldots$, hold. The equation (11.14) has n_0 linearly independent solutions and if (11.14), and (11.15$_1$), $\ldots$, (11.15$_{k-1}$) hold, then the solutions of (11.15$_k$) are of the form $z_0 + z_k$, where $y_k = z_0$ is a particular solution of (11.15$_k$) and z_k varies over the n_k-dimensional linear manifold of solutions of $Bz - kz = 0$. This proves (11.34).

In the last part of the lemma, it is supposed that $\lambda = 0$ is an eigenvalue of B, but $\lambda = 1, 2, \ldots$ are not. Then (11.14) has n_0 linearly independent solutions y_0 and if (11.14) and (11.15$_1$), $\ldots$, (11.15$_{k-1}$) hold, then (11.15$_k$) has a unique solution. Thus for a given y_0, the vectors $y_1, y_2, \ldots$ are uniquely determined. The corresponding series (11.13) is convergent for $|t| < a$ and is a solution of (11.1) by Theorem 11.3. This proves (11.35) and completes the proof of Lemma 11.4.

Proof of Theorem 11.4. In view of (11.34) in Lemma 11.4, only the first inequality in (11.32) remains to be proved. Since $N_\lambda \geqq N_{\lambda+1} \geqq \cdots$, this inequality will be proved if it is shown that

$$(11.36) \qquad N_\lambda \geqq n_\lambda.$$

As in the last proof, there is no loss of generality in supposing that $\lambda = 0$ in (11.35). It can be supposed that $n_0 > 0$ and, in view of (11.35) in Lemma 11.4, it can also be supposed that B has an eigenvalue which is a positive integer.

The proof of Lemma 11.3 makes it clear that there exists a change of variables (11.25) such that $U(t)$ is a polynomial in t and that (11.1) becomes (11.26) in which the eigenvalues of C are the same as those of B except that any integral eigenvalue $\lambda = n > 0$ of B has been replaced by $\lambda = 0$. Let $n_\lambda(C)$ have the same meaning for C as n_λ does for B. It will be shown that

$$(11.37) \qquad\qquad n_0(C) \geqq n_0.$$

To this end, it suffices to re-examine the proof of Lemma 11.3 and show that the analogue of (11.37) holds at each step. Thus, if the first step leads from (11.1) to (11.29), it suffices to show that

$$(11.38) \qquad\qquad n_0(C^1) \geqq n_0.$$

Let B be in a Jordan normal form, $B = \mathrm{diag}\,[J(1), \ldots, J(g)]$ and let $\lambda(1) > 0$ be an integer. Make the change of variables (11.28) transforming (11.1) into (11.29), where (11.30) holds. Since $\lambda(1) \neq 0$ and B_2 is in a Jordan normal form, it is clear that n_0 rows of B_2 contain only zero elements. Hence the rank of C^1 in (11.30) is at most $d - n_0$, so that (11.38) holds. Consequently (11.37) follows.

Since $\lambda = 1, 2, \ldots$ are not eigenvalues of C, the last part of Lemma 11.4 implies that (11.26) has $n_0(C)$ linearly independent solutions $\eta(t)$ which are regular at $t = 0$. Since $U(t)$ is a polynomial in t, (11.25) shows that (11.1) has at least $n_0(C)$ linearly independent solutions $y(t)$ regular at $t = 0$. In view of (11.37), the inequality (11.36) follows for $\lambda = 0$. This completes the proof of Theorem 11.4.

Exercise 11.5. (a) Let $A(t) = (a_{jk}(t))$ be a $d \times d$ matrix of functions analytic for $|t| < a$. Let $\alpha(j) = 0$ or 1 and $\alpha < d$ if $\alpha = \alpha(1) + \cdots + \alpha(d)$. Then the system

$$t^{\alpha(j)} y^{j'} = \sum_{k=1}^{d} a_{jk}(t) y^k \qquad \text{for } j = 1, \ldots, d$$

has at least $d - \alpha$ linearly independent solutions analytic at $t = 0$. (b) Let $a_0(t)$, $a_1(t)$ be analytic for $|t| < a$, then the differential equation

$$tu'' + a_1(t)u' + a_0(t)u = 0$$

has at least one solution $u(t) \not\equiv 0$ analytic at $t = 0$. For a generalization of this exercise, see § 13.

12. Higher Order Equations

Consider a differential equation of dth order for a function u,

$$(12.1) \qquad u^{(d)} + p_{d-1}(t)u^{(d-1)} + \cdots + p_1(t)u' + p_0(t)u = 0,$$

in which the coefficient functions are single-valued and analytic on a punctured disc $0 < |t| < a$. Instead of writing (12.1) as a first order system in the standard way, transform it into a system for the vector $y = (y^1, \ldots, y^d)$, where

$$(12.2) \qquad y^j = t^{j-1}u^{(j-1)} \qquad \text{for } j = 1, \ldots, d.$$

Then

$$
\begin{aligned}
(12.3) \qquad & ty^{j'} = (j-1)y^j + y^{j+1} \qquad \text{for } j = 1, \ldots, d-1, \\
& ty^{d'} = (d-1)y^d - \sum_{k=1}^{d} t^{d-(k-1)}p_{k-1}(t)y^k.
\end{aligned}
$$

Thus $t = 0$ is a simple singularity for this system if $t^{d-k}p_k(t)$ is analytic at $t = 0$ for $k = 0, \ldots, d-1$; i.e., if $p_{d-1}(t)$ has at most a pole of the first order, $p_{d-2}(t)$ has at most a pole of the second order, $\ldots, p_0(t)$ has at most a pole of the dth order. In this case, let

$$(12.4) \quad a_k(t) \equiv t^{d-k}p_k(t) = b_k + \sum_{n=1}^{\infty} p_{kn}t^n \qquad \text{for } k = 0, \ldots, d-1,$$

where b_k and p_{kn} are constants and the series in (12.4) is convergent for $|t| < a$. Then (12.1) is of the form

$$(12.5) \quad u^{(d)} + t^{-1}a_{d-1}(t)u^{(d-1)} + \cdots + t^{-(d-1)}a_1(t)u' + t^{-d}a_0(t)u = 0,$$

where $a_0(t), \ldots, a_{d-1}(t)$ are analytic at $t = 0$. Correspondingly, (12.3) is of the form (11.1), where B is the constant matrix:

$$(12.6) \quad B = \begin{pmatrix}
0 & 1 & 0 & 0 & \cdots & & 0 \\
0 & 1 & 1 & 0 & \cdots & & 0 \\
0 & 0 & 2 & 1 & \cdots & & 0 \\
0 & 0 & 0 & 3 & \cdots & & 0 \\
\cdot & \cdot & \cdot & \cdot & \cdot & \cdot & \cdot \\
0 & 0 & 0 & 0 & \cdots & & 1 \\
-b_0 & -b_1 & -b_2 & -b_3 & \cdots & & (d-1) - b_{d-1}
\end{pmatrix}$$

The coefficient matrix on the right of (12.3) reduces to the constant matrix B if $p_{kn} = 0$ for $k = 0, \ldots, d - 1$ and $n = 1, 2, \ldots$. This is the case when (12.1) is the differential equation

$$(12.7) \qquad u^{(d)} + t^{-1}b_{d-1}u^{(d-1)} + \cdots + t^{-(d-1)}b_1 u' + t^{-d}b_0 u = 0,$$

in which $b_0, \ldots, b_{d-1}$ are constants. This is called *Euler's differential equation*. The solutions of (12.7) are easily determined from the fact that a fundamental matrix for the corresponding system (12.3) is $Y(t) = t^B$; cf. the remark following (11.3). Thus the solutions of (12.7) are linear combinations of functions of the form $t^\lambda (\log t)^k$.

The numbers λ and permissible values of k are determined by the Jordan normal form of B. The equation (12.7) obviously has a solution of the form $u = t^\lambda$ if and only if λ is an eigenvalue of B. Substituting $u = t^\lambda$ into (12.6), it is seen that this is the case if and only if $F(\lambda) = 0$, where

$$(12.8) \qquad F(\lambda) = \sum_{j=0}^{d} b_{d-j}\lambda(\lambda - 1) \ldots (\lambda - j + 1) \quad \text{and} \quad b_d = 1.$$

The equation $F(\lambda) = 0$ is called the *indicial equation for* (12.7).

Let $\lambda(1), \ldots, \lambda(g)$ be the distinct solutions of $F(\lambda) = 0$ with the respective multiplicities $h(1), \ldots, h(g)$, where $h(1) + \cdots + h(g) = d$. Then a linearly independent set of solutions of (12.7) is $t^{\lambda(j)}(\log t)^k$, where $j = 1, \ldots, g$ and $k = 0, \ldots, h(j) - 1$.

Exercise 12.1.   (a) Verify this last statement by the type of argument in § 8(vi) following Exercise 8.1. (b) The remarks concerning (11.1) and (11.3) show that the change of variables $t = e^s$ reduces the system (12.3) belonging to (12.7) to one with constant coefficients. Verify directly that the substitution $t = e^s$ reduces (12.7) to an equation with constant coefficients and, hence, that (a) follows directly from § 8(vi).

Returning to the general equation (12.1) and its corresponding system (12.3), the following theorem will be proved:

Theorem 12.1 (Fuchs). *Let $p_k(t)$ be single-valued and analytic for $0 < |t| < a$. Then $t = 0$ is a regular singular point for (12.3) if and only if $t = 0$ is a simple singularity for (12.3) (i.e., if and only if (12.4) holds with convergent series on the right).*

It is clear that $t = 0$ is a regular singular point for the system (12.3) if and only if the solutions of (12.1) are linear combinations of functions of the form $t^\lambda (\log t)^k \alpha(t)$, where $\alpha(t)$ is analytic for $|t| < a$.

Proof.   The "if" portion of the theorem is a consequence of Theorem 11.1. Thus it is sufficient to prove the "only if" portion.

It follows from Theorem 10.1 that (12.1) has at least one solution of the form $u_1(t) = t^\lambda \alpha_1(t)$, where $\alpha_1(t)$ is analytic for $0 < |t| < a$. If it is assumed that $t = 0$ is a regular singular point for (12.3), then $\alpha_1(t)$ has at most a pole at $t = 0$. In fact, by changing λ, it can be supposed that $\alpha_1(t)$ is analytic at $t = 0$. In particular $\alpha_1(t) \neq 0$ for small $|t| > 0$.

The proof proceeds by induction on the order d. Consider first an equation of order $d = 1$,

$$u' + p_0(t)u = 0,$$

with a solution $u_1(t) = t^\lambda \alpha_1(t)$, $\alpha_1(t)$ analytic at $t = 0$. It is clear that $p_0(t) = -u_1'(t)/u_1(t)$ has at most a pole of order $d = 1$ at $t = 0$.

Assume $d > 1$ and that the theorem is correct for equations of order $d - 1$. Let $u = u_1(t)$ be the type of solution described above and introduce the new dependent variable $v = u/u_1$ for small $t > 0$. Then (12.1) is transformed into an equation of the form

$$(12.9) \qquad v^{(d)} + q_{d-2}(t)v^{(d-1)} + \cdots + q_0(t)v' = 0,$$

which is an equation order $d - 1$ for v'. It is readily seen that

$$q_{d-2}(t) = C_{d1}(u_1'/u_1) + p_{d-1}(t)$$
$$(12.10) \quad q_{d-3}(t) = C_{d2}(u_1''/u_1) + C_{d-1,1}p_{d-1}(u_1'/u_1) + p_{d-2}$$
$$\cdot \quad \cdot \quad \cdot \quad \cdot \quad \cdot \quad \cdot \quad \cdot \quad \cdot \quad \cdot \quad \cdot \quad \cdot \quad \cdot$$
$$q_0(t) = C_{d,d-1}(u_1^{(d-1)}/u_1) + C_{d-1,d-2}p_{d-1}(u_1^{(d-2)}/u_1) + \cdots + p_1(t),$$

and since $u = u_1(t)$ is a solution of (12.1),

$$(12.11) \qquad 0 = u_1^{(d)}/u_1 + p_{d-1}(u_1^{(d-1)}/u_1) + \cdots + p_0(t),$$

where $C_{jk} = j!/k!(j - k)!$ are binomial coefficients.

The equation (12.9), as an equation of order $d - 1$ for v', has the solutions $v' = (u/u_1)'$ for arbitrary solutions u of (12.1). Consequently, $t = 0$ is a regular singular point for the system associated with (12.9). Hence, by the induction hypotheses, $t^{d-1-k}q_k(t)$ is analytic at $t = 0$ for $k = 0, \ldots, d - 2$. Also, $u_1^{(k)}/u_1$ has a pole of at most the order k at $t = 0$. It follows from (12.10) and (12.11) that p_{d-1} has at most a pole of order 1 at $t = 0$, p_{d-2} has a pole of order at most 2, etc. This proves the theorem.

Exercise 12.2. Let $p_0(t), \ldots, p_{d-1}(t)$ be analytic for $a < |t| < \infty$. The point $t = \infty$ is called a simple singularity [or regular singular point] for (12.1) if it is a simple singularity [or regular singular point] for (12.3); cf. Exercise 11.2. (a) Necessary and sufficient that $t = \infty$ be a simple singularity for (12.1) is that $t^{d-k-1}p_k(t) \to 0$ as $|t| \to \infty$ for $k = 0, \ldots, d - 1$. (b) Let $p_0(t), \ldots, p_{d-1}(t)$ be analytic for $t \neq t_1, \ldots, t_n, \infty$. Necessary and sufficient that $t = t_1, \ldots, t_n, \infty$ be simple singularities for (12.1) is that $p_{d-k}(t)$ be of the form $p_{d-k}(t) = (t - t_1)^{-k} \ldots (t - t_n)^{-k}a_k(t)$, for $k = 1, \ldots, d$, where $a_k(t)$ is a polynomial of degree at most $k(d - 1)$. (Differential equations with this property are said to be of Fuchs' type.)

For a second order $(d = 2)$ equation (12.5) having a regular singular point at $t = 0$, it is comparatively simple to discuss the behavior of a set of fundamental solutions. We attempt first to find a solution of the form

$$(12.12) \qquad u = t^\lambda \sum_{k=0}^{\infty} c_k t^k$$

by the method of undetermined coefficients. If the roots λ_1, λ_2 of the indicial equation $F(\lambda) = 0$ [cf. (12.4) and (12.8)] are such that $\lambda_1 - \lambda_2$ is not an integer, then we obtain two solutions (12.12) with $\lambda = \lambda_1$, λ_2 in this way; see Corollary 11.2. If $\lambda_1 - \lambda_2 \geq 0$ is an integer, we can still obtain in this way a solution $u(t)$ of the form (12.12) with $\lambda = \lambda_1$; see Lemma 11.4. A second linearly independent solution $v(t)$ can be obtained from the fact that, by § 8(iv),

$$(v/u)' = u^{-2}(t) \exp - \int^t s^{-1} a_1(s) \, ds.$$

Exercise 12.3. Discuss the nature of the solutions at the (finite) singular points of the equations: (a) $t^2 u'' + t u' + (t^2 - \mu^2)u = 0$ (Bessel); (b) $(1 - t^2)u'' - 2tu' + n(n + 1)u = 0$ (Legendre); (c) $(1 - t^2)u'' - 2tu' + [n(n + 1) - m^2(1 - t^2)^{-1}]u = 0$ (associated Legendre).

13. A Nonsimple Singularity

In this section, we shall prove a theorem about the number of analytic solutions of a particular type of linear homogeneous system for which $t = 0$ is a singularity, but not necessarily a simple singularity.

Theorem 13.1 (Lettenmeyer). *Let $A(t) = (a_{jk}(t))$ be a $d \times d$ matrix of functions analytic at $t = 0$. Let $\alpha(j) \geq 0$ be an integer for $j = 1, \ldots, d$ and let $\alpha < d$, where $\alpha = \alpha(1) + \cdots + \alpha(d)$. Then the system*

$$(13.1) \qquad t^{\alpha(j)} y^{j\prime} = \sum_{k=1}^{d} a_{jk}(t) y^k \qquad \text{for } j = 1, \ldots, d$$

has at least $d - \alpha$ linearly independent solutions analytic at $t = 0$.

This theorem generalizes Theorem 11.4 which corresponds to the case $\alpha(j) = 0$ or 1; cf. Exercise 11.5.

Proof. The solutions $y(t)$ of (13.1) analytic at $t = 0$ will be determined by the method of undetermined coefficients. Let the function $a_{jk}(t)$ have the expansion

$$(31.2) \qquad a_{jk}(t) = \sum_{m=0}^{\infty} a_{jk,m} t^m.$$

Consider a solution $y(t) = (y^1(t), \ldots, y^d(t))$ of (13.1) with convergent expansions

(13.3)
$$y^j(t) = \sum_{m=0}^{\infty} y_m{}^j t^m.$$

Then the numbers $y_m{}^j$ satisfy

(13.4$_{jm}$)
$$[m - \alpha(j) + 1]y^j_{m-\alpha(j)+1} = \sum_{k=1}^{d} \sum_{n=0}^{m} a_{jk,m-n} y_n{}^k$$

for $j = 1, \ldots, d$ and $m = 0, 1, \ldots$. It is understood that $y_m{}^j = 0$ if $m < 0$. Conversely, if a set of numbers $y_n{}^j$ satisfies (13.4) and if (13.3) is convergent for small $|t|$ and $j = 1, \ldots, d$, then $y(t) = (y^1(t), \ldots, y^d(t))$ is a solution of (13.1) analytic at $t = 0$.

Let N be a large fixed integer to be specified below. Divide the system of equations (13.4$_{jm}$) into two systems

(13.5) $\sum_1$: (13.4$_{jm}$) for $j = 1, \ldots, d$ and $0 \leq m \leq N + \alpha(j) - 2$,

(13.6) $\sum_2$: (13.4$_{jm}$) for $j = 1, \ldots, d$ and $m \geq N + \alpha(j) - 1$.

Since it is assumed that the series in (13.2) is convergent, there exist positive numbers, c and ρ, such that

(13.7) $\quad |a_{jk,m}| \leq \dfrac{c}{\rho^m}$ \quad for \quad $j, k = 1, \ldots, d$ \quad and \quad $m = 0, 1, \ldots$.

Let $0 < \theta < 1$ and define

(13.8)
$$z_{jm} = (\theta\rho)^m y_m{}^j,$$

(13.9)
$$c_{jm,kn} = -\frac{a_{jk,m+\alpha(j)-n-1}(\theta\rho)^{m-n}}{m}.$$

The last set of relations is equivalent to

$$c_{jh,kn} = -\frac{a_{jk,m-n}(\theta\rho)^{m-\alpha(j)-n+1}}{m - \alpha(j) + 1}, \quad \text{where } h = m - \alpha(j) + 1.$$

Thus the system $\sum_2$ of equations can be written as

$$z_{jh} + \sum_{k=1}^{d} \sum_{n=0}^{m} c_{jh,kn} z_{kn} = 0 \quad \text{for} \quad m \geq N + \alpha(j) - 1,$$

where $h = m - \alpha(j) + 1$, or in the form

$$z_{jm} + \sum_{k=1}^{d} \sum_{n=0}^{m+\alpha(j)-1} c_{jm,kn} z_{kn} = 0 \quad \text{for} \quad m \geq N,$$

or, finally, as

(13.10) $\quad z_{jm} + \sum_{k=1}^{d} \sum_{n=N}^{m+\alpha(j)-1} c_{jm,kn} z_{kn} = -\sum_{p=1}^{d} \sum_{q=0}^{N-1} c_{jm,pq} z_{pq}$ \quad for \quad $m \geq N$.

It is understood that the inner sum on the left over $N \leq n \leq m + \alpha(j) - 1$ is 0 if $m + \alpha(j) = N$.

Let p, q be fixed integers satisfying $1 \leq p \leq d$ and $0 \leq q \leq N - 1$. Instead of (13.10) consider the set of equations

$$(13.11) \qquad z_{jm} + \sum_{k=1}^{d} \sum_{n=N}^{m+\alpha(j)-1} c_{jm,kn} z_{kn} = -c_{jm,pq},$$

for $j = 1, \ldots, d$ and $m \geq N$. It will be shown that if N is sufficiently large, then (13.11) has a solution $z_{jm} = z_{jm}^{pq}$ satisfying

$$(13.12) \qquad |z_{jm}^{pq}| \leq 2 \quad \text{for} \quad j = 1, \ldots, d \quad \text{and} \quad m \geq N.$$

To this end, note that, by (13.7) and (13.9),

$$|c_{jm,kn}| \leq \frac{c\theta^{m-n} \rho^{-\alpha(j)+1}}{m}.$$

Thus, if c', c'' are suitable constants (independent of p, q, N), then

$$\sum_{m=N}^{\infty} |c_{jm,pq}|^2 \leq \frac{c'}{N^2} \theta^{-2N} \sum_{m=N}^{\infty} \theta^{2m} = \frac{c'}{N^2(1 - \theta^2)},$$

$$\sum_{n=N}^{m+\alpha(j)-1} |c_{jm,kn}|^2 \leq \frac{c'}{m^2} \sum_{n=N}^{m+\alpha(j)-1} \theta^{2(m-n)} \leq \frac{c'}{m^2} \sum_{n=1-\alpha(j)}^{\infty} \theta^{2n} \leq \frac{c''}{m^2}.$$

Hence, if N is sufficiently large,

$$(13.13) \qquad \sum_{j=1}^{d} \sum_{m=N}^{\infty} |c_{jm,pq}|^2 \leq \frac{c''}{N^2} \leq 1,$$

$$(13.14) \qquad \sum_{j=1}^{d} \sum_{k=1}^{d} \sum_{m=N}^{\infty} \sum_{n=N}^{m+\alpha(j)-1} |c_{jm,kn}|^2 \leq d^2 c'' \sum_{m=N}^{\infty} \frac{1}{m^2} \leq \frac{1}{4},$$

for all choices of p, q.

From now on, it is supposed that N is fixed so large that (13.13) and (13.14) hold. Let $\mu > N$ denote a fixed integer and replace the infinite system (13.11), where $j = 1, \ldots, d$ and $m \geq N$, by the finite system of equations,

$$(13.15) \qquad z_{jm}^{\mu} + \sum_{k=1}^{d} \sum_{n=N}^{v} c_{jm,kn} z_{kn}^{\mu} = -c_{jm,pq}$$

for $j = 1, \ldots, d$ and $N \leq m \leq \mu$, where v in (13.15) is

$$(13.16) \qquad v \equiv v(j, m, \mu) \equiv \min [\mu, m + \alpha(j) - 1].$$

The system (13.15) can be written in the form

$$(13.17) \qquad \xi + C\xi = \eta,$$

where ξ is a vector of dimension $e = d(\mu - N + 1)$ with components

z_{jm}^μ for $j = 1, \ldots, d$ and $m = N, \ldots, \mu$; η is a vector of dimension e with components $-c_{jm, pq}$ for $j = 1, \ldots, d$ and $m = N, \ldots, \mu$; and C is an $e \times e$ matrix. Using Euclidean norms, it is clear from (13.13) and (13.14) that

$$\|\eta\| \leq 1 \quad \text{and} \quad \|C\| \leq 1/2.$$

Hence (13.17) has a unique solution ξ satisfying $\|\xi\| \leq 2$, since $\|\eta\| = \|\xi + C\xi\| \geq \|\xi\| - \|C\xi\| \geq \frac{1}{2} \|\xi\|$. Thus the finite system (13.15) has a unique solution z_{jn}^μ satisfying

$$|z_{jm}^\mu| \leq 2 \quad \text{for} \quad j = 1, \ldots, d \quad \text{and} \quad m = N, \ldots, \mu.$$

Then by the Cantor Selection Theorem I 2.1, there exists a sequence of integers $(N <) \mu(1) < \mu(2) < \cdots$ such that

$$z_{jm}^{pq} = \lim_{h \to \infty} z_{jm}^{\mu(h)} \quad \text{exists for} \quad j = 1, \ldots, d \quad \text{and} \quad m \geq N.$$

Note that ν in (13.15) and (13.16) satisfies $\nu(j, m, \mu) \to m + \alpha(j) - 1$ as $\mu \to \infty$. Hence, by letting $\mu = \mu(h) \to \infty$ in (13.15), it follows that z_{jm}^{pq} is a solution of (13.11) for $j = 1, \ldots, d$ and $m \geq N$ satisfying (13.12).

Consequently, if z_{pq} are any dN numbers for $p = 1, \ldots, d$ and $q = 0, \ldots, N - 1$, then a solution of (13.10) for $j = 1, \ldots, d$ and $m \geq N$ is given by

$$(13.18_{jm}) \qquad\qquad z_{jm} = \sum_{p=1}^{d} \sum_{q=0}^{N-1} z_{jm}^{pq} z_{pq}$$

for $j = 1, \ldots, d$ and $m \geq N$. In other words, the equations of the system $\sum_2$ in (13.6) [i.e., the equations in (13.10) for $j = 1, \ldots, d$ and $m \geq N$] are satisfied if (13.18_{jm}) hold for $j = 1, \ldots, d$ and $m \geq N$.

Thus the original system of equations (13.4_{jm}) for the $y_m{}^j$ or, equivalently, for the z_{jm} are satisfied if $\sum_1$ and (13.18) hold. If $\alpha^* = \max [\alpha(1), \ldots, \alpha(d)]$, the system $\sum_1$ involves the $d(N + \alpha^* - 1)$ unknown z_{jm} for $j = 1, \ldots, d$ and $m = 0, \ldots, N + \alpha^* - 2$; cf. (13.4_{jm}) for $j = 1, \ldots, d$ and $0 \leq m \leq N + \alpha(j) - 2$. Also the system $\sum_1$ consists of $[N - 1 + \alpha(1)] + \cdots + [N - 1 + \alpha(d)] = d(N - 1) + \alpha$ equations, where $\alpha = \alpha(1) + \cdots + \alpha(d)$. Add to the system $\sum_1$, the (possibly vacuous) set of equations

$$\sum_3: (13.18_{jm}) \quad \text{for} \quad j = 1, \ldots, d \quad \text{and} \quad N \leq m \leq N + \alpha^* - 2.$$

The system $\sum_3$ involves the same set of $d(N + \alpha^* - 1)$ variables that occur in $\sum_1$ and consists of $d(\alpha^* - 1)$ equations. Thus, the combined systems, $\sum_1$ and $\sum_3$, has at least $d(N + \alpha^* - 1) - [d(N - 1) + \alpha + d(\alpha^* - 1)] = d - \alpha$ linearly independent solutions.

Corresponding to any solution of $\sum_1$ and $\sum_3$, the equations (13.18_{jm}) for $m > N + \alpha^* - 2$ (together with the equations of $\sum_3$) give a solution of $\sum_2$. The resulting set of numbers $y_m{}^j = z_{jm}/(\theta\rho)^m$, for $j = 1, \ldots, d$ and $m = 0, 1, \ldots$, is a solution of $\sum_1$ and $\sum_2$. In view of (13.12) and (13.18), there is a constant c_0 such that $|z_{jm}| \leqq c_0$; hence $|y_m{}^j| \leqq c_0/(\theta\rho)^m$ for $j = 1, \ldots, d$ and $m \geqq 0$. Consequently, (13.3) is convergent for $|t| < \theta\rho$. This proves Theorem 13.1.

Exercise 13.1 (*Perron*). Let $a_j(t)$ be analytic for $|t| < a$, $j = 0, \ldots, d$. Let α be an integer, $0 \leqq \alpha < d$. Show that

$$t^\alpha u^{(d)} + a_{d-1}(t)u^{(d-1)} + \cdots + a_1(t)u' + a_0(t)u = 0$$

has at least $d - \alpha$ linearly independent solutions analytic at $t = 0$.

Exercise 13.2 (*Lettenmeyer*). (*a*) Let $A(t)$, $X(t)$ be $d \times d$ matrices of functions analytic at $t = 0$. Let $\det X(t) \not\equiv 0$ and $\det X(t)$ have a zero of order α, $0 \leqq \alpha < d$, at $t = 0$. Then the system

(13.19) $$X(t)y' = A(t)y$$

has at least $d - \alpha$ linearly independent solutions analytic at $t = 0$. (*b*) Let $X(t)$, $A(t)$ be analytic in a simply connected t-domain E such that $\det X(t) \not\equiv 0$ and $\det X(t)$ has exactly α zeros in E, counting multiplicities. Let $\alpha < d$. Then (13.19) has at least $d - \alpha$ linearly independent solutions $y(t)$ analytic on E.

Exercise 13.3. Let $X(t)$ be as in Exercise 13.2(*a*). Let $f(t, y) = f(t, y^1, \ldots, y^d)$ be a d-dimensional vector, each component of which is an analytic function of (i.e., a convergent power series in) $t, y^1, \ldots, y^d$. Then

$$X(t)y' = f(t, y)$$

has a $d - \alpha$ parameter family of solutions $y(t)$ analytic at $t = 0$. See Bass [1].

Notes

SECTION 1. Equation (1.5) in Theorem 1.2 as well as the special case (8.4) are given by Liouville [2] in 1838, although some authors refer to (8.4) as Liouville's formula and (1.5) as Jacobi's. Theorem 1.2 was also given by Jacobi [1, IV, p. 403] in 1845; for the particular case when the system (1.1) is replaced by one linear equation of the second order, the corresponding formula (8.4) occurs in a paper (1827) by Abel [1, I, p. 251]. The notion of a fundamental set of solutions is due to Lagrange (circa 1765); see [2, I, p. 473]. The term "fundamental set of solutions" was introduced by Fuchs [4, p. 117] in 1866.

SECTION 2. The method of variation of constants (and Corollary 2.1) is essentially due to Lagrange (1774, 1775); see [2, IV, p. 9 and p. 159]. The result of Exercise 2.2 goes back to Perron [11]. It has also been proved by Diliberto [1], [2]. The proof given in Hints, using Exercise 2.1, is that of Reid [4].

SECTION 3. The possibility of reducing the order when a solution is known goes back (1762–1765) to d'Alembert for a linear equation of order d. Corollary 3.1 when (1.1) is replaced by a single equation of order d [cf. (8.2) and (8.19)] is implicit in a result (1833) of Libri [1, pp. 185–194] (who also makes remarks about the possibility of extending the result to systems). Corollary 3.1 for the case of an equation of dth order is given explicitly (1874) by Frobenius [1].

SECTION 5. For the analogous case of a linear dth order equation, the general solution [§ 8 (vi)] is due to Euler (1743).

SECTION 6. Theorem 6.1 for the case of a dth order equation was given (1883) by Floquet [1].

SECTION 7. The adjoint of a linear dth order equation was given by Lagrange about 1765, see [2, I, p. 471]. The adjoint system (7.1) was defined in 1837 by Jacobi; cf. [1, IV, p. 403]. The term "adjoint" was introduced by Fuchs [1, p. 422] in 1873.

SECTION 8. See comments on §§ 1–7. Exercise 8.3 contains results*of Pólya [1]; cf. comments of Hartman [15].

SECTION 9. The type of formalism discussed in connection with (9.1)–(9.6) was used extensively by G. D. Birkhoff [3]. Exercise 9.1 is based on considerations used by Cesari [1]; see also Levinson [3].

APPENDIX. For a historical survey and references, see Schlesinger [2j, Forsyth [1], and the encyclopedia article of Hilb [1]; for later references, see A. Schmidt [1].

SECTION 10. Problems arising out of (10.2) were initiated by Riemann in work dated 1857 (cf. [1, pp. 379–390]) and independently by Fuchs in 1865 (see [1, p. 124]). Theorem 10.1 is essentially due to M. Hamburger [1]; the proof in the text follows Coddington and Levinson [2]. Although the paper by Wintner [5] contains many misstatements, including the main theorem, it has a number of good ideas including the suggestion to view (10.5) as a system (10.6) of d^2 equations rather than as a matrix equation; for applications, see § 11 [cf. the proof of Theorem 11.1 which may be new].

SECTION 11. Theorem 11.1 is due to Sauvage [1]; cf. Hilb [1] for references to Horn and Schlesinger. The Lemma 11.1 is a result of Birkhoff [1]. The partial converse, Theorem 11.2, was given by Sauvage and Koenigsberger; the first complete proof is due to Horn [1]. The proof in the text is that of Schlesinger [1, pp. 141–162], who attributes the arguments in the proof of Lemma 11.2 to Kronecker. For similar proofs of this lemma, see Lettenmeyer [1] and Moser [2]; generalizations have been given by Hilbert, J. Plemelj, and G. D. Birkhoff, see Birkhoff [2] for references. Corollary 11.2 and Lemma 11.3 are in Rasch [1, p. 113]. Their use in connection with (11.25)–(11.27) is that of A. Schmidt [1]. Exercise 11.5 is a special case of Lettenmeyer's [1] result, Theorem 13.1.

SECTION 12. Theorem 12.1 is due to Fuchs in 1868 [1, p. 212]. It was first proved in special cases by Riemann [1, pp. 379–390]. The proof in the text for the "only if" part is that of Thomé [1]; see Hilb [1] for references to Frobenius.

SECTION 13. Theorem 13.1, which is due to Lettenmeyer [1], generalizes the result of Perron [1] in Exercise 13.1 dealing with one equation of dth order. The proof in the text is a modification of Lettenmeyer's which, in turn, is based on Hilb's proof [2] of Perron's theorem. For Exercise 13.2, see Lettenmeyer [1].

---

*For related results, see Mammana [S1].

Chapter V

Dependence on Initial Conditions and Parameters

1. Preliminaries

Let $f(t, y)$ be defined on an open (t, y)-set E with the property that if $(t_0, y_0) \in E$, then the initial value problem

$$(1.1) \qquad y' = f(t, y) \quad \text{and} \quad y(t_0) = y_0$$

has a unique solution $y(t) = \eta(t, t_0, y_0)$ which is then defined on a maximal t-interval (ω_-, ω_+), where $\omega_\pm$ depends on (t_0, y_0). In this chapter, the problem of the smoothness (i.e., of the continuity or differentiability properties) of $\eta(t, t_0, y_0)$ will be considered.

Often, a more general situation is encountered in which (1.1) is replaced by a family of initial value problems depending on a set of parameters $z = (z^1, \ldots, z^e)$,

$$(1.2) \qquad y' = f(t, y, z) \quad \text{and} \quad y(t_0) = y_0,$$

where for each fixed z, (1.2) has a unique solution $y(t) = \eta(t, t_0, y_0, z)$. In most cases, the question of the dependence of solutions of (1.1) on t and initial conditions can be reduced to the question of the dependence on t, z of solutions of a family of initial value problems (1.2) for *fixed* initial conditions $y(t_0) = y_0$; conversely, the question of the dependence of solutions of (1.2) on t, t_0, y_0, z can be reduced to the question of smoothness of solutions of initial value problems in which extra parameters z do not occur. The first reduction is accomplished by the change of variables $t, y \to t - t_0, y - y_0$ which changes (1.1) to

$$(1.3) \qquad y' = f(t - t_0, y - y_0) \quad \text{and} \quad y(0) = 0,$$

in which $z = (t_0, y_0) = (t_0, y_0^1, \ldots, y_0^d)$ can be considered as a set of parameters (and the initial condition $y(0) = 0$ is fixed). The second reduction is obtained by replacing (1.2) by an initial value problem for a $(d + e)$-dimensional vector (y, z), in which no extra parameters occur:

$$(1.4) \qquad y' = f(t, y, z), z' = 0 \quad \text{and} \quad y(t_0) = y_0, z(t_0) = z_0,$$

where (t_0, y_0, z_0) denotes any of the possible choices for (t, y, z). For this reason, some of the theorems which follow will be stated for (1.2) but proved for (1.1).

Below $x, y, \eta, f \in \mathbf{R}^d$ and $z \in \mathbf{R}^e$, where $d, e \geqq 1$.

2. Continuity

The assumption of uniqueness implies the continuity of the general solution $y = \eta(t, t_0, y_0, z)$ of $y' = f(t, y, z)$:

Theorem 2.1. *Let $f(t, y, z)$ be continuous on an open (t, y, z)-set E with the property that for every $(t_0, y_0, z) \in E$, the initial value problem (1.2), with z fixed, has a unique solution $y(t) \equiv \eta(t, t_0, y_0, z)$. Let $\omega_- < t < \omega_+$ be the maximal interval of existence of $y(t) = \eta(t, t_0, y_0, z)$. Then $\omega_+ = \omega_+(t_0, y_0, z)$ [or $\omega_- = \omega_-(t_0, y_0, z)$] is a lower [or upper] semicontinuous function of $(t_0, y_0, z) \in E$ and $\eta(t, t_0, y_0, z)$ is continuous on the set $\omega_- < t < \omega_+$, $(t_0, y_0, z) \in E$.*

It is understood that $\omega_+[\omega_-]$ can assume the value $+\infty$ $[-\infty]$. The lower semicontinuity of ω_+ at (t_1, y_1, z_1) means that if $t^0 < \omega_+(t_1, y_1, z_1)$, then $\omega_+(t_0, y_0, z) \geqq t^0$ for all (t_0, y_0, z) near (t_1, y_1, z_1); i.e., $\omega_+(t_1, y_1, z_1) \leqq$ lim inf $\omega_+(t_0, y_0, z_0)$ as $(t_0, y_0, z_0) \to (t_1, y_1, z_1)$. The upper semicontinuity of ω_- is similarly defined.

It is easy to see that $\omega_+(t_0, y_0, z)$ need not be continuous. For suppose that $(t_1, y_1, z_1) \in E$ and $t_1 < t^0 < \omega_+(t_1, y_1, z_1)$. If E is replaced by the set obtained from E by deleting the point $(t, y, z) = (t^0, \eta(t^0, t_1, y_1, z_1), z_1)$, then $\omega_+(t_1, y_1, z_1)$ now takes the value t^0, but ω_+ is not altered for all points (t_0, y_0, z) near (t_1, y_1, z_1).

Remark. Let $f(t, y, z)$, $\eta(t, t_0, y_0, z)$ be as in Theorem 2.1. For fixed (t, t_0, z), the relation $y = \eta(t, t_0, y_0, z)$ can be considered as a map carrying y_0 into y. The assumption that the solution of (1.2), for $(t_0, y_0, z) \in E$, is unique implies that this map is one-to-one. In fact the inverse map is given by $y_0 = \eta(t_0, t, y, z)$. A consequence of Theorem 2.1 is that the map $y_0 \to y$ is continuous.

Proof. Since (1.2) can be replaced by (1.4) and (y, z) by y, there is no loss of generality in supposing that f does not depend on z. Thus, it will be supposed that $f(t, y)$ is defined on an open (t, y)-set E and that (1.1) has a unique solution $y(t) = \eta(t, t_0, y_0)$ on a maximal interval of existence $\omega_- < t < \omega_+$, where $\omega_\pm = \omega_\pm(t_0, y_0)$. It will be shown that, in this form, Theorem 2.1 is merely a corollary of Theorem II 3.2 for the case $f_n(t, y) \equiv f(t, y)$, $n = 1, 2, \ldots$.

In order to verify that $\omega_+(t_0, y_0)$ is lower semicontinuous, choose a sequence of points $(t_1, y_{10}), (t_2, y_{20}), \ldots$ in E such that $(t_n, y_{n0}) \to (t_0, y_0) \in E$ and $\omega_+(t_n, y_{n0}) \to c(\leqq \infty)$ as $n \to \infty$, where $c = $ lim inf $\omega_+(t^*, y^*)$ as

$(t^*, y^*) \to (t_0, y_0)$. Since the solution of (1.1) is unique, it follows from Theorem II 3.2 that $c \geqq \omega_+ = \omega_+(t_0, y_0)$; i.e., $\omega_+(t_0, y_0)$ is lower semi-continuous. The proof of the upper semi-continuity of $\omega_-(t_0, y_0)$ is the same.

Note that, for the case where $f_n = f$, $n = 1, 2, \ldots$, and the solution of (1.1) is unique, a selection of a subsequence in Theorem II 3.2 is unnecessary. Thus it follows that $\eta(t, t_0, y_0)$ is a continuous function of (t_0, y_0) for every fixed t, $\omega_-(t_0, y_0) < t < \omega_+(t_0, y_0)$; in fact, this continuity is uniform for $t^1 \leqq t \leqq t^2$, $\omega_-(t_0, y_0) < t^1 < t^2 < \omega_+(t_0, y_0)$. In other words, if $\epsilon > 0$, there exists a $\delta_{\epsilon 0} > 0$, depending on (t_0, y_0, t^1, t^2), such that

$$|\eta(t, t_0, y_0) - \eta(t, t_1, y_1)| < \epsilon \quad \text{if} \quad |t_0 - t_1|, |y_0 - y_1| < \delta_{\epsilon 0}$$

for $t^1 \leqq t \leqq t^2$. But since $\eta(t, t_0, y_0)$ is a continuous function of t for fixed (t_0, y_0), there is a $\delta_{\epsilon 1} > 0$, depending on (t_0, y_0, t^1, t^2), such that

$$|\eta(t, t_0, y_0) - \eta(s, t_0, y_0)| < \epsilon \quad \text{if} \quad |t - s| < \delta_{\epsilon 1}, \quad t^1 \leqq s, t \leqq t^2.$$

Hence, if $\delta_\epsilon = \min(\delta_{\epsilon 0}, \delta_{\epsilon 1})$ and $|t - s|, |t_0 - t_1|, |y_0 - y_1| < \delta_\epsilon$, then

$$|\eta(s, t_0, y_0) - \eta(t, t_1, y_1)| < 2\epsilon.$$

This completes the proof of Theorem 2.1.

3. Differentiability

If it is assumed that $f(t, y, z)$ is of class C^1, it follows that the general solution $y = \eta(t, t_0, y_0, z)$ of (1.2) is of class C^1. In fact, even more is contained in the following theorem.

Theorem 3.1 (Peano). *Let $f(t, y, z)$ be continuous on an open (t, y, z)-set E and possess continuous first order partials $\partial f/\partial y^k$, $\partial f/\partial z^j$ with respect to the components of y and z: (i) Then the unique solution $y = \eta(t, t_0, y_0, z)$ of (1.2) is of class C^1 on its open domain of definition $\omega_- < t < \omega_+$, $(t_0, y_0, z) \in E$, where $\omega_\pm = \omega_\pm(t_0, y_0, z)$. (ii) Furthermore, if $J(t) = J(t, t_0, y_0, z)$ is the Jacobian matrix $(\partial f/\partial y)$ of $f(t, y, z)$ with respect to y at $y = \eta(t, t_0, y_0, z)$,*

$$(3.1) \qquad J(t) = J(t, t_0, y_0, z) = \left(\frac{\partial f}{\partial y}\right) \qquad \text{at } y = \eta(t, t_0, y_0, z),$$

then $x = \partial\eta(t, t_0, y_0, z)/\partial y_0{}^k$ is the solution of the initial value problem,

$$(3.2) \qquad\qquad x' = J(t)x, \qquad x(t_0) = e_k,$$

where $e_k = (e_k{}^1, \ldots, e_k{}^d)$ with $e_k{}^j = 0$ if $j \neq k$ and $e_k{}^k = 1$; $x = \partial\eta(t, t_0, y_0, z)/\partial z^j$ is the solution of

$$(3.3) \qquad\qquad x' = J(t)x + g_j(t), \qquad x(t_0) = 0,$$

where $g_j(t) = g_j(t, t_0, y_0, z)$ is the vector $\partial f(t, y, z)/\partial z^j$ at $y = \eta(t, t_0, y_0, z)$; and $\partial\eta(t, t_0, y_0, z)/\partial t_0$ is given by

$$(3.4) \qquad \frac{\partial\eta}{\partial t_0} = -\sum_{k=1}^{d} \frac{\partial\eta}{\partial y_0^{k}} f^k(t_0, y_0, z) \ .$$

The uniqueness of the solution of (1.2) is assured, e.g., by Theorem II 1.1. Note that the assertions concerning $\partial\eta/\partial y_0^{k}$ and (3.2) or $\partial\eta/\partial z^j$ and (3.3) result on "formally" differentiating both equations in (1.2), i.e., both equations in

$$\eta'(t, t_0, y_0, z) = f(t, \eta, z), \qquad \eta(t_0, t_0, y_0, z) = y_0$$

with respect to y_0^{k} or z^j. Similarly, differentiating these equations formally with respect to t_0 shows that $x = \partial\eta/\partial t_0$ is also a solution of $x' = J(t)x$ satisfying the initial condition $x(t_0) = -\eta'(t_0, t_0, y_0, z) = -f(t_0, y_0, z)$. Writing $f(t_0, y_0, z) = \Sigma f^k(t_0, y_0, z)e_k$, it follows that (3.4) is a formal consequence of (3.2) and the principle of superposition (Theorem IV 1.1) for the linear system $x' = J(t)x$.

More generally, if $y(t, s)$ is a 1-parameter family of solutions of $y' = f(t, y, z)$ for fixed z, if $y(t_0, s_0) = y_0$, and if $y(t, s)$ is of class C^1 in (t, s), then the partial derivative $x = \partial y(t, s)/\partial s$ at $s = s_0$ is also a solution of the system $x' = J(t)x$. For this reason $x' = J(t)x$ is called the *equation of variation* of (1.2) along the solution $y = \eta(t, t_0, y_0, z)$.

The assertion concerning $x = \partial\eta/\partial y_0^{k}$ and (3.2) shows that the Jacobian matrix $(\partial\eta/\partial y_0)$ is the fundamental matrix for $x' = J(t)x$ which reduces to the identity for $t = t_0$. In particular, Theorem IV 1.2 implies

Corollary 3.1. *Under the conditions of Theorem 3.1,*

$$(3.5) \qquad \det\left(\frac{\partial\eta(t, t_0, y_0, z)}{\partial y_0}\right) = \exp\int_{t_0}^{t} \sum_{j=1}^{d} \frac{\partial f^{j}}{\partial y^{j}}\, ds$$

for $\omega_- < t < \omega_+$, where the argument of the integrand is $(s, \eta(s, t_0, y_0, z), z)$.

Remark. By (3.5), $\det(\partial\eta/\partial y_0) \neq 0$. Thus, the continuous one-to-one map $y_0 \to y = \eta(t, t_0, y_0, z)$ for fixed (t, t_0, z), considered in the Remark after Theorem 2.1, is of class C^1 and has an inverse of class C^1 with respect to (t, t_0, y, z). This statement about the inverse is also clear from the explicit formula $y \to y_0 = \eta(t_0, t, y, z)$.

Exercise 3.1 *(Liouville).* Let $f(y)$ be of class C^1 on an open set E and let $y = \eta(t, y_0)$ be the unique solution of the initial value problem $y' = f(y)$, $y(t_0) = y_0$ for $(t_0, y_0) \in E$. Show that the set of maps $y_0 \to y$ defined by $y = \eta(t, y_0)$ for fixed t are volume preserving if and only if $\operatorname{div} f(y) \equiv \Sigma\, \partial f^k/\partial y^k$ is 0.

Note that the assertion that $x = \partial\eta/\partial y_0^{k}$ is a solution of (3.2) implies that the iterated derivative $\partial(\partial\eta/\partial y_0^{k})/\partial t$ exists and is $J(t)\, \partial\eta/\partial y_0^{k}$. The last expression is a continuous function of t, t_0, y_0, z; hence, Schwarz's theorem implies that $\partial(\partial\eta/\partial t)/\partial y_0{}^k$ exists and is $\partial(\partial\eta/\partial y_0{}^k)/\partial t$. A similar remark applies to $\partial\eta/\partial z^j$. Also, note that on the right side of (3.4) the variable t only occurs in $\partial\eta/\partial y_0{}^k$.

Corollary 3.2. *Under the conditions of Theorem* 3.1., *the second mixed derivatives* $\partial^2\eta/\partial y_0{}^k\,\partial t = \partial^2\eta/\partial t\,\partial y_0{}^k$, $\partial^2\eta/\partial z^j\,\partial t = \partial^2\eta/\partial t\,\partial z^j$, $\partial^2\eta/\partial t_0\,\partial t = \partial^2\eta/\partial t\,\partial t_0$ *exist and are continuous.*

In order to avoid an interruption to the proof of Theorem 3.1, the following simple lemma will be proved first. This lemma is a convenient substitute for the mean value theorem of differential calculus when dealing with vectors, for it avoids some awkwardness in the fact that θ_k depends on k in $y(b) - y(a) = (b - a)(y^{1\prime}(\theta_1), \ldots, y^{d\prime}(\theta_d))$, where $a < \theta_k < b$.

Lemma 3.1. *Let* $f(t, y)$ *be continuous on a product set* $(a, b) \times K$, *where K is an open convex y-set, and let f have continuous partials* $\partial f/\partial y^k$ *with respect to the components of y. Then there exist continuous functions* $f_k(t, y_1, y_2)$, $k = 1, \ldots, d$, *on the product set* $(a, b) \times K \times K$ *such that*

$$(3.6) \qquad f_k(t, y, y) = \frac{\partial f(t, y)}{\partial y^k}$$

and that if $(t, y_1, y_2) \in (a, b) \times K \times K$, *then*

$$(3.7) \qquad f(t, y_2) - f(t, y_1) = \sum_{k=1}^{d} f_k(t, y_1, y_2)(y_2{}^k - y_1{}^k).$$

In fact, $f_k(t, y_1, y_2)$ *is given by*

$$(3.8) \qquad f_k(t, y_1, y_2) = \int_0^1 \frac{\partial f(t, sy_2 + (1 - s)y_1)}{\partial y^k}\,ds.$$

Proof. Put $F(s) = f(t, sy_2 + (1 - s)y_1)$ for $0 \leq s \leq 1$. The convexity of K implies that $F(s)$ is defined. Then $dF/ds = \Sigma\,(y_2{}^k - y_1{}^k)\,\partial f(t, sy_2 + (1 - s)y_1)/\partial y^k$. Hence $F(1) - F(0)$ is the right side of (3.7) if f_k is defined by (3.8). Since $F(1) = f(t, y_2)$ and $F(0) = f(t, y_1)$, the lemma follows.

Proof of Theorem 3.1. Since (1.2) can be replaced by (1.4) and (y, z) by y, there is no loss of generality in supposing that f does not depend on z when proving the existence and continuity of the partial derivatives of η. Thus the initial value problem (1.1) having the solution $y = \eta(t, t_0, y_0)$ on $\omega_- < t < \omega_+$ is under consideration.

In order to simplify the domain on which the function η must be considered, let a, b be arbitrary numbers satisfying $\omega_- < a < b < \omega_+$, $\omega_\pm = \omega_\pm(t_1, y_1)$. Then, by Theorem 2.1, $\eta(t, t_0, y_0)$ is defined and continuous for $a \leq t \leq b$ and (t_0, y_0) near (t_1, y_1). In the following only such (t, t_0, y_0) will be considered. Since the assertions of Theorem 3.1 are "local," it clearly suffices to prove the assertions on the interior of such a (t, t_0, y_0)-set.

(a) In order to prove the existence of $\partial\eta/\partial y_0{}^k$, let h be a scalar, e_k the vector in (3.2) and, for small $|h|$,

$$(3.9) \qquad y_h(t) = \eta(t, t_0, y_0 + he_k).$$

This is defined on $a \leq t \leq b$ and, by Theorem 2.1,

$$(3.10) \qquad y_h(t) \to y_0(t) \qquad \text{as } h \to 0$$

uniformly on $[a, b]$. By (1.1), $(y_h(t) - y_0(t))' = f(t, y_h(t)) - f(t, y_0(t))$. Applying Lemma 3.1 with $y_2 = y_h(t)$, $y_1 = y_0(t)$,

$$(3.11) \qquad [y_h(t) - y_0(t)]' = \sum_{k=1}^{d} f_k(t, y_0(t), y_h(t))[y_h{}^k(t) - y_0{}^k(t)].$$

Introduce the abbreviation

$$(3.12) \qquad x_h = \frac{y_h(t) - y_0(t)}{h}, \qquad h \neq 0.$$

The existence of $\partial\eta(t, t_0, y_0)/\partial y_0{}^k$ is equivalent to the existence of $\lim x_h(t)$ as $h \to 0$.

By (1.1) and (3.9), $y_h(t_0) = y_0 + he_k$, and so $x_h(t_0) = e_k$. Thus, by (3.11) and (3.12), $x = x_h(t)$ is the solution of the initial value problem

$$(3.13) \qquad x' = J(t; h)x, \qquad x(t_0) = e_k,$$

where $J(t; h)$ is a $d \times d$ matrix in which the kth column is the vector $f_k(t, y_0(t), y_h(t))$. By (3.6), the continuity of $f_k(t, y_1, y_2)$ and (3.10), it follows that $J(t; h) \to J(t; 0)$ as $h \to 0$ uniformly on $[a, b]$, where $J(t; 0) = J(t)$ is the matrix defined by (3.1).

Consider (3.13) to be a family of initial value problem depending on a parameter h, where the right side $J(t; h)x$ of the differential equation is continuous on the open-set $a < t < b$, $|h|$ small, x arbitrary. Since the solutions of (3.13) are unique, Theorem 2.1 implies that the general solution is a continuous function of h [for fixed (t, t_0)]. In particular, $x(t) = \lim x_h(t)$, $h \to 0$, exists and is the solution of (3.2) on $a < t < b$. Hence $\partial\eta(t, t_0, y_0)/\partial y_0{}^k$ exists.

In order to verify that this partial derivative is continuous with respect to all of its arguments, rewrite (3.2) as

$$(3.14) \qquad x' = J(t, t_0, y_0)x, \qquad x(t_0) = e_k,$$

a family of initial value problems depending on parameters (t_0, y_0). Since $J(t, t_0, y_0)$ is a continuous function of $(t, \quad t_0, y_0)$ and initial value problems associated with   linear differential equations have unique solutions, Theorem 2.1 implies that the solution $x = \partial\eta(t, t_0, y_0)/\partial y_0{}^k$ of (3.14) is a continuous function of its arguments.

(b) The existence and continuity of $\partial \eta(t, t_0, y_0)/\partial t_0$ will now be considered. Put

$$x_h(t) = \frac{\eta(t, t_0 + h, y_0) - \eta(t, t_0, y_0)}{h}, \qquad h \neq 0.$$

The solution of the initial value problem $y' = f$, $y(t_0) = \eta(t_0, t_0 + h, y_0)$ is the same as the solution of $y' = f$, $y(t_0 + h) = y_0$; i.e., $\eta(t, t_0 + h, y_0) = \eta(t, t_0, \eta(t_0, t_0 + h, y_0))$. Thus

$$hx_h(t) = \eta(t, t_0, \eta(t_0, t_0 + h, y_0)) - \eta(t, t_0, y_0).$$

Since $\eta(t, t_0, y_0)$ has continuous partial derivatives with respect to the components of y_0 and $\eta(t_0, t_0 + h, y_0) \to \eta(t_0, t_0, y_0) = y_0$ as $h \to 0$, it follows that

$$hx_h(t) = \sum \left[\frac{\partial \eta(t, t_0, y_0)}{\partial y_0{}^k} + o(1) \right] [\eta^k(t_0, t_0 + h, y_0) - y_0{}^k]$$

as $h \to 0$. By the mean value theorem of differential calculus and $y_0 = \eta(t_0 + h, t_0 + h, y_0)$, there is a $\theta = \theta_k$ such that

$$\eta^k(t_0, t_0 + h, y_0) - y_0{}^k = -h\eta^{k'}(t_0 + \theta h, t_0 + h, y_0), 0 < \theta < 1.$$

Note that $\eta^{k'}(t_0 + \theta h, t_0 + h, y_0) = f^k(t_0 + \theta h, \eta(t_0 + \theta h, t_0 + h, y_0))$ is $f^k(t_0, y_0) + o(1)$ as $h \to 0$. Thus, as $h \to 0$,

$$x_h(t) = -\sum \left[\frac{\partial \eta(t, t_0, y_0)}{\partial y_0{}^k} + o(1) \right] [f^k(t_0, y_0) + o(1)].$$

This shows that $\partial \eta/\partial t_0 = \lim x_h(t)$ exists as $h \to 0$ and satisfies the analogue of the relation (3.4). This relation implies that $\partial \eta/\partial t_0$ is a continuous function of (t, t_0, y_0).

(c) Returning to (1.2), so that f can depend on z and $\eta = \eta(t, t_0, y_0, z)$, it follows that η is of class C^1. By applying the results just proved for (1.1) to (1.4), it is readily verified that the assertions concerning (3.2), (3.3), and (3.4) hold. This verification will be left to the reader.

The proof of Theorem 3.1 has the following consequence.

Corollary 3.3. *Let $f(t, y, z, z^*)$ be a continuous function on an open (t, y, z, z^*)-set E, where z^* is a vector of any dimension. Suppose that f has continuous first order partial derivatives with respect to the components of y and z. Then*

$$(3.15) \qquad y' = f(t, y, z, z^*), \qquad y(t_0) = y_0$$

has a unique solution $\eta = \eta(t, t_0, y_0, z, z^*)$ for fixed z, z^* with $(t_0, y_0, z, z^*) \in E$; η has first order partials with respect to t, t_0, the components of y and of z, and the second order partials $\partial^2 \eta/\partial t\, \partial t_0$, $\partial^2 \eta/\partial t\, \partial y_0{}^k$, $\partial^2 \eta/\partial t\, \partial z^j$; finally, these partials of η are continuous with respect to (t, t_0, y_0, z, z^*).

Proof. Since the partial derivatives of η involved in this statement are calculated with z^* fixed, their existence follows from Theorem 3.1. Their continuity with respect to (t, t_0, y_0, z, z^*) follows, as in the proof of Theorem 3.1, by the use of analogues of (3.1), (3.2), Theorem 2.1, and an analogue of (3.4).

4. Higher Order Differentiability

The question of higher order differentiability of the general solution is easily settled by the use of Theorem 3.1 and Corollary 3.3.

Theorem 4.1. *Let $f(t, y, z, z^*)$ be a continuous function on an open (t, y, z, z^*)-set E such that f has continuous partial derivatives of all orders not exceeding m, $m \geq 1$, with respect to the components of y and z. Then*

$$(4.1) \qquad y' = f(t, y, z, z^*), \qquad y(t_0) = y_0$$

has a unique solution $\eta = \eta(t, t_0, y_0, z, z^*)$, for fixed z, z^* with $(t_0, y_0, z, z^*) \in E$, and η has all continuous partial derivatives of the form

$$(4.2) \qquad \frac{\partial^{i+i_0+\alpha_1+\cdots+\alpha_d+\beta_1+\cdots+\beta_e}\eta}{\partial t^i\, \partial t_0^{i_0}\, \partial(y_0^1)^{\alpha_1} \ldots \partial(y_0^d)^{\alpha_d}\, \partial(z^1)^{\beta_1} \ldots \partial(z^e)^{\beta_e}},$$

where $i \leq 1$, $i_0 \leq 1$ and $i_0 + \Sigma\beta_k + \Sigma\alpha_j \leq m$.

Proof. The proof will be given first with $i_0 = 0$ by induction on m. The case $m = 1$ is correct by Corollary 3.3 for z^* of any dimension. Assume the validity of the theorem if m is replaced by $m - 1 (\geq 1)$.

Consider the analogue of (3.2),

$$(4.3) \qquad x' = J(t, t_0, y_0, z, z^*)x, \qquad x(t_0) = e_k,$$

where $J = (\partial f/\partial y)$ at $y = \eta(t, t_0, y_0, z, z^*)$. By the assumption on f and by the induction hypothesis, the right side, $J(t, t_0, y_0, z, z^*)y$, of the differential equation in (4.3) has continuous partial derivatives of order $\leq m - 1$ with respect to the components of y, y_0, and z. Hence, by the induction hypothesis, the solution $x = \partial\eta(t, t_0, y_0, z, z^*)/\partial y_0^k$ of (4.3) has continuous partial derivatives of all orders $\leq m - 1$ with respect to the components of y_0 and each of these partials has a continuous partial derivative with respect to t. Similarly, the analogue of (3.3) shows that $\partial\eta/\partial z^j$ has continuous partial derivatives of all orders $\leq m - 1$ with respect to the components of y_0, and z, each of these partials has a continuous partial derivative with respect to t. This completes the induction and shows that $\eta(t, t_0, y_0, z, z^*)$ has continuous partial derivatives of the form (4.2) with $i_0 = 0$, $i \leq 1$, $\Sigma\alpha_k + \Sigma\beta_j \leq m$.

The existence and continuity of derivatives of the form (4.2) with $i_0 = 1$, $i \leq 1$, $\Sigma\alpha_k + \Sigma\beta_j \leq m - 1$ follows from the analogue of (3.4). This completes the proof.

Corollary 4.1. *Let $f(t, y, z)$ be of class C^m, $m \geq 1$, on an open (t, y, z)-set. Then the solution $y = \eta(t, t_0, y_0, z)$ of (1.2) is of class C^m on its domain of existence.*

The proof of this useful corollary will be left as an exercise.

5. Exterior Derivatives

Several useful concepts will be introduced in this section. All concepts are of a "local" nature.

By a (piece of) 2-dimensional surface S of class C^m, $m \geq 1$, in a Euclidean y-space $\mathbf{R}^d$ is meant a set S of points y in $\mathbf{R}^d$ which can be put into one-to-one correspondence with an open set D of points (u, v) in a Euclidean plane by a function $y = y(u, v)$ of class C^m on D such that the two vectors $\partial y/\partial u$, $\partial y/\partial v$ are linearly independent at every point of D. The function $y = y(u, v)$ is called an admissible parametrization of S.

If $y = y(u, v)$ is any given function of class C^m, $m \geq 1$, on an open (u, v)-set D such that $\partial y/\partial u$, $\partial y/\partial v$ are linearly independent at a point, hence near a point (u_0, v_0) of D, then the set of points $y = y(u, v)$ for (u, v) near (u_0, v_0) is a piece of surface. For by the implicit function theorem, the map $(u, v) \to y$ is one-to-one for (u, v) near (u_0, v_0).

Consider a piece of surface S_0 of class C^1 with an admissible parametrization $y = y(u, v)$ defined on a simply connected, bounded open set D_0 and a piecewise C^1 Jordan curve C in D_0 bounding an open subset D of D_0. Let S, J be the y-image of D, C, respectively. This situation will be described briefly by saying "a piece of C^1 surface S bounded by piecewise C^1 Jordan curve J."

A differential r-form on an open set E is a formal expression

$$\omega = \sum_{i_1=1}^{d} \cdots \sum_{i_r=1}^{d} p_{i_1 \ldots i_r}(y) \, dy^{i_1} \wedge \cdots \wedge dy^{i_r}$$

with real-valued coefficients defined on E, where $p_{i_1 \ldots i_r}(y) = \pm p_{j_1 \ldots j_r}(y)$ according as $(j_1, \ldots, j_r)$ is an even or odd permutation of $(i_1, \ldots, i_r)$. In particular, $p_{i_1 \ldots i_r}(y) = 0$ if two of the indices $i_1, \ldots, i_r$ are equal. The form ω is called continuous [or of class C^m or 0] if its coefficients are continuous [or of class C^m or identically 0] on E. A sequence of differential r-forms on E is said to be uniformly bounded [or uniformly convergent] if the sequences of the corresponding coefficients are uniformly bounded [or uniformly convergent]. Differential r-forms can be added in the obvious way. Differential r- and s-forms can be multiplied to give an $(r + s)$-form by the usual associative, distributive laws and anti-commutative law $dy^i \wedge dy^j = -dy^j \wedge dy^i$; see § VI 2.

A continuous linear differential form (1-form or Pfaffian)

$$(5.1) \qquad \omega = \sum_{j=1}^{d} p_j(y) \, dy^j$$

on E is said to possess a *continuous exterior derivative* $d\omega$ if there exists a continuous differential 2-form

$$(5.2) \qquad d\omega = \sum_{i=1}^{d} \sum_{j=1}^{d} p_{ij}(y) \, dy^i \wedge dy^j, \qquad p_{ij} = -p_{ji},$$

on E such that Stokes' formula

$$(5.3) \qquad \int_J \omega = \iint_S d\omega$$

holds for every piece of C^1 surface S in E bounded by a C^1 piecewise Jordan curve J in E.

It is clear that if S is the image of D on the surface S_0: $y = y(u, v)$ for $(u, v) \in D$, and J is the image of the Jordan curve C, then (5.3) means that

$$(5.4) \qquad \int_C \sum_{j=1}^{d} p_j(y(u, v)) \, dy^j(u, v) = \int_D \sum_{i=1}^{d} \sum_{j=1}^{d} p_{ij}(y(u, v)) \frac{\partial(y^i, y^j)}{\partial(u, v)} \, du \, dv,$$

with the usual convention as to orientation of C around D.

If the coefficients $p_j(y)$ of (5.1) are of class C^1, then ω has a continuous exterior derivative with $d\omega = \Sigma \, dp_j(y) \wedge dy^j$ or

$$(5.5) \qquad d\omega = \sum_{i=1}^{d} \sum_{j=1}^{d} \frac{1}{2} \left(\frac{\partial p_j}{\partial y^i} - \frac{\partial p_i}{\partial y^j} \right) dy^i \wedge dy^j.$$

There are cases, however, where (5.1) has a continuous exterior derivative when the coefficients of (5.1) are only continuous. Consider, for example, the case that there exists a real-valued function $U(y)$ of class C^1 such that $\omega = dU$ [i.e., $p_j(y) = \partial U/\partial y^j$], then ω has the exterior derivative $d\omega = 0$.

The fundamental lemma about the existence of continuous exterior derivatives is the following:

Lemma 5.1. *Let* (5.1) *be a continuous linear differential 1-form on an open set E. Then* (5.1) *has a continuous exterior derivative* (5.2) *on E if and only if, on every open subset E^0 with compact closure $\bar{E}^0 \subset E$, there exists a sequence of 1-forms $\omega^1, \omega^2, \ldots$ of class C^1 such that $\omega^n \to \omega$ as $n \to \infty$ uniformly on E^0 and $d\omega^1, d\omega^2, \ldots$ is uniformly convergent on E^0 (in which case, $d\omega^n \to d\omega$ uniformly on E^0 as $n \to \infty$).*

Proof. If a sequence $\omega^1, \omega^2, \ldots$ of the specified type exists on E^0 and if the case $\omega = \omega^n$ of (5.3) is written in the form (5.4), an obvious term-by-term integration gives (5.3). Thus, it is seen that the existence of sequences $\omega^1, \omega^2, \ldots$ is sufficient for the existence of a continuous $d\omega$.

Conversely, if (5.1) is a continuous 1-form on E with a continuous exterior derivative (5.2), approximate the coefficients of ω, $d\omega$ by the method in § I 3: Let $\varphi(t)$ be as in § I 3 and put, for $\epsilon = 1/n$,

$$p_i^{(n)}(y) = c\epsilon^{-d}\int_{-\infty}^{+\infty}\ldots\int_{-\infty}^{+\infty} p_i(y - \eta)\varphi(\epsilon^{-2}\|\eta\|^2)\, d\eta^1 \ldots d\eta^d,$$

$$p_{ij}^{(n)}(y) = c\epsilon^{-d}\int_{-\infty}^{\infty}\ldots\int_{-\infty}^{\infty} p_{ij}(y - \eta)\varphi(\epsilon^{-2}\|\eta\|^2)\, d\eta^1 \ldots d\eta^d.$$

Since the integrals are actually integrals over spheres $\|\eta\| \leqq \epsilon$, $p_i^{(n)}$ and $p_{ij}^{(n)}$ are defined on the sets E_n consisting of points y whose (Euclidean) distance from the boundary of E exceeds $\epsilon = 1/n$. In particular, they are defined on E^0 for large n and tend uniformly to p_i, p_{ij}, respectively, on E^0, as $n \to \infty$.

Define the C^∞ forms $\omega^n = \Sigma p_j^{(n)}(y)\, dy^j$ and $\alpha^n = \Sigma\Sigma p_{ij}^{(n)}(y)\, dy^i \wedge dy^j$ on E^0 for large n. Let S be a piece of C^1 surface in E^0 bounded by a C^1 piecewise Jordan arc J in E^0. Then if $\epsilon = 1/n$ is sufficiently small and $\|\eta\| \leqq \epsilon$, the translation $S(\eta)$ of S by the vector $-\eta$ is in E and (5.3) is valid if S is replaced by $S(\eta)$. This can be written in a form analogous to (5.4),

$$\int_C \Sigma\, p_j(y(u, v) - \eta)\, dy^j(u, v) = \iint_D \Sigma\Sigma\, p_{ij}(y(u, v) - \eta)\, \frac{\partial(y^i, y^j)}{\partial(u, v)}\, du\, dv.$$

Let this relation be multiplied by $c\epsilon^{-d}\varphi(\epsilon^{-2}\|\eta\|^2)$ and integrated over $\|\eta\| \leqq \epsilon$ with respect to $d\eta^1 \ldots d\eta^d$. An obvious change of the order of integration shows that the result can be interpreted as the Stokes' relation $\int_J \omega^n = \iint_S \alpha^n$. Thus ω^n has the continuous exterior derivative $d\omega^n = \alpha^n$ in E_0. This completes the proof.

Remark. In deciding whether or not a continuous 1-form (5.1) has the continuous exterior derivative (5.2), it suffices to verify Stokes' formula (5.3) for rectangles S on coordinate 2-planes $y^i = $ const. for $i \neq j, k$, where $1 \leqq j < k \leqq d$. This is a consequence of the following exercise.

Exercise 5.1. A continuous differential 1-form (5.1) on an open set E has a continuous exterior derivative if and only if there exists a continuous differential 2-form (5.2) such that for every pair j, k $(1 \leqq j < k \leqq d)$ and fixed y^i, with $i \neq j$, $i \neq k$, the 1-form $p_j(y)\, dy^j + p_k(y)\, dy^k$ has the continuous exterior derivative $p_{jk}\, dy^j \wedge dy^k + p_{kj}\, dy^k \wedge dy^j$; in fact, if and only if Stokes' formula (5.3) holds for all rectangle S on 2-planes $y^i = $ const. for $i \neq j, k$ with $S \subset E$.

Exercise 5.2.   Let the continuous differential 1-form (5.1) possess a continuous exterior derivative and let $p_1(y)$ have a continuous derivative with respect to y^j for a fixed $j \neq 1$. Show that $p_j(y)$ possesses a continuous partial derivative with respect to y^1.

Exercise 5.3.   Let $p_j(t)$, where $j = 1, \ldots, d$, be continuous functions on E such that $p_j(y)$ has continuous partial derivatives with respect to the components y^k, $k \neq j$, of y. Show that (5.1) has a continuous exterior derivative.

For the sake of brevity, "vector" and "matrix" notation will be used in connection with 1-forms and their exterior derivatives. For example, an ordered set of e 1-forms $\omega_1, \ldots, \omega_e$ will be abbreviated $\omega = (\omega_1, \ldots, \omega_e)$; analogously if these forms have continuous exterior derivatives, $d\omega$ denotes the ordered set of 2-forms $d\omega = (d\omega_1, \ldots, d\omega_e)$. Finally, if $A = (a_{ij}(y))$ is an $e \times d$ matrix function on E, by $\omega = A(y)\ dy$ will be meant the ordered set of 1-forms $\omega = (\omega_1, \ldots, \omega_e)$, where $\omega_i = \sum\limits_{j=1}^{d} a_{ij}(y)\ dy^j$ for $i = 1, \ldots, e$.

6. Another Differentiability Theorem

The main result (Theorem 3.1) on the differentiability of general solutions has the following generalization.

Theorem 6.1.   *Let $f(t, y, z)$ be continuous on an open (t, y, z)-set E. A necessary and sufficient condition that the initial value problem*

$$(6.1) \qquad y' = f(t, y, z), \qquad y(t_0) = y_0$$

have a unique solution $y = \eta(t, t_0, y_0, z)$ for all $(t_0, y_0, z) \in E$ which is of class C^1 with respect to (t, t_0, y_0, z) on its domain of definition is that every point of E have an open neighborhood E^0 on which there exist a continuous nonsingular $d \times d$ matrix $A(t, y, z)$ and a continuous $d \times e$ matrix $C(t, y, z)$ such that the d differential 1-forms

$$(6.2) \qquad \omega = A(dy - f\ dt) + C\ dz$$

in the variables $dt, dy^1, \ldots, dy^d, dz^1, \ldots, dz^e$ have continuous exterior derivatives on E^0.

In contrast to Theorem 3.1, the conditions of Theorem 6.1 are invariant under C^1 changes of the variables t, y, z.

It is understood that if $A = (a_{ij}(t, y, z))$ and $C = (c_{ik}(t, y, z))$, then (6.2) represents an ordered set of 1-forms, the ith one of which is

$$(6.3) \qquad \omega_i = \sum_{j=1}^{d} a_{ij}\ dy^j - \left(\sum_{k=1}^{d} a_{ik} f^k \right) dt + \sum_{k=1}^{e} c_{ik}\ dz^k,$$

$i = 1, \ldots, d$. If this has a continuous exterior derivative, the latter is a differential 2-form of the type

$$(6.4) \quad d\omega_i = \sum_{j=1}^{d} \sum_{k=1}^{d} \alpha_{ijk}\, dy^j \wedge dy^k + \sum_{j=1}^{d} \beta_{ij}\, dt \wedge dy^j + \sum_{k=1}^{e} \gamma_{ik}\, dt \wedge dz^k$$

$$+ \sum_{j=1}^{d} \sum_{k=1}^{e} \delta_{ijk}\, dz^k \wedge dy^j + \sum_{j=1}^{e} \sum_{k=1}^{e} \epsilon_{ijk}\, dz^j \wedge dz^k,$$

where $\alpha_{ijk} = -\alpha_{ikj}$, β_{ij}, γ_{ik}, δ_{ijk}, $\epsilon_{ijk} = -\epsilon_{ikj}$ are continuous function of (t, y, z). In this case, define a $d \times d$ matrix $F(t, y, z) = (f_{ij}(t, y, z))$ and a $d \times e$ matrix $N(t, y, z) = (n_{ij}(t, y, z))$ by

$$(6.5) \qquad f_{ij} = \beta_{ij} - 2\sum_{k=1}^{d} \alpha_{ijk} f^k$$

$$(6.6) \qquad n_{ij} = \gamma_{ij} + \sum_{k=1}^{d} \delta_{ikj} f^k$$

Theorem 6.1 will be proved in § 11 below. The proof of Theorem 6.1 will have the following consequence.

Corollary 6.1. *Let $f(t, y, z)$ be as in Theorem 6.1, let $A(t, y, z)$, $C(t, y, z)$ exist on E^0 as specified, and consider only $(t, y, z) \in E^0$. Then $x = \partial\eta/\partial y_0{}^k$ is the solution of*

$$(6.7) \qquad [A(t, \eta, z)x]' = F(t, \eta, z)x, \qquad x(t_0) = e_k$$

and $x = \partial\eta/\partial z^i$ is the solution of

$$(6.8) \quad [A(t, \eta, z)x + c_i(t, \eta, z)]' = F(t, \eta, z)x + n_i(t, \eta, z), \qquad x(t_0) = 0,$$

where $c_i(t, y, z)$, $n_i(t, y, z)$ are the ith columns of $C(t, y, z)$, $N(t, y, z)$, respectively, for $i = 1, \ldots, e$ and $\eta = \eta(t, t_0, y_0, z)$.

Note that a solution $x = x(t)$ of (6.7) does not necessarily have a derivative, but $A(x, \eta, z)x(t)$ has a derivative satisfying (6.7). An obvious change of variables reduces the linear equations in (6.7) and (6.8) to the type considered in Chapter IV; cf. (11.1)–(11.3).

The statements concerning (6.7) and (6.8) can be written more conveniently as matrix equations

$$(6.9) \quad \left[A(t, \eta, z) \frac{\partial\eta}{\partial y_0} \right]' = F(t, \eta, z) \frac{\partial\eta}{\partial y_0}, \qquad \frac{\partial\eta}{\partial y_0} = I \text{ at } t = t_0,$$

$$(6.10) \quad \left[A(t, \eta, z) \frac{\partial\eta}{\partial z} + C(t, \eta, z) \right]'$$

$$= F(t, \eta, z) \frac{\partial\eta}{\partial z} + N(t, \eta, z), \qquad \frac{\partial\eta}{\partial z} = 0 \text{ at } t = t_0,$$

where $\partial\eta/\partial y_0$, $\partial\eta/\partial z$ denote Jacobian matrices.

Exercise 6.1. Let $f(t, y, y')$ be a continuous d-dimensional vector defined on an open (t, y, y')-set E. Let $(i_1, \ldots, i_d)$ be any set of d integers, $0 \leq i_j \leq d$, such that no integer except 0 occurs more than once. Let $t = y^0$ and suppose that $f^k(y^0, y, y')$ has continuous partial derivatives with respect to each of its arguments except possibly y^{i_k}. Then the initial value problem

$$y'' = f(t, y, y'), \qquad y(t_0) = y_0, \qquad y'(t_0) = y_0',$$

where $y_0'^k \neq 0$ if $i_k \neq 0$, has a unique solution $y = \eta(t, t_0, y_0, y_0')$ and $\eta(t, t_0, y_0, y_0')$, $\eta'(t, t_0, y_0, y_0')$ are of class C^1.

The following two exercises are applications of Theorem 6.1 and Corollary 6.1 to differential geometry.

Exercise 6.2. (*a*) Let $(g_{jk}(x))$, where $x = (x^1, \ldots, x^d)$, be a $d \times d$ nonsingular symmetric matrix with real entries which are functions of class C^1 for small $\|x\|$ and let $(g^{jk}(x))$ be the inverse matrix. Consider the initial value problem

$$(6.11) \qquad x^{i''} + \sum_{j=1}^{d} \sum_{k=1}^{d} \Gamma_{jk}^i x^{j'} x^{k'} = 0, \qquad x(0) = x_0 \text{ and } x'(0) = x_0'$$

for the geodesics of $ds^2 = \Sigma\Sigma\, g_{jk}\, dx^j\, dx^k$, where $\Gamma_{jk}^i = \Gamma_{jk}^i(x)$ are the Christoffel symbols of the second kind defined by

$$\Gamma_{jk}^i = \tfrac{1}{2} \sum_{m=1}^{d} g^{im} \left[\frac{\partial g_{jm}}{\partial x^k} + \frac{\partial g_{km}}{\partial x^j} - \frac{\partial g_{jk}}{\partial x^m} \right].$$

Assume that "ds^2 has a continuous Riemann curvature tensor" in the sense that each of the d^2 differential 1-forms $\omega_j{}^i = \Sigma_k\, \Gamma_{jk}^i\, dx^k$ has a continuous exterior derivative. Show that (6.11) has a unique solution $x = \xi(t, x_0, x_0')$ for small $|t|$, $|x_0|$ and arbitrary x_0' and that $\xi(t, x_0, x_0')$, $\xi'(t, x_0, x_0')$ are of class C^1 as functions of their $2d + 1$ variables. (*b*) Let $z = (z^1, \ldots, z^e)$ and $x = (x^1, \ldots, x^d)$, where $e \geq d$, and let $z = z(x)$ be a function of class C^2 for small $\|x\|$ with a Jacobian matrix $(\partial z^j/\partial x^k)$ of rank d. Show that (*a*) is applicable to $ds^2 = \|dz\|^2 = \Sigma\Sigma\, g_{jk}\, dx^j\, dx^k$, where g_{jk} is the scalar product $(\partial z/\partial x^j) \cdot (\partial z/\partial x^k)$. (*c*) Show that $ds^2 = [1 + 9(x^2)^{4/3}]\, [(dx^1)^2 + (dx^2)^2]$, where $d = 2$, has more than one geodesic through the point $x_0 = 0$ in the direction $x_0' = (1, 0)$.

Exercise 6.3. Let $(h_{jk}(y))$, where $y = (y^1, y^2)$, be a 2×2 symmetric matrix of real-valued functions of class C^1 for small $\|y\|$ such that det $(h_{jk}) < 0$. Let α denote the indefinite quadratic form

$$\alpha = \sum_{j=1}^{2} \sum_{k=1}^{2} h_{jk}\, dy^j\, dy^k.$$

Then α has factorizations $\alpha = 2\omega_1\omega_2$, where ω_1, ω_2 are linearly independent differential 1-forms. (Note that α is not a differential 2-form and

$\alpha = 2\omega_1\omega_2$ is an ordinary, not exterior, product.) (a) There exists a unique continuous differential 1-form ω_{12} such that

$$d\omega_1 = \omega_{12} \wedge \omega_2 \quad \text{and} \quad d\omega_2 = \omega_1 \wedge \omega_{12}.$$

(b) Assume that α has a continuous curvature $K = K(y)$ in the sense that the factors ω_1, ω_2 can be chosen so that ω_{12} has a continuous exterior derivative, in which case, K is defined by

$$d\omega_{12} = K\,\omega_1 \wedge \omega_2.$$

Show that there exist functions $y(u) = y(u^1, u^2)$ of class C^1 for small $\|u\|$ such that $y(0) = 0$ and $y = y(u)$ transforms α into the form

$$\alpha = 2T\,du^1\,du^2, \quad \text{where } T = T(u) > 0$$

is of class C^1 and has a continuous second mixed derivative such that

$$\frac{\partial^2(\log T)}{\partial u^1\,\partial u^2} + KT = 0.$$

(It can be shown that $y = y(u)$ is of class C^2; Hartman [16].) (c) Show that (b) is applicable if $\alpha = 0$ is the differential equation for the asymptotic lines on a piece of surface of class C^3 of negative curvature in Euclidean 3-space. (d) Show that (b) is applicable if $\alpha = 0$ is the differential equation for the lines of curvature on a piece of surface of class C^3 without umbilical points in Euclidean 3-space.

7. S- and L-Lipschitz Continuity

The proof of sufficiency in Theorem 6.1 falls into two parts: uniqueness and differentiability. It turns out that the necessary and sufficient condition of Theorem 6.1 can be lightened considerably for "uniqueness" alone. Consider the case in which no parameters z occur,

(7.1) $$y' = f(t, y), \qquad y(t_0) = y_0.$$

Correspondingly, $A = A(t, y)$ and the analogue of (6.2) is

(7.2) $$\omega = A(dy - f\,dt).$$

Here, $\omega = (\omega_1, \ldots, \omega_d)$, where

(7.3) $$\omega_i = \sum_{j=1}^{d} a_{ij}\,dy^j - \left(\sum_{j=1}^{d} a_{ij}f^j\right) dt$$

and when $d\omega$ exists, it is of the form

(7.4) $$d\omega_i = \sum_{j=1}^{d}\sum_{k=1}^{d}\alpha_{ijk}\,dy^j \wedge dy^k + \sum_{j=1}^{d}\beta_{ij}\,dt \wedge dy^j,$$

where $\alpha_{ijk} = -\alpha_{ikj}$, β_{ij} are continuous functions of (t, y). Correspondingly

(7.5) $\qquad F = (f_{ij}) \qquad$ where $\quad f_{ij} = \beta_{ij} - 2\sum_{k=1}^{d} \alpha_{ijk} f^k.$

The notion of a differential 1-form ω with a continuous exterior derivative generalizes the notion of a C^1 form ω. Lemma 5.1 suggests the following generalization of a 1-form ω with uniformly Lipschitz continuous coefficients:

A continuous linear differential form (5.1) on a domain E will be said to be *S-Lipschitz continuous on E* if there exists a sequence of 1-forms ω^1, ω^2, ... of class C^1 on E such that $\omega^n \to \omega$ as $n \to \infty$ uniformly on E and $d\omega^1$, $d\omega^2$, ... are uniformly bounded on E.

Exercise 7.1.   Show that if the coefficients of (5.1) are uniformly Lipschitz continuous on E, then (5.1) is S-Lipschitz continuous on E.

Exercise 7.2.   Consider the case of dimension $d = 2$. Let

$$\omega = p_1(y)\, dy^1 + p_2(y)\, dy^2$$

be continuous on a simply connected, open (y^1, y^2)-set E with the property that there exists a bounded, measurable function $p_{12}(y)$ on E such that for any subset S of E bounded by a C^1 piecewise Jordan curve J in E,

$$\int_J p_1(y)\, dy^1 + p_2(y)\, dy^2 = \iint_S p_{12}(y)\, dy^1\, dy^2.$$

Show that ω is S-Lipschitz continuous on every open subset E_0 with compact closure $\bar{E}_0 \subset E$. (It is understood that "measurable" means measurable with respect to plane Lebesgue measure.) One might say that ω has a "bounded exterior derivative." This notion does not generalize readily to arbitrary dimensions d, for a piece of (2-dimensional) surface S is of d-dimensional measure zero if $d > 2$.

The condition that each of the d forms of (7.2) is S-Lipschitz continuous can be generalized as follows: Let $f(t, y)$ be continuous and $A(t, y)$ be continuous and nonsingular on E. The form (7.2) is said to be *L-Lipschitz continuous on E* if there exists a sequence of forms ω^1, ω^2, ... of the type

(7.6) $\qquad\qquad \omega^n = A_n(t, y)\, [dy - f_n(t, y)\, dt]$

of class C^1 on E such that $\omega^n \to \omega$ uniformly on E as $n \to \infty$, and that there exist constants c_0, c satisfying

(7.7) $\qquad\qquad c_0 I \leqq A_n{}^*F_n \leqq cI \qquad$ for $n = 1, 2, \ldots$

on E. Here $F_n = F_n(t, y)$ is the matrix belonging to (7.6) defined by the analogue of formulae (7.3)–(7.5). The inequalities in (7.7) have the following meaning: if B, C are two $d \times d$ matrices, $B \leqq C$ means that the corresponding quadratic forms satisfy $\xi \cdot B\xi \leqq \xi \cdot C\xi$ for all real d-vectors ξ, where the dot indicates scalar multiplication. $B \leqq C$ is equivalent to $B^H \leqq C^H$, where $B^H = \frac{1}{2}(B + B^*)$ is the Hermitian part of B.

The form (7.2) is called *upper L-Lipschitz continuous on E* if (7.7) is replaced by

(7.8) $A_n{}^*F_n \leqq cI$ for $n = 1, 2, \ldots$.

Consider these conditions when A, f are of class C^1. In this case, the exterior derivative of (7.2) can be calculated formally from $d\omega = (dA) \wedge (dy - f\,dt) - A\,df \wedge dt$ which gives

(7.9)
$$\alpha_{ijk} = \frac{1}{2}\left(\frac{\partial a_{ik}}{\partial y^j} - \frac{\partial a_{ij}}{\partial y^k}\right),$$
$$\beta_{ij} = \frac{\partial a_{ij}}{\partial t} + \sum_{k=1}^{d}\left(f^k\frac{\partial a_{ik}}{\partial y^j} + a_{ik}\frac{\partial f^k}{\partial y^j}\right).$$

Hence, (7.5) shows that

(7.10) $$f_{ij} = \frac{\partial a_{ij}}{\partial t} + \sum_{k=1}^{d} a_{ik}\frac{\partial f^k}{\partial y^j} + \sum_{k=1}^{d} f^k\frac{\partial a_{ij}}{\partial y^k}.$$

If $B = A^*F = (b_{ij})$ and $G = A^*A$, then

(7.11) $$b_{ij} = \frac{1}{2}\frac{\partial g_{ij}}{\partial t} + \sum_{k=1}^{d} g_{ik}\frac{\partial f^k}{\partial y^j} + \frac{1}{2}\sum_{k=1}^{d} f^k\frac{\partial g_{ij}}{\partial y^k} + h_{ij},$$

where $H = (h_{ij})$ is skew-symmetric. This can be seen as follows: the partial derivative of $G = A^*A$ with respect to t is $G' = A^*A' + A^{*\prime}A$, so that the Hermitian part of A^*A' is $\frac{1}{2}G'$. This accounts for the replacement of the first term of (7.10) by that in (7.11). The same remark applies to the third terms involving differentiation with respect to y^k instead of t. For other applications of (7.11), see § XIV 12.

Exercise 7.3.   Let $f(t, y)$ be continuous on E and satisfy $[f(t, y_2) - f(t, y_1)] \cdot (y_2 - y_1) \leqq 0$. Show that $\omega = dy - f(t, y)\,dt$ is upper L-Lipschitz continuous with $A_n = I$ in (7.6) and $c = 0$ in (7.8).

8. Uniqueness Theorem

A generalization of the uniqueness theorem contained in Theorem 6.1 is the following:

Theorem 8.1. *Let $f(t, y)$ be continuous on an open (t, y)-set E and let there exist a continuous, nonsingular matrix $A(t, y)$ on E such that the 1-forms (7.2) are S-Lipschitz continuous on E or, more generally, that (7.2) is L-Lipschitz continuous on E. Then (7.1) has a unique solution $y = \eta(t, t_0, y_0)$ for all $(t_0, y_0) \in E$. Furthermore, $\eta(t, t_0, y_0)$ is uniformly Lipschitz continuous on compact subsets of its domain of definition.*

This theorem will be proved in § 10 below. It is possible to formulate a one-sided uniqueness theorem analogous to Theorem III 6.2. This is given by the following exercise.

Exercise 8.1.   Let f, A be as in Theorem 8.1 except that (7.2) is supposed to be only upper L-Lipschitz continuous on E. Then (7.1) has a unique solution $y = \eta(t, t_0, y_0)$ to the right $(t \geq t_0)$ of t_0 for all $(t_0, y_0) \in E$. Furthermore, an inequality of the type

$$|\eta(t, t_1, y_1) - \eta(t, t_2, y_2)| \leqq \text{const.} \, |\eta(t^*, t_1, y_1) - \eta(t^*, t_2, y_2)|$$

holds for $t \geq t^* \geq \max(t_1, t_2)$ on compact subsets of the domain of definition of $\eta(t, t_0, y_0)$.

Exercise 8.2 (*Another One-Sided Generalization of Corollary III* 6.1). Let $f(t, y)$ be continuous on R: $0 \leqq t \leqq a$, $|y| \leqq b$. On R, let there exist a continuous, nonsingular $A(t, y)$ such that (7.2) is upper L-Lipschitz continuous on R_ϵ: $(0 <) \epsilon \leqq t \leqq a$, $|y| < b$ with (7.8) replaced by

$$F_n A_n^{-1} \leqq \frac{I}{t}, \quad \text{i.e.,} \quad A_n{}^* F_n \leqq \frac{A_n{}^* A_n}{t} \quad \text{on} \quad R_\epsilon$$

for $n = 1, 2, \ldots$. Then (7.1) has a unique solution for $t_0 = 0$, $|y_0| < b$.

9. A Lemma

The proof of Theorem 8.1 will depend on the Uniqueness Theorem III 7.1 and the following lemma.

Lemma 9.1.   *Let* $f(t, y)$ *be of class* C^1 *and let* $A(t, y)$ *be a nonsingular matrix of class* C^1 *on an open set* E *such that*

$$(9.1) \qquad\qquad A^* F \leqq \mu(t) A^* A,$$

where $\mu(t)$ *is a continuous function. Let* $y = \eta(t) = \eta(t, t_0, y_0)$ *be a solution of* (7.1) *and* $J(t, y)$ *the Jacobian matrix* $(\partial f/\partial y)$. *Then a solution* $x(t)$ *of the linear "equations of variation"*

$$(9.2) \qquad\qquad x' = J(t, \eta(t))x$$

satisfies, for $t \geq t_0$,

$$(9.3) \qquad \|A(t, \eta)x(t)\| \leqq \|A(t_0, y_0)x(t_0)\| \exp \int_{t_0}^{t} \mu(s) \, ds.$$

Proof.   A differentiation of $A(t, \eta(t))x(t)$, using (9.2) and (7.10), shows that (9.2) implies that

$$(9.4) \qquad\qquad (A^0 x)' = F^0 x,$$

where the superscript 0 indicates that the argument is $(t, y) = (t, \eta)$ with $\eta = \eta(t) = \eta(t, t_0, y_0)$. Thus

(9.5) $$(A^0 x) \cdot (A^0 x)' = x \cdot (A^0)^* F^0 x,$$

since $A^0 x \cdot F^0 x = x \cdot (A^0)^* F^0 x$. Note that $\|A^0 x\|^2 = A^0 x \cdot A^0 x = x \cdot (A^0)^* A^0 x$. Hence (9.1) and (9.5) imply that $d \|A^0 x\|^2 / dt \leqq 2\mu(t) \|A^0 x\|^2$ and so a quadrature gives (9.3); cf. Lemma IV 4.2.

Note that if

(9.6) $$-cI \leqq A^* F \leqq cI, \qquad c \geqq 0,$$

(9.7) $$m^{-1} I \leqq A^* A \leqq mI, \qquad m > 1,$$

then $-cm A^* A \leqq A^* F \leqq cm A^* A$. In this case, (9.3) and the corresponding inequality for $t \leqq t_0$ show that

(9.8) $$\|A^0 x(t)\| \leqq \|A(t_0, y_0) x(t_0)\| \exp cm |t - t_0|$$

where $\eta(t)$, hence $x(t)$, is defined. In particular, since $x = \partial \eta(t, t_0, y_0) / \partial y_0{}^k$ is a solution of (9.2) by Theorem 3.1, (9.6) and (9.7) imply that

(9.9) $$\left\| \frac{\partial \eta(t, t_0, y_0)}{\partial y_0{}^k} \right\| \leqq m \exp cm |t - t_0|.$$

Estimates for $\partial \eta(t, t_0, y_0) / \partial t_0$ follow from the analogue of (3.4),

(9.10) $$\frac{\partial \eta(t, t_0, y_0)}{\partial t_0} = -\sum_{k=1}^{d} \frac{\partial \eta(t, t_0, y_0)}{\partial y_0{}^k} f^k(t_0, y_0).$$

10. Proof of Theorem 8.1

Consider first the case that f, A are of class C^1. Let $(t_1, y_1) \in E$ and let E^0 be a convex open neighborhood of (t_1, y_1) with compact closure $\bar{E}^0 \subset E$. Then, by Peano's existence theorem (Corollary II 2.1), there exists an open set $E_0 \subset E^0$ and numbers a, b ($> t_1 > a$) such that if $(t_0, y_0) \in \bar{E}_0$, then $\eta(t, t_0, y_0)$ exists for $a \leqq t \leqq b$ and $(t, \eta) \in E^0$. Note that E_0, a, b depend only on E^0 and a bound for $|f|$ on E^0.

Suppose, in addition, that (9.6) and (9.7) hold on E^0. Then (9.9) and (9.10) show that there exists a constant K (depending only on E^0, a bound for $|f|$ on E^0 and on m, c) such that

(10.1) $$|\eta(t, t_0, y_0) - \eta(t, t_0, y^0)| \leqq K |y_0 - y^0|$$

for $a \leqq t \leqq b$, $(t_0, y_0) \in \bar{E}_0$, $(t, y^0) \in \bar{E}_0$;

(10.2) $$|\eta(t, t_0, y_0) - \eta(t, t^0, y_0)| \leqq K |t_0 - t^0|$$

for $a \leqq t \leqq b$, $(t_0, y_0) \in \bar{E}_0$, $(t^0, y_0) \in \bar{E}_0$; and

(10.3) $$|\eta(t, t_0, y_0) - \eta(t^*, t_0, y_0)| \leqq K |t - t^*|$$

for $a \leq t, t^* \leq b$, $(t_0, y_0) \in \bar{E}_0$. Finally, interchanging t and t_0 in (10.1) gives

(10.4) $$|y_0 - y^0| \leq K |\eta(t, t_0, y_0) - \eta(t, t_0, y^0)|,$$

provided that $a \leq t \leq b$, $(t, \eta(t, t_0, y_0)) \in \bar{E}_0$, $(t, \eta(t, t_0, y^0)) \in \bar{E}_0$. There is a neighborhood $E_{00} \subset E_0$ of (t_1, y_1) such that the last proviso [for (10.4)] is satisfied if the interval $[a, b]$ is sufficiently small and $(t_0, y_0), (t_0, y^0) \in E_{00}$.

It must be emphasized that $[a, b]$, E_0, E_{00} and the constant K in (10.1)–(10.4) depend only on E^0, a bound for $|f|$ in E^0, and the validity of (9.6)– and (9.7) on E^0.

Return to the case that A, f are not necessarily of class C^1; instead A, f are only continuous, A is nonsingular and (7.2) is L-Lipschitz continuous on every open set E^0 with compact closure $\bar{E}^0 \subset E$. Then there exists a sequence of d ordered 1-forms

$$\omega^{(n)} = A_n(t, y) \, dy - h_n(t, y) \, dt,$$

on E^0 with C^1 coefficients such that $A_n \to A$, $h_n \to Af$ uniformly on E^0 as $n \to \infty$ and (7.7) holds on E^0. Since A is nonsingular, $A_n(t, y)$ is non-singular on E^0 for large n, say, for all n, and $\omega^{(n)}$ can be written as

(10.5) $$\omega^{(n)} = A_n(t, y) [dy - f_n(t, y) \, dt],$$

where $f_n = A_n^{-1} h_n \to f$ uniformly on E^0 as $n \to \infty$. It can be supposed that (7.7) holds on E^0 with $c_0 = -c \leq 0$; also that there exists an $m > 1$ such that

(10.6) $$m^{-1}I \leq A_n{}^*A_n \leq mI, \qquad m > 1, \qquad \text{for } n = 1, 2, \ldots$$

on E^0.

For $(t_1, y_1) \in E$, let E^0 in the last paragraph be any convex open neighborhood of (t_1, y_1) with compact closure $\bar{E}^0 \subset E$. Since $f, f_1, f_2, \ldots$ are uniformly bounded for $(t, y) \in E^0$, there is an open neighborhood E_0 of (t_1, y_1) such that any solution $y = y(t)$ of (7.1) or of

(10.7) $$y' = f_n(t, y), \qquad y(t_0) = y_0$$

for $(t_0, y_0) \in \bar{E}_0$ exists and $(t, y(t)) \in E^0$ for $a \leq t \leq b$, where $a, b \, (>t_1 > a)$ are independent of n and the solution $y = y(t)$. Thus there exists a K, independent of n, such that if $y = \eta_n(t, t_0, y_0)$ is the solution of (10.7), then (10.1)–(10.4) with $\eta = \eta_n$. In particular, the sequence $\eta_1(t, t_0, y_0)$, $\eta_2(t, t_0, y_0), \ldots$ is uniformly bounded and equicontinuous for $a \leq t \leq b$, $(t_0, y_0) \in \bar{E}_0$. Thus there exists a subsequence, which after renumbering can be taken to be the full sequence, such that

(10.8) $$\eta(t, t_0, y_0) = \lim_{n \to \infty} \eta_n(t, t_0, y_0)$$

exists uniformly for $a \leq t \leq b$, $(t_0, y_0) \in \bar{E}_0$.

By Theorem I 2.4, $y = \eta(t, t_0, y_0)$ is a solution of (7.1) for $a \leqq t \leqq b$. Also, (10.1) and (10.4) hold under the conditions specified on t, t_0, y_0, y^0. The inequality (10.4) implies that if (t, t_0, y_0) is sufficiently near to (t_1, t_1, y_1), then no two distinct arcs $y = \eta(t, t_0, y_0)$ pass through the same point (t, y). Hence, by Theorem III 7.1, $y = \eta(t, t_1, y_1)$ is the only solution of (7.1) with $(t_0, y_0) = (t_1, y_1)$ on small intervals $[t_1 - \epsilon, t_1]$, $[t_1, t_1 + \epsilon]$. Since (t_1, y_1) is an arbitrary point on E, Theorem 8.1 follows.

11. Proof of Theorem 6.1

Sufficiency. It is assumed that there exist continuous A, C such that (6.2) has a continuous exterior derivative and det $A \neq 0$ in a vicinity of a point of E. It will first be shown that it is sufficient to consider the case that f does not depend on z. To this end, write (6.1) as (1.4) and let (y, z) be replaced by y_*, so that correspondingly $(f, 0)$ is replaced by f_*. Let $A_*(t, y_*)$ be the matrix

$$\begin{pmatrix} A & C \\ 0 & I \end{pmatrix}$$

and $\omega_* = A_*(dy_* - f_* \, dt)$. Then $\omega_* = (\omega, dz)$, where ω is given by (6.2). Hence ω_* has continuous exterior derivatives. Thus, if the asterisks are omitted from y_*, f_*, A_*; it is seen that it is sufficient to consider the case that $f = f(t, y)$ does not depend on z.

Let (t_1, y_1), E^0, E_0, a, b, (10.5), $\eta_n(t, t_0, y_0)$ be as in the proof of Theorem 8.1. Then (10.8) holds uniformly for $a \leqq t \leqq b$, $(t_0, y_0) \in \bar{E}_0$. Also, in obvious notation, $x = \partial \eta_n / \partial y_0{}^k$ is the solution of

$$(11.1) \qquad [A_n(t, \eta_n)x]' = F_n(t, \eta_n)x, \qquad x(t_0) = e_k;$$

cf. the derivation of (9.4) from (9.2). Introducing the new variables

$$(11.2) \qquad\qquad x^* = A_n(t, \eta_n)x$$

shows that

$$(11.3) \qquad x^{*\prime} = F_n(t, \eta_n)A_n^{-1}(t, \eta_n)x^*, \qquad x^*(t_0) = A_n(t_0, y_0)e_k.$$

Since $A_n(t, \eta_n) \to A(t, \eta)$, $F_n(t, \eta_n) \to F(t, \eta)$, $A_n^{-1}(t, \eta_n) \to A^{-1}(t, \eta)$ as $n \to \infty$ uniformly for $a \leqq t \leqq b$, $(t_0, y_0) \in \bar{E}_0$, Corollary IV 4.1 shows that, for fixed $(t_0, y_0) \in \bar{E}_0$, $\lim A_n(t, \eta_n) \, \partial \eta_n / \partial y_0{}^k$ exists uniformly for $a \leqq t \leqq b$ as $n \to \infty$ and is the solution of

$$(11.4) \qquad x^{*\prime} = F(t, \eta)A^{-1}(t, \eta)x^*, \qquad x^*(t_0) = A(t_0, y_0)e_k.$$

Actually this limit is uniform for $a \leqq t \leqq b$, $(t_0, y_0) \in \bar{E}_0$. This can be seen, e.g., by Theorem 2.1, by constructing a family of linear differential equations $x^{*\prime} = H(t, t_0, y_0, \epsilon)x^*$, where $H(t, t_0, y_0, \epsilon)$ is a matrix continuous in (t, t_0, y_0, ϵ) which becomes $F_n(t, \eta_n)A_n^{-1}(t, \eta_n)$ for $\epsilon = 1/n$ and $F(t, \eta)$ $A^{-1}(t, \eta)$ for $\epsilon = 0$.

If follows that

$$(11.5) \qquad\qquad \lim_{n \to \infty} \frac{\partial \eta_n}{\partial y_0{}^k} \quad \text{exists uniformly}$$

for $a \leqq t \leqq b$, $(t_0, y_0) \in \bar{E}_0$ and is the solution of

$$(11.6) \qquad [A(t, \eta)x]' = F(t, \eta)x, \qquad x(t_0) = e_k.$$

Consequently, a standard term-by-term differentiation theorem implies that $\partial \eta / \partial y_0{}^k$ exists for $a \leqq t \leqq b$, $(t_0, y_0) \in \bar{E}_0$ and is the limit (11.5). As in the proof of Theorem 3.1, it is seen that $\partial \eta / \partial t_0$ exists and is given by the analogue (9.10) of (3.4).

This proves that $\eta(t, t_0, y_0)$ is of class C^1 if (t, t_0, y_0) is sufficiently near to (t_1, t_1, y_1), where (t_1, y_1) is an arbitrary point of E. If now a and b are chosen arbitrarily, subject only to the condition that $\eta(t, t_1, y_1)$ exists for $a \leqq t \leqq b$, then a finite number of applications of formulae of the type $\eta(t, t_0, y_0) = \eta(t, t^*, \eta(t^*, t_0, y_0))$ shows that $\eta(t, t_0, y_0)$ is of class C^1 on its domain of existence.

All assertions of Corollary 6.1 except that concerning (6.8) also have been verified. The verification of this assertion will be left as an exercise.

Necessity. Assume that (6.1) has a unique solution $\eta = \eta(t, t_0, y_0, z)$ which is of class C^1. Let $(t_1, y_1, z_1) \in E$ be fixed. Then there is a neighborhood E^0 of (t_1, y_1, z_1) such that $\eta(t, t_0, y_0, z)$ exists on an interval containing t_0, t_1 if $(t_0, y_0, z) \in E^0$. Also, since the Jacobian matrix $\partial \eta / \partial y_0$ is the identity matrix at $(t, t_0, y_0, z) = (t_1, t_1, y_1, z_1)$, it can be supposed that E^0 is so small that $\det (\partial \eta(t_1, t_0, y_0, z_0)/\partial y_0) \neq 0$ for $(t_0, y_0, z_0) \in E^0$.

Consider the function $\eta(t_1, t, y, z)$ of $(t, y, z) \in E^0$ for fixed t_1. Then

$$(11.7) \qquad d\eta(t_1, t, y, z) = \left(\frac{\partial \eta}{\partial y_0}\right)[dy - f(t, y, z)\, dt] + \left(\frac{\partial \eta}{\partial z}\right) dz,$$

where (3.4) has been used with (t_0, y_0) replaced by (t, y). Thus, if $A(t, y, z)$ $= (\partial \eta / \partial y_0)$ at $(t, t_0, y_0, z) = (t_1, t, y, z)$ and $C(t, y, z) = (\partial \eta(t_1, t, y, z)/\partial z)$, then ω in (6.2) becomes $\omega = d\eta(t_1, t, y, z)$ which has the continuous exterior derivative $d\omega = 0$. Also, $\det A \neq 0$. This completes the proof.

12. First Integrals

Consider a system of differential equations

$$(12.1) \qquad\qquad y' = f(t, y)$$

in which f is continuous on an open set E. A real-valued function $u(t, y)$ defined on an open subset E_0 of E is called a *first integral* of (12.1) if it is constant along solutions of (12.1). i.e., if $y = y(t)$ is any solution of (12.1) on a t-interval (a, b) such that $(t, y(t)) \in E_0$ for $a < t < b$, then $u(t, y(t))$ is independent of t.

Lemma 12.1. *Let $u(t, y)$ be a function of class C^1 on an open set $E_0 \subseteq E$. Then $u(t, y)$ is a first integral of (12.1) if and only if it is a solution of the linear partial differential equation*

$$(12.2) \qquad \frac{\partial u}{\partial t} + \sum_{k=1}^{d} \frac{\partial u}{\partial y^k} f^k(t, y) = 0.$$

In fact, (12.2) is equivalent to $du(t, y(t))/dt = 0$ for all solutions $y = y(t)$ of (12.1) such that $(t, y(t)) \in E_0$.

Theorem 12.1. *Let $u = \eta^1(t, y)$, $\eta^2(t, y), \ldots, \eta^d(t, y)$ be first integrals of (12.1) of class C^1 on an open subset $E_0 \subset E$ such that the Jacobian matrix $(\partial \eta / \partial y)$ is nonsingular, where $\eta = (\eta^1, \ldots, \eta^d)$. Let $(t_0, y_0) \in E_0$, $\eta_0 = \eta(t_0, y_0)$, and $y = y(t, \eta)$ the function inverse to $\eta = \eta(t, y)$ for (t, η) near (t_0, η_0). Then, for fixed η, $y = y(t, \eta)$ is a solution of (12.1). Furthermore, if $(t_0, y_0) \in E_0$ and $u(t, y)$ is of class C^1 for (t, y) near (t_0, y_0), then $u(t, y)$ is a first integral for (12.1) if and only if there exists a function $U = U(\eta)$ of class C^1 for η near η_0 such that $u(t, y) = U(\eta(t, y))$ for (t, y) near (t_0, y_0).*

Proof. Since $u = \eta^i(t, y)$ for $i = 1, \ldots, d$ is a first integral of (12.1), it follows from (12.2) that

$$(12.3) \qquad \frac{\partial \eta}{\partial t} + \left(\frac{\partial \eta}{\partial y} \right) f = 0.$$

Hence $(\partial \eta / \partial y)^{-1} (\partial \eta / \partial t) + f = 0$. Since $y = y(t, \eta)$ is inverse to $\eta = \eta(t, y)$, it is seen that $y' \equiv \partial y / \partial t = -(\partial \eta / \partial y)^{-1}(\partial \eta / \partial t)$, so that $y = y(t, \eta)$ is a solution of (12.1) for fixed η. This proves the first part of the theorem.

If $U(\eta)$ is of class C^1, it follows readily from the criterion (12.2) that $u(t, y) = U(\eta(t, y))$ is a first integral. Conversely, let $u(t, y)$ be a first integral for (t, y) near (t_0, y_0) and put $U(\eta, t) = u(t, y(t, \eta))$. Clearly, $u(t, y) = U(\eta(t, y), t)$. Thus it suffices to verify that $U(\eta, t)$ is independent of t. But $\partial U / \partial t = \partial u / \partial t + \Sigma(\partial u / \partial y^k) y^{k'}$ which is 0 by (12.1) and (12.2). This proves the theorem.

Theorem 12.2. *Let $f(t, y)$ be continuous on an open set E. Then, for any $(t_0, y_0) \in E$, (12.1) has d first integrals $\eta = \eta(t, y)$ of class C^1 on a neighborhood of (t_0, y_0) satisfying $\det (\partial \eta / \partial y) \neq 0$ if and only if the initial value problem $y' = f$, $y(t_0) = y_0$ has a unique solution $y = \eta(t, t_0, y_0)$ of class C^1 (with respect to all of its variables.)*

Proof. If $y = \eta(t, t_0, y_0)$ exists and is of class C^1, put $\eta(t, y) = \eta(t_0, t, y)$ for (t, y) near (t_0, y_0) and fixed t_0. Then each component of $\eta(t, y)$ is a first integral, for $\eta(t, y(t))$ is the constant $y(t_0)$. Also $(\partial \eta(t, y) / \partial y)$ is the unit matrix at $t = t_0$, hence nonsingular for t near t_0.

Conversely, if the components of $\eta = \eta(t, y)$ are first integrals of class C^1 on a neighborhood of a point (t_0, y_0) of E, $\det(\partial\eta/\partial y) \neq 0$ and $y = y(t, \eta)$ is the inverse function, put $\eta(t, t_0, y_0) = y(t, \eta(t_0, y_0))$. This is a solution of the initial value problem $y' = f$, $y(t_0) = y_0$ and is of class C^1. For fixed t_0, it follows that $d\eta(t_0, t, y) = (\partial\eta(t_0, t, y)/\partial y_0)[dy - f(t, y)\,dt]$; cf. (11.7). Thus Theorem 6.1 implies that the solution $y = \eta(t, t_0, y_0)$ of $y' = f$, $y(t_0) = y_0$ is unique and of class C^1.

Notes

SECTION 3. Theorem 3.1 was first proved for $d = 1$ by the method of successive approximations independently by Picard (see Darboux [1, p. 363]) and Bendixson [1]. (It had been proved earlier by Nicoletti [1] assuming an additional Lipschitz condition on the partial derivatives of f.) Theorem 3.1, without the parameters z, was proved by Peano [3] using a method similar to that in the text (except that, instead of using Theorem 2.1, he employed an estimate of the type $|x_h(t)| \leq \exp c\, |t - t_0|$ for (3.12) where c is related to bounds for $|\partial f^i/\partial y^j|$; cf. Lemma IV 4.1). The result was rediscovered by von Escherich [2], using the method of successive approximations, and by Lindelöf [2], using a method similar to Peano's. In [1], Hadamard indicates a proof similar to that in the text; Lemma 3.1 is given in Hadamard [3, pp. 351–352] with a different proof.

SECTION 5. The definition of a continuous exterior derivative was given by E. Cartan [1, pp. 65–71]. Lemma 5.1 is due to Gillis [2] (and, in a more general form, was used by him to answer affirmatively the question of E. Cartan whether $\omega_r \wedge \omega_s$ has a continuous exterior derivative if the r-form ω_r and s-form ω_s have continuous exterior derivatives). The proof of Lemma 5.1 in the text is that of H. Cartan [1, pp. 62–63].

SECTION 6. Theorem 6.1 is in Hartman [17]; the proof in the text follows Hartman [26]; cf. [14] for the case $d = 1$. For Exercise 6.2 and generalizations to extremals, see Hartman [17]. For Exercise 6.3, see Hartman [14].

SECTIONS 7 AND 8. See Hartman [26].

SECTION 9. See Hartman [26]. Lemma 9.1 is essentially in Lewis [2] (cf. Opial [8]); for a generalization see Lewis [3].

SECTION 12. Theorem 12.1 goes back to Lagrange's work in 1779; cf. [2, IV pp. 624–634].

Chapter VI

Total and Partial Differential Equations

This chapter treats certain problems involving partial differential equations which can be solved by the use of the theory of ordinary differential equations. Parts I and II are independent of each other.

PART I. A THEOREM OF FROBENIUS

1. Total Differential Equations

Let $H(y, z)$ be a continuous $d \times e$ matrix on a $(d + e)$-dimensional open set E, say $H = (h_{ij})$. Consider the set of *total differential equations*

$$(1.1) \quad dy - H(y, z)\, dz = 0 \quad \text{or} \quad dy^i - \sum_{j=1}^{e} h_{ij}(y, z)\, dz^j = 0, \quad i = 1, \ldots, d,$$

and initial condition

$$(1.2) \qquad\qquad y(z_0) = y_0$$

for some $(y_0, z_0) \in E$. Equation (1.1) is an abbreviation for the set of partial differential equations

$$(1.3) \quad \frac{\partial y^i}{\partial z^j} = h_{ij}(y, z) \quad \text{for} \quad i = 1, \ldots, d \text{ and } j = 1, \ldots, e.$$

If $e = 1$, then (1.3) is a set of ordinary differential equations; existence and uniqueness theorems for corresponding initial value problems are supplied by earlier chapters. If $e > 1$, then in general we cannot expect (1.1) to have solutions. For example, suppose that $H(y, z)$ is of class C^1. If a C^1 solution $y = y(z)$ of (1.1) exists, then (1.3) makes it clear that $y(z)$ is of class C^2 [since the right side of (1.3) is of class C^1]. But then $\partial^2 y^i / \partial z^j\, \partial z^m = \partial^2 y^i / \partial z^m\, \partial z^j$ and this leads to the condition

$$(1.4) \qquad \frac{\partial h_{ij}}{\partial z^m} + \sum_{k=1}^{d} \frac{\partial h_{ij}}{\partial y^k} h_{km} = \frac{\partial h_{im}}{\partial z^j} + \sum_{k=1}^{d} \frac{\partial h_{im}}{\partial y^k} h_{kj},$$

for $i = 1, \ldots, d$ and $j, m = 1, \ldots, e$, which must hold along the solution $(y, z) = (y(z), z)$. Thus, in this case, a necessary condition that (1.1)–(1.2)

117

have a solution $y(z) = y(z, z_0, y_0)$ for *arbitrary* points (y_0, z_0) of E, is that the "integrability conditions" (1.4) hold as an identity on E. When H is of class C^1, it will be seen that this necessary condition is also sufficient for the existence of $y = y(z, z_0, y_0)$ and, furthermore, in this case the solution of (1.1)–(1.2) is unique, and $y(z, z_0, y_0)$ is of class C^1 with respect to all of its arguments.

Instead of dealing with (1.1)–(1.2) directly, it is more convenient to consider

$$(1.5) \qquad \omega = A(y, z)[dy - H(y, z)\, dz],$$

where $A(y, z)$ is a continuous, nonsingular $d \times d$ matrix, and to pose the following problem: When does there exist a function

$$(1.6) \qquad y = y(\eta, z)$$

of class C^1 on a $(d + e)$-dimensional neighborhood of a point (η_0, z_0) such that

$$(1.7) \qquad y(\eta_0, z_0) = y_0, \qquad \text{where} \quad (y_0, z_0) \in E,$$

$$(1.8) \qquad \det \left(\frac{\partial y(\eta, z)}{\partial \eta} \right) \neq 0,$$

and that (1.6) transforms (1.5) into a differential form

$$(1.9) \qquad \omega = D(\eta, z)\, d\eta$$

in $d\eta = (d\eta^1, \ldots, d\eta^d)$ with coefficients depending, of course, on (η, z). (The insertion of $A(y, z)$ in (1.5) is for convenience only and does not affect the problem.) When $y = y(\eta, z)$ exists, the system (1.5) will be said to be *completely integrable* at (y_0, z_0).

If $y = y(\eta, z)$ is of class C^1 and $dy = (\partial y/\partial \eta)\, d\eta + (\partial y/\partial z)\, dz$ is inserted into (1.5), it is seen that the result is of the form (1.9) if and only if

$$(1.10) \qquad \left(\frac{\partial y}{\partial z} \right) - H(y, z) = 0,$$

in which case

$$(1.11) \qquad A(y, z) \left(\frac{\partial y}{\partial \eta} \right) = D(\eta, z).$$

In particular, (1.8) implies that $D(\eta, z)$ is nonsingular.

The reduction of (1.5) to a form (1.9) is equivalent to satisfying (1.10), i.e., (1.1) or (1.3). Hence, the complete integrability of (1.1) or (1.5) is equivalent to the existence of a family of solutions $y = y(\eta, z)$ of (1.1) depending on parameters $\eta = (\eta^1, \ldots, \eta^d)$ and satisfying (1.7), (1.8).

The question of the existence of (1.6) can be viewed from a slightly different point of view: When does ω in (1.5) possess a local "integrating factor", i.e., when does there exist a nonsingular continuous matrix $E(y, z)$ such that $E(y, z)\omega$ is a total differential $d\eta(y, z)$, $E(y, z)\omega = d\eta$. It is clear that $E(y, z)$ exists if and only if the problem just posed has a solution; in which case, $E(y, z) = D^{-1}(\eta(y, z),z)$ where $\eta(y, z)$ is the inverse function of $y = y(\eta, z)$.

Finally, the question of the existence of (1.6) can be dealt with in still another way. Consider the problem of finding real-valued functions $u = u(y, z)$ of $d + e$ independent variables satisfying e simultaneous linear partial differential equations

$$(1.12) \qquad X_j[u] \equiv \frac{\partial u}{\partial z^j} + \sum_{i=1}^{d} h_{ij}(y, z) \frac{\partial u}{\partial y^i} = 0, \qquad j = 1, \dots, e,$$

where $H(y, z)$ is a continuous $d \times e$ matrix on an open set E. The system (1.12) is called the *system adjoint to* (1.1).

The system (1.12) is said to be *complete* on E if there exist d solutions $u = \eta^1(y, z), \dots, \eta^d(y, z)$ on E such that the rank of the Jacobian matrix $(\partial\eta/\partial y, \partial\eta/\partial z)$ is maximal (i.e., d) at every point. Actually, (1.12) shows that this rank condition holds if and only if $(\partial\eta(y, z)/\partial y)$ is nonsingular. If d such solutions $\eta^1, \dots, \eta^d$ exist, the system (1.12) can be written as

$$(1.13) \qquad \left(\frac{\partial\eta}{\partial z}\right) + \left(\frac{\partial\eta}{\partial y}\right) H(y, z) = 0.$$

Multiplication by $(\partial\eta/\partial y)^{-1}$ shows that this is equivalent to

$$(1.14) \qquad \left(\frac{\partial\eta}{\partial y}\right)^{-1} \left(\frac{\partial\eta}{\partial z}\right) + H(y, z) = 0.$$

The condition $\det(\partial\eta/\partial y) \neq 0$ implies that $\eta = \eta(y, z)$ has a local inverse $y = y(\eta, z)$ of class C^1 and so, $\partial y/\partial z = -(\partial\eta/\partial y)^{-1}(\partial\eta/\partial z)$. Thus (1.14) is equivalent to (1.10), i.e., (1.3). This argument can be reversed. Hence the complete integrability of (1.1) or (1.5) at (y_0, z_0) is equivalent to the completeness of (1.12) on a neighborhood of (y_0, z_0). In particular, when $e > 1$ and $H(y, z)$ is of class C^1, (1.4) is therefore necessary and sufficient for completeness of (1.12) on a vicinity of every point $(y_0, z_0) \in E$. The condition (1.4) can be written as

$$(1.15) \quad X_m[h_{ij}] = X_j[h_{im}] \qquad \text{for} \quad i = 1, \dots, d \quad \text{and} \quad m, j = 1, \dots, e$$

by virtue of (1.12)

The arguments of § V 12 show that if (1.1)–(1.2) has solutions for all $(y_0, z_0) \in E$, then a function $u(y, z)$ of class C^1 on $E_0 \subset E$ is a solution of (1.12) if and only if it is a "first integral" of (1.1), i.e., $u(y(z), z) = \text{const.}$ for every solution $y = y(z)$ of (1.1) for which $(y(z), z) \in E_0$.

A system (1.1) or equivalent system (1.10) for which the integrability conditions (1.15) are satisfied is called a *Jacobi system*. Theorem 3.2 is an existence and uniqueness theorem for such systems. In Theorem 3.2, it is not assumed that $H(y, z)$ is of class C^1, so that integrability conditions take a form different from (1.15), in fact, a more convenient algebraic form [so that no calculations involve the rather formidable relations (1.4)]. This theorem will be deduced from Theorem 3.1 which treats the question of the complete integrability of (1.1) or (1.5).

Exercise 1.1. (a) Let $b_{1k}(y), \ldots, b_{dk}(y)$ for $k = 1, 2$ be $2d$ functions of class C^1. Define the partial differential operators $X_k[u] = \Sigma_j b_{jk}(y) \, \partial u / \partial y^j$ for $k = 1, 2$ and the operator $X_{12}[u] = \Sigma_j (X_2[b_{1j}] - X_1[b_{2j}]) \, \partial u / \partial y^j$. Show that X_{12} is the commutator of X_1, X_2 in the sense that if $u(y)$ is of class C^2, then $X_{12}[u] = X_1[X_2(u)] - X_2[X_1(u)]$. (b) Show that if $u(y)$ is of class C^1 and $X_k[u] = 0$ for $k = 1, 2$. Then $X_{12}[u] = 0$. See Exercise 8.2(b) for a generalization.

Exercise 1.2. Let $B(x) = (b_{ij}(x))$, where $i = 1, \ldots, d + e$ and $j = 1, \ldots, e$, be a $(d + e) \times e$ matrix of rank e, continuous on an open set E in the $(d + e)$-dimensional x-space, $x = (x^1, \ldots, x^{d+e})$. Define the differential operators $Y_j[u] = \Sigma_i b_{ij}(x) \, \partial u / \partial x^i$ for $j = 1, \ldots, e$. The system (*) $Y_j[u] = 0, j = 1, \ldots, e$, is called *complete* on E if there exist d solutions $u = \eta^1(x), \ldots, \eta^d(x)$ of class C^1 such that the $(d + e) \times d$ Jacobian matrix $(\partial \eta^k / \partial x_j)$ is of rank d. Using Theorem 3.1 show that if $B(x)$ is of class C^1, then the system (*) is complete on a neighboorhood of every point $x_0 \in E$ if and only if the commutator Y_{jk} of Y_j, Y_k (cf. Exercise 1.1) is a linear combination of $Y_1, \ldots, Y_e$ [i.e., if and only if there exist functions $c_{jkm}(x)$, for $j, k, m = 1, \ldots, e$, such that $Y_{jk} = \Sigma_m c_{jkm} Y_m$ or, equivalently, $Y_k[b_{ij}] - Y_j[b_{ik}] \equiv \Sigma_m c_{jkm}(x) b_{im}(x)$ for $j, k = 1, \ldots, e$ and $i = 1, \ldots, d + e$].

2. Algebra of Exterior Forms

In order to be able to write integrability conditions in a convenient form, some simple facts about exterior forms will be recalled.

Let $0 < r \leq d$ and $C_{rd} = d! / r! (d - r)!$. Consider a vector space W over the real field of dimension C_{rd} with basis elements $e_{i_1 \ldots i_r}$, where $1 \leq i_1 < i_2 < \cdots < i_r \leq d$. Thus any vector ω in the space W has a unique representation

$$\omega = r! \sum_{1 \leq i_1 < i_2 < \cdots < i_r \leq d} \cdots \sum c_{i_1 \ldots i_r} e_{i_1 \ldots i_r},$$

where $c_{i_1 \ldots i_r}$ are real numbers. Introduce the symbols $e_{j_1 \ldots j_r}$ for $j_1, \ldots, j_r = 1, \ldots, d$, where $e_{j_1 \ldots j_r} = 0$ if two indices $j_l, \ldots, j_r$ are equal and $e_{j_1 \ldots j_r} = \pm e_{i_1 \ldots i_r}$ according as $(j_1, \ldots, j_r)$ is an even or odd permutation of $(i_1, \ldots, i_r)$, $1 \leq i_1 < \cdots < i_r \leq d$. Then any vector ω has a unique representation of the form

$$\omega = \sum_{j_1=1}^{d} \cdots \sum_{j_r=1}^{d} c_{j_1 \ldots j_r} e_{j_1 \ldots j_r},$$

subject to the conditions

(2.1) $c_{j_1 \ldots j_r} = \pm c_{i_1 \ldots i_r},$

where $(j_1, \ldots, j_r)$ is an even or odd permutation of $(i_1, \ldots, i_r)$ and $1 \le i_1 < \cdots < i_r \le d$. In particular, $c_{j_1 \ldots j_r} = 0$ if two of the indices $j_1, \ldots, j_r$ are equal.

Change the notation for the "basis elements" $e_{i_1 \ldots i_r}$ again, writing $e_{j_1 \ldots j_r} = dy^{j_1} \wedge \cdots \wedge dy^{j_r}$. Correspondingly, a vector is a differential r-form

(2.2) $\omega = \sum c_{j_1 \ldots j_r} dy^{j_1} \wedge \cdots \wedge dy^{j_r}$

with constant coefficients subject to (2.1).

As in the last chapter, multiplication of a differential r-form ω^r and an s-form ω^s is defined as a differential $(r + s)$-form obtained by the usual associative and distributive laws and the anticommutative law $dy^i \wedge dy^j = -dy^j \wedge dy^i$; so that $\omega^r \wedge \omega^s = (-1)^{rs} \omega^s \wedge \omega^r$. This type of multiplication will be referred to as "exterior" multiplication.

We can obtain a "change of basis" for the vector space W of differential r-forms in the following way: Let $T = (t_{ij})$ be a nonsingular $d \times d$ matrix and let

(2.3) $dy^i = \sum_{j=1}^{d} t_{ij} d\eta^j, \qquad \det (t_{ij}) \ne 0,$

and, more generally, let the "basis" elements be the exterior product

$$dy^{j_1} \wedge \cdots \wedge dy^{j_r} = \left(\sum_{j=1}^{d} t_{j_1 j} d\eta^j \right) \wedge \cdots \wedge \left(\sum_{j=1}^{d} t_{j_r j} d\eta^j \right).$$

Then (2.2) becomes a differential r-form of the type

(2.4) $\omega = \sum \gamma_{j_1 \ldots j_r} d\eta^{j_1} \wedge \cdots \wedge d\eta^{j_r},$

where the convention analogous to (2.1) is observed.

In order to see that this is the usual type of change of basis for the space W, it is necessary to prove the following:

Lemma 2.1. *Let the "change of basis" (2.3) transform (2.2) into (2.4). Then (2.2) is 0 (i.e., all $c_{j_1 \ldots j_r} = 0$) if and only if (2.4) is 0 (i.e., all $\gamma_{j_1 \ldots j_r} = 0$).*

This follows form the fact that "changes of basis" (2.3) are associative, i.e., if

$$d\eta^i = \sum_{j=1}^{d} s_{ij} d\zeta^j, \qquad \text{where} \quad \det (s_{ij}) \ne 0,$$

transforms (2.4) into

(2.5) $$\omega = \sum \delta_{j_1 \ldots j_r} \, d\zeta^{j_1} \wedge \cdots \wedge d\zeta^{j_r},$$

then

$$dy^i = \sum_{j=1}^{d} \left(\sum_{k=1}^{d} t_{ik} s_{kj} \right) d\zeta^j$$

transforms (2.2) into (2.5). Thus the choice $(s_{ij}) = (t_{ij})^{-1}$ gives Lemma 2.1.

If certain of the dy^k are linear combinations of the others (e.g., if $dy^{h+1}, \ldots, dy^d$ are linear combinations of $dy^1, \ldots, dy^h$ with constant coefficients), then (2.2) becomes a differential r-form in $dy^1, \ldots, dy^h$. This notion is involved in the following lemma:

Lemma 2.2. *Let $\omega_1, \ldots, \omega_s$ be linearly independent differential 1-forms and ω a differential r-form with constant coefficients. Then the exterior product $\omega_1 \wedge \cdots \wedge \omega_s \wedge \omega$ is 0 if and only if the relations $\omega_1 = \cdots = \omega_s = 0$ imply that $\omega = 0$.*

Let the forms $\omega_1, \ldots, \omega_s$ be given by

(2.6) $$\omega_i = \sum_{j=1}^{d} s_{ij} \, dy^j, \qquad i = 1, \ldots, s.$$

The assumption of linear independence means that $\omega_1, \ldots, \omega_s$ considered as vectors are linearly independent; i.e., rank $(s_{ij}) = s$ where $i = 1, \ldots, s$ and $j = 1, \ldots, d$. The relations $\omega_1 = \cdots = \omega_s = 0$ mean that s of the dy^j are expressed as linear combination of the other $d - s$ dy^k, in which case ω becomes an r-form in the latter. The statement of the lemma is to the effect that this r-form is 0 (i.e., all coefficients are 0) if and only if $\omega_1 \wedge \cdots \wedge \omega_s \wedge \omega = 0$.

The proof will show that the conclusion of Lemma 2.2 can be stated as follows: $\omega_1 \wedge \cdots \wedge \omega_s \wedge \omega = 0$ if and only if there exist s differential $(r - 1)$-forms $\alpha_1, \ldots, \alpha_s$ such that $\omega = \alpha_1 \wedge \omega_1 + \cdots + \alpha_s \wedge \omega_s$.

Proof. Join $d - s$ rows to the $d \times s$ matrix (s_{ij}) to obtain a nonsingular square matrix. Consider the change of basis (2.3), where $(t_{ij}) = (s_{ij})^{-1}$. Then, with respect to the new basis, $\omega_i = d\eta^i$ for $1 \leq i \leq s$ and, say, ω is given by (2.4). The product $\omega_1 \wedge \cdots \wedge \omega_s \wedge \omega = 0$ if and only if every nonzero term of ω contains a factor $\omega_i = d\eta^i$ for $1 \leq i \leq s$, i.e., if and only if $\omega = 0$ when $\omega_1 = \cdots = \omega_s = 0$. Thus the assertion is correct with respect to the $(d\eta^1, \ldots, d\eta^d)$-basis and, by Lemma 2.1, with respect to the $(dy^1, \ldots, dy^d)$-basis.

3. A Theorem of Frobenius

The following theorem of Frobenius is the main theorem concerning (1.5).

Theorem 3.1. *Let $H(y, z)$ be a continuous $d \times e$ matrix on an open set E. A necessary and sufficient condition for the complete integrability of $\omega_0 = dy - (H(y, z) \, dz$ at (y_0, z_0) is that there exists a continuous, nonsingular $d \times d$ matrix $A(y, z)$ on a neighborhood of (y_0, z_0) such that if $\omega = (\omega_1, \ldots, \omega_d)$ is defined by $\omega = A\omega_0$, then ω has a continuous exterior derivative satisfying*

$$(3.1) \qquad (\omega_1 \wedge \cdots \wedge \omega_d) \wedge d\omega_i = 0 \qquad \text{for } i = 1, \ldots, d.$$

Conditions (3.1) represent the integrability conditions. The expression on the left is the exterior product of d differential 1-forms $\omega_1, \ldots, \omega_d$ and the differential 2-form $d\omega_i$. Condition (3.1) will be used in the equivalent form (Lemma 2.2): $\omega = 0$ (i.e., $\omega_1 = \cdots = \omega_d = 0$) implies that $d\omega = 0$ (i.e., $d\omega_1 = \cdots = d\omega_d = 0$).

Exercise 3.1. Show that conditions (3.1) reduce to (1.4) if $A(y, z) = I$ and $H(y, z)$ is of class C^1.

Theorem 3.1 should be completed by the following:

Lemma 3.1. *Let $H(y, z)$ be a continuous $d \times e$ matrix on an open set E. Let $\Omega = \Omega(E)$ be the (possibly empty) set of continuous, nonsingular $d \times d$ matrices $A(y, z)$ on E such that $\omega = A[dy - H \, dz]$ has a continuous exterior derivative. Then the integrability conditions (3.1) hold for all $A \in \Omega$ or for no $A \in \Omega$.*

For example, if $H(y, z)$ is of class C^1, so that $dy - H \, dz$ has a continuous exterior derivative, and if (1.4) does not hold, then (3.1) does not hold for any choice of continuous nonsingular A.

Exercise 3.2 (*A Simplified Version of Lemma* 3.1). Let $A(y, z)$ and $H(y, z)$ be continuous, det $A \neq 0$, and $\omega = A[dy - H(y, z) \, dz]$ have a continuous exterior derivative satisfying the integrability conditions (3.1). Let $A_0(y, z)$ be a C^1, nonsingular $d \times d$ matrix. Show that $A_0(y, z)\omega$ has a continuous exterior derivative given by $d(A_0\omega) = A_0 \, d\omega + (dA_0) \wedge \omega$, and hence that the form $A_0\omega$ satisfies integrability conditions analogous to (3.1).

Exercise 3.3. Let x, y, z be real variables; $P(x, y, z)$, $Q(x, y, z)$, $R(x, y, z)$ real-valued functions of class C^1; and $P^2 + Q^2 + R^2 \neq 0$. Show that the integrability condition $\omega \wedge d\omega = 0$ for the existence of local integrating factors for $\omega = P \, dx + Q \, dy + R \, dz$ is that $P(R_y - Q_z) + Q(P_z - R_x) + R(Q_y - P_x) = 0$.

Exercise 3.4. Show that if $H(y, z)$ is continuous on E and there exists a continuous nonsingular $A(y, z)$ on E such that $\omega = A[dy - H(y, z) \, dz]$ has a continuous exterior derivative satisfying the integrability conditions (3.1), then every point $(y_0, z_0) \in E$ has a neighborhood E_0 on which there is defined a sequence of C^1 1-forms $\omega^1, \omega^2, \ldots$ such that $\omega^n \to \omega$ and $d\omega^n \to d\omega$ uniformly as $n \to \infty$ and ω^n satisfies the integrability conditions.

Exercise 3.5. Let $H(y, z)$ be continuous on E. Show that $\omega = dy - H(y, z)\, dz$ has a continuous exterior derivative $d\omega$ satisfying the integrability conditions (3.1) if and only if $H(y, z)$ has continuous partial derivatives with respect to the components of y (cf. Exercise V 5.1) and, for $1 \leq j < m \leq e$ and $i = 1, \ldots, d$,

$$(3.2) \quad \int_J h_{ij}\, dz^j + h_{im}\, dz^m = \iint_S \sum_{k=1}^d \left(\frac{\partial h_{ij}}{\partial y^k} h_{km} - \frac{\partial h_{im}}{\partial y^k} h_{kj} \right) dz^j\, dz^m$$

for all rectangles S with boundaries J, $\bar{S} \subset E$, in the 2-planes $y = $ const. and $z^k = $ const. for $k \neq j, m$; cf. (1.4).

Theorem 3.2. *Let $H(y, z)$ be continuous on an open set E. A necessary and sufficient condition for (1.12) to be complete on a neighborhood of a point $(y_0, z_0) \in E$ is the existence of a continuous nonsingular matrix $A(y, z)$ on a neighborhood of (y_0, z_0) as in Theorem 3.1. In particular, if $e > 1$ and $H(y, z)$ is of class C^1, then (1.4) [or (1.15)] is necessary and sufficient for the (local) completeness of (1.12).*

The concept of completeness leads at once to:

Corollary 3.1. *Let $H(y, z)$ be continuous on a neighborhood of a point (y_0, z_0) and let (1.12) be complete [i.e., possess d solutions $u = \eta^1(y, z), \ldots, \eta^d(y, z)$ such that $\eta = (\eta^1, \ldots, \eta^d)$ satisfies $\det (\partial \eta / \partial y) \neq 0$]. Put $\eta_0 = \eta(y_0, z_0)$. Let $u(y, z)$ be a real-valued function of class C^1 on a neighborhood of (y_0, z_0). Then $u(y, z)$ is a solution of (1.12) if and only if there exists a function $U(\eta)$ of class C^1 for η near η_0 such that $u(y, z) \equiv U(\eta(y, z))$ for (y, z) near (y_0, z_0).*

The proof of Theorem 3.1 will be given in §4 and that of Lemma 3.1 in §5. Theorem 3.2 follows from Theorem 3.1 and the considerations of §1. The proof of Corollary 3.1 is similar to that of Theorem V 12.1 and will be omitted.

4. Proof of Theorem 3.1

Necessity. Let there exist a function $y = y(\eta, z)$ of class C^1 in a vicinity of a point (η_0, z_0) satisfying (1.7), (1.8) and transforming

$$(4.1) \qquad \omega_0 = dy - H(y, z)\, dz$$

into a form

$$(4.2) \qquad \omega_0 = D_0(\eta, z)\, d\eta.$$

Thus $D_0(\eta, z) = (\partial y / \partial \eta)$ is nonsingular; cf. (1.11). By condition (1.8), (1.6) has an inverse $\eta = \eta(y, z)$ of class C^1 on a vicinity of (y_0, z_0). Let $A(y, z) = D_0^{-1}(\eta(y, z), z)$; so that A is continuous and nonsingular, furthermore, $\omega = A(dy - H\, dz) = d\eta$ has the continuous exterior derivative $d\omega = d(d\eta) = 0$. This proves the "necessity."

The proof of "sufficiency" will depend on the following lemma which exhibits the role of the integrability conditions (3.1).

Lemma 4.1. *Let $f(t, y, z)$ be continuous on a neighborhood of a point (t_0, y_0, z_0) and let there exist a continuous, nonsingular $d \times d$ matrix $A(t, y, z)$ and a continuous $d \times e$ matrix $C(t, y, z)$ such that*

(4.3) $\omega = A(dy - f\,dt - C\,dz)$ *has a continuous exterior derivative*

satisfying (3.1). *Let* $y = \eta(t, t_0, y_1, z)$ *be the solution of*

(4.4) $$y' = f(t, y, z), \qquad y(t_0) = y_1.$$

Then the matrix

(4.5) $$\left(\frac{\partial \eta}{\partial y_1}\right)^{-1}\left[\left(\frac{\partial \eta}{\partial z}\right) - C(t, \eta, z)\right]$$

is independent of t.

Proof of Lemma 4.1. By Theorem V 6.1, (4.4) has a unique solution $y = \eta(t, t_0, y_1, z)$ of class C^1. In the notation of Corollary V 6.1 [cf. (V 6.9)–(V 6.10)], $Y = A(t, \eta, z)\,(\partial\eta/\partial y_1)$ is a fundamental matrix for the linear system

(4.6) $$Y' = F(t, \eta, z)A^{-1}(t, \eta, z)Y$$

and $Y = A(t, \eta, z)[(\partial\eta/\partial z) - C(t, \eta, z)]$ is a solution of

(4.7) $$Y' = F(t, \eta, z)A^{-1}(t, \eta, z)Y + N(t, \eta, z) + F(t, \eta, z)C(t, \eta, z).$$

Note some differences in notations here and in § V 6; here y_1 plays the role of y_0, and $-AC$ in (4.3) that of C in (V 6.2).

For t_0 fixed, the change of variables $(t, y, z) \to (t, y_1, z)$, where $y = \eta(t, t_0, y_1, z)$, transforms (4.3) into

(4.8) $$\omega = A(t, \eta, z)\left\{\left(\frac{\partial \eta}{\partial y_1}\right) dy_1 + \left[\left(\frac{\partial \eta}{\partial z}\right) - C(t, \eta, z)\right] dz\right\}.$$

Since the coefficients in (4.8) possess continuous derivatives with respect to t, the proof of Lemma V 5.1 shows that

$$d\omega = \left[A\left(\frac{\partial \eta}{\partial y_1}\right)\right]' dt \wedge dy_1 + \left\{A\left[\left(\frac{\partial \eta}{\partial z}\right) - C\right]\right\}' dt \wedge dz + \cdots,$$

where the omitted terms involve $dy_1{}^j \wedge dy_1{}^k$, $dy_1{}^j \wedge dz^k$, and $dz^j \wedge dz^k$. In view of the remarks concerning (4.6) and (4.7), this is

(4.9) $$d\omega = F\left(\frac{\partial \eta}{\partial y_1}\right) dt \wedge dy_1 + \left[F\left(\frac{\partial \eta}{\partial z}\right) + N\right] dt \wedge dz + \cdots,$$

where the argument of A, C, F, N is (t, η, z) and $\eta = \eta(t, t_0, y_1, z)$.

At a fixed (t, y_1, z), choose dy_1 to make $\omega = 0$, i.e.,

$$(4.10) \qquad dy_1 = -\left(\frac{\partial \eta}{\partial y_1}\right)^{-1}\left[\left(\frac{\partial \eta}{\partial z}\right) - C(t, \eta, z)\right] dz;$$

then (4.9) has the form $d\omega = \{\cdots\} dt \wedge dz + $ (terms in $dz^j \wedge dz^k$), where $\{\cdots\}$ is given by

$$\{\cdots\} = -F\left[\left(\frac{\partial \eta}{\partial z}\right) - C\right] + F\left(\frac{\partial \eta}{\partial z}\right) + N = N + FC$$

Since (4.10) makes $\omega = 0$, Lemma 2.2 and the integrability conditions (3.1) imply that $d\omega = 0$. In particular $\{\cdots\} = 0$, i.e., $N + FC = 0$. In this case, (4.7) reduces to (4.6). Thus $Y = A\,(\partial \eta/\partial y_1)$ is a fundamental solution of (4.6) and $Y = A[(\partial \eta/\partial z) - C]$ is a matrix solution of the same system (4.6). Consequently there is a matrix C^0 independent of t such that

$$A\left[\left(\frac{\partial \eta}{\partial z}\right) - C\right] = A\left(\frac{\partial \eta}{\partial y_1}\right)C^0;$$

cf. § IV 1. Since the matrix (4.5) is C^0, the lemma is proved.

Proof of "Sufficiency" in Theorem 3.1. By assumption, there is a continuous, nonsingular $A(y, z)$ satisfying the conditions of the theorem. Change notation as follows: let $t = z^1$; if $e = 1$, let $z_2 = 0$ and if $e > 1$, let z_2 be the $(e - 1)$-dimensional vector $(z^2, \ldots, z^e)$. Then (1.1) can be written as

$$\omega = A(y, t, z_2)[dy - f\,dt - C_2\,dz_2],$$

where f is the first column of the matrix H and C_2 is the matrix consisting of the other columns of H. Let $y = \eta(t, z_2, y_1)$ be the solution of $y' = f(y, t, z_2)$, $y(z_0^1) = y_1$ for fixed z_0^1. The change of variables $(y, t, z_2) \to (y_1, t, z_2)$ transforms ω into the form

$$\omega = A(\eta, z)\left\{\left(\frac{\partial \eta}{\partial y_1}\right) dy_1 + \left[\left(\frac{\partial \eta}{\partial z_2}\right) - C_2(\eta, t, z_2)\right] dz_2\right\}.$$

This can be written as

$$\omega = A_1(y_1, z)[dy_1 - H_2(y_1, z_2)\,dz_2],$$

where H_2 does not depend on $t = z^1$ by Lemma 4.1; cf. (4.8) and (4.5). The form ω, of course, satisfies the integrability conditions in the new variables (y_1, z).

If $e = 1$, so that $z_2 = 0$, the theorem is proved. If $e > 1$, change notation by letting $z^2 = t$ and $z_3 = 0$ or $z_3 = (z^3, \ldots, z^e)$ according as $e = 2$ or $e > 2$ and write

$$\omega = A_1(y_1, z^1, t, z_3)[dy_1 - f_1\,dt - C_3\,dz_3],$$

where $f_1 = f_1(y_1, t, z_3)$ is the first column of H_2 and C_3 is the matrix consisting of the other columns. Let $y_1 = \eta_1(t, z_3, y_2)$ be the solution of $y_1' = f_1(y_1, t, z_3), y_1(z_0^2) = y_2$. Thus η_1 does not depend on z^1. The change of variables $(y_1, z^1, t, z_3) \rightarrow (y_2, z^1, t, z_3)$ transforms ω into a form of the type

$$\omega = A_2(y_2, z)[dy_2 - H_3 \, dz_3],$$

where, by Lemma 4.1, $H_3 = H_3(y_2, z_3)$ does not depend on z^2.

If $e = 2$, so that $z_3 = 0$, the theorem is proved. It is clear that the process can be continued to obtain a proof of the theorem for any $e > 2$.

5. Proof of Lemma 3.1

Since the integrability conditions are local conditions, it can be supposed that E is the neighborhood of a point (y_0, z_0). Suppose that $\Omega(E)$ is not empty and that the integrability conditions are satisfied for some element of $\Omega(E)$. Then, by Theorem 3.1, there exists a $y = y(\eta, z)$ satisfying (1.7)–(1.8) and transforming

$$(5.1) \qquad \omega_0 = dy - H(y, z) \, dz$$

into

$$(5.2) \qquad \omega_0 = \left(\frac{\partial y}{\partial \eta}\right) d\eta.$$

Let $A(y, z)$ be an arbitrary element of $\Omega(E)$. Then

$$(5.3) \qquad \omega = A(dy - H \, dz)$$

is transformed into

$$(5.4) \qquad \omega = D(\eta, z) \, d\eta, \qquad \text{where} \quad D = A(y, z)\left(\frac{\partial y}{\partial \eta}\right).$$

Since the property of having a continuous exterior derivative is not lost under a C^1 change of variables $y, z \rightarrow \eta, z$ [see the definition of continuous exterior derivative in § V 5], $A \in \Omega(E)$ implies that (5.4) has a continuous exterior derivative. It is clear from the proof of Lemma V 5.1 that $d\omega$ is of the form

$$(5.5) \qquad d\omega = \sum_{j=1}^{d} \sum_{k=1}^{d} A_{jk}(\eta, z) \, d\eta^j \wedge d\eta^k + \sum_{j=1}^{d} \sum_{k=1}^{d} B_{jk}(\eta, z) \, d\eta^j \wedge dz^k,$$

where $A_{jk} = -A_{kj}$, B_{jk} are continuous d-dimensional vectors.

As $D(\eta, z)$ is nonsingular, $\omega = 0$ is equivalent to $d\eta = 0$, in which case $d\omega = 0$. Thus (5.4) satisfies the integrability condition. But then (5.3), which results from (5.4) by a C^1 change of variables $\eta, z \rightarrow y, z$, also satisfies the integrability condition. This proves the lemma.

6. The System (1.1)

The theorems of § 3 give necessary and sufficient conditions for the existence of local solutions of initial value problems (1.1)–(1.2).

Theorem 6.1. *Let $H(y, z)$ be a continuous $d \times e$ matrix on an open set E. A necessary and sufficient condition that (1.1)–(1.2) have a unique solution $y = y(z, z_0, y_0)$ for z near z_0 and all $(y_0, z_0) \in E$ such that $y(z, z_0, y_0)$ is of class C^1 is the complete integrability of (1.5) at every $(y_0, z_0) \in E$, i.e., the existence of an $A(y, z)$ for each $(y_0, z_0) \in E$ as in Theorem 3.1. In this case, if the product of the two Euclidean spheres R: $(\|y - y_0\| \leqq b) \times (\|z - z_0\| \leqq a)$ is in E and if $|\eta \cdot H(y, z)\zeta| \leqq M$ for all Euclidean unit d-dimensional vectors η, e-dimensional vectors ζ and $(y, z) \in R$, then $y(z) = y(z, z_0, y_0)$ exists on the sphere $\|z - z_0\| \leqq \min (a, b/M)$.*

It follows, as remarked in § 1, that if $e > 1$ and $H(y, z)$ is of class C^1, then (1.4) is necessary and sufficient for the existence of local solutions (1.1)–(1.2) with (y_0, z_0) arbitrary.

Particularly important cases of (1.1) are those in which $H(y, z)$ is linear in y. Let $H_0(z), H_1(z), \ldots, H_d(z)$ be continuous $d \times e$ matrices on an open z-set D and let

$$(6.1) \qquad H(y, z) = H_0(z) + \sum_{k=1}^{d} H_k(z)y^k,$$

so that (1.1)–(1.2) becomes

$$\cdot(6.2) \qquad dy - \left[H_0(z) + \sum_{k=1}^{d} H_k(z)y^k \right] dz = 0, \qquad y(z_0) = y_0.$$

Corollary 6.1. *Let $H_k(z) = (h_{kij}(z))$, where $k = 0, \ldots, d$, be continuous $d \times e$ matrices $(i = 1, \ldots, d$ and $j = 1, \ldots, e)$ on an open z-set D. Necessary and sufficient for (6.2) to have a solution $y = y(z, z_0, y_0)$ for all $z_0 \in D$, y_0 arbitrary is that for $1 \leqq j < m \leqq d$, and $r = 0, 1, \ldots, d$,*

$$(6.3) \qquad \int_J h_{rij} \, dz^j + h_{rim} \, dz^m = \sum_{k=1}^{d} \int_S (h_{kij}h_{rkm} - h_{kim}h_{rkj}) \, dz^j \, dz^m$$

for every rectangle S with boundary J, $\bar{S} \subset D$, on coordinate 2-planes $z^h = $ const. for $h \neq j, m$. In this case, the solution $y(z, z_0, y_0)$ of (6.2) is unique and of class C^1 with respect to all of its variables. When $H_0(z), \ldots, H_d(z)$ are of class C^1, conditions (6.3) are equivalent to

$$(6.4) \qquad \frac{\partial h_{rij}}{\partial z^m} + \sum_{k=1}^{d} h_{kij}h_{rkm} = \frac{\partial h_{rim}}{\partial z^j} + \sum_{k=1}^{d} h_{kim}h_{rkj}.$$

This corollary is a consequence of Theorem 6.1 and the following exercises.

Exercise 6.1. (*a*) Let $H_0, \ldots, H_d$ be of class C^1, show that conditions (6.4) are the integrability conditions for $\omega = dy - H(y, z) \, dz$ in the case (6.1). (*b*) Let $H_0, \ldots, H_d$ be continuous. Show that $\omega = dy - H(y, z) \, dz$ in case (6.1) has a continuous exterior derivative satisfying the integrability conditions if and only if the conditions on (6.3) of Corollary 6.1 are satisfied.

Exercise 6.2 (*Continuation*). Show that if (6.1) is continuous, then there exists a continuous nonsingular $A(y, z)$ such that $A(dy - H \, dz)$ has a continuous exterior derivative satisfying (3.1) if and only if the same is true of $dy - H \, dz$.

The proof of Theorem 6.1 combined with the usual arguments of the monodromy theorem will be used to show that, in the linear cases (6.1), the solutions exist in the large.

Corollary 6.2. *Let $H(y, z)$ be continuous for $z \in D$ and all y, where D is a simply connected open set. Assume that the sufficient conditions of Theorem 6.1 for the local solvability of (1.1)–(1.2) are satisfied. For every compact subset D_0 of D, let there exist a constant $K = K(D_0)$ such that $\| H(y, z) \| \leq K(\| y \| + 1)$ for $z \in D_0$ and all y. Then the solution $y(z) = y(z, z_0, y_0)$ of (1.1)–(1.2) exists for all $z \in D$.*

It will be clear that the condition $\| H(y, z) \| \leq K(\| y \| + 1)$ can be refined along the lines of Theorem III 5.1.

Proof of Theorem 6.1. *Necessity.* Let (1.1)–(1.2) have a unique solution $y = y(z, z_0, y_0)$ for z near z_0 of class C^1. The Jacobian matrix $(\partial y / \partial y_0)$ is the identity matrix at $z = z_0$ and hence is nonsingular for z near z_0. For fixed z_0, put $y(z, \eta) = y(z, z_0, \eta)$. This is of class C^1 near the point $(\eta_0, z_0) = (y_0, z_0)$; it satisfies (1.7), (1.8) and transforms (4.1) into a form (4.2) since, for fixed η, $y(z) = y(z, \eta)$ is a solution of (1.1), i.e. of (1.10). Thus "necessity" in Theorem 6.1 follows from that in Theorem 3.1.

Uniqueness. Let $y = y(z)$ be a solution of (1.1)–(1.2), say on the Euclidean sphere $\| z - z_0 \| \leq a$. For a fixed e-dimensional vector ζ of (Euclidean) length one, consider the values $y_1(t) = y(z_0 + t\zeta)$ of $y(z)$ on the line segment $z = z_0 + t\zeta$, $0 \leq t \leq a$. By (1.1)–(1.2),

$$(6.5) \qquad \frac{dy_1}{dt} = H(y_1, z_0 + t\zeta)\zeta, \qquad y_1(0) = y_0.$$

If $H(y, z)$ is smooth (e.g., is uniformly Lipschitz continuous in y), then the initial value problem (6.5) has a unique solution which is necessarily $y_1(t) = y(z_0 + t\zeta)$ for $0 \leq t \leq a$. The same can be concluded under the conditions imposed here; namely, that A, H are continuous, $\det A \neq 0$, and (1.5) has a continuous exterior derivative. For then the forms

$$(6.6) \qquad A(y_1, z_0 + t\zeta)[dy_1 - H(y_1, z_0 + t\zeta)\zeta \, dt]$$

in dy_1, dt for fixed z_0, ζ have continuous exterior derivatives. (This follows at once from the definition of "exterior derivative" in § V 5.) Hence Theorem V 6.1 implies the uniqueness of the solution of (6.5).

Existence. By Theorem 3.1 there exists a $y(\eta, z)$ of class C^1 satisfying (1.7), (1.8) and transforming (1.5) into (1.9). For $\eta = \eta_0$, $y(\eta_0, z)$ is a solution of (1.1)–(1.2); i.e., $y(z, z_0, y_0) = y(\eta_0, z)$ exists. It only remains to verify that $y(z, z_0, y_0)$ is of class C^1 in all of its variables. By (1.8), (1.6) has a C^1 inverse $\eta = \eta(y, z)$ for (y, z) near (y_0, z_0). But $y(\eta(y_1, z_1), z)$ for fixed (y_1, z_1), hence fixed $\eta = \eta(y_1, z_1)$, is a solution of (1.1) reducing to y_1 when $z = z_1$. In other words, $y(z, z_1, y_1) = y(\eta(y_1, z_1), z)$ which shows that $y(z, z_1, y_1)$ is of class C^1.

Domain of Existence. It is readily verified that the conditions on H in the last part of Theorem 6.1 imply that the solution $y_1(t) = y_1(t, \zeta)$ of (6.5) exists for $0 \leqq t \leqq \alpha$, where $\alpha = \min(a, b/M)$, for all unit vectors ζ. In fact, if $\|y_1\|'$ is a right or left derivative, (6.5) and the conditions on H imply that $| \|y_1\|'| \leqq M$; cf. the proof of Lemma III 4.2. The existence and uniqueness of solutions of (1.1)–(1.2) for all (t_0, y_0) imply that $y_1(t, \zeta)$ is a function $y(z)$ of $z = z_0 + t\zeta$ and that $y = y(z)$ is a solution of (1.1)–(1.2). This proves Theorem 6.1.

Exercise 6.3. Assume that $H(y, z)$ is of class C^1 and that (1.4) holds. Prove the existence of a solution of (1.1)–(1.2) by the use of (6.5).

Exercise 6.4. Assume that $H(y, z)$ is of class C^1 and that (1.4) holds. Let $z_0 = 0$. Prove the existence of a solution of (1.1)–(1.2) in the following manner: Let $h_j(y, z)$ be the d-dimensional vector which is the jth column of $H(y, z)$. Define $y = y_1(z^1)$ as the solution of $dy/dz^1 = h_1(y, z^1, 0, \ldots, 0)$, $y(0) = y_0$. If $y_{j-1}(z^1, \ldots, z^{j-1})$ has been defined, let $y_j(z^1, \ldots, z^j)$ be the solution of $dy/dz^j = h_j(y, z^1, \ldots, z^j, 0, \ldots, 0)$, $y(0) = y_{j-1}(z^1, \ldots, z^{j-1})$. Show that $y = y_e(z^1, \ldots, z^e)$ is the desired solution of (1.1)–(1.2).

Proof of Corollary 6.2. It is clear from Theorem III 5.1 and the conditions of Corollary 6.2 that, for a fixed ζ, the solution $y_1 = y_1(t)$ of (6.5) exists on any t-interval J containing $t = 0$ on which $z = z_0 + t\zeta$ is in D. By the proof of uniqueness of solutions $y(z) = y(z, z_0, y_0)$ of (1.1)–(1.2), a pair of solutions $y(z, z_{01}, y_{01})$, $y(z, z_{02}, y_{02})$ of (1.1) on z-spheres about z_{01}, z_{02}, where $z_{0i} = z_0 + t_i\zeta$, $y_{0i} = y_1(t_i)$, and $t_1, t_2 \in J$, coincide on any common domain of definition. Consequently, the solution $y = y(z, z_0, y_0)$ can be defined on an open subset of D containing the line segment $z = z_0 + t\zeta$, $t \in J$.

These arguments show that the same is true if the line segment on $z = z_0 + t\zeta$ is replaced by any polygonal path P in D which begins at z_0 and has no self-intersections. If two such polygonal paths P_1, P_2 which begin at z_0 and end at z_1 are considered, the two solutions $y(z, z_0, y_0)$ defined on neighborhoods of P_1, P_2 agree at $z = z_1$. First, this is clear if the paths P_1, P_2 are sufficiently near to each other. Second, it then follows without this nearness condition on P_1, P_2 by virtue of the simply connectedness of D. Hence the solution $y(z) = y(z, z_0, y_0)$ can be defined (as a single-valued function of z) on D, as was to be proved.

PART II. CAUCHY'S METHOD OF CHARACTERISTICS

7. A Nonlinear Partial Differential Equation

Consider a partial differential equation

$$(7.1) \qquad F\left(u, y^1, \ldots, y^d, \frac{\partial u}{\partial y^1}, \ldots, \frac{\partial u}{\partial y^d}\right) = 0$$

for a real-valued function $u = u(y)$ of d independent real variables, where $F(u, y, p)$ is a real-valued function of $1 + d + d$ variables on an open-set E_{2d+1}. A solution of (7.1) is a function $u = u(y)$ of class C^1 on an open y-set E_d such that $(u(y), y, u_y(y)) \in E_{2d+1}$ for $y \in E_d$ and (7.1) becomes an identity in y. Here $u_y(y) = (\partial u/\partial y^1, \ldots, \partial u/\partial y^d)$ is the gradient of u.

In general, solutions of initial value problems are sought, i.e., solutions of (7.1) which take given values on a piece of a hypersurface S. To be more explicit, let S be a piece of C^1-hypersurface in y-space, i.e., let S be a set of points

$$(7.2) \qquad S: y = \zeta(\gamma), \qquad \text{where } \gamma = (\gamma^1, \ldots, \gamma^{d-1}),$$

$\zeta(\gamma)$ is of class C^1 in a vicinity of $\gamma = \gamma_0$ and rank $(\partial \zeta/\partial \gamma) = d - 1$. Let φ be a given function on S or, equivalently, let $\varphi = \varphi(\gamma)$ be a given function of γ for γ near γ_0. Then the "initial condition" is the requirement that the solution $u = u(y)$ reduces to φ on S, i.e.,

$$(7.3) \qquad u(\zeta(\gamma)) = \varphi(\gamma).$$

The existence theorems to be obtained are local in the sense that only solutions $u = u(y)$ defined for y near $y_0 = \zeta(\gamma_0)$ will be obtained. The method to be used is that of Cauchy which reduces the problem to the theory of ordinary differential equations and is called *Cauchy's method of characteristics*. There is no analogue of this method for *systems* of first order partial differential equations.

The following abbreviations $F_u = \partial F/\partial u$, $F_y = (\partial F/\partial y^1, \ldots, \partial F/\partial y^d)$ and $F_p = (\partial F/\partial p^1, \ldots, \partial F/\partial p^d)$ will be employed. A dot denotes the usual scalar product of d-dimensional vectors.

In order to motivate the method to be employed, consider the following heuristic arguments in the special case of (7.1) which is a linear partial differential equation of the form

$$(7.4) \qquad F \equiv \sum_{k=1}^{d} f^k(y) \frac{\partial u}{\partial y^k} = 0;$$

i.e., $F(u, y, p)$ does not depend on u and is of the form $F(u, y, p) = \Sigma f^k(y)p^k = F_p(y) \cdot p$. If $u = u(y)$ is a solution of (7.4) and $y = y(t)$ is a solution of the system of ordinary differential equations

$$(7.5) \qquad\qquad y' = f = F_p,$$

then (7.4) shows that $u(y(t))$ is a constant. Solutions $y = y(t)$ of (7.5) are called *characteristics* of the partial differential equation (7.4).

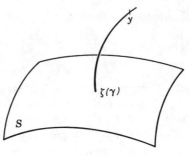

Figure 1.

Suppose that no characteristic is tangent to S. This condition can be expressed by

$$(7.6) \qquad\qquad \det\left[\left(\frac{\partial\zeta(\gamma)}{\partial\gamma}\right), F_p(y(\gamma))\right] \neq 0$$

if S is given by (7.2). In (7.6), $[(\partial\zeta/\partial\gamma), F_p]$ is a $d \times d$ matrix, the first $d - 1$ column of which constitute the $d \times (d - 1)$ Jacobian matrix $(\partial\zeta(\gamma)/\partial\gamma)$ and the last column is the vector F_p.

In this case, S is said to be noncharacteristic and the characteristics with initial points on S fill up a small piece E_d of y-space. The value $u(y)$ of a solution u at a point $y \in E_d$ must be the same as the given value of $\varphi(\gamma)$ at the initial point $\zeta(\gamma)$ on S of the characteristic through y; see Figure 1. Conversely, it is to be expected that a function $u(y)$ defined on E_d in this fashion is a solution of (7.3)–(7.4). Under suitable smoothness conditions on F this turns out to be the case; cf. § V 12 for the relationship between solutions of (7.4) and first integrals of (7.5).

Instead of a linear equation (7.4), consider a somewhat more complicated equation, say, a quasi-linear equation, i.e., an equation in which the highest order ($=$first) partial derivatives occur linearly,

$$(7.7) \qquad\qquad F \equiv \sum_{k=1}^{d} f^k(u, y)\frac{\partial u}{\partial y^k} + U(u, y) = 0.$$

If a solution $u = u(y)$ is known and we consider a solution $y(t)$ of (7.5), where the right side is $f(u, y) = f(u(y), y)$, then the equation $F = 0$ implies that $du(y(t))/dt = -U(u(y(t)), y(t))$. This leads to the set of ordinary (autonomous) differential equations

(7.8) $$y' = f, \qquad u' = -U$$

in which the right sides are functions of u and y, but not of the independent variable t. It will turn out that problems for the quasi-linear equation $F = 0$ can be reduced to problems for this system of ordinary differential equation.

Returning to the general nonlinear case (7.1), characteristics will be defined. These are not generally level curves of a solution as in the linear case. Assume that $F(u, y, p)$ is of class C^1 on some open (u, y, p)-set and assume that $u = u(y)$ in (7.1) is of class C^2. We can reduce the "nonlinearity" of (7.1) by differentiating (7.1) with respect to a fixed component y^m of y to obtain a second order partial differential equation for u which is quasi-linear, i.e., linear in the second order partials of u. This equation can be formally written as a first order, quasi-linear equation for $p^m = \partial u / \partial y^m$,

$$\sum_{j=1}^{d} \frac{\partial F}{\partial p^j} \frac{\partial p^m}{\partial y^j} + \frac{\partial F}{\partial y^m} + F_u p^m = 0.$$

Thus, in analogy to the above, we are led to the ordinary differential equations

$$\frac{dy^j}{dt} = \frac{\partial F}{\partial p^j} \quad \text{for} \quad j = 1, \dots, d, \qquad \frac{dp^m}{dt} = -\left(\frac{\partial F}{\partial y^m} + F_u p^m \right),$$

to which we can add $du/dt = \Sigma(\partial u / \partial y^j) y^{j\prime}$ or

$$\frac{du}{dt} = \sum_{j=1}^{d} p^j \frac{\partial F}{\partial p^j}.$$

The differential equations for y^j and u do not depend on m. Letting $m = 1, \dots, d$ gives a set of autonomous ordinary differential equations for u, y, p which can be written as

(7.9) $$y' = F_p, \qquad p' = -F_y - pF_u, \qquad u' = p \cdot F_p,$$

where the argument of F_u, F_y, F_p is (u, y, p) and the independent variable t does not occur. A solution $y = y(t)$, $p = p(t)$, $u = u(t)$ of (7.9) is called a *characteristic strip* and the projection $u(t)$, $y(t)$ of a solution into the (u, y)-space is called a *characteristic*. A condition of the type

(7.10) $$F_p(u, y, p) \neq 0$$

prevents a characteristic from reducing to a point. The "derivation" of (7.9) will be stated as a formal result in Lemma 8.2.

A solution of the initial value problem (7.1), (7.3) cannot exist unless there exists a function $p = p(\gamma)$ on S which is the gradient of $u(y)$ at $y = \zeta(\gamma)$ and satisfies

$$(7.11) \qquad F(\varphi(\gamma), \zeta(\gamma), p) = 0$$

$$(7.12) \qquad p \cdot \frac{\partial \zeta(\gamma)}{\partial \gamma^i} = \frac{\partial \varphi(\gamma)}{\partial \gamma^i} \qquad \text{for} \quad i = 1, \dots, d - 1.$$

The last condition results upon differentiation of (7.3) with respect to γ^i. In particular, there is a vector $p_0 = p(\gamma_0)$ such that

$$(7.13) \qquad F(u_0, y_0, p_0) = 0$$

$$(7.14) \qquad p_0 \cdot \frac{\partial \zeta(\gamma_0)}{\partial \gamma^i} = \frac{\partial \varphi(\gamma_0)}{\partial \gamma^i} \qquad \text{for} \quad i = 1, \dots, d - 1.$$

Assume that the "initial data is noncharacteristic at $\gamma = \gamma_0$," i.e., that

$$(7.15) \qquad \det \left[\left(\frac{\partial \zeta(\gamma_0)}{\partial \gamma} \right), F_p(u_0, y_0, p_0) \right] \neq 0,$$

where the first $d - 1$ columns of the matrix in (7.15) are the vectors $\partial \zeta(\gamma_0)/\partial \gamma^i$, $i = 1, \dots, d - 1$, and the last is $F_p(u_0, y_0, p_0)$. Then, by the implicit function Theorem I 2.5, (7.13), (7.14), (7.15) and $y = \zeta(\gamma) \in C^2$, $F \in C^1$, $\varphi \in C^2$ imply that (7.11)–(7.12) has a unique solution $p = p(\gamma)$ of class C^1 in a vicinity of $\gamma = \gamma_0$ which satisfies $p(\gamma_0) = p_0$.

Note that in the nonlinear case, we cannot speak of "S being non-characteristic" but only of "the initial data being noncharacteristic." The initial data consists of S: $y = \zeta(\gamma)$, the function $\varphi(\gamma)$, the vector p_0, and, when (7.15) holds, the implicitly determined $p(\gamma)$. By continuity, (7.15) implies that

$$(7.16) \qquad \det \left[\left(\frac{\partial \zeta(\gamma)}{\partial \gamma} \right), F_p(\varphi(\gamma), \zeta(\gamma), p(\gamma)) \right] \neq 0$$

for γ near γ_0. When (7.16) holds, the initial data is called *noncharacteristic*.

Exercise 7.1. Suppose that F in (7.1) does not depend on u. What is the form of the system (7.9) for the characteristic strips? Note that the determination of u in (7.9) reduces to a quadrature.

The initial value problem (7.1), (7.3) is often considered in another form, to be obtained now. This form of the problem will be used in § 10. Suppose that S in (7.2) is of class C^1. Without loss of generality, it can be supposed that $y_0 = 0$ and that S in (7.2) is given in the form $y^d = \psi(y^1, \dots, y^{d-1})$, where ψ is of class C^2 for $(y^1, \dots, y^{d-1})$ near 0 and $\psi(0) = 0$. If

$$(y^1, \dots, y^{d-1}, y^d - \psi(y^1, \dots, y^d))$$

are introduced as new coordinates, again called y, then in the new coordinates S is a piece of the hyperplane $y^d = 0$ near $y = 0$. The partial differential equation (7.1) is transformed into another of the same type, although if, e.g., the original F is of class C^2, the new F has continuous first and second derivatives except possibly for those of the type $\partial^2 F/\partial y^j \, \partial y^k$. The condition (7.15) in the new coordinates system becomes $\partial F(u_0, y_0, p_0)/\partial p^d \neq 0$. Thus, if (7.13) holds, the equation $F(u, y, p) = 0$ can be solved for p^d in terms of $u, y, p^1, \ldots, p^{d-1}$, say $p^d = -H(u, y, p^1, \ldots, p^{d-1})$, and (7.1) is equivalent to $p^d + H(u, y, p^1, \ldots, p^{d-1}) = 0$ for (u, y, p) near (u_0, y_0, p_0).

Thus, if the notation is changed by replacing $d - 1$ by d and y by (y, t), then the initial value problem takes the form

$$(7.17) \qquad\qquad u_t + H(u, t, y, u_y) = 0,$$

$$(7.18) \qquad\qquad u(0, y) = \varphi(y),$$

where $u = u(t, y)$ is the unknown function, $\varphi(y)$ is the given initial function, $u_t = \partial u/\partial t$, and $u_y = (\partial u/\partial y^1, \ldots, \partial u/\partial y^d)$.

Exercise 7.2. (a) Write H as $H(u, t, y, q)$, where $y = (y^1, \ldots, y^d)$ and $q = (q^1, \ldots, q^d)$. Find the differential equations for the characteristic strips for the partial differential equation (7.17), using t as the independent variable. (b) Simplify the result of part (a), assuming that $H(t, y, q)$ is independent of u. [Note that in this case, when H is a Hamiltonian function, (7.17) is the Hamilton-Jacobi partial differential equation and the nontrivial parts of the equations for the characteristic strips are the equation of motion in the Hamiltonian form.]

8. Characteristics

The relationships between solutions of (7.1) and characteristic strips or characteristics will now be determined.

Lemma 8.1. *Let $F(u, y, p)$ be of class C^1. Then $F(u, y, p)$ is a first integral of the system* (7.9); *i.e., F is constant along any solution of* (7.9).

Proof. It suffices to verify that if $y(t), p(t), u(t)$ is a solution of (7.9), then the derivative of $F(u(t), y(t), p(t))$ is 0. This is equivalent to

$$F_y \cdot F_p + F_p \cdot (-F_y - pF_u) + F_u(p \cdot F_p) = 0.$$

Lemma 8.2. *Let $F(u, y, p)$ be of class C^1 on an open set E_{2d+1} and let $u = u(y)$ be a solution of* (7.1) *of class C^2 on an open set E_d. For $y_0 \in E_d$, there exists a characteristic strip $y(t), p(t), u(t)$ for small $|t|$, such that $u(t) = u(y(t)), p(t) = u_y(y(t))$ and $y(0) = y_0$.*

In particular, in the (u, y)-space, the arc $(u, y) = (u(t), y(t))$ lies on the hypersurface $u = u(y)$ and $(-1, p(t))$ is a normal vector to this hypersurface at the point $(u, y) = (u(t), y(t))$. Thus if $u = u(y)$ is a solution of

(7.1) of class C^2, the hypersurface $u = u(y)$ can be considered to be made up of characteristics.

Corollary 8.1. *If the solutions of initial value problems associated with* (7.9) *are unique* (e.g., *if $F \in C^2$*) *and $u = u_1(y)$, $u_2(y)$ are two solutions of* (7.1) *of class C^2 which "touch" at $y = y_0$* [i.e., $u_1(y_0) = u_2(y_0)$, $u_{1y}(y_0) = u_{2y}(y_0)$], *then they "touch" along a characteristic arc $y = y(t)$.*

Proof of Lemma 8.2. [This is a repetition of the "derivation" of (7.9).] Consider a solution $y = y(t)$ of the initial value problem

$$(8.1) \qquad y' = F_p(u(y), y, u_y(y)), \qquad y(0) = y_0.$$

Differentiating (7.1) with respect to y^m gives

$$(8.2) \qquad F_u \frac{\partial u}{\partial y^m} + \frac{\partial F}{\partial y^m} + \sum_{j=1}^{d} \frac{\partial F}{\partial p^j} \frac{\partial^2 u}{\partial y^j \, \partial y^m} = 0$$

Put $p(t) = u_y(y(t))$. Then (8.1) implies that (8.2) can be written as the mth component of $F_u p + F_y + p' = 0$, where the argument of F_u, F_y is $(u(y(t)), y(t), p(t))$. Also, if $u = u(y(t))$, then $u' = u_y(y(t)) \cdot y'$ is $p \cdot F_p$ by (8.1). Thus $y = y(t)$, $p = u_y(y(t))$, $u = u(y(t))$ is a solution of (7.9). This proves the lemma.

Remark. It remains unknown whether Lemma 8.2 is valid if it is only assumed that $u(y) \in C^1$ and $d > 2$. In this direction, there is a partial result for $d > 2$ and a complete result for $d = 2$:

Exercise 8.1. Let $F(u, y, p)$ be of class C^1, $u(y)$ of class C^1 in a neighborhood of y_0 and a solution of (7.1). (a) Let m be fixed, $1 \leqq m \leqq d$. Show that there exists a solution $y(t)$ of the initial value problem

$$y' = F_p(u(y), y, u_y(y)), \qquad y(0) = y_0$$

such that $p^m(t) = [\partial u(y)/\partial y^m]_{y=y(t)}$ has a continuous derivative with respect to t satisfying $p^{m'} = -\partial F/\partial y^m - F_u p^m$, where the argument of $\partial F/\partial y^m$, F_u is $[u(y(t)), y(t), u_y(y(t))]$. (b) In particular, if the solution of the initial value problem $y' = F_p(u(y), y, u_y(y))$, $y(0) = y_0$ is unique [so that $y(t)$ in part (a) does not depend on m], then the conclusion of Lemma 8.2 is valid. (This is applicable, e.g., if $F(u, y, p)$ depends linearly on p.) (c) If $F_p \neq 0$ at $(u, y, p) = (u(y_0), y_0, u_y(y_0))$ and $d = 2$, then the conclusion of Lemma 8.2 is valid (under the assumptions $F \in C^1$, $u \in C^1$).

Exercise 8.2. Let $F(u, y, p)$ and $G(u, y, p)$ be of class C^1 on an open (u, y, p)-domain. Define the function $H(u, y, p)$ by

$$H = F_p \cdot (G_y + pG_u) - G_p \cdot (F_y + pF_u).$$

(If F, G are linear in p and independent of u, H corresponds to the "commutator" of F, G defined in Exercise 1.1.) Let $u(y)$ be a solution of class C^1 of both $F = 0$ and $G = 0$ on some y-domain D. (a) Show that if, in addition, $u(y)$ is of class C^2, then $u(y)$ is a solution of $H = 0$. (b) Show that if $u(y) \in C^1$ and if through each point $(u, y, p) = (u(y_0), y_0, u_y(y_0))$, there passes a characteristic strip for $F = 0$ as in Lemma 8.2 [e.g., if Exercise 8.1(b) is applicable], then $u(y)$ is a solution of $H = 0$.

9. Existence and Uniqueness Theorem

The main theorem on (7.1) is the following.

Theorem 9.1. *Let $F(u, y, p)$ be of class C^2 on an open domain E_{2d+1}. Let $(u_0, y_0, p_0) \in E_{2d+1}$. Let (7.2) be a piece of hypersurface of class C^2 defined for γ near γ_0 and $\zeta(\gamma_0) = y_0$. Let $\varphi(\gamma)$ be a function of class C^2 for γ near γ_0 and $\varphi(\gamma_0) = u_0$. Finally, let (7.13), (7.14), and (7.15) hold. Then, on a neighborhood E_d of $y = y_0$, there exists a unique solution $u = u(y)$ of class C^2 of the initial value problem (7.1)–(7.3).*

Note that (7.15) implies that $F_p(u_0, y_0, p_0) \neq 0$ and that rank $(\partial \zeta(\gamma)/\partial \gamma)$ is $d - 1$. The condition $F_p(u_0, y_0, p_0) \neq 0$ implies the existence of hypersurfaces S satisfying (7.15). For example, if $\partial F/\partial p^d \neq 0$ at (u_0, y_0, p_0), then the hyperplane $y^d = y_0{}^d$ is an admissible S.

Proof. By the argument in §7, there is a unique function $p = p(\gamma)$ of class C^1 for γ near γ_0 satisfying (7.11)–(7.12) and $p(\gamma_0) = p_0$. Let $y = Y(t, \gamma)$, $p = P(t, \gamma)$, $u = U(t, \gamma)$ be the solution of (7.9) satisfying the initial condition

$$(9.1) \qquad Y(0, \gamma) = \zeta(\gamma), \qquad P(0, \gamma) = p(\gamma), \qquad U(0, \gamma) = \varphi(\gamma).$$

By Theorem V 3.1, this solution is unique and $Y(t, \gamma), P(t, \gamma), U(t, \gamma)$, $Y'(t, \gamma), P'(t, \gamma), U'(t, \gamma)$ are of class C^1 for small $|t|$ and γ near γ_0. Since F is a first integral for (7.9),

$$(9.2) \qquad F(U(t, \gamma), Y(t, \gamma), P(t, \gamma)) = 0.$$

In fact, the function in (9.2) does not depend on t by Lemma 8.1 and, at $t = 0$, (9.2) reduces to (7.11).

By the first part of (7.9) and by (9.1), at $(t, \gamma) = (0, \gamma_0)$, the Jacobian $\partial Y(t, \gamma)/\partial(t, \gamma)$ at $t = 0, \gamma = \gamma_0$ is

$$\frac{\partial(y^1, \ldots, y^d)}{\partial(\gamma^1, \ldots, \gamma^{d-1}, t)} = \det \left[\left(\frac{\partial \zeta(\gamma_0)}{\partial \gamma} \right), F_p(u_0, y_0, p_0) \right].$$

Thus assumption (7.15) shows that this Jacobian determinant is not 0. Hence, there exists a unique map

$$(9.3) \qquad t = t(y), \qquad \gamma = \gamma(y)$$

for y near y_0 inverse to

$$(9.4) \qquad y = Y(t, \gamma).$$

The map (9.3) is of class C^1. Put

(9.5) $u(y) = U(t(y), \gamma(y))$

for y near y_0. Then (9.2) becomes

$$F(u(y), y, P[t(y), \gamma(y)]) = 0.$$

Thus the existence assertion will be proved if it is shown that

(9.6) $u_y(y) = P(t(y), \gamma(y)),$

i.e., $du(y) = P(t(y), \gamma(y)) \cdot dy$. Under the change of variables (9.4), $y \rightarrow (t, \gamma)$, with nonvanishing Jacobian, this is equivalent to

$$dU(t, \gamma) = \sum_{j=1}^{d-1} P(t, \gamma) \cdot \frac{\partial Y(t, \gamma)}{\partial \gamma^j} \, dy^j + P(t, \gamma) \cdot Y'(t, \gamma) \, dt$$

or to

(9.7) $\dfrac{\partial U(t, \gamma)}{\partial \gamma^j} - P(t, \gamma) \cdot \dfrac{\partial Y(t, \gamma)}{\partial \gamma^j} = 0, \qquad j = 1, \dots, d - 1,$

(9.8) $U'(t, \gamma) - P(t, \gamma) \cdot Y'(t, \gamma) = 0.$

The relation (9.8) follows from (7.9). Thus only (9.7) remains to be verified.

For a fixed γ, let $\lambda_j(t)$ denote the expression on the left of (9.7). Thus it suffices to show that $\lambda_j(t) \equiv 0$. Note that, by (7.12) and (9.1),

(9.9) $\lambda_j(0) = 0.$

In what follows, let $F = F(u, y, p)$ and $u = U(t, \gamma), y = Y(t, \gamma), p = P(t, \gamma)$. Differentiating the left side of (9.7) with respect to t gives

$$\lambda_j' = \frac{\partial u'}{\partial \gamma^j} - p' \cdot \frac{\partial y}{\partial \gamma^j} - p \cdot \frac{\partial y'}{\partial \gamma^j}.$$

The change of order of differentiation is permitted since y', p', u' are of class C^1. Using (7.9), the last relation becomes

$$\lambda_j' = \frac{\partial(p \cdot F_p)}{\partial \gamma^j} + (F_y + pF_u) \cdot \frac{\partial y}{\partial \gamma^j} - p \cdot \frac{\partial F_p}{\partial \gamma_j}$$

$$= F_p \cdot \frac{\partial p}{\partial \gamma^j} + F_y \cdot \frac{\partial y}{\partial \gamma^j} + F_u p \cdot \frac{\partial y}{\partial \gamma^j}.$$

If (9.2) is differentiated with respect to γ^j, it is seen that the sum of the first two terms on the right is $-F_u \, \partial u / \partial \gamma^j$. Hence,

$$\lambda_j' = -F_u \left(\frac{\partial u}{\partial \gamma^j} - p \cdot \frac{\partial y}{\partial \gamma^j} \right) = -F_u \lambda_j.$$

Since $\lambda_j(t)$ satisfies a linear homogeneous differential equation and the initial condition (9.9), it follows that $\lambda_j(t) \equiv 0$.

Hence (9.5) is a solution of (7.1)–(7.3). Also $u(y)$ is of class C^2 since its gradient (9.6) is of class C^1. Finally, when $F \in C^2$, so that solutions of (7.9) are uniquely determined by initial conditions, the uniqueness of C^2 solutions of (7.1)–(7.3) follows from Lemma 8.2, the remarks following it, and the existence proof just completed. (For another uniqueness proof, see the next section.)

It should be mentioned that when the initial data is not "noncharacteristic," in general there will not exist a solution.

In a sense, the existence Theorem 9.1 is unsatisfactory for it produces solutions of class C^2 when it is natural to ask for solutions only of class C^1. It is reasonable to inquire as to whether or not the differentiability conditions can be lightened in Theorem 9.1 and still obtain solutions of class C^1. To some extent, this question can be answered in the negative.

Exercise 9.1. Let x, y be real variables and $f(x)$ a continuous nowhere differentiable function of x. Show that $u_x - u_y + f(x + y) = 0, u(0, y) = 0$ has no C^1 solution. Thus continuity of F is not sufficient to assure the existence of solutions.

Exercise 9.2. Even $F \in C^1$ and analytic initial data is insufficient to assure existence of solutions. Let x, y, q be real variables. Let $f(q)$ be a real-valued function of class C^1 for small $|q|$ such that $\partial f/\partial q$ is not Lipschitz continuous at $q = 0$. Show that, on the one hand, the procedure in the proof of Theorem 9.1 does not lead to a solution of $u_x = f(u_y)$, $u(0, y) = \frac{1}{2}y^2$. (The difficulty arises from the fact that the analogue of the map (9.4) has no inverse.) On the other hand, Exercise 8.1(c) implies that if a solution exists, then it is obtainable by such a procedure. Hence there is no C^1 solution.

Exercise 9.3. The last exercise shows that, in a sense, the following is the "best" theorem: Theorem 9.1 remains correct if "$F, \zeta(\gamma), \varphi, u(y) \in C^2$" is replaced by "$F, \zeta(\gamma), \varphi, u(y)$ are of class C^1 with uniformly Lipschitz continuous partial derivatives." See Ważewski [3]. This can be proved by a suitable modification of the proof of Theorem 9.1 using, e.g., the fact that uniformly Lipschitz continuous functions possess total differentials almost everywhere. For a different proof, see Digel [2].

10. Haar's Lemma and Uniqueness

Let (7.1), (7.3) be replaced by (7.17) and (7.18), i.e., by

$$(10.1) \qquad u_t + H(u, t, y, u_y) = 0,$$

$$(10.2) \qquad u(0, y) = \varphi(y).$$

It follows from Theorem 9.1 that if $H(u, t, y, q)$ is of class C^2 on an open set E_{2+2d} containing the point $(u, t, y, q) = (\varphi(0), 0, 0, \varphi_y(0))$ and $\varphi(y)$ is of class C^2 for y near $y = 0$, then (10.1), (10.2) has a unique solution $u = u(t, y)$ of class C^2 for small $|t|, |y|$. The situation as to uniqueness for solutions of (10.1)–(10.2) is very simple.

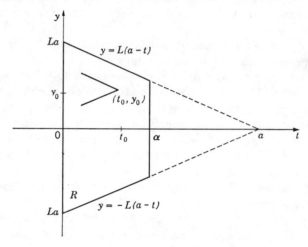

Figure 2.   The case $d = \dim y$ is 1.

Theorem 10.1.   *Let $H(u, t, y, q)$ be defined on an open set E_{2+2d} containing the point $(u, t, y, q) = 0$ and satisfy a uniform Lipschitz condition with respect to (u, q). Let $\varphi(y)$ be a function of class C^1 satisfying $\varphi(0) = 0$, $\varphi_y(0) = 0$. Then (10.1)–(10.2) has at most one solution of class C^1 on a neighborhood E_d of $y = 0$.*

This follows by applying the following lemma (with $C = N = 0$) to the difference $v = u_2(t, y) - u_1(t, y)$ of two solutions $u_1(t, y)$, $u_2(t, y)$.

Lemma 10.1.   *Let $v = v(t, y)$ be a real-valued function of class C^1 on a set R: $0 \leqq t \leqq \alpha(<a)$, $|y^i| \leqq L(a - t)$ for $i = 1, \ldots, d$ and satisfy*

$$(10.3) \qquad\qquad |v(0, y)| \leqq C,$$

$$(10.4) \qquad\qquad |v_t| \leqq L \sum_{k=1}^{d} \left| \frac{\partial v}{\partial y^k} \right| + M |v| + N,$$

where $L, M > 0$ and $C, N \geqq 0$ are constants. Then, on R,

$$(10.5) \qquad\qquad |v(t, y)| \leqq C e^{Mt} + N \frac{e^{Mt} - 1}{M} .$$

See Figure 2.

Proof.   Let C', N' be arbitrary constants satisfying

(10.6) $C < C'$, $N < N'$

and put

(10.7) $$u(t, y) = C'e^{Mt} + N'\frac{e^{Mt} - 1}{M}$$

so that

(10.8) $$u_t = L\sum_{k=1}^{d}\frac{\partial u}{\partial y^k} + Mu + N' = Mu + N'.$$

It will be shown that

(10.9) $$u(t, y) - v(t, y) > 0$$

on R, so that letting $C' \to C$ and $N' \to N$ gives

(10.10) $$v(t, y) \leqq Ce^{Mt} + N\frac{e^{Mt} - 1}{M}$$

Replacing v by $-v$ in this argument gives (10.5).

It is clear from (10.3), (10.6), and (10.7) that $u - v > 0$ for small $t > 0$. If (10.9) does not hold on R, there is a point (t_0, y_0) of R such that $0 < t_0 \leqq \alpha$, (10.9) holds on that portion of R where $0 \leqq t < t_0$, and equality holds at (t_0, y_0).

For any of the 2^d choices of $\pm$, the points of the line segments

(10.11) $(t, y) = (t, \pm L(t_0 - t) + y_0^1, \ldots, \pm L(t_0 - t) + y_0^d)$

are in R for $0 \leqq t \leqq t_0$, for $|\pm L(t_0 - t) + y_0^k| \leqq L(a - t)$ since $|y_0^k| \leqq L(a - t_0)$; see Figure 2. The difference $u - v$ at the point (10.11) is positive for $0 \leqq t < t_0$ and 0 at $t = t_0$. Consequently, the derivative of $u - v$ along the line (10.11) is nonpositive at $t = t_0$. This gives

(10.12) $$u_t - v_t + \sum_{k=1}^{d}(\pm L)\frac{\partial(u - v)}{\partial y^k} \leqq 0$$ at (t_0, y_0).

From (10.8), $u_t = Mu + N'$, so that $u_t = Mv + N' = M|v| + N'$ at $(t, y) = (t_0, y_0)$. This fact and $u_y = 0$ give

$$v_t \geqq M|v| + N' + L\sum_{k=1}^{d}\left(\pm\frac{\partial v}{\partial y^k}\right)$$ at (t_0, y_0).

If $\pm$ is chosen so that $\pm\partial v/\partial y^k = |\partial v/\partial y^k|$ at (t_0, y_0), the resulting inequality contradicts (10.4) since $N' > N$. This implies (10.9) on R and proves the lemma.

Exercise 10.1.   (a) Let B be the (t, y)-set $B = \{(t, y) : 0 \leqq t < a$, $c_k + L_k t \leqq y_k \leqq d_k -, L_k t$ for $k = 1, \ldots, d\}$, where $L_k \geqq 0$, $c_k < d_k$, and

$2L_k a \leqq d_k - c_k$. Let $u(t, y)$ be a real-valued function of class C^1 on B and let $m(s) = \max u(s, y)$ taken over the set $B_s = \{(t, y):(t, y) \in B, t = s\}$. Then $m(t), 0 \leqq t < a$, has a right derivative $D_R m(t)$ and there exists a point $(t, y_0) \in B_t$ such that $m(t) = u(t, y_0)$ and $D_R m(t) = \partial u/\partial t - \sum\limits_{k=1}^{d} |\partial u/\partial y^k| \, L_k$ evaluated at (t, y_0). (b) Let $\omega(t, u)$ be continuous for $0 < t < a, u \geqq 0$ and such that the only solution of $u' = \omega(t, u)$ defined for $0 < t \leqq \epsilon \, (<a)$ and satisfying $u(t) \to 0$ and $u(t)/t \to 0$, as $t \to +0$, is $u(t) \equiv 0$. Let $H(u, t, y, q)$ be continuous for (u, t, y, q) near $(u, t, y, q) = 0$ and $|H(u_1, t, y, q_1) - H(u_2, t, y, q_2)| \leqq \Sigma \, L_k \, |q_1{}^k - q_2{}^k| + \omega(t, |u_1 - u_2|)$. Let $\varphi(y)$ be of class C^1 for small $|y|$ and satisfy $\varphi(0) = 0, \varphi_y(0) = 0$. Then (10.1), (10.2) has at most one solution on the set B.

Exercise 10.2. (a) Let B denote a bounded (t, y)-set defined by inequalities: $0 \leqq t < a, b_j(t, y) \geqq 0$ for $j = 1, \ldots, m$, where $b_j(t, y)$ is a real-valued function of class C^1. It is assumed that every boundary point of B lies on either $t = 0, t = a$, or k of the m hypersurfaces $b_j(t, y) = 0$, $1 \leqq k \leqq m$; also, if k of the hypersurfaces $b_j(t, y) = 0$, say $j = j_1, \ldots, j_k$, have a point (t, y) in common, then the k differential 1-forms $\sum\limits_{i=1}^{d} (\partial b_j/\partial y^i) \, dy^i$, where $j = j_1, \ldots, j_k$, are linearly independent at (t, y). Let $H(u, t, y, q)$ be defined on a $(1 + 1 + d + d)$-dimensional domain E with a projection on the (t, y)-space containing B. Let $u(t, y), v(t, y)$ be real-valued functions of class C^1 on B such that $(u, t, y, u_y) \in E, (v, t, y, v_y) \in E$. Suppose that $u_t > H(u, t, y, u_y), v_t \leqq H(v, t, y, v_y)$ on B and that $u(0, y) > v(0, y)$. Finally, suppose that, at every boundary point (t, y) of B common to k hypersurfaces $b_j(t, y) = 0$ for $j = j_1, \ldots, j_k$, we have the inequality

$$H(u, t, y, u_y) - H\left(u, t, y, \left[u - \sum_{i=1}^{k} \lambda_i b_{j_i}\right]_y\right) \geqq \sum_{i=1}^{k} \lambda_i \frac{\partial b_{j_i}}{\partial t}$$

for all non-negative numbers $\lambda_1, \ldots, \lambda_k$ such that $(u, t, y, [u - \Sigma \, \lambda_i b_{j_i}]_y) \in B$. Then $u(t, y) > v(t, y)$ on B. See Nagumo [3]. (b) Let B be the same as the (t, y)-set in Exercise 10.1(a). Let $H(u, t, y, q)$ be defined on a $(1 + 1 + d + d)$-dimensional set E with a projection on the (t, y)-space containing B and satisfying $|H(u, t, y, q_1) - H(u, t, y, q_2)| \leqq \Sigma \, L_k |q_1{}^k - q_2{}^k|$. Let $u(t, y), v(t, y)$ be C^1 functions on B such that $(u, t, y, u_y) \in E, (v, t, y, v_y) \in E$; that $u_t > H(u, t, y, u_y)$ and $v_t \leqq H(v, t, y, v_y)$ on E; and that $u(0, y) > v(0, y)$. Then $u(t, y) > v(t, y)$ on B. (c) Deduce Lemma 10.1 from part (b).

Notes

SECTION 1. Connections between the systems (1.1) and (1.12) were considered by Boole in 1862; see E. A. Weber [1] for historical remarks and early references. Discussions of integrability conditions and existence theorems for Jacobi systems go back (1862) to Jacobi [1, V, p. 39] and to Clebsch [2] who introduced the concept of "complete systems" (1.12). (See also the reference to A. Mayer [1] in connection with Exercise 6.3.) The result in Exercise 1.1(b) is due to E. Schmidt. For references to Schmidt, Perron, and Gillis and for generalizations of part (b) see Ostrowski [1]. Further generalizations, based on Pliś [2], are given by Hartman [13]; see Exercise 8.2.

SECTION 2. See E. Cartan [1, pp. 49–64].

SECTION 3. Under analyticity assumptions, Theorem 3.1 is due to Frobenius [2]. The statement of the theorem in the text, avoiding differentiability assumptions, is due to Hartman [17]. The proof in the text is adapted from E. Cartan [1, pp. 99–100]. For a related but somewhat less general theorem on Jacobi systems, see Gillis [1].

SECTION 6. The arguments in the "sufficiency" proof and the existence proof for (1.1)–(1.2) suggested in Exercise 6.3 go back to A. Mayer [1]; cf. Caratheodory [1, pp. 26–30]. The proof outlined in Exercise 6.4 was used by Weyl [2, pp. 64–68]. Still another proof, using successive approximations, is given by Nikliborc [1]; cf. also Gillis [1].

SECTIONS 7–9. The initial value problem considered in § 7 is the simplest example of the type called "Cauchy's problem" in the theory of partial differential equations. The differential equations (7.9) for the characteristic strips, Theorem 9.1, and its proof are due to Cauchy (about 1819) under conditions of analyticity; see, e.g., [1, pp. 423–470]. Actually, a few years earlier, Pfaff [1] had considered the problem of finding solutions of (7.1) and also introduced, in a cumbersome manner, the system (7.9) for the characteristic strips in order to reduce the problem to the theory of ordinary differential equations. A treatment of nonanalytic equations (7.1) awaited, of course, a knowledge of Theorem V 3.1. In fact, both Picard's and Bendixson's work in 1896, mentioned in connection with Theorem V 3.1, were written from the point of view of solving a nonanalytic linear partial differential equation. A theorem similar to Theorem 9.1 was given by Gross [1]. For fuller treatments of this problem, see Caratheodory [1] and Kamke [5]. For a theorem involving slightly less differentiability conditions, see Exercise 9.3 and the reference to Wazewski [3] and Digel [2]. Regions of existence for the solutions were investigated by Kamke and also by Ważewski [2]. For Exercise 8.1, see Pliś [2]. The example in Exercise 9.1 was given by Perron [1].

SECTION 10. Lemma 10.1 was given by Haar [1] for the purpose of proving the uniqueness Theorem 10.1. (This paper contains a wrong proof for Lemma 8.2 under the assumption that $u \in C^1$.) For Exercise 10.1, see Ważewski [1] (and Turski [1] for a generalization which is contained in Exercise 10.2). For Exercise 10.2, see Nagumo [3].

Chapter VII

The Poincaré-Bendixson Theory

The main part (§§ 2–9) of this chapter deals with the geometry of solutions of differential equations on a plane ($d = 2$). The restriction to a plane appears essential since the arguments will make repeated use of the Jordan curve theorem. In § 10, the results obtained are applied to certain non-linear second order differential equations.

Recent extensions of the Poincaré-Bendixson theory from planes to 2-dimensional manifolds are presented in the Appendix in § 12. The last section (§ 14) concerns the behavior of solutions of differential equations on a torus.

1. Autonomous Systems

A system of differential equations in which the independent variable t does not occur explicitly,

(1.1) $$y' = f(y),$$

is called *autonomous*. A trivial but important property of such systems is the fact that if $y = y(t)$, $\alpha < t < \beta$, is a solution of (1.1), then $y = y(t + t_0)$ is also a solution for $\alpha - t_0 < t < \beta - t_0$ for any constant t_0. An *orbit* will mean a set of points y on a solution $y = y(t)$ of (1.1) without reference to a parametrization.

Any system $y' = f(t, y)$ can be considered autonomous if the dependent variable y is replaced by the $(d + 1)$-vector (t, y) and the system $y' = f$ is replaced by $t' = 1$, $y' = f(t, y)$, where the prime denotes differentiation with respect to a new independent variable. For most purposes, however, this remark is not useful.

A point y_0 is called a *stationary* or singular point of (1.1) if $f(y_0) = 0$ and a *regular* point if $f(y_0) \neq 0$. The stationary points y_0 are characterized by the fact that the constant $y(t) \equiv y_0$ is a solution of (1.1). When solutions of (1.1) are uniquely determined by initial conditions, $f(y_0) = 0$ and $y(t_0) = y_0$ for some t_0 imply $y(t) \equiv y_0$. This need not be the case in general.

144

If (1.1) has a solution $C^+ : y = y(t)$ defined on a half-line $t \geqq t_0$, its set $\Omega(C^+)$ of ω-limit points is the (possibly empty) set of points y_0 for which there exists a sequence $t_0 < t_1 < \ldots$ such that $t_n \to \infty$ and $y(t_n) \to y_0$ as $n \to \infty$. Correspondingly, if $C^- : y = y(t)$ is defined for $t \leqq t_0$, we define the set $A(C^-)$ of α-limit points and if $C : y = y(t)$ is defined for $-\infty < t < \infty$, its set of limit points is defined to be $A(C) \cup \Omega(C)$.

Remark 1. $\Omega(C^+)$ is contained in the closure of the set of points $C^+ : y = y(t), t \geqq t_0$.

Theorem 1.1. *Assume that $f(y)$ is continuous on an open y-set E and that $C^+ : y = y_+(t)$ is a solution of (1.1) for $t \geqq 0$. Then $\Omega(C^+)$ is closed. If C^+ has a compact closure in E, then $\Omega(C^+)$ is connected.*

Proof. The verification that $\Omega(C^+)$ is closed is trivial. In order to prove the last part, note that by Remark 1, $\Omega(C^+)$ is a compact set. Suppose that $\Omega(C^+)$ is not connected, then it has a decomposition into the union of two closed (hence, compact) sets C_1, C_2 such that dist $(C_1, C_2) = \delta > 0$. It is clear that there exists a sequence $0 < t_1 < t_2 < \ldots$ of t-values satisfying dist $(y_+(t_{2n+1}), C_1) \to 0$, dist $(y_+(t_{2n}), C_2) \to 0$ as $n \to \infty$. Hence, for large n, there is a point $t = t_n{}^*$ such that $t_n < t_n{}^* < t_{n+1}$, dist $(y_+(t_n{}^*), C_i) \geqq \delta/4$ for $i = 1, 2$. The sequence $y_+(t_1{}^*), y_+(t_2{}^*), \ldots$ has a cluster point y_0, since C^+ has compact closure. Clearly, $y_0 \in \Omega(C^+)$ and dist $(y_0, C_i) \geqq \delta/4$ for $i = 1, 2$. This contradiction proves the assertion.

Theorem 1.2. *Let f, C^+ be as in Theorem 1.1 and $y_0 \in E \cap \Omega(C^+)$. Then*

$$(1.2) \qquad y' = f(y), \qquad y(0) = y_0$$

has at least one solution $y = y_0(t)$ on a maximal interval (ω_-, ω_+) such that $y_0(t) \in \Omega(C^+)$ for $\omega_- < t < \omega_+$. In particular, when C^+ has a compact closure in E, then $C_0 : y = y_0(t)$ exists on $(-\infty, \infty)$ and $C_0 \cup A(C_0) \cup \Omega(C_0) \subset \Omega(C^+)$.

An orbit $C_0 : y = y_0(t)$, $\omega_- < t < \omega_+$, which is contained in some $\Omega(C^+)$, $C^+ \not\subset C_0$, is called an (ω-) *limit orbit*. If, in addition, $y = y_0(t)$ is periodic, $y_0(t + p) = y_0(t)$ for all t and some $p > 0$, the orbit $C_0 : y = y_0(t)$ is called an (ω-) *limit cycle*. (The condition $C^+ \not\subset C_0$ assures that not every periodic $C_0 : y = y_0(t)$ is a limit cycle; cf. the case of a family of closed orbits.)

Proof. Let $t_0 < t_1 < \ldots$ and $t_n \to \infty$, $y_n \to y_0$ as $n \to \infty$, where $y_n = y_+(t_n)$. Then $y_n(t) = y_+(t + t_n)$ is a solution of

$$(1.3) \qquad y' = f(y), \qquad y(0) = y_n.$$

It follows therefore from Theorem II 3.2, where $f_n(t, y) = f(y)$ for $n = 1, 2, \ldots$, that (1.2) has a solution $y_0(t)$ on a maximal interval (ω_-, ω_+) and that there exists a sequence of positive integers $n(1) < n(2) < \ldots$

such that

(1.4) $$y_0(t) = \lim_{k \to \infty} y_{n(k)}(t) = \lim_{k \to \infty} y_+(t + t_{n(k)})$$

holds uniformly on compact intervals of $\omega_- < t < \omega_+$. It is clear that $y_0(t) \in \Omega(C^+)$ for $\omega_- < t < \omega_+$. This proves the first part of the theorem.

The second part concerning existence on $(-\infty, \infty)$ follows at once from Theorem II 3.1 which implies that the right maximal interval $[0, \omega_+)$ for $y = y_0(t)$ is either $[0, \infty)$ or $y_0(t)$ tends to ∂E as $t \to \omega_+ < \infty$.

The last part concerning $A(C_0)$ and $\Omega(C_0)$ follows from (1.4) and the fact that $\Omega(C^+)$ is closed.

Remark 2.   If solutions of all initial value problems associated with (1.1) are unique, then "selection" in the proof of the theorem is unnecessary; thus $y_n = y_+(t_n) \to y_0$ as $n \to \infty$ implies that

(1.5) $$y_0(t) = \lim_{n \to \infty} y_+(t + t_n)$$

holds uniformly on every closed, bounded interval in (ω_-, ω_+).

Corollary 1.1.   *If $\Omega(C^+)$ consists of a single point $y_0 \in E$, then y_0 is a stationary point and $y_+(t) \to y_0$ as $t \to \infty$.*

2. Umlaufsatz

In the plane $(d = 2)$, where the Jordan curve theorem is available, the notions of the last section can be carried much further to give the Poincaré-Bendixson theory. This will be done in §§ 4–6. In order to avoid an interruption of the proofs, the idea of the index of a plane stationary point will first be discussed.

Recall that a Jordan curve J is defined as a topological image of a circle; in other words, J is a y-set of points $y = y(t)$, $a \leqq t \leqq b$, where $y(t)$ is continuous, $y(a) = y(b)$, and $y(s) \neq y(t)$ for $a \leqq s < t < b$. The Jordan curve theorem will be stated here for reference. For a proof, the reader is referred, e.g., to Newman [1, p. 115].

Jordan Curve Theorem.   *If J is a plane Jordan curve, then its complement in the plane is the union of two disjoint connected open sets, E_1 and E_2, each having J as its boundary, $\partial E_1 = \partial E_2 = J$.*

One of the sets E_1 or E_2 is bounded and is called the interior of J; furthermore the interior of J is simply connected.

Consider a continuous arc $J : y = y(t)$, $a \leqq t \leqq b$, in the $y = (y^1, y^2)$ plane. Let $\eta = \eta(t) \neq 0$, $a \leqq t \leqq b$, be a continuous 2-dimensional vector attached to the point $y(t)$, i.e., $\eta \neq 0$ is a vector field on J. Consider an angle $\varphi = \varphi(t)$ from the positive y^1-direction $(1, 0)$ to $\eta(t)$, so that $\cos \varphi = \eta^1 / \|\eta\|$, $\sin \varphi = \eta^2 / \|\eta\|$, where $\|\eta\|^2 = (\eta^1)^2 + (\eta^2)^2$. These formulae determine $\varphi(t)$ up to an integral multiple of 2π but if $\varphi(t)$ is fixed at some point, say $t = a$, then $\varphi(t)$ is uniquely determined as a continuous function. By $\varphi(t)$ below is always meant such a continuous determination. Define $j_\eta(J)$ by

$$2\pi j_\eta(J) = \varphi(b) - \varphi(a).$$

For example, if $\eta(t)$ is continuously differentiable,

$$2\pi j_\eta(J) = \int_a^b \frac{\eta^1\,d\eta^2 - \eta^2\,d\eta^1}{\|\eta\|^2}.$$

If $J = J_1 + J_2$ in the sense that $J:y = y(t)$, $a \leqq t \leqq b$, and $a < c < b$, $J_1:y = y(t)$, $a \leqq t \leqq c$ and $J_2:y = y(t)$, $c \leqq t \leqq b$, then

$$j_\eta(J_1 + J_2) = j_\eta(J_1) + j_\eta(J_2).$$

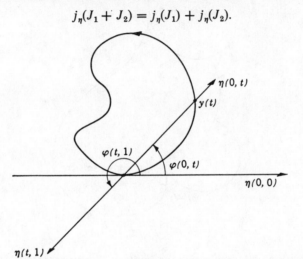

Figure 1.

Actually, if $\eta(t)$ is given, $j_\eta(J)$ has nothing to do with J but, in applications, $\eta(t)$ will be a "vector beginning at the point $y = y(t)$" of J.

The main interest below will be in the case that J is a Jordan curve, in which case it will always be assumed that J is positively oriented and $\eta \neq 0$ is a continuous vector on J [so that $y(a) = y(b)$ and $\eta(a) = \eta(b)$]. [Only Jordan curves $J:y = y(t)$ which are piecewise of class C^1 will occur, so that the positive orientation means that the normal vector $(-dy^2/dt, dy^1/dt) \neq 0$, defined except at corners of J, points into the interior of J.] It is clear that $j(J_\eta)$ is an integer. It is called the *index of η with respect to J*.

Theorem 2.1 (Umlaufsatz). *Let $J : y = y(t)$, $0 \leqq t \leqq 1$, be a positively oriented Jordan curve of class C^1 and $\eta(t) = dy/dt$ $(\neq 0)$ the tangent vector field on J. Then $j_\eta(J) = 1$.*

Proof. On the triangle $\Delta : 0 \leqq s \leqq t \leqq 1$, define $\eta(s, t) = [y(t) - y(s)] /\|\,y(t) - y(s)\|$ if $s \neq t$ or $(s, t) \neq (0, 1)$, $\eta(t, t) = y'(t)/\|y'(t)\|$, and

$\eta(0, 1) = -\eta(0, 0)$. It is clear that $\eta(s, t)$ is continuous and $\eta(s, t) \neq 0$ on Δ. Note that $\eta(0, t)$ and $\eta(t, 1)$ are oppositely oriented vectors; see Figure 1.

Suppose that the point $y = y(0)$ on J is chosen so that the tangent line through $y(0)$ is parallel to the y^1-axis and no part of J lies below this tangent line. Since Δ is simply connected, it is possible to define (uniquely) a continuous function $\varphi(s, t)$ such that $\varphi(0, 0) = 0$ and $\varphi(s, t)$ is an angle from the positive y^1-direction to $\eta(s, t)$. Then $2\pi j_\eta(J) = \varphi(1, 1) - \varphi(0, 0)$, as can be seen by considering $\varphi(t, t)$.

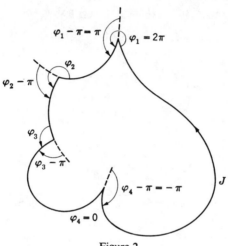

Figure 2.

The position of J implies that $0 \leq \varphi(0, t) \leq \pi$ and that $\varphi(0, 1)$ is an odd multiple of π, hence $\varphi(0, 1) = \pi$. Similarly, a consideration of $\varphi(s, 1) - \varphi(0, 1) = \varphi(s, 1) - \pi$ for $0 \leq s \leq 1$ shows that $\varphi(1, 1) - \pi = \pi$. Consequently, $\varphi(1, 1) = 2\pi$. Since $2\pi j_\eta(J) = \varphi(1, 1) - \varphi(0, 0) = \varphi(1, 1)$, the theorem is proved.

The "rounding off" of corners in the case of a J which is piecewise of class C^1 gives the following:

Corollary 2.1. *Let $J : y = y(t)$, $0 \leq t \leq 1$, be a positively oriented Jordan curve which is piecewise of class C^1 with corners at the t-values $(0 <) t_1 < \cdots < t_n (< 1)$ and $\eta(t) = dy/dt$ for $t \neq t_k$. Let $J_k : y = y(t)$, $t_{k-1} \leq t \leq t_k$, for $k = 1, \ldots, n + 1$ with $t_0 = 0$ and $t_{n+1} = 1$. Then $2\pi \sum_{k=1}^{n+1} j_\eta(J_k) + \sum_{k=1}^{n} (\varphi_k - \pi) = 2\pi$, where φ_k is the exterior angle $0 \leq \varphi_k \leq 2\pi$ at $y(t_k)$.*

Note that $\varphi_k - \pi$ is an angle, $-\pi \leq \varphi_k - \pi \leq \pi$, from $\eta(t_k - 0)$ to $\eta(t_k + 0)$; see Figure 2.

The essential idea in the proof of Theorem 2.1 is contained in:

Lemma 2.1. *Let $J : y = y(t)$, $a \leq t \leq b$, be a Jordan curve, $\xi(t)$ and $\eta(t)$ two vector fields on J which can be deformed into one another without vanishing. Then $j_\xi(J) = j_\eta(J)$.*

The possibility of a deformation without vanishing means the existence of a continuous vector $\eta = \eta(t, s)$ for $a \leq t \leq b$, $0 \leq s \leq 1$, such that $\eta(t, 0) = \xi(t)$, $\eta(t, 1) = \eta(t)$, $\eta(a, s) = \eta(b, s)$, and that $\eta(t, s) \neq 0$. For example, $\eta(t, s) = (1 - s)\xi(t) + s\eta(t)$ is such a deformation if $\xi(t), \eta(t)$ are not in opposite directions for any t.

Proof. Let $j(s)$ be the index of $\eta(t, s)$ for a fixed s. It is clear that $j(s)$ is a continuous function of s. Since $j(s)$ is an integer, it is constant. In particular, $j(0) = j(1)$.

3. Index of a Stationary Point

In what follows, $f(y) = f(y^1, y^2)$ is continuous on an open plane set E. As before, a point where $f = 0$ is called a stationary point and a point where $f \neq 0$ is a regular point.

Let $J : y = y(t)$, $a \leq t \leq b$, be an arc in E on which $f(y) \neq 0$. Define $j_f(J)$ to be $j_\eta(J)$, where $\eta(t) = f(y(t))$. For example, if $f(y), y(t)$ are of class C^1, then $j_f(J)$ is given by the line integral

$$(3.1) \qquad 2\pi j_f(J) = \int_J \frac{f^1 df^2 - f^2 df^1}{\|f\|^2}, \qquad \text{where} \quad f = (f^1, f^2).$$

When J is a positively oriented Jordan curve in E on which $f \neq 0$, the integer $j_f(J)$ is called the *index of f with respect to J*.

Lemma 3.1. *Let J_0 and J_1 be two Jordan curves in E which can be deformed into one another in E without passing through a stationary point. Then $j_f(J_0) = j_f(J_1)$.*

The assumption here means the existence of a continuous $y(t, s)$, $a \leq t \leq b$, $0 \leq s \leq 1$, such that (i) for a fixed s, $J(s) : y = y(t, s)$ is a Jordan curve in E; (ii) $J(0) = J_0$, $J(1) = J_1$; and (iii) $f(y(t, s)) \neq 0$. The proof is the same as that of Lemma 2.1.

Corollary 3.1. *Let J be a positively oriented Jordan curve in E such that the interior of J is in E and that $f(y) \neq 0$ on and inside J. Then $j_f(J) = 0$.*

Proof. Since the interior of a Jordan curve is simply connected, J can be deformed (in its interior) to a small circle J_1 around a point y_0 of its interior. Since $f(y_0) \neq 0$, it is clear that if the circle J_1 is sufficiently small, the change of the angle between $f(y)$ and the y^1-direction around J_1 is small. Since $j_f(J_1)$ is an integer, $j_f(J_1) = 0$. By Lemma 3.1, $j_f(J) = 0$.

Let $y_0 \in E$. Lemma 3.1 shows that the integer $j_f(J)$ is independent of the Jordan curve J in the class of curves J in E with interiors in E containing no stationary point except possibly y_0. This integer $j_f(J)$ is called the *index* $j_f(y_0)$ *of* y_0 with respect to f. By Corollary 3.1, $j_f(y_0) = 0$ if y_0 is a regular point. For this reason, only the *indices of isolated stationary points* y_0 are considered.

Corollary 3.2. *Let J be a positively oriented Jordan curve in E on which* $f(y) \neq 0$ *and let the interior of J be in E and contain only a finite number of stationary points* $y_1, \ldots, y_n$. *Then* $j_f(J) = j_f(y_1) + \cdots + j_f(y_n)$.

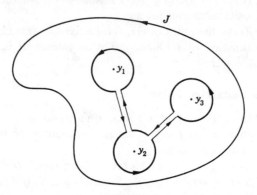

Figure 3.

For J can be deformed into a path consisting of circles around each stationary point and "cuts" between the circles traced in both directions; see Figure 3.

Exercise 3.1. Show that according as $ad - bc > 0$ or $ad - bc < 0$, the index of the origin with respect to $f_0(y) = (ay^1 + by^2,\ cy^1 + dy^2)$ is $+1$ or -1.

Exercise 3.2 (*Continuation*). Let $f_0(y)$ be as in the last exercise and $f_1(y)$ a continuous function defined for small $\|y\|$ such that $f_1(y)/\|y\| \to 0$ as $y \to 0$. Show that if $f(y) = f_0(y) + f_1(y)$, then $y = 0$ is an isolated stationary point and the index $j_f(0) = \pm 1$ according as $ad - bc \gtrless 0$.

Exercise 3.3. Let $f(y)$ be of class C^1 on an open set E with a Jacobian determinant $\det(\partial f/\partial y) = \partial(f^1, f^2)/\partial(y^1, y^2)$ different from 0 wherever $f = 0$. Let J be a positively oriented Jordan curve in E with interior I in E and $f(y) \neq 0$ on J. Show that there are at most a finite number of stationary points $y_1, \ldots, y_k$ in I and that $j_f(J) = n_+ - n_-$, where n_+ or n_- is the number of these points at which $\det(\partial f/\partial y) > 0$ or $\det(\partial f/\partial y) < 0$.

Theorem 3.1. *Let* $f(y)$ *be continuous on an open set E and let* $y = y_p(t)$ *be a solution of* $y' = f(y)$ *of period p,* $y_p(t + p) = y_p(t)$ *for* $-\infty < t < \infty$. *Let* $y = y_p(t)$, $0 \leq t \leq p$, *be a Jordan curve with an interior I contained in E and* $f(y_p(t)) \neq 0$. *Then I contains a stationary point.*

Proof.   Let J be the Jordan curve, $y = y_p(t)$, $0 \leq t \leq p$, with a positive orientation.   By Theorem 2.1, $j_f(J) = 1 \neq 0$.   Thus the theorem follows from Corollary 3.1.

A return to the study of stationary points and their indices will be made in § 6 below.

4. The Poincaré-Bendixson Theorem

The discussion of the differential equation $y' = f(y)$ in § 1 will now be continued for the plane case $(d = 2)$ with the aid of the Jordan curve theorem. The main result is the following theorem of Poincaré-Bendixson.

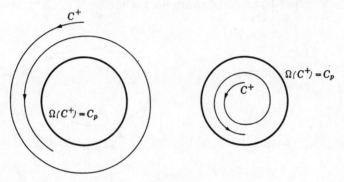

Figure 4.

Theorem 4.1.   *Let $f(y) = f(y^1, y^2)$ be continuous on an open plane set E and let $C^+ : y = y_+(t)$ be a solution of*

$$(4.1) \qquad\qquad y' = f(y)$$

for $t \geq 0$ with a compact closure in E. In addition, suppose that $y_+(t_1) \neq y_+(t_2)$ for $0 \leq t_1 < t_2 < \infty$ and that $\Omega(C^+)$ contains no stationary points. Then $\Omega(C^+)$ is the set of points y on a periodic solution $C_p : y = y_p(t)$ of (4.1). Furthermore, if $p > 0$ is the smallest period of $y_p(t)$, then $y_p(t_1) \neq y_p(t_2)$ for $0 \leq t_1 < t_2 < p$; i.e., $J : y = y_p(t)$, $0 \leq t \leq p$, is a Jordan curve.

In the case that initial value problems associated with (4.1) have unique solutions, either $y_+(t_1) \neq y_+(t_2)$ for $0 \leq t_1 < t_2 < \infty$ or $y_+(t)$ is periodic [i.e., $y_+(t + p) = y_+(t)$ for all t for some fixed positive number p]. In the latter case (excluded in Theorem 4.1), $\Omega(C^+)$ coincides with the set of points on C^+.

The proof of Theorem 4.1 will show that C^+ is a spiral which tends to the closed curve $\Omega(C^+) : y = y_p(t)$ either from the exterior or from the interior; see Figure 4. It will have the following consequence:

Corollary 4.1. *Assume the conditions of Theorem 4.1 and let $p > 0$ be a period of $y = y_p(t)$. Then there exists a sequence $(0 \leqq) t_1 < t_2 < \dots$ such that*

(4.2) $y_+(t + t_n) \to y_p(t)$ as $n \to \infty$

uniformly for $0 \leqq t \leqq p$ and

(4.3) $t_{n+1} - t_n \to p$ as $n \to \infty$.

Proof of Theorem 4.1.   A closed, bounded line segment L in E is called *transversal* to (4.1) if $f(y) \neq 0$ for $y \in L$ and the direction of $f(y)$ at points $y \in L$ are not parallel to L. All crossings of L by a solution $y = y(t)$ of $y' = f$ are in the same direction with increasing t.

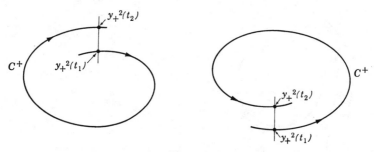

Figure 5.

The proof will be divided into steps (a)–(e).

(a) Let $y_0 \in E$, $f(y_0) \neq 0$, L a transversal through y_0. Then Peano's existence theorem implies that there is a small neighborhood E_0 of y_0 and an $\epsilon > 0$ such that any solution $y = y^0(t)$ of the initial value problem $y' = f$, $y(0) = y^0$ for $y^0 \in E_0$ exists for $|t| \leqq \epsilon$ and crosses L exactly once for $|t| \leqq \epsilon$. In fact, if $\delta > 0$ is arbitrary, E_0 and ϵ can be chosen so that $y^0(t)$ exists and differs from $y^0 + tf(y_0)$ by at most $\delta |t|$ for $|t| \leqq \epsilon$. Thus if E_0 is sufficiently small, $y = y^0(t)$ crosses L at least once, but can cross L at most once for $|t| \leqq \epsilon$ since crossings of L are in the same direction.

In particular, it follows that if $y = y^0(t)$ is a solution of $y' = f$ on a closed bounded interval, then $y = y^0(t)$ has at most a finite number of crossings of L.

(b) Let L be a transversal which, without loss of generality, can be supposed to be on the y^2-axis, where $y = (y^1, y^2)$. Suppose that $y = y_+(t)$ crosses L at t-values $t_1 < t_2 < \dots$, then $y_+^2(t_n)$ is strictly monotone in n.

In order to see this, suppose without loss of generality that crossings of L occur with increasing y^1 (i.e., y^1 changes from negative to positive values at crossings). Consider the case that $y_+^2(t_1) < y_+^2(t_2)$; see Figure 5.

The set consisting of the arc $y = y_+(t)$, $t_1 \leq t \leq t_2$, and the line segment $y_+{}^2(t_1) \leq y^2 \leq y_+{}^2(t_2)$ on the y^2-axis forms a Jordan curve J. For all $t > t_2$, $y = y_+(t)$ is in the exterior of J or in the interior of J, by the assumption on $y_+(t)$ and the fact that crossings of L occur only in one direction. This makes it clear that $y_+{}^2(t_2) < y_+{}^2(t_3)$ and the argument can bé repeated.

(c) It will now be verified that if L is a transversal, $\Omega(C^+)$ contains at most one point on L. For if $y_0 \in L \cap \Omega(C^+)$, part (a) implies that $y = y_+(t)$ crosses L infinitely many times [in fact, whenever $y_+(t)$ comes near y_0]. With increasing t, the intersections of $y = y_+(t)$ and L tend monotonously along L to y_0, by part (b). Thus $L \cap \Omega(C^+)$ cannot contain any other point $y = y_0$.

(d) Since C^+ is bounded, $\Omega(C^+)$ is not empty. Let $y_0 \in \Omega(C^+)$. By Theorem 1.2, $y' = f, y(0) = y_0$ has a solution $C_0 : y = y_0(t)$, $-\infty < t < \infty$, contained in $\Omega(C^+)$; thus $\Omega(C_0) \subset \Omega(C^+)$.

$\Omega(C_0)$ is not empty. Let $y^0 \in \Omega(C_0)$, so that y^0 is a regular point since $\Omega(C^+)$ contains no stationary points. Thus there is a transversal L^0 through y^0 and $y = y_0(t)$ has infinitely many crossings of L^0 near y^0, but y^0 and every such crossing is a point of $\Omega(C^+)$. By (c), these points coincide. In particular, there exist points $t_1 < t_2$ such that $y^0 = y_0(t_1) = y_0(t_2)$. It follows that (4.1) has a periodic solution $y = y_p(t)$ of period $p = t_2 - t_1$ such that $y_p(t) = y_0(t)$ for $t_1 \leq t \leq t_2$. Since $y_0(t)$ is not constant on any t-interval, it can be supposed that $y_p(t^0) \neq y_p(t_0)$ for $0 \leq t_0 < t^0 < p$.

(e) It must be shown that $\Omega(C^+)$ coincides with its subset $C_p : y = y_p(t)$, $-\infty < t < \infty$. If not, $\Omega(C^+) - C_p$ is not empty. Then C_p contains a point y_1 which is a cluster point of $\Omega(C^+) - C_p$, since $\Omega(C^+)$ is connected by Theorem 1.1. Let L_1 be a transversal through y_1. Any small sphere about y_1 contains points $y_2 \in \Omega(C^+) - C_p$. For any such y_2, $y' = f$ has a solution $y = y_2(t)$, $-\infty < t < \infty$, such that $y_2(0) = y_2$ and $y_2(t)$ is contained in $\Omega(C^+)$, by Theorem 1.2. If y_2 is sufficiently close to y_1, then $y_2(t)$ crosses the transversal L_1. The crossing is necessarily at the point y_1 by part (c).

Since $y_2 \notin C_p$, this is impossible when solutions of initial value problems belonging to (4.1) are unique. That it is impossible in the general case can be seen as follows: Let $y_2(t_p) \in C_p$, while $y_2(t) \notin C_p$ for t between 0 and t_p. Since $y_2(t_p)$ is a regular point, there is a transversal L_p through $y_2(t_p)$. Then a small translation of L_p in a suitable direction is a transversal L_{p0} which meets C_p and $y = y_2(t)$ in two distinct points; see Figure 6. This contradicts part (c) and proves the theorem.

Remark. For subsequent use, note that the argument in part (e) shows that whenever $C^+ : y = y_+(t)$, $t \geq 0$, has the property that $y_+(t_1) \neq y_+(t_2)$ for $t_1 \neq t_2$ and $y_0 \in \Omega(C^+) \cap E$ is a regular point, then there is a neighborhood E_0 of y_0 such that the solution of $y' = f, y(0) = y_0$ in $\Omega(C^+) \cap E_0$ is unique. If, in addition, $\Omega(C^+)$ is connected and there is a periodic orbit $C_p: y = y_p(t)$ in $\Omega(C^+)$ consisting only of regular points, then $\Omega(C^+) = C_p$.

Proof of Corollary 4.1. Let $y^0 = y_p(0)$ and let L^0 be a transversal through y^0. Let the successive crossings of L^0 by $y = y_+(t)$ occur at $(0 \leqq)$ $t_1 < t_2 < \ldots$ Then $y_+(t_n)$ tends monotonically along L^0 to y^0. Since $y = y_p(t)$ is the unique solution of $y' = f, y(0) = y^0$ in $\Omega(C^+)$, an analogue of Remark 2 following Theorem 1.2 shows that (4.2) holds uniformly for bounded t-intervals, in particular for $0 \leqq t \leqq p$.

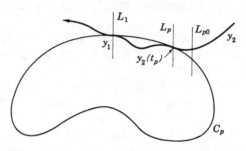

Figure 6.

Note that $y_+(t_n + p) \to y_p(p) = y^0, n \to \infty$. Thus if $\epsilon > 0$ and n is large, $y_+(t)$ crosses L^0 in the interval $[t_n + p - \epsilon, t_n + p + \epsilon]$. Hence $t_{n+1} \leqq t_n + p + \epsilon$. Also $\|y_+(t_n + t) - y_p(t)\|$ is small for large n, $0 < \epsilon \leqq t \leqq p - \epsilon < p$, which implies that there is a $\delta > 0$ such that $\|y_+(t_n + t) - y^0\| \geqq \delta$ for $0 < \epsilon \leqq t \leqq p - \epsilon$. In particular, there is no crossing of L^0 for $\epsilon \leqq t \leqq p - \epsilon$. Hence $t_{n+1} \geqq t_n + p - \epsilon$ for large n. This proves the corollary.

Theorem 4.2. *Let f, C^+ be as in Theorem 4.1 except that $\Omega(C^+)$ contains a finite number n of stationary points of (4.1). If $n = 0$, Theorem 4.1 applies. If $n = 1$ and $\Omega(C^+)$ is a point, Corollary 1.1 applies. If $1 \leqq n < \infty$ and $\Omega(C^+)$ is not a point, then $\Omega(C^+)$ consists of stationary points $y_1, y_2, \ldots,$ y_n and a finite or infinite sequence of orbits $C_0: y = y_0(t), -\infty \leqq \alpha_- < t < \alpha_+ \leqq \infty$, which do not pass through a stationary point but $y_0(\alpha_{\pm}) = \lim y(t)$, as $t \to \alpha_{\pm}$, exist and are among the set $y_1, \ldots, y_n$.*

It is possible that $y_0(\alpha_+) = y_0(\alpha_-)$. It is not claimed that (α_-, α_+) is the maximal interval of existence of $y = y_0(t)$. But when initial conditions uniquely determine solutions of (4.1), so that the only solution of (4.1) through a stationary point y_k is $y(t) \equiv y_k$, then it follows that $\alpha_- = -\infty$ and $\alpha_+ = \infty$; cf. Lemma II 3.1.

Proof. Consider the case that $n \geqq 1$ and $\Omega(C^+)$ is not a point. Since $\Omega(C^+)$ is connected by Theorem 1.1, it contains regular points y_0. For any such point y_0, there exist solutions $C_*:y = y_*(t)$, $-\infty < t < \infty$, of $y' = f$, $y(0) = y_0$ in $\Omega(C^+)$. Let C_* denote any such solution.

Consider only $C_*^+:y = y_*(t)$ for $t \geqq 0$. The treatment of $t \leqq 0$ is similar. There are two cases: (i) there is a first positive $t = \alpha_+$, where $y_*(\alpha_+)$ is a stationary point (hence one of the points $y_1, \ldots, y_n$) or (ii) $y_*(t)$ is not a stationary point for any finite $t \geqq 0$.

Consider case (ii). If $\Omega(C_*^+)$ contains a regular point y^0, then, by part (d) in the proof of Theorem 4.1, C_*^+ contains a periodic solution path $C_p:y = y_p(t)$. But then C_p contains a stationary point y_i; otherwise, by part (e) and the Remark at the end of the proof of Theorem 4.1, $\Omega(C^+) = C_p$. This is impossible since $y_*(t) \not\equiv y_i$ for $t \geqq 0$. Hence in case (ii), $\Omega(C_*^+)$ can contain only stationary points and, since it is connected, only one stationary point y_*.

Thus, in case (ii), $\Omega(C_*^+)$ is a stationary point y_* and $y_*(t) \to y_*$ as $t \to \infty$ by Corollary 1.1. Thus any regular point $y_0 \in \Omega(C^+)$ is on an arc $C_0:y = y_0(t)$, $\alpha_- < t < \alpha_+$, in $\Omega(C^+)$ of the type specified.

It remains to show that the set of such arcs C_0 is at most denumerable. Note that if $y_0 \in \Omega(C^+)$ is a regular point, the solution $C_*:y = y_*(t)$ above is unique on a sufficiently small interval $|t| < \epsilon$; cf. part (e) and the Remark following the proof of Theorem 4.1. Thus, no two of the arcs C_0 can meet.

Since $y_+(t_1) \neq y_+(t_2)$ for $t_2 > t_1$, it can be supposed that $y_+(t)$ is not one of the stationary points $y_1, \ldots, y_n$ for $t \geqq 0$. Otherwise, $y_+(t)$, $t \geqq 0$, is replaced by $y_+(t)$, $t \geqq t_0$, for a suitable $t_0 > 0$. Also if $y_0 \in \Omega(C^+)$ is a regular point, then $y_+(t) \neq y_0$ for $t \geqq 0$ for the sequence of intersections of $y = y_+(t)$ with a transversal through y_0 tends to y_0 in a strictly monotone sense; cf. part (b) of the proof of Theorem 4.1. Thus C^+ and C_0 have no points in common.

Suppose, if possible, that there exists a nondenumerable set of C_0 which can then be assumed to join the same, not necessarily distinct, pair of stationary points. Any one or two of these joined together forms a Jordan curve J. Since there is at most a denumerable set of J with pairwise disjoint interiors, there exist three such distinct J, say J_1, J_2, J_3, such that J_3 is contained in the closure of the interior I_n of J_n for $n = 1, 2$. This is impossible, for since C^+ does not meet J_n, C^+ is either between J_1 and J_2 or J_2 and J_3.

In the next theorem, the assumption "$y_+(t_1) \neq y_+(t_2)$ for $t_1 \neq t_2$" is omitted.

Theorem 4.3. *Let $f(y) = f(y^1, y^2)$ be continuous on an open plane set E and $C^+:y = y_+(t)$ a solution of (4.1) for $t \geqq 0$ with compact closure in E. Then $\Omega(C^+)$ contains a closed (periodic) orbit $C_p:y = y_p(t)$ of (4.1) which can reduce to a stationary point $y_p(t) \equiv y_0$.*

Proof. Suppose that $\Omega(C^+)$ does not contain a stationary point. Let $y_0 \in \Omega(C^+)$ and $C_0^+ : y = y_0(t)$, $0 \leq t < \infty$, be a solution supplied by Theorem 1.2, so that $C_0^+ \subset \Omega(C^+)$. Since $\Omega(C^+)$ is closed, $\Omega(C_0^+) \subset \Omega(C^+)$. If $y_0(t_1) \neq y_0(t_2)$ for $0 \leq t_1 < t_2 < \infty$, then $\Omega(C_0^+)$ is a closed orbit $y = y_p(t)$ by Theorem 4.1. If $y_0(t_1) = y_0(t_2)$ for certain t_1, t_2 with $0 \leq t_1 < t_2 < \infty$, then the orbit C_0^+ contains the periodic solution path $y = y_p(t)$ of period $p = t_2 - t_1$ which coincides with $y_0(t)$ for $t_1 \leq t \leq t_2$. In either case $\Omega(C^+)$ contains a closed orbit $y = y_p(t)$.

Theorem 4.4. *Let $f(y)$ be continuous on an open, simply connected plane set E where $f(y) \neq 0$ and $y = y(t)$, a solution of (4.1) on its maximal interval of existence (ω_-, ω_+). Then $y = y(t)$ does not remain in any compact subset E_0 of E as $t \to \omega_+$ [or $t \to \omega_-$].*

Proof. If, e.g., $C^+ : y = y(t)$, $t_0 \leq t < \omega_+$, is in a compact set E_0 in E for some t_0, then $\omega_+ = \infty$ by Theorem II 3.1 and $\Omega(C^+)$ contains a periodic solution $y = y_p(t)$ by the last theorem. Since $f \neq 0$ on E, $y_p(t)$ does not reduce to a constant on any t-interval. Let t_1 be the first $t > t_0$, where $y_p(t_1) = y_p(t_0)$. Then $y = y_p(t)$, $t_0 \leq t \leq t_1$, is a Jordan curve J. Thus (4.1) has a periodic solution $y = y_0(t)$ of period $t_1 - t_0$ such that $y_0(t) \equiv y_p(t)$ for $t_0 \leq t \leq t_1$. Since E is simply connected, the interior of $J : y = y_0(t)$, $t_0 \leq t \leq t_1$, is contained in E. By Theorem 3.1, this interior contains a stationary point. This contradicts $f \neq 0$ and proves the theorem.

5. Stability of Periodic Solutions

Return to the situation in Theorem 4.1.

Theorem 5.1. *Let f, C^+, C_p be as in Theorem 4.1 Then there exists an $\epsilon > 0$ such that if y_0 is within a distance ϵ of $\Omega(C^+) = C_p : y = y_p(t)$ and on the same side (interior or exterior) of C_p as C^+, then*

$$(5.1) \qquad y' = f(y), \qquad y(0) = y_0$$

has a solution $C_0 : y = y_0(t)$ for $t \geq 0$ such that $y_0(t_1) \neq y_0(t_2)$ for $t_1 \neq t_2$ and $\Omega(C_0^+) = C_p$.

Proof For sake of definiteness, consider the case that C^+ is exterior to C_p. Let $y^0 = y_p(0)$, L^0 a transversal through y^0, the successive crossings of L^0 by $y_+(t)$ occur at $(0 \leq) t_1 < t_2 < \dots$. Thus $y_+(t_n)$ tends monotonously along L^0 to y^0.

Let J_n be the Jordan curve consisting of the arc $y = y_+(t)$, $t_n \leq t \leq t_{n+1}$, and the open segment I_n of L^0 joining $y_+(t_n)$, $y_+(t_{n+1})$. Let D_n be the interior of J_n and $E_n = D_n - \bar{D}_{n+1}$. Note that E_n is a simply connected open set in E since J_{n+1}, except for the point $y_+(t_{n+1})$, is interior to J_n; see Figure 7.

It is clear that if $\epsilon > 0$ is sufficiently small, then $\cup E_n$ contains all points $y_0 \notin C^+ \cup C_p$ within a distance ϵ of C_p on the same side as C^+. It is also clear from Corollary 4.1 that the union $\cup E_n$, $n \geqq N$, for large N is within a distance ϵ of C_p. Since $f(y) \neq 0$ for $y \in C_p$, it can be supposed that $\cup E_n$ contains no stationary points; otherwise, $t_1, t_2, \ldots$ is replaced by $t_N, t_{N+1}, \ldots$.

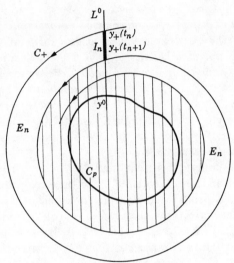

Figure 7.   Shaded area is D_{n+1}.

In the proof, it is sufficient to consider only those $y_0 \notin C^+ \cup C_p$, so that $y_0 \in E_n$ for some n. Let $y = y_0(t)$ be a solution of (5.1) for t near 0. By Theorem 4.4, a continuation of this solution for increasing t meets the boundary ∂E_n of E at a finite t-value. Let t_0 be the first $t_0 > 0$ at which $y_0(t_0) \in \partial E_n$, where $\partial E_n \subset C^+ \cup I_n \cup I_{n+1}$. It can be supposed that $y_0(t^1) \neq y_0(t^2)$ for $0 \leqq t^1 < t^2 \leqq t_0$. Because of the direction of crossings of solutions on I_n, $y_0(t_0) \notin I_n$ so that $y_0(t_0) \in C^+ \cup I_{n+1}$. If $y_0(t_0) \in C^+$, say $y_0(t_0) = y^+(t^0)$, then $y_0(t)$ can be defined for $t \geqq t_0$ to be $y_0(t) = y^+(t + t^0 - t_0)$. If $y_0(t_0) \in I_{n+1}$, then $y_0(t)$ exists for $t(> t_0)$ near t_0 and is E_{n+1}.

Continuing this procedure, we obtain $y_0(t)$ defined for $t \geqq 0$ so that either $y_0(t) = y^+(t + a)$ for some a and large t or the solution $y = y_0(t)$ with increasing t successively passes through $E_n, E_{n+1}, \ldots$. This completes the proof.

Let $C_p : y = y_p(t)$ be a periodic solution of $y' = f$ of period $p > 0$ such that $J : y = y_p(t)$, $0 \leqq t \leqq p$, is a Jordan curve. C_p is called *orbitally stable from the exterior as* $t \to +\infty$ if, for every $\epsilon > 0$, there exists a $\delta = \delta_\epsilon > 0$

with property that if y_0 is exterior to but within a distance δ of J, then all solutions $C_0^+ : y = y_0(t)$ of (5.1) exist and remain within a distance ϵ of J for $t \geqq 0$. C_p is called *asymptotically orbitally stable from the exterior as* $t \to \infty$, if there exists a $\delta > 0$ such that if y_0 is exterior to but within a distance δ of J, then all solutions $C_0^+ : y = y_0(t)$ of (5.1) exist for $t \geqq 0$ and $\Omega(C_0^+) = J$. Similar definitions hold with "exterior" replaced by "interior" and/or "$t \to \infty$" by "$t \to -\infty$".

Theorem 5.2. *Let* $f(y)$ *be continuous on an open plane set* E *with the property that initial values determine unique solutions of* $y' = f$. *Let* $C_p : y = y_p(t)$ *be a periodic solution of* $y' = f$ *with a least positive period* p. (i) *Then* C_p *is asymptotically orbitally stable from the exterior as* $t \to \infty$ *if and only if the orbit* C_p *is* $\Omega(C_0^+)$ *for some solution* $C_0^+ : y = y_0(t)$, $t \geqq 0$, *on the exterior of* C_p. (ii) *Then* C_p *is orbitally stable from the exterior as* $t \to \infty$ *if and only if either the orbit* C_p *is* $\Omega(C_0^+)$ *for some solution* C_0^+ *exterior to* C_p *or, for every* $\epsilon > 0$, *there is a periodic solution of* (1.1) *exterior to and within a distance* ϵ *of* C_p.

Proof. In (i), "only if" is trivial and "if" follows from Theorem 5.1. In (ii), "if" is clear from Theorem 5.1. In order to prove the converse, suppose that C_p is orbitally stable from the exterior. Since solutions of initial value problems (5.1) are unique, $f(y) \neq 0$ on C_p and, hence, in some ϵ_0-vicinity of C_p. Let $C_0^+ : y = y_0(t)$, $t \geqq 0$, be a solution of $y' = f$, exterior to but within a distance $\epsilon < \epsilon_0$ of C_p. Then C_0^+ has a compact closure in E and $\Omega(C_0^+)$ contains no stationary points, so that $\Omega(C_0^+)$ is a periodic solution path of $y' = f$ by Theorem 4.1. Thus either $\Omega(C_0^+) = C_p$ or $\Omega(C_0^+)$ is periodic solution path exterior to and within a distance ϵ of C_p.

6. Rotation Points

In §§ 6–9, the behavior of solutions of

$$(6.1) \qquad y' = f(y), \qquad y(0) = y_0$$

near an isolated stationary point will be considered. Let $f(y)$ be defined for small $\|y\|$, say, on an open set containing $\|y\| \leqq b$, and let

$$(6.2) \qquad f(0) = 0 \quad \text{and} \quad f(y) \neq 0 \quad \text{if} \quad y \neq 0.$$

Note that if (6.1) has a unique solution for all y_0, then Theorems 4.1 and 4.2 imply that if $0 < \|y_0\| < b$ and $C_0^+ : y = y_0(t)$ is the solution of (6.1) on the maximal interval $[0, \omega_+)$, then only the following (not mutually exclusive) cases can occur: (i) there is a least t_0, $0 < t_0 < \omega_+$, such that $\|y_0(t_0)\| = b$; (ii) $\omega_+ = \infty$ and the solution path C_0^+ is a Jordan curve with $y = 0$ in its interior; (iii) $\omega_+ = \infty$ and C_0^+ is a spiral approaching such a closed orbit; (iv) $\omega_+ = \infty$ and $y_0(t) \to 0$ as $t \to \infty$; and (v) $\omega_+ = \infty$, C_0^+ is a spiral around $y = 0$ and $\Omega(C_0^+)$ consists of a finite or infinite sequence of orbits $y(t)$, $-\infty < t < \infty$, such that $y(t) \to 0$ as $t \to \pm \infty$.

By a *spiral* $y = y_0(t)$, $0 \leqq t < \omega_+$, around $y = 0$ is meant an arc $y_0(t) \neq 0$ such that a continuous determination of arc $\tan y_0^2(t)/y_0^1(t)$ tends to either ∞ or $-\infty$ as $t \to \omega_+$.

If every neighborhood of $y = 0$ contains closed orbits surrounding $y = 0$, the stationary point $y = 0$ is called a *rotation point*.

When $y = 0$ is a rotation point and solutions of arbitrary initial value problems (6.1) are unique, the set of solutions of $y' = f$ in a neighborhood of $y = 0$ can be described as follows: there is a neighborhood E_0 of $y = 0$ such that the solution $C_0 : y = y_0(t)$ of (6.1) for every $y_0 \in E_0$ is either a closed orbit surrounding $y = 0$ or is spiral such that $\Omega(C_0)$, $A(C_0)$ are closed orbits surrounding $y = 0$.

This is illustrated by the following: Consider the differential equations

$$(6.3) \qquad y^{1'} = y^1 u(r) - y^2 v(r), \qquad y^{2'} = y^1 v(r) + y^2 u(r),$$

where u, v are continuous, real-valued functions of $r = \|y\|$ for small $r \geqq 0$. In polar coordinates, these equations become

$$(6.4) \qquad\qquad r' = ru(r), \qquad \theta' = v(r).$$

Example 1. Suppose that $u(r) = 0$, $v(r) \equiv \beta \neq 0$. Then (6.3) is

$$(6.5) \qquad\qquad y^{1'} = -\beta y^2, \qquad y^{2'} = \beta y^1$$

and (6.4) is $r' = 0$, $\theta' = \beta$, so that all orbits ($y \not\equiv 0$) are circles.

Example 2. Suppose that $u(r) = r \sin(1/r)$, $v(r) = 1$. Then (6.4) is $r' = r^2 \sin(1/r)$, $\theta' = 1$. Thus, besides the trivial solution $y \equiv 0$, there are closed orbits $r = 1/n\pi$, $n = 1, 2, \ldots$. Between the orbits, $r = 1/n\pi$ and $r = 1/(n + 1)\pi$, $(-1)^n r' > 0$ and so the corresponding orbits are spirals which tend to the circles $r = 1/n\pi$, $r = 1/(n + 1)\pi$ as $t \to \infty$, $t \to -\infty$ or $t \to \infty$, $t \to -\infty$, depending on the parity of n.

Example 3. In the last example, $u(r)$ can be redefined to be 0, say, between $r = 1/n\pi$ and $r = 1/(n + 1)\pi$ for a finite or infinite sequence of n-values $1, 2, \ldots$. Correspondingly, the spiral orbits between $r = 1/n\pi$ and $r = 1/(n + 1)\pi$ are replaced by circular orbits.

Exercise 6.1. Let C be a closed set on $0 \leqq r \leqq 1$. Show that there exist functions $u(r)$, $v(r)$ which are uniformly Lipschitz continuous on $0 \leqq r \leqq 2$, $u^2(r) + v^2(r) \neq 0$ for $r \neq 0$, and the solution of (6.3) with initial condition $y(0) = y_0$ is a closed orbit if $0 < \|y_0\| \in C$ and is a spiral if $\|y_0\| \notin C$, $0 < \|y_0\| \leqq 1$.

A rotation point $y = 0$ such that all orbits, except $y \equiv 0$, in a vicinity of $y = 0$ are closed curves is called a *center*. The simplest illustration of a center is the linear system (6.5) in Example 1.

7. Foci, Nodes, and Saddle Points

Assume that the only solution of $y' = f(y)$, $y(0) = 0$ is $y \equiv 0$ [so that no solution $y(t) \not\equiv 0$ can tend to 0 as t tends to a finite value].

The simplest nonrotation points are called "attractors." The isolated stationary point $y = 0$ is called an *attractor* for $t = \infty$ [or $t = -\infty$] if all solutions $y = y_0(t)$ of (6.1) for small $\|y_0\|$ exist for $t \geqq 0$ [or $t \leqq 0$] and $y_0(t) \to 0$ as $t \to \infty$ [or $t \to -\infty$]. If, in addition, all orbits $y_0(t) \not\equiv 0$ are

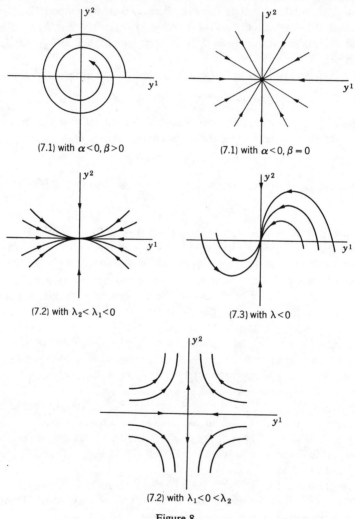

(7.1) with $\alpha < 0, \beta > 0$

(7.1) with $\alpha < 0, \beta = 0$

(7.2) with $\lambda_2 < \lambda_1 < 0$

(7.3) with $\lambda < 0$

(7.2) with $\lambda_1 < 0 < \lambda_2$

Figure 8.

spirals, then the attractor $y = 0$ is called a *focus*. If all orbits, $y_0(t) \not\equiv 0$ have a tangent at $y = 0$; i.e., if a continuous determination of $\theta(t) =$ arc tan $y_0^2(t)/y_0^1(t)$ tends to a limit θ_0, $-\infty < \theta_0 < \infty$, then the attractor $y = 0$ is called a *node*. A node is called a *proper node* if for every θ_0 mod 2π, there is a unique solution $y = y_0(t)$ such that $\theta(t) \to \theta_0$; otherwise it is called an *improper node*.

Illustration of these cases of attractors are given by real linear equations; see Figure 8. The system

(7.1) $$y^{1\prime} = \alpha y^1 - \beta y^2, \qquad y^{2\prime} = \beta y^1 + \alpha y^2$$

is an attractor for $t = \infty$ [or $t = -\infty$] if $\alpha < 0$ [or $\alpha > 0$]. It is a *focus* if $\alpha, \beta \neq 0$ and a *proper node* if $\alpha \neq 0$, $\beta = 0$. In the case,

(7.2) $$y^{1\prime} = \lambda_1 y^1, \qquad y^{2\prime} = \lambda_2 y^2$$

$y = 0$ is an *improper node* if $\lambda_1 \lambda_2 > 0$, $\lambda_1 \neq \lambda_2$. The case

(7.3) $$y^{1\prime} = \lambda y^1, \qquad y^{2\prime} = y^1 + \lambda y^2,$$

where $\lambda \neq 0$ is also an improper node.

There are nonrotation points which are not attractors and attractors which are neither foci nor nodes. The simplest example of a stationary point which is not an attractor is a *saddle point*. This is a stationary point $y = 0$ with the property that only a finite number of orbits tend to 0 as $t \to +\infty$ or $t \to -\infty$. This is illustrated by (7.2) where λ_1, λ_2 are real and $\lambda_1 \lambda_2 < 0$.

Exercise 7.1.   Verify the statements just made about (7.1)–(7.3).

Exercise 7.2.   Consider the linear system $y' = Ay$, where A is a real constant 2×2 matrix with det $A \neq 0$, so that $y = 0$ is the only stationary point. Let λ_1, λ_2 be the characteristic values for A. Show that $y = 0$ is an attractor for $t = \infty$ [or $t = -\infty$] if and only if Re $\lambda_k < 0$ [or > 0] for $k = 1, 2$; $y = 0$ is a center if and only if Re $\lambda_1 = $ Re $\lambda_2 = 0$; $y = 0$ is a focus if and only if λ_1, λ_2 are complex conjugates, but not real or purely imaginary; $y = 0$ is a proper node if $\lambda_1 = \lambda_2$ and the elementary divisors are simple; $y = 0$ is an improper node if $\lambda_1, \lambda_2 > 0$ but either $\lambda_1 \neq \lambda_2$ or $\lambda_1 = \lambda_2$ and the elementary divisor is multiple.

8. Sectors

The general nonrotation point will now be considered. It will be convenient to have the following terminology: A solution $y = y(t) \not\equiv 0$ of $y' = f$ defined on an interval $[0, \omega)$ (or an interval $(-\omega, 0]$) for $0 < \omega \leq \infty$ is called a *positive* [or *negative*] *null solution* if $y(t) \to 0$ as $t \to \omega$ [or $-\omega$]. When the solution of $y' = f, y(0) = 0$ is unique, then necessarily $\omega = \infty$.

Lemma 8.1. *Let $f(y)$ be continuous for small $\|y\|$, $f(0) = 0$, and $f(y) \neq 0$ for $y \neq 0$. Suppose that $y = 0$ is not a rotation point. Then there exists at least one null solution.*

Proof. Suppose that $\epsilon > 0$ is so small that there is no closed orbit in $\|y\| \leq \epsilon$ surrounding $y = 0$ and suppose, if possible, that there is no null solution in $\|y\| \leq \epsilon$.

Then there is no solution $C_0 : y = y_0(t) \neq 0$ of $y' = f$ defined and satisfying $\|y_0(t)\| \leq \epsilon$ for $t \geq 0$. For otherwise $y_0(t_1) \neq y_0(t_2)$ when $t_1 \neq t_2$ (since there are no closed orbits in $\|y\| \leq \epsilon$) and Theorem 4.2 implies that there exists at least one positive null curve. Similarly, no solution $y = y_0(t) \neq 0$ if $y' = f$ exists and satisfies $\|y_0(t)\| \leq \epsilon$ for $t \leq 0$.

Thus, if $\|y_0\| < \epsilon$ and $y_0(t)$ is a solution of (6.1), there is a bounded interval $-s \leq t \leq 0$ such that $y_0(0) = y_0$, $\|y_0(t)\| < \epsilon$ for $-s < t \leq 0$ and $\|y_0(-s)\| = \epsilon$. Correspondingly, the solution $y = y_0(t - s)$ is defined for $0 \leq t \leq s$, $\|y_0(t - s)\| < \epsilon$ for $0 < t \leq s$, $\|y_0(t - s)\| = \epsilon$ for $t = 0$ and $y_0(t - s) = y_0$ for $t = s$.

By considering a sequence of points $y_0 = y_1, y_2, \ldots$ tending to the origin, we obtain a sequence of solutions $y = y_n(t)$, $0 \leq t \leq s_n$, such that $\|y_n(0)\| = \epsilon$, $\|y_n(t)\| < \epsilon$ for $0 < t \leq s_n$, $\|y_n(s_n)\| \to 0$ as $n \to \infty$. After a selection of a subsequence and renumbering, it can be supposed that $y_0 = \lim y_n(0)$ and $\omega = \lim s_n$, $n \to \infty$, exist, where $\|y_0\| = \epsilon$ and $0 \leq \omega \leq \infty$. Also, if $\omega > 0$, it can be supposed that $y_0(t) = \lim y_n(t)$, $n \to \infty$, exists uniformly on every closed bounded interval of $[0, \omega)$ and is a solution of $y' = f$.

It is clear that $\omega > 0$. For otherwise, for large n, $y = y_n(t)$ is in a small vicinity of y_0 for $0 \leq t \leq s_n$; thus $y_n(s_n) \to 0$, $n \to \infty$, is impossible. If $0 < \omega < \infty$, then Peano's existence theorem shows that, for large n, $y_n(t)$ can be defined on an interval containing $0 \leq t \leq \omega$, and $y_0(t) = \lim y_n(t)$ exists uniformly for $0 \leq t \leq \omega$; thus, $y_0(\omega) = \lim y_n(s_n) = 0$. Finally, if $\omega = \infty$, then $y_0(t)$ is defined and $\|y_0(t)\| \leq \epsilon$ for $t \geq 0$. This is a contradiction and proves the lemma.

Hypothesis. *In what follows, assume that solutions of arbitrary initial value problems $y' = f(y)$, $y(0) = y_0$ are unique.*

Let C be a positively oriented Jordan curve surrounding $y = 0$. A solution $y = y(t)$ of $y' = f(y)$ is called a positive or negative *base solution* for C if $y(t)$ is defined for either $t \geq 0$ or $t \leq 0$, $y(0) \in C$, $y(t)$ is interior to C for $t \neq 0$, and $y(t)$ is a null solution.

Let $y = y_1(t)$, $y_2(t)$ be base solutions for C. The open subset S of the interior of C with boundary consisting of $y = 0$, the arcs $y = y_1(t)$, $y_2(t)$ and the (oriented closed) subarc C_{12} from $y_1(0)$ to $y_2(0)$ will be called the *sector* of C [determined by the ordered pair $y = y_1(t)$, $y = y_2(t)$]. It is not excluded that $y_1(0) = y_2(0)$ so that C_{12} can be C or reduce to a point.

Consider the case that there exists a solution $y = y_0(t)$, $-\infty < t < \infty$, of $y' = f$ which is interior or on C for all t and $y_0(t + t_1) \equiv y_1(t)$ for $t \geqq 0$, $y_0(t + t_2) = y_2(t)$ for $t \leqq 0$ for some $t_1, t_2 (\leqq t_1)$; see Figure 9. The point $y = 0$ and the arc $y = y_0(t)$, $-\infty < t < \infty$, form a Jordan

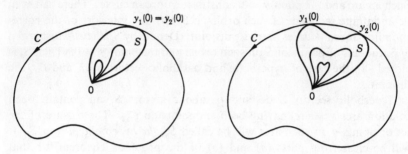

Figure 9.   Elliptic sectors.

curve J with interior I. If S contains I, then it is called an *elliptic sector*. When $t_1 = t_2$ [so that $y_1(0) = y_2(0) = y_0(t_1)$], and C_{12} reduces to the point $y_0(t_1)$, then S is elliptic and coincides with I. When $t_1 \neq t_2$, S can contain points not in I. By considering the possibilities (i)–(v) mentioned after (6.2), it is seen that if $y = y(t)$ is a solution of (6.1) with $y(0) \in I$, then $y(t)$ exists for $-\infty < t < \infty$ and $y(t) \to 0$ as $t \to \pm\infty$.

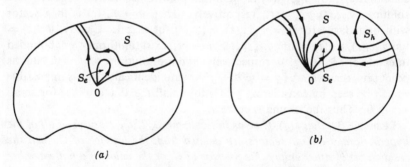

Figure 10.   (*a*) Hyperbolic sector. (*b*) Parabolic sector.

A sector S with the properties that it is not an elliptic sector and that $S \cup C_{12}$ contains no base solution is called a *hyperbolic sector*; see Figure 10(*a*). Part (*a*) of the proof of Theorem 9.1 below implies that one of the boundary arcs $y = y_1(t), y_2(t)$ of a hyperbolic sector is a positive and the other a negative base solution.

A sector S with the properties that both boundary arcs $y_1(t), y_2(t)$ are positive [or negative] base solutions and that the closure of S contains no negative [or positive] base solution is called a positive [or negative] *parabolic sector*; see Figure 10(*b*).

Note that any type of sector S (elliptic, hyperbolic, or parabolic) can contain solutions $y = y(t)$, $-\infty < t < \infty$, such that $y(t) \to 0$ as $t \to \pm \infty$. Such an arc and the point $y = 0$ constitute a Jordan curve. There can even be an infinite sequence of such orbits in S with the interiors of the corresponding Jordan curves pairwise disjoint. Denote by S_e, the *elliptic portion of S*, the sets of points of $\bar{S}$ on such orbits and the point $y = 0$. Then S_e is closed. In the case of hyperbolic and parabolic sectors S, S_e and C_{12} are disjoint.

Hyperbolic sectors S do and parabolic sectors S can contain open solution arcs $y = y(t)$ having both endpoints on C_{12}. The closure of the set of points y on such arcs will be called S_h, *the hyperbolic part of S*. It will be clear from parts (*a*) and (*b*) of the proof of Theorem 9.1 that $y = 0 \in S_h$ or $y = 0 \notin S_h$ according as S is hyperbolic or parabolic.

Lemma 8.2. *Let $f(y)$ be continuous on a simply connected open set E containing $y = 0$ such that $f(0) = 0$, $f(y) \neq 0$ if $y \neq 0$, and that the solutions of initial value problems $y' = f(y)$, $y(0) = y_0$ are unique. Let C be a Jordan curve in E surrounding $y = 0$. Then there is at most a finite number of elliptic and hyperbolic sectors in C.*

Proof. If the lemma is false, then there is a point $y_0 \in C$ and a sequence of points $y_{(1)}(0)$, $y_{(2)}(0)$, ... of C tending monotonously to y_0 along C such that $y_{(2n)}(0)$, $y_{(2n+1)}(0)$ is the initial point of a positive, negative base solution, $y_{(2n)}(t)$, $y_{(2n+1)}(t)$, respectively. Then $y = y_{(n)}(t)$ is in a sector with boundary arcs $y = y_{(n-1)}(t)$, $y_{(n+1)}(t)$ for $n \geq 1$. Clearly, $y^{(0)}(t) = \lim y_{(2n)}(t)$, $y^{(1)}(t) = \lim y_{(2n+1)}(t)$, $n \to \infty$, exist uniformly on bounded intervals of $t \geq 0$, $t \leq 0$, respectively, and are solutions of $y' = f$. But as point sets, the two arcs $y = y^{(0)}(t)$, $y^{(1)}(t)$ are identical. This is impossible as can be seen by considering $y^{(0)}(t)$ for small $t \geq 0$ and $y^{(1)}(t)$ for small $-t \geq 0$. Thus the lemma is correct.

Lemma 8.3. *Let $f(y)$, C be as in Lemma 8.2. If the closures of all of the hyperbolic and elliptic sectors are deleted from the interior of C, then the residual set is either empty, the interior of C, or the union of a finite number of pairwise disjoint parabolic sectors.*

Proof. It is sufficient to consider the case when there exist hyperbolic and/or elliptic sectors and that the residual set is the union of a finite number of disjoint sectors. It has to be shown that these sectors are parabolic.

Let S be a sector not containing any hyperbolic or elliptic sectors. Suppose first that of the two boundary base solutions $y = y_1(t)$, $y_2(t)$ of S, one is positive and the other negative. It will be shown that this leads to a contradiction.

For the sake of definiteness, let $y_1(t)$ be a positive, $y_2(t)$ a negative base solution; see Figure 11. Moving on the boundary arc C_{12} from $y_1(0)$ to $y_2(0)$, there is a last point $y_1{}^*$ [possibly $y_1(0)$] such that the solution $y = y_1{}^*(t)$ of $y' = f, y(0) = y_1{}^*$ exists and is in S for $t \geq 0$ (and hence is a positive null solution). Then $y_1{}^* \neq y_2(0)$, since S contains no elliptic sectors. Moving on C_{12} from $y_2(0)$ toward $y_1{}^*$, there is a last point $y_2{}^*$ such that the solution $y = y_2{}^*(t)$ of $y' = f, y(0) = y_2{}^*$ exists and is in S for $t \leq 0$. Then $y_2{}^* \neq y_1{}^*$ since S contains no elliptic sector.

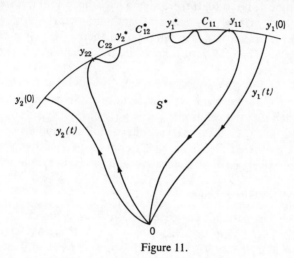

Figure 11.

Let C_{12}^* be the subarc of C_{12} joining $y_1{}^*$ and $y_2{}^*$. The solution $y = y_1{}^*(t)$ [or $y_2{}^*(t)$] with t increasing [or decreasing] from 0 has a last point y_{11} [or y_{22}] on C, where y_{11} [or y_{22}] can coincide with $y_1{}^*$ [or $y_2{}^*$] and is on the arc (or point) of C from $y_1(0)$ to $y_1{}^*$ [or $y_2{}^*$ to y_2]. Let C_{11} [or C_{22}] be the subarc on C from y_{11} to $y_1{}^*$ [or $y_2{}^*$ to y_{22}] and $C^{12} = C_{11} \cup C_{12}^* \cup C_{22}$. Let $y = y_{11}(t)$ [or $y_{22}(t)$] be the solution of $y' = f, y(0) = y_{11}$ [or y_{22}] for $t \geq 0$ [or $t \leq 0$]. Then there is a sector S^* with boundary consisting of $y = 0$, the base solutions $y = y_{11}(t), y_{22}(t)$, and the arc C^{12}.

Since S^* is subset of S, it is not an elliptic sector. No solution $y = y_0(t)$ of (6.1) with y_0 an interior point of C^{12} is in $S^* \cup C^{12}$ for $t \geq 0$ or $t \leq 0$. This is clear if $y_0 \in C_{12}^*$ by the definition of the endpoints of $y_1{}^*, y_2{}^*$ of C_{12}^*. It is also clear if $y_0 \subset C_{11} \cup C_{22}$, for the solution arc beginning at such a point y_0 either leaves S or is a part of the solution arcs $y = y_1{}^*(t), y_2{}^*(t)$ which for large $|t|$ are not in S^* (but on the boundary of S^*).

Thus S^* is a hyperbolic sector in S. This contradiction shows that the boundary base solutions $y = y_1(t), y_2(t)$ cannot be of opposite type (i.e., positive and negative).

Consider the case that both boundary base solutions $y = y_1(t), y_2(t)$ of S are positive [or negative]. Then if S is not parabolic, it contains a negative [or positive] base solution $y = y_3(t)$. But then $y_1(t), y_3(t)$ and $y_2(t), y_3(t)$ define sectors of the type just discussed. This is impossible, hence S is parabolic and the lemma is proved.

In order to avoid a consideration of special cases, the following convention will be adopted: If $y = 0$ is a nonrotation point, then, by Lemma 8.1, there exists at least one base solution $y = y_1(t)$ if C is in a sufficiently small neighborhood of $y = 0$. If there are no hyperbolic or elliptic sectors, the arcs $y_1(t)$ and $y_2(t) = y_1(t)$ define a parabolic sector with $C_{12} = C$. Thus, for small C around a nonrotation point, there is always a decomposition of the interior of C into a finite number of elliptic, hyperbolic, and parabolic sectors.

Exercise 8.1.   Let f, C be as in Lemma 8.2. Suppose that the closure of the interior I of C contains no periodic solutions. Show that there is a point $y_0 \in C$ such that the solution $C_0 : y = y_0(t)$ of (6.1) exists and is in I for either $t \geqq 0$ or $t \leqq 0$. Thus $y_0(t)$ is either a null solution or is a spiral around $y = 0$ such that $\Omega(C_0)$ or $A(C_0)$ contains a solution $y = y(t)$, $-\infty < t < \infty$, which is both a negative and a positive null solution.

9. The General Stationary Point

The object of this section is to prove the following theorem:

Theorem 9.1.   *Let $f(y)$ be continuous on a simply connected open set E containing $y = 0$ such that $f(0) = 0$ and $f(y) \neq 0$ for $y \neq 0$ and that solutions of initial value problems (6.1) are unique. Let C be a Jordan curve in a sufficiently small neighborhood of $y = 0$ surrounding $y = 0$, n_e the number of elliptic and n_h the number of hyperbolic sectors in C. Then the index $j_f(0)$ of the point $y = 0$ is given by*

$$(9.1) \qquad\qquad 2j_f(0) = 2 + n_e - n_h.$$

It is clear that for a rotation point $y = 0$, $n_e = n_h = 0$ and $j_f(0) = 1$, so that (9.1) holds and such points need not be considered. Thus it can be supposed that there is a decomposition of the interior of C into elliptic, hyperbolic and parabolic sectors.

Proof. C will be replaced by a piecewise C^1 Jordan curve C', around $y = 0$, made up of solution arcs and orthogonal trajectories, as shown in Figure 12, where E, H, P represent elliptic, hyperbolic, and parabolic sectors. If η is the tangent vector to C' and J is a subarc of C', with endpoints which are not corners, then, in this proof $2\pi j_\eta(J)$ will represent the contribution of J to $2\pi j_\eta(C') = 2\pi$, i.e., the variation of the turning of η along J taking into account discontinuities of η in accordance with Corollary 2.1. Furthermore $j_f(0) = j_f(C')$.

If L is an arc joining two points A and B, the notation $[AB]$, (AB), etc. will be used to denote the corresponding closed arc, open arc, etc. For any sector with boundary base solutions $y = y_1(t)$, $y_2(t)$, let $A_i = y_i(\pm \epsilon_0)$ for a fixed $\epsilon_0 > 0$, where $\pm$ is chosen according as $y_i(t)$ is a positive or negative base solution.

(a) *Hyperbolic Sectors.* Let S be a hyperbolic sector determined by $y = y_1(t)$, $y_2(t)$. In order to fix ideas, suppose that $y = y_1(t)$ is a positive base solution. Thus $A_1 = y_1(+\epsilon_0)$, where $\epsilon_0 > 0$. It is clear that $A_1 \notin S_e$, the elliptic part of S. Since S_e is closed, points y of S near A_1 are not in S_e.

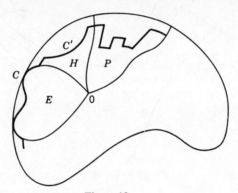

Figure 12.

Consider the differential equation

(9.2) $$y' = g(y), \quad \text{where} \quad g = (-f^2(y), f^1(y)),$$

for the orthogonal trajectories of solutions of $y' = f$. Let $L_i = [A_i B_i]$, $i = 1, 2$, be a solution arc of (9.2) with initial point A_i such that $(A_i B_i]$ is in $S - S_e$; see Figure 13. It will be shown that if $L_1 = [A_1 B_1]$ is sufficiently short and $y_0 \in (A_1 B_1]$, then the solution of $y = y_0(t)$ of $y' = f$, $y(\epsilon_0) = y_0$ exists and $y_0(t) \in S$ on an interval $\epsilon_0 \leq t \leq t^*$, $y_0(t^*)$ is on and is the only point of $y = y_0(t)$ on $L_2 = (A_2 B_2]$, and $y_0(t^*) \to A_2$ as $y_0 \to A_1$.

Let $T > 0$, $\epsilon > 0$. Then, by Theorem V 2.1, there exists a $\delta = \delta(\epsilon, T) > 0$ such that $\|y_0 - A_1\| < \delta$ implies that $y_0(t)$ exists and satisfies $\|y_0(t) - y_1(t)\| < \epsilon$ for $\epsilon_0 \leq t \leq T$. In particular, $y_0(t) \in S$ for $\epsilon_0 \leq t \leq T$ if T and $1/\epsilon$ are sufficiently large. Let $t^1 = t^1(y_0)$ be the least $t > \epsilon_0$ such that $y_0(t^1) \in C_{12}$, so that $t^1 > T$. The fact that S is hyperbolic and that $y_0 \notin S_e$ implies the existence of t^1. It is clear that $t^1(y_0) \to \infty$ as $y_0 \to A_1$.

Choose $y_{01}, y_{02}, \ldots$ such that $y_{0n} \in (A_1 B_1]$, $A_1 = \lim y_{0n}$, $y^0 = \lim y_{0n}(t^1)$ exists on C_{12} as $n \to \infty$, where $y_{0n}(t)$ is the solution belonging to $y_0 = y_{0n}$ and $t^1 = t^1(y_{0n})$. Thus, the solution $y_{-n}(t) = y_{0n}(t + t^1)$ exists and is in $S \cup C_{12}$ for $-t^1 \leq t \leq 0$ and $y_{-n}(0) \to y^0$, $n \to \infty$. By Theorem

V 2.1, $y_{-\infty}(t) = \lim y_{-n}(t)$ exists uniformly on every bounded interval $-T \leqq t < 0$, is the solution of $y' = f$, $y(0) = y^0(0) \in C_{12}$, and $y_{-\infty}(t) \in \bar{S}$ for $t \leqq 0$. Since S is a hyperbolic sector, $y_{-\infty}(t)$ is on the boundary of S for large $-t$; i.e., $y_{-\infty}(t - t_0) = y_2(t)$ for $t \leqq 0$ and $y_{-\infty}(-t_0)$ is the last point of $y_{-\infty}(t) \in C_{12}$ when t decreases from 0.

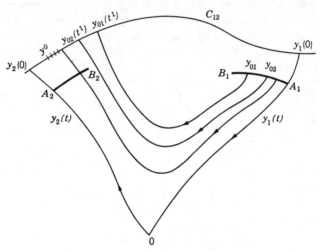

Figure 13.

It follows that according as $t_0 > 0$ or $t_0 = 0$, $y_0(t^1 - t_0)$ or $y_0(t^1)$ tends to $y_2(0)$ as $y_0 \to A_1$. In particular, if y_0 is sufficiently near to A_1, there exists a least $t^* > \epsilon_0$, $t^* = t^*(y_0)$, such that $y_0(t^*) \in L_2 = (A_2 B_2]$. Also $y_0(t^*) \to A_2$ as $y_0 \to A_1$.

Let $L_1 = [A_1 B_1]$ be so small that $t^*(y_0)$ exists for all $y_0 \in L_1$. Choose B_2 to be $y_0(t^*)$ for $y_0 = B_1$; see Figure 13. Let C_i be an interior point of $[A_i B_i]$ and J the arc $C_1 B_1 B_2 C_2$ consisting a piece of the orthogonal trajectory L_1, the solution arc $y = y_0(t)$ joining B_1 and B_2, and the piece $[C_2 B_2]$ of L_2. It is easy to see that

$$(9.3) \qquad\qquad 2\pi j_f(J) = 2\pi j_\eta(J) - \pi,$$

for the angle from η to f is $\tfrac{1}{2}\pi$ on $[C_1 B_1]$. It changes to 0 on going through B_1 and is 0 on $(B_1 B_2)$. It jumps to $-\pi/2$ on passing through B_2 and is $-\pi/2$ on $(B_2 C_2)$. This gives $2\pi[j_f(J) - j_\eta(J)] = -\pi/2 - \pi/2 = -\pi$, i.e., (9.3). If $y_1(t)$ is a negative and $y_2(t)$ a positive base solution, the formula (9.3) is still valid.

(b) *Parabolic Sectors.* Let S be a parabolic sector. In order to fix ideas, let S be a *positive* parabolic sector, so that A_i is a point $y_i(\epsilon_0)$ where $\epsilon_0 > 0$.

Note that $y = 0 \notin S_h$, the hyperbolic part of S. For otherwise, the arguments of part (a) show that there exists a negative base solution in $S \cup C_{12}$. Also, if $y \neq 0$ is a point of S_e, then $y \neq S_h$ by the definitions of S_e, S_h. Hence $S_e \cap S_h$ is empty, so that dist $(S_e, S_h) > 0$. It is also clear that $S_e \cap C_{12}$ is empty. Furthermore, $y_i(t) \notin S_e$ for $i = 1, 2$ and $t \geq 0$, for otherwise $y_i(t) \in S_e$ for $i = 1$ or 2 and all $t \geq 0$ and, in particular, $y_i(0) \in C_{12} \cap S_e$, which is impossible. Similarly, $y_i(t) \notin S_h$ for $i = 1, 2$ and $t > 0$, otherwise $y_i(t) \in S_h$ for $i = 1$ or 2 and all $t \geq 0$ and, in particular, the limit point $y = 0 \in S_h$.

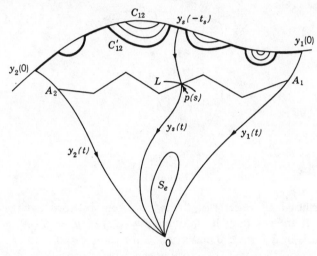

Figure 14.

If S_h is not empty, there are points $y_0 \in S$ for which a solution $y = y_0(t)$ of (6.1) on some interval $\alpha_- \leq t \leq \alpha_+$ is such that $y_0(t) \in S$ for $\alpha_- < t < \alpha_+$ and $y_0(\alpha_{\pm}) \in C_{12}$. The arc $y = y_0(t)$, $\alpha_- \leq t \leq \alpha_+$ and the corresponding subarc of C_{12} joining $y_0(\alpha_{\pm})$ form a Jordan curve. There is a finite or infinite sequence of such maximal Jordan curves $J_1, J_2, \ldots$ in the sense that the interiors of $J_1, J_2, \ldots$ are pairwise disjoint and the union of $J_1, J_2, \ldots$ and their interiors contains $\bar{S} \cap S_h$; see Figure 14. The set of points on C_{12} not on any J_n together with the points on the closures of the arcs $J_n \cap S$ form a Jordan arc C_{12}' in $S \cup C_{12}$ joining $y_1(0)$, $y_2(0)$. Let S' be the interior of the Jordan curve consisting of C_{12}', $y = 0$, and the arcs $y = y_i(t)$ for $i = 1, 2$ and $t \geq 0$.

If S_h is empty, let $S = S'$ and $C_{12} = C_{12}'$. Then, whether or not S_h is empty, $S_e \subset S'$, and so $S_e \cap C_{12}'$ is empty. Hence there is a polygonal path P; $y = p(s)$, $0 \leq s \leq 1$, which joins $A_1 = p(0)$, $A_2 = p(1)$ and $p(s) \in S' - S_e$ for $0 < s < 1$.

If $0 < t < 1$, the solution $y = y_s(t)$ of $y' = f$, $y(0) = p(s)$ is such that $y_s(t) \in S$ for $t \geqq 0$ and $y_s(-t_s) \in C_{12}$ for some $t_s > 0$. Through each point $p(s) \in P$, draw an open orthogonal trajectory arc L; i.e., a solution arc of (9.2) through $y = p(s)$ such that the closure of L is in $S' - S_e$. In the cases $p(0) = A_1$ and $p(1) = A_2$, let the corresponding L be half-closed and have A_i as an endpoint instead of an interior point.

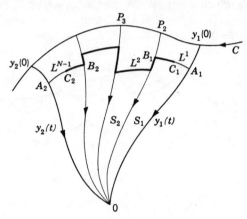

Figure 15.

The orthogonal trajectories L can be taken so short that they lie in $S' - S_e$. It follows that the solution $y = y_0(t)$ of $y' = f$, $y(0) = y_0 \in L$ exists and is in S for $t \geqq 0$ and meets C for some $t < 0$.

The set of orthogonal trajectory arcs L can be considered to form an "open covering" of the s-set $0 \leqq s \leqq 1$ in the sense that s is "contained" in an L if an arc $y = y_s(t)$, $0 \leqq t \leqq t^s$ or $-t^s \leqq t \leqq 0$ is in S and contains a point of L. If an s_0 is "in L," then s near s_0 is "in L." Thus, by the theorem of Heine-Borel, there is a finite set $L_1 = [A_1, B^1)$, $L_2 = [A_2, B^2)$, $L_3 = (A^3, B^3)$, . . . of these arcs such that every solution $y = y_s(t)$ meets at least one of the $L_1, L_2, \ldots$ at some t^s ($\gtreqless 0$); in which case $y_s(t) \in S$ for $t \geqq t^s$.

Let $A_1 = y_{(1)}, y_{(2)}, \ldots, y_{(n)} = A_2$ be the endpoints of the arcs L_1, $L_2, \ldots$. The solution $y_{(k)}(t)$ of $y' = f(y)$, $y(0) = y_{(k)}$ exists and is in S for $t \geqq 0$. Also, since $y_{(k)} \in S' - S_e$, there is a least $t = t_k > 0$ such that $P_k \equiv y_{(k)}(-t_k) \in C$. Thus $y_{(k)}(t - t_k)$, $t \geqq 0$, is a positive base solution. After a suitable change of enumeration, it can be supposed that $P_1 = y_1(0), P_2, \ldots, P_n = y_2(0)$ are ordered on C and that $P_i \neq P_j$ for $i \neq j$; see Figure 15.

Each solution pair $y = y_{(k)}(t - t_k)$, $y = y_{(k+1)}(t - t_{k+1})$ defines a sector S_k which is a subsector of S. Let s_0 be the largest s-value, $0 \leqq s \leqq 1$, such that $p(s)$ meets the arc $y = y_{(k)}(t - t_k)$ and s_1 the least s-value $s_1 > s_0$ such that $p(s_1)$ is on the arc $y = y_{(k+1)}(t - t_{k+1})$. Now $y = y_s(t)$, for s fixed in $s_0 < s < s_1$, is in S_k and meets at least one of the selected $L_1, L_2, \ldots$ Since the arcs $L_1, L_2, \ldots$ have no endpoints in S_k, there is at least one of the arcs $L_1, L_2, \ldots$ which meets every solution arc $y = y_s(t)$ for $s_0 < s < s_1$. Thus, there are closed orthogonal trajectory arcs $L^1, \ldots, L^{N-1}$ each of which is a subarc of one of the closed arcs $\bar{L}_1, \bar{L}_2, \ldots$, such that L^j joins $y = y_{(k)}(t - t_k)$ and $y = y_{(k+1)}(t - t_{k+1})$. It can be supposed that L^1, L^{N-1} begin and terminate, respectively, at A_1, A_2; say $L^1 = [A_1 B_1]$, $L^{N-1} = [B_2 A_2]$.

Let C_1, C_2 be interior points of L^1, L^{N-1} and J the arc joining C_1, C_2 consisting successively of subarcs of L^1, $y = y_{(2)}(t - t_2)$, L^2, $y = y_{(3)}(t - t_3), \ldots, y = y_{(N-1)}(t - t_{N-1})$, L^{N-1}. It will be verified that

$$(9.4) \qquad\qquad 2\pi j_f(J) = 2\pi j_\eta(J).$$

Note that the tangent vector η to J at a point $y \in L^k$ is in the direction $\pm g(y)$, where $g(y)$ occurs in (9.2) and $\pm$ is independent of k. (That the $\pm$ is independent of k can be seen as follows: Suppose that $g(y)$ points into S_1 at A_1, then it clearly points into S_2 at B_1. By continuity it points into S_2 along the solution arc $y = y_{(2)}(t)$ and, hence, at the end-point of L_2 on $y = y_{(2)}(t)$. Similarly, it points into S_3 at the endpoint of L_2 on $y = y_3(t)$. This argument can be continued and shows that $\pm$ does not depend on k.)

Thus the sign of the sine of the angle from η to f is ∓ 1 and, consequently, the angle is of the form $2n\pi \mp \frac{1}{2}\pi$, where n is an integer and $\mp$ is independent of k. On $L^1 = [C_1 B_1]$, the angle from η to f is $\frac{1}{2}\pi$ (so that on $L^2, \ldots$, L^{N-1}, it is of the form $2n\pi + \frac{1}{2}\pi$). On the part of J consisting of the solution arc $y = y_{(2)}(t - t_2)$, the angle from η to f becomes 0 or π. Hence on L^2, it is $\frac{1}{2}\pi$. Continuing this argument, it is seen that the angle from η to f is $\frac{1}{2}\pi$ on every L^j. Thus $2\pi[j_f(J) - j_\eta(J)] = \frac{1}{2}\pi - \frac{1}{2}\pi = 0$; i.e., (9.4) holds.

It is readily verified that if S is a negative parabolic sector, a similar construction also leads to (9.4).

(c) *Elliptic Sectors.* Let S be an elliptic sector with boundary solutions $y = y_1(t), y_2(t)$ which are subarcs of some $y = y_0(t)$, $-\infty < t < \infty$, in S. Suppose that the constructions just described have been made on all of the hyperbolic and parabolic sectors. Since S is adjacent to such sectors, there is a solution arc $[A_2 A_1]$ on the arc $y = y_0(t)$ [containing the points $y_1(0)$, $y_2(0)$ in its interior] and two orthogonal trajectory arcs $[C_2 A_2)$, $(A_1 C_1]$ in the interiors of the adjacent sectors, respectively; see Figures 16 and 17. If J is the arc $[C_2 C_1]$ consisting of the orthogonal trajectory $[C_2 A_2)$, the solution arc $[A_2 A_1]$, and $(A_1 C_1]$, then

$$(9.5) \qquad\qquad 2\pi j_f(J) = 2\pi j_\eta(J) + \pi.$$

For, on $[C_2 A_2)$ in Figure 16 [or in Figure 17], the angle from η to f is $-\frac{1}{2}\pi$ [or $\frac{1}{2}\pi$]; on $(A_2 A_1)$, it is 0 [or π]; and finally, on $(A_1 C_1]$, it becomes $\frac{1}{2}\pi$ [or $3\pi/2$]. Thus $2\pi[j_f(J) - j_\eta(J)]$ is $\frac{1}{2}\pi - (-\frac{1}{2}\pi) = \pi$ [or $3\pi/2 - \frac{1}{2}\pi = \pi$]; so that (9.5) holds.

   (d) Completion of the Proof.   In the constructions in parts (a) and (b), the same number $\epsilon_0 > 0$ has been used for all of the hyperbolic and parabolic sectors. Thus, if a base solution $y = y(t)$ is on the boundary of

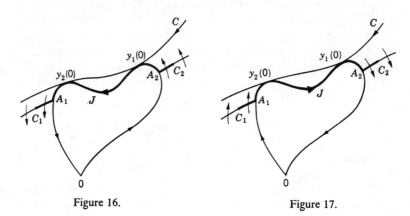

Figure 16.   Figure 17.

two such adjacent regions, there is an orthogonal trajectory arc $[C_2 C_1]$ cutting $y = y(t)$ at a point A. For the arc $J = [C_2 C_1]$, it is clear that

$$(9.6) \qquad\qquad j_f(J) = j_\eta(J).$$

Thus if the relations (9.3), (9.4), (9.5), and (9.6) are added for all arcs J, we obtain $2\pi j_f(0) = 2\pi - \pi n_h + \pi n_e$, where the 2π on the right is $2\pi \Sigma j_\eta(J)$ by the (Umlaufsatz) Theorem 2.1 and its Corollary 2.1. This proves (9.2).

   Remark.   For the purpose of the next exercise, note that the assumption in Theorem 9.1 that "the solution of $y' = f(y)$, $y(0) = y_0$ is unique" can be relaxed to the assumption that "the solution of $y' = f(y)$, $y(0) = y_0$ is unique when $y_0 \neq 0$." (This involves an obvious modification of the definitions of "base solution," "elliptic sector," "hyperbolic sector," and "parabolic sector.") For $y' = f(y)$ can be replaced by $y' = h(y)$, where $h(y) = \|y\| f(y)$. The two indices $j_f(0)$, $j_h(0)$ are obviously equal. It is clear that an arc $y = y(t)$, where $y(t) \neq 0$, is a solution of $y' = f(y)$ if and only if it becomes a solution of $dy/ds = h(y)$ after the change of parameters $t \to s$ where $ds = dt/\|y(t)\|$. Thus "n_e, n_h" are the same for both systems $y' = f(y)$, $y' = h(y)$. Finally, since $y' = h$ implies that $\|y'\| \leq \|y\|$ for small $\|y\|$, it follows that $y \equiv 0$ is the only solution of $y' = h(y)$, $y(0) = 0$.

Exercise 9.1.   Let $U(y)$ be a real-valued function of class C^1 for $\|y\| < b$ such that $U(0) = 0$ and the gradient $g(y) = (\partial U/\partial y^1, \partial U/\partial y^2)$ vanishes only for $y = 0$. Thus if $f(y) = (\partial U/\partial y^2, -\partial U/\partial y^1)$, then U is constant on solutions of $y' = f$. Show that either (i) there is an $\epsilon > 0$ such that $U(y) \neq 0$ in $0 < \|y\| < \epsilon$ or (ii) the set of arcs in $0 < \|y\| < b$ which join $y = 0$ and $\|y\| = \epsilon$ and on which $U = 0$ consists of a finite (even) number $2n$ of arcs, $2n > 0$; furthermore, $j_f(0) = 1 - n$. [Note: The initial value problems $y' = f(y), y(0) = y_0 \neq 0$ have unique solutions. (Why?) Case (i) occurs only if $y = 0$ is a rotation point, in which case it is a center. This happens only if U has a strict local maximum or minimum at $y = 0$. Case (ii) occurs if $y = 0$ is a nonrotation point, in which case there are no elliptic or parabolic sectors for any Jordan curve C surrounding $y = 0$.]

Theorem 9.2.   *Let $f(y)$ be continuous on a simply connected open set E such that solutions of initial value problems (6.1) are unique. Let C be a positively oriented Jordan curve of class C^1 in E with the property that $f(y) \neq 0$ on C and that $f(y)$ is tangent to C at only a finite number of points $y_1, \ldots, y_n$ of C. Let n^e, n^h be the number of these points y_i where the solution arc $y = y(t)$ of $y' = f, y(0) = y_i$ for small $|t|$ is internally, externally tangent to C at y_i (so that $n^e + n^h \leq n$). Then $2j_f(C) = 2 + n^e - n^h$.*

The solution arc $y(t)$ of $y' = f$, $y(0) = y_i \in C$ is said to be internally [or externally] tangent to C at y_i if there exists an $\epsilon > 0$ such that $y(t)$ is interior [or exterior] to C for $0 < |t| \leq \epsilon$.

Proof.   Let η denote the positively oriented tangent vector on C. If $n^e = n^h = 0$, then it is clear that the angle from η to f along C does not pass through a value 0 mod π. Hence the two integers $j_f(C), j_\eta(C)$ are equal. Since $j_\eta(C) = 1$ by Theorem 2.1, it follows that $2j_f(C) = 2$ if $n^e = n^h = 0$.

Hence it can be supposed that not both n^e and n^h are zero. A point y_0 of C will be called an elliptic [or hyperbolic] point if the solution arc through y_0 is internally [or externally] tangent to C at y_0. Thus n^e [or n^h] is the number of elliptic [or hyperbolic] points.

Let $J = [AB]$ be a subarc of C such that A, B are elliptic or hyperbolic points but no interior point of J is elliptic or hyperbolic. Then at a point y_0 interior to J, the solution arc through y_0 crosses C in a direction (from interior to exterior or from exterior to interior of C) independent of y_0. Thus by considering the angle from η to f, it is seen that $2\pi j_f(J) - 2\pi j_\eta(J)$ is $-\pi$, 0, or π.

If one of the points A, B is elliptic and the other hyperbolic, then $f(y)$ and the tangent vector η have the same orientation at both A, B or have the opposite orientation at both A, B; see Figure 18a. In this case, $2\pi j_f(J) - 2\pi j_\eta(J) = 0$.

If both points A, B are elliptic or both are hyperbolic, then $f(y)$ and the tangent vector η have the same direction at one of the points A, B and have the opposite orientation at the other point. In this case, $2\pi j_r(J) - 2\pi j_\eta(J) = \pm\pi$. It will be left to the reader to verify that $2\pi j_r(J) - 2\pi j_\eta(J)$ is π or $-\pi$ according as both A, B are elliptic or both are hyperbolic; cf. Figure 18b.

Thus, if $n^e(J)$ or $n^h(J)$ denotes the number (0, 1, or 2) of the endpoints of J which are elliptic or hyperbolic, then

$$2\pi j_r(J) = 2\pi j_\eta(J) + \tfrac{1}{2}\pi[n^e(J) - n^h(J)]$$

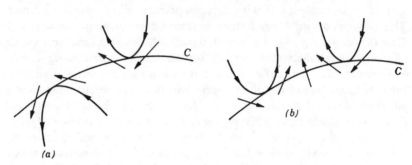

(a)

(b)

Figure 18.

holds in all cases. Summing this relation for all subarcs J gives the desired result since

$$\sum_J j_\eta(J) = 1, \qquad \tfrac{1}{2}\sum_J n^e(J) = n^e, \qquad \tfrac{1}{2}\sum_J n^h(J) = n^h.$$

10. A Second Order Equation

An application of the theorem of Poincaré-Bendixson will be given in this section in Corollary 10.1. It concerns the second order equation

$$(10.1) \qquad u'' + h(u, u')u' + g(u) = 0$$

for a real-valued function u or the equivalent autonomous first order system

$$(10.2) \qquad u' = v, \qquad v' = -h(u, v)v - g(u).$$

If, in (10.1), $g(u)u > 0$, then the term $g(u)$ is a "restoring force" (as for the harmonic oscillator $u'' + u = 0$). If $h > 0$, then the "frictional" term hu' tends to decrease the speed $|u'|$ (as in the equation $u'' + hu' = 0$ with $h > 0$ a constant). The next theorem can be interpreted as saying that if the restoring force and frictional term are not too small, then the solutions of (10.1) are bounded.

Theorem 10.1. *Let* $g(u)$, $h(u, v)$ *be real-valued, continuous functions for all* u, v *with the properties:* (i) *solutions of* (10.2) *are uniquely determined by initial conditions;* (ii) *there exists a number* $a > 0$ *such that*

$$(10.3) \qquad\qquad g(u)u > 0 \qquad \text{for} \quad |u| \geqq a,$$

$$(10.4) \qquad\qquad G(u) \equiv \int_0^u g(r) \, dr \to \infty \qquad \text{as} \quad |u| \to \infty;$$

(iii) *there exists a number* $m > 0$ *such that*

$$(10.5) \qquad\qquad h(u, v) \geqq -m, \qquad \text{for} \quad |u| \leqq a, \ -\infty < v < \infty;$$

(iv) *if* $h_0(u) = \inf h(u, v)$ *for* $-\infty < v < \infty$, *then*

$$(10.6) \qquad\qquad h_0(u) > 0 \qquad \text{for} \quad |u| \geqq a,$$

$$(10.7) \qquad\qquad H(u) \equiv \int_a^u h_0(r) \, dr \to \infty \qquad \text{as} \quad u \to \infty.$$

Then there exists a Jordan curve C *bounding a domain* E, *containing the origin* $(u, v) = (0, 0)$, *such that no point of* C *is an egress point for* E *and that if* $u(t)$, $v(t)$ *is a solution of* (10.2) *starting, say, at* $t = 0$, *then* $u(t)$, $v(t)$ *exists for* $t \geqq 0$ *and* $(u(t), v(t)) \in E$ *for large* t.

A sufficient condition for (iv) is that there exists a number $M > 0$ such that $h(u, v) \geqq M > 0$ for $|u| \geqq a$, $-\infty < v < \infty$. Actually, the proof below does not use the full force of (iv) but only that $h(u, v) > 0$ for $|u| \geqq a$, that $h^0(u) > 0$ for $u \geqq a$ if $h^0(u) = \inf h(u, v)$ for $v \leqq -v_0$, and that $h^1(u) \geqq 0$ for $u \geqq a$ and $\int_a^u h^1(r) \, dr \to \infty$ as $u \to \infty$ if $h^1(u) = \inf h(u, v)$ for $v \geqq v_0 > 0$ for a fixed v_0.

Proof. Along a solution arc $u(t)$, $v(t)$ of (10.2),

$$(10.8) \qquad\qquad \frac{dv}{du} = -h(u, v) - \frac{g(u)}{v}.$$

Below u', v' refer to the derivatives in (10.2) and dv/du refers to (10.8). Let $|g(u)| \leqq b$ for $|u| \leqq a$. Then, by (10.5) and (10.8),

$$(10.9) \qquad\qquad \frac{dv}{du} \leqq m + \frac{b}{|v|} \leqq 2m \qquad \text{if} \quad |u| \leqq a, \ |v| \geqq \frac{b}{m}.$$

Let $\varphi(u)$ be a positive, continuous increasing function of $u \geqq a$ such that

$$(10.10) \qquad\qquad \varphi(u) > \frac{g(u)}{h_0(u)} \qquad \text{and} \qquad \varphi(u) \to \infty \qquad \text{as } u \to \infty.$$

Then $v \leqq -\varphi(u) < -\dfrac{g(u)}{h_0(u)}$ implies that $v < 0$, $h(u, v) \geqq h_0(u) >$
$-\dfrac{g(u)}{v}$; so that

(10.11) $\dfrac{dv}{du} < 0$ if $u \geqq a$, $v \leqq -\varphi(u)$.

Consider the arcs

(10.12) $C_\alpha: \tfrac{1}{2}v^2 + G(u) = \alpha$ for $u| \geqq |a$

for a large constant $\alpha > 0$. These arcs are symmetric with respect to the
u-axis. A portion of these arcs and a segment of the line $u = a$ [or $u = -a$]

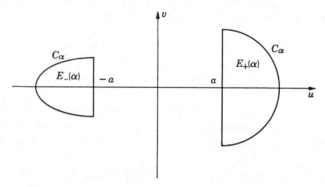

Figure 19.

form a Jordan curve bounding a domain $E_+(\alpha)$ [or $E_-(\alpha)$]; see Figure 19.
Along a solution $u(t), v(t)$ of (10.2), the quantity $\psi(t) = \tfrac{1}{2}v^2(t) + G(u(t))$
has a derivative $\psi' = v(-hv - g) + gv = -h(u, v)v^2 < 0$ if $|u(t)| \geqq a$,
$v(t) \neq 0$. Thus $\psi(t)$ is decreasing and so the arc $u = u(t), v = v(t)$ enters
$E_\pm(\alpha)$ with increasing t as soon as it meets the curved boundary C_α of $E_\pm(\alpha)$
and remains in $E_\pm(\alpha)$ as long as $|u(t)| \geqq a$.

For convenience, the construction of C will refer to Figure 20. A letter
on this diagram denotes either a point or one of its coordinates; e.g., v_2
denotes either the point (a, v_2) or the ordinate v_2.

Choose $\sigma > a$ so large that

(10.13) $\varphi(\sigma) \geqq \dfrac{b}{m}$ and $\dfrac{H(\sigma)}{4a} > 2m$.

Choose α so that C_α passes through the point $A = (\sigma, -\varphi(\sigma))$; i.e.,
$\alpha = \tfrac{1}{2}\varphi^2(\sigma) + G(\sigma)$. In particular, C_α passes through the point $(u, v) =$
$(\sigma, \varphi(\sigma))$. Let $\tau > 0$ denote the point where C_α meets the u-axis.

Let $u^0(t), v^0(t)$ denote the solution of (10.2) determined by the initial
condition $(u_0(0), v_0(0)) = (\tau, 0)$. As t decreases from 0, $u^0(t)$ decreases and

$v^0(t)$ increases until $u^0(t)$ takes the value a at some point $t = t_2 < 0$. This is clear from $v' = -hv - g \leqq -g < 0$, hence $u' = v > 0$, as long as $u \geqq a$. Let C^0 denote the arc $(u, v) = (u^0(t), v^0(t))$. In view of the remarks above, the part of C^0 for $t_2 \leqq t \leqq 0$ has only the point τ in common with C_α. In particular, if $v^0(t) = v_1$ when $u^0(t) = \sigma$ and if $v_0 = \varphi(\sigma)$, then

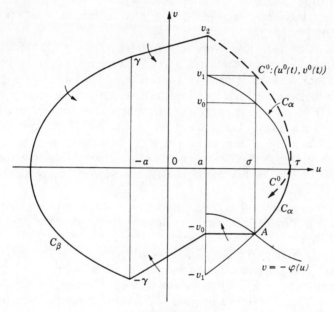

Figure 20.

$v_1 > v_0$. Also, by (10.8), $dv/du \leqq -h(u, v) \leqq -h_0(u)$ on C^0 for $a \leqq u \leqq \sigma$. Hence, if $v_2 = v^0(t_2)$, then

$$(10.14) \qquad v_2 \geqq v_1 + \int_a^\sigma h_0(r)\, dr = v_1 + H(\sigma) > v_0 + H(\sigma).$$

Put $\gamma = v_0 + \tfrac{1}{2}H(\sigma) < v_2 - \tfrac{1}{2}H(\sigma)$. Thus the slopes of the line segments γ to v_2 and $-\gamma$ to $-v_0$ are at least $\tfrac{1}{2}H(\sigma)/2a$ and thus exceed $2m$, by (10.13). Define β by $\beta = \tfrac{1}{2}\gamma^2 + G(a)$, so the arcs C_β pass through the points $\pm\gamma$.

Let C denote the Jordan curve consisting of the arc C^0 from τ to v_2, the line segment from v_2 to γ, the arc C_β from γ to $-\gamma$, the line segment from $-\gamma$ to $-v_0$, the horizontal line segment from $-v_0$ to A, and C_α from A to τ. Let E denote the interior of C.

It will first be verified that the points of C, except for the points on C^0, are strict ingress points of E. To this end, it is sufficient to verify that the solution arcs of (10.2) reaching the lines segments on C cross C as indicated in Figure 20. This is clear along the horizontal segment from $-v_0$ to A, for $u' = v < 0$ and $dv/du < 0$ by (10.11) and the monotony of $\varphi(u)$ for $u \geqq a$. The slope of the segment from $-\gamma$ to $-v_0$ is $H(\sigma)/4a > 2m$ while, along this segment, $u' = v < 0$, $dv/du \leqq 2m$ by (10.9) since $v_0 = \varphi(\sigma) \geqq b/m$ by (10.13). Similarly, the slope of the segment from γ to v_2 exceeds $2m$ and on this segment $v \geqq \gamma > v_0 \geqq b/m$ so that $u' = v > 0$ and $dv/du \leqq 2m$.

It remains to show that every solution $u(t), v(t)$ of (10.2) starting at $t = 0$ exists for $t \geqq 0$ and that $(u(t), v(t)) \in E$ for large t. For this purpose, rename σ to σ_0 and the Jordan curve C to $C(\sigma_0)$. Then, for each $\sigma \geqq \sigma_0$, the above construction leads to a Jordan curve $C(\sigma)$ and its interior $E(\sigma)$. The sets $E(\sigma)$ are increasing with σ and the union $U \equiv \bigcup C(\sigma)$ for $\sigma > \sigma_0$ is the exterior of $C(\sigma_0)$. Let $u(t), v(t)$ be a solution of (10.2) starting at a point of U for $t = 0$. Then, as long as $u(t), v(t)$ remains in U, there is a unique $\sigma = \sigma(t)$ such that $(u(t), v(t)) \in C(\sigma(t))$ and $\sigma(t)$ is a nonincreasing function of t. This implies that $(u(t), v(t))$ exists for $t \geqq 0$; cf. Corollary. II 3.2.

Suppose, if possible, that $u(t), v(t)$ does not enter $E(\sigma_0)$ eventually, so that $\sigma(t) \geqq \sigma_0$ for all $t \geqq 0$. Let $\sigma_1 = \lim \sigma(t)$ as $t \to \infty$. Thus $(u(t), v(t))$ is arbitrarily near to $C(\sigma_1)$ for large t-values. It is clear in this case, however, that $(u(t), v(t)) \in E(\sigma_1)$ for arbitrarily large t. But then $\sigma(t) < \sigma_1$ for large t. This is a contradiction and proves the theorem.

Corollary 10.1 *In addition to the assumptions of Theorem 10.1, suppose that $g(u) \neq 0$ for $u \neq 0$ [so that $g(u)u > 0$ for $u \neq 0$ and the origin is the only stationary point for (10.2)] and suppose that the origin is not an ω-limit point for every solution of (10.2). Then (10.2) has a periodic solution $(u_0(t), v_0(t)) \not\equiv 0$ which is asymptotically orbitally stable from the exterior as $t \to \infty$ and the interior of the Jordan curve $u = u_0(t)$, $v = v_0(t)$ contains the origin.*

Proof. If some solution $(u(t), v(t))$ does not have the origin as an ω-limit point, then the theorem of Poincaré-Bendixson (Theorem 4.1) implies that its set of ω-limit points is a periodic solution $(u_0(t), v_0(t))$, since $(u(t), v(t)) \in E$ for large t.

The Jordan curve $u = u_0(t)$, $v = v_0(t)$ surrounds the origin by Theorem 3.1. It also follows that if there are two periodic solutions, then one of the. corresponding Jordan curves is contained in the interior of the other. Since all periodic orbits are contained in the compact set $E \cup C$, it follows that there exists a unique periodic solution $u_0(t), v_0(t)$ such that the Jordan curve $C_0 : u = u_0(t)$, $v = v_0(t)$ contains all other periodic orbits in its interior.

The periodic solution $u_0(t), v_0(t)$ is asymptotically orbitally stable from the exterior as $t \to \infty$. In order to see this, consider a solution $u(t), v(t)$ starting for $t = 0$ at a point exterior to C_0. Since the origin is not an ω-limit point for $u(t), v(t)$, its set of ω-limit points is a periodic orbit (Theorem 4.1) which is necessarily C_0. Thus the assertion follows from Theorem 5.2.

Exercise 10.1.   (*a*) Let $g(u), h(u, v)$ be continuous for all u, v and let $e(t)$ be continuous for all t. Suppose that solutions of

$$(10.15) \qquad u'' + h(u, u')u' + g(u) = e(t)$$

are uniquely determined by initial conditions. Let there exist positive constants a, e_0, m, M such that (i) $h(u, v) \geqq M$ for $|u| \geqq a$, $|v| \geqq a$; (ii) $h(u, v) \geqq -m$ for all u, v; (iii) $|e(t)| \leqq e_0$ for all t; and (iv) $ug(u) > 0$ for large $|u|$ and $\lim \inf |g(u)| > ma + e_0$ as $|u| \to \infty$. Then there exists a Jordan curve C in the (u, v)-plane such that if $u(t)$ is a solution of (10.15) starting at $t = t_0$, then $u(t)$ exists for $t \geqq t_0$, $(u, v) = (u(t), u'(t))$ is in the interior E of C for large t and the $(u(t), u'(t))$ cannot leave E with increasing t. (*b*) If, in addition, $e(t)$ is periodic of period $p > 0$, then (10.15) has a solution of period p. See Opial [7].

Exercise 10.2.   Show that Theorem 10.1 remains correct if condition (iii) is strengthened to $h(u, v) \geqq m > 0$ for $|u| \leqq a$, $-\infty < v < \infty$, and (iv) is relaxed to $h(u, v) \geqq 0$ for $|u| \geqq a$.

A particular case of (10.1) is the equation

$$(10.16) \qquad u'' + h(u)u' + g(u) = 0.$$

If $g(u) = u$, (10.16) is called Liénard's equation. An elegant simple argument shows, under certain conditions, the existence of a periodic solution which is unique (up to translations of the independent variable t). The equation (10.16) will be treated as the first order system

$$(10.17) \qquad u' = v - H(u), \qquad v' = -g(u),$$

where

$$(10.18) \qquad H(u) = \int_0^u h(r) \, dr \qquad \text{and} \qquad G(u) = \int_0^u g(r) \, dr.$$

Theorem 10.2.   *Let $h(u), g(u)$ be continuous for all u with the properties: (i) that solutions of (10.17) are uniquely determined by initial conditions; (ii) that $h(u) = h(-u)$ is even, $H(u) < 0$ for $0 < u < a$, $H(u) > 0$ and is increasing for $u > a$, $H(u) \to \infty$ as $u \to \infty$; finally, (iii) that $g(u) = -g(-u)$ is odd and $ug(u) > 0$ for $u \neq 0$. Then (10.17) has exactly one periodic solution $(u_0(t), v_0(t)) \not\equiv 0$, up to translations of t, and this solution is asymptotically, orbitally stable (from the exterior and interior) as $t \to \infty$.*

Of course, $(u(t), v(t))$ is a [periodic] solution of (10.17) if and only if $u(t)$ is a [periodic] solution of (10.16).

Proof. The functions on the right of (10.17) are odd functions of (u, v); so that if $u(t), v(t)$ is a solution, so is $-u(t), -v(t)$. The tangent to a solution arc $u(t), v(t)$ is horizontal at a point (u, v) if and only if $u = 0$ and is vertical if and only if $v = H(u)$.

The arguments to follow refer to Figure 21. Along a solution starting at $v_1 > 0$, u increases and v decreases until $v = H(u)$, say, at the point γ. Then u decreases, v continues to decrease and the solution arc remains

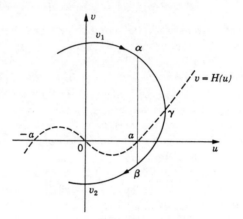

Figure 21.

below the curve $v = H(u)$ until the solution meets the v-axis, say, at $v = v_2$. Otherwise the solution would have a horizontal tangent at a point where $u \neq 0$ or v would tend to $-\infty$ as u tends to a finite value, which is impossible by

$$(10.19) \qquad \frac{dv}{du} = -\frac{g(u)}{v - H(u)}.$$

By symmetry, it is clear that a continuation of the solution is closed if and only if $v_2 = -v_1$. It is clear that we can start at any point γ, i.e., $(\gamma, H(\gamma))$, and determine the points v_1, v_2 by moving along the corresponding solution arc with decreasing, increasing t. Denote $\frac{1}{2}(v_2{}^2 - v_1{}^2)$ by $\varphi(\gamma)$. Thus if $\psi(u, v) = \frac{1}{2}v^2 + G(u)$, then

$$\varphi(\gamma) = \int_{v_1 \gamma v_2} d\psi,$$

where the integral denotes a line integral along the solution arc from v_1 to γ to v_2.

If $0 < \gamma \leq a$, then $\varphi(\gamma) > 0$, for along a solution arc, $d\psi/dt = vv' + g(u)u' = -g(u)H(u) > 0$. If $\gamma > a$, let α and β denote the points where the

The Poincaré-Bendixson Theory   181

solution arc meets the line $u = a$; see Figure 21. Write

$$\varphi(\gamma) = \varphi_1(\gamma) + \varphi_2(\gamma), \qquad \text{where} \quad \varphi_1 = \int_{v_1\alpha} + \int_{\beta v_2}, \quad \varphi_2 = \int_{\alpha\gamma\beta}$$

Along the solution arcs contributing to φ_1, $d\psi/du = v(dv/du) + g(u) = -g(u)H(u)/(v - H(u))$, by (10.19). Thus $d\psi/du > 0$, $du > 0$ on $v_1\alpha$ and $d\psi/du < 0$, $du < 0$ on βv_2, so that $\varphi_1(\gamma) > 0$. As γ increases, the arc $v_1\alpha$ is raised, so that $g(u)|H(u)|/(v - H(u))$ decreases; and βv_2 is lowered, so that $g(u)|H(u)|/|v - H(u)|$ decreases. Hence $\varphi_1(\gamma)$ decreases as γ increases.

On the solution arc $\alpha\gamma\beta$, $dv < 0$ and $d\psi/dv = v + g(u) \, du/dv = H(u) > 0$. Thus

$$\varphi_2(\gamma) = \int_{\alpha\gamma\beta} H(u) \, dv < 0 \qquad \text{for} \quad \gamma > a.$$

If $\alpha\gamma\beta$ has a parametric representation $u = u(v) = u(v, \gamma)$, it is clear that $u(v, \gamma)$, hence $H(u(v, \gamma))$ is an increasing function of γ. Thus $dv < 0$ implies that $\varphi_2(\gamma)$ decreases as γ increases. If $u \geqq a + \delta > \alpha$, then $H(u) \geqq \epsilon > 0$ for some constant ϵ. This assures that $\varphi_2(\gamma) \to -\infty$ as $\gamma \to \infty$.

Thus, $\varphi(\gamma) > 0$ for $0 < \gamma \leqq a$, $\varphi(\gamma) = \varphi_1(\gamma) + \varphi_2(\gamma)$ is a decreasing function of $\gamma > a$, and $\varphi(\gamma) \to -\infty$ as $\gamma \to \infty$ [since $\varphi_2(\gamma) \to -\infty$ and $\varphi_1(\gamma)$ is decreasing]. Hence, there exists a unique γ_0 such that $\varphi(\gamma_0) = 0$. The corresponding solution of (10.17) is a nontrivial periodic solution, which is unique up to translations of the parameter t.

The fact that $\varphi(\gamma) < 0$ for $\gamma > \gamma_0$ [$\varphi(\gamma) > 0$ for $0 < \gamma < \gamma_0$] makes it clear that this periodic solution is asymptotically orbitally stable from the exterior [interior] as $t \to \infty$.

Exercise 10.3. (a) Verify that Theorem 10.2 applies to van der Pol's equation

$$(10.20) \qquad u'' + \mu(u^2 - 1)u' + u = 0,$$

where $\mu > 0$ is a constant. (b) If (10.17) is the system corresponding to (10.20), what is the nature of the stationary point $(u, v) = (0, 0)$?

Exercise 10.4. Show that if the condition "$H(u) > 0$ and is increasing for $u \geqq a$" is omitted in Theorem 10.2, then the conclusions of Theorem 10.2 are valid with "exactly one" replaced by "at least one" and "exterior and interior" replaced by either "exterior" or "interior."

Exercise 10.5. Let h, g satisfy the conditions of Theorem 10.2; in addition, let $g(u)$ be increasing for $u > 0$; and let $\mu > 0$ be a constant. Let $\Gamma(\mu)$ denote the limit cycle in the (u, v)-plane belonging to

$$(10.21) \qquad u'' + \mu h(u)u' + g(u) = 0;$$

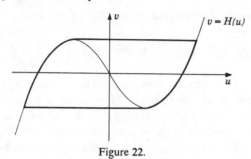

Figure 22.

i.e., the unique nontrivial periodic solution arc of

(10.22) $u' = v - \mu H(u), \quad v' = -g(u).$

Show that, as $\mu \to \infty$, $\Gamma(\mu)$ tends to the closed curve consisting of two horizontal line segments and two arcs on $v = H(u)$ as in Figure 22. Compare Lefschetz [1, pp. 342–346].

APPENDIX: POINCARÉ-BENDIXSON THEORY ON 2-MANIFOLDS

Although the proof of the Poincaré-Bendixson Theorem 4.1 for plane autonomous systems depended on the validity of the Jordan curve theorem, it turns out that analogous results hold under suitable smoothness assumptions on arbitrary 2-dimensional differentiable manifolds. These analogous results will be the main object of this appendix.

11. Preliminaries

The objects of study of this appendix will be flows on 2-manifolds.
Definition. 2-manifold of class C^k. Let M be a connected, Hausdorff topological space for which (i) there is given an open covering $M = \bigcup U_\alpha$, where $\alpha \in A$ and A is some index set, and for each α, a continuous, one-to-one map g_α of U_α onto an open (plane) square such that (ii) if $U_\alpha \cap U_\beta$ is not empty, then $g_\beta(g_\alpha^{-1})$ is a map of $g_\alpha(U_\alpha \cap U_\beta)$ onto $g_\beta(U_\alpha \cap U_\beta)$ which is of class C^k (so that, if $k \geq 1$, this map has a nonvanishing Jacobian). Then $\{M; U_\alpha; g_\alpha\}$ is called a 2-manifold of class C^k.

If m denotes a point of M, then $y_\alpha = g_\alpha(m)$ is a function of $m \in U_\alpha$, the values of which are binary (real) vectors $y_\alpha = (y_\alpha^1, y_\alpha^2)$. As m varies over U_α, y_α varies over a square; e.g., $|y_\alpha^1|, |y_\alpha^2| < 1$. U_α is called a coordinate neighborhood of any point $m \in U_\alpha$ and y_α the local coordinates of $m = g_\alpha^{-1}(y_\alpha)$. Condition (ii) concerns the map $y_\beta = g_\beta(g_\alpha^{-1}(y_\alpha))$. (The parenthetical part concerning the Jacobian $\partial(y_\beta^1, y_\beta^2)/\partial(y_\alpha^1, y_\alpha^2)$ is redundant since the inverse maps $g_\beta(g_\alpha^{-1})$ and $g_\alpha(g_\beta^{-1})$ are assumed of class C^k, $k \geqq 1$.)

It is important to emphasize that the 2-manifold $\{M; U_\alpha; g_\alpha\}$ consists of the space M, the given covering $\bigcup U_\alpha$, and the given set of homeomorphisms g_α. Usually, U_α and g_α are fixed and $\{M; U_\alpha; g_\alpha\}$ will be abbreviated to M.

Definition. Let M be a 2-manifold of class C^k, $k \geqq 0$. Let U, g be an open set of M and $y = g(m)$ a homeomorphism of U onto an open square. Then U is called an admissible coordinate neighborhood on M and $y = g(m)$ (admissible) local coordinates of $m \in U$ provided that $\{M; U_\alpha$ and $U; g_\alpha$ and $g\}$ satisfies conditions (i) and (ii) of the last definition.

Definition. Let M be a 2-manifold of class C^k, $k \geqq 0$. By a flow

(11.1) $$m^t = T^t m \quad \text{or} \quad m^t = \mu(t, m)$$

of class C^k is meant a function $\mu(t, m)$ defined for $-\infty < t < \infty$, $m \in M$ with values in M such that (i) for a fixed t, $T^t : M \to M$ is a homeomorphism of M onto M; (ii) T^t is a group of maps

(11.2) $$T^t T^s = T^{t+s}, \quad \text{i.e.,} \quad \mu(t, \mu(s, m)) = \mu(t + s, m),$$

in particular, $\mu(0, m) = m$; (iii) $\mu(t, m)$ is a continuous function of (t, m); finally, (iv) if $k \geqq 1$, then $\mu(t, m)$ is of class C^k as a function of (t, m).

The last two conditions have the following meaning: Consider any given (t_0, m_0), let U_α be a coordinate neighborhood containing m_0, $y_\alpha = g_\alpha(m)$ local coordinates on U_α, and let $\mu(t_0, m_0) \in U_\beta$. Then (iii) means that $\mu(t, m) \in U_\beta$ for (t, m) near (t_0, m_0) and (iv) means that

(11.3) $$y_\beta = g_\beta(\mu[t, g_\alpha^{-1}(y_\alpha)])$$

is of class C^k as a function of (t, y_α) on the open (t, y_α)-set on which the right side is defined.

Lemma 11.1. *Let $k \geqq 1$. Then, for a fixed t, the Jacobian of the map $S_{\alpha\beta}^t : y_\alpha \to y_\beta$ given by (11.3) does not vanish (wherever $S_{\alpha\beta}^t$ is defined).*

Proof. For, by (ii), $S_{\alpha\beta}^t$ has the inverse

$$S_{\beta\alpha}^{-t} : y_\alpha = g_\alpha(\mu[-t, g_\beta^{-1}(y_\beta)]),$$

which by assumption is also of class C^k. This implies the lemma.

A "flow" is a generalization of the concept of "general solution of autonomous differential equations on M"; cf., e.g., § 14 or § IX 2.

For any given point m_0, the subset $C(m_0) = \{m_0^t = \mu(t, m_0), -\infty < t < \infty\}$ of M is called the orbit through m_0. Since each point $m_0 \in M$ uniquely determines its orbit, it is clear from the group property (ii) that two orbits are either identical or have no points in common.

When the orbit $C(m_0)$ reduces to the point m_0 [i.e., $\mu(t, m_0) = m_0$ for $-\infty < t < \infty$], then m_0 is called a *stationary point*.

Lemma 11.2. *Let $k \geqq 1$ and $m_0 \in U_\alpha$. Then m_0 is a stationary point if and only if $y = g_\alpha(\mu(t, m_0))$ has the derivative 0 at $t = 0$.*

Exercise 11.1.   Verify this lemma.

A subset M_1 of M is called an *invariant set* if $T^t M_1 = M_1$ for every t, $-\infty < t < \infty$. Thus M_1 is an invariant set if and only if $T^t M_1 \subset M_1$ for $-\infty < t < \infty$ or, equivalently, if and only if $C(m_1) \subset M_1$ whenever $m_1 \in M_1$. In particular, if M_1 is a closed invariant set and $m_1 \in M_1$, then $\bar{C}(m_1)$, the closure of the orbit through m_1, is contained in M_1.

A subset N of M is called a *minimal set* if (i) N is a closed invariant set and (ii) N contains no proper subset which is closed and invariant.

For example, if m_0 is a stationary point, then the set consisting of m_0 is a minimal set. More generally, if $\mu(t, m_0)$ is periodic [i.e., if there exists a number $p > 0$ such that $\mu(t + p, m_0) = \mu(t, m_0)$ for $-\infty < t < \infty$], then $C(m_0)$ is a minimal set.

Exercise 11.2.   Let M be a 2-manifold and T^t a flow on M. (a) If M_1, M_2 are invariant sets, then $M_1 \cup M_2$, $M_1 \cap M_2$ and $M_1 - M_2$ are invariant sets. (b) If M_1 is an invariant set and ∂M_1, $M_1{}^0 = M_1 - \partial M_1$ are its boundary and set of interior points, then ∂M_1 and $M_1{}^0$ are invariant sets. (c) If N is a minimal set, then either $N = M$ or N is a nowhere dense set.

Let $m_0 \in M$ and let $C^+(m_0)$ denote the semi-orbit $C^+(m_0) = \{\mu(t, m_0)$ for $t \geqq 0\}$ starting at m_0. A point $m \in M$ is called an *ω-limit point* of $C^+(m_0)$ if there exists a sequence of t-values $0 < t_1 < t_2 < \ldots$ such that $t_n \to \infty$ and $\mu(t_n, m_0) \to m$ as $n \to \infty$. The set of ω-limit points of $C^+(m_0)$ will be denoted by $\Omega(m_0)$. The analogues of Theorems 1.1 and 1.2 are valid.

Exercise 11.3.   Let M be a 2-manifold and T^t a flow on M of class C^0. Let $m_0 \in M$. (a) Let $C^j(m_0) = \{m = \mu(t, m_0)$ for $t \geqq j\} = C^+(\mu(j, m_0))$. Then $\Omega(m_0) = \bar{C}^1(m_0) \cap \bar{C}^2(m_0) \cap \ldots$ and $\Omega(m_0)$ is closed. (b) If $m_1 \in \Omega(m_0)$, then $C(m_1) \subset \Omega(m_0)$; i.e., $\Omega(m_0)$ is an invariant set. (c) If, in addition, $C^+(m_0)$ has a compact closure, then $\Omega(m_0)$ is connected.

Exercise 11.4.   Let M, T^t be as in Exercise 11.3, N a minimal set, and $m_0 \in N$. (a) Then $\Omega(m_0) \subset \bar{C}(m_0) = N$. (b) If, in addition, $\Omega(m_0)$ is not empty (e.g., if N is compact), then $\Omega(m_0) = N$.

Let M, T^t be of class C^k, $k \geqq 1$. Let U_α be a coordinate neighborhood on M and $y = g_\alpha(m)$ local coordinates of $m \in U_\alpha$; thus $g_\alpha : U_\alpha \to S$ is a homeomorphism of U_α onto a square S, say, $S : |y^1|, |y^2| < 1$. A closed [or open] arc $\Gamma : y = y_0(s)$ for $|s| \leqq a$ [or $|s| < a$] of class C^k in S or its image $\tilde{\Gamma} : m_0(s) = g_\alpha^{-1}(y_0(s))$ in U_α is called a *transversal* of the flow T^t if no orbit is tangent to $\tilde{\Gamma}$, i.e., if the differentiable arc $y = g_\alpha(\mu(t, m_0(s)])$, which is defined for small $|t|$ and starts at $y_0(s)$ when $t = 0$, is not tangent to Γ for any s on $|s| \leqq a$ [or $|s| < a$].

Lemma 11.3. *Let $k \geqq 1$, $m_0 \in U_\alpha$ a nonstationary point of the flow T^t*

and $y = g_\alpha(m)$ local coordinates on U_α. Then there exists a C^k transversal arc $\Gamma : y = y_0(s)$ or $\tilde{\Gamma} : m_0(s) = g_\alpha^{-1}(y_0(s))$ for $|s| \leqq a_0$ satisfying $m_0(0) = m_0$. Furthermore, for any such transversal and sufficiently small $a > 0$, the set $U = \{m = \mu(t, m_0(s)),$ where $|t| < a, |s| < a\}$ and $y = g(m) \equiv (s, t)$ for $m \in U$ are an admissible coordinate neighborhood and local coordinates on U, respectively.

Exercise 11.5. Verify this lemma.

12. Analogue of the Poincaré-Bendixson Theorem

The main object of this section is to prove:

Theorem 12.1 (A. J. Schwartz). *Let M be a 2-manifold of class C^2 and T^t a flow on M of class C^2. Let N be a nonempty compact minimal set. Then N is either* (i) *a stationary point m_0; or* (ii) *a periodic orbit (which is homeomorphic to a circle); or* (iii) *$N = M$.*

In case (iii), $M = N$ is compact and has a flow T^t on it without stationary points. It follows from the Euler-Poincaré formula (relating genus and the sum of indices of singular points of a vector field on M) that the genus of M is 1; thus M is homeomorphic to a torus or a Klein bottle. Actually, a flow on a Klein bottle without stationary points necessarily has a periodic orbit. Thus, in case (iii), M is a torus. For a discussion of these remarks, see H. Kneser [2].

The assumptions concerning the C^2-property of T^t cannot be reduced to C^1 even if M is of class C^∞; see Exercise 14.3.

Proof of Theorem 12.1. In view of Exercise 11.2(c), $N = M$ or N is a nowhere dense set. Suppose that the theorem is false. Thus $N \neq M$ is a compact, nowhere dense, minimal set in M which does not contain a stationary point or a periodic orbit. It will be shown that this is impossible.

Let $m_0 \in N$ and let $\tilde{U}$ and $y = g(m)$, where $y \equiv (s, t)$, be an admissible neighborhood and local coordinates as furnished by Lemma 11.3. Thus, if $m = g^{-1}(y) \equiv g^{-1}(s, t)$, where $|s|, |t| < a$, then $m_0 = g^{-1}(0, 0)$, $\tilde{\Gamma} : m_0(s) = g^{-1}(s, 0)$ is a transversal arc of class C^2 and, for a fixed s, $g^{-1}(s, t) = \mu(t, m_0(s))$ is part of the orbit $C(m_0(s))$.

A point m and its local coordinates (s, t) will be identified; e.g., m_0 will be referred to as the point $(0, 0)$. Let Γ represent the open line segment $|s| < a$, $t = 0$, which is the transversal arc $\Gamma : y = g(m_0(s)) \equiv (s, 0)$. The point $(s, 0)$ of Γ will be referred to simply as s.

Since N contains no interior points, it is clear that $K = N \cap \Gamma$ contains no s-interval, so that K is a nonempty, nowhere dense set, closed relative to Γ. Let

(12.1) $$W = (-a, a) - K,$$

so that W is an open s-set and has a decomposition $W = \bigcup(\alpha_j, \beta_j)$ into pairwise disjoint, open s-intervals (α_1, β_1), (α_2, β_2), ...

Let U be the set of s-values on $|s| < a$ such that the semi-orbit $C^+(m_0(s))$ starting for $t = 0$ at the point $m_0(s) \in \tilde{\Gamma}$ meets $\tilde{\Gamma}$ for some $t = t(s)$ and does not pass through an endpoint of $\tilde{\Gamma}$ for $0 < t < t(s)$. It is clear that U is an open s-set on $|s| < a$. For $s \in U$, let $t(s)$ be the least positive t-value [in fact, $t(s) \geqq a$] such that $\mu(t, m_0(s)) \in \tilde{\Gamma}$ and let $f(s)$ denote the s-coordinate of $\mu(t(s), m_0(s))$; i.e.,

(12.2) $(f(s), 0) = g(\mu[t(s), m_0(s)])$ for $s \in U$.

We can obtain $t(s)$ and $f(s)$ in another way. Let $s_0 \in U$ and define the real-valued functions σ, τ of (s, t) by

$$(\sigma, \tau) = g(\mu[t, m_0(s)]) = g(\mu[t, g^{-1}(s, 0)])$$

for small $|s - s_0|$, $|t - t(s_0)|$. Thus, on the orbit starting at $m_0(s)$, (σ, τ) are coordinates of the point corresponding to the time t. In particular $\sigma(s, t(s)) = f(s)$ and $\tau(s, t(s)) = 0$. Then σ, τ are of class C^2 with a non-vanishing Jacobian $\partial(\sigma, \tau)/\partial(s, t)$ by Lemma 11.1.

Since Γ is a transversal, $\partial\tau(s, t)/\partial t \neq 0$ at $t = t(s)$, hence $t(s)$ is of class C^2 by the Implicit Function Theorem. Hence

(12.3) $f(s)$ is of class C^2.

Note that $\sigma(s, t)$, $\tau(s, \tau)$ satisfy

$$(\sigma(s, t + t(s)), \tau(s, t + t(s))) = (\sigma(s, t + t(s)), t)$$

for small $|s - s_0|$, $|t|$. Since $\partial(\sigma, \tau)/\partial(s, t) \neq 0$, it follows that the Jacobian of $(\sigma(s, t + t(s)), t)$ with respect to (s, t), which is the product of the Jacobians $\partial(\sigma, \tau)/\partial(s, t)$ and $\partial(s, t + t(s))/\partial(s, t)$, does not vanish. Hence the partial derivative of $\sigma(s, t + t(s))$ with respect to s is not 0. Since $f(s) = \sigma(s, t + t(s))$ for $t = 0$,   it follows that

(12.4) $Df(s) \neq 0$ for $s \in U$ if $D = d/ds$.

Since N is a compact minimal set, $m_1 \in N$ implies that $\bar{C}(m_1) = N$; cf. Exercise 11.4. This shows that if $s_1 \in K = N \cap \Gamma$, then $s_1 \in U$, i.e., that $K \subset U$. Let V be an open subset of U such that

(12.5) $K \subset V \subset \bar{V} \subset U$.

In particular, (12.3) and (12.4) imply that there exist constants R and C such that

(12.6) $0 < 1/R \leqq |Df(s)| \leqq R$ for $s \in V$ and $R > 1$,

(12.7) $|D^2f(s)| \leqq C$ for $s \in V$ and $C > 1$.

It is clear that

(12.8) $$f: U \to (-a, a)$$

is a one-to-one map and has an inverse $f^{-1}(s)$. Below $f^k(s)$, $k = 0, \pm 1, \ldots$, denote the iterates of f and f^{-1}; e.g., $f^0(s) = s$ and if $s \in U$ and $f(s) \in U$, then $f^2(s) = f(f(s))$. In particular, if $k > 0$, then $f^k(s)$ is defined on $U \cap f^{-1}(U) \cap \cdots \cap f^{-(k-1)}(U)$; a similar remark applies to $k < 0$.

In addition to the properties (12.6), (12.7), the function $f(s)$ has the properties that

(12.9) $$f(K) = K$$

since N is invariant; that

(12.10) $f^k(s) = s$ for some $s \in K$ implies that $k = 0$,

since N contains no periodic orbits; and finally, that

(12.11) K is a nonempty minimal set with respect to f

in the sense that K contains no proper subsets K_0 such that K_0 is closed with respect to K and that $f(K_0) = K_0$ [or equivalently, $f^{\pm 1}(K_0) \subset K_0$].

The remainder of the proof consists in proving the assertion:

Lemma 12.1. *If K is a closed set and V, U are open sets of $|s| < a$ satisfying (12.5), then there cannot exist a function $f(s)$ defined on U satisfying (12.3)–(12.11).*

Proof. (a) Assume that this lemma is false and that $f(s)$ exists. Put

(12.12) $\epsilon = \text{dist}\,(K, (-a, a) - V)$,   so that $0 < \epsilon < a$.

Note that if (α_j, β_j) is an interval of $W = (-a, a) - K$, then

(12.13) $\beta_j - \alpha_j < \epsilon$ implies that $[\alpha_j, \beta_j] \subset V \subset U$

since $\alpha_j, \beta_j \in K$.

(b) It is easy to see that

(12.14) $f(K_1) = K_1$ if $K_1 = \{\alpha_1, \beta_1, \alpha_2, \beta_2, \ldots\} - \{-a, a\}$

are the endpoints $\alpha_j, \beta_j \in K$. For $s_0 \in K - K_1$ if and only if s_0 is a limit point of both $K \cap (-a, s_0)$ and $K \cap (s_0, a)$. Thus (12.14) follows from (12.4) and (12.9).

(c) It will be shown that there exists an integer i such that $\alpha = \alpha_i$, $\beta = \beta_i$ satisfy

(12.15) $f^k([\alpha, \beta]) \subset V$ for $k = 0, 1, \ldots$.

Let Q be the finite set of endpoints α_j, β_j of intervals (α_j, β_j) of W such that $\beta_j - \alpha_j \geqq \epsilon$. In view of (12.10), there is an integer n such that

$f^k(\alpha_1) \notin Q$ for $k \geqq n$. Then, by (12.14), $f^n(\alpha_1) = \alpha_i$ or $f^n(\alpha_1) = \beta_i$ for some i and $|f^k(\beta_1) - f^k(\alpha_1)| < \epsilon$ for $n \geqq k$. Thus (12.15) follows from (12.13).

(d) Let $n \geqq 0$ be an integer and $[p, q]$ a closed s-interval such that $f^k([p, q]) \subset V$ for $0 \leqq k \leqq n$. Then

$$(12.16) \qquad \left| \frac{Df^{k+1}(s)}{Df^{k+1}(r)} \right| \leqq \exp CR \sum_{j=0}^{k} |f^j(p) - f^j(q)|$$

for $0 \leqq k \leqq n$ and $p \leqq r, s \leqq q$. In order to verify this, note that $f^{k+1}(s) = f(f^k(s))$. Thus if $Df \circ f^k(s)$ denotes the derivative of f evaluated at $f^k(s)$, then we have

$$(12.17) \quad Df^{k+1}(s) = [Df \circ f^k(s)] Df^k(s), \qquad \text{hence} \qquad Df^{k+1}(s) = \prod_{j=0}^{k} Df \circ f^j(s).$$

Consequently,

$$\left| \log \left| \frac{Df^{k+1}(s)}{Df^{k+1}(r)} \right| \right| \leqq \sum_{j=0}^{k} \left| \log |Df \circ f^j(s)| - \log |Df \circ f^j(r)| \right|$$

and, by the mean value theorem of differential calculus,

$$\left| \log \left| \frac{Df^{k+1}(s)}{Df^{k+1}(r)} \right| \right| \leqq \sum_{j=0}^{k} |Df(\theta_j)|^{-1} |D^2 f(\theta_j)| \cdot |f^j(s) - f^j(r)|$$

for suitable θ_j between $f^j(r)$ and $f^j(s)$. Thus (12.16) follows from (12.6), (12.7).

(e) Let α, β be as in (12.15) in (c). Then

$$(12.18) \quad \delta \equiv \sum_{k=0}^{\infty} |Df^k(\alpha)| \qquad \text{satisfies} \quad 1 \leqq \delta \leqq 2a(\beta - \alpha)^{-1} e^{2aCR}$$

In order to see this, note first that $Df^0(s) = 1$, so $\delta \geqq 1$. There is a $\theta_k \subset (\alpha, \beta)$ such that

$$|f^k(\alpha) - f^k(\beta)| = |Df^k(\theta_k)| (\beta - \alpha).$$

Since the intervals with endpoints $f^k(\alpha), f^k(\beta)$ are contained in $(-a, a)$ and are pairwise disjoint for $k \geqq 0$ by (12.10), it follows that

$$(\beta - \alpha) \sum_{k=0}^{\infty} |Df^k(\theta_k)| \leqq \sum_{k=0}^{\infty} |f^k(\alpha) - f^k(\beta)| \leqq 2a.$$

In addition, by (12.16),

$$|Df^k(\alpha)| \leqq |Df^k(\theta_k)| e^{2aCR} \qquad \text{for} \quad k = 0, 1, \ldots.$$

This proves (12.18).

(f) Let d denote the number

$$(12.19) \qquad d = \epsilon/3CR\delta(1 + a); \quad \text{in particular } 0 < d < \epsilon.$$

It will be shown that

(12.20k) $f^k([\alpha - d, \alpha + d]) \subset V,$

(12.21k) $|f^k(s) - f^k(\alpha)| < \epsilon$ for $|s - \alpha| \leq d,$

(12.22k) $|Df^k(s)| \leq 3 |Df^k(\alpha)|$ for $|s - \alpha| \leq d$

hold for $k = 0, 1, \ldots$. In view of $f^k(\alpha) \in K$ and the definition (12.12) of ϵ, it is seen that (12.20k) is a consequence of (12.21k); cf. (12.13).

The relations (12.21k), (12.22k) are trivial for $k = 0$. Assume that (12.21k), (12.22k) hold for $k = 0, \ldots, n$. Then, by (12.16) in (d),

$$|Df^{n+1}(s)| \leq |Df^{n+1}(\alpha)| \exp CR \sum_{j=0}^{n} |f^j(s) - f^j(\alpha)| .$$

By the mean value theorem and (12.22j), for $j = 0, \ldots, n$,

$$\sum_{j=0}^{n} |f^j(s) - f^j(\alpha)| \leq |s - \alpha| \sum_{j=0}^{n} |Df^j(\theta_j)| \leq 3 |s - \alpha| \sum_{j=0}^{n} |Df^j(\alpha)| ,$$

where θ_j is between α and s. Thus (12.18) implies that

$$|Df^{n+1}(s)| \leq |Df^{n+1}(\alpha)| e^{3CRd\delta}$$

But since $3CRd\delta < \epsilon/a \leq 1$ and $e < 3$, the inequality (12.22 n + 1) holds. This inequality shows that

$$|f^{n+1}(s) - f^{n+1}(\alpha)| \leq |s - \alpha| \cdot |Df^{n+1}(\theta)| \leq 3d |Df^{n+1}(\alpha)| \leq 3d\delta,$$

where $|\alpha - \theta| < |s - \alpha|$. Since $C > 1$ and $R > 1$ imply that $3d\delta < \epsilon$, the inequality (12.21 n + 1) holds. This proves (12.20k)–(12.22k) for $k = 0, 1, \ldots$.

(g) In view of (12.18) and (12.22k),

(12.23) $\lim_{n \to \infty} Df^n(s) = 0$ uniformly for $|\alpha - s| \leq d.$

By the minimality property of K, the closure of the sequence of points $f^{j+n}(\alpha)$ for $n = 0, 1, \ldots$ (and j fixed) is the set K. Hence there exists a large value of k (>0) such that

$$|f^k(\alpha) - \alpha| \leq \frac{d}{2} \quad \text{and} \quad |Df^k(s)| < \tfrac{1}{2} \quad \text{for} \quad |s - \alpha| \leq d.$$

Thus $|f^k(\alpha \pm d) - \alpha| < d$, and so $f^k(s) - s$ has opposite signs at $s = \alpha \pm d$. Hence there is an s-value s_0 such that $f^k(s_0) = s_0$ and $|s_0 - \alpha| < d$. In addition, $f^{nk}(s_0) = s_0$ for $n = 1, 2, \ldots$ and, by (12.23), $f^{nk}(\alpha) \to s_0$ as $n \to \infty$. Since K is closed and $\alpha \in K$ implies that $f^{nk}(\alpha) \in K$, we have $s_0 \in K$. This contradicts (12.10) and proves Lemma 12.1 and Theorem 12.1.

Lemma 12.2. *Let M be a 2-manifold and T^t a flow on M of class C^0. Let M_1 be a nonempty, compact, invariant subset. Then M_1 contains at least one nonempty, minimal set.*

Since the intersection of invariant sets is an invariant set, this lemma is an immediate consequence of Zorn's lemma. (For the statement and proof of the latter, see Kelley [1, p. 33]). Theorem 12.1 and Lemma 12.1 give an analogue of the Poincaré-Bendixson theorem:

Theorem 12.2. *Let M be an orientable 2-manifold of class C^2, T^t a flow on M of class C^2, and $m_0 \in M$. Suppose that $\Omega(m_0) \neq M$ and that $\Omega(m_0)$ is a nonempty compact set which contains no stationary points. Then $\Omega(m_0)$ is a Jordan curve and $C^+(m_0)$ spirals toward $\Omega(m_0)$.*

When N is a Jordan curve on M containing no stationary points, then $C^+(m_0)$ is said to spiral toward N if, for every $m \in N$, there is a transversal arc $\Gamma = \Gamma(m)$ through m such that successive intersections of $C^+(m_0) = \{m_0{}^t = \mu(t, m_0), t \geqq 0\}$ and Γ tend monotonically to m.

The concept of the "orientability" of M is used here to mean the following: Let N be a Jordan curve on M of class C^2. Then there exist an open set V containing N and a C^2-diffeomorphism of the cylinder $\{(s, y^1, y^2): |s| < 1, (y^1)^2 + (y^2)^2 = 1\}$ onto V such that the circle $s = 0$ is mapped onto N. Thus either (s, y^1) or (s, y^2) are local coordinates.

Proof. By Lemma 12.2, $\Omega(m^0)$ contains a nonempty, compact minimal set N and, by Theorem 12.1, N is a Jordan curve (containing no stationary points). Let V be a neighborhood of N described above. By considering the image of the flow on the cylindrical image of V, it is clear that the arguments in parts (*b*) and (*c*) of the proof of the Poincaré-Bendixson Theorem 4.1 are valid. Hence $C^+(m_0)$ spirals toward N. This implies, in particular, that $N = \Omega(m_0)$ and completes the proof of the theorem.

13. Flow on a Closed Curve

In view of the remarks following Theorem 12.1, it is seen that the torus is an exceptional 2-manifold M in that it can admit flows for which M is a minimal set. It seems therefore of interest to examine flows on a torus. This section is a preparation for such an examination and concerns a "flow" on a Jordan curve Γ or, equivalently, a topological map of Γ onto itself.

Let Γ be a Jordan curve and $S : \Gamma \to \Gamma$ an orientation preserving homeomorphism of Γ onto itself. The discrete group of homeomorphisms $S^n, n = 0, \pm 1, \ldots,$ of Γ onto itself will be called a flow on Γ. As in § 11, we can define the orbit $C(\gamma) : \{S^n\gamma : n = 0, \pm 1, \ldots\}$ through a point $\gamma \in \Gamma$, semi-orbit $C^+(\gamma) = \{S^n\gamma : n = 0, 1, \ldots\}$, the set $\Omega(\gamma)$ of ω-limit points of $C^+(\gamma)$, invariant sets, minimal sets, etc.

The points of the curve Γ can be considered to be parametrized, say $\gamma = \gamma(y)$, where $0 \leq y \leq 1$, and the points $\gamma \in \Gamma$ corresponding to $y = 0$ and $y = 1$ are identical. Or, more conveniently, Γ can be considered as a line, $-\infty < y < \infty$, on which we identify any two points y_1, y_2 for which $y_1 - y_2$ is an integer. Then $\gamma_1 = S\gamma$ can be represented as real-valued function $y_1 = f(y)$ satisfying

(i) $f(y)$ is continuous and strictly increasing,
(ii) $f(y + 1) = f(y) + 1$, so that $f(y) - y$ has the period 1.

The fact that f is nondecreasing merely reflects the fact that S is orientation preserving. The condition that f is strictly increasing and satisfies (ii) is a consequence of the fact that S is a homeomorphism. Conversely, any function $f(y)$ satisfying (i) and (ii) induces an orientation preserving homeomorphism S of Γ onto itself.

Let $f^0(y) = y$ and $f^{-1}(y)$ be the function inverse to $f(y)$. Let $f^2(y) = f(f(y))$, $f^3(y) = f(f^2(y))$, ... and similarly $f^{-2}(y) = f^{-1}(f^{-1}(y))$, $f^{-3}(y) = f^{-1}(f^{-2}(y))$, ..., so that $f^n(y)$ corresponds to S^n for $n = 0, \pm 1, \ldots$ and each $f^n(y)$ satisfies (i) and (ii).

Lemma 13.1. *Let* $f(y)$ *be a continuous function for* $-\infty < y < \infty$ *satisfying* (i) *and* (ii)*, then there exists a number* α *such that*

(13.1) $$f^n(y)/n \to \alpha \quad \text{as} \quad |n| \to \infty \quad \text{for all } y;$$

in fact, if δ_n*,* ϵ_n *are defined by*

$$n\alpha - \delta_n = \min (f^n(y) - y), \quad n\alpha + \epsilon_n = \max (f^n(y) - y)$$

for $-\infty < y < \infty$*, so that*

$$-\delta_n \leq f^n(y) - y - n\alpha \leq \epsilon_n,$$

then

(13.2) $$\delta_n \geq 0, \quad \epsilon_n \geq 0 \quad \text{and} \quad \delta_n + \epsilon_n < 1.$$

The number α is called the *rotation number* of the map S or of the flow S^n on Γ.

Proof. (*a*) The continuous function

(13.3) $$\varphi_m(y) = f^m(y) - y.$$

has the period 1, by the analogue of (ii), and satisfies

(13.4) $|m| (\beta^m - \beta_m) < 1$ if β^m [or β_m] = max [or min] $\varphi_m(y)/|m|$.

For suppose that (13.4) does not hold. Then $\varphi_m(y_1) - \varphi_m(y_2) = 1$ for some points y_1, y_2. Since φ_m is periodic, it can be supposed that $y_2 < y_1 < y_2 + 1$. Then, by (13.3), $f^m(y_1) - f^m(y_2) = 1 + y_1 - y_2 > 1$; so that $f^m(y_1) > 1 + f^m(y_2) = f^m(1 + y_2)$. But this contradicts the increasing character of f^m.

(b) Let $k \geq 1$ be an integer. Then (13.3) shows that

$$f^{jk}(y) = f^k(f^{(j-1)k}(y)) = f^{(j-1)k}(y) + \varphi_k(f^{(j-1)k}(y)).$$

Hence

$$k\beta_k \leq f^{jk}(y) - f^{(j-1)k}(y) \leq k\beta^k \quad \text{for} \quad j = 1, 2, \ldots;$$

in particular, for $k = 1$,

$$\beta_1 \leq f^i(y) - f^{i-1}(y) \leq \beta^1 \quad \text{for} \quad i = 1, 2, \ldots.$$

Then if $n > 0$ and $n = kd + r$, where $1 \leq r < k$, a sum of these relations for $j = 1, \ldots, d$ and $i = kd + 1, \ldots, kd + r$ gives

$$dk\beta_k + r\beta_1 \leq f^n(y) - y \leq dk\beta^k + r\beta^1.$$

By the definition of β_n, β^n, and $\varphi_n(y)$, this implies that

$$dk\beta_k + r\beta_1 \leq n\beta_n \leq n\beta^n \leq dk\beta^k + r\beta^1.$$

Consequently

(13.5) $$\beta_k \leq \liminf_{n \to \infty} \beta_n \leq \limsup_{n \to \infty} \beta^n \leq \beta^k.$$

But since $\beta^k - \beta_k \to 0$, $k \to \infty$, by (13.4), it follows that these upper and lower limits are equal, say, to α. Thus $\varphi_n(y)/n \to \alpha$ as $n \to \infty$ uniformly in y. This gives the part of (13.1) concerning $n \to +\infty$.

By (13.4) and (13.5), $k\beta_k \leq \varphi_k(y) \leq k\beta^k$ and $k\beta_k \leq k\alpha \leq k\beta^k$. Hence if $\delta_k = k\alpha - k\beta_k$ and $\epsilon_k = k\beta^k - k\alpha$, then (13.2) holds for $n = 0, 1, \ldots$.

Note that $f^{-n}(y) - y = z - f^n(z)$ if $z = f^{-n}(y)$ and so (13.2) holds with $\delta_{-n} = \epsilon_n$ and $\epsilon_{-n} = \delta_n$ for $n = 1, 2, \ldots$. This completes the proof.

Lemma 13.2. *The rotation number α of S is a topological invariant, i.e., is independent of the parametrization of Γ.*

Proof. Let S determine $f(y)$. A change of parameters $z = g(y)$ on Γ, preserving orientation, is given by a function $g(y)$ satisfying the analogue of (i) and (ii). If $R:z = g(y)$, then, in the z-parametrization, S becomes $RSR^{-1}:z_1 = h(z)$, where $h(z) = g(f(g^{-1}(z)))$ satisfies (i) and (ii). Since $(RSR^{-1})^n = RS^nR^{-1}:z_n = h^n(z) \equiv g(f^n(g^{-1}(z)))$, it follows that $h^n(z) = g(g^{-1}(z) + n\alpha + r_n)$ where $-\delta_n \leq r_n \leq \epsilon_n$ by (13.2). Let $\psi(z) = g(z) - z$, so that $\psi(z)$ has the period 1. Then, as $n \to \infty$,

$$\frac{h^n(z)}{n} = \frac{g^{-1}(z) + n\alpha + r_n + \psi(g^{-1}(z) + n\alpha + r_n)}{n} \to \alpha.$$

This proves the lemma.

Theorem 13.1. *Let S^n be a flow on a Jordan curve Γ. Then the rotation number α of S is rational if and only if S^k has a fixed point γ_0, $S^k\gamma_0 = \gamma_0$, for some integer $k > 0$.*

Proof. Let S belong to $f(y)$. Suppose that $S^k\gamma_0 = \gamma_0$ for some $k > 0$, i.e., that there is a number y_0 and an integer r such that $f^k(y_0) = y_0 + r$.

Hence

$$f^{nk}(y_0) = y_0 + nr \quad \text{and} \quad \alpha = \lim \frac{f^{nk}(y_0)}{nk} = \frac{r}{k}$$

is rational.

Conversely, suppose that $\alpha = r/k$ is rational, where $k > 0$. Then $f^k(y) - y - r$ attains non-negative and non-positive values by (13.2). Hence there is a y_0 such that $f^k(y_0) - y_0 - r = 0$; i.e., $S^k\gamma_0 = \gamma_0$ for the point $\gamma_0 \in \Gamma$ corresponding to y_0. This proves the lemma.

Lemma 13.3. *Let S have an irrational rotation number α. Let $\gamma_0 \in \Gamma$ be fixed, $\gamma_n = S^n\gamma_0$, j and k fixed integers, and Γ_0 either of the closed arcs on Γ bounded by γ_j, γ_k. Then there exists an integer $n > 0$ such that $\Gamma \subset \Gamma^n$, where*

$$\Gamma^n = \Gamma_0 \cup S^{-|j-k|}\Gamma_0 \cup \cdots \cup S^{-n|j-k|}\Gamma_0.$$

Hence, if $\gamma \in \Gamma$, there is an $i = i(\gamma)$, $0 \leq i \leq n$, such that $S^{i|j-k|}\gamma \in \Gamma_0$.

Proof. For the sake of definiteness, let Γ_0 be the oriented subarc of Γ from γ_k to γ_j and let $\Gamma^{(n)}$ be the union of arcs

$$(13.6) \qquad \Gamma^{(n)} = \Gamma_0 \cup S^{-(j-k)}\Gamma_0 \cup \cdots \cup S^{-n(j-k)}\Gamma_0.$$

Thus $\Gamma^n = \Gamma^{(n)}$ or $\Gamma^n = S^{n(j-k)}\Gamma^{(n)}$ according as $j > k$ or $j < k$ and so, $\Gamma \subset \Gamma^n$ if and only if $\Gamma \subset \Gamma^{(n)}$. Thus it suffices to show that if $n > 0$ is large, then $\Gamma \subset \Gamma^{(n)}$. The mth and $(m + 1)$st arc on the right of (13.6) abut at a common endpoint $S^{-m(j-k)}\gamma_k$. Hence if $\Gamma \not\subset \Gamma^{(n)}$ for $n = 1, 2, \ldots$, the endpoints $S^{-n(j-k)}\gamma_k$ tend monotonously to a point γ^0 on Γ as $n \to \infty$. But then $\gamma^0 = \lim S^{-n(j-k)}\gamma_k = \lim S^{-(n-1)(j-k)}\gamma_k = S^{j-k}\gamma^0$ whereas, by Theorem 13.1, S^{j-k} has no fixed point. This contradiction shows that $\Gamma \subset \Gamma^{(n)}$ for large n.

Thus, if $\gamma \in \Gamma$, then $\gamma \in S^{-m|j-k|}\Gamma_0$ or $S^{m|j-k|}\gamma \in \Gamma_0$ for some $m = m(\gamma)$, $0 \leq m \leq n$. This proves the lemma.

If $S^k\gamma_0 = \gamma_0$ for some $k > 0$, the orbit $C(\gamma_0)$ consists of the points $\gamma_0, S\gamma_0, \ldots, S^k\gamma_0$. If no S^k, $k \neq 0$, has a fixed point, the situation is as follows:

Theorem 13.2. *Let S^n be a flow on a Jordan curve Γ having an irrational rotation number α. For any $\gamma \in \Gamma$, let $\Omega(\gamma)$ be the set of ω-limit points of the semi-orbit $C^+(\gamma)$. Then $N_0 = \Omega(\gamma)$ is independent of γ and hence is a unique minimal set of S^n. Furthermore $N_0 = \Gamma$ or N_0 is a perfect nowhere dense set on Γ.*

When $N_0 = \Gamma$, the flow S^n is said to be *ergodic*.

Remark. Since N_0 is a unique minimal set of the flow S^n, as well as S^{-n}, it is clear that S^n is ergodic if and only if S^{-n} is ergodic. Thus the set N_0 is the set of limit points of any semi-orbit $C^+(\gamma) = \{S^n\gamma, n = 0, 1, \ldots\}$ or of any semi-orbit $C^-(\gamma) = \{S^{-n}\gamma, n = 0, 1, \ldots\}$ or of any orbit $C(\gamma) = \{S^n\gamma, n = 0, \pm 1, \ldots\}$.

Proof. Consider the semi-orbit $C^+(\gamma_0)$ and its set $\Omega(\gamma_0)$ of ω-limit points. Let $\gamma^0 \in \Omega(\gamma_0)$ and γ any point of Γ. Since there are points γ_j, $\gamma_k \in C^+(\gamma_0)$ arbitrarily near to γ^0 and the smaller of the arcs Γ_0 bounded by them contains a point $S^i\gamma$ for some $i \geq 0$, it follows that $\gamma^0 \in \Omega(\gamma)$; i.e., $\Omega(\gamma_0) \subset \Omega(\gamma)$. Interchanging the roles of γ and γ_0 shows that $N_0 = \Omega(\gamma)$ is independent of γ.

The set N_0 is closed and invariant. If $\gamma \in N_0$, then γ is a limit point of $C^+(\gamma) \subset N_0$, hence a limit point of N_0. Thus N_0 is perfect. Clearly N_0 is minimal. Hence $N_0 = \Gamma$ or N_0 is nowhere dense; cf. Exercise 1*.2(c). This proves Theorem 13.2.

Lemma 13.4. *Let S have an irrational rotation number α. For a given y_0, the function $g(y_n + k) = n\alpha + k$, where $y_n = f^n(y_0)$ and $n, k = 0, \pm 1, \ldots$ is an increasing function on the sequence of numbers $\{y_n + k\}$.*

Proof. It has to be shown that

$$(13.7) \quad f^n(y_0) + k < f^m(y_0) + j \quad \text{implies that} \quad n\alpha + k < m\alpha + j.$$

Applying f^{-m} to the first inequality in (13.7) gives

$$f^{n-m}(y_0) + k < y_0 + j \quad \text{or} \quad f^{n-m}(y_0) - y_0 < j - k.$$

Since $f^{n-m}(y) - y$ cannot be an integer, $\max (f^{n-m}(y) - y) < j - k$ and so (13.2) implies that $(n - m)\alpha < j - k$. This proves (13.7) and the lemma.

For the next theorem, the following simple lemma will be needed.

Lemma 13.5 (Kronecker). *Let α be an irrational number and S^n the flow on a Jordan curve Γ such that the corresponding $f(y)$ is $f(y) = y + \alpha$. Then S^n is ergodic (i.e., $N_0 = \Gamma$).*

In view of the Remark following Theorem 13.2, this is equivalent to the statement that the set of points $\{y = n\alpha + k$, where $k, n = 0, \pm 1, \ldots\}$ is dense on $-\infty < y < \infty$. In other words, if $[y]$ denotes the largest integer not exceeding y and $\gamma_n = n\alpha - [n\alpha]$ is the fractional part of $n\alpha$, then the lemma is equivalent to the statement that the set $\{y = \gamma_n$, where $n = 0, \pm 1, \ldots\}$ is dense on $0 \leq \gamma \leq 1$.

Exercise 13.1. Prove Lemma 13.5.

Theorem 13.3. *Let S^n be a flow on a Jordan curve Γ having an irrational rotation number α. Then S^n is ergodic (i.e., $N_0 = \Gamma$) if and only if S is topologically equivalent to a "rotation," i.e., if and only if there exists an orientation preserving change of parameters $R : z = g(y)$ such that*

$$(13.8) \quad RSR^{-1} : z_1 = z + \alpha,$$

Proof. Let there exist an R satisfying (13.8). The minimal set of the map RSR^{-1} is the set N_0 of limit points of $\alpha, 2\alpha, \ldots$ on Γ by Theorem

13.2. Since α is irrational, this minimal set N_0 is Γ. Hence the minimal set for S is Γ.

Conversely, let $N_0 = \Gamma$ be the minimal set for S. Let $y_0 = 0$ and $y_n = f^n(0)$ in Lemma 13.4. Then the increasing function $z = g(y)$ is defined on the sequence $\{y_n + k, n, k = 0, \pm 1, \ldots\}$ which is dense on $-\infty < y < \infty$. It is continuous since $\{n\alpha + k, \text{ for } n, k = 0, \pm 1, \ldots\}$ is dense on $-\infty < z < \infty$ and, hence has a unique continuous extension to an increasing function for $-\infty < y < \infty$. Furthermore, $g(y + 1) = g(y) + 1$. Thus $R{:}z = g(y)$ defines a change of parameters on Γ.

The relation $g(f^n(0) + k) = n\alpha + k$ implies that if $y = f^n(0) + k$, then $g(f(y)) = g(f^{n+1}(0) + k) = (n + 1)\alpha + k = g(y) + \alpha$. Since the set of the points $\{f^n(0) + k\}$ is dense on $-\infty < y < \infty$, it follows that $g(f(y)) = g(y) + \alpha$ for all y. That is, $RSR^{-1}{:}z_1 = g(f(g^{-1}(z))) = z + \alpha$. This proves the theorem.

Exercise 13.2.   Let Γ be a circle and N_0 any perfect, nowhere dense set on Γ. There exists an orientation preserving homeomorphism S of Γ onto itself having (an irrational rotation number α and having) N_0 as its unique minimal set. See Denjoy [1].

Lemma 12.1 occurring in the proof of Theorem 12.1 has the following consequence.

Theorem 13.4.   *Let* S^n *be a flow on* Γ *having an irrational rotation number* α. *Let* S *belong to* $f(y)$ *and suppose that* $f(y), f^{-1}(y)$ *are of class* C^2. *Then* S *is ergodic (i.e.,* $N_0 = \Gamma$).

This assertion can be improved slightly as follows:

Exercise 13.3.   (a) Let the condition that "$f(y), f^{-1}(y)$ are of class C^2" be relaxed to "$f(y), f^{-1}(y)$ are of class C^1 and df/dy is of bounded variation for $0 \leq y \leq 1$." Then S is ergodic. Denjoy; see van Kampen [1]. (b) This assertion is false if the condition that "df/dy is of bounded variation for $0 \leq y \leq 1$" is omitted. See Denjoy [1].

14. Flow on a Torus

A torus M will be viewed as a square $0 \leq x \leq 1, 0 \leq y \leq 1$ in the (x, y)-plane in which the points $(x, 0), (x, 1)$ or $(0, y), (1, y)$ on opposite sides of the square are identified or, even more conveniently, as the entire (x, y)-plane in which pairs of points $(x_1, y_1), (x_2, y_2)$ are considered identical if and only if $x_1 - x_2, y_1 - y_2$ are integers.

Consider a continuous flow

$$(14.1) \qquad T^t{:}x^t = \xi(t, x, y), \qquad y^t = \eta(t, x, y)$$

on the torus M, so that ξ, η are continuous for $-\infty < t, x, y < \infty$ and T^t is a group of homeomorphisms of the (x, y)-plane onto itself. Further-more, since $(x, y), (x + 1, y), (x, y + 1)$ are considered identical points of the torus M

$$
\begin{aligned}
\xi(t, x + 1, y) &= \xi(t, x, y) + 1 \quad \text{and} \quad \eta(t, x + 1, y) = \eta(t, x, y), \\
\xi(t, x, y + 1) &= \xi(t, x, y) \quad \text{and} \quad \eta(t, x, y + 1) = \eta(t, x, y) + 1.
\end{aligned}
$$
(14.2)

[The first line means, e.g., that if the point (x, y) is translated to $(x + 1, y)$, then its orbit is merely translated a distance 1 in the x-direction.] The group property $T^{s+t} = T^s T^t$ is equivalent to

(14.3) $\zeta(s + t, x, y) = \zeta(s, \xi(t, x, y), \eta(t, x, y))$ for $\zeta = \xi$ or $\zeta = \eta$.

Suppose that ξ, η have continuous derivatives with respect to t. Then $F_1(x, y) = [\partial \xi(t, x, y)/\partial t]_{t=0}$, $F_2(x, y) = [\partial \eta(t, x, y)/\partial t]_{t=0}$ satisfy

(14.4) $F_j(x, y) = F_j(x + 1, y) = F_j(x, y + 1)$ for $j = 1, 2$

and $x = \xi(t, x_0, y_0)$, $y = \eta(t, x_0, y_0)$ is a solution of the initial value problem

(14.5) $x' = F_1(x, y), \qquad y' = F_2(x, y),$

(14.6) $x(0) = x_0, \qquad y(0) = y_0,$

by (14.3).

If, e.g., ξ and η are of class C^1, then it follows from Theorem III 7.1 that (14.5), (14.6) has the unique solution $x = \xi(t, x_0, y_0)$, $y = \eta(t, x_0, y_0)$.

Lemma 14.1. *Let M be a torus and (14.1) a flow on M of class C^k, $k \geq 1$ [or a continuous flow such that ξ, η have continuous partials with respect to t]. Suppose that T^t has no stationary point [or that $F_1{}^2 + F_2{}^2 \neq 0$]. Then there exists a Jordan curve Γ of class C^k [or of class C^1] on M which is transversal to the flow. Furthermore, such a transversal curve Γ cannot be contracted to a point on M, so that Γ does not bound a 2-cell (i.e., does not bound a subset of M homeomorphic to a disc).*

For example, suppose that $F_1(x, y)$ does not vanish and let $F(x, y) = F_2/F_1$, so that (14.5) has the same solution paths as

(14.7) $\dfrac{dy}{dx} = F(x, y),$ i.e., $x' = 1, \quad y' = F(x, y).$

Here every circle $x = $ const. on M is a transversal curve Γ.

Proof. Consider the differential equations for the orthogonal trajectories

(14.8) $x' = F_2(x, y), \qquad y' = -F_1(x, y).$

Let $x_0(t), y_0(t)$ be a solution of (14.8), so that $(x_0(t), y_0(t)) \not\equiv$ const. since

$F_1^2 + F_2^2 \neq 0$. Also $(x_0(t), y_0(t))$ is of class C^k [or of class C^1] since (F_1, F_2) is of class C^{k-1} [or class C^0].

Suppose first that $x = x_0(t)$, $y = y_0(t)$ is a closed curve Γ on M, i.e., let there exist a least number $p > 0$ and integers r_0, s_0 such that $x_0(t + p) = x_0(t) + r_0$, $y_0(t + p) = y_0(t) + s_0$. It is clear that Γ is a transversal curve.

If $x_0(t)$, $y_0(t)$ is not a closed curve on M, then the half-trajectory $x_0(t)$, $y_0(t)$ for $t \geqq 0$, viewed as a path on M, has at least one ω-limit point, say, (x_1, y_1). The point (x_1, y_1) is contained in arbitrarily small curvilinear rectangles $R:ABCD$ on M, in which the arcs AB, CD are solution arcs of (14.5) and BC, AD are solution arcs of (14.8). The point $x_0(t)$, $y_0(t)$ is inside R for some large $t = t_0$ and leaves at some point P_1 on CD [or AB] at a first time $t_1 > t_0$ and then meets AB [or CD] at a point P_2 for some first $t_2 > t_1$. It is clear that if R is small enough that there exists an arc $P_2 P_1$ in R which together with the arc $x = x_0(t)$, $y = y_0(t)$ for $t_1 \leqq t \leqq t_2$ constitutes a transversal curve Γ of class C^k.

Suppose, if possible, that a transversal curve $\tilde{\Gamma}$ on M can be contracted to a point. Then it has an image Γ in the (x, y)-plane which is a Jordan curve of class C^1 bounding an open set Ω^0. Clearly the points of Γ are all egress or all ingress points of Ω^0 for (14.5). Then the index of Γ relative to (14.5) is $+1$ or -1; cf. § 2. Hence Ω^0 contains at least one stationary point by Corollary 3.1. This is a contradiction and completes the proof of Lemma 14.1.

In what follows, it is supposed that

(H_1) M is a torus, T^t is a flow on M of class C^k, $k \geqq 1$, without stationary points.

Let Γ be a transversal (Jordan) curve on M. After a suitable C^k homeomorphism of the plane, it can be supposed that

(H_2)   The circle $\Gamma : x = 0$ is a transversal curve. In particular, $F_1(n, y) \neq 0$ if $n = 0, \pm 1, \ldots$. Without loss of generality it can be supposed that

$$(14.9) \qquad F_1(n, y) > 0 \qquad \text{for} \quad n = 0, \pm 1, \ldots ;$$

for otherwise t is replaced by $-t$.

Let m_0 be a point of Γ on M and suppose that the semi-orbit $C^+(m_0)$ through m_0 meets Γ again for some $t_1 > 0$. Let m_1 be the first such point. In other words, if $m_0 = (0, y_0)$, then there is a unique $t_1 > 0$ such that $\xi(t_1, 0, y_0) = 1$ [cf. (14.9)], so that $m_1 = (1, \eta(t_1, 0, y_0))$. Put $f(y_0) = \eta(t_1, 0, y_0)$. The set U of points y_0 where U is defined is open and of period 1 (in the sense that $y_0 \in U$ if and only if $y_0 + 1 \in U$). It will be supposed that

(H_3)   Every orbit meets Γ and $y_1 = f(y)$ is defined for all y, $-\infty < y < \infty$.

The hypothesis (H_3) holds, e.g., if $F_1(x, y)$ does not vanish [so that (14.5) is "equivalent" to (14.6)] or if there are no closed orbits.

Exercise 14.1.   Verify the assertions of the last paragraph. The function $f(y)$ has the following properties: (i) $f(y)$ is strictly increasing; (ii) $f(y + 1) = f(y) + 1$, so that $f(y) - y$ has the period 1; (iii) $f(y)$ is of class C^k. The property (i) follows from the fact that two orbits are either identical or have no points in common. The property (ii) follows from the last line of (14.2); for if $x = 0$ and $t = t_1$, then $\xi(t_1, 0, y) = 1$ and $f(y) = \eta(t_1, 0, y)$. The proof of property (iii) is similar to (but simpler than) the proof of (12.3) in § 12.

The results of § 13 can now be transcribed to results about certain flows on a torus. Let the number α of Lemma 13.1 for the function $f(y)$ be called the *rotation number of the flow* T^t. Thus Lemma 13.2 implies the assertion:

Theorem 14.1.   *Let* T^t *be a flow of class* C^1 *on a torus M satisfying* (H_1), (H_2), *and* (H_3). *Then there exists a periodic* (*closed*) *orbit if and only if the rotation number* α *of* T^t *is rational.*

Exercise 14.2.   Let T^t be as in Theorem 14.1 and let the rotation number α be rational. Show that every semi-orbit $C^+(m_0)$ on M is either a Jordan curve or that $\Omega(m_0)$ is a Jordan curve and $C^+(m_0)$ spirals toward $\Omega(m_0)$.

Theorems 13.2 and 13.4 give

Theorem 14.2.   *Let* T^t *be a flow of class* C^2 *on a torus M satisfying* (H_1), (H_2), *and* (H_3) *and let its rotation number* α *be irrational. Then M is a minimal set* (*and every semi-orbit* $C^+(m_0)$ *is dense on M*).

When M is a minimal set, the flow T^t is called *ergodic*.

Remark.   This theorem is false if the condition "T^t is of class C^2" is relaxed to "T^t is of class C^1." The situation becomes even worse if it is only assumed that "T^t is continuous, ξ and η have continuous partials with respect to t, and $F_1{}^2 + F_2{}^2 \neq 0$," although the other results above have analogues in this case. This is indicated by the next exercise which is an extension of the results of Exercise 13.3.

Exercise 14.3   Let points (x, y), $(x + 1, y)$, and $(x, y + 1)$ be identified so that the plane becomes a torus M and the line $x = 0$ a circle Γ. (*a*) Let N_0 be an arbitrary, perfect, nowhere dense set on Γ. There exists a continuous function $F(x, y) = F(x + 1, y) = F(x, y + 1)$ such that initial value problems belonging to (14.7) have unique solutions and such that if $T^t : x^t = t + x_0$, $y^t = \eta(t, x_0, y_0)$ is the flow induced by (14.7) on M and S is the corresponding homeomorphism on Γ, then (the rotation number α is irrational and) N_0 is the unique, nonempty, minimal set of S. (*b*) There exist functions $F(x, y) = F(x + 1, y) = F(x, y + 1)$ of class C^1 such that the corresponding flow T^t of class C^1 has an irrational rotation number α but T^t is not ergodic (i.e., M is not a minimal set). See Denjoy [1].

The simplest flow on M is given by

(14.10) $$x^t = t + x, \qquad y^t = \alpha t + y$$

arising from the differential equations

$$x' = 1, \qquad y' = \alpha,$$

where α is a constant (in fact, the rotation number of the flow). Consider a flow of the form

(14.11) $$x^t = t + x, \qquad y^t = \eta(t, x, y)$$

arising, e.g., from a system of differential equations of the form $x' = 1$, $y' = F(x, y)$. The function $f(y)$ belonging to (14.11) is

(14.12) $$f(y) = \eta(1, 0, y).$$

In examining the orbits of (14.11), it is sufficient to consider $\eta(t, 0, y)$, i.e., the orbits beginning for $t = 0$ at a point $(0, y)$ of the y-axis. For the orbit starting at (x, y) meets the x-axis when $t = -x$, and so

(14.13) $$\eta(t, x, y) = \eta(t + x, 0, \eta(-x, x, y)),$$

since $T^t = T^{t+x}T^{-x}$.

It will be verified that

(14.14) $$\eta(t, 0, y) - \alpha t - y \qquad \text{is bounded}$$

for $-\infty < t, y < \infty$. To this end, note that $f^n(y) = \eta(n, 0, y)$, so that

$$\eta(n, 0, y) - n\alpha - y \qquad \text{is bounded}$$

by Lemma 13.1. If $t = n + \theta_1$, where $0 \le \theta_1 < 1$, then $\eta(t, 0, y) = \eta(n, 0, \eta(\theta_1, 0, y))$ satisfies

$$\eta(t, 0, y) - n\alpha - \eta(\theta_1, 0, y) \qquad \text{is bounded, where } n\alpha = \alpha t - \theta_1 \alpha.$$

Also, $\eta(t, 0, y + 1) = 1 + \eta(t, 0, y)$ shows that if $y = j + \theta_2$, where $0 \le \theta_2 < 1$, then

$$\eta(\theta_1, 0, y) - j = \eta(\theta_1, 0, \theta_2) \qquad \text{is bounded, where } j = y - \theta_2.$$

The last two relations give (14.14). The assertion (14.14) can be greatly improved in the ergodic case. This is the essence of the following result.

Theorem 14.3 (Bohl). *Let T^t be a continuous flow on the torus M of the form* (14.11) *such that T^t is ergodic and has the (irrational) rotation number α. Then there exist continuous functions $\psi(y)$, $G(t, z)$ such that*

(14.15) $$\psi(y + 1) = \psi(y) \quad \text{and} \quad G(t, z) = G(t + 1, z) = G(t, z + 1)$$

and that

(14.16) $$\eta(t, 0, y) = \alpha t + y + \psi(y) + G(t, \alpha t + y + \psi(y)).$$

Note that for a fixed y [or t], $\eta(t, 0, y) - \alpha t - y$ is an almost periodic function of t [or periodic function of y].

Proof. Let $z = g(y)$ be the function supplied by Theorem 13.3 and let $\psi(y)$, $\psi_0(y)$ be the functions of period 1 defined by

$$(14.17) \qquad g(y) = y + \psi(y), \qquad g^{-1}(z) = z + \psi_0(z).$$

Make the change of variables $R:(x, z) = (x, g(y))$. Then T^t becomes

$$(14.18) \qquad RT^tR^{-1}:x^t = t + x, \; z^t = \zeta(t, x, z),$$

where

$$(14.19) \qquad \zeta(t, x, z) = g(\eta[t, x, g^{-1}(z)]).$$

By the analogue of (14.2)

$$(14.20) \quad \zeta(t, x + 1, z) = \zeta(t, x, z) \quad \text{and} \quad \zeta(t, x, z + 1) = 1 + \zeta(t, x, z)$$

and, by the group property of RT^tR^{-1} [cf. (14.3)],

$$(14.21) \quad \zeta(t + 1, x, z) = \zeta(t, x + 1, \zeta(1, x, z)) = \zeta(t, x, \zeta(1, x, z)).$$

Introduce the abbreviations

$$(14.22) \qquad \eta(t, y) = \eta(t, 0, y) \quad \text{and} \quad \zeta(t, z) = \zeta(t, 0, z).$$

Note that, since $f(y) = \eta(1, y)$, Theorem 13.3 gives $\zeta(1, z) = z + \alpha$. Hence (14.20)–(14.22) imply that

$$\zeta(t, z + 1) = 1 + \zeta(t, z) \quad \text{and} \quad \zeta(t + 1, z) = \zeta(t, \zeta(1, z)) = \zeta(t, z + \alpha).$$

Thus the continuous function

$$\psi_1(t, z) = \zeta(t, z - \alpha t) - z$$

has the period 1 with respect to t [or z] for fixed z [or t]. Write the last relation as

$$(14.23) \qquad \zeta(t, z) = \alpha t + z + \psi_1(t, \alpha t + z).$$

In view of (14.19), $\eta(t, y) = g^{-1}(\zeta[t, g(y)])$, so that, by (14.17),

$$\eta(t, y) = \zeta(t, g(y)) + \psi_0(\zeta[t, g(y)]).$$

Consequently (14.23) shows that

$$(14.24) \qquad \eta(t, y) = \alpha t + g(y) + G(t, \alpha t + g(y)),$$

if $G(t, z)$ is defined by

$$G(t, z) = \psi_1(t, z) + \psi_0(z + \psi_1(t, z)).$$

It is clear that G satisfies the last part of (14.15). Finally, (14.24) and the first part of (14.17) give (14.16). This proves the theorem.

Notes

Most of §§ 1–9 of this chapter is contained in (or related to) the four-part memoir of Poincaré [3]; see also Bendixson [2]. L. E. J. Brouwer [1] considers similar problems without analyticity assumptions and even without the assumption that there is only one solution through a given point.

SECTION 1. The terminology "α- and ω-limit points" is due to G. D. Birkhoff [3]; "limit cycle" to Poincaré [3].

SECTION 2. The Umlaufsatz (Theorem 2.1) goes back to Riemann [1, pp. 106–107]. It was first given a formal statement and proof by G. N. Watson [1]. The proof in the text is that of H. Hopf [1]; cf. also van Kampen [4].

SECTION 3. The definition of "index" was given by Poincaré [3, I].

SECTION 4. See Poincaré [3] (in particular, part II) and Bendixson [2].

SECTION 5. See Poincaré [3].

SECTIONS 6–7. See Poincaré [3].

SECTIONS 8–9. For Lemma 8.1, cf. Bendixson [2, p. 26]. In the analytic case, Theorems 9.1 and 9.2 are due to Poincaré [3] (in particular, part I) and to Bendixson [2]. The exposition in the text of Theorem 9.1 is an adaptation and simplification of Brouwer's treatment [1, II] (which is complicated by the fact that initial conditions do not determine a unique solution). For Exercise 9.1, see Schilt [1].

SECTION 10. Problems of the type considered here go back to van der Pol [1] and to Liénard [1]. Theorem 10.1 is a variant of a result of Levinson and Smith [1]. An analogous result concerning (10.15) was given by Levinson [1], improved by Langenhop [1], and further improved by Opial [7] as in Exercise 10.1. Theorem 10.2 is a generalization of a result of Liénard [1]. For Exercise 10.5, see, e.g., Lefschetz [1, pp. 342–346] and references to Flanders and Stoker [1], LaSalle [2], and Stoker [1]. For fuller treatments of related problems and for references to the work of Cartwright, Littlewood, Duff, Massera, Reuter, Sansone, Schimizu, etc.; see Andronow and Chaikin [1], Bogolyubov and Mitropol'ski [1], Bogolyubov and Krylov [1], Lefschetz [1], Minorsky [1], Conti and Sansone [1], and Stoker [1].

APPENDIX. This type of investigation was initiated by Poincaré [2], [3, III] for the case of differential equations on a torus; see notes on §§ 13–14.

SECTION 11. The ideas in this section are due to Poincaré.

SECTION 12. The results and methods of this section are those of A. J. Schwartz [1]; for a generalization, see Sacksteder [1]. The analogue of Theorem 12.1 was known earlier for the torus; Denjoy [1]. See notes on §§ 13–14.

SECTIONS 13–14. Except for Theorem 13.4 and Exercises 13.1, 13.2 in § 13 and Theorems 14.2, 14.3, and Exercise 14.3 in § 14, the results and methods of these sections are essentially in Poincaré [3, III]. The presentation in the text follows the completed arguments and simplifications introduced by Denjoy [1] and van Kampen [1]; cf. also Siegel [1]. (After Poincaré, proofs for the main part of Lemma 13.1 have been given by E. E. Levi, Bohl, H. Kneser, Nielsen, and Denjoy; see Bohl [1] and van Kampen [1] for references.) Theorem 13.4 and its consequence, Theorem 14.2, were conjectured by Poincaré [3, III] for the analytic case. They were first proved by Denjoy [1] in a slightly stronger form (cf. Exercise 13.3). A very simple proof of Denjoy's result has been given by van Kampen [1]. A similar proof which avoids, however, the use of the rotation number has been given by Siegel [1]. Denjoy [1] has given examples (cf. Exercises 13.2 and 14.3) showing that the results are false if the smoothness assumptions are lightened. Theorem 14.3 is a result of Bohl [1].

Chapter VIII

Plane Stationary Points

This chapter continues the discussion of the behavior of solutions of plane autonomous systems. The basic existence theorems will be proved in the first section for autonomous systems of arbitrary dimension.

1. Existence Theorems

In this section, there will be considered an autonomous system

$$(1.1) \qquad\qquad z' = f(z)$$

for a real d-dimensional vector $z = (z^1, \ldots, z^d)$. By a half-trajectory of (1.1) will be meant a solution arc in the z-space: C^+: $z = z(t)$, $0 \leqq t < \omega_+ (\leqq \infty)$, or C^-: $z = z(t)$, $0 \geqq t > \omega_- (\geqq -\infty)$, defined on a right or left maximal interval of existence. Correspondingly, Γ will be called of type C^+ or C^-. The point $z(0)$ will be called the endpoint of Γ.

Theorem 1.1. *Let $f(z)$ be continuous on an open z-set Ω. Let Ω_0 be an open subset of Ω such that the part of its boundary in Ω (i.e., $\partial\Omega_0 \cap \Omega$) is the union of two disjoint sets $L \cup R$, where R is compact, the points of L are egress points, and $\partial\Omega_0 \cap \Omega = L \cup R$ is not compact. Then $\Omega_0 \cup L \cup R$ contains at least one half-trajectory Γ of (1.1) with endpoint in L or R.*

For the definition of egress point, cf. § III 8. It is not assumed in this theorem that L exhausts the set of egress points of Ω_0. Since the boundary $\partial\Omega_0$ is closed relative to Ω, the condition that $L \cup R$ is not compact implies that either $L \cup R$ is unbounded or has a limit point z_∞ which is not in Ω.

It will be clear from the proof that either there exists a half-trajectory Γ of type C^- in $\Omega_0 \cup L$ with endpoint on L or there exists a half-trajectory of type C^+ in $\Omega_0 \cup L \cup R$ with endpoint on R. Sufficient conditions for the latter case are given by the next theorem. The conclusion of this theorem does not specify, however, whether Γ is of type C^+ or C^-.

Theorem 1.2. *In Theorem 1.1, assume that solutions of (1.1) are uniquely determined by initial conditions, that R is not empty, and that $L \cup R$ is* *connected, then there exists a half-trajectory in* $\Omega_0 \cup L \cup R$, *endpoint on* R.

The condition that "$L \cup R$ is connected" can be relaxed to "$L_0 \cup R$ is connected, where $L_0 \subset L$ and $L_0 \cup R$ is either unbounded or has a limit point not in $L \cup R$ (i.e., not in Ω)."

Remark. In Theorem 1.2, the condition that solutions of (1.1) are uniquely determined by initial conditions can be omitted if there exist

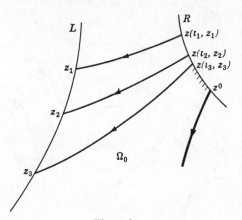

Figure 1.

smooth functions $f_n(z)$, $n = 1, 2, \ldots$, on Ω which approximate $f(z)$ uniformly on compact subsets of Ω and if Theorem 1.2 is applicable to $z' = f_n(z)$. This will be illustrated in Corollary 1.1 below.

Proof of Theorem 1.1. For brevity, we write $\omega_-(z_0)$, $\omega_+(z_0)$, $\tau(z_0)$, $\ldots$, although solutions are not determined by the initial point z_0, so that ω_-, ω_+, τ, $\ldots$ depend on z_0 *and* the selected solution. Let $z(t, z_0)$ be any solution of (1.1) determined by $z(0) = z_0$ and let its maximal interval of existence be $\omega_-(z_0) < t < \omega_+(z_0)$. If $z_0 \in L$, then $z(t, z_0) \in \Omega_0$ for small $-t > 0$. If there exists a $z_0 \in L$ having a half-trajectory $z(t, z_0)$ in $\Omega_0 \cup L$ on its left maximal interval $\omega_-(z_0) < t \leqq 0$, then the conclusion of Theorem 1.1 holds.

Suppose therefore that, for every $z_0 \in L$ and solution $z(t, z_0)$, there exists a $\tau = \tau(z_0)$, $\omega_-(z_0) < \tau(z_0) < 0$, such that $z(t, z_0) \in \Omega_0 \cup L$, $\tau(z) < t < 0$, but $z(\tau, z_0) \notin \Omega_0 \cup L$. Then $z(\tau, z_0) \in R$; so that, R is not empty.

Since R is compact and $L \cup R$ is not compact, there exists a sequence of points $z_1, z_2, \ldots$ of L such that if $\tau_m = \tau(z_m)$, then, as $m \to \infty$, $z^0 = \lim z(\tau_m, z_m) \in R$ exists and either $\|z_m\| \to \infty$ or $z_\infty = \lim z_m$ exists and $z_\infty \notin \Omega$; see Figure 1.

Put $z_m(t) = z(t, z(\tau_m, z_m))$, so that $z_m(0) = z(\tau_m, z_m) \in R$, $z_m(-\tau_m) = z_m \in L$, $z_m(t) \in \Omega_0 \cup L$, $0 < t < -\tau_m$. Since $z_m(0) \to z^0$ as $m \to \infty$, Theorem II 3.2 implies that if the sequence $z_1, z_2, \ldots$ is suitably chosen, it can be supposed that there exists a solution $z^0(t)$ of (1.1) satisfying $z(0) = z^0$, having a maximal interval of existence (ω_0, ω^0), and such that

$$(1.2) \qquad z_m(t) \to z^0(t) \qquad \text{uniformly as} \quad m \to \infty$$

on any t-interval $[t_*, t^*]$ in (ω_0, ω^0). In particular, $\omega_-(z_m) < t_* < t^* < \omega_+(z_m)$ for large m.

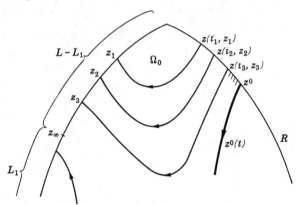

Figure 2. $z^0(t) \in \Omega_0 \cup R$ for $0 \leqq t < \omega_+(z^0)$.

Suppose, if possible, that $z^0(t) \notin \Omega_0 \cup L \cup R, 0 < t < \omega^0$. Then $z^0(t) \notin \bar{\Omega}_0 \cap \Omega = \Omega_0 \cup L \cup R$ for $0 \leqq t < \omega^0$ and, hence there is a $t = t_1$, $0 < t_1 < \omega^0$ such that $z^0(t_1) \notin \bar{\Omega}_0$. Let $t_1 < t_2 < \omega^0$. It follows that, for large m, $z_m(t)$ is defined for $0 \leqq t \leqq t_2$ and that (1.2) holds for $0 \leqq t \leqq t_2$. But then $z_m(t_1) \notin \bar{\Omega}_0$ for large m. Consequently, $0 < -\tau_m < t_1$ for large m. In this case, $z^0(-\tau_m) - z_m(-\tau_m) \to 0$, $m \to \infty$, by (1.2). Since $z_m(-\tau_m) = z_m$ and either $\|z_m\| \to \infty$ or $z_m \to z_\infty$ as $t \to \infty$, it follows that either $\|z^0(-\tau_m)\| \to \infty$ or $z^0(-\tau_m) \to z_\infty \notin \Omega$ as $m \to \infty$. This is impossible as $-\tau_m < t_1 < \omega^0$. Hence Theorem 1.1 is proved.

Proof of Theorem 1.2.   In view of the proof of Theorem 1.1, it suffices to consider the case that there exist points $z_0 \in L$ such that (the unique) $z(t, z_0)$ is in $\Omega_0 \cup L$, hence in $\Omega_0 \cup L \cup R$, for $\omega_-(z_0) < t < 0$. Let L_1 be the (nonempty) subset of L of such points. Clearly L_1 is closed relative to L, since L consists of egress points; cf. Theorem II 3.2.

If there is a point $z^0 \in R$ which is a limit point of L_1, then $C^- : z(t, z^0) \in \Omega_0 \cup L \cup R, \omega_-(z^0) < t \leqq 0$. Suppose therefore that L_1 has no limit point in R. Thus L_1 is closed relative to $L \cup R$.

Consequently $L - L_1$ has a limit point z_∞ in L_1. Otherwise $L \cup R$

has a decomposition into nonempty, disjoint sets $(L - L_1) \cup R$ and L_1, which are closed relative to $L \cup R$. But $L \cup R$ is connected.

Thus the constructions of the last proof can be repeated with the modifications that $z_m \in L - L_1$ and $z_m \to z_\infty \in L_1$. If $z^0(t) \in \Omega_0 \cup L \cup R$, $0 \leqq t < \omega_+(z_0)$, the proof is complete; cf. Figure 2. If not, it becomes necessary to examine the conclusions at the end of the proof to the effect that $-\tau_m < t_1 < \omega^0$ and $z^0(-\tau_m) \to z_\infty$ as $m \to \infty$.

Hence there exists a $t = \tau_\infty (\leqq 0)$ which is a (finite) limit point of the sequence $\tau_1, \tau_2, \ldots$. Then $z^0(-\tau_\infty) = z_\infty$. It is clear that $\tau_\infty < 0$ for

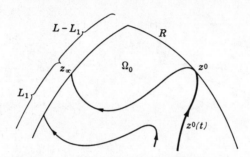

Figure 3. $z^0(t) \in \Omega_0 \cup R$ for $\omega_-(z^0) < (t) \leqq 0$.

$z^0(0) \in R$, $z_\infty \in L_1$ (and L, R are disjoint). Since $z_\infty \in L_1$ and $z(t, z_\infty)$ is uniquely determined by z_∞, it follows that $z(t, z_\infty) \in \Omega_0 \cup L \cup R, \omega_-(z_\infty) < t < 0$; see Figure 3. [This is the critical point where the assumption that solutions of (1.1) are uniquely determined by initial conditions is used.]

In particular, the half-trajectory $z^0(t - \tau_\infty) = z(t, z_\infty) \in \Omega_0 \cup L \cup R$, $\omega_-(z_\infty) < t \leqq \tau_\infty$. Since the endpoint of this trajectory is $z(\tau_\infty, z_\infty) = z^0 \in R$, the theorem is proved: $z^0(t) \in \Omega_0 \cup L \cup R$ for $\omega_-(z^0) < t \leqq 0$.

Corollary 1.1. *Let* dim $z \geqq 2$, $f(z)$ *be continuous on an open set* Ω *containing the closure of the spherical sector*

$$(1.3) \qquad \Omega_0 = \left\{ z : 0 < \|z\| < \delta, \ \left\| \frac{z}{\|z\|} - z^* \right\| < \eta < 2 \right\},$$

where δ, η *are positive numbers and* $\|z^*\| = 1$, *and let*

$$(1.4) \qquad\qquad\qquad f(0) = 0.$$

Let L, R *be the lateral and spherical parts of the boundary of* Ω_0,

$$(1.5) \qquad L = \left\{ z : 0 < \|z\| < \delta, \ \left\| \frac{z}{\|z\|} - z^* \right\| = \eta \right\},$$

$$(1.6) \qquad R = \left\{ z : \|z\| = \delta, \ \left\| \frac{z}{\|z\|} - z^* \right\| \leqq \eta \right\},$$

respectively. Suppose that every point of L is an egress point for Ω_0. Then (1.1) *has at least one half-trajectory Γ in the set $\bar{\Omega}_0$ with endpoint in R (and defined on a half-line $t \geq 0$ or $t \leq 0$); see Figure 4.*

Proof.   Suppose first that the solutions of (1.1) are uniquely determined by initial conditions, then Theorem 1.2 is applicable if the open set Ω of Corollary 1.1 is replaced by the open set obtained by deleting $z = 0$ from Ω, so that $\partial\Omega_0 \cap \Omega = L \cup R$. Then there exists a half-trajectory, e.g., $C^+ : z(t)$, $0 \leq t < \omega_+ (\leq \infty)$, in $\Omega_0 \cup L \cup R$, $z(0) \in R$. If $\omega_+ < \infty$, then

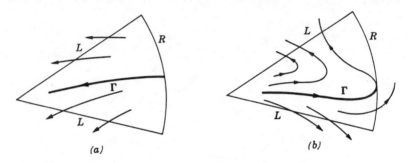

Figure 4.   (a) Γ of type C^+.   (b) Γ of type C^-.

$z(t) \to 0$ as $t \to \omega_+$; cf. Lemma II 3.1.   But, in this case, the definition of $z(t)$ can be extended to $0 \leq t < \infty$ by defining $z(t) = 0$ for $t \geq \omega_+$. [Actually, this situation cannot arise when the solutions of (1.1) are uniquely determined by initial conditions, since $z(t_0) = 0$ for some t_0 implies that $z(t) \equiv 0$. It can arise, however, in the general case to be considered now.]

It has to be shown that Corollary 1.1 is valid if it is not assumed that the solutions of (1.1) are uniquely determined by initial conditions. Since $z \in L$ is an egress point, if follows that the trajectory derivative of $u(z) = \|z/\|z\| - z^*\|^2 - \eta^2$ is non-negative at $z \in L$; i.e.,

$$(1.7) \qquad (z \cdot z^*)(z \cdot f) - \|z\|^2(z^* \cdot f) \geq 0$$

for $0 < \|z\| \leq \delta, \|z/\|z\| - z^*\| = \eta$; cf. § III 8.   Thus, if $f_\epsilon(z) = f(z) - \epsilon z^*$, then

$$(1.8) \qquad (z \cdot z^*)(z \cdot f_\epsilon) - \|z\|^2(z^* \cdot f_\epsilon) \geq \epsilon(\|z\|^2 - (z \cdot z^*)^2),$$

which exceeds a positive constant $\epsilon c_n > 0$ when

$$(1.9) \qquad \frac{1}{n} \leq \|z\| \leq \delta, \qquad \left\|\frac{z}{\|z\|} - z^*\right\| = \eta$$

for large n.

Let $f_1(z), f_2(z), \ldots$ be a sequence of smooth functions such that $f_m(z) \to f(z)$ as $m \to \infty$ uniformly on an open set $\Omega^1 \supset \bar{\Omega}_0$. Let the integer $n > 0$ be so large that $1/n < \delta$. Then if $f_m(z)$ is replaced by $f_m(z) - z^*/n$, if necessary, it can be supposed that there is an $m = m(n)$ such that

$$(1.10) \qquad (z \cdot z^*)(z \cdot f_m) - \|z\|^2(z^* \cdot f_m) \geqq \frac{c_n}{2n} > 0,$$

when (1.9) holds, and that $m(n) \to \infty$ as $n \to \infty$.

Let Ω_n be the open set obtained by deleting the sphere $\|z\| \leqq 1/n$ from Ω^1. Let

$$\Omega_{0n} = \left\{ z : \frac{1}{n} < \|z\| < \delta, \ \left\| \frac{z}{\|z\|} - z^* \right\| < \eta < 2 \right\}$$

and let L_n, R be the lateral and spherical parts of the boundary of Ω_{0n} in Ω_n. By (1.10), $z \in L_n$ is an egress point of Ω_{0n} with respect to the differential equation

$$(1.11) \qquad z' = f_m(z).$$

Thus, by Theorem 1.2, (1.11) has a half-trajectory Γ_m in $\Omega_{0n} \cup L_n \cup R$, endpoint on R. For the sake of definiteness, let $\Gamma_m = C_m^+ : z = z_m(t)$, $0 \leqq t < \tau_m \ (\leqq \infty)$, where $z_m(0) \in R$.

Here $0 \leqq t < \tau_m$ is the right maximal interval of existence if (1.11) is considered only on Ω_n. Let the right maximal interval of existence of $z_m(t)$ be $0 \leqq t < \omega_m \ (\leqq \infty)$ if (1.11) is considered on Ω^1. Thus $\tau_m \leqq \omega_m$ and $\tau_m < \omega_m$ implies that $\|z_m(\tau_m)\| = 1/n$.

By choosing a subsequence, if necessary, it can be supposed that $z_0 = \lim z_m(0) \in R$ exists as $m = m(n) \to \infty$. By Theorem II 3.2, it can also be supposed that (1.1) has a solution $z_0(t)$ satisfying $z_0(0) = z_0$ and that $z_m(t) \to z_0(t)$ as $m = m(n) \to \infty$ uniformly on every compact interval of the right maximal interval $[0, \omega_0)$ of existence of $z_0(t)$ relative to Ω^1.

Suppose that the half-trajectory $\Gamma : z_0(t)$, $0 \leqq t < \omega_0$, is not in the set $\bar{\Omega}_0$. Thus $z_0(t_1) \notin \bar{\Omega}_0$ for some t_1, $0 < t_1 < \omega_0$. If $\epsilon > 0$, then $0 < \tau_m < t_1 + \epsilon$ and $\tau_m < \omega_m$ for large m. Consequently,

$$\|z_0(\tau_m) - z_m(\tau_m)\| \to 0$$

and $\|z_m(\tau_m)\| = 1/n \to 0$ as $m(n) \to \infty$. Thus if τ_∞ is a limit point of $\tau_1, \tau_2, \ldots$, then $z_0(\tau_\infty) = 0$. Here, it is possible to change the definition of $z_0(t)$ as follows: if $t_0 > 0$ is the least t-value where $z_0(t) = 0$, put $z_0(t) = 0$ for $t \geqq t_0$. Now $\Gamma : z_0(t)$ is defined for $0 \leqq t < \infty$ and $z_m(t) \to z_0(t)$, $m(n) \to \infty$, uniformly for $0 \leqq t \leqq t_0 < \infty$. Repeating the argument just concluded, it follows that $z_0(t)$ is in the set $\bar{\Omega}_0$ for $t \geqq 0$. This completes the proof.

Corollary 1.2. *Let* dim $z = 2$. *In addition to the assumptions of Corollary* 1.1, *suppose that*

$$(1.12) \qquad\qquad f(z) \neq 0 \text{ for } 0 \neq z \in \bar{\Omega}_0.$$

Then the half-trajectory $\Gamma : z(t)$ *in Corollary* 1.1 *is defined for* $t \geqq 0$ *or* $t \leqq 0$ *and satisfies* $z(t) \to 0$ *as* $t \to \infty$ *or* $t \to -\infty$.

Proof. If this corollary is false, then $\Gamma : z(t)$ for $t \geqq 0$ or $t \leqq 0$ remains at a positive distance from $z = 0$; cf. Lemma II 3.1. In view of (1.12) this is impossible by the general theory of plane autonomous systems; cf. Theorem VII 4.4.

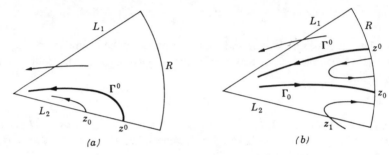

Figure 5.

Corollary 1.3. *Let the assumptions of Corollary* 1.2 *hold.* (i) *If, in addition,* $z \cdot f(z) < 0$ *for* $z \in R$, *then any half-trajectory* Γ *in the set* $\bar{\Omega}_0$ *with endpoint on* R *is defined for* $t \geqq 0$. (ii) *If* $z \cdot f(z) > 0$ *for* $z \in R$ *and* $z(t)$ *is any solution of* (1.1) *with* $z(0)$ *interior to* R [*i.e.,* $\|z(0)\| = \delta$, $\|z(0)/ \|z(0)\| - z^*\| < \eta$], *then* $z(t)$ *can be defined for* $t \leqq 0$ *so that* $z(t) \in \bar{\Omega}_0$ *and* $z(t) \to 0$ *as* $t \to -\infty$.

Exercise 1.1.   Prove Corollary 1.3.

A different situation is considered in the next result.

Corollary 1.4. *Let* dim $z = 2$, *so that* L *in* (1.5) *is the union of two disjoint open line segments* L_1, L_2. *Let the conditions of Corollary* 1.2 *hold but, instead of assuming that every point* z *of* L *is an egress point for* Ω_0, *assume that every* $z \in L_1$ *is an egress point for* Ω_0 *and that every* $z \in L_2$ *is an ingress point for* Ω_0. *Then there exists no, or at least two, half-trajectories in* $\bar{\Omega}_0$ *with endpoint on* $L \cup R$.

Proof. Assume that there exists a half-trajectory Γ^0 in Ω_0 with endpoint $z^0 \in L \cup R$. Suppose that Γ^0 is of type C^+ (otherwise, replace t by $-t$ and interchange L_1 and L_2). Then $z^0 \in L_2 \cup R$.

Suppose that $z^0 \in \bar{L}_2$. Then every point z_0 of the segment of $\bar{L}_2$ from 0 to z^0 is the endpoint of a half-trajectory Γ^0 of type C^+ in $\Omega_0 \cup L_2$; cf. Figure 5(a).

Suppose that z^0 is an interior point of R and that no point z_1 of L_2 is an endpoint of a half-trajectory in $\Omega_0 \cup L_2$. Then a solution arc of (1.1) through z_1 does not meet Γ^0 for increasing t, but does meet R; cf. Figure 5(b). The arguments used in the proof of Theorem 1.1 show that there exists a half-trajectory Γ_0 of type C^- in $\Omega_0 \cup R$ with endpoint $z_0 \in R$. This proves Corollary 1.4.

Exercise 1.2. In Corollary 1.4, assume that $z \cdot f(z) \neq 0$ for $z \in R$. Show that there exists no, or infinitely many, half-trajectories in Ω_0 with endpoints on $L \cup R$.

For arbitrary $d = \dim z$, Corollary 1.3 has the following analogue.

Corollary 1.5. *Let $d \geqq 2$. Let the assumptions of Corollary 1.1 hold and, in addition, $z \cdot f(z) \neq 0$ for $z \in \Omega_0 \cup R$. Then any half-trajectory $\Gamma : z(t)$ in the set $\bar{\Omega}_0$ with endpoint in R is defined on a half-line $t \geqq 0$ or $t \leqq 0$ and tends to $z = 0$ as $t \to \infty$ or $t \to -\infty$ according as $z \cdot f(z) < 0$ or $z \cdot f(z) > 0$.*

This is clear because $\|z\|$ is decreasing or increasing with t according as $z \cdot f(z) < 0$ or $z \cdot f(z) > 0$.

2. Characteristic Directions

In this section, $\dim z = 2$. Write $z = (x, y)$ and (1.1) as

$$(2.1) \qquad x' = X(x, y), \qquad y' = Y(x, y).$$

It will be supposed that X, Y are continuous for small $|x|$, $|y|$ and that

$$(2.2) \qquad X(0, 0) = Y(0, 0).$$

Introducing polar coordinates $x = r \cos \theta$, $y = r \sin \theta$ transforms (2.1) into

$$(2.3) \quad \begin{aligned} r' &= X(r \cos \theta, r \sin \theta) \cos \theta + Y(r \cos \theta, r \sin \theta) \sin \theta, \\ r\theta' &= Y(r \cos \theta, r \sin \theta) \cos \theta - X(r \cos \theta, r \sin \theta) \sin \theta. \end{aligned}$$

A direction $\theta = \theta_0$ at the origin is called *characteristic* for (2.1) if there exists a sequence (r_1, θ_1), (r_2, θ_2), $\ldots$ such that $0 < r_n \to 0$ and $\theta_n \to \theta_0$ as $n \to \infty$; $(X_n, Y_n) = (X, Y)$ at $(x, y) = (r_n \cos \theta_n, r_n \sin \theta_n)$ is not $(0, 0)$, and

$$(2.4) \qquad \frac{Y_n \cos \theta_0 - X_n \sin \theta_0}{(X_n{}^2 + Y_n{}^2)^{1/2}} \to 0 \qquad \text{as} \quad n \to \infty.$$

The condition (2.4) means that the angle (mod π) between the vectors (X_n, Y_n) and $(\cos \theta_0, \sin \theta_0)$ tends to 0 as $n \to \infty$.

Lemma 2.1. *Let $X(x, y)$, $Y(x, y)$ be continuous for small $|x|$, $|y|$ and*

$X^2 + Y^2 \geqq 0$ *according as* $x^2 + y^2 \geqq 0$. *Let* (2.1) *possess a solution* $(x(t), y(t))$ *for* $0 \leqq t < \omega$ $(\leqq \infty)$ *such that*

(2.5) $0 < x^2(t) + y^2(t) \to 0$ as $t \to \omega$.

Let $r(t) = (x^2(t) + y^2(t))^{\frac{1}{2}} > 0$ *and* $\theta(t)$ *a continuous determination of* arc tan $y(t)/x(t)$. *Let* $\theta = \theta_0$ *be a noncharacteristic direction. Then either* $\theta'(t) > 0$ *or* $\theta'(t) < 0$ *for all t near* ω *for which* $\theta(t) = \theta_0$ mod 2π.

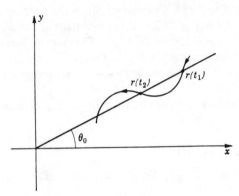

Figure 6.

Proof.   Clearly $\theta'(t) \neq 0$ for all t near ω for which $\theta(t) = \theta_0$ mod 2π. Otherwise there exists a sequence $t_1 < t_2 < \ldots$ such that $t_n \to \omega$, $\theta(t_n) = \theta_0$ (mod 2π) and $\theta'(t_n) = 0$. But then (2.4) holds with $(r_n, \theta_n) = (r(t_n), \theta_0)$ because the expression in (2.4) is zero by (2.3). This is impossible since $\theta = \theta_0$ is noncharacteristic.

Suppose if possible that the lemma is false. Then there exists $t_1 < t_2 < \ldots$ such that $t_n \to \omega$, $r(t_1) > r(t_2) > \ldots$, $(-1)^n \theta'(t_n) > 0$ and $\theta(t_n) = \theta_0$ (mod 2π); see Figure 6. Let $\Theta(r, \theta)$ denote the right side of the second equation in (2.3), so that $(-1)^n \Theta(r(t_n), \theta_0) > 0$. By the continuity of $\Theta(r, \theta_0)$ with respect to $r > 0$, it follows that there exists an r_n such that $r(t_n) > r_n > r(t_{n+1})$ and $\Theta(r_n, \theta_0) = 0$. Since $X^2 + Y^2 \neq 0$ if $x^2 + y^2 \neq 0$, it follows that (2.4) holds with $(r_n, \theta_n) = (r_n, \theta_0)$; i.e., θ_0 is characteristic. This is a contradiction and proves the lemma.

Theorem 2.1.   *Let* $X(x, y)$, $Y(x, y)$ *and* $(x(t), y(t))$ *be as in Lemma* 2.1. *Suppose that every θ-interval, $\alpha < \theta < \beta$, contains a noncharacteristic direction. Then either*

(2.6) $\theta_0 = \lim\limits_{t \to \omega} \theta(t)$ exists (and is finite) *or $(x(t), y(t))$ is a spiral; i.e.,*

(2.7) $$|\theta(t)| \rightarrow \infty \quad \text{as} \quad t \rightarrow \omega.$$

In the case (2.6), $\theta = \theta_0$ is a characteristic direction.

Proof. Suppose, if possible, that lim $\theta(t)$, as $t \rightarrow \omega$, does not exist either as a finite or infinite value. Then there exist numbers α, β such that, as $t \rightarrow \omega$,

$$(-\infty) \leq \liminf \theta(t) < \alpha < \beta < \limsup \theta(t) \; (\leq \infty).$$

By assumption, there is a noncharacteristic θ_0 satisfying $\alpha < \theta_0 < \beta$. Since $\theta(t)$ is continuous, there exist t-values arbitrarily close to ω where $\theta(t) = \alpha < \theta_0$ and t-values close to ω where $\theta(t) = \beta > \theta_0$. It follows that there are t-values arbitrarily close to ω where $\theta(t) = \theta_0$, $\theta'(t) \geq 0$ and other t-values where $\theta(t) = \theta_0$, $\theta'(t) \leq 0$. This is impossible by the last lemma. Hence lim $\theta(t)$, $t \rightarrow \omega$, exists as a finite or infinite value. Since the last assertion of Theorem 2.1 is clear from the definition of "characteristic direction," the proof is complete.

Exercise 2.1. Let X, Y be continuous for small $|x|$, $|y|$ and satisfy $X = Y = 0$ at $(0, 0)$. Let $\psi(r)$ be a positive continuous function for small $r > 0$ such that $\psi(+0) = 0$. Suppose that the limits

$$p(\theta) = \lim_{r \to 0} \frac{X(r \cos \theta, r \sin \theta)}{\psi(r)}, \qquad q(\theta) = \lim_{r \to 0} \frac{Y(r \cos \theta, r \sin \theta)}{\psi(r)}$$

exist uniformly for θ near θ_0 and that $p^2(\theta) + q^2(\theta) \neq 0$. Show that $\theta = \theta_0$ is a characteristic direction if and only if $q(\theta) \cos \theta - p(\theta) \sin \theta = 0$ at $\theta = \theta_0$.

Theorem 2.2. *Let $X(x, y)$, $Y(x, y)$ be continuous on the triangle T: $0 < x \leq a$, $|y| \leq \eta x$ and such that $X \neq 0$. Let $\omega(x, u)$ be a non-negative, continuous function for $0 < x \leq a$, $0 \leq u \leq 2\eta x$ with the properties that $\omega(x, 0) \equiv 0$ and that the only solution $u(x)$ of*

(2.8) $$\frac{du}{dx} = \omega(x, u)$$

for small $x > 0$ satisfying

(2.9) $$u(x), \frac{u(x)}{x} \rightarrow 0 \quad \text{as} \quad x \rightarrow +0$$

is $u \equiv 0$. Let the function

(2.10) $$U(x, y) = \frac{Y(x, y)}{X(x; y)}$$

satisfy

(2.11) $U(x, y_2) - U(x, y_1) \leqq \omega(x, y_2 - y_1)$ for $y_1 \leqq y_2$

and (x, y_1), $(x, y_2) \in T$. *Then, up to replacement of the parameter t by t +* const., *(2.1) has at most one solution $(x(t), y(t))$ which for large t [or $-t$] satisfies*

(2.12) $x^2(t) + y^2(t) \to 0$ and $\dfrac{y(t)}{x(t)} \to 0$

as t [or $-t$] $\to \infty$.

Exercise 2.2.   Prove Theorem 2.2. Note that (2.1) is equivalent to $dy/dx = U(x, y)$.

Exercise 2.3.   Show that the conclusion of Theorem 2.2 is valid if (2.11) is replaced by $U(x, y_2) - U(x, y_1) \leqq (y_2 - y_1)/x$ for $-\eta x \leqq y_1 \leqq y_2 \leqq \eta x$ and $0 < x \leqq a$. In particular, this is the case if $X > 0$ and X, Y have continuous partial derivatives X_y, Y_y with respect to y satisfying $XY_y - X_yY \leqq X^2/x$ on T.

Exercise 2.4.   Replace T in Theorem 2.2 by $R_0 : 0 < x \leqq a$, $|y| \leqq b$; also, let $\omega(x, u)$ be continuous on $0 < x \leqq a$, $|u| \leqq 2b$. Show that an analogue of Theorem 2.2 is valid if "$u(x)/x \to 0$" and "$y(t)/x(t) \to 0$" are deleted from (2.9) and (2.12), respectively.

3. Perturbed Linear Systems

The results of §§ 1 and 2 will be applied to a 2-dimensional system obtained by perturbing a linear system

(3.1) $z' = Ez$,

in which $z = (x, y)$ is a real 2-dimensional vector and E is a constant matrix with real entries. Unless otherwise specified, it will be assumed that

(3.2) $\det E \neq 0$.

Let λ_1, λ_2 be the eigenvalues of E; so that λ_1, λ_2 are real or are complex conjugates since E is real. If $\lambda = \lambda_1$ or $\lambda = \lambda_2$ is a simple eigenvalue or if $\lambda = \lambda_1 = \lambda_2$ and E has a double elementary divisor, then, up to factors ± 1, there is only one (real) unit eigenvector z^* satisfying $Ez^* = \lambda z^*$.

Recall from Exercises VII 7.1 and 7.2 that if λ_1, $\lambda_2 = \pm i\beta$ are purely imaginary (with $\beta \neq 0$), then $z = 0$ is a center; that if λ_1, $\lambda_2 = \alpha \pm i\beta$ are complex conjugates but not real or imaginary (α, β real, $\neq 0$), then $z = 0$ is a focus (at $t = \pm\infty$ according as $\alpha \lessgtr 0$); that if λ_1, λ_2 are real and det $E = \lambda_1\lambda_2 > 0$, then $z = 0$ is an attractor, in fact, a node for $t = \pm\infty$ according as $\lambda_1, \lambda_2 \lessgtr 0$ (and a proper node only if $\lambda_1 = \lambda_2 < 0$ and E has simple elementary divisors); finally, if λ_1, λ_2 are real and $\lambda_1\lambda_2 = \det E < 0$, then $z = 0$ is a saddle point.

After a real linear change of variables, it can be supposed, when convenient, that E is in one of the normal forms

(3.3-1)

$$E = \begin{pmatrix} 0 & -\beta \\ \beta & 0 \end{pmatrix}$$

$\lambda_1, \lambda_2 = \pm i\beta$

(3.3-2)

$$E = \begin{pmatrix} \alpha & -\beta \\ \beta & \alpha \end{pmatrix}$$

$\lambda_1, \lambda_2 = \alpha \pm i\beta$

(3.3-3)

$$E = \begin{pmatrix} \lambda_1 & 0 \\ 0 & \lambda_2 \end{pmatrix}$$

λ_1, λ_2 real; $\lambda_1 \neq \lambda_2$

(3.3-4)

$$E = \begin{pmatrix} \lambda & 0 \\ 0 & \lambda \end{pmatrix}$$

$\lambda = \lambda_1 = \lambda_2$ real
(simple elementary divisors)

(3.3-5)

$$E = \begin{pmatrix} \lambda & 0 \\ \epsilon & \lambda \end{pmatrix}$$

$\lambda = \lambda_1 = \lambda_2, \quad \epsilon \neq 0$
(double elementary divisor)

The system to be considered in this section is of the form

(3.4) $$z' = Ez + F(z),$$

where $F(z)$ is continuous for small $\|z\|$ and satisfies

(3.5) $$\frac{F(z)}{\|z\|} \to 0 \quad \text{as} \quad 0 \neq z \to 0.$$

Theorem 3.1. *Assume* (3.2) *and that the continuous* $F(z)$ *satisfies* (3.5). *Let* $z = 0$ *be an attractor for* (3.1) *at* $t = \infty$, *so that* $\alpha_k = \text{Re } \lambda_k < 0$ *for* $k = 1, 2$.

(i) *In this case,* $z = 0$ *is an attractor for* (3.4) *at* $t = \infty$. *More generally, if* $\alpha_k < -c < 0$ *for* $k = 1, 2$, *then there exists a constant* $M = M(c)$ *such that if* $\|z_0\| \neq 0$ *is sufficiently small, every solution of* (3.4) *satisfying the initial condition* $z(0) = z_0$ *exists for* $t \geq 0$ *and satisfies*

(3.6) $$\|z(t)\| \leq M \|z_0\| e^{-ct} \quad \text{for} \quad t \geq 0;$$

(3.7) $$t^{-1} \log \|z(t)\| \to \alpha \quad \text{as} \quad t \to \infty,$$

where $\alpha = \alpha_1$ *or* $\alpha = \alpha_2$.

(ii) *If* $\alpha_1 < \alpha_2 < 0$, *so that* $\lambda_1 = \alpha_1, \lambda_2 = \alpha_2$ *are real, and if* (3.7) *holds, then*

(3.8) $$\lim_{t \to \infty} \frac{z(t)}{\|z(t)\|} = z^0$$

exists and is an eigenvector of E belonging to α $(=\lambda_1$ *or* λ_2). *In particular, if E is in the normal form* (3.3-3), *then*

(3.9) $$\frac{y(t)}{x(t)} \to 0 \quad \text{or} \quad \frac{x(t)}{y(t)} \to 0 \qquad \text{as} \quad t \to \infty$$

according as $\alpha = \lambda_1$ *or* $\alpha = \lambda_2$.

(iii) *Let* $\alpha_1 < \alpha_2 < 0$. *If* z^0 *is either of the two real unit eigenvectors of E belonging to* $\lambda = \alpha_1$, *then* (3.4) *has at least one solution* $z(t)$ *satisfying* (3.8) *and* (3.7) *with* $\alpha = \alpha_1$. *If* z^0 *is either of the two real unit eigenvectors of E belonging to* $\lambda = \alpha_2$ *and if* $\|z_0\| \neq 0$ *and* $\|z_0/\|z_0\| - z^0\|$ *are sufficiently small, then any solution of* (3.4) *determined by* $z(0) = z_0$ *exists for* $t \geq 0$ *and satisfies* (3.8) *and* (3.7) *with* $\alpha = \alpha_2$.

Proof of (i).   After a real linear change of variables, it can be supposed that E is in one of the real normal forms (3.3). In (3.3-5), it can also be supposed that $\epsilon > 0$ is so small that Re $\lambda + \frac{1}{2}\epsilon < -c$. It is then readily verified that if $r = \|z\|$ and $r \neq 0$ is small, then $r' \leq -cr$. This implies that along a solution $z(t)$, $r(t) \leq r(0)e^{-ct}$ for small $t > 0$. Consequently if $r(0) > 0$ is sufficiently small, then $z(t)$ exist for $t \geq 0$ and satisfies (3.6) with $M = 1$.

When E is not in a normal form and L is a nonsingular matrix such that $L^{-1}EL$ is in the form just used, then (3.6) holds with $M = 1$ if z is replaced by $\eta = Lz$. In this case, (3.6) holds if $M = \|L\| \cdot \|L^{-1}\|$.

When $\alpha_1 = \alpha_2 < 0$ and E is in a normal form (3.3-2), (3.3-4), or (3.3-5), it is easy to see that $0 \neq z(t) \to 0$ as $t \to \infty$ implies (3.7). This completes the proof of (i) for this case. [The cases $\alpha_1 \neq \alpha_2$ will be considered in the proofs of (ii) and (iii).]

Proof of (ii) and (iii).   Assume that E is in the normal form (3.3-3), so that the unit eigenvectors of E are $(\pm 1, 0)$ for $\lambda = \lambda_1$ and $(0, \pm 1)$ for $\lambda = \lambda_2$. On introducing polar coordinates $z = (r \cos \theta, r \sin \theta)$, (3.4) takes the form, as $r \to 0$,

(3.10)
$$r' = [\lambda_1 \cos^2 \theta + \lambda_2 \sin^2 \theta + o(1)]r,$$
$$\theta' = (\lambda_2 - \lambda_1) \sin \theta \cos \theta + o(1).$$

It follows that the only characteristic directions (mod 2π) are $\theta = 0$, $\pi/2$, π, $3\pi/2$; cf. Exercise 2.1. Hence, by Theorem 2.1, if $0 \neq z(t) \to 0$ as $t \to \infty$, then $z = z(t)$ is a spiral or (3.9) holds.

Let Ω_0 be a wedge $\Omega_0 : 0 < \|z\| < \delta$, $\|z/\|z\| - z^0\| < \eta$, where $z^0 = (0, \pm 1)$ and δ, η are small. It is seen that if a solution $z(t)$ starts in Ω_0 or enters in Ω_0, it remains in Ω_0, for the boundary points are strict ingress points for Ω_0; cf. Corollary 1.3 (ii) with t replaced by $-t$. Thus such solutions satisfy $z(t) \to 0$ as $t \to \infty$ and the second part of (3.9), i.e., (3.8).

The first equation of (3.10) implies (3.7) with $\alpha = \lambda_2$. In particular, no solution $z = z(t) \not\equiv 0$ tending to 0 as $t \to \infty$ is a spiral.

If, in the definition of the wedge Ω_0, z^0 is taken to be $(\pm 1, 0)$, the existence of solutions $z(t) \neq 0$ satisfying (3.8) follows from Corollary 1.3(i). As above, such solutions satisfy (3.7) with $\alpha = \lambda_1$. This proves Theorem 3.1.

Theorem 3.2. *Let the eigenvalues of E be $\alpha \pm i\beta$, where $\alpha \leq 0$ and $\beta \neq 0$ are real; let (3.5) hold; let $z(t)$ be a solution of (3.4) such that $0 < \|z(t)\| < \delta_0$ for all t and $\theta(t)$ a continuous determination of arc tan $y(t)/x(t)$. Let $0 < \epsilon < |\beta|$. Then there exists a $\delta_\epsilon > 0$ such that if $\delta_0 \leq \delta_\epsilon$, then*

$$(3.11) \qquad |\theta(t) - \beta t| < \epsilon t \qquad \text{for large } t.$$

In particular, $\theta(t) \to \pm \infty$ as $t \to \infty$ according as $\beta \gtrless 0$. If, in addition, $\alpha < 0$, then

$$(3.12) \qquad \theta'(t) \to \beta, \qquad \text{hence} \quad t^{-1}\theta(t) \to \beta, \qquad \text{as} \quad t \to \infty.$$

Thus, if $z = 0$ is a center for (3.1), it is a center or focus for (3.4) and if $z = 0$ is a focus for (3.1), then it is a focus for (3.4).

Proof. It is readily verified that neither the assumptions nor conclusions are affected if z is subjected to a real linear transformation. Hence, it can be supposed that E is in the normal form (3.3-2), with $\alpha \leq 0$. Let $F(z) = (F^1, F^2)$ and write (3.4) as

$$(3.13) \quad x' = \alpha x - \beta y + F^1(x, y), \qquad y' = \beta x + \alpha y + F^2(x, y).$$

Introducing polar coordinates gives

$$(3.14) \qquad r' = \alpha r + R(r, \theta), \qquad \theta' = \beta + S(r, \theta),$$

where

$$(3.15) \quad rR = F^1 \cos \theta + F^2 \sin \theta, \qquad rS = F^2 \cos \theta - F^1 \sin \theta;$$

so that $R(r, \theta), S(r, \theta) \to 0$ as $r \to +0$. Thus, there exists a $\delta_\epsilon > 0$ such that $|S(r, \theta)| < \epsilon$ if $0 < r \leq \delta_\epsilon$. Hence $|\theta' - \beta| < \epsilon$ if $0 < \|z(t)\| \leq \delta_\epsilon$, and (3.11) holds. If $\alpha < 0$, then, by Theorem 3.1, $r(t) \to 0$ as $t \to \infty$, and so $S(r(t), \theta(t)) \to 0$ as $t \to \infty$. In this case, (3.12) follows. This proves Theorem 3.2.

As can be expected, the property that $z = 0$ is a center (i.e., that a solution starting at any point $z_0 \neq 0$ at $t = 0$ returns to exactly the point z_0 at some positive t) is very sensitive to perturbations. This is illustrated by the following exercise which shows that no condition of smallness on $F(z) \not\equiv 0$ at $z = 0$ can assure that if $z = 0$ is a center for (3.1), then it is a center for (3.4).

Exercise 3.1.   Let $h(r)$ be a continuous function for $0 \leqq r \leqq 1$ such that $h(r) \to 0$ as $r \to 0$. Consider a system (3.4) of the form

$$(3.16) \qquad x' = -\beta y + xh(r), \qquad y' = \beta x + yh(r),$$

where $r = (x^2 + y^2)^{1/2}$, the function $F(z) = (xh(r), yh(r))$ is continuous for $\|z\| \leqq 1$, and $\|F(z)\|/\|z\| = |h(r)| \to 0$ as $r \to 0$. If $h(r) \equiv 0$, i.e., $F = 0$, then $z = 0$ is a center. Show that if $h(r) < 0$ for $0 < r \leqq 1$, then $z = 0$ is a focus (at $t = \infty$) for (3.16).

It might also be guessed that the other cases, $\lambda_1 = \lambda_2$, determined by equalities (rather than by inequalities) are sensitive to perturbations. This turns out to be the case. For example, the next two exercises show that if $z = 0$ is a node for (3.1) with $\lambda_1 = \lambda_2 < 0$, then, even if (3.5) holds, $z = 0$ can be a focus for (3.4), whether or not E has simple or double elementary divisors. However, as will be shown in Theorems 3.5 and 3.6, suitable conditions of smallness on F at $z = 0$, more stringent than (3.5), preserve the character of this type of stationary point.

Exercise 3.2.   Let $E = \text{diag } [\lambda, \lambda]$, $\lambda < 0$. Show that there exist continuous functions $F(z)$ for $\|z\| < \delta$ satisfying (3.5) and such that (*a*) $z = 0$ is a focus for (3.4); and (*b*) the equation (3.4) has a solution $z(t) \to 0$ as $t \to \infty$ satisfying any one of the seven possibilities compatible with

$$-\infty \leqq \liminf_{t \to \infty} \theta(t) \leqq \limsup_{t \to \infty} \theta(t) \leqq \infty.$$

Exercise 3.3.   Let E be as in (3.3–5) with $\lambda < 0$ and $\epsilon = 1$, so that E has a double elementary divisor and is in a Jordan normal form. Show that there exist continuous $F(z)$ for $\|z\| \leqq \delta$ satisfying (3.5) such that (*a*) all, (*b*) some but not all, (*c*) no solutions $z(t)$ of (3.4) which tend to 0 as $t \to \infty$ are spirals (i.e., $|\theta(t)| \to \infty$ as $t \to \infty$). [Case (*b*) cannot occur if the solutions of (3.4) are uniquely determined by initial conditions]. See Theorem 3.3.

Theorem 3.3.   *Let E be as in (3.3–5) with $\lambda < 0$ and $\epsilon = 1$. Let $F(z)$ be continuous for small $\|z\|$ and satisfy (3.5) and let $z(t) \neq 0$ be a solution of (3.4) for large t satisfying $z(t) \to 0$ as $t \to \infty$. Then either $z = z(t) = (x(t), y(t))$ is a spiral (i.e., $|\theta(t)| \to \infty$ as $t \to \infty$) or $\theta(t) \to 0$ (mod π) as $t \to \infty$.*

Exercise 3.4.   Prove Theorem 3.3.

Conditions which assure that all or that no solutions in Theorem 3.3 are spirals will be considered subsequently; cf. Exercise 4.5.

Theorem 3.4.   *Let $E = \text{diag } (\lambda_1, \lambda_2)$, where $\lambda_1 < \min (0, \lambda_2)$; let $F(z)$ be continuous and satisfy (3.5). If $z^0 = (1, 0)$ or $(-1, 0)$, then (3.4) has at least one solution $z(t)$, $t \geqq 0$, satisfying (3.8) and (3.7) with $\alpha = \lambda_1$; furthermore, if $\lambda_2 > 0$ and $z(t)$ is a solution of (3.4) for large t such that $z(t) \to 0$ as $t \to \infty$, then (3.7) holds with $\alpha = \lambda_1$ and $y(t)/x(t) \to 0$ as $t \to \infty$.*

Exercise 3.5.   Deduce Theorem 3.4 from Corollary 1.3 (i).

Exercise 3.6.   Let $E = \text{diag}\ (\lambda_1, \lambda_2)$, $\lambda_1 < \min\ (0, \lambda_2)$, and let $F(z)$ be continuous and satisfy (3.5) and

$$(3.17) \qquad \frac{\|F(z_1) - F(z_2)\|}{\|z_1 - z_2\|} \to 0 \qquad \text{as} \quad z_1, z_2 \to 0$$

(with $z_1 \neq z_2$). Then, up to reparametrizations (i.e., replacements of t by $t + \text{const.}$), (3.4) has unique pair of solutions $z_\pm(t)$ for large t such that $0 \not\equiv z_\pm(t) \to 0$ as $t \to \infty$. These solutions satisfy $z_\pm(t)/\|z_\pm(t)\| \to (\pm 1, 0)$ as $t \to \infty$, and, hence (3.7) with $\alpha = \lambda_1$.

Exercise 3.7.   Let $E = \text{diag}\ (\lambda_1, \lambda_2)$ with $\lambda_1 < \min\ (0, \lambda_2)$ and let $F(z)$ be continuous for small $\|z\|$ and satisfy (3.5). Use Theorem 2.2 and/or Exercise 2.3 to find conditions, more general than (3.17), to assure that (3.1) has at most one solution (up to changes of the parameter) satisfying $z(t) \to 0$ and $z(t)/\|z(t)\| \to (1, 0)$ as $t \to \infty$.

This completes the discussion of (3.4) under the assumption (3.5). Except in the case of a center, assumptions slightly stronger then (3.5) suffice to preserve the character of the stationary point $z = 0$ in passing from the linear system (3.1) to the perturbed system (3.4). Results of this type are consequences of general theorems in Chapter X (in particular, in § X 16). Some will be stated here for the sake of completeness. The deduction of these theorems from results in Chapter X will be given as Exercises 3.7–3.11; cf. also Theorem X 13.1 and its corollaries in § X 16.

The first condition to be imposed on $F(z)$ will involve the function

$$(3.18) \qquad \varphi_0(r) = \max \|F(z)\| \qquad \text{for} \quad \|z\| \leqq r$$

[so that $\varphi_0(r)$ is a continuous, nondecreasing function for small $r \geqq 0$ and $\varphi_0(0) = 0$] and the condition

$$(3.19) \qquad \int_{+0}^r r^{-2} \varphi_0(r)\ dr < \infty.$$

This last condition is satisfied if, e.g.,

$$(3.20) \qquad \frac{\|F(z)\|}{\|z\|^{1+\epsilon}} \to 0 \qquad \text{as} \quad z \to 0$$

for some $\epsilon > 0$, since (3.20) implies that $\varphi_0(r)/r^{1+\epsilon} \to 0$ as $r \to 0$. Conditions of the type (3.18), (3.19), or (3.20) are invariant under linear changes of the variables, $z \to Lz$ where L is a constant matrix, so that, in the theorems to follow, the assumption that the matrix E is in a normal form is no loss of generality. This does not apply, e.g., if (3.18)˙ is replaced by $\varphi_0(r) = \max \|F(z)\|$ for $\|z\| = r$ and it is not assumed that $\varphi_0(r)$ is monotone.

Theorem 3.5. *Let $F(z)$ be continuous for small $\|z\|$ and satisfy (3.18)–(3.19) and let $z(t)$ be a solution of (3.4) satisfying $0 \not\equiv z(t) \to 0$ as $t \to \infty$.*
(i) *Let E be as in (3.3–2), where $\alpha < 0$, $\beta \neq 0$. Then there exist constants $c > 0$, θ_0 such that*

$$(3.21) \qquad z(t) = ce^{\alpha t}(\cos [\beta t + \theta_0 + o(1)], \sin [\beta t + \theta_0 + o(1)])$$

as $t \to \infty$; conversely if $c > 0$, θ_0 are given constants, there exists a solution $z(t)$ of (3.4) satisfying (3.21).
  (ii) *Let $E = \mathrm{diag}\ [\lambda_1, \lambda_2]$ with $\lambda_1 < \lambda_2 < 0$. Then there exists a constant $c_1 \neq 0$ or a constant $c_2 \neq 0$ such that either*

$$(3.22) \qquad z(t) = e^{\lambda_1 t}(c_1 + o(1), o(1)) \qquad as \quad t \to \infty$$

or

$$(3.23) \qquad z(t) = e^{\lambda_2 t}(o(1), c_2 + o(1)) \qquad as \quad t \to \infty;$$

conversely if $c_1 \neq 0$ and $c_2 \neq 0$, there exist solutions $z(t)$ of (3.4) satisfying (3.22) and (3.23), respectively.
  (iii) *Let $E = \mathrm{diag}\ [\lambda_1, \lambda_2]$ with $\lambda_1 < 0 < \lambda_2$. Then there exists a constant $c_1 \neq 0$ such that (3.22) holds; conversely, if $c_1 \neq 0$, there is a solution $z(t)$ of (3.4) satisfying (3.22).*
  (iv) *Let $E = \mathrm{diag}\ [\lambda, \lambda]$ with $\lambda < 0$. Then there exist constants c_1, c_2, not both 0, such that*

$$(3.24) \qquad z(t) = e^{\lambda t}(c_1 + o(1), c_2 + o(1)) \qquad as \quad t \to \infty;$$

conversely, if c_1, c_2 are given constants, not both 0, then (3.4) has a solution satisfying (3.24).
  Exercise 3.8.   Denote by Theorem 3.5* the analogue of Theorem 3.5 in which the hypothesis (3.18)–(3.19) is replaced by the slightly heavier condition:

$$(3.25) \qquad \varphi(r) = \sup \frac{\|F(z)\|}{\|z\|} \qquad for \quad 0 < \|z\| \leqq r,$$

$$(3.26) \qquad \int_{+0}^{r} r^{-1}\varphi(r)\, dr < \infty.$$

(*a*) Deduce parts (i), (ii) of Theorem 3.5* from Theorem X 1.1 (i.e., from variants of Corollary X 1.2). (*b*) Deduce parts (iii), (iv) of Theorem 3.5* from Lemma X 4.3 (i.e., from Corollary X 4.2).
  Exercise 3.9.   Using an analogue of the Remark 2 following Lemma X 4.3 and the result of Exercise 3.8, prove Theorem 3.5.
  Theorem 3.6.   *Let the condition (3.18), (3.19) of Theorem 3.5 be replaced by the assumption that there exists a non-negative, non-decreasing, continuous*

function $\varphi(r)$ for small $r \geqq 0$ such that

(3.27) $\|F(z_1) - F(z_2)\| \leqq \varphi(r) \|z_1 - z_2\|$ for $\|z_1\|, \|z_2\| \leqq r$

and that (3.26) *holds. Then the constants* c, θ_0 *in* (3.21), *the constant* c_1 *in* (3.22) *in both* (ii) *and* (iii), *and the constants* c_1, c_2 *in* (3.24) *uniquely determine the solution* $z(t)$. (*In particular, in case* (iv), $z = 0$ *is a proper node.*)

Exercise 3.10. (*a*) Deduce the assertions concerning (3.21) and (3.24) from Theorem X 1.1 (i.e., from variants of Corollary X 1.2). (*b*) Deduce the assertion concerning (3.22) in both parts (ii), (iii) from Exercise 3.6.

In the case of a multiple elementary divisor for E, the condition (3.19) of Theorem 3.5 has to be strengthened to

(3.28) $$\int_{+0} r^{-2} |\log r| \, \varphi_0(r) \, dr < \infty.$$

This condition is also satisfied if (3.20) holds.

Theorem 3.7. *Let* $F(z)$ *be continuous for small* $\|z\|$ *and satisfy* (3.18), (3.28). *Let* E *be as in* (3.3–5) *with* $\lambda < 0$ *and* $\epsilon = 1$. *Let* $z(t)$ *be a solution of* (3.4) *with small* $\|z(0)\| \neq 0$. *Then* $z(t)$ *exists for* $t \geqq 0$ *and either there exists a constant* $c_1 \neq 0$ *such that*

(3.29) $z(t) = c_1 e^{\lambda t}(1 + o(1), t + o(t))$ as $t \to \infty$

or there exists a constant $c_2 \neq 0$ *such that*

(3.30) $z(t) = c_2 e^{\lambda t}(o(1/t), 1 + o(1))$ as $t \to \infty$;

conversely, if $c_1 \neq 0$ *and* $c_2 \neq 0$ *are given, then there exist solutions* $z(t)$ *satisfying* (3.29) *and* (3.30), *respectively.*

Exercise 3.11. Deduce Theorem 3.7 from Corollary X 4.1. In order to obtain the assertions concerning (3.29) [or (3.30)], make the change of dependent variables $z = (x, y) \to (u, v)$ defined by $x = e^{\lambda t}u$, $y = e^{\lambda t}t(u + v)$ [or $x = e^{\lambda t}u/t$, $y = e^{\lambda t}(u + v)$] and the change of independent variable $t = e^s$. This deduction is more straightforward if the condition (3.18), (3.28) is strengthened to (3.25),

(3.31) $$\int_{+0} r^{-1} |\log r| \, \varphi(r) \, dr < \infty.$$

Otherwise, the necessary arguments involve the analogue of Remark 2 following Lemma X 4.2; see also Corollary X 16.3 and Exercise X 16.2.

Theorem 3.8. *Let the condition* (3.18), (3.28) *of Theorem 3.7 be replaced by the assumption that there exists a non-negative, non-decreasing continuous function* $\varphi(r)$ *for small* $r \geqq 0$ *satisfying* (3.27), (3.31). *Then the constant* c_2 *in* (3.30) *uniquely determines the solution* $z(t)$.

Exercise 3.12. Deduce Theorem 3.8 from the changes of variables in Exercise 3.11 and from Theorem X 8.2.

4. More General Stationary Point

A discussion similar to that of the last section will be given for a plane autonomous system of the form

(4.1) $x' = P(x, y) + p(x, y)$, $y' = Q(x, y) + q(x, y)$,

where P, Q are homogeneous polynomials of degree $m \geqq 1$ and

(4.2) $p^2(x, y) + q^2(x, y) = o(r^{2m})$ as $r^2 = x^2 + y^2 \to 0$,

(4.3) $(P + p)^2 + (Q + q)^2 \geqq 0$ according as $x^2 + y^2 \geqq 0$.

In terms of polar coordinates, $x = r \cos \theta$ and $y = r \sin \theta$, define

(4.4) $R(\theta) = r^{-m}(P \cos \theta + Q \sin \theta)$,

(4.5) $S(\theta) = r^{-m}(Q \cos \theta - P \sin \theta)$;

thus R, S are homogeneous polynomials of $\sin \theta, \cos \theta$ of degree $m + 1$.

In terms of polar coordinates, (4.1) can be written as

(4.6) $r' = r^m[R(\theta) + \rho(r, \theta)]$, $r\theta' = r^m[S(\theta) + \sigma(r, \theta)]$,

where

(4.7) $\rho(r, \theta) = r^{-m}(p \cos \theta + q \sin \theta)$, $\sigma(r, \theta) = r^{-m}(q \cos \theta - p \sin \theta)$

tend to 0 as $r \to 0$ uniformly in θ.

If $S(\theta) \not\equiv 0$ and $R(\theta) \not\equiv 0$, then a necessary condition for $\theta = \theta_0$ to be characteristic is that $S(\theta_0) = 0$ and a sufficient condition is that $S(\theta_0) = 0$, $R(\theta_0) \neq 0$; cf., e.g., Exercise 2.1. If $S(\theta) \not\equiv 0$, it has only a finite number of zeros (mod 2π). Theorem 2.1 implies the following:

Theorem 4.1.   *Assume* (4.2), (4.3) *and* $S(\theta) \not\equiv 0$. *If* $(x(t), y(t))$ *is a solution of* (4.1) *for large* $t > 0$ *[or* $-t > 0$*] satisfying*

(4.8) $0 < x^2(t) + y^2(t) \to 0$ *as* $t \to \infty$ *[or* $t \to -\infty$*]*,

then a continuous determination of $\theta(t) = \arctan y(t)/x(t)$ *satisfies either*

(4.9) $\theta_0 = \lim \theta(t)$ *exists (and is finite)*

and $S(\theta_0) = 0$ *or*

(4.10) $|\theta(t)| \to \infty$ *as* $t \to \infty$ *[or* $t \to -\infty$*]*.

The question to be considered first is the following: If $S(\theta_0) = 0$, do there exist solutions of (4.1) satisfying (4.8), (4.9)? After a rotation of the (x, y)-plane, it can be supposed that $\theta_0 = 0$. Suppose $\theta = 0$ is a zero of degree $k > 0$ for $S(\theta)$,

(4.11) $S(\theta) = c_0\theta^k + o(|\theta|^k), \theta \to 0$ and $c_0 \neq 0, k > 0$.

In order to state the next result, introduce the sector

(4.12) $\Omega_0(\delta, \eta) = \{(x, y) : 0 < r \leq \delta, |\theta| < \eta < \tfrac{1}{2}\pi\}.$

Theorem 4.2. *Assume* (4.2), (4.3), (4.11) *and that k is an odd integer. Then if $\delta, \eta > 0$ are sufficiently small,* (4.1) *has a half-trajectory Γ in $\Omega_0(\delta, \eta)$ with endpoint on $r = \delta$. For any such half-trajectory,* (4.8) *and* (4.9) *hold with $\theta_0 = 0$. If, in addition, $R(0) \neq 0$, then Γ is defined for large $t > 0$ or $-t > 0$ according as $R(0) \lessgtr 0$.*

Exercise 4.1. (a) Using Theorem 2.2 (and Exercise 2.3), obtain sufficient conditions to assure that Γ in Theorem 4.2 is unique (up to the replacement of t by $t +$ const.). (b) In addition to the conditions of Theorem 4.2, assume that $R(0) \neq 0$. Using Exercise 2.4, applied to (4.6) rather than (4.1), deduce sufficient conditions for the uniqueness of Γ. For example, show that if

$$\Phi(r, \theta) = \frac{S(\theta) + \sigma(r, \theta)}{R(\theta) + \rho(r, \theta)}$$

satisfies

$$\Phi(r, \theta_2) - \Phi(r, \theta_1) \leq \psi(r)(\theta_2 - \theta_1) \quad \text{for} \quad -\epsilon \leq \theta_1 \leq \theta_2 \leq \epsilon$$

and small $r > 0$, where $\psi(r) > 0$ is continuous and satisfies $\int_{+0} \psi(r) \, dr/r < \infty$, then Γ is unique.

Proof. If $0 < r \leq \delta$ and $\theta = \pm \eta$, where $\delta, \eta > 0$ are sufficiently small, then, by (4.11), $c_0 \theta' \gtrless 0$. Thus, if $c_0 > 0$, then the lateral boundaries $\theta = \pm \eta$ of Ω_0 are strict egress points and Corollary 1.1 is applicable. If $c_0 < 0$, this corollary becomes applicable if t is replaced by $-t$. Also, (4.3) implies that Corollary 1.2 can be used. This gives the existence of $\Gamma : (x(t), y(t))$ satisfying (4.8). Also, (4.9) follows from Theorem 4.1 since $\eta > 0$ can be taken so small that $\theta = 0$ is the only characteristic direction in $|\theta| \leq \eta$. This gives the first part of Theorem 4.2. The second part of Theorem 4.2 is much simpler since Corollary 1.3 (or even Corollary 1.5) can be used in place of Corollary 1.1.

In order to obtain refinements of Theorem 4.2 and to deal with the case that $k > 0$ is even, suppose that

(4.13) $R(\theta) = d_0 \theta^j + o(|\theta|^j), \theta \to 0 \quad \text{and} \quad d_0 \neq 0, j \geq 0.$

Theorem 4.3. *Assume* (4.2), (4.3), (4.11) *with $k > 0$ an even integer and $R(0) \neq 0$ [i.e.,* (4.13) *with $j = 0$]. Then, if $\delta, \eta > 0$ are sufficiently small,* (4.1) *has no or infinitely many half-trajectories Γ in $\Omega_0(\delta, \eta)$. For any such half-trajectory,* (4.8) *and* (4.9) *hold with $\theta_0 = 0$.*

The first part of this assertion follows from Exercise 1.2; the second part from Theorem 4.1. See Theorem 4.5 and Exercises 4.6, 4.7 for criteria for the alternatives in Theorem 4.3.

When $j = 0$ [so that $R(0) = d_0 \neq 0$], k is odd, and

(4.14) $c_0 d_0 > 0$,

then (4.1) has infinitely many half-trajectories Γ satisfying (4.8) and (4.9) with $\theta_0 = 0$; cf. Corollary 1.3(ii). In the next theorem, there will be no assumption on the parity of j, k; instead, it will be supposed that

(4.15) $k > j + 1$ and $c_0 d_0 > 0$.

[It can be mentioned that if both j, k are even, then condition (4.14) involves no loss of generality. For the substitution $\theta \to -\theta$ (i.e., $y \to -y$) changes the sign of c_0 if k is even (cf. (4.6)) but not that of d_0 if j is even.] Whenever (4.14) holds, it can also be supposed that

(4.16) $c_0 < 0$ and $d_0 < 0$,

otherwise t is replaced by $-t$.

Exercise 4.2. Show that there are examples of P, Q with $c_0 d_0 < 0$, $k > 0$ even, $j > 0$ odd, $k > j + 1$ such that no condition of smallness on p, q assures that (4.1) has a half-trajectory $(x(t), y(t))$ satisfying (4.8).

Theorem 4.4. *Assume* (4.2), (4.3), (4.11), (4.13), (4.15), (4.16). *Then there exists a positive* $\epsilon_0 = \epsilon_0(c_0, d_0, j, k)$ *such that if*

(4.17) $|p(x, y)|, |q(x, y)| \leqq \dfrac{\epsilon_0 r^m}{(\log 1/r)^{k/(k-j-1)}}$

holds for small $r > 0$ *[e.g., if* $p, q = o(r^{m+\epsilon})$ *as* $r \to +0$ *for some* $\epsilon > 0$*], then* (4.1) *possesses infinitely many half-trajectories defined for* $t \geqq 0$ *satisfying* (4.8) *and* (4.9) *with* $\theta_0 = 0$.

In order to prove this, introduce the following notation: Let $\delta, \eta > 0$; c_1 and d_1 are positive constants satisfying

(4.18) $-S(\theta) \leqq c_1 \theta^k$, $-R(\theta) \geqq d_1 \theta^j$ for $0 \leqq \theta \leqq \eta$.

Let $\psi_1(r)$, $\psi_2(r)$ be positive continuous functions for $0 < r \leqq \delta$ such that the functions (4.7) satisfy

(4.19)
$$\psi_1(r) \geqq \max [-\sigma(r, \theta)] \quad \text{for} \quad 0 \leqq \theta \leqq \eta,$$
$$\psi_2(r) \geqq \max [\rho(r, \theta)] \quad \text{for} \quad 0 \leqq \theta \leqq \eta,$$

and that

(4.20) $\psi_1(r), \psi_2(r) \to 0$ as $r + 0$.

It follows from (4.6) that if $\eta > 0$ is small and $\delta = \delta(\eta) > 0$ is sufficiently small, then

(4.21) $\theta' < 0$ for $0 < r \leqq \delta$, $\theta = \eta$.

Let Ω_1 be the set

(4.22) $\Omega_1 = \{(x, y) : 0 < r \leqq \delta, 0 < \theta \leqq \eta, d_2\theta^j \geqq \psi_2(r)\}$,

where d_2 is any fixed constant, $0 < d_2 < d_1$; see Figure 7. Then (4.6) implies that, on Ω_1,

(4.23) $-r' \geqq d_3 r^m \theta^j > 0$,   where $d_3 = d_1 - d_2 > 0$,

(4.24) $-r\theta' \leqq [c_1\theta^k + \psi_1(r)]r^m$.

Thus, along a solution of (4.1) in Ω_1, $r' < 0$ and it is permissible to

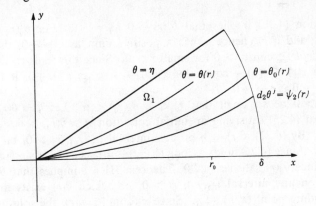

Figure 7.

introduce r as as independent variable, so that

(4.25) $$r\,\frac{d\theta}{dr} = \frac{S(\theta) + \sigma(r, \theta)}{R(\theta) + \rho(r, \theta)},$$

(4.26) $$d_3\theta^j r\,\frac{d\theta}{dr} \leqq c_1\theta^k + \psi_1(r).$$

In addition, by (4.21) and (4.23),

(4.27) $$\frac{d\theta}{dr} > 0 \text{ for } 0 < r \leqq \delta, \qquad \theta = \eta.$$

Theorem 4.4 will be deduced from the following:

Lemma 4.1. *If there exists a continuously differentiable function* $\theta = \theta_0(r)$, $0 < r \leqq \delta$, *satisfying the differential inequality*

(4.28) $$d_3\theta^j r\,\frac{d\theta}{dr} \geqq c_1\theta^k + \psi_1(r)$$

and $d_2\theta_0{}^j \geqq \psi_2(r)$, *then* (4.1) *has infinitely many half-trajectories defined for* $t \geqq 0$ *satisfying* (4.8) *and* (4.9) *with* $\theta_0 = 0$.

Proof of the Lemma. If the new variable

(4.29) $$v = \theta^{j+1}$$

is introduced, (4.28) becomes

(4.30) $$r \frac{dv}{dr} \geq c^* v^\lambda + \psi^*(r),$$

where

(4.31) $\lambda = \dfrac{k}{j+1} > 1, \qquad c^* = \dfrac{c_1(j+1)}{d_3}, \qquad \psi^* = \dfrac{\psi_1(r)(j+1)}{d_3}.$

The relation (4.30) implies that $\theta_0(r) \to 0$ as $r \to 0$. For $\theta_0(r) \geq 0$ is increasing and if θ_0, hence v, has a positive limit as $r \to +0$, then, by (4.30), $r \, dv/dr \geq$ const. > 0 for small $r > 0$. Since this leads to the contradiction $v(r) \leq v(\eta) -$ const. $\log (\eta/r) \to -\infty$ as $r \to +0$, it follows that $\theta_0(r) \to 0$ as $r \to +0$.

Let $r_0 > 0$ be so small that $r_0 < \delta$, $\theta_0(r_0) < \eta$. Let $\theta = \theta(r)$ be a solution of (4.25) satisfying an initial condition $\eta > \theta(r_0) > \theta_0(r_0)$; see Figure 7. By (4.27), $\theta(r) < \eta$ on any interval $[r_1, r_0)$, $r_1 > 0$, on which $\theta(r)$ exists. Since (4.26) holds as long as $(x, y) = (r \cos \theta(r), r \sin \theta(r))$ is in Ω_1 and since $\theta_0(r)$ satisfies (4.28), Theorem III 4.1 implies that $\theta_0(r) \leq \theta(r) < \eta$ on any interval $[r_1, r_0)$, $r_1 > 0$, on which $\theta(r)$ exists and the corresponding point $(x, y) \in \Omega_1$. Since $d_2 \theta_0{}'(r) \geq \psi_2(r)$, the solution $\theta(r)$ can be defined on $(0, r_0]$ and the corresponding point $(x, y) \in \Omega_1$. This implies the lemma.

Proof of Theorem 4.4. In view of (4.17) and (4.19), it is possible to choose

$$\psi_1(r) = \psi_2(r) = \frac{2\epsilon_0}{(\log 1/r)^{k/(k-j-1)}}.$$

For a constant $\epsilon > 0$ to be specified, put

$$v_0 = \frac{\epsilon}{(\log 1/r)^{(j+1)/(k-j-1)}},$$

so that

$$r \frac{dv_0}{dr} = \frac{\epsilon(j+1)}{(k-j-1)(\log 1/r)^{k/(k-j-1)}}, \qquad v_0{}^\lambda = \frac{\epsilon^\lambda}{(\log 1/r)^{k/(k-j-1)}}.$$

The inequality (4.28) or (4.30) is equivalent to

$$\frac{\epsilon(j+1)}{(k-j-1)} \geq c^* \epsilon^\lambda + \frac{2\epsilon_0(j+1)}{d_3}.$$

Since $\lambda > 1$, it is clear that if $\epsilon_0 > 0$ is sufficiently small, it is possible to choose an $\epsilon > 0$ satisfying this inequality. Finally, the condition

$d_2\theta_0{}^j(r) \geqq \psi_2(r)$ becomes, by (4.29),

$$\frac{d_2\epsilon^{j/(j+1)}}{(\log 1/r)^{j/(k-j-1)}} \geqq \frac{2\epsilon_0}{(\log 1/r)^{k/(k-j-1)}}\,.$$

But if $d_2 > 0$ and $\epsilon > 0$, this holds for small r since $k > j + 1 > j$. Thus Theorem 4.4 follows from Lemma 4.1.

When $j = 0$ in Theorem 4.4, the result can be sharpened somewhat.

Theorem 4.5. *Assume* (4.2), (4.3), (4.11) *with* $c_0 < 0$, $R(0) = d_0 < 0$, *and* $k > 0$ *even. Put*

(4.32)
$$\epsilon^* = \frac{(-d_0)^{k/(k-1)}}{[-c_0 k^k (k-1)]^{1/(k-1)}}\,.$$

(i) *Let* $0 < \epsilon_0 < \epsilon^*$, $\delta > 0$, $\eta > 0$. *Suppose that* $\sigma(r, \theta)$ *in* (4.7) *satisfies*

(4.33) $-\sigma(r, \theta) \leqq \dfrac{\epsilon_0}{(\log 1/r)^{k/(k-1)}}$ *for* $0 < r \leqq \delta$, $0 < \theta \leqq \eta$.

Then (4.1) *has infinitely many half-trajectories defined for* $t \geqq 0$ *satisfying* (4.8) *and* (4.9) *with* $\theta_0 = 0$. (ii) *Let* $\epsilon^* < \epsilon^0$, $\delta > 0$, $\eta > 0$. *Suppose that*

(4.34) $-\sigma(r, \theta) \geqq \dfrac{\epsilon^0}{(\log 1/r)^{k/(k-1)}}$ *for* $0 < r \leqq \delta$, $|\theta| \leqq \eta$.

Then no half-trajectory satisfies (4.8) *and* (4.9) *with* $\theta_0 = 0$.

The proof of (i) is similar to that of Theorem 4.4. Let $0 < -c_0 < c_1$, $0 < d_1 < -d_0$, $\psi_1(r)$ be defined as in (4.19) and

$$\Omega_1 = \{(x, y): 0 < r \leqq \delta, \ 0 < \theta \leqq \eta\}.$$

It is clear from (4.25) that if δ, $\eta > 0$ are sufficiently small, then $(x, y) \in \Omega_1$ implies that

(4.35)
$$d_1 r \frac{d\theta}{dr} \leqq c_1 \theta^k + \psi_1(r)$$

and (4.27). The required analogue of Lemma 4.1 is the following:

Lemma 4.2. *Let* $c = c_1/d_1$, $\psi(r) = \psi_1(r)/d_1$. *If there exists a continuously differentiable function* $\theta_0(r) > 0$, $0 < r \leqq \eta$ *satisfying*

(4.36)
$$r \frac{d\theta}{dr} \geqq c\theta^k + \psi(r),$$

then the conclusion of (i) *in Theorem 4.5 holds.*

Exercise 4.3. (*a*) Prove Lemma 4.2. (*b*) Deduce Theorem 4.5(i) from it.

For the proof of Theorem 4.5 (ii), let $0 < c_1 < -c_0$, $-d_0 < d_1$, and

$$0 < \psi^1(r) \leqq \min\,[-\sigma(r, \theta)]   \text{for} |\theta| \leqq \eta.$$

Then if δ, η are sufficiently small,

$$(4.37) \qquad\qquad d_1 r \frac{d\theta}{dr} \geqq c_1 \theta^k + \psi^1(r).$$

Lemma 4.3. *Let* $c = c_1/d_1$, $\psi(r) = \psi^1(r)/d_1$. *If, for every small* $\eta > 0$ *and* $r_0 > 0$, *the solution of*

$$(4.38) \qquad\qquad r \frac{d\theta}{dr} = c\theta^k + \psi(r), \qquad \theta(r_0) = \eta,$$

satisfies $\theta(r_1) = -\eta$ *at some* $r_1, 0 < r_1 < r_0$, *then the conclusion of Theorem 4.5(ii) holds.*

Exercise 4.4.   (a) Prove Lemma 4.3. (b) Deduce Theorem 4.5(ii) from it.

Exercise 4.5.   Apply Theorem 4.5 to the case $m = 1$, $P(x, y) = \lambda x$, $Q(x, y) = x + \lambda y$, and $\lambda < 0$ [so that (4.1) is the system considered in Theorem 3.3].

Exercise 4.6.   The proof of Theorem 4.5(i) makes it clear that if $\psi(r) \geqq 0$ is any continuous function for $0 \leqq r \leqq \eta$, $\psi(0) = 0$, and if

$$(4.39) \qquad\qquad r \frac{d\theta}{dr} = c\theta^k + \psi(r),$$

$c > 0$, has a solution $\theta_0(r) > 0$ for $0 < r \leqq \eta$, then we can obtain analogues of Theorem 4.5(i) by replacing (4.33) by $-\sigma(r, \theta) \leqq \epsilon\psi(r)$ for a suitable $\epsilon > 0$. This exercise deals with conditions on $\psi \geqq 0$ to assure that (4.39) has positive solutions for $0 < r \leqq \eta$. Introduce the new independent variable t defined by $r = e^{-t/c}$, so that $dr/r = -dt/c$ and $r = 0$ corresponds to $t = \infty$. Writing $\varphi(t) = \psi(e^{-t/c})/c$ and $\lambda = k$ transforms (4.39) into

$$(4.40) \qquad\qquad \theta' = -\theta^\lambda - \varphi(t), \qquad \text{where} \quad \theta' = \frac{d\theta}{dt},$$

$\lambda > 1$, and $\varphi(t) \geqq 0$ is continuous for large t. The problem is to find conditions on continuous $\varphi(t) \geqq 0$ to assure that (4.40) has a positive solution for large t; cf. § XI 7 for the case $\lambda = 2$. For brevity, a function $\varphi(t) \geqq 0$ continuous for large t for which (4.40) has a positive solution for large t will be called of class N_λ.

(a) Show that if $\varphi(t)$ is of class N_λ, then $\int^\infty \varphi(t)\, dt < \infty$. (b) Show that $\varphi(t) \in N_\lambda$ if and only if there exists a continuously differentiable positive function $\theta = \theta_0(t)$ for large t such that

$$(4.41) \qquad\qquad \theta' + \theta^\lambda \leqq -\varphi;$$

hence if $\varphi_0(t) \in N_\lambda$ and $0 \leqq \varphi(t) \leqq \varphi_0(t)$, then $\varphi(t) \in N_\lambda$. (c) If $\mu = \lambda/(\lambda - 1)$, $\epsilon^* = \max (u - u^\lambda)/(\lambda - 1)^{\lambda/(\lambda-1)}$ for $u > 0$, and $0 \leqq \varphi(t) \leqq$

ϵ^*/t^μ, then $\varphi(t) \in N_\lambda$. (d) If $\displaystyle\int^\infty \varphi(t)\, dt < \infty$ and

$$(4.42) \qquad \int_t^\infty \varphi(s)\, ds \leqq \frac{\epsilon_*}{t^{\mu-1}}, \qquad \text{where} \quad \epsilon_* = \frac{(\lambda - 1)^{(\lambda-1)/(\lambda-2)}}{\lambda^{\lambda/(\lambda-1)}}$$

is max $[u - (\lambda - 1)u^\lambda]$ for $u > 0$, then $\varphi(t) \in N_\lambda$. (e) If $\displaystyle\int^\infty t^{1/(\lambda-1)}\varphi(t)$
$dt < \infty$, then $\varphi(t) \in N_\lambda$.

Exercise 4.7. Formulate analogues of Theorem 4.5 (i) using parts (d) and (e) of the last exercise.

Notes

SECTION 1. Theorems 1.1 and 1.2 may be new and are suggested by the result of Hartman and Wintner [1] refining a paper of Perron [3]. The results of this section have the advantage of permitting the treatment, e.g., of some cases of (4.6) when $R(\theta)$, $S(\theta)$ have a common zero.

SECTION 2. The main result (Theorem 2.1) of this section goes back to Bendixson [2] under conditions of analyticity. The treatment in the text follows that of Nemytskiĭ and Stepanov [1]; cf. Hartman and Wintner [11] and Kowalski [1]. For results related to Theorem 2.2, see Hoheisel [1], Hartman and Wintner [1], Hartman [1], and Keil [1].

SECTION 3. For early references on the subject of this section, see the encyclopedia articles of Painlevé [1] and Liebmann [1]; see also Dulac [1], [2]. Investigations on the questions considered here were begun by Briot and Bouquet [1] for equations of the form $x\, dy/dx = ax + by + \ldots$ which, because of § 2, contain most of the cases of (3.4) when the eigenvalues of E are real. Poincaré [1] initiated the discussion of solutions of (3.4), under conditions of analyticity, when (3.5) holds. Perron [3], [5] was the first to systematically investigate the questions of § 3 for nonanalytic differential equations. He obtained existence and uniqueness theorems of the type Theorems 3.5–3.8 but with much heavier conditions. Weyl [4] obtained existence and uniqueness theorems for the cases of real eigenvalues under conditions similar to those of Theorems 3.6, 3.8; cf. also Hoheisel [1]. Wintner was the first to omit a Lipschitz condition of the type occurring in Theorems 3.6 and 3.8 in considering questions of existence. Theorems 3.5, 3.7 are due to Wintner [6], [11] (and are based on his papers [3], [8]). This type of result has been generalized for nonautonomous systems of arbitrary dimension by Hartman and Wintner [19], see § X 13. Examples of the type occurring in Exercise 3.1 and 3.2(a) were given by Perron [5, I] and were modified by Hartman and Wintner [11] to obtain all the assertions of Exercises 3.2 and 3.3.

SECTION 4. Many of the papers mentioned in connection with § 3 are relevant here in some cases of $m = 1$. Most papers in the literature on the problems of this section invoke the hypothesis $S(\theta_0) = 0$, $R(\theta_0) \neq 0$ (i.e., deal with the cases $k > 0, j = 0$); see, e.g., Frommer [1], Forster [1], Lonn [2], Grobman and Vinograd [1], and Nemytskiĭ and Stepanov [1]. The particular uniqueness criterion involving $\psi(r)$ in Exercise 4.1(b) is due to Lonn [1]; for other criteria, see Vinograd and Grobman [1]. Theorem 4.4 may be new. Theorem 4.5 is a result of Lonn [2]; for the case $k = 2$ corresponding to (4.1) of the type $x' = \lambda x + p$, $y' = x + \lambda y + q$ as in Exercise 4.4, the number ϵ^* in (4.32) is $-d_0^2/4c_0$ and part (i), but a less sharp form of part (ii) was given earlier by Lonn [1].

Chapter IX

Invariant Manifolds and Linearizations

This chapter concerns the behavior of solutions of an autonomous system (of arbitrary dimension) in the vicinity of a simple type of stationary point or of a periodic solution. Many results to be obtained will be extended to nonautonomous systems in the next chapter by very different methods; cf., e.g., § IX 6 and §§ X 8, 11. The lemmas of this chapter, however, dealing with local maps from one Euclidean space to another have intrinsic interest, give some insight which is not furnished by other methods, and are applicable to the study of both stationary points and periodic solutions.

1. Invariant Manifolds

For every (real) t, let $T^t : \xi \to \xi_t$ be a continuous mapping of a neighborhood D_t of $\xi = 0$ in a Euclidean ξ-space into a neighborhood of $\xi = 0$ in the same space, with $T^t(0) = 0$. A set S is called *invariant* with respect to the family of maps $\{T^t\}$ if $T^t(D_t \cap S) \subset S$ for all t. A set S is called *locally invariant* with respect to $\{T^t\}$ if there exists an $\epsilon > 0$ such that $\xi \in S$ implies that $T^{t_0}\xi \in S$ for all t_0 for which $\|T^t\xi\| < \epsilon$ on the t-interval joining 0 and t_0.

The problem of the behavior of solutions of a smooth autonomous system near a stationary point can, in some cases, be viewed as the comparison of the solutions of a linear system with constant coefficients

$$(1.1) \qquad \xi' = E\xi$$

and solutions of a perturbed system

$$(1.2) \qquad \xi' = E\xi + F(\xi).$$

Unless the contrary is stated, it will be supposed that $F(\xi)$ is of class C^1 for small $\|\xi\|$ and that

$$(1.3) \qquad F(\xi) = o(\|\xi\|) \qquad \text{as} \quad \xi \to 0;$$

or, equivalently,

(1.4) $$F(0) = 0, \qquad \partial_\xi F(0) = 0$$

where $\partial_\xi F$ is the Jacobian matrix of F with respect to ξ.

Let $\xi_t = \eta(t, \xi_0)$ be the solution of (1.2) satisfying the initial condition $\eta(0, \xi_0) = \xi_0$. For a fixed t, consider $\xi_t = \eta(t, \xi_0)$ as a map $T^t : \xi_0 \to \xi_t$ of a neighborhood D_t of $\xi = 0$ in the ξ-space into a neighborhood of $\xi = 0$ in the same space. The map T^t is defined on the set D_t of points ξ_0 for which the solution $\eta(t, \xi_0)$ is defined on a t-interval containing 0 and t. [The maps T^t are the germ of a group; cf. (2.2).]

A set S in the ξ-space which is [locally] invariant with respect to the family of maps T^t will be called [locally] invariant with respect to (1.2). S is invariant [or locally invariant] with respect to (1.2) if and only if it has the property that $\xi_0 \in S$ implies that $\eta(t, \xi_0) \in S$ for all t on the maximal interval of existence of the solution $\eta(t, \xi_0)$ [or for some $\epsilon > 0$, $\eta(t_0, \xi_0) \in S$ whenever $\|\eta(t, \xi_0)\| < \epsilon$ on the t-interval joining 0 and t_0].

If S is an invariant set, then the intersection of S and a sphere $\|\xi\| < \epsilon$ is locally invariant. Conversely, if S is a locally invariant set, then $S_0 = \bigcup T^t(S \cap D_t)$ is an invariant set. Thus the investigation of invariant sets can be reduced to the study of locally invariant sets and vice versa. This is convenient by virtue of the following remark: If $F(\xi)$ is altered outside of a small sphere, $\|\xi\| < \epsilon$, and an invariant set S_0 is determined for the new differential equation, then the intersection of S_0 and the sphere $\|\xi\| < \epsilon$ is a locally invariant set for the original differential equation (1.2).

Locally invariant sets are convenient for another reason. The conditions imposed on F are of "local" nature and it is not reasonable to expect that invariant sets, involving notions "in the large," should be simple sets. For example, suppose that dim $\xi = 2$ and that (1.2) has a pair of solutions $\xi = \xi_1(t), \xi_2(t)$ for $-\infty < t < \infty$, such that $\xi_1(t), \xi_2(t) \to 0$ as $t \to \pm\infty$ as in Figure 1. Then the set S_0 consisting of $\xi = 0$ and the points $\xi = \xi_j(t)$, $-\infty < t < \infty$ and $j = 1, 2$, is an invariant set S_0. S_0 is a curve with a self-intersection. But each of the sets $S_1 = \{\xi = (\xi^1, 0), |\xi^1| < \epsilon\}$ or $S_2 = \{\xi = (0, \xi^2), |\xi^2| < \epsilon\}$, for a sufficiently small $\epsilon > 0$, is a locally invariant set and is a C^1-arc.

After a linear change of variables, with a constant matrix N,

(1.5) $$\xi = N\zeta, \qquad \det N \neq 0,$$

the equation (1.2) becomes

(1.6) $$\zeta' = N^{-1}EN\zeta + N^{-1}F(N\zeta).$$

Suppose that N is chosen so that

(1.7) $$N^{-1}EN = \operatorname{diag}[P, Q],$$

where P is a $d \times d$ and Q an $e \times e$ matrix with eigenvalues $p_1, \ldots, p_d$ and $q_1, \ldots, q_e$, respectively, where $d > 0$, $e \geqq 0$ and

(1.8) $\operatorname{Re} p_j \leqq \alpha < 0$ and $\operatorname{Re} q_k \geqq \beta > \alpha.$

Write $\zeta = (y, z)$, where y is a d-dimensional vector, z an e-dimensional vector and $(N^{-1}EN)\zeta = (Py, Qz)$. Thus, in ζ-coordinates, the linear equation (1.1) becomes

(1.9) $y' = Py, \qquad z' = Qz.$

The solution $(y(t), z(t))$ of (1.9) with an initial point $(y(0), z(0)) = (y(0), 0)$ satisfies $z(t) \equiv 0$ and $\|y(t)\| \leqq$ const. $t^j e^{\alpha t} \leqq$ const. $e^{(\alpha + \epsilon)t}$ for some integer

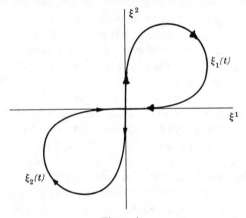

Figure 1.

j and for arbitrary $\epsilon > 0$ and large t; cf. § IV 5. In addition, if $(y(t), z(t))$ is a solution of (1.9) for which $\|(y(t), z(t))\| \leqq e^{(\beta - \epsilon)t}$ for some $\epsilon > 0$ and large $t \geqq 0$, then $z(t) \equiv 0$. Thus the d-dimensional flat $z = 0$ in the $\zeta = (y, z)$-space is invariant with respect to (1.9) and is made up of all solutions $(y(t), z(t))$ satisfying $\|(y(t), z(t))\| \leqq e^{(\beta - \epsilon)t}$ for some $\epsilon > 0$ and large t.

The first question concerning (1.2) to be considered is whether or not an analogous situation holds for (1.2). More precisely, for the system (1.2) or (1.6) written as

(1.10) $y' = Py + F_1(y, z), \qquad z' = Qy + F_2(y, z),$

where F_1, F_2 are of class C^1 for small $\|y\|, \|z\|$,

(1.11) $F_1, F_2 = o(\|y\| + \|z\|)$ as $(y, z) \to 0,$

is there a d-dimensional locally invariant manifold S of the form $S : z = g(y)$ defined for small $\|y\|$ which is made up of all solutions $(y(t), z(t))$ of

(1.10) in a neighborhood of $(y, z) = 0$ for large t satisfying $\|(y(t), z(t))\| \leq e^{(\beta - \epsilon)t}$ for some $\epsilon > 0$. It will be shown in § 6 that the answer is in the affirmative.

2. The Maps T^t

(i) Consider the unique solution $\xi = \eta(t, \xi_0)$ of the initial value problem

$$(2.1) \qquad \xi' = E\xi + F(\xi), \qquad \xi(0) = \xi_0.$$

Since the solution $\eta(t, 0) = 0$ for $\xi_0 = 0$ exists for all t, $\eta(t, \xi_0)$ exists on an arbitrarily large interval $|t| \leq t_0$ if $\|\xi_0\|$ is sufficiently small; cf. Theorem V 2.1.

For a fixed t, consider $\xi_t = \eta(t, \xi_0)$ as a map $T^t: \xi_0 \to \xi_t$ from the ξ-space into itself. The set of maps T^t behaves like an Abelian group in the sense that if $\|\xi_0\|$ is so small that $\xi_t = \eta(t, \xi_0)$ is defined on a t-interval containing $t = 0$, t_1, t_2, and $t_1 + t_2$, then

$$(2.2) \qquad T^{t_1 + t_2} = T^{t_1} T^{t_2}$$

for the given ξ_0, i.e., $\eta(t_1 + t_2, \xi_0) = \eta(t_1, \eta(t_2, \xi_0))$ since the solution of (2.1) is unique.

(ii) Consider a change of variables $R: \zeta = Z_0(\xi)$ which together with its inverse $R^{-1}: \xi = X_0(\zeta)$ is of class C^1. Then (2.1) becomes of the form

$$(2.3) \qquad \zeta' = (N^{-1}EN)\zeta + G(\zeta), \qquad \zeta(0) = \zeta_0$$

where $G(\zeta) = o(\|\zeta\|)$ as $\zeta \to 0$ and N is the Jacobian matrix $N = (\partial X_0/\partial \zeta)_{\zeta = 0} = \partial_\zeta X_0(0)$. In general, $G(\zeta)$ is not of class C^1. Solutions $\zeta_t = \zeta(t, \zeta_0)$ of (2.3) are, of course, unique since solutions of (2.1) are unique and the map R is one-to-one. The map $\zeta_0 \to \zeta_t$, is given by

$$(2.4) \qquad RT^tR^{-1}: \zeta_t = \zeta(t, \zeta_0).$$

This can be seen by considering the action of RT^tR^{-1}; thus $R^{-1}\zeta_0$ is a point $\xi_0 = X_0(\zeta_0)$, $T^tR^{-1}\zeta_0$ is the solution $\xi_t = \eta(t, \xi_0)$ of (2.1) for fixed ξ_0 and $R(T^tR^{-1}\zeta_0)$ is therefore the solution $\zeta_t = \zeta(t, \zeta_0)$ of (2.3) for fixed ζ_0.

(iii) By Theorem V 3.1, $\eta(t, \xi_0)$ is of class C^1 and its Jacobian $H(t, \xi_0) = \partial_{\xi_0}\eta$ with respect to ξ_0 satisfies the linear initial value problem

$$(2.5) \qquad H'(t, \xi_0) = [E + \partial_\xi F(\eta)]H(t, \xi_0), \qquad H(0, \xi_0) = I.$$

In particular, if $\xi_0 = 0$,

$$(2.6) \qquad H'(t, 0) = EH(t, 0), \qquad H(0, 0) = I;$$

thus

$$(2.7) \qquad H(t, 0) = e^{Et}.$$

Therefore, the expansion for $\eta(t, \xi_0)$, for a fixed t, in terms of linear terms in ξ_0 and higher order terms is of the form

$$(2.8) \qquad \eta(t, \xi_0) = e^{Et}\xi_0 + \Xi(t, \xi_0),$$

where

$$(2.9) \qquad \Xi(t, 0) = 0 \quad \text{and} \quad \partial_{\xi_0}\Xi(t, 0) = 0.$$

3. Modification of $F(\xi)$

In order to avoid technical difficulties (as, e.g., the fact that the domain D_t of the map $T^t : \xi_0 \to \xi_t$ depends on t), it will be convenient to replace $F(\xi)$ in (2.1) by a function which is defined for all ξ, is identical with $F(\xi)$ for small $\|\xi\|$, say, for $\|\xi\| \leq \frac{1}{2}s$, and vanishes for $\|\xi\| \geq s > 0$. If the new function is called $F(\xi)$ again, then the solution $\xi = \eta(t, \xi_0)$ of (2.1) is defined for all ξ. Thus, for every t, the domain of $T^t : \xi_0 \to \xi_t$ is the entire ξ_0-space and the set of maps T^t is indeed a group.

Lemma 3.1. *Let $F(\xi)$ be a vector function of class C^1 for small $\|\xi\|$ satisfying $F(0) = 0$, $\partial_\xi F(0) = 0$. Let $\theta > 0$ be arbitrary. Then there exists a number $s = s(\theta) > 0$ (which tends to 0 with θ) and a function $G(\xi)$ of class C^1 defined for all ξ satisfying $G(\xi) = F(\xi)$ for $\|\xi\| \leq \frac{1}{2}s$, $G(\xi) = 0$ for $\|\xi\| \geq s$, and $\|\partial_\xi G\| \leq \theta$ for all ξ.*

In this lemma, F and ξ need not be of the same dimension. Here and in the remainder of this chapter, the norm $\|A\|$ of a rectangular matrix A is the norm of A as a linear operator from one Euclidean space into another, i.e., the least constant c such that $\|Ay\| \leq c \|y\|$ for all y.

Proof. Let $s > 0$ be so small that $\|\partial_\xi F(\xi)\| \leq \theta/8$, in particular $\|F(\xi)\| \leq \theta \|\xi\|/8$, for $\|\xi\| \leq s$. Let $\varphi(t)$ be a smooth real-valued function of t for $t \geq 0$ such that $\varphi(t) = 1$ for $t \leq (\frac{1}{2}s)^2$, $0 < \varphi(t) < 1$ for $(\frac{1}{2}s)^2 < t < s^2$, $\varphi(t) = 0$ for $t > s^2$ and $0 \leq -d\varphi/dt \leq 2/s^2$ for all $t \geq 0$. Put $G(\xi) = F(\xi)\varphi(\|\xi\|^2)$ or $G(\xi) = 0$ according as $\|\xi\| \leq$ or $\geq s$. Then $\partial_\xi G = 0$ for $\|\xi\| > s$. For $\|\xi\| \leq s$, $\partial_\xi G = (\partial F^i/\partial\xi^j)\varphi + 2(F^i\xi^j) \, d\varphi/dt$ and so $\|\partial_\xi G\| \leq (\theta/8) + 2(\theta \|\xi\|^2/8)(2/s^2) \leq \theta$. This proves the lemma.

Thus, in dealing with solutions of (2.1) only in a small neighborhood of $\xi = 0$, Lemma 3.1 shows that there is no loss of generality in supposing that $F(\xi)$ is of class C^1 for all ξ,

$$(3.1) \qquad \|\partial_\xi F(\xi)\| \leq \theta$$

for all ξ, and

$$(3.2) \qquad F(\xi) = 0 \quad \text{for} \quad \|\xi\| \geq s,$$

where $s = s(\theta)$.

It will now be verified that there exist $s_0 = s_0(s, \theta) > 0$, $\theta_0 = \theta_0(s, \theta)$ such that $s_0, \theta_0 \to 0$ as $s, \theta \to 0$ and that if the solution $\xi = \eta(t, \xi_0)$ of

(2.1) is written as (2.8), then

(3.3) $\Xi(t, \xi_0) = 0$ for $0 \leq t \leq 1$, $\|\xi_0\| \geq s_0$

(3.4) $\|\partial_\xi \Xi(t, \xi_0)\| \leq \theta_0$ for $0 \leq t \leq 1$, arbitrary ξ_0.

In order to see this, note that (3.1) implies that $\|F(\xi)\| \leq \theta \|\xi\|$, thus a solution of (2.1) satisfies $\|\xi'\| \leq c_0 \|\xi\|$ for $c_0 = \|E\| + \theta$. Hence the solution $\xi = \xi(t)$ of (2.1) satisfies $\|\xi(t)\| \geq \|\xi_0\| \exp(-c_0 t)$; cf. Lemma IV 4.1. Thus, if $\|\xi_0\| \geq s_0$, where $s_0 = s \exp c_0$, then $\|\xi(t)\| \geq s$ for $0 \leq t \leq 1$. In this case, (2.1) reduces to $\xi' = E\xi$, $\xi(0) = \xi_0$ for $0 \leq t \leq 1$ and the solution $\xi(t)$ is $e^{Et}\xi_0$; i.e., in (2.8), $\Xi(t, \xi_0) = 0$ for $0 \leq t \leq 1$ and $\|\xi_0\| \geq s_0$.

The relation $\Xi(t, \xi_0) = \eta(t, \xi_0) - e^{Et}\xi_0$ implies that $\partial_{\xi_0}\Xi(t, \xi_0) = H(t, \xi_0) - e^{Et}$ or

$$\partial_{\xi_0}\Xi(t, \xi_0) = e^{Et}[K(t, \xi_0) - I], \quad \text{where} \quad K(t, \xi_0) = e^{-Et}H(t, \xi_0).$$

The matrix $K(t, \xi_0)$ has the derivative $K' = e^{-Et}(H' - EH)$ or, by (2.5),

$$K'(t, \xi_0) = e^{-Et} \partial_\xi F(\eta)e^{Et}K(t, \xi_0), \quad K(0, \xi_0) = I.$$

Since $\|e^{-Et} \partial_\xi F(\eta)e^{Et}\|$ has the bound $c_1\theta$ for $0 \leq t \leq 1$, where $c_1 = (e^{\|E\|})^2$, it follows from Lemma IV 4.1 that $\|K(t, \xi)\|$ has a bound of the form $\exp c_1\theta$ for $0 \leq t \leq 1$. Hence $\|K'\| \leq (c_1\theta) \exp c_1\theta$, and so $\|K(t, \xi_0) - I\| \leq (c_1\theta) \exp c_1\theta$ for $0 \leq t \leq 1$. Consequently,

$$\|\partial_{\xi_0}\Xi(t, \xi_0)\| \leq e^{\|E\|} (c_1\theta) \exp c_1\theta \quad \text{for} \quad 0 \leq t \leq 1;$$

i.e., (3.4) holds with $\theta_0 = e^{\|E\|}(c_1\theta) \exp c_1\theta$.

4. Normalizations

After a linear change of variables, $\xi = N\zeta$, (2.1) can be written as

(4.1) $y' = Py + F_1(y, z)$, $z' = Qz + F_2(y, z)$ and $y(0) = y_0, z(0) = z_0$,

where $N^{-1}EN = \text{diag}[P, Q]$. It is supposed that the eigenvalues p_j, q_k of P, Q satisfy

(4.2) $\text{Re } p_j \leq \alpha < 0$ and $\text{Re } q_k \geq \beta > \alpha$.

The eigenvalues of the nonsingular matrices

(4.3) $A = e^P$, $C = e^Q$

are e^{p_j}, e^{q_k}, respectively, where $0 < |e^{p_j}| \leq e^\alpha < 1, |e^{q_k}| \geq e^\beta > e^\alpha$. Thus if $\epsilon > 0$ is arbitrary, there exist real nonsingular matrices N_1, N_2 such that $N_1^{-1}AN_1 = \exp(N_1^{-1}PN_1)$ has a norm $\leq e^{\alpha+\epsilon}$ and that $N_2^{-1}C^{-1}N_2 = \exp(N_2^{-1}C^{-1}N_2)$ has a norm $\leq e^{-\beta+\epsilon}$. This follows by considering the "real" analogue of the Jordan normal forms in which the usual 1 on the subdiagonal is replaced by an arbitrarily small ϵ; cf. § IV 9.

Since diag $[A, C] = $ exp diag $[P, Q]$, it can be supposed that N is replaced by the product N diag $[N_1, N_2]$, so that

(4.4) $$\|A\| \leqq e^{\alpha+\epsilon}; \quad \|C^{-1}\| \leqq e^{-\beta+\epsilon}.$$

It will be supposed that $\epsilon > 0$ is so small that

(4.5) $$a = \|A\|, \quad 1/c = \|C^{-1}\|$$

satisfy

(4.6) $$a < c, \quad a < 1.$$

It will also be supposed that F_1, F_2 are of class C^1 and

(4.7) $F_i, \partial_y F_i, \partial_z F_i$ vanish at $(y, z) = (0, 0)$ for $i = 1, 2,$

(4.8) $\|\partial_y F_i\|, \|\partial_z F_i\| \leqq \theta$ for all (y, z) and $i = 1, 2,$

(4.9) $F_i(y, z) = 0$ for $\|y\|^2 + \|z\|^2 \geqq s^2 > 0$, $i = 1, 2.$

Correspondingly, the general solution of (4.1) defines, for fixed t, a map T^t from (y_0, z_0) to the point $(y, z) \equiv (y_t, z_t)$ such that T^t is of the form

(4.10) $T^t: y_t = e^{Pt} y_0 + Y(t, y_0, z_0), \quad z_t = e^{Qt} z_0 + Z(t, y_0, z_0),$

where

(4.11) Y, Z and the Jacobian matrices $\partial_{y_0, z_0} Y, Z$ vanish at $(y_0, z_0) = 0$

for all t,

(4.12) $\|\partial_{y_0} Y\|, \|\partial_{z_0} Y\|, \|\partial_{y_0} Z\|, \|\partial_{z_0} Z\| \leqq \theta_0$ for all y_0, z_0

and $0 \leqq t \leqq 1$,

(4.13) $Y = 0, \quad Z = 0$ if $\|y_0\|^2 + \|z_0\|^2 \geqq s_0^2$

and $0 \leqq t \leqq 1$. In (4.12)–(4.13), θ_0, s_0 depend on (θ, s) in such a way that $\theta_0, s_0 \to 0$ as $\theta, s \to 0$. Finally, the set of maps T^t form a group: $T^{t_1+t_2} = T^{t_1} T^{t_2}$.

5. Invariant Manifolds of a Map

One of the basic results to be proved concerns one map $T: (y_0, z_0) \to (y_1, z_1)$ rather than a group of maps T^t. In the application of this result, $T = T^1$.

Lemma 5.1. *Let A be a $d \times d$ matrix, C an $e \times e$ nonsingular matrix such that (4.5), (4.6) hold. Let $T: (y_0, z_0) \to (y_1, z_1)$ be a map of the form*

(5.1) $T: y_1 = Ay_0 + Y(y_0, z_0), \quad z_1 = Cz_0 + Z(y_0, z_0),$

where Y, Z are of class C^1 for small $\|y_0\|, \|z_0\|$ and satisfy (4.11). Then there *exists an e-dimensional vector function $z = g(y)$ of class C^1 for small $\|y\|$ such that*

(5.2) $$g(0) = 0, \qquad \partial_y g(0) = 0$$

and that the maps

(5.3) $R: u = y, \quad v = z - g(y)$ and $R^{-1}: y = u, \quad z = v + g(u)$

transform T into the form

(5.4) $RTR^{-1}: u_1 = Au_0 + U(u_0, v_0), \qquad v_1 = Cv_0 + V(u_0, v_0),$

where

(5.5)   *U, V and their Jacobian matrices vanish at $(u_0, v_0) = 0$*

and

(5.6) $V(u_0, 0) = 0.$

Condition (5.6) means that the set of points (u_0, v_0) near the origin on the flat $v_0 = 0$ is invariant under the map (5.4); i.e., the manifold $z = g(y)$ is locally invariant for (5.1). In applications of Lemma 5.1, the following two remarks will often be used. Remark 2 will be used in §§ 8–9.

Remark 1. In view of (4.11) and Lemma 2.1, it can be supposed that Y, Z are of class C^1 for all (y_0, z_0) and satisfy (4.12)–(4.13), where θ_0, s_0 are arbitrarily small positive numbers. Let θ_0 satisfy

(5.7) $$0 < \theta_0 < \min\left(\frac{c-a}{4}, \frac{1-a}{2}\right).$$

It will be shown, in this case, that $g(y)$ can be defined for all y [so that RTR^{-1} is defined for all (u_0, v_0)] and (5.6) holds for all u_0. Furthermore, there is a constant $\sigma = \sigma(\theta_0)$ such that

(5.8) $$\|\partial_y g(y)\| \leqq \sigma < 1$$

and $\sigma \to 0$ as $\theta_0 \to 0$.

Remark 2. If, in addition, it is assumed that $c > 1$, then $g(y) \to 0$ as $\|y\| \to \infty$.

Lemma 5.1 will now be proved by the method of successive approximations.* Another proof will be given in Exercises 5.3 and 5.4 at the end of this section.

Proof of Lemma 5.1 and Remark 2. Assume for a moment that R [i.e., $g(y)$] is known. Then (5.1), (5.3) show that

(5.9) $RTR^{-1}: \begin{aligned} u_1 &= Au_0 + Y(u_0, v_0 + g(u_0)), \\ v_1 &= Cv_0 + Cg(u_0) + Z(u_0, v_0 + g(u_0)) \\ &\qquad - g(Au_0 + Y[u_0, v_0 + g(u_0)]). \end{aligned}$

---

*See also Appendix, page 271.

Hence, by (5.4)

(5.10) $V(u, v) = Cg(u) + Z(u, v + g(u)) - g(Au + Y[u, v + g(u)])$,

and (5.6) holds if and only if

(5.11) $g(u) = C^{-1}\{g(Au + Y[u, g(u)]) - Z(u, g(u))\}.$

Hence, it must be shown that the functional equation (5.11) for $g(u)$ has a solution of class C^1 satisfying (5.2)–(5.8).

The equation (5.11) will be solved by successive approximations. Let

(5.12) $g_0(u) \equiv 0$

and, if $g_{n-1}(u)$ has been defined, put

(5.13) $g_n(u) = C^{-1}\{g_{n-1}(Au + Y[u, g_{n-1}(u)]) - Z(u, g_{n-1}(u))\}.$

Below let $g_{n-1} = g_{n-1}(u)$, $g_{n-1}^0 = g_{n-1}(Au + Y^0)$, where $Y^0 = Y(u, g_{n-1}(u))$, $Z^0 = Z(u, g_{n-1}(u))$. It is clear that $g_0, g_1, \ldots$ are defined and of class C^1 for all u. In addition, if ∂g_n is the Jacobian matrix of g_n, then

(5.14) $\partial g_n = C^{-1}\{(\partial g_{n-1}^0)[A + \partial_y Y^0 + (\partial_z Y^0)(\partial g_{n-1})]$
$$- [\partial_y Z^0 + (\partial_z Z^0)(\partial g_{n-1})]\},$$

where, e.g., $\partial_y Y^0 = \partial_y Y(y, z)$ at $(y, z) = (u, g_{n-1}(u))$.

Define the number σ by

(5.15) $\sigma = \dfrac{\theta_0}{c - a - 3\theta_0},$ so that $0 < \sigma < 1$

by (5.7). It will be shown by induction that

(5.16) $\|\partial g_n(u)\| \leqq \sigma$ for all u.

It is clear that (5.16) holds for $n = 0$. Assume that (5.16) holds when n is replaced by $n - 1$. Then, by (5.14) and $\sigma < 1$,

$$\|\partial g_n\| \leqq c^{-1}\{\sigma[a + \theta_0 + \theta_0\sigma] + [\theta_0 + \theta_0\sigma]\} \leqq c^{-1}[\sigma(a + 3\theta_0) + \theta_0].$$

Since $c^{-1}[\sigma(a + 3\theta_0) + \theta_0] = \sigma$, (5.16) follows and the induction is complete.

It will now be verified that $\partial g_0, \partial g_1, \ldots$ are equicontinuous. For any function $f = f(u)$ or $f = f(y, z)$, let $\Delta f = f(u + \Delta u) - f(u)$ or $\Delta f = f(y + \Delta y, z + \Delta z) - f(y, z)$. Put

(5.17) $h_1(\delta) = \sup \|\Delta \partial_{y,z} Y, Z\|$ for $\|\Delta y\|, \|\Delta z\| \leqq \delta$,

where $\partial_{y,z} Y, Z$ means any of the four Jacobian matrices $\partial_y Y$, $\partial_z Y$, $\partial_y Z$, $\partial_z Z$ of $Y(y, z)$, $Z(y, z)$. It will be shown by induction that

(5.18) $\|\Delta \partial g_n\| \leqq h(\delta)$ for $\|\Delta u\| \leqq \delta < 1$, where

(5.19)
$$h(\delta) = \frac{4h_1(\delta)}{c - a - 4\theta_0}.$$

It is clear that (5.18) holds for $n = 0$. Assume its validity if n is replaced by $n - 1$. Note that, by (5.16),

$$\|\Delta g_{n-1}(u)\| \leq \sigma \|\Delta u\| \leq \|\Delta u\|,$$

hence

(5.20)
$$\|\Delta \partial_{y,z} Y^0, Z^0\| \leq h_1(\|\Delta u\|),$$

$$\|\Delta[Au + Y(u, g_{n-1}(u))]\| \leq (a + 2\theta_0) \|\Delta u\| \leq \|\Delta u\|,$$

where the last two inequalities follow from (4.12) and (5.7). Using the analogue of $\Delta[f_1(u)f_2(u)] = f_1(u + \Delta u) \Delta f_2 + (\Delta f_1)f_2(u)$ and $\sigma < 1$, it follows from (5.14) that if $\|\Delta u\| \leq \delta < 1$, then

(5.21) $\|\Delta \partial g_n\| \leq c^{-1}\{h(\delta)[a + 2\theta_0] + [h_1(\delta) + h_1(\delta) + \theta_0 h(\delta)]$

$$+ [h_1(\delta) + h_1(\delta) + \theta_0 h(\delta)]\} = c^{-1}\{h(\delta)(a + 4\theta_0) + 4h_1(\delta)\},$$

where the right side is $h(\delta)$ by (5.19).

Next, it will be shown that the sequence $g_0, g_1, \ldots$ converges uniformly on every bounded u-set. This is true if there exist constants M, r such that $0 < r < 1$ and for $n = 1, 2, \ldots,$

(5.22)
$$\|g_n(u) - g_{n-1}(u)\| \leq M \|u\| r^n.$$

This inequality holds for $n = 1$, if M and r are chosen subject to $Mr = \sigma$. Assume the validity of (5.22) when n is replaced by $n - 1$. By (5.13), $c \|g_n(u) - g_{n-1}(u)\|$ is at most

$$\|g_{n-1}(Au + Y[u, g_{n-1}]) - g_{n-2}(Au + Y[u, g_{n-2}])\|$$
$$+ \|Z(u, g_{n-1}) - Z(u, g_{n-2})\|.$$

The first term is majorized by

$$\|g_{n-1}(Au + Y[u, g_{n-1}]) - g_{n-2}(Au + Y[u, g_{n-1}])\|$$
$$+ \|g_{n-2}(Au + Y[u, g_{n-1}]) - g_{n-2}(Au + Y[u, g_{n-2}])\|.$$

Hence $c \|g_n(u) - g_{n-1}(u)\|$ is not greater than

$$M \|Au + Y(u, g_{n-1})\| r^{n-1} + \sigma \theta_0 M \|u\| r^{n-1} + \theta_0 M \|u\| r^{n-1},$$

which is at most $Mr^{n-1} \|u\| (a + 4\theta_0)$. Thus, if $r = (a + 4\theta_0)/c$ and $M = \sigma/r$, then (5.22) holds and $r < 1$ by (5.7).

Consequently, $g(u) = \lim g_n(u)$ exists uniformly on every bounded u-set. In view of (5.13), this limit function $g(u)$ satisfies the functional equation (5.11). Finally, since the sequence $\partial g_0, \partial g_1, \ldots$ is uniformly bounded and equicontinuous, there is a subsequence which is uniformly convergent on every bounded u-set. It follows that $g(u)$ is of class C^1. This completes the proof.

Proof of Remark 2. Let $M = \max \|g(u)\|$ for $\|u\| \leq s_0$. By (4.13) and (5.11), $g(u) = C^{-1}g(Au)$ if $\|u\| \geq s_0$ and $g(u) = C^{-n}g(A^n u)$ if $\|A^{n-1}u\| \geq s_0$. Thus if $\|A^{n-1}u\| \geq s_0$ but $\|A^n u\| \leq s_0$, then $\|g(u)\| \leq Mc^{-n}$. This implies Remark 2 since $c > 1$ and, for large $\|u\|$, there exixts a unique integer $n = n(u)$ satisfying $\|A^{n-1}u\| \geq s_0 > \|A^n u\|$ and $n(u) \to \infty$ as $\|u\| \to \infty$.

Exercise 5.1. (*a*) This part of the exercise concerns variants of Lemma 5.1 under various smoothness assumptions on Y, Z. Instead of the assumption that Y, Z is of class C^1 and satisfies (4.11)–(4.12), suppose that Y, Z satisfies one of the following hypotheses: (i) $Y = 0$, $Z = 0$ at $(y, z) = 0$ and Y, Z is uniformly Lipschitz continuous with an arbitrarily small Lipschitz constant for small $\|y\|$, $\|z\|$ [i.e., if $\epsilon > 0$ is arbitrary, then $\|\Delta Y\| + \|\Delta Z\| \leq \epsilon(\|\Delta y\| + \|\Delta z\|)$ for sufficiently small ($\|y\|$, $\|y + \Delta y\|$, $\|z\|$, $\|z + \Delta z\|$)]; (ii) Y, Z is of class C^m, $1 \leq m \leq \infty$, and satisfies (4.11); (iii) Y, Z satisfies (ii) with $1 \leq m < \infty$ and its partial derivatives of order m have a degree of continuity majorized by a (constant times a) monotone, non-negative function $h_m(\delta) \to 0$, $\delta \to + 0$; (iv) Y, Z are analytic and satisfy (4.11). Then the analogue of Lemma 5.1 holds with a $g(y)$ having the corresponding property (i), (ii), (iii) or (iv), instead of being of class C^1.

(*b*) Verify that if F_1, F_2 in (4.1) have the analogues of property (i), (ii), (iii) or (iv), then $Y = Y(t, y_0, z_0)$, $Z = Z(t, y_0, z_0)$ in (4.10) have the corresponding property (i), (ii), (iii) or (iv) with respect (y_0, z_0) uniformly for $0 \leq t \leq 1$.

Exercise 5.2. Show that the restriction $a < 1$ in Lemma 5.1 is not needed. [Note that the condition $a + 2\theta_0 < 1$ was used in the proof only in connection with (5.20), (5.21).]

Corollary 5.1. *Let $T, g(y), \theta_0$ be as in Lemma 5.1 and Remark 1 following it. For a given (y_0, z_0), put $(y_1, z_1) = T(y_0, z_0)$, $(y_2, z_2) = T(y_1, z_1)$, ... Then, on the one hand, $z_0 = g(y_0)$ implies that $\|(y_n, z_n)\| = O((a + \theta_0)^n)$ as $n \to \infty$ (in fact, if (4.2), (4.3) hold and $y_0 \neq 0$, then $y_n \neq 0$ for all n, $\|z_n\|/\|y_n\| \to 0$ and $\limsup n^{-1} \log \|(y_n, z_n)\| \leq \alpha$ as $n \to \infty$); and, on the other hand, $z_0 \neq g(y_0)$ implies that $(c - 2\theta_0)^n = O(\|(y_n, z_n)\|)$ as $n \to \infty$.*

Remark 3. If $c > 1$ (so that $a < 1 < c$) then the manifold $z = g(y)$ in a neighborhood of $(y, z) = (0, 0)$ can be described as the set of points (y_0, z_0) such that $(y_n, z_n) = T^n(y_0, z_0)$ satisfy $\|(y_n, z_n)\| \to 0$ exponentially as $n \to \infty$ and/or $\|(y_n, z_n)\| \to 0$ as $n \to \infty$ and/or (y_n, z_n) remains in a neighborhood of $(0, 0)$ for $n = 0, 1, 2, \ldots$. In the case $a < 1 < c$, the manifold $z = g(y)$ is called the *stable manifold* of (5.1) as $n \to \infty$; the corresponding manifold for $n \to -\infty$ is called the *unstable manifold* of (5.1).

Proof of Corollary 5.1. Note that $z_0 = g(y_0)$ is equivalent to $v_0 = 0$. In this case, $v_0 = v_1 = \cdots = 0$ by (5.4), (5.6). Correspondingly, $u_n = Au_{n-1} + U(u_{n-1}, 0)$, so that $\|u_n\| \leq (a + \theta_0) \|u_{n-1}\|$ and $\|u_n\| \leq (a + \theta_0)^n \|u_0\| \to 0$ as $n \to \infty$. Thus, if $\epsilon > 0$ is arbitrary, there is an $N = N_\epsilon$ such that $\|u_n\| \leq (a + \epsilon) \|u_{n-1}\|$ for $n \geq N$ and $\|u_{n+N}\| \leq (a + \epsilon)^n \|u_N\|$ for $n \geq 0$. Since $y_n = u_n$, $z_n = g(u_n) = o(\|u_n\|)$ as $n \to \infty$, it is seen that $\|(y_n, z_n)\| \leq (1 + \sigma) \|u_n\|$ and the first assertion follows. Also $\limsup n^{-1} \log \|(y_n, z_n)\| \leq \log a$. Since a linear transformation of the y-variables can bring $\log a$ arbitrarily near to α, it follows that $\limsup n^{-1} \log \|(y_n, z_n)\| \leq \alpha$.

By virtue of (5.10), we have the relation
$$\partial_v V(u, v) = \partial_z Z(u, v + g(u)) - \partial g(Au + Y[u, v + g(u)]) \, \partial_z Y(u, v + g(u)),$$
so that $\|\partial_v V\| \leq \theta_0 + \sigma\theta_0 \leq 2\theta_0$, and by (5.6), $\|V(u, v)\| \leq 2\theta_0\|v\|$. Thus, $v_n = Cv_{n-1} + V(u_{n-1}, v_{n-1})$ implies that $\|v_n\| \geq (c - 2\theta_0) \|v_{n-1}\|$ or $\|v_n\| \geq (c - 2\theta_0)^n \|v_0\|$. Also $\|(y_n, z_n)\| \geq \|(u_n, v_n)\| - \|g(u_n)\| \geq (1 - \sigma) \|(u_n, v_n)\|$ which implies the last assertion.

Theorem 5.1. *In the map $T: \xi_0 \to \xi_1$,*

$$(5.23) \qquad T: \quad \xi_1 = \Gamma\xi_0 + \Xi(\xi_0),$$

let $\Xi(\xi_0)$ be of class C^1 for small $\|\xi_0\|$ and satisfy $\Xi(0) = 0$, $\partial_{\xi_0}\Xi(0) = 0$; Γ a constant, nonsingular matrix having d, e_0, e eigenvalues of absolute value less than 1, equal to 1, greater than 1, respectively, where $d, e_0, e \geq 0$. Then there exists a map R of a neighborhood of $\xi_0 = 0$ onto a neighborhood of the origin in the Euclidean (u_0, v_0, w_0)-space such that R is of class C^1 with a nonvanishing Jacobian, and RTR^{-1} is of the form

$$(5.24) \qquad RTR^{-1}: \quad \begin{aligned} u_1 &= Au_0 + U(u_0, v_0, w_0), \\ w_1 &= Bw_0 + W(u_0, v_0, w_0), \\ v_1 &= Cv_0 + V(u_0, v_0, w_0), \end{aligned}$$

where A, B, C is a square $d \times d$, $e_0 \times e_0$, $e \times e$ matrix with eigenvalues of absolute value less than 1, equal to 1, greater than 1, respectively; U, V, W and their partial derivatives vanish at the origin; and

$$(5.25) \qquad V = 0, \; W = 0 \quad \text{if} \quad v_0 = 0, \, w_0 = 0,$$

$$(5.26) \qquad U = 0, \; W = 0 \quad \text{if} \quad u_0 = 0, \, w_0 = 0.$$

The condition (5.25) [or (5.26)] means that the plane $v_0 = 0$, $w_0 = 0$ of dimension d [or $u_0 = 0$, $w_0 = 0$ of dimension e] is a locally invariant manifold. When Γ has no eigenvalues of absolute value 1, so that dim $\xi_0 = d + e$, then the variables w_0, w_1 are absent in (5.24).

Proof. (Details will be left to the reader.) By Lemma 5.1, there is a map R_0: $\xi_0 \to (u_0, v_0, w_0)$ of class C^1 with nonvanishing Jacobian such that if $R_0 T R_0^{-1}$ is given by the right side of (5.24), then (5.25) holds. If Lemma 5.1 is applied to $(R_0 T R_0^{-1})^{-1} = R_0 T^{-1} R_0^{-1}$, a new map R_1 results and the desired map R in Theorem 5.1 is given by $R = R_1 R_0$.

The results of this section are applicable to differential equations by virtue of the arguments used to obtain the following corollary of Lemma 5.1 and Corollary 5.1.

Corollary 5.2. *Let* (4.10) *be a group of maps* T^t *of class* C^1 *for all* (y_0, z_0) *satisfying* (4.11) *and* (4.12) *for* $0 \leqq t \leqq 1$, *where* P, Q *are constant matrices such that* (4.2), (4.3), (4.5), *and* (4.6) *hold, and* θ_0 *satisfies* (5.7). *Let* $g(y)$ *be the function furnished by Lemma 5.1 and the Remark 1 following it when* $T = T^1$. *Then* $RT^t R^{-1}$ *is of the form*

$$(5.27)\quad RT^t R^{-1}: \; u_t = e^{Pt} u_0 + U(t, u_0, v_0), \qquad v_t = e^{Qt} v_0 + V(t, u_0, v_0),$$

where

$$(5.28)\qquad\qquad V(t, u_0, 0) = 0 \qquad \text{for all} \quad t, u_0.$$

Furthermore, if $y_0 \neq 0$ *and* $z_0 = g(y_0)$, *then* $z_t = g(y_t)$ *for all* t, $y_t \neq 0$ *for all* t, $\|z_t\|/\|y_t\| \to 0$ *and* $\lim \sup t^{-1} \log \|y_t\| \leqq \alpha$ *as* $t \to \infty$; *if* $z_0 \neq g(y_0)$, *then* $(c - 2\theta_0)^t = O(\|(y_t, z_t)\|)$ *as* $t \to \infty$.

If $c > 1 > a$, a remark similar to that following Corollary 5.1 is applicable here.

Proof. It will first be verified that if $n \leqq t \leqq n + 1$, then there exist positive constants c_1, c_2 such that

$$(5.29)\qquad\qquad c_1\|(y_n, z_n)\| \leqq \|(y_t, z_t)\| \leqq c_2\|(y_n, z_n)\|.$$

In order to see this, note that $T^t = T^{t-n}T^n$. Thus (4.10) and (4.11) for $0 \leqq t - n \leqq 1$ give

$$(5.30)\qquad\qquad \|y_t - e^{P(t-n)} y_n\| \leqq \theta_0(\|y_n\| + \|z_n\|)$$

and a similar inequality for z_t. These inequalities imply (5.29).

Let $z_0 = g(y_0)$, then the behavior of (y_n, z_n) for large n is described by Corollary 5.1. The last part of (5.29) gives $\lim \sup t^{-1} \log \|(y_t, z_t)\| \leqq \alpha$ as $t \to \infty$. Suppose, if possible, that $z_t \neq g(y_t)$ for some t, say $t = t_0$, then $(c - 2\theta_0)^n = O(\|(y_{n+t_0}, z_{n+t_0})\|)$ as $n \to \infty$ by Corollary 5.1. But this is a contradiction. Hence $z_t = g(y_t)$ for all t; i.e., $v_t = 0$ for all t so that (5.28) holds.

Note that if $y_t = 0$ for some t, then $z_t = g(y_t)$ implies $\|z_t\| \leqq \sigma \|y_t\| = 0$ for the same t. But then $(y_t, z_t) \equiv 0$ for all t by the group property of T^t. The remaining assertions of Corollary 5.2 follow from those of Corollary 5.1 and from (5.29).

The following two exercises give another proof of Lemma 5.1 based on the methods of §§ X 8–10 for nonautonomous differential equations.[*]

_____

[*]See Appendix, page 271, for simplification and extension of these exercises.

The main part of the proof is Exercise 5.4(*b*), which leads to a comparatively simple proof of Lemma 5.1 because it deals with maps of the form $T_n = S_n \circ S_{n-1} \circ \cdots \circ S_1$, where S_k depends on k, rather than with $T^n = T \circ \cdots \circ T$; cf. parts (*d*) and (*e*) of Exercise 5.4.

Exercise 5.3. (*a*) For $n = 1, 2, \ldots$, let $\xi \to S_n \xi$ be a continuous map of the $\xi = (\xi^1, \ldots, \xi^d)$ space R $=$ R^d into itself and $T_n = S_n \circ S_{n-1} \circ \cdots \circ S_1$. Let S be a compact and $K_1, K_2, \ldots$ closed ξ-sets such that $S_n(\text{R} - K_{n-1}) \subset \text{R} - K_n$, and $K_n \cap T_n(S)$ is not empty for $n = 2, 3, \ldots$. Then there exists a point $\xi_0 \in S$ such that $T_n \xi_0 \in K_n$ for $n = 1, 2, \ldots$. (*b*) Let E be a nonsingular constant $d \times d$ matrix, $F(\xi)$ a continuous vector-valued function for all ξ such that $F(\xi) = 0$ for large $\|\xi\|$. Then $\xi \to E\xi + F(\xi)$ maps the ξ-space onto itself (i.e., the equation $E\xi + F(\xi) = \eta$ has at least one solution ξ for every $\eta \in \text{R}^d$).

Exercise 5.4. (*a*) Let A, C be matrices as in Lemma 5.1. For $n = 1, 2, \ldots$, let $Y_n(y, z)$, $Z_n(y, z)$ be continuous functions for all (y, z) which vanish for large $\|y\| + \|z\|$. Let S_n denote the map

$$S_n: (y, z) \to (Ay + Y_n(y, z), Cz + Z_n(y, z)),$$

T_n the map $T_n = S_n \circ S_{n-1} \circ \cdots \circ S_1$, and P the projection $P(y, z) = z$ of the (y, z)-space onto the z-space. For a fixed y_0, show that

$$z \to P \circ T_n(y_0, z)$$

is a map onto the z-space. (*b*) Let Y_n, Z_n satisfy

$$\| Y_n(y, z)\|, \|Z_n(y, z)\| \leqq (\|y\| + \|z\|)\delta,$$

where $0 \leqq 4\delta < c - a$. Let K be the cone $\|z\| \leqq \|y\|$ and, for a fixed y_0, S the sphere $S = \{(y, z): y = y_0, \|z\| \leqq \|y_0\|\}$. Show that $S_n(\text{R} - K) \subset \text{R} - K$ and that there exists a $z = z_{(n)}$ such that $P \circ T_n(y_0, z_{(n)}) = 0$, hence $(y_0, z_{(n)}) \in S$ and $K \cap T_n(S)$ is not empty. Consequently, there exists a $(y_0, z_0) \in S$ such that $(y_n, z_n) = T_n(y_0, z_0) \in K$ for $n = 0, 1, \ldots$. (*c*) Show that if $(y_n, z_n) = T_n(y_0, z_0) \in K$ for $n = 0, 1, \ldots$, then $\|z_n\| \leqq \|y_n\| \leqq (a + 2\delta)^n \|y_0\|$, where $a + 2\delta < \frac{1}{2}(a + c)$; but if

$$(y_n, z_n) = T_n(y_0, z_0) \notin K$$

for some n (hence for all large n), then, for large n,

$$\|y_n\| > \|z_n\| \geqq (\text{const.})(c - 2\delta)^n > 0, \quad \text{where} \quad c - 2\delta > \tfrac{1}{2}(a + c).$$

(*d*) In addition to the conditions of parts (*a*), (*b*), and (*c*), assume that $Q_n = Y_n, Z_n$ satisfy

$$\|Q_n(y, z) - Q_n(y^*, z^*)\| \leqq (\|y - y^*\| + \|z - z^*\|)\delta$$

for all y, z, y^*, z^*. In terms of a sequence $(y_n, z_n) = T_n(y_0, z_0)$, introduce the maps

$$S_n^*: (y, z) \to (Ay + Y_n^*(y, z), Cz + Z_n^*(y, z)) \quad \text{for} \quad n = 1, 2, \ldots,$$

where

$$Y_n{}^*(y, z) = Y_n(y + y_{n-1}, z + z_{n-1}) - Y_n(y_{n-1}, z_{n-1});$$
$$Z_n{}^*(y, z) = Z_n(y + y_{n-1}, z + z_{n-1}) - Z_n(y_{n-1}, z_{n-1});$$

so that if $(y_n{}^*, z_n{}^*) = T_n(y_0{}^*, z_0{}^*)$, then $S_{n+1}{}^*(y_n{}^* - y_n, z_n{}^* - z_n) = (y_{n+1}^* - y_{n+1}, z_{n+1}^* - z_{n+1})$. Show that y_0 in part (b) uniquely determines z_0; in fact, if $(y_n, z_n), (y_n{}^*, z_n{}^*) \in K$ for $n = 0, 1, \ldots$, then

$$\|z_n{}^* - z_n\| \leqq \|y_n{}^* - y_n\| \quad \text{for} \quad n = 0, 1, \ldots.$$

(e) Assume the conditions of (a), (b), (c), (d) and, in addition, that $Y_n(y, z)$, $Z_n(y, z)$ are of class C^1. Let $z_0 = g(y_0)$ be the unique z_0 in (b). Show that $g(y_0)$ is of class C^1 [and that the partial derivatives of g vanish at $y_0 = 0$ if the partials of Y_n, Z_n vanish at $(y, z) = (0, 0)$]. In fact, let $T_n{}^{**} = S_n{}^{**} \circ S_{n-1}{}^{**} \circ \cdots \circ S_1{}^{**}$, where $S_n{}^{**}$ is the linear map

$$(u, v) \to (Au + (\partial_y Y_n)u + (\partial_z Y_n)v, \quad Cv + (\partial_y Z_n)u + (\partial_z Z_n)v)$$

with the Jacobian matrices $\partial_{y,z} Y_n, Z_n$ evaluated at (y_{n-1}, z_{n-1}); let $e_j = (0, \ldots, 0, 1, 0, \ldots, 0)$ be the vector with the kth component $e_j{}^k = 0$ or $e_j{}^k = 1$ according as $k \neq j$ or $k = j$. Then $(u_n, v_n) = (\partial y_n(y_0)/\partial y_0{}^j, \partial z_n(y_0)/\partial y_0{}^j)$, $n = 0, 1, \ldots$, exists and is the unique sequence satisfying $(u_n, v_n) = T_n{}^{**}(u_0, v_0)$, $u_0 = e_j$, and $\|v_n\| \leqq \|u_n\|$ for $n = 0, 1, \ldots$. (f) Deduce Lemma 5.1 from part (e) with the choice $S_1 = S_2 = \cdots = T$.

6. Existence of Invariant Manifolds

A consequence of Corollary 5.2 is the following:

Theorem 6.1. *In the differential equation*

$$(6.1) \qquad \xi' = E\xi + F(\xi),$$

let $F(\xi)$ be of class C^1 and $F(0) = 0$, $\partial_\xi F(0) = 0$. Let the constant matrix E possess d (>0) eigenvalues having negative real parts, say, d_i eigenvalues with real parts equal to α_i, where $\alpha_1 < \cdots < \alpha_r < 0$ and $d_1 + \cdots + d_r = d$, whereas the other eigenvalues, if any, have non-negative real parts. If $0 < \epsilon < -\alpha_r$, then (6.1) has solutions $\xi = \xi(t) \neq 0$ satisfying

$$(6.2) \qquad \|\xi(t)\| \, e^{\epsilon t} \to 0 \qquad \text{as} \quad t \to \infty$$

and any such solution satisfies

$$(6.3) \qquad \lim t^{-1} \log \|\xi(t)\| = \alpha_i \qquad \text{for some} \quad i.$$

Furthermore, for sufficiently small $\epsilon > 0$, the point $\xi = 0$ and the set of points ξ on solutions $\xi(t)$ satisfying $\lim t^{-1} \log \|\xi(t)\| \leqq \alpha_i$ for a fixed i [or $\limsup t^{-1} \log \|\xi(t)\| < 0$] as $t \to \infty$ constitute a locally invariant C^1 manifold S_i [or S_r] of dimension $d_1 + \cdots + d_i$ [or $d_1 + \cdots + d_r = d$].

It will be clear that the proof has the following consequence.

Corollary 6.1. *Let $F(\xi)$ be of class C^1 for small $\|\xi\|$, $F(0) = 0$, and $\partial_\xi F(0) = 0$. Then (6.1) has a solution $\xi = \xi(t) \not\equiv 0$ satisfying (6.2) for some $\epsilon > 0$ if and only if E has at least one eigenvalue with negative real part.*

For another proof of Theorem 6.1 and a generalization to nonautonomous systems, see §§ X 8, 11.

Proof. If "lim" is replaced by "lim sup," the last part of Theorem 6.1 follows from the normalizations of § 4 and from Corollary 5.2 with $\alpha = \alpha_i$, $\beta = \alpha_{i+1}$ [or $\alpha = \alpha_r$, $\beta = 0$]. This argument also shows that, as $t \to \infty$,

$$\liminf t^{-1} \log \|\xi(t)\| < \alpha_{i+1} \quad \text{implies that} \quad \limsup t^{-1} \log \|\xi(t)\| \leq \alpha_i,$$

for $i = 1, \ldots, r$ with α_{r+1} interpreted as 0. Hence $\limsup t^{-1} \log \|\xi(t)\| = \alpha_i$ for some i implies that $\lim \inf = \lim \sup$.

Remark. We have similar results for the solutions $\xi(t) \not\equiv 0$ satisfying $\|\xi(t)\| \, e^{-\epsilon t} \to 0$ as $t \to -\infty$. This follows by replacing t by the new variable $-t$, so that (6.1) becomes

$$\xi' = -E\xi - F(\xi)$$

and applying Theorem 6.1 to this equation.

The arguments used to obtain Theorem 6.1 and Corollary 5.2 give

Theorem 6.2. *Let E, $F(\xi)$ be as in the last theorem. In addition, let E have $e\ (>0)$ eigenvalues with positive real parts. Let $\xi_t = \xi(t, \xi_0)$ be the solution of (6.1) satisfying $\xi(0, \xi_0) = \xi_0$ and T^t the corresponding map T^t: $\xi_t = \xi(t, \xi_0)$. Let $\epsilon > 0$. Then there exists a map R of a neighborhood of $\xi = 0$ in the ξ-space onto a neighborhood of the origin in the Euclidean (u, v, w)-space, where $\dim u = d$, $\dim v = e$, $\dim u + \dim v + \dim w = \dim \xi$, such that R is of class C^1 with a nonvanishing Jacobian and $RT^t R^{-1}$ has the form*

$$(6.4) \qquad RT^t R^{-1}: \begin{aligned} u_t &= e^{Pt}u_0 + U(t, u_0, v_0, w_0), \\ w_t &= e^{P_0 t}w_0 + W(t, u_0, v_0, w_0), \\ v_t &= e^{Qt}v_0 + V(t, u_0, v_0, w_0); \end{aligned}$$

U, V, W and their partial derivatives with respect to u_0, v_0, w_0 vanish at $(u_0, v_0, w_0) = 0$. Furthermore, $V = 0$, $W = 0$ if $v_0 = 0$, $w_0 = 0$; and $U = 0$, $W = 0$ if $u_0 = 0$, $w_0 = 0$; finally $\|e^P\| < 1$, $\|e^{-Q}\| < 1$ and the eigenvalues of e^{P_0} are of absolute value 1.

It can be remarked that the change of variables $R: \xi \to (u, v, w)$ transforms (6.1) into differential equations of the form

$$(6.5) \qquad u' = Pu + F_1(u, v, w), \qquad w' = P_0 w + F_2(u, v, w),$$
$$v' = Qv + F_3(u, v, w),$$

where F_1, F_2, F_3 are $o(\|u\| + \|v\| + \|w\|)$ as $(u, v, w) \to 0$ (but F_1, F_2, F_3 need not be of class C^1); $F_2 = 0$, $F_3 = 0$ if $v = 0$, $w = 0$; and $F_1 = 0$, $F_2 = 0$ if $u = 0$, $w = 0$.

The condition $V = 0$, $W = 0$ when $v_0 = 0$, $w_0 = 0$ [or $U = 0$, $W = 0$ when $u_0 = 0$, $w_0 = 0$] means that the d-dimensional plane $v_0 = 0$, $w_0 = 0$ [or e-dimensional plane $u_0 = 0$, $w_0 = 0$] are locally invariant manifolds. When E has no eigenvalues with real part 0, then the variables w, w_0 are absent; in this case, the manifold $u_0 = 0$ [or $v_0 = 0$] that consists of the solution arcs which tend to 0 as $t \to \infty$ [or $t \to -\infty$] is called the *stable* [or *unstable*] manifold of (6.1) through $\xi = 0$.

7. Linearizations

In the differential equation

(7.1) $$\xi' = E\xi + F(\xi),$$

suppose that no eigenvalue of E has a vanishing real part. The remarks concerning (6.5) suggest the question as to whether or not there is a C^1 change of variables $R : \xi \to \zeta$ with nonvanishing Jacobian in a neighborhood of $\xi = 0$ which transforms (7.1) into the linear system

(7.2) $$\zeta' = E\zeta$$

in a neighborhood of $\zeta = 0$. In general the answer is in the negative if $\dim \xi > 2$; see Exercises 7.1 and 8.1–8.2. A discussion of this problem is given in the Appendix of this chapter.

Exercise 7.1. Let ξ, η, ζ be real variables and consider the system of three differential equations:

$$\xi' = \alpha\xi, \qquad \eta' = (\alpha - \gamma)\eta + \epsilon\xi\zeta, \qquad \zeta' = -\gamma\zeta,$$

where $\alpha > \gamma > 0$ and $\epsilon \neq 0$. Show that there is no map $R : (\xi, \eta, \zeta) \to (u, v, w)$ of class C^1 with nonvanishing Jacobian from a neighborhood of $(\xi, \eta, \zeta) = 0$ onto a neighborhood of $(u, v, w) = 0$ transforming the given differential equations into the linear system

$$u' = \alpha u, \qquad v' = (\alpha - \gamma)v, \qquad w' = -\gamma w.$$

See Hartman [21].

For a topological, rather than a C^1, map R, we have:

Theorem 7.1. *Suppose that no eigenvalue of E has a vanishing real part and that $F(\xi)$ is of class C^1 for small $\|\xi\|$, $F(0) = 0, \partial_\xi F(0) = 0$. Let $T^t : \xi_t = \eta(t, \xi_0)$ and $L^t : \zeta_t = e^{Et}\zeta_0$ be the general solution of (7.1) and (7.2), respectively. Then there exists a continuous one-to-one map of a neighborhood of $\xi = 0$ onto a neighborhood of $\zeta = 0$ such that $RT^tR^{-1} = L^t$; in particular, $R : \xi \to \zeta$ maps solutions of (7.1) near $\xi = 0$ onto solutions of (7.2) preserving parametrizations.*

Thus the topological structure of the set of solutions of (7.1) in a neighborhood of $\xi = 0$ is identical with that of the solutions of (7.2) near $\zeta = 0$. This is no longer true if some eigenvalues of E have vanishing real parts. For in this case, (7.2) has closed solutions paths arbitrarily near $\zeta = 0$, but (7.1) need not have closed solution paths near $\xi = 0$; cf. Exercise VIII 3.1. Theorem 7.1 will be proved in § 9.

8. Linearization of a Map

Instead of the problem of linearizing a group of maps T^t, the corresponding question involving one map T will be considered first.

Lemma 8.1. *Let A, C be non-singular constant matrices, where A is a $d \times d$ matrix, C an $e \times e$ matrix, and*

$$(8.1) \qquad a = \|A\| < 1 \quad \text{and} \quad 1/c = \|C^{-1}\| < 1.$$

Let $T:(y_0, z_0) \to (y_1, z_1)$ be a map of the form

$$(8.2) \qquad T: y_1 = Ay_0 + Y(y_0, z_0), \qquad z_1 = Cz_0 + Z(y_0, z_0),$$

where Y, Z are functions of class C^1 for small $\|y_0\|$, $\|z_0\|$ which vanish together with their Jacobian matrices at $(y_0, z_0) = 0$. Then there exists a continuous, one-to-one map

$$(8.3) \qquad R: u = \Phi(y, z), \qquad v = \Psi(y, z)$$

of a neighborhood of $(y, z) = 0$ onto a neighborhood of $(u, v) = 0$ such that R transforms T into the linear map

$$(8.4) \qquad RTR^{-1} = L: u_1 = Au_0, \qquad v_1 = Cv_0.$$

Remark. In view of Lemma 2.1, it can be supposed that Y, Z are of class C^1 for all (y_0, z_0) and satisfy (4.12)–(4.13), where θ_0, s_0 are arbitrarily small positive numbers. It will be shown in this case that if $\theta_0 > 0$ is sufficiently small, then R can be chosen so that it is a continuous, one-to-one map of the (y, z)-space onto the (u, v)-space, and that (8.4) holds for all (u_0, v_0).

The following exercises give positive and negative results concerning the existence of linearizing maps R which are smoother than (8.3) in Lemma 8.1; see also the Appendix.

Exercise 8.1. (a) Using the example $T: x^1 = ax$, $y^1 = ac(y + \epsilon xz)$, $z^1 = cz$, where $0 < c < 1 < a$, $ac > 1$, and $\epsilon > 0$, show that if R is any linearizing map, then R and R^{-1} are not of class C^1. See Hartman [21]. (b) Let dim $y = $ dim $u \leq 2$. Let the map $T: y_1 = Ay + Y(y)$ be of class C^2 for small $\|y\|$; Y and its first order partials vanish at $y = 0$; A is a constant matrix having no eigenvalue of absolute value 0 or 1. Show that there exists a map $R:y \to u$ of class C^1 of a neighborhood of $y = 0$ onto a neighborhood of $u = 0$ with nonvanishing Jacobian such that RTR^{-1} is the linear map $RTR^{-1}:u_1 = Au$. See Hartman [20, Part 3]. (c) Using the example $T: x_1 = \epsilon^2 x + y^2$, $y_1 = \epsilon y$, where x, y are real variables and $\epsilon > 0$ is fixed, show that R in part (b) cannot always be chosen of class C^2 even if T is analytic and $\|A\| < 1$. See Sternberg [3, p. 812].

Exercise 8.2. *Contractions.* (a) Let dim $y = $ dim $u = d$ be arbitrary. Consider a map $T:y_1 = Ay + Y(y)$, where A is a nonsingular $d \times d$ matrix such that $\|A\| < 1$ and $Y(y)$ is of class C^2 for small $\|y\|$ with $Y(0) = 0$, $\partial_y Y(0) = 0$. Show that there exists a map $R:y \to u$ of a neighborhood of $y = 0$ onto a neighborhood of $u = 0$ such that $R(0) = 0$, R is of class C^1 and has a nonvanishing Jacobian, and RTR^{-1} is the linear map $RTR^{-1}: u_1 = Au$. See Hartman [20]. [Note, that by Exercise 8.1 (c), there may not exist an R of class C^2 even if $Y(y)$ is analytic in y.] (b) In part (a), let $\alpha_1, \ldots, \alpha_d$ be the eigenvalues of A and suppose that (*) $\alpha_j \neq \alpha_1^{m_1} \alpha_2^{m_2} \ldots \alpha_d^{m_d}$ for any set of non-negative integers $(m_1, \ldots, m_d)$ satisfying $1 < \Sigma m_k \leq n$, where n is an integer such that $n > \log |\alpha_j|/\log |\alpha_k|$ for $1 \leq j, k \leq d$. Let $Y(y)$ be of class C^n. Then R in part (a) can be chosen of class C^n. Also, if (*) holds for all sets of non-negative integers $(m_1, \ldots, m_d)$ satisfying $\Sigma m_k > 1$ and $Y(y)$ is of class C^∞ [or analytic], then R can be chosen of class C^∞ [or analytic]. See Sternberg [3] and Appendix to this chapter.

Proof of Lemma 8.1.   In order to prove Lemma 8.1, we prove two simple lemmas

Lemma 8.2.   *Let B be a nonsingular $m \times m$ matrix and let $b_1 = \|B^{-1}\|$. Let $S:x_0 \to x_1$ be a map of the form*

$$(8.5) \qquad\qquad S:x_1 = Bx_0 + X(x_0),$$

where $X(x_0)$ is defined for all x_0 and satisfies a Lipschitz condition

$$(8.6) \qquad \|X(x_0) - X(x_0 + \Delta x_0)\| \leq \theta_1 \|\Delta x_0\| ,$$

where $\theta_1 b_1 < 1$. Then S is one-to-one and onto the x_1-space. If, in addition, $\|X(x_0)\| \leq c^0$ for all x_0 and

$$S^{-1}:x_0 = B^{-1}x_1 + X^1(x_1),$$

then $\|X^1(x_1)\| \leq b_1 c^0$ for all x_1.

Proof.   Note that
$$(8.7) \qquad \|B^{-1}[X(x_0) - X(x_0 + \Delta x_0)]\| \leq b_1 \theta_1 \|\Delta x_0\| .$$
In order to see that S is one-to-one, it suffices to show that

(8.8) $$B^{-1}S{:}x_1 = x_0 + B^{-1}X(x_0)$$

is one-to-one. But this is clear from $1 - b_1\,\theta_1 > 0$ and

$$\| [x_0 + B^{-1}X(x_0)] - [x_0 + \Delta x_0 + B^{-1}X(x_0 + \Delta x_0)] \|$$

$$\geq (1 - b_1\theta_1)\|\Delta x_0\| \,.$$

In order to show that S is onto, it suffices to show that $B^{-1}S$ is onto; i.e., that if x_1 is given, then there exists an x_0 satisfying the equation in (8.8). Note that (8.8) can be written as

(8.9) $$x_0 = x_1 - B^{-1}X(x_0) \,.$$

We show the existence of a solution x_0 by using successive approximations defined by $x^0 = 0$ and

(8.10) $$x^n = x_1 - B^{-1}X(x^{n-1}) \quad \text{for } n \geq 1.$$

By (8.7), for $n \geq 2$,

$$\| x^n - x^{n-1} \| = \| B^{-1}[X(x^{n-1}) - X(x^{n-2})] \| \leq b_1\theta_1 \| x^{n-1} - x^{n-2} \|.$$

A simple induction gives $\| x^n - x^{n-1} \| \leq (b_1\,\theta_1)^{n-1} \| x^1 - x^0 \|$ for $n \geq 1$. Since $0 < b_1\theta_1 < 1$, the series $\Sigma\ (x^n - x^{n-1})$ is convergent; i.e., the sequence $x^0, x^1, \dots$ has a limit, say x_0, as $n \to \infty$. The last part of (8.10) shows that the limit x_0 satisfies (8.9), hence the equation in (8.8).

Finally, the last part of the lemma follows from $X^1(x_1) = B^{-1}(Bx_0 - x_1) = - B^{-1}X(x_0)$. This completes the proof.

In order to formulate the next lemma, we introduce some notation and terminology. Since A in Lemma 8.1 is non-singular, it has an inverse. Let

(8.11) $$a_1 = \|A^{-1}\|$$

Let b_1, θ_1, θ be constants satisfying

(8.12) $$b_1 = \max(a_1, 1/c) > 0, \qquad 0 < b_1\,\theta_1 < 1,$$

$$\theta = \theta_1(1+c) + \max(a,c) < 1.$$

If $c_0 > 0$ is a constant, let $\Omega\ (\theta_1, c_0)$ denote the set of pairs of functions $(\ Y(y_0, z_0),\ Z(y_0, z_0))$ defined and satisfying, for all (y_0, z_0),

(8.13) $$Y(0,0) = 0,\ Z(0,0) = 0,$$

(8.14) $$\|Y(y_0, z_0)\| + \|Z(y_0, z_0)\| \leq c_0,$$

(8.15) $\| \Delta Y \|$, $\| \Delta Z \| \leq \theta_1 (\| \Delta y_0 \| + \| \Delta z_0 \|)/2$,

where $\Delta Y = Y(y_0 + \Delta y_0, z_0 + \Delta z_0) - Y(y_0, z_0)$, etc.

It is clear that Lemma 8.1 is contained in the case $Y_1 = 0$, $Z_1 = 0$ (so that $U = L$) of the following:

Lemma 8.3.   *Let A,C be as in Lemma 8.1, (Y,Z) and (Y_1, Z_1) a pair of elements of $\Omega(\theta_1, c_0)$ and*

(8.16) $T : y_1 = Ay_0 + Y(y_0, z_0)$, $z_1 = Cz_0 + Z(y_0, z_0)$,

(8.17) $U : y_1 = Ay_0 + Y_1(y_0, z_0)$, $z_1 = Cz_0 + Z_1(y_0, z_0)$.

Then there exists a unique continuous map

(8.18) $R_0 : u = y + \Lambda (y,z)$, $v = z + \Theta(y,z)$

defined for all (y,z) such that $\Lambda (0,0) = 0$, $\Theta(0,0) = 0$, Λ and Θ are bounded,

(8.19) $R_0 T = UR_0$.

Furthermore, R_0 is one-to-one and onto the (u,v)-space.

Proof *(a)* It follows from Lemma 8.2 that T has an inverse defined for all (y_1, z_1), say,

(8.20) $T^{-1} : y_0 = A^{-1}y_1 + Y^1(y_1, z_1)$, $z_0 = C^{-1}z_1 + Z^1(y_1, z_1)$,

and that

(8.21) $\| Y^1 (y_1, z_1) \|$, $\| Z^1(y_1, z_1) \| \leq b_1 c_0$,

(b) The equation (8.19) is equivalent to the pair of equations
$$Ay + Y + \Lambda (Ay + Y, Cz + Z) = A(y + \Lambda) + Y_1(y + \Lambda , z + \Theta),$$
$$Cz + Z + \Theta (Ay + Y, Cz + Z) = C(z + \Theta) + Z_1(y + \Lambda , z + \Theta).$$
where the argument of Y, Z, Λ, Θ is (y,z). Using (8.20), the first of these equations can be written
$$y + \Lambda = A[A^{-1}y + Y^1 + \Lambda (T^{-1})]$$
$$+ Y_1(A^{-1}y + Y^1 + \Lambda (T^{-1}), C^{-1}z + Z^1 + \Theta(T^{-1})),$$
where $\Lambda = \Lambda(y,z)$ and $\Lambda(T^{-1}) = \Lambda \circ T^{-1}$ with argument (y,z) Consequently, (8.19) is equivalent to the last two equations which can be written
(8.22) $\Theta = C^{-1} [Z - Z_1(y + \Lambda, z + \Theta) + \Theta(Ay + Y, Cz + Z)]$,
(8.23) $\Lambda = A[Y^1 + \Lambda (T^{-1})]$
$$+ Y_1(A^{-1}y + Y^1 + \Lambda (T^{-1}), C^{-1}z + Z^1 + \Theta(T^{-1})).$$

(c) *Existence of R_0.* We prove the existence[*] of R_0 by showing the existence of a solution (Λ, Θ) of the functional equations (8.22),(8.23), using successive approximations. These are defined by $\Lambda^0 = 0$, $\Theta^0 = 0$ and

$$(8.24) \quad \Theta^n = C^{-1}[Z - Z_1(y + \Lambda^{n-1}, z + \Theta^{n-1}) + \Theta^{n-1}(Ay + Y, Cz + Z)],$$

$$(8.25) \quad \Lambda^n = A[Y^1 + \Lambda^{n-1}(T^{-1})] + Y_1(A^{-1}y + Y^1 + \Lambda^{n-1}(T^{-1}), C^{-1}z + Z^1 + \Theta^{n-1}(T^{-1})],$$

for $n = 1,2,\dots$. Thus Λ^n, Θ^n are defined and continuous for all (y,z). They are also bounded, for it is clear that Λ^0, Θ^0 and Λ^1, Θ^1 are bounded and, if we put

$$r_n = |||\Lambda^n - \Lambda^{n-1}||| + |||\Theta^n - \Theta^{n-1}||| \quad \text{for } n = 1,2,\dots,$$

where $|||...||| = \sup ||...||$, then (8.24),(8.25) give

$$||| \Theta^n - \Theta^{n-1} ||| \leq c[\theta_1 r^{n-1} + |||\Theta^{n-1} - \Theta^{n-2}|||],$$

$$||| \Lambda^n - \Lambda^{n-1} ||| \leq [a|||\Lambda^{n-1} - \Lambda^{n-2}||| + \theta_1 r_{n-1}].$$

On adding, we get

$$r_n \leq [\theta_1(c + 1) + \max(a,c)] r_{n-1} = \theta r_{n-1},$$

by the last part of (8.12). Hence, $r_n < r_1 \theta^{n-1}$ for $n = 1,2,\dots$,

so that the series $\Sigma(\Lambda^n - \Lambda^{n-1})$, $\Sigma(\Theta^n - \Theta^{n-1})$ have the convergent majorant $r_1 \Sigma \theta^{n-1}$, independent of (y,z). Consequently the limits $\Lambda = \lim \Lambda^n$, $\Theta = \lim \Theta^n$ exist uniformly on the *(y,z)*-space, as $n \to \infty$. These limits are continuous and bounded. The definitions (8.24),(8.25) imply (8.22),(8.23) which are equivalent to (8.19).

(d) *Uniqueness.* This follows in the usual way.

(e) *R_0 is one-to-one and onto.* Denote the unique R_0 satisfying (8.19) by R_{TU}, so that $R_{TU}T = UR_{TU}$. If we interchange the roles of T and U, it follows that there exists a unique R_{UT} satisfying the prescribed conditions and $R_{UT}U = TR_{UT}$. Hence

$$R_{TU}R_{UT}U = T_{TU}TR_{UT} = UR_{TU}R_{UT},$$
$$TR_{UT}R_{TU} = R_{UT}UR_{TU} = R_{UT}R_{TU}T.$$

---

[*] The existence of R_0 also follows from the Banach space analogue of Lemma 8.2.

By virtue of uniqueness , $R_{TU}R_{UT} = R_{UU} = I$ and $R_{UT}R_{TU} = R_{TT} = I$, where I is the identity map. Hence both $R_0 = R_{TU}$ and R_{UT} are one-to-one and onto. This completes the proof.

9. Proof of Theorem 7.1

In the proof of the theorem, it will be supposed that E has d eigenvalues with negative real parts and e eigenvalues with positive real parts. (The case when $d = 0$ or $e = 0$ is easier and is contained in the general case by the addition of dummy components to ξ.) After preliminary normalizations, it can be supposed that T^t is of the form (4.10) for all (t, y_0, z_0), where (4.11) holds, (4.12) and (4.13) hold for $0 \leq t \leq 1$, and $A = e^P$, $C = e^Q$ satisfy the conditions in Lemma 8.1.

It will also be supposed that θ_0 is so small that Lemma 8.1 and the Remark following it are applicable to $T = T^1$. Denote by R_0 the corresponding map supplied by Lemma 8.1, so that $R_0 T^1 R_0^{-1} = L$. Put

$$(9.1) \qquad R = \int_0^1 L^{-s} R_0 T^s \, ds.$$

Then

$$(9.2) \qquad L^t R = \left(\int_0^1 L^{t-s} R_0 T^{s-t} \, ds \right) T^t$$

If $s - t$ is introduced as a new integration variable, say s, the last integral becomes

$$\left\{ \int_{-t}^0 + \int_0^{1-t} \right\} L^{-s} R_0 T^s \, ds.$$

In the first of these integrals, the integrand can be written as

$$L^{-s} R_0 T^s = L^{-1-s} R_0 T^{s+1}$$

since $L = R_0 T^1 R_0^{-1}$. Thus (9.2) becomes

$$(9.3) \qquad L^t R = \left(\int_0^1 L^{-s} R_0 T^s \, ds \right) T^t = R T^t.$$

Thus, in order to complete the proof, it is sufficient to verify that $R = R_0$. To this end, let $U = L$ in (8.18) and R_0 is given by (8.19) in Lemma 8.3. The definition (9.1) of R implies that R has the form prescribed in Lemma 8.3. Hence $R = R_0$ follows from $t = 1$ in (9.3) and the uniqueness of R_0.

10. Periodic Solution

Lemmas 5.1, 8.1 will now be applied to the study of solutions in the neighborhood of a periodic solution of an autonomous system:

$$(10.1) \qquad \xi' = f(\xi).$$

Lemma 10.1. *Let $f(\xi)$ be of class C^1 on an open set containing $\xi = 0$. Let $\xi = \eta(t, \xi_0)$ be the solution of* (10.1) *satisfying $\eta(0, \xi_0) = \xi_0$. Suppose that $\gamma(t) = \eta(t, 0)$ is periodic of least period $p > 0$. [Thus $\gamma(t) \not\equiv$ const., $f(\gamma(t)) \neq 0$ and $\eta(t, \xi_0)$ exists on an open t-interval containing $[0, p]$ if $\|\xi_0\|$ is sufficiently small.] Let π be the hyperplane π: $\xi \cdot f(0) = 0$ orthogonal to the curve $\mathscr{C}$: $\xi = \gamma(t)$ at $\xi = 0$. Then there exists a unique (real-valued) function $t = \tau(\xi_0)$ of class C^1 for small $\|\xi_0\|$ such that $\tau(0) = p$ and $\eta(t, \xi_0) \in \pi$ when $t = \tau(\xi_0)$; i.e.,*

$$(10.2) \qquad \eta(\tau(\xi_0), \xi_0) \cdot f(0) = 0.$$

Roughly speaking, if a solution starts at ξ_0 near 0, then at a time $t = \tau(\xi_0)$ near p, the solution meets the hyperplane π; see Figure 2.

Proof of Lemma 10.1. This is an immediate consequence of the implicit function theorem. The equation $\eta(t, \xi_0) \cdot f(0) = 0$ is satisfied if $t = p$, $\xi_0 = 0$. Also the derivative $\eta'(t, \xi_0) \cdot f(0) = f(\eta(t, \xi_0)) \cdot f(0)$ at $\xi_0 = 0$ is $f(\gamma(t)) \cdot f(0)$. At $t = p$, it becomes $|f(0)|^2 \neq 0$ since $\gamma(p) = \gamma(0) = 0$. This gives the lemma.

If we consider small $\|\xi_0\|$, $\xi_0 \in \pi$, then

$$(10.3) \qquad T: \xi_1 = \eta(\tau(\xi_0), \xi_0)$$

is a map from one neighborhood of $\xi = 0$ on π into another. The meaning of T is clear; the solution $\xi = \eta(t, \xi_0)$ starts for $t = 0$ at $\xi_0 \in \pi$ and $\xi_1 = T(\xi_0)$ is the first point $(t > 0)$ where the solution $\xi = \eta(t, \xi_0)$ again meets π. The applicability and consequences of Lemmas 5.1, 8.1 will be considered. Roughly, we can expect the following type of results: On the one hand, if the Jacobian matrix of this map at $\xi_0 = 0$ has a norm less than 1, then $\xi = \gamma(t)$ is orbitally asymptotically stable (in the sense that if ξ^0 is sufficiently near a point of the curve $\mathscr{C}$: $\xi = \gamma(t)$, then the solution

$\xi = \eta(t, \xi^0)$ "tends" to $\mathscr{C}$ as $t \to \infty$, i.e., is in an arbitrarily small neighborhood of $\mathscr{C}$ for large t). On the other hand, if dim $\xi = m$ and the Jacobian matrix at $\xi_0 = 0$ of this map has only $d\, (< m - 1)$ eigenvalues with absolute value less than 1 and $(m - 1) - d$ with absolute value greater than 1, then the set of $\xi_0 \in \pi$ for which $\xi = \eta(t, \xi_0)$ "tends" to $\mathscr{C}$, as $t \to \infty$, constitutes a d-dimensional manifold S.

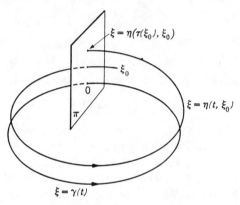

Figure 2.

In order to calculate the eigenvalues of the Jacobian matrix of the map $T : \xi_0 \to \xi_1$ at $\xi_0 = 0$, consider ξ_0 arbitrary for a moment (i.e., not subject to $\xi_0 \in \pi$). The matrix $H(t, \xi_0) = \partial_{\xi_0}\eta(t, \xi_0)$ satisfies

$$(10.4) \qquad H'(t, \xi_0) = \partial_\xi f(\eta(t, \xi_0))H(t, \xi_0), \qquad H(0, \xi_0) = I.$$

In particular, for $\xi_0 = 0$,

$$(10.5) \qquad H'(t, 0) = \partial_\xi f(\gamma(t))H(t, 0), \qquad H(0, 0) = I.$$

Note that $H(t, 0)$ is a fundamental matrix of (10.5) and that the coefficient matrix, $\partial_\xi f(\gamma(t))$, is periodic of period p. It follows from the Floquet theory in § IV 6 that $H(t, 0)$ has a representation of the form

$$(10.6) \qquad H(t, 0) = K(t)e^{Dt}, \qquad \text{where} \quad K(t + p) = K(t)$$

is a periodic matrix function and D is a constant matrix. In particular, $\tau(0) = p$ and $K(p) = K(0) = I$ imply that

$$(10.7) \qquad H(\tau(0), 0) = H(p, 0) = e^{Dp}.$$

The characteristic roots (= eigenvalues) $e_1, e_2, \ldots, e_m$ of the matrix $H(p, 0) = e^{Dp}$ are called the *characteristic roots* of the periodic solution $\xi = \gamma(t)$. Note that $e_k \neq 0$ since $H(p, 0)$ is nonsingular. Correspondingly, $p^{-1} \log e_1,\ p^{-1} \log e_2, \ldots$ are called the *characteristic exponents*, so that only the real parts of the characteristic exponents of $\xi = \gamma(t)$ are uniquely defined. The characteristic exponents, modulo $2\pi i$, are eigenvalues of D (and D is not unique).

When ξ is subjected to a linear change of coordinates, $\xi = N\zeta$ it is readily verified that $H(t, 0)$ is replaced by $NH(t, 0)N^{-1}$, so that the set of characteristic roots are not changed.

Lemma 10.2. *Let $f(\xi)$ be as in Lemma 10.1, dim $\xi = m$, and T the map $\xi_0 \to \xi_1$ in (10.3), where $\xi_0, \xi_1 \in \pi$. Let $e_1, \ldots, e_m$ be the characteristic roots of $\xi = \gamma(t)$. Then one of these, say, e_m is 1 and $e_1, \ldots, e_{m-1}$ are the eigenvalues of the Jacobian matrix of the map T at $\xi_0 = 0 \in \pi$. In fact, if the coordinates in the ξ-space are chosen so that*

$$(10.8) \qquad\qquad f(0) = (0, \ldots, 0, 1)$$

and $\pi: \xi^m = 0;$ then the last column of $H(p, 0)$ is $(0, \ldots, 0, 1)$ and the $(m-1) \times (m-1)$ matrix obtained by deleting the last row and column from $H(p, 0) = e^{Dp}$ is the Jacobian matrix of the map T at $\xi_0 = 0$.

Proof. It will first be verified that $H(p, 0)$ satisfies

$$(10.9) \qquad\qquad H(p, 0)f(0) = f(0);$$

i.e., $\lambda = 1$ is an eigenvalue of $H(p, 0)$ and $f(0)$ is a corresponding eigenvector. Note that $\gamma'(t) = f(\gamma(t))$. If this relation is differentiated with respect to t, it is seen that $\xi = \gamma'(t)$ is a solution of linear initial value problem

$$(10.10) \qquad \xi' = \partial_\xi f(\gamma(t))\xi, \qquad \xi(0) = \gamma'(0) = f(0).$$

Since $H(t, 0)$ is a fundamental matrix for this linear system reducing to I at $t = 0$, $\gamma'(t) = H(t, 0)f(0)$; cf. § IV 1. For $t = p$, this relation becomes (10.9).

Thus, if (10.8) holds, then

$$(10.11) \qquad \text{the last column of} \quad H(p, 0) \quad \text{is} \quad (0, \ldots, 0, 1).$$

The Jacobian matrix of the map (10.3), without the restriction $\xi_0 \in \pi$, is

$$\partial_{\xi_0}[\eta(\tau(\xi_0), \xi_0)] = \eta'(\tau, \xi_0)\,\partial_{\xi_0}\tau + H(\tau, \xi_0).$$

At $\xi_0 = 0$, this becomes

$$\partial_{\xi_0}[\eta(\tau(\xi_0), \xi_0)]_{\xi_0=0} = f(0)\,\partial_{\xi_0}\tau(0) + H(p, 0).$$

The first term on the right is the matrix $(f^i(0)\,\partial\tau(0)/\partial\xi_0{}^j)$, so that the first $m-1$ rows are 0 by virtue of (10.8). The last term is $H(p, 0)$. Consequently, the lemma follows.

11. Limit Cycles

The remarks of the last section make it clear that, in a study of the solutions of (10.1) near $\xi = \gamma(t)$, the real parts of the (nontrivial) characteristic exponents of $\xi = \gamma(t)$ play a role similar to that of the real parts of the eigenvalues of E in a study of solutions of (1.2) near $\xi = 0$.

Theorem 11.1 *Let $f(\xi)$ be of class C^1 on an open set and let* (10.1) *possess a periodic solution $\xi = \gamma(t)$ of (least) period $p > 0$. Let $\dim \xi = m$ and let the real parts of $m - 1$ characteristic exponents of $\xi = \gamma(t)$ be negative, say, less than $\alpha/p < 0$. Then there exists a $\delta > 0$ and a constant L with the property that for each ξ^0 on the open set $\operatorname{dist}(\xi^0, \mathscr{C}) < \delta$, where $\mathscr{C}: \xi = \gamma(t)$, $0 \le t \le p$, there is an asymptotic phase t_0 such that the solution $\xi = \eta(t, \xi^0)$ of*

$$(11.1) \qquad\qquad \xi' = f(\xi), \qquad \xi(0) = \xi^0$$

satisfies

$$(11.2) \qquad \|\eta(t + t_0, \xi^0) - \gamma(t)\| \le L e^{\alpha t/p} \qquad \text{for } t \ge 0.$$

In particular, $\xi = \gamma(t)$ is a limit cycle and is asymptotically, orbitally stable.

Proof. As before, let $\gamma(0) = 0$. The assumptions make it clear that after a linear change of the variables ξ, it can be supposed that the map (10.3) is of the form

$$T: \xi_1 = A\xi_0 + \Xi(\xi_0), \qquad \xi_0 \in \pi,$$

where $\Xi(\xi_0)$ is of class C^1 for small $\|\xi_0\|$ and vanishes together with its Jacobian matrix at $\xi_0 = 0$ and that $a \equiv \|A\| < e^\alpha$. Thus, if $\|\xi_0\|$ is sufficiently small, say, $\|\xi_0\| < \epsilon$, then $\|\xi_1\| \le e^\alpha \|\xi_0\|$. Hence $\xi_n = T^n \xi_0$ is defined for $n = 1, 2, \ldots$ and $\|\xi_n\| \le e^{n\alpha} \|\xi_0\| \to 0$ as $n \to \infty$.

Since the solutions of (11.1) are unique, it follows from Theorem V 2.1 that if $\epsilon > 0$ is arbitrary, then there exists a $\delta = \delta_\epsilon > 0$ with the property that if $\operatorname{dist}(\mathscr{C}, \xi^0) < \delta$, then there exists a least positive value $t = \tau^0(\xi^0)$ such that $\eta(t, \xi^0)$ exists for $0 \le t \le \tau^0$ and $\eta(\tau^0, \xi^0) \in \pi$, $\|\eta(\tau^0, \xi^0)\| < \epsilon$.

Let $\xi_0 = \eta(\tau^0, \xi^0) \in \pi$. Put $\tau^1 = \tau(\xi_0) + \tau^0$ and, for $n = 2, 3, \ldots,$ $\tau^n = \tau(\xi_{n-1}) + \tau^{n-1}$, so that $\xi_n = \eta(\tau^n, \xi^0)$. Clearly, $\tau(0) = p$ implies that $\tau(\xi_n) \to p$ and $\tau^n/np \to 1$ as $n \to \infty$. Actually, $t_0 = \lim (\tau^n - np)$ exists as $n \to \infty$ and there is a constant L_1 such that

$$(11.3) \qquad\qquad |\tau^n - (np + t_0)| \le L_1 e^{\alpha n} \|\xi_0\| \, .$$

In order to see this, note that

$$|(\tau^n - np) - (\tau^{n-1} - (n-1)p)| = |\tau(\xi_{n-1}) - p|.$$

Since $\tau(0) = p$ and $\tau(\xi)$ is of class C^1, $|\tau(\xi_{n-1}) - p|$ is majorized by $L_0 \|\xi_{n-1}\| \le L_0 e^{\alpha(n-1)} \|\xi_0\|$ for a suitable constant L_0. Thus the existence of a t_0 satisfying (11.3) with $L_1 = L_0/(1 - e^\alpha)$ follows.

Since $\eta(t, \xi^0)$ is of class C^1, $\|\eta(t + \tau^n, \xi^0) - \gamma(t)\| = \|\eta(t, \xi_n) - \eta(t, 0)\| \le L_2 \|\xi_n\| \le L_2 e^{\alpha n} \|\xi_0\|$ for some constant L_2 and $0 \le t \le p$.

Also, the boundedness of $f(\xi)$ in a vicinity of $\mathscr{C}$ implies that $\|\eta(t + \tau^n, \xi^0) - \eta(t + np + t_0, \xi^0)\| \leq L_3 e^{\alpha n} \|\xi_0\|$ by virtue of (11.3). Hence

$$\|\eta(t + np + t_0, \xi^0) - \gamma(t)\| \leq (L_2 + L_3)e^{\alpha n} \|\xi_0\| \qquad \text{for} \quad 0 \leq t \leq p.$$

If $t + np$ is replaced by t, it follows that

$$\|\eta(t + t_0, \xi^0) - \gamma(t)\| \leq (L_2 + L_3)e^{\alpha t/p}\|\xi_0\|$$

for $np \leq t \leq (n + 1)p$ and $t = 0, 1, \ldots$. This proves the theorem.

Theorem 11.2. *Let* dim $\xi = m, f(\xi)$ *be of class C^1 on an open set, and let* (10.1) *possess a periodic solution $\xi = \gamma(t)$ of least period $p > 0$. (i) Then there exists a solution $\xi = \xi(t) \not\equiv \gamma(t)$ defined for large t and satisfying*

$$(11.4) \qquad \text{dist } (\mathscr{C}, \xi(t))e^{\epsilon t} \to 0 \qquad \text{as} \quad t \to \infty,$$

for some $\epsilon > 0$, where $\mathscr{C} : \xi = \gamma(t)$, $0 \leq t \leq p$, if and only if $\xi = \gamma(t)$ has at least one characteristic exponent with negative real part. (ii) If $\xi = \gamma(t)$ has exactly d ($\leq m - 1$) characteristic exponents with negative real parts, then the set of points ξ near $\mathscr{C}$ on solutions $\xi = \xi(t)$ of (10.1) satisfying (11.4) for some $\epsilon > 0$ constitutes a $(d + 1)$-dimensional C^1 manifold S. (iii) If at least one characteristic exponent has positive real part, then $\xi = \gamma(t)$ is not orbitally stable. (iv) If $0 < d < m - 1$ in (ii) and $(m - 1) - d$ characteristic exponents have positive real parts, then S can also be described as the set of points ξ near $\mathscr{C}$ on solutions $\xi = \xi(t)$ satisfying

$$(11.5) \qquad \text{dist } (\mathscr{C}, \xi(t)) \to 0, \qquad \text{as} \quad t \to \infty$$

and/or

$$(11.6) \qquad \text{dist } (\mathscr{C}, \xi(t)) \leq \epsilon \qquad \text{for} \quad 0 \leq t < \infty$$

for some sufficiently small $\epsilon > 0$.

It will be clear from the proof and from Exercise 5.1 that the C^1 assumption on f and assertion on S can be weakened somewhat or strengthened either to analyticity or to C^m, $2 \leq m \leq \infty$. In case (iv), there is, of course, an analogous $(m - d)$-dimensional manifold corresponding to $t \to -\infty$ which intersects S transversally along $\mathscr{C}$; S and the corresponding $(m - d)$-dimensional manifold are called the *stable* and *unstable manifolds of $\mathscr{C}$.*

Proof. Only the proof of (ii), when $0 < d < m - 1$, will be indicated. It can be supposed that $\gamma(0) = 0$ and that (10.8) holds. After a linear change of variables in π [i.e., in the $(\xi^1, \ldots, \xi^{m-1})$-subspace], it can be supposed that T in (10.3) is of the form (5.1), where Y, Z are of class C^1 for small $\|y_0\|, \|z_0\|$, and Y, Z and their Jacobian matrices vanish at $(y_0, z_0) = 0$. Here dim $y_0 = d$, $(y_0, z_0) = (\xi^1, \ldots, \xi^{m-1})$, $a = \|A\| < 1$ and $1/c = \|C^{-1}\| \leq 1 + \theta$, where $\theta > 0$ is arbitrarily small. Let $y = g(z)$

be the manifold supplied by Lemma 5.1. Then $T^n(y_0, z_0)$ is defined for $n = 1, 2, \ldots$ and $\| T^n(y_0, z_0) \|$ tends to 0 exponentially as $n \to \infty$ if and only if $z_0 = g(y_0)$.

If $\| \xi_0 \|$ is small and $\xi_0 \in \pi$, there is a (y_0, z_0) such that $\xi_0 = (y_0, z_0, 0)$, where the last 0 is the real number 0. Consider the set of ξ-points given by Σ: $\xi = \eta(t, \xi_0)$, where $\xi_0 = (y_0, g(y_0), 0) \in \pi$ and $0 \leq t \leq \tau(\xi_0)$.

It will be left to the reader to verify that the subset S of Σ in a small open vicinity of $\mathscr{C}$ satisfies the assertion of the theorem.

Remark. From Lemma 8.1, we can deduce the topological nature of the set of solutions of (10.1) near $\mathscr{C}$ when the real parts of the nontrivial $m - 1$ characteristic exponents of $\xi = \gamma(t)$ are not 0.

Exercise 11.1.   Under the conditions of Theorem 11.2(i), let $\xi(t)$ be a solution of (10.1) satisfying (11.4) for some $\epsilon > 0$. Show the existence of numbers t_0 and $c > 0$ such that $\| \xi(t + t_0) - \gamma(t) \| e^{ct} \to 0$ as $t \to \infty$.

Theorem 11.3.   *Let $f(\xi)$ be of class C^1 on an open set and such that* (10.1) *possesses a periodic solution $\xi = \gamma(t)$ of (least) period $p > 0$. Put*

$$(11.7) \qquad \Delta = \int_0^p \operatorname{tr} \partial_\xi f(\gamma(s)) \, ds.$$

If $\Delta > 0$, then $\xi = \gamma(t)$ is not orbitally stable. If $\dim \xi = 2$ and $\Delta < 0$, then $\xi = \gamma(t)$ is exponentially, asymptotically, orbitally stable.

Proof.   For arbitrary $m = \dim \xi$, (10.5) implies that

$$\det H(t, 0) = \exp \int_0^t \operatorname{tr} \partial_\xi f(\gamma(s)) \, ds;$$

cf. Theorem IV 1.2. Thus

$$(11.8) \qquad \det H(p, 0) = e^\Delta,$$

so that e^Δ is the product $e_1 e_2 \ldots e_m$. When $\Delta > 0$, then $|e_k| > 1$ for some k and when $m = 2$, it can be supposed that $e_2 = 1$ and $e_1 = e^\Delta$. Thus Theorem 11.3 follows from Theorems 11.1, 11.2.

Exercise 11.2.   Let $\dim \xi = 2$, $f(\xi)$ be of class C^1 on a simply connected open set E, and $\operatorname{tr} \partial_\xi f(\xi) \equiv \operatorname{div} f(\xi) \neq 0$. Then (10.1) does not possess a periodic solution ($\neq$ const.) nor a solution $\xi = \xi(t)$, $-\infty < t < \infty$, such that $\xi(\pm \infty) = \lim \xi(t)$ as $t \to \pm \infty$ exist, are equal, and $\xi(\pm \infty) \in E$.

APPENDIX: SMOOTH EQUIVALENCE MAPS

12. Smooth Linearizations

As pointed out earlier, Theorem 7.1 and Lemma 8.1 become false if "the continuity of R and R^{-1}" in the assertions is replaced by the assertion that 'R, R^{-1} are of class C^1.'' This and the following two sections concern the existence of smooth linearizing maps R, R^{-1} under additional hypotheses.

Theorem 12.1. *Let $n > 0$ be an integer [or $n = \infty$]. Then there exists an integer $N = N(n) \geqq 2$ [or $N = \infty$] with the following properties: If Γ is a real, constant, non-singular $d \times d$ matrix with eigenvalues $\gamma_1, \ldots, \gamma_d$ such that*

$$(12.1) \quad \gamma_1^{m_1}\gamma_2^{m_2} \ldots \gamma_d^{m_d} \neq \gamma_k \quad \text{for } k = 1, \ldots, d \quad \text{and } 2 \leqq \sum_{j=1}^{d} m_j \leqq N$$

for all sets of non-negative integers $m_1, \ldots, m_d$ and if, in the map

$$(12.2) \quad\quad\quad T: \xi_1 = \Gamma\xi + \Xi(\xi),$$

$\Xi(\xi)$ is of class C^N for small $\|\xi\|$ satisfying $\Xi(0) = 0$, $\partial_\xi\Xi(0) = 0$, then there exists a map R: $\zeta = Z_0(\xi)$ of class C^n for small $\|\xi\|$ such that

$$(12.3) \quad\quad\quad Z_0(0) = 0, \quad\quad \partial_\xi Z_0(0) = I,$$

$$(12.4) \quad\quad\quad RTR^{-1}: \zeta_1 = \Gamma\zeta.$$

Note that (12.1) implies that $|\gamma_k| \neq 1$ for $k = 1, \ldots, d$ for, since Γ is real, its eigenvalues are real-valued or occur in pairs of complex conjugates e.g., if $|\gamma_1| = 1$ and $\gamma_2 = \bar{\gamma}_1 = 1/\gamma_2$, then $\gamma_1\gamma_2^2 = \gamma_1$). When T^t is the "group" of maps associated with the differential equation (7.1) and T in (12.2) is $T = T^1$, then $\Gamma = e^E$; so that if $e_1, \ldots, e_d$ are the eigenvalues of E, then $\gamma_j = \exp e_j$ and (12.1) is replaced by $m_1e_1 + \cdots + m_de_d \neq e_k$.

Exercise 12.1. Formulate an analogous theorem concerning the linearization of the differential equation (7.1) and prove it by using Theorem 12.1 and the device involving (9.1) in obtaining Theorem 7.1 from Lemma 8.1.

Remark. The proof of Theorem 12.2 supplies a choice of $N(n)$ or, equivalently, of $\lambda = N - n$ (probably, far from the best choice): Let $0 < \alpha < a < 1$ be such that the eigenvalues of Γ satisfy $0 < \alpha < \min (|\gamma_k|, 1/|\gamma_k|) < a < 1$ for $k = 1, \ldots, d$. (In particular, in a suitable coordinate system, the norms of Γ, Γ^{-1} are less than $c = 1/\alpha$; furthermore $\Gamma = A$ or $\Gamma = \text{diag} [A, B]$, where the norms of A and B^{-1} satisfy $\alpha < |A|$, $|B^{-1}| < a$.) Then N can be chosen to be $n + \lambda$ if λ is an integer such that

$$\frac{A_{n,d} d^2 a^\lambda}{\alpha^{n+2}} < 1, \quad \text{where} \quad A_{n,d} = \frac{(d + n - 1)!}{n!\,(d-1)!}$$

is the number of partial derivatives of order n of a function of d variables; cf. (14.29) and steps (l) and (m) in the proof of Theorem 12.2 in §14.

Theorem 12.1 will be obtained from a more general result, Theorem 12.2. The fact that Theorem 12.1 is contained in Theorem 12.2 is implied by the following lemma which depends on a simple formal calculation.

Lemma 12.1. *Let $N \geq 2$ be an integer [or $N = \infty$]. Let Γ be a real, constant, nonsingular matrix with eigenvalues satisfying* (12.1) *and let $\Xi(\xi)$ be of class C^N for small $\|\xi\|$ satisfying $\Xi(0) = 0, \partial_\xi \Xi(0) = 0$. Then there exists a map $R_1: \zeta = Z_1(\xi)$ of class C^N for small $\|\xi\|$ satisfying $Z_1(0) = 0, \partial_\xi Z_1(0) = I;$ and $\Xi_1(\xi)$ in*

$$(12.5) \qquad\qquad T_1: \zeta_1 = \Gamma\zeta + \Xi_1(\zeta),$$

where $T_1 = R_1 T R_1^{-1}$, has the property that all partial derivatives of $\Xi_1(\zeta)$ of order $\leq N$ vanish at $\zeta = 0$.

A generalization of Theorem 12.1 involves the "equivalence" of two maps T, T_1 without the assumption that either is linear:

Theorem 12.2. *Let* (12.5) *be a map of class C^N for small $\|\xi\|$, where $2 \leq N \leq \infty$ and $\Xi_1(0) = 0, \partial_\zeta \Xi_1(0) = 0$, and suppose that the eigenvalues γ_k of Γ satisfy $|\gamma_k| \neq 0, 1$. Let $n > 0$ be an integer. Then there exists an integer $\lambda = \lambda(n) > 0$ depending only on n and Γ with the following property: If $N \geq n + \lambda(n)$ and T in* (12.2) *is of class C^N such that all partial derivatives of $\Xi(\xi) - \Xi_1(\xi)$ of order $\leq N$ vanish at $\xi = 0$, then there exists a map $R: \zeta = Z_0(\xi)$ of class C^n for small $\|\xi\|$ satisfying* (12.3) *and*

$$(12.6) \qquad\qquad RTR^{-1} = T_1.$$

In the proof of this theorem, the definition of R will not depend on n, thus R is of class C^n for every applicable n. In particular, if $N = \infty$, then $R \in C^\infty$. Also, the proof will show that for a given $n > 0$, the assumption that Ξ, Ξ_1 are of class C^N can be relaxed to the assumption that Ξ, Ξ_1 are of class C^{n+1} and that each partial derivative of $\Xi(\xi) - \Xi_1(\xi)$ of order $k \leq n + 1$ is majorized by const. $\|\xi\|^{N_0 - k}$ for a fixed $N_0 \geq n + \lambda(n)$, in which case $R \in C^n$ is such that each partial derivative of $R\xi - \xi$ of order $j \leq n$ is majorized by const. $\|\xi\|^{N_0 - j}$.

Lemma 12.1 will be proved in the next section and Theorem 12.2 in § 14. The proof of the following theorem which is the analogue of Theorem 12.2 for differential equations depends on a simple modification of the proof of Theorem 12.2 and will be left as an exercise; see Exercise 14.1.

Theorem 12.3. *In the differential equation*

$$(12.7) \qquad\qquad \zeta' = E\zeta + F_1(\zeta),$$

let $F_1(\zeta)$ be of class C^N for small $\|\zeta\|$, where $2 \leq N \leq \infty$ and $F_1(0) = 0, \partial_\zeta F_1(0) = 0$, and suppose that no eigenvalue of E has a vanishing real part. Let $n > 0$ be an integer. Then there exists an integer $\lambda = \lambda(n) > 0$ depending only on n and E with the following property: If $N \geq n + \lambda(n)$ and $F(\xi)$ in

$$(12.8) \qquad\qquad \xi' = E\xi + F(\xi)$$

is of class C^N such that all partial derivatives of $F(\xi) - F_1(\xi)$ of order $\leqq N$ vanish at $\xi = 0$, then there exists a map $R: \zeta = Z_0(\xi)$ of class C^n for small $\|\xi\|$ satisfying (12.3) and transforming (12.7) into (12.8).

The remarks following Theorem 12.2 on the smoothness of Ξ, Ξ_1 have analogues concerning the smoothness of F, F_1. In particular, the analogue of the last part of the remark concerning R implies the following result on asymptotic integration:

Corollary 12.1. *Under the conditions of Theorem 12.3, there is a one-to-one correspondence between solutions $\xi(t)$ of (12.8) and solutions $\zeta(t) = R\xi(t)$ of (12.7) satisfying $\xi(t)$, $\zeta(t) \to 0$ as $t \to \infty$ [or $-\infty$]; furthermore $\|\xi(t) - \zeta(t)\| \leqq$ const. $\|\zeta(t)\|^N$ as $t \to \infty$ [or $-\infty$].*

Exercise 12.2. (a) Let T and T_1 in (12.2) and (12.5) be maps of class C^∞ for small $\|\xi\|$ such that the eigenvalues γ_k of Γ satisfy $|\gamma_k| \neq 0$, 1 and that $\Xi(0) = 0$, $\partial_\xi\Xi(0) = 0$, $\Xi_1(0) = 0$, $\partial_\zeta\Xi_1(0) = 0$. Let T_0, T_{10} denote the (not necessarily convergent) Taylor expansion for $T\xi$, $T_1\xi$ at the origin. Then there exists a map $R: \zeta = Z_0(\xi)$ of class C^∞ for small $\|\xi\|$ satisfying (12.3) and $T_1 = RTR^{-1}$ if and only if there exists a formal power series map $R_0: \zeta = \xi + \cdots$ such that formally $T_{10}R_0 = R_0T_0$. (This assertion is a consequence of Theorem 12.2 and Lemma 13.1) The question of the existence of R_0 depends on the solvability of certain linear equations; cf., e.g., the proof of Lemma 12.1. *(b)* Formulate the analogue of *(a)* in which the maps T, T_1 are replaced by the differential equations (12.7), (12.8).

13. Proof of Lemma 12.1

Case 1 $(2 \leqq N < \infty)$. After a suitable linear change of variables, it can be supposed that Γ is in a form similar to a Jordan normal form except that the subdiagonal elements are 0 or $\epsilon > 0$, where ϵ will be specified below; cf. § IV 9. Thus the transformation (12.2) can be written in terms of components as follows:

$$(13.1) \qquad T: \xi_1{}^j = \gamma_j\xi^j + \epsilon_j\xi^{j-1} + \sum_{2 \leqq |i| \leqq N} \cdots \sum c^j_{(i)}\xi^{(i)} + o(\|\xi\|^N),$$

where $j = 1, \ldots, d$ and ϵ_j is 0 or ϵ. In (13.1) and below, (i) represents a d-tuple $(i_1, \ldots, i_d)$ of non-negative integers; $|i| = i_1 + \cdots + i_d$; and $\xi^{(i)}$ is the product $\xi^{(i)} = (\xi^1)^{i_1} \ldots (\xi^d)^{i_d}$.

The map R_1 will be determined so that each component of $R_1\xi$ is a polynomial in $(\xi^1, \ldots, \xi^d)$, say,

$$(13.2) \qquad R_1: \zeta^j = \xi^j + \sum_{2 \leqq |m| \leqq N} \cdots \sum r^j_{(m)}\xi^{(m)} \quad \text{for} \quad j = 1, \ldots, d.$$

The object is to determine R_1 so that $R_1TR_1^{-1}\xi = \Gamma\xi + o(\|\xi\|^N)$, or, equivalently,

$$(13.3) \qquad R_1T\xi = \Gamma R_1\xi + o(\|\xi\|^N) \qquad \text{as} \quad \xi \to 0.$$

Up to terms of order $o(\|\xi\|^N)$, the j-th component of the left side of (13.3) is

$$\gamma_j \xi^j + \epsilon_j \xi^{j-1} + \sum_{(i)} c^j_{(i)} \xi^{(i)} + \sum_{(m)} r^j_{(m)} \prod_{\alpha=1}^d (\gamma_\alpha \xi^\alpha + \epsilon_\alpha \xi^{\alpha-1} + \sum_{(i)} c^\alpha_{(i)} \xi^{(i)})^{m_\alpha},$$

while that of the right side is

$$\gamma_j \xi^j + \gamma_j \sum_{(m)} r^j_{(m)} \xi^{(m)} + \epsilon_j \xi^{j-1} + \epsilon_j \sum_{(m)} r^{j-1}_{(m)} \xi^{(m)}.$$

Thus (13.3) is equivalent to

$$\sum_{2 \le |m| \le N} r^j_{(m)} \left[\gamma_j \xi^{(m)} - \prod_{\alpha=1}^d \left(\gamma_\alpha \xi^\alpha + \epsilon_\alpha \xi^{\alpha-1} + \sum_{(k)} c^\alpha_{(k)} \xi^{(k)} \right)^{m_\alpha} \right]$$
$$+ \epsilon_j \sum_{(m)} r^{j-1}_{(m)} \xi^{(m)} = \sum_{2 \le |i| \le N} c^j_{(i)} \xi^{(i)} + o(\|\xi\|^N).$$

Comparing coefficients gives a linear system of equations for $r^j_{(m)}$.

It is easy to see that in view of (12.1) these linear equations uniquely determine the numbers $r^j_{(m)}$ for $j = 1, \ldots, d$ and $2 \le |m| \le N$ if $\epsilon_1 = \cdots \epsilon_d = 0$. In fact, the main part of the factor of $r^j_{(m)}$ on the left (i.e., the part of lowest order in ξ) is $(\gamma_j - \prod \gamma_\alpha^{m_\alpha}) \xi^{(m)}$, so that we can successively determine $r^j_{(m)}$ first for $|m| = 2$, then $|m| = 3$, etc. This shows that if $\epsilon_1 = \cdots = \epsilon_d = 0$, then the determinant of the matrix of coefficients of the unknown $r^j_{(m)}$ is not zero.

It follows that if $\epsilon > 0$ is sufficiently small, where ϵ_j is 0 or ϵ, then the matrix of coefficients is nonsingular. This proves Lemma 12.1 when $2 \le N < \infty$.

Remark. For the purpose of treating the case $N = \infty$, note the following corollary of the proof just completed: if $2 \le N < \infty$, then there exists a *unique* map R_1 of the form (13.2) satisfying the conclusions of Lemma 12.1. Actually, the proof shows that R_1 is unique after a certain linear change of variables (which leaves the form (13.2) unchanged) and hence is unique before the linear change of variables.

Case 2 ($N = \infty$). Since T is of class C^∞, it has a formal (not necessarily convergent) Taylor expansion at $\xi = 0$;

$$T: \xi_1^j = \sum_{k=1}^d \gamma_{jk} \xi^k + \sum_{|i| \ge 2} c^j_{(i)} \xi^{(i)} \quad \text{for} \quad j = 1, \ldots, d,$$

where $\Gamma = (\gamma_{jk})$. The proof just completed and the remark following it show that there is a unique formal power series map

$$(13.4) \qquad R_1: \zeta^j = \xi^j + \sum_{|m| \ge 2} r^j_{(m)} \xi^{(m)} \quad \text{for } \ j = 1, \ldots, d$$

such that $R_1 T = \Gamma R_1$ formally.

In order to complete the proof, grant the following fact for a moment:

Lemma 13.1.   *Corresponding to any formal power series*

$$(13.5) \qquad \sum_{|m|=0}^{\infty} r_{(m)} \xi^{(m)}$$

with real coefficients, there is a function $r(\xi)$ of class C^{∞} having (13.5) as its formal Taylor development.

For if Lemma 13.1 holds, a desired C^{∞} map R_1 is obtained as $R_1\xi = (r^1(\xi), \ldots, r^d(\xi))$ in which $r^j(\xi)$ is of class C^{∞} and has the formal Taylor series on the right of (13.4). Thus in order to complete the proof of Lemma 12.1, it only remains to verify Lemma 13.1.

Proof of Lemma 13.1.   Let $\Phi(t)$ be a real-valued function of class C^{∞} such that $\Phi(t) = 1$ for $|t| \leq 1/4$, $0 < \Phi(t) < 1$ for $1/4 < |t| < 3/4$, and $\Phi(t) = 0$ for $|t| \geq 3/4$. It is easy to see that

$$(13.6) \qquad r(\xi) = \sum_{|m|=0}^{\infty} r_{(m)} \xi^{(m)} \Phi\left(m_1! \ldots m_d! \left(\sum_{|i|=|m|} |r_{(i)}|\right) \|\xi\|^2\right)$$

is a uniformly convergent series for $\|\xi\| \leq 1$ and represents a C^{∞} function with the desired property. The (m)-th term is zero unless

$$m_1! \ldots m_d! \left(\sum |r_{(i)}|\right) \|\xi\|^2 \leq 3/4;$$

in which case the (m)-th term of (13.6) is majorized by

$$|r_{(m)}| \cdot \|\xi^{(m)}\| \leq \left(\sum |r_{(i)}|\right) \|\xi\|^2 \leq \frac{3}{4 m_1! \ldots m_d!}$$

if $|m| \geq 2$. Thus the series in (13.6) is uniformly convergent for $\|\xi\| \leq 1$ and $r(0) = r_{(0)}$. Similarly, the series in (13.6) can be differentiated formally any number of times to give a uniformly convergent series and $\partial^{|m|} r(\xi)/\partial(\xi^1)^{m_1} \ldots \partial(\xi^d)^{m_d} = r_{(m)} m_1! \ldots m_d!$ at $\xi = 0$. This proves Lemma 13.1.

14. Proof of Theorem 12.2

In what follows it is supposed that T, T_1 are of class C^N and that there exists a constant c_N such that

$$(14.1) \qquad \|D_{(m)}(T - T_1)\xi\| \leq c_N \|\xi\|^{N-|m|} \quad \text{for small} \quad \|\xi\|$$

and $|m| = 0, 1, \ldots, N$, where $D_{(m)} = \partial^{|m|}/\partial(\xi^1)^{m_1} \ldots \partial(\xi^d)^{m_d}$. In the following only $\|\xi\| \leq r_0 < 1$ is considered, so that $\|\xi\|^j \leq \|\xi\|^k$ if $j \geq k$.

In view of the normalizations of § 4 and of the results of Exercise 5.1, it is no loss of generality to suppose that T, T_1 are defined on the entire ξ-space, are one-to-one, and reduce to the linear map $\Gamma\xi$ for large $\|\xi\|$; that $\Gamma = \text{diag}\,[A, B]$ and $\xi = (\xi_-, \xi_+)$; that T is of the form

$$T: \xi_{1-} = A\xi_- + Y(\xi_-, \xi_+), \qquad \xi_{1+} = B\xi_+ + Z(\xi_-, \xi_+);$$

that $Y(0, \xi_+) = 0$ and $Z(\xi_-, 0) = 0$, i.e., that

(14.2) $M^+: \xi_- = 0$ and $M^-: \xi_+ = 0$

are invariant manifolds; and that there exist constants

(14.3) $0 < \alpha < a < 1$

such that

(14.4) $\alpha \|\xi\|_- \leq \|T\xi\|_- \leq a \|\xi\|_-,$ $\alpha \|T\xi\|_+ \leq \|\xi\|_+ \leq a \|T\xi\|_+,$

where $\|\xi_\pm\| = \|\xi\|_\pm$. These inequalities will also be used in the equivalent form

$\alpha \|T^{-1}\xi\|_- \leq \|\xi\|_- \leq a \|T^{-1}\xi\|_-,$ $\alpha \|\xi\|_+ \leq \|T^{-1}\xi\|_+ \leq a \|\xi\|_+.$

Both sets of inequalities are illustrated in Figure 3 when $\dim \xi_- = \dim \xi_+ = 1$.

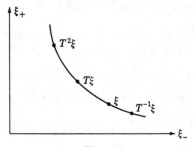

Figure 3.

Let K^+, K^0, K^- denote the ξ-sets

(14.5) $K^+: \|\xi\|_- \leq \|\xi\|_+;$ $K^0: \|\xi\|_- = \|\xi\|_+;$ $K^-: \|\xi\|_- > \|\xi\|_+.$

Thus K^0 is a conical hypersurface. Let $K^j = T^j K^0$ for $j = 0, \pm 1, \ldots,$ so that $T^{-j}K^j = K^0$ and, in view of (14.4), $K^j \subset K^+$ if $j \geq 0$ and $K^j \subset K^-$ if $j < 0$. Let Q^0 denote the ξ-set between K^0 and K^1 including K^0 but not K^1, i.e.,

$$Q^0 = \{\xi: \xi \in K^+, \quad T^{-1}\xi \notin K^+ - K^0\}.$$

In addition, let $Q^j = T^j Q^0$ the corresponding set between K^j and K^{j+1} including K^j but not K^{j+1}, so that

$$Q^j = \{\xi: T^{-j}\xi \in K^+, \quad T^{-j-1}\xi \notin K^+ - K^0\};$$

cf. Figure 4. It is clear that

$$K^j \to M^\pm \qquad \text{as} \quad j \to \pm\infty,$$

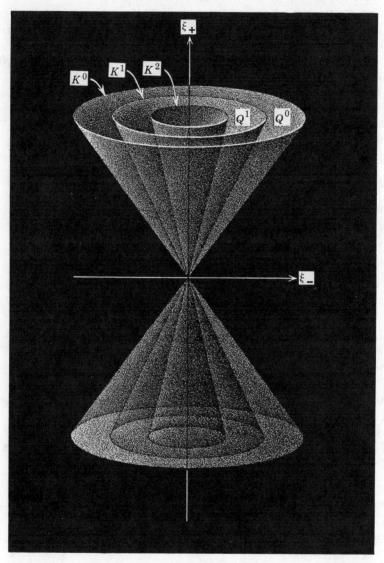

Figure 4.

that the sets Q^j are pairwise disjoint, that $K^+ - M^+$ is the union of Q^j for $j \geqq 0$, and that $K^- - M^-$ is the union of Q^j for $j < 0$. For $\xi \notin M^+ \cup M^-$, let $k(\xi)$ denote the unique integer k such that $\xi \in Q^k$.

For $r > 0$, let $S(r)$ denote the sphere $\|\xi\| \leqq r$, $K^{\pm}(r) = K^{\pm} \cap S(r)$, $K^0(r) = K^0 \cap S(r)$, and $Q^j(r) = Q^j \cap S(r)$. For a given $r_0 > 0$, there exists an $r_1 = r_1(r_0) > 0$ such that $0 < r_1 < r_0$ and that if $\xi \in Q^k(r_1)$ for $k > 0$ [or $k < 0$], then $T^{-j}\xi \in S(r_0)$ for $j = 0, 1, \ldots, k$ [or $-j = 0, 1, \ldots, k$].

The proof will be divided into steps (a)–(m). For brevity, most discussions will be given for $\xi \in K^+$; the obvious analogous statements and discussions for $\xi \in K^-$ will not be given, but will be used.

(a) *If $k(\xi) \geqq 0$, then, for $j = 0, \ldots, k(\xi)$,*

(14.6) $$\alpha^j \|\xi\|_+ \leqq \|T^{-j}\xi\|_+ \leqq a^j \|\xi\|_+,$$

(14.7) $$\|T^{-j}\xi\| \leqq 2 \|T^{-j}\xi\|_+ \leqq 2a^j \|\xi\|_+ \leqq 2a^j \|\xi\|.$$

This is clear from (14.4) and from the fact that $T^{-j}\xi \in K^+$ for $j = 0, \ldots, k(\xi)$.

(b) *Let $k(\xi) \geqq 0$. Then*

(14.8) $$a^{-2k(\xi)} \leqq \|\xi\|_+/\|\xi\|_- \leqq \alpha^{-2k(\xi)-2}$$

In order to verify this, let $\eta = T^{-k(\xi)}\xi$, so that $\eta \in Q^0 \subset K^+$ but $T^{-1}\eta \notin K^-$. Then, by (14.4),

$$\|\eta\|_+ \leqq a^k \|\xi\|_+, \qquad a^{-k} \|\xi\|_- \leqq \|\eta\|_-,$$

$$\alpha^{k+1} \|\xi\|_+ \leqq \|T^{-1}\eta\|_+, \qquad \|T^{-1}\eta\|_- \leqq \alpha^{-k-1} \|\xi\|_-,$$

where $k = k(\xi)$; so that (14.8) follows from $\|\eta\|_- \leqq \|\eta\|_+$ and $\|T^{-1}\eta\|_- > \|T^{-1}\eta\|_+$.

(c) *There exist constants $c < 1$ and $\kappa > 1$ such that if $k(\xi) > \kappa$, then*

(14.9) $$\|T^{-1}\xi\| \leqq c \|\xi\|.$$

[*In fact, c can be chosen arbitrarily close to a if κ is sufficiently large.*] By (14.4) and (14.8),

$$\|T^{-1}\xi\|^2 \leqq \alpha^{-2} \|\xi\|_-^2 + a^2 \|\xi\|_+^2 \leqq (\alpha^{-2}a^{4k(\xi)} + a^2) \|\xi\|_+^2.$$

Thus if κ is so large that $\alpha^{-2}a^{4\kappa} + a^2 < 1$, the result follows with $c = (\alpha^{-2}a^{4\kappa} + a^2)^{1/2}$.

(d) *Let $\|T_1\xi\| \leqq C \|\xi\|$ for $\|\xi\| \leqq r_0$ and let $T_1{}^j\xi, T_1{}^j\eta \in S(r_0)$ for $j = 0, 1, \ldots, h$. Then*

(14.10) $$\|T_1{}^j\xi - T_1{}^j\eta\| \leqq C^j \|\xi - \eta\| \qquad \text{for } j = 0, \ldots, h.$$

Put $\zeta(j) = T_1{}^j\xi - T_1{}^j\eta$, so that $\zeta(j) = T_1\zeta(j-1)$ for $j = 1, \ldots, h$. Hence $\|\zeta(j)\| \leqq C \|\zeta(j-1)\|$ and (14.10) follows.

(e) *Let $r_0 > 0$ be fixed, $r_1 = r_1(r_0) > 0$ as before part (a). Let $C \geq 1$ be a constant such that $\|T\xi\|$, $\|T_1\xi\| \leq C \|\xi\|$ for $\|\xi\| \leq r_0$. Let λ, N be integers such that*

$$(14.11) \qquad \lambda \geq 2, \quad Ca^\lambda < 1 \quad \text{and} \quad N \geq \lambda$$

and let there exist a constant c_N such that (14.1) holds for $\|\xi\| \leq r_0$. Then there exists an $r_2 = r_2(\lambda)$, $0 < r_2 < r_1$, such that (i) for $j = 0, \ldots, k(\xi)$,

$$(14.12) \qquad \|T_1^j T^{-k(\xi)}\xi\| \leq r_1 \quad \text{for} \quad \xi \in K^+(r_2) - M^+;$$

and (ii) there exists a constant C_N such that

$$(14.13) \qquad \|T_1^{k(\xi)} T^{-k(\xi)}\xi - \xi\| \leq C_N \|\xi\|^N \quad \text{for} \quad \xi \in K^+(r_2) - M^+.$$

Proof. Let $\eta \in K^+(r_1)$ and suppose that $T^j\eta$, $T_1^j\eta$ are defined for $j = 0, 1, \ldots, h$ for some $h \leq k(\xi)$. Put

$$(14.14) \qquad \zeta(j) = T_1^j\eta - T^j\eta.$$

Then $\zeta(j) = T_1[T_1^{j-1}\eta - T^{j-1}\eta] + (T_1 - T)T^{j-1}\eta$; i.e.,

$$\zeta(j) = T_1\zeta(j - 1) + (T_1 - T)T^{j-1}\eta.$$

Since $\zeta(0) = 0$, an easy induction gives

$$\zeta(j) = \sum_{i=0}^{j-1} T_1^{j-1-i}(T_1 - T)T^i\eta \quad \text{for} \quad j = 1, \ldots, h.$$

Hence, by (14.1),

$$\|\zeta(j)\| \leq c_N \sum_{i=0}^{j-1} C^{j-1-i} \|T^i\eta\|^N.$$

Choose $\eta = T^{-k(\xi)}\xi$, so that

$$\|\zeta(j)\| \leq c_N \sum_{i=0}^{j-1} C^{j-1-i} \|T^{i-k}\xi\|^N, \qquad k = k(\xi).$$

By (14.7), for $j = 0, \ldots, h$,

$$\|\zeta(j)\| \leq 2^N c_N \|\xi\|^N C^{-1} a^{(k-j)N} \sum_{i=0}^{j-1} (Ca^N)^{j-i}.$$

Hence, if $Ca^\lambda < 1$,

$$\|\zeta(j)\| \leq \frac{2^N c_N a^{(k-j)N}}{C(1 - Ca^N)} \|\xi\|^N,$$

so that

$$(14.15) \qquad \|\zeta(j)\| \leq C_N \|\xi\|^N \quad \text{if} \quad C_N = \frac{2^N c_N}{C(1 - Ca^N)}$$

for $j = 0, \ldots, h$.

From the definition of $\zeta(j)$ and η,

$$T_1^j T^{-k(\xi)}\xi = \zeta(j) + T^{j-k(\xi)}\xi;$$

hence

$$\|T_1^j T^{-k(\xi)}\xi\| \leq \|\zeta(j)\| + \|T^{j-k(\xi)}\xi\|.$$

This inequality, (14.7), and (14.15) make assertion (i) clear. The definition of $\zeta(j)$ and the choice $j = k(\xi)$ in (14.15) give (14.13).

(f) *Definition of* R. Let $R^{(0)}$ be a map of the closure of Q^0 into the ξ-space such that $R^{(0)}$ is of class C^N, reduces to the identity I on K^0 and to $T_1 T^{-1}$ on K^1, and satisfies an inequality of the form

$$(14.16) \qquad \| D_{(m)}(R^{(0)}\xi - \xi) \| \leqq d_N \| \xi \|^{N - |m|} \qquad \text{for} \quad m = 0, \ldots, N,$$

small $\| \xi \|$, and some constant d_N. Define a map $R^{(k)}$ on $\bar{Q}^k$, the closure of Q^k, by putting

$$(14.17) \qquad R^{(k)}\xi = T_1^k R^{(0)} T^{-k}\xi \qquad \text{for} \quad \xi \in \bar{Q}^k$$

and $k = 0, \pm 1, \ldots$. If $\xi \notin M^{\pm}$, put

$$(14.18) \qquad R\xi = R^{(k)}\xi \qquad \text{for} \quad \xi \in \bar{Q}^k.$$

The conditions $R^{(0)} = T_1 T^{-1}$ on K^1, $R^{(0)} = I$ on K^0 imply that R is continuous at $\xi \in K^1$ and hence for all $\xi \notin M^{\pm}$. It will also be supposed that $R^{(0)}$ is chosen so that

$$(14.19) \qquad R\xi \quad \text{is of class } C^N \qquad \text{for} \quad \xi \notin M^{\pm}.$$

Such a choice of $R^{(0)}$ is clearly possible; it suffices to choose $R^{(0)}\xi = \xi$ for $0 \neq \xi$ near K^0, and $R^{(0)}\xi = T_1 T^{-1}\xi$ for $0 \neq \xi$ near K^1.

A possible construction of $R^{(0)}$ is as follows: Note that if $0 \neq \xi \in K^1 = TK^0$, then $\alpha^2 \leqq \| \xi \|_- / \| \xi \|_+ \leqq a^2$ and if $0 \neq \xi \in Q^0$, then $\alpha^2 \leqq \| \xi \|_- / \| \xi \|_+ \leqq 1$. Let $0 < a^2 < \epsilon_1 < \epsilon_2 < 1$, so that the set $\{\xi: \epsilon_1 \| \xi \|_+ \leqq \| \xi \|_- \leqq \epsilon_2 \| \xi \|_+, \xi \neq 0\}$ is interior to Q^0 while the set $\{\xi: \alpha^2 \| \xi \|_+ \leqq \| \xi \|_- \leqq \| \xi \|_+\}$ contains $\bar{Q}^0$. Let $\Phi(t)$ be of class C^∞ for $\alpha^2 \leqq t \leqq 1$ such that $\Phi(t) = 0$ for $\alpha^2 \leqq t \leqq \epsilon_1$ and $\Phi(t) = 1$ for $\epsilon_2 \leqq t \leqq 1$. For $0 \neq \xi \in \bar{Q}^{(0)}$, put

$$R^{(0)}\xi = \Phi\left(\frac{\| \xi \|_-}{\| \xi \|_+}\right)\xi + \left[1 - \Phi\left(\frac{\| \xi \|_-}{\| \xi \|_+}\right)\right] T_1 T^{-1}\xi$$

and $R^{(0)}\xi = 0$ if $\xi = 0$. Note that $R^{(0)} = I$ on K^0, $R^{(0)} = T_1 T^{-1}$ on K^1 and

$$R^{(0)}\xi - \xi = \left[1 - \Phi\left(\frac{\| \xi \|_-}{\| \xi \|_+}\right)\right](T_1 - T)T^{-1}\xi,$$

from which (14.16) readily follows.

(g) *The map* R *satisfies, for* $\xi \notin M^{\pm}$,

$$(14.20) \qquad RT\xi = T_1 R\xi$$

This is clear from (14.17) and (14.18), for if $\xi \in \bar{Q}^k$, then $T\xi \in \bar{Q}^{k+1}$ and $RT\xi = T_1^{k+1} R^{(0)} T^{-k}\xi = T_1 R\xi$.

(h) Assertion (e) remains valid if (14.12), (14.13) *are replaced by*

$$\| T_1{}^j R^{(0)} T^{-k(\xi)} \xi \| \leqq r_1 \quad \text{for} \quad \xi \in K^+(r_2) - M^+,$$

(14.21)
$$\| R\xi - \xi \| \leqq C_N \| \xi \|^N \quad \text{for} \quad \xi \in K^+(r_2) - M^+,$$

respectively. Note that

$$\| T_1{}^j R^{(0)} T^{-k(\xi)} \xi - T_1{}^j T^{-k(\xi)} \xi \| \leqq C^j \| R^{(0)} T^{-k(\xi)} \xi - T^{-k(\xi)} \xi \|.$$

The right side is majorized by $C^j d_N \| T^{-k(\xi)} \xi \|^N$ in view of (14.16) and hence by $2^N d_N C^j \| \xi \|^N a^{k(\xi)N}$ in view of (14.7). Assuming that $C \geqq 1$, it follows that, for $j = 0, \ldots, k(\xi)$,

$$\| T_1{}^j R^{(0)} T^{-k(\xi)} \xi - T_1{}^j T^{-k(\xi)} \xi \| \leqq 2^N d_N (Ca^N)^{k(\xi)} \| \xi \|^N \leqq 2^N d_N \| \xi \|^N.$$

In view of *(e)*, this makes the validity of *(h)* clear.

The object of the remainder of the proof is to obtain estimates for the derivatives of $R\xi - \xi$ for $\xi \notin M^\pm$. To this end introduce the following notation:

To avoid confusion with other superscripts, the coordinate index of a vector will be written before, instead of after, the vector symbol; e.g., $f = ({}^1f, \ldots, {}^df)$ and $\xi = ({}^1\xi, \ldots, {}^d\xi)$. For a function $f(\xi)$, let $f_j = \partial f / \partial ({}^j\xi)$. If $(\alpha) = (\alpha_1, \ldots, \alpha_d)$ is a d-tuple of non-negative integers, write $f_{(\alpha)} = D_{(\alpha)} f = \partial^{|\alpha|} f / \partial ({}^1\xi)^{\alpha_1} \ldots \partial ({}^d\xi)^{\alpha_d}$. Let $f_{(\alpha)} \circ \eta, f_j \circ \eta$ denote the value of the corresponding derivative of f evaluated at the point $\xi = \eta$. Similarly for a map, say T, the abbreviations $T_j\xi$ or $T_{(\alpha)}\xi$ mean $\partial (T\xi) / \partial ({}^j\xi)$ or $(T\xi)_{(\alpha)}$.

It will be shown that if $\lambda = \lambda(n)$ in *(e)* is sufficiently large, then for a suitable constant $C_N(|\alpha|)$

(14.22)
$$\| (R\xi - \xi)_{(\alpha)} \| \leqq C_N(|\alpha|) \| \xi \|^{N-|\alpha|}$$

for $\xi \in K^+(r_2) - M^+$ and $|\alpha| \leqq n$. The proof will be by induction on $|\alpha|$. The relation (14.21) corresponds to the case $|\alpha| = 0$.

For $0 \neq \xi \in Q^k$, define

(14.23)
$$S^{(k)}\xi = R^{(k)}\xi - \xi.$$

By (14.17),

$$R^{(k+1)}\xi = T_1 R^{(k)} T^{-1} \xi \quad \text{for} \quad 0 \neq \xi \in Q^{k+1};$$

hence

$$S^{(k+1)}\xi = T_1 R^{(k)} T^{-1} \xi - T T^{-1} \xi$$

or

(14.24)
$$S^{(k+1)}\xi = T_1 R^{(k)} T^{-1} \xi - T_1 T^{-1} \xi + F\xi \quad \text{for} \quad 0 \neq \xi \in Q^{k+1},$$

where

(14.25)
$$F = (T_1 - T) T^{-1}.$$

The chain rule of differentiation gives

$$R_j^{(k+1)}\xi = \sum_{m=1}^{d} \sum_{h=1}^{d} (T_{1m} \circ R^{(k)}T^{-1}\xi)(^m R_h^{(k)} \circ T^{-1}\xi)[^h(T^{-1})_j\xi].$$

Repeated differentiation leads to a formula of the type

$$(14.26) \quad {}^iR_{(\alpha)}^{(k+1)}\xi = \sum_{m=1}^{d} \sum_{|\beta|=|\alpha|} (^iT_{1m} \circ R^{(k)}T^{-1}\xi)(^m R_{(\beta)}^{(k)} \circ T^{-1}\xi)\Phi^{\alpha,\beta} + \Psi^{i,\alpha,k},$$

where the second summation is over all d-tuples $\beta = (\beta_1, \ldots, \beta_d)$ with $|\beta| = |\alpha|$; $\Phi^{\alpha,\beta}$ is a product of $|\alpha|$ factors of the form $^h(T^{-1})_j\xi$; $\Psi^{i,\alpha,k}$ is a polynomial (independent of k) in the components of $T_{1(\beta)} \circ R^{(k)}T^{-1}\xi$, $(T^{-1})_{(\beta)}\xi$ for $|\beta| \leqq |\alpha|$ and $R_{(\gamma)}^{(k)} \circ T^{-1}\xi$ for $|\gamma| < |\alpha|$. Note that the polynomial in $\Psi^{i,\alpha,k}$ does not depend on k; the dependence on k arises from its arguments $T_{1(\beta)} \circ R^{(k)}T^{-1}\xi$ and $R_{(\gamma)}^{(k)} \circ T^{-1}\xi$.

If the second term on the right of (14.24) is written as $T_1T^{-1}\xi = T_1 IT^{-1}\xi$, where I is the identity operator, it follows that the analogue of (14.26) holds, i.e.,

$$^i(T_1T^{-1})_{(\alpha)}\xi = \sum_{m=1}^{d} \sum_{|\beta|=|\alpha|} (^iT_{1m} \circ T^{-1}\xi)(^m I_{(\beta)} \circ T^{-1}\xi)\Phi^{\alpha,\beta} + \Psi^{i,\alpha},$$

where $\Psi^{i,\alpha}$ is obtained by replacing $T_{1(\beta)} \circ R^{(k)}T^{-1}\xi$, $R_{(\gamma)}^k \circ T^{-1}\xi$ by $T_{1(\beta)} \circ T^{-1}\xi$, $I_{(\gamma)} \circ T^{-1}\xi$, respectively, in $\Psi^{i,\alpha,k}$. Consequently, (14.24) implies that

$$(14.27) \qquad {}^iS_{(\alpha)}^{(k+1)} = \sum_{\mathrm{I}} + \sum_{\mathrm{II}} + \Psi^{i,\alpha,k} - \Psi^{i,\alpha} + {}^iF_{(\alpha)}\xi$$

where

$$(14.28) \quad \begin{aligned} \sum_{\mathrm{I}} &= \sum_{m=1}^{d} \sum_{|\beta|=|\alpha|} (^iT_{1m} \circ R^{(k)}T^{-1}\xi)(^m S_{(\beta)}^{(k)} \circ T^{-1}\xi)\Phi^{\alpha,\beta}, \\ \sum_{\mathrm{II}} &= \sum_{m=1}^{d} \sum_{|\beta|=|\alpha|} (^iT_{1m} \circ R^{(k)}T^{-1}\xi - {}^iT_{1m} \circ T^{-1}\xi)(^m I_{(\beta)} \circ T^{-1}\xi)\Phi^{\alpha,\beta}, \end{aligned}$$

(i) *The sum*

$$\sum_{h=1}^{d} \sum_{m=1}^{d} \sum_{|\beta|=|\alpha|} |^hT_{1m} \circ R^{(k)}T^{-1}\xi| \cdot |\Phi^{\alpha,\beta}|$$

which is defined for $0 \neq \xi \in Q^{k+1}(r_2)$ *has a bound* $L_{|\alpha|}$ *independent of* $k = 0, 1, \ldots$. For, if C is a bound for the first order derivatives of $T_1\xi$, $T^{-1}\xi$ on $\|\xi\| \leqq r_0$, then $|\Phi^{\alpha,\beta}| \leqq C^{|\alpha|}$ and $\|T_{1m} \circ R^{(k)}T^{-1}\xi\| \leqq C$. Hence a majorant for the sum in (i) is

$$(14.29) \quad L_{|\alpha|} = C^{1+|\alpha|}d^2 \sum_{|\beta|=|\alpha|} 1, \quad \text{where} \quad \sum_{|\beta|=|\alpha|} 1 \equiv A_{|\alpha|,d} = \frac{(d + |\alpha| - 1)!}{|\alpha|!\,(d-1)!}.$$

Remark. By considering sufficiently small r_0, the number C, hence $L_{|\alpha|}$, can be made to depend only on the norms of Γ and Γ^{-1}.

(j) *There exists a number C_0 such that*

$$\sum_{h=1}^{d} \sum_{|\beta|=|\alpha|} |^h F_{(\beta)}\xi| \leqq C_0 \|T^{-1}\xi\|^{N-|\alpha|} \quad \text{for} \ \ 0 \leqq |\alpha| \leqq N \ \ \text{and} \ \ \|\xi\| \geqq r_0.$$

By (14.25), a differentiation of $F\xi$ gives

$$F_j\xi = \sum_{m=1}^{d} [(T_1 - T)_m \circ T^{-1}\xi](^m T^{-1})_j\xi.$$

Repeated differentiations show that $^h F_{(\beta)}\xi$ is a polynomial in $^h(T_1 - T)_{(\gamma)}°.$ $T^{-1}\xi$ and $^m(T^{-1})_{(\gamma)}$ for $|\gamma| \leqq |\beta|$ and $m = 1, \ldots, d$, and that each term contains a factor of the first type. Since there exists a constant C_0', such that

$$|^h(T_1 - T)_{(\gamma)} \circ T^{-1}\xi| \leqq C_0' \|T^{-1}\xi\|^{N-|\gamma|}$$

for $0 \leqq |\gamma| \leqq N$ and small $\|\xi\|$, the assertion (j) follows.

(k) *There exists a number C_0 such that*

$$|\Sigma_{\mathrm{II}}| \leqq C_0 \|T^{-1}\xi\|^N \quad \text{for} \ \ 0 \leqq |\alpha| \leqq N, \ \ \|\xi\| \leqq r_0, \ \ \text{and} \ \ k = 0, 1, \ldots.$$

If C' is a bound for the second order derivatives of $T_1\xi$ for small $\|\xi\|$, the mean value theorem of differential calculus shows that

$$|\Sigma_{\mathrm{II}}| \leqq C'C^{|\alpha|} d\Big(\sum_{|\beta|=|\alpha|} 1\Big) \|R^{(k)}T^{-1}\xi - T^{-1}\xi\|.$$

In view of (14.21), the assertion (k) holds with $C_0 = C'C^{|\alpha|} d(\Sigma\, 1)C_N$.

(l) *Let $1 \leqq n \leqq N - \lambda$. Let the conditions of (e) hold and, in addition, let λ be so large that $L_\nu a^\lambda < 1$ for $\nu = 1, \ldots, n$, where L_ν is given in part (i). Then there exists a constant $C_N(|\alpha|)$ such that (14.22) holds for $\xi \in K^+(r_2) - M^+$ if $0 \leqq |\alpha| \leqq n.$*

Proof. Introduce the abbreviation

$$(14.30) \qquad S^{(k,\mu)}\xi = \sum_{m=1}^{d} \sum_{|\beta|=\mu} |^m S_{(\beta)}^{(k)}\xi|.$$

The assertion to be proved can be cast in the form

$$(14.31) \qquad S^{(k,\mu)}\xi \leqq C_N(\mu) \|\xi\|^{N-\mu} \quad \text{for} \ \mu \leqq n,$$

$$0 \neq \xi \in Q^k(r_2), \ \ \text{and} \ \ k = 0, 1, \ldots.$$

The proof will be by induction on n. The case $n = 0$ is contained in (14.21) in assertion (h). Assume $0 < n \leqq N - \lambda$ and the desired result for $0, 1, \ldots, n - 1$.

It will first be shown that there is a constant $C_{N0}(n)$, independent of k, such that

$$(14.32) \qquad |\Psi^{i,\alpha,k} - \tilde{\Psi}^{i,\alpha}| \leqq C_{N0}(n) \|T^{-1}\xi\|^{N-n} \quad \text{for} \ \ |\alpha| = n$$

and $0 \neq \xi \in Q^{k+1}(r_2)$. Note that the descriptions of $\Psi^{i,\alpha,k}$ and $\Psi^{i,\alpha}$ following (14.26) imply that the expression on the left of (14.32) is majorized by

$$\text{const.} \left\{ \sum_{|\beta| \le n} \| T_{1(\beta)} \circ R^{(k)} T^{-1} \xi - T_{1(\beta)} \circ T^{-1} \xi \| + \sum_{|\gamma| < n} \| S^{(k)}_{(\gamma)} \circ T^{-1} \xi \| \right\},$$

where "const." is independent of k. Since $N > n$, the argument used in part (k) shows that there exists a constant $K_N(n)$, depending on n but independent of k, such that the first sum does not exceed $K_N(n) \| T^{-1} \xi \|^N$. Hence the existence of $C_{N0}(n)$ in (14.32) follows from the induction hypothesis.

In view of (14.27) and (i), (j), (k), (14.32), and (14.30),

$$(14.33) \quad S^{(k+1,n)} \xi \le L_n S^{(k,n)} T^{-1} \xi + [C_{N0}(n) + 2C_0] \| T^{-1} \xi \|^{N-n}.$$

It is clear from (14.16) that there exists a constant such that

$$S^{(0,n)} \xi \le \text{const.} \| \xi \|^{N-n} \quad \text{for} \quad 0 \neq \xi \in Q^0(r_2).$$

It follows from (14.33) and a simple induction on k that similar inequalities hold if $S^{(0,n)}$, Q^0 are replaced by $S^{(k,n)}$, Q^k. Thus if κ is given in step (c), then, for the finite set of k-values, $k = 0, 1, \ldots \kappa$, there exists a constant $C_N(n)$ such that

$$(14.34) \quad S^{(k,n)} \xi \le C_N(n) \| \xi \|^{N-n} \quad \text{for} \quad 0 \neq \xi \in Q^k(r_2),$$

$$(14.35) \quad C_N(n) \ge \frac{[C_{N0}(n) + 2C_0] c^{N-n}}{1 - L_n c^{N-n}}.$$

Note that, by (c), $L_n c^{N-n} \le L_n c^\lambda < 1$ if κ is (fixed) sufficiently large.

An induction on k will now be used to show that (14.34) holds for all $k = 0, 1, \ldots$. Assume that (14.34) holds for some $k \ge \kappa$. Then, by (14.33) for $0 \neq \xi \in Q^{k+1}(r_2)$,

$$S^{(k+1,n)} \xi \le [L_n C_N(n) + C_{N0}(n) + 2C_0] \| T^{-1} \xi \|^{N-n}.$$

But $\| T^{-1} \xi \| \le c \| \xi \|$ by (c), so that the right side is majorized by $[L_n C_N(n) + C_{N0}(n) + 2C_0] c^{N-n} \| \xi \|^{N-n}$. Since (14.35) implies that the factor of $\| \xi \|^{N-n}$ is at most $C_N(n)$, the inequality (14.34) holds if k is replaced by $k + 1$. This completes the induction on k and also on n.

(m) *Let the conditions of (e) and (l) hold. Then R can be defined on $M^{\pm}$ so that R is of class C^{n-1} on $\| \xi \| \le r_2$.*

Proof. It has been shown that R, considered on $K^+(r_2) - M^+$, is of class C^N and that its partial derivatives of order $\le n$ are bounded if $N - n \ge \lambda$. Without loss of generality, it can be supposed dim $M^+ < d - 1$ (e.g., by increasing d, by adding dummy coordinates to ξ_+). Thus points of $K^+(r_2) - M^+$ which are near can be joined by short rectifiable paths in $K^+(r_2) - M^+$. It follows that R and its partial derivatives of order $\leqq n - 1$ are uniformly continuous in $K^+(r_2) - M^+$. Hence R has an extension to $K^+(r_2)$ of class C^{n-1}. This proves (m).

In view of (g) and the continuity of R, (14.20) holds on $\|\xi\| \leqq r_2$. Since R is close to the identity for small $\|\xi\|$, R has an inverse of class C^{n-1}. This proves Theorem 12.2. [We should remark that λ depends only on Γ; cf. (c) and the Remark following (i).]

Exercise 14.1. Prove Theorem 12.3. To this end, treat the "group" of maps T^t, T_1^t instead of the differential equations and the problem of finding a suitable R satisfying $RT^t = T_1^t R$. For $\xi \notin M^{\pm}$, define $t(\xi)$ so that $T^{-t(\xi)}\xi \in K^0$. Put $R\xi = T_1^{t(\xi)}T^{-t(\xi)}\xi$. The restriction of this R to $\bar{Q}^0$ gives an $R^{(0)}$ as in (f). It is only necessary to verify (14.16); but this can be deduced from (e).

APPENDIX: SMOOTHNESS OF STABLE MANIFOLDS*

In this Appendix, we give another proof of Lemma 5.1 which has the advantages that (1) it is valid for infinite dimensional Banach spaces and (2) it yields easy proofs for additional smoothness as in Exercise 5.1(a),(ii). Additional differentiability is deduced from "continuity" if one allows parameters; cf. Chapter V §§ 3-4. The simplification of the proof of, say, differentiability results by dealing with a more general situation: first, the iterates T^n of T are replaced by products $T^n = T_n \circ \ldots \circ T_1$ of maps $T_1, T_2, \ldots$ and, second, the introduction of terms $U_n(w)$ as in (1.2) below. The first generalization permits "changes of dependent variables" and the second permits the analogue of the "interchange of parameters and initial conditions" as in Chapter V § 1.

---

*This Appendix is reprinted with permission of Academic Press from Hartman [S3] in J. Differential Equations 9(1971)361-372. For other treatments, see Hirsch, Pugh and Shub [S1] and references there.

1. NOTATION AND HYPOTHESES

Let $W, X, Y, Z = X \times Y = XY$, $E = W \times X \times Y = WXY$ be Banach spaces. Norms for $w \in W$, $x \in X$, $y \in Y$, $z = (x, y) \in Z$, $e = (w, x, y) \in E$ will be denoted by

$$\| x \|, \| y \|, \| z \| = \max(\| x \|, \| y \|),$$
$$\| e \| = \max(\| w \|, \| x \|, \| y \|).$$

Finally, let $X_r(x_0) = \{x \in X : \| x - x_0 \| < r\}$.

DEFINITION STABILITY SET.   Let $X^0 \subset X$ be open; $T_n' : X^0 \to X$ a map for $n = 1, 2,...$; $\mathscr{D}(T_n' \circ \cdots \circ T_1')$ the (possibly empty) open set where $T_n' \circ \cdots \circ T_1'$ is defined. The (possibly empty) subset $\mathscr{D} = \mathscr{D}(\{T_n'\}, X^0) = \bigcap \mathscr{D}(T_n' \circ \cdots \circ T_1')$ of X^0 will be called the *stability set* of the sequence $\{T_n'\}$, with respect to X^0.

We shall be interested in stability sets of sequences of maps $T_n' : Z_1(0) \to Z$ of the form

$$T_n'(z) = (A_n x, B_n y) + (F_n'(z) + x^n, G_n'(z) + y^n), \qquad (1.0)$$

where $A_n \in L(X, X)$, $B_n \in L(Y, Y)$ are bounded linear operators, B_n is invertible, $\| A_n \| < 1, \| B_n^{-1} \| < 1; F_n'$ and G_n' are maps from $Z_1(0)$ to X and Y, respectively, such that $F_n'(0) = 0$, $G_n'(0) = 0$; $\| F_n'(z_1) - F_n'(z_2) \|$, $\| G_n'(z_1) - G_n'(z_2) \| \leqslant \epsilon \| z_1 - z_2 \|$; and $\| x^n \|, \| y^n \| \leqslant \epsilon$; where $\epsilon > 0$ is a fixed small number. In order to reduce (1.0) to the case $x^n = 0$, $y^n = 0$, we can replace Z, T_n' by $R^1 Z, T_n''$, where

$$T_n''(t, z) = (t, A_n x, B_n y) + (0, F_n'(z) + 2tx^n, G_n'(z) + 2ty^n). \quad (1.1)$$

It is clear that the stable set $\mathscr{D}(\{T_n'\}, Z_1(0))$ will be determined if $\mathscr{D}(\{T_n''\}, (R^1 Z)_1(0))$ is known; in fact,

$$(x, y) \in \mathscr{D}(\{T_n'\}, Z_1(0)) \Leftrightarrow (1/2, x, y) \in \mathscr{D}(\{T_n''\}, (R^1 Z)_1(0)).$$

This reduction to the case $t^n = 0$, $x^n = 0$, $y^n = 0$ has been accomplished at the cost of replacing the linear map $x \mapsto A_n x$, having a norm $a < 1$, by a map $(t, x) \mapsto (t, A_n x)$, having norm 1. It will turn out that, in view of the form of the latter, nothing is lost and, indeed, there are other advantages.

For these "methodical" reasons, we shall not consider maps (1.0) below, but maps $T_n : E_r(0) \to E = W \times X \times Y = WXY$ of the form

$$T_n(e) = (U_n(w), A_n x, B_n y) + (0, F_n(e), G_n(e)), \qquad (1.2)$$

where r is fixed $(0 < r < \infty)$; $A_n \in L(X, X)$ and $B_n \in L(Y, Y)$ are linear bounded operators, B_n is invertible,

$$\| A_n \| \leqslant a < 1, \qquad \| B_n^{-1} \| \leqslant 1/b < 1; \tag{1.3}$$

$$0 < \delta < 1/10, \qquad 0 < a < 1 - 2\delta < 1 + 5\delta < b; \tag{1.4}$$

$$U_n(0) = 0, \qquad F_n(0) = 0, \qquad G_n(0) = 0; \tag{1.5}$$

$U_n : W_r(0) \to W$; F_n, $G_n : E_r(0) \to X$, Y; and

$$\| U_n(w_1) - U_n(w_2) \| \leqslant \| w_1 - w_2 \|, \tag{1.6}$$

$$\| F_n(e_1) - F_n(e_2) \|, \| G_n(e_1) - G_n(e_2) \| \leqslant \delta^2 \| e_1 - e_2 \|. \tag{1.7}$$

The stability set of $\{T_n\}$ with respect to $\{E_r(0)\}$ is

$$\mathscr{D} = \bigcap_{n=1}^{\infty} \mathscr{D}_n, \qquad \mathscr{D}_n = \mathscr{D}(S_n), \qquad S_n = T_n \circ \cdots \circ T_1. \tag{1.8}$$

The first result (Section 2) will give, under suitable assumptions on U_n, F_n and G_n, the existence of a continuous function $y_0 : (WX)_r(0) \to Y_r(0)$ such that

$$\mathscr{D} = \{e = (w, x, y) : y = y_0(w, x) \text{ on } (WX)_r(0)\}. \tag{1.9}$$

In such a case, we shall speak of the *stable manifold* $\mathscr{D} : y = y_0(w, x)$ of the sequence $\{T_n\}$, with respect to $E_r(0)$. Actually, the assumptions on U_n, F_n, and G_n will imply that $y_0 \in C^1$, but this and other smoothness properties will be deduced from the existence proof for a continuous $y_0(w, x) = y_0(w, x, \sigma)$ carried out for the case that T_n depends on parameters σ.

HYPOTHESIS (H).   Let Σ be a metric space and $\sigma \in \Sigma$. Let $T_n \in C^0(E_r(0) \times \Sigma, E)$. For each $\sigma \in \Sigma$, assume that T_n satisfies (1.2)–(1.7), where a, b, δ are independent of σ. Let T_n have a Fréchet derivative $D_e T_n \in C^0(E_r(0) \times \Sigma)$.

HYPOTHESIS (H$_0$).   Let $\Sigma = \Lambda \times U \times T$, where Λ, U are metric spaces and $T = [0, \epsilon]$ is a real t-interval. We say that T_1, T_2,... satisfies (H$_0$) if (H) holds, $T_{n0} = (T_n)_{t=0}$ is independent of $u \in U$, and T_n, $D_e T_n \to T_{n0}$, $D_e T_{n0}$ uniformly on U, for fixed (e, λ), as $t \to +0$.

By (1.4), we have

$$\alpha \equiv b - \delta/(1 - 2\delta) > (1 + \delta)/(1 - 2\delta) > 1; \tag{1.10}$$

also, if

$$c_n = \sum_{k=0}^{n} (a + \delta^2)^k \quad \text{and} \quad c = \lim_{h \to \infty} c_n = 1/(1 - a - \delta^2), \quad (1.11)$$

then $a < 1 - 2\delta < 1 - \delta - \delta^2$ implies that

$$\delta^2 c_n < \delta^2 c < \delta. \qquad (1.12)$$

Generally, we shall not exhibit the dependence of functions on parameters, unless convenient to do so.

2. Main Lemma

In this section, we prove

Lemma 2.1(a). *Assume hypothesis* (H). *Then there exists* $y_0(w, x) = y_0(w, x, \sigma) \in C^0((WX)_r(0) \times \Sigma, Y_r(0))$ *such that, for a fixed* σ, *the stability set* $\mathscr{D} = \mathscr{D}(\{T_n\}, E_r(0))$ *is given by* (1.9); *also if* $e_0 = (w_0, x_0, y_0(w_0, x_0))$ *and* $e_k = S_k(e_0) = (w_k, x_k, y_k)$, *then, for* $k \geqslant 0$, $e_k = 0$ *if* $(w_0, x_0) = 0$,

$$\| y_k(w^0, x^0) - y_k(w_0, x_0) \|$$
$$\leqslant (1 - 2\delta)(\delta \| w^0 - w_0 \| + \| x_k(w^0, x^0) - x_k(w_0, x_0) \|), \qquad (2.1)$$

$$\| x_k(w^0, x^0) - x_k(w_0, x_0) \| \leqslant \delta \| w^0 - w_0 \| + (a + 2\delta^2)^k \| x^0 - x_0 \|; \qquad (2.2)$$

and the manifold $y = y_0(w, x)$ *is "invariant," that is,*

$$y_k(w_0, x_0) \equiv y_0(w_k(w_0, x_0), x_k(w_0, x_0)) \quad on \quad (WX)_r(0). \qquad (2.3)$$

(b) *If, in addition,* (H_0) *holds and* $\sigma = (\lambda, u, t)$, *then* $y_0(w, x, \lambda, u, 0)$ *is independent of* u *and* $y_0(w, x, \lambda, u, t) \to y_0(w, x, \lambda, u, 0)$ *uniformly on* U, *for fixed* (w, x, λ), *as* $t \to +0$.

Note that (2.1), (2.2) imply

$$\| y_k(w, x) \| \leqslant (1 - 2\delta)(\delta \| w \| + \| x_k(w, x) \|) < (1 - 4\delta^2)r \text{ for } k \geqslant 0, \quad (2.1')$$

$$\| x_k(w, x) \| \leqslant \delta \| w \| + (a + 2\delta^2)^k \| x \| < (a + 2\delta)r \quad \text{for} \quad k \geqslant 1, \quad (2.2')$$

where $w = w_0$, $x = x_0$. But if $y \neq y_0(w, x)$ and $e_0 = (w, x, y) \in \mathscr{D}_n$ for some $n > 0$, then $e_k = S_k(e_0) = (w_k, x_k, y_k)$ satisfies

$$\| y_k \| \geqslant \alpha^k \| y_{0k}(w, x) - y \| \quad \text{for} \quad k = 0, ..., n, \qquad (2.2'')$$

where (2.18) holds; cf. the proof of Proposition 2.5. In particular, if $\alpha^N \| y_{0N}(w, x) - y \| \geqslant r$, then $(w, x, y) \notin \mathscr{D}_N$.

PROPOSITION 2.1. *Let Γ be a complete metric space, and Σ as in* (H); *$C = C(\gamma, \sigma) : \Gamma \times \Sigma \to \Gamma$ continuous and, for fixed σ, $C_\sigma = C(\cdot, \sigma) : \Gamma \to \Gamma$ a contraction*

$$\operatorname{dist}(C_\sigma(\gamma^0), C_\sigma(\gamma_0)) \leqslant \theta \operatorname{dist}(\gamma^0, \gamma_0), \qquad 0 \leqslant \theta < 1,$$

with θ independent of σ. Then, for each $\sigma \in \Sigma$, the map $C_\sigma : \Gamma \to \Gamma$ has a unique fixed point $\gamma = \gamma(\sigma)$ and $\gamma(\sigma) \in C^0(\Sigma, \Gamma)$. If $\Sigma = \Lambda \times U \times T$ as in (H_0), $C(\gamma, \lambda, u, 0)$ is independent of $u \in U$, and $C(\gamma, \lambda, u, t) \to C(\gamma, \lambda, u, 0)$, as $t \to +0$, uniformly on U for fixed (γ, λ), then $\gamma(\lambda, u, 0)$ is independent of u and $\gamma(\lambda, u, t) \to \gamma(\lambda, u, 0)$ uniformly on U, for fixed λ, as $t \to +0$.

For, by the proof of the contraction principle, if $\gamma_0 \in \Gamma$, then $\gamma(\sigma) = \lim C_\sigma{}^n(\gamma_0)$, as $n \to \infty$, and $\operatorname{dist}(\gamma(\sigma), C_\sigma{}^n(\gamma_0)) \leqslant \theta^n \operatorname{dist}(C_\sigma(\gamma_0), \gamma_0)/(1 - \theta)$.

PROPOSITION 2.2. *Let $n > 0$; $e_0, e^0 \in \mathscr{D}_n = \mathscr{D}(S_n)$; and $e_k = S_k(e_0)$, $e^k = S_k(e^0)$ for $k = 0,..., n$.*

(a) *The inequality*

$$\| y^m - y_m \| \geqslant (1 - 2\delta)(\delta \| w^m - w_m \| + \| x^m - x_m \|) \quad \text{for some } m,$$
$$0 \leqslant m < n \quad (2.4)$$

(for example, $w^m = w_m$, $x^m = x_m$), implies that, for $k = m + 1,..., n$,

$$\| y^k - y_k \| \geqslant \delta \| w^k - w_k \| + \| x^j - x_k \|,$$
$$\| y^k - y_k \| \geqslant \alpha^{k-m} \| y^m - y_m \|, \quad (2.5)$$

where $\alpha > 1$; cf. (1.10).

(b) *The inequality*

$$\| y^n - y_n \| < (1 - 2\delta)(\delta \| w^n - w_n \| + \| x^n - x_n \|) \quad (2.6)$$

(for example, $y^n = y_n$) and (1.11), (1.12) *imply that, for $k = 0,..., n$,*

$$\| y^k - y_k \| \leqslant (1 - 2\delta)(\delta \| w^k - w_k \| + \| x^k - x_k \|), \quad (2.7)$$
$$\| x^k - x_k \| \leqslant \delta^2 c_k \| w^0 - w_0 \| + (a + \delta^2)^k \| x^0 - x_0 \|, \ \delta^2 c_k < \delta. \quad (2.8)$$

Proof (a). Since $\| e \| = \max(\| w \|, \| x \|, \| y \|)$, (2.4) implies that

$$\delta \| e^m - e_m \| \leqslant \| y^m - y_m \|/(1 - 2\delta). \quad (2.9)$$

Hence, by (1.6) and (1.7), $\| w^{m+1} - w_{m+1} \| \leqslant \| w^m - w_m \|$ and

$$\| x^{m+1} - x_{m+1} \| \leqslant a \| x^m - x_m \| + \delta \| y^m - y_m \|/(1 - 2\delta);$$

so that, by (2.4),

$$\delta \| w^{m+1} - w_{m+1} \| + \| x^{m+1} - x_{m+1} \| \leqslant (1 + \delta) \| y^m - y_m \|/(1 - 2\delta).$$

Also, (2.9) and (1.6), (1.7) give

$$\| y^{m+1} - y_{m+1} \| \geqslant [b - \delta/(1 - 2\delta)] \| y^m - y_m \|. \tag{2.10}$$

The last two inequalities and (1.10) give (2.5) for $k = m + 1$, and an induction gives it for $k = m + 1, ..., n$.

Proof (b). The inequalities (2.7) follow from part (a). In order to obtain (2.8), note that (2.7) implies that

$$\| e^k - e_k \| \leqslant \| w^k - w_k \| + \| x^k - x_k \| \leqslant \| w^0 - w_0 \| + \| x^k - x_k \|.$$

Hence, by (1.7),

$$\| x^{k+1} - x_{k+1} \| \leqslant \delta^2 \| w^0 - w_0 \| + (a + \delta^2) \| x^k - x_k \|.$$

An induction gives (2.8).

PROPOSITION 2.3.   *Write* $S_n(e) = (P_n(e), Q_n(e), R_n(e))$, *where* $e \in \mathscr{D}_n$ *and* $P_n \in W$, $Q_n \in X$, $R_n \in Y$. *For a fixed* σ *and* $n \geqslant 1$, *there exists a function*

$$y_{0n}(w, x) = y_{0n}(w, x, \sigma) \in C^0((WX)_r(0) \times \Sigma, Y_r(0))$$

such that $(w, x, y_{0n}(w, x)) \in \mathscr{D}_n$; $y_{0n} = 0$ *if* $(w, x) = 0$; *and*

$$R_n(w, x, y) = 0 \ \text{if and only if} \ \ y = y_{0n}(w, x). \tag{2.11}$$

If, in addition (H_0) *holds, then* y_{0n} *is independent of* u *when* $t = 0$ *and* $y_{0n}(w, x, \lambda, u, t) \to y_{0n}(w, x, \lambda, u, 0)$, *uniformly on* U *for fixed* (w, x, λ), *as* $t \to +0$.

Proof. We shall give the proof for a fixed σ, but with some detail, so that the continuity and uniformity assertions will be clear from Proposition 2.1.
   Let $w^0 = w_0 = w$, $x^0 = x_0 = x$ and $e_0 = (w_0, x_0, y_0)$, $e^0 = (w^0, x^0, y^0) \in \mathscr{D}_n$. Thus (2.4) holds for $m = 0$, so that by Proposition 2.2(a),

$$\| R_n(w, x, y^0) - R_n(w, x, y_0) \| \geqslant \alpha^n \| y^0 - y_0 \|. \tag{2.12}$$

Thus, for fixed (w, x), the equation

$$R_n(w, x, y) = 0 \qquad (2.13)$$

has at most one solution y. Define the set

$$J_n = \{(w, x) : (2.13) \text{ has a solution } y, (w, x, y) \in \mathcal{D}_n\}.$$

In particular $(w, x) = 0 \in J_n \neq \varnothing$.

We shall show, by the implicit function theorem, that J_n is open. Let $(w_0, x_0) \in J_n$ and $R_n(w_0, x_0, y_0) = 0$. By (2.12), $D_y R_n(w, x, y)$ is invertible[*] if $(x, y) \in \mathcal{D}_n$, also $\|[D_y R_n]^{-1}\| \leqslant 1/\alpha^n$. Put $D_0 = D_y R_n(w_0, x_0, y_0)$ and write (2.13) as $y - D_0^{-1} R_n(w, x, y) = y$, so that a solution y of (2.13) is a fixed point of the map $y \mapsto y - D_0^{-1} R_n(w, x, y) \equiv C_{w,x}(y)$, depending on parameters w, x. Since $D_y C_{w,x} = 0$ at (w_0, x_0, y_0), $\|D_y C_{w,x}\| \leqslant \theta < 1$ for (w, x, y) near (w_0, x_0, y_0). It follows that there are positive numbers ϵ, s such that, for fixed $(w, x) \in W_s(w_0) \times X_s(x_0)$, $C_{w,x}$ is a contraction mapping of $Y_\epsilon(y_0)$ into $Y_\epsilon(y_0)$, with contracting factor θ. Hence, $W_s(w_0) \times X_s(x_0) \in J_n$, and J_n is open.

Let $(w, x) \in J_n$, so that $e_{0n}(w, x) = (w, x, y_{0n}(w, x)) \in \mathcal{D}_n$ and put $e_{kn}(w, x) = S_n(e_{0n}(w, x))$ for $k = 0, 1, \ldots, n$. Thus, $y_{nn}(w, x) = R_n(e_{0n}(w, x)) = 0$; so that if $(w^0, x^0), (w_0, x_0) \in J_n$, Proposition 2.2(b) gives

$$\|y_{kn}(w^0, x^0) - y_{kn}(w_0, x_0)\|$$
$$\leqslant (1 - 2\delta)(\delta \| w^0 - w_0 \| + \| x_{kn}(w^0, x^0) - x_{kn}(w_0, x_0)\|), \qquad (2.14)$$

$$\| x_{kn}(w^0, x^0) - x_{kn}(w_0, x_0)\| \leqslant \delta \| w^0 - w_0 \| + (a + \delta^2)^k \| x^0 - x_0 \|, \qquad (2.15)$$

for $k = 0, \ldots, n$. In particular, for $0 \leqslant k \leqslant n$ and $1 \leqslant m \leqslant n$,

$$\| x_{mn}(w, x)\| \leqslant \delta \| w \| + (a + \delta^2) \| x \| < (a + 2\delta)r < r, \qquad (2.16)$$

$$\| y_{kn}(w, x)\| \leqslant (1 - 2\delta)(2\delta \| w \| + \| x \|) < (1 - 4\delta^2)r < r. \qquad (2.17)$$

By the uniform continuity (2.14) of $y_{0n}(w, x)$ on J_n, $y_{0n}(w, x)$ and $e_0(w, x) = (w, x, y_{0n}(w, x))$ have unique continuous extensions to $\bar{J}_n \cap (WX)_r(0)$ satisfying (2.14)–(2.17). In particular, $e_0(w, x) \in \mathcal{D}_n$ and (2.11) holds on $\bar{J}_n \cap (WX)_r(0)$. Thus, J_n is both open and closed relative to $(WX)_r(0)$ and so, $J_n = (WX)_r(0)$. This completes the proof of Proposition 2.3.

PROPOSITION 2.4.   *Let $y_{0n}(w, x)$ be given by Proposition 2.3. Then*

$$y_0(w, x) = \lim_{n \to \infty} y_{0n}(w, x) \text{ exists uniformly} \qquad (2.18)$$

---

[*]When dim $Y = \infty$, $\mathcal{D}_n$ in the definition of J_n should be replaced by its connected component containing $e = 0$.

on $(WX)_r(0) \times \Sigma$. Also, the stability set (1.8) satisfies

$$\mathscr{D} \supset \{(w, x, y) : y = y_0(w, x) \text{ on } (WX)_r(0)\}, \tag{2.19}$$

and (2.1), (2.2) hold.

Proof. Use the notation of the last proof, and let $1 \leqslant k \leqslant n$. Since $w_{0n}(w, x) = w_{0k}(w, x) = w$ and $x_{0n}(w, x) = x_{0k}(w, x) = x$, Proposition 2.2(a), with $m = 0$ in (2.4), and $y_{kk} = 0$ give

$$r > \| y_{kn} \| = \| y_{kn} - y_{kk} \| \geqslant \alpha^k \| y_{0n}(w, x) - y_{0k}(w, x)\|.$$

This proves the statement concerning (2.18), since $\alpha > 1$.

Keeping k fixed and letting $n \to \infty$ in (2.14), (2.15) gives (2.1), (2.2), hence $\| y_k \| \leqslant (1 - 4\delta^2)r < r$ for $k \geqslant 0$, and $\| x_k \| \leqslant (a + 2\delta)r < r$ for $k \geqslant 1$. This implies (2.19).

PROPOSITION 2.5. *The relations* (1.9) *and* (2.3) *hold.*

Proof. On (1.9). It only remains to prove the reverse inclusion to (2.19). Suppose that $(w, x, y) \in \mathscr{D}$ and $y \neq y_0(w, x)$. Let $e_{kn}(w, x) = \dot{S}_k(w, x, y_{0n}(w, x))$ for $k = 0, ..., n$ and $e_k(w, x) = S_k(w, x, y)$ for $k = 0, 1, ...$. Since $w_{0n} = w_0 = w$, $x_{0n} = x_0 = x$, Proposition 2.2(a) and $y_{nn} = 0$ give $\| y_n \| = \| y_{nn} - y_n \| \geqslant \alpha^n \| y_{0n}(w, x) - y \| \sim \alpha^n \| y_0(w, x) - y \| \to \infty$, as $n \to \infty$. This contradicts $(w, x, y) \in \mathscr{D}$.

On (2.3). In view of what has been proved, for every $(w, x) \in (WX)_r(0)$, there is one and only one y such that $e = (w, x, y)$ is in the stability set $\mathscr{D}(\{T_n\}, E_r(0))$. The relation (2.3) follows by applying this uniqueness statement to the sequence $T_k, T_{k+1}, ...$.

3. SMOOTHNESS OF STABLE MANIFOLDS

Let $e_0(w_0, x_0) = (w_0, x_0, y_0(w_0, x_0)) \in \mathscr{D}$; $e_n(w_0, x_0) = S_n(e_0)$ for $n = 1, 2, ...$; and

$$\Omega = L(W, W), \quad \Xi = L(X, X), \quad H = L(W \times X, Y) \tag{3.1}$$

Banach spaces of bounded linear operators. Define a sequence of linear maps of $\Omega\Xi H$ into itself by

$$\tau_{n0}(\omega, \xi, \eta) \tag{3.2}$$

$$= (K_{n0}\omega, A_n\xi + M_{n0}\omega + N_{n0}\xi + O_{n0}\eta, B_n\xi + P_{n0}\omega + Q_{n0}\xi + R_{n0}\eta),$$

where $n = 1, 2,...$ and

$$K_{n0} = D_w U_n(w_{n-1}(w_0, x_0));$$

$$M_{n0}, N_{n0}, O_{n0} = D_w, D_x, D_y F_n(e_{n-1}(w_0, x_0)); \qquad (3.3)$$

$$P_{n0}, Q_{n0}, R_{n0} = D_w, D_x, D_y G_n(e_{n-1}(w_0, x_0));$$

so that τ_{n0} depends on parameters (w_0, x_0, σ). By Lemma 2.1, there exists a function

$$\eta = \gamma(\omega, \xi) = \gamma(\omega, \xi, w_0, x_0, \sigma) \qquad (3.4)$$

in $C^0(\Omega\Xi \times (WX)_r(0) \times \Sigma)$, such that for a fixed (w_0, x_0, σ) and $\rho > 0$, the restriction $\gamma \mid (\Omega\Xi)_\rho(0)$ gives the stable manifold $\mathscr{D}(\{\tau_{n0}\}, (\Omega\Xi H)_\rho(0))$.

THEOREM 3.1. *Under assumption* (H), *the stability set* $\mathscr{D}(\{T_n\}, E_r(0))$ *is a manifold* $y = y_0(w, x) = y_0(w, x, \sigma)$, *where* y_0 *satisfies the conclusions of Lemma* 2.1(a) *and has a Fréchet derivative* $D_{(w,x)} y_0$, *continuous in* (w, x, σ), *given by*

$$D_{(w,x)} y_0 = \gamma(id_W, id_X, w, x, \sigma). \qquad (3.5)$$

This will be deduced from Lemma 2.1 in the next section. In the case (1.0) or equivalently (1.1), we can obtain the existence of derivatives of y_0 with respect to some of the parameters on which it depends, and also higher order differentiability. In the next theorem, we assume that the parameter space Σ is of the form

$$\sigma = (v, \lambda) \in \Sigma = V^0 \times \Lambda, \qquad (3.6)$$

where V^0 is an open ball in a Banach space V and Λ is a metric space. We shall also assume that

$$U_n(w) = K_n w, \qquad K_n \in L(W, W), \qquad \| K_n \| \leqslant 1, \qquad (3.7)$$

$$K_n = K_n(\lambda) \text{ does not depend on } v. \qquad (3.8)$$

THEOREM 3.2_k. *Assume* (H), (3.6), (3.7), *and* (3.8). *Suppose that the map* T_n *in* (1.2) *is of class* C^k, $k \geqslant 1$, *as function of* (w, x, y, v), *that the derivatives of* A_n, B_n, F_n, G_n *are continuous functions of* (w, x, y, v, λ), *that the first order derivatives are bounded on* $\{small\ v\text{-}balls\} \times E_s(0) \times \Lambda \times \{n = 1, 2,...\}$ *for all* s, $0 < s < r$, *and that all derivatives of order* $\leqslant k$ *are bounded on* $\{small\ (v, w, x, y)\text{-}balls\} \times \Lambda \times \{n = 1, 2,...\}$. *Then* $y_0(w, x, v, \lambda)$ *is of class* C^k *with respect to* (w, x, v) *and its derivatives are continuous functions of* (w, x, v, λ).

Furthermore, if $e_0 = (w, x, y_0(w, x, v, \lambda)) \in E$ and $e_n = S_n(e_0)$ for $n = 1, 2,...$, then the first order derivatives $D_{(w,x,v)}e_n$ are bounded on

$$\{small\ v\text{-}balls\} \times E_s(0) \times \Lambda \times \{n = 0, 1,...\}, \quad 0 < s < r,$$

and all derivatives of order $\leqslant k$ are bounded on

$$\{small\ (w, x, v)\text{-}balls\} \times \Lambda \times \{n = 0, 1,...\}.$$

4. Proof of Theorem 3.1

Let $\epsilon > 0$ be arbitrarily small and $s = r - \epsilon$. We shall prove Theorem 3.1 on $E_s(0)$. From now on, assume that $(w_0, x_0) \in (WX)_s(0)$ and $(w, x) \in (WX)_1(0)$, so that $(w_0 + tw, x_0 + tx) \in (WX)_r(0)$ if $t \in T = [0, \epsilon]$. Using the notations of Section 2, let $e_0 = (w_0, x_0, y_0(w_0, x_0))$ and $e_n = S_n(e_0)$ for $n = 1, 2,...$. For $0 < t \leqslant \epsilon$, put

$$\bar{e}_n \equiv (\bar{w}_n, \bar{x}_n, \bar{y}_n) = [e_n(w_0 + tw, x_0 + tx) - e_n(w_0, x_0)]/t. \quad (4.1_n)$$

By (2.1), (2.2) in Lemma 2.1,

$$\| \bar{e}_n \| \leqslant 1 \quad \text{for} \quad n = 0, 1,... . \quad (4.2)$$

We can write

$$\bar{e}_n = T_n{}'(\bar{e}_{n-1}), \quad (4.3)$$

where $T_n{}'$ is the linear map

$$T_n{}'(\bar{w}, \bar{x}, \bar{y}) \quad (4.4)$$
$$= (K_n\bar{w}, A_n\bar{x} + M_n\bar{w} + N_n\bar{x} + O_n\bar{y}, B_n\bar{y} + P_n\bar{w} + Q_n\bar{x} + R_n\bar{y}),$$

in which

$$K_n = \int_0^1 D_w U_n\, d\theta; \qquad M_n, N_n, O_n = \int_0^1 D_w, D_x, D_y F_n\, d\theta;$$

$$P_n, Q_n, R_n = \int_0^1 D_w, D_x, D_y G_n\, d\theta; \quad (4.5)$$

and the argument of the integrands is

$$\theta e_n(w_0 + tw, x_0 + tx) + (1 - \theta)\, e_n(w_0, x_0).$$

Although $t > 0$ is required in (4.1), (4.2), we allow $t = 0$ in (4.5); so that the operators there do not depend on (w, x) at $t = 0$ and tend, as $t \to +0$, to their values (3.3) at $t = 0$ uniformly for $(w, x) \in (WX)_1(0)$, with (w_0 , x_0 , σ) fixed.

By Lemma 2.1, there is a function

$$\bar{y} = h(\bar{w}, \bar{x}) = h(\bar{w}, \bar{x}, w_0 , x_0 , \sigma, w, x, t)$$

defined on

$$WX \times (WX)_s(0) \times \varSigma \times (WX)_1(0) \times [0, \epsilon]$$

such that, for a fixed set of parameters,

$$\mathscr{D}(\{T_n{}'\}, E_\rho(0)) = \{(\bar{w}, \bar{x}, \bar{y}) : \bar{y} = h(\bar{w}, \bar{x}) \text{ on } (WX)_\rho(0)\}$$

for all $\rho > 0$.

One can associate with (4.4) a linear map of $\varOmega\varXi H$ into itself,

$$\tau_n(\omega, \xi, \eta) = (K_n\omega, A_n\xi + M_n\omega + N_n\xi + O_n\eta, B_n\eta + P_n\omega + Q_n\xi + R_n\eta). \tag{4.6}$$

Again, by Lemma 2.1, there is a function $\eta = \gamma(\omega, \xi)$, defined on $\varOmega\varXi$ and depending continuously on parameters $(w_0 , x_0 , \sigma, w, x, t)$, such that the stable manifold $\mathscr{D}(\{T_n\}, (\varOmega\varXi H)_\rho(0))$ is $\eta = \gamma \mid (\varOmega\varXi)_\rho(0)$ for any $\rho > 0$. At $t = 0$, (4.6) reduces to (3.2), (3.3) and is independent of (w, x). By Lemma 2.1,

$$\eta = \gamma(\omega, \xi) = \gamma(\omega, \xi, w_0 , x_0 , \sigma, w, x, t),$$

as an $H = L(WX, Y)$-valued function, is continuous in all of its variables and, as $t \to +0$,

$$\gamma(\omega, \xi, w_0 , x_0 , \sigma, w, x, t) \to \gamma(\omega, \xi, w_0 , x_0 , \sigma) = \gamma(\omega, \xi, w_0 , x_0 , \sigma, \omega, x, 0)$$

uniformly for $(w, x) \in (WX)_1(0)$, for fixed $(\omega, \xi, w_0 , x_0 , \sigma)$. In particular, if $(\omega, \xi) = (id_W , id_X)$ is fixed, then

$$\gamma(id_W , id_X , w_0 , x_0 , \sigma, w, x, t)(\bar{w}, \bar{x}) \to \gamma(id_W , id_X , w_0 , x_0 , \sigma)(\bar{w}, \bar{x}), \quad (4.7)$$

as $t \to +0$, uniformly for $(w, x), (\bar{w}, \bar{x}) \in (WX)_1(0)$, and fixed (w_0 , x_0 , σ).

On the one hand, it is clear that

$$h(\bar{w}, \bar{x}, w_0 , x_0 , \sigma, w, x, t) = \gamma(id_W , id_X , w_0 , x_0 , \sigma, w, x, t)(\bar{w}, \bar{x}). \quad (4.8)$$

On the other hand, (4.2), (4.3) imply that $\bar{e}_0 \in \mathscr{D}(\{T_n{}'\}, E_\rho(0)\}$ for $\rho > 1$; so that, by (4.1_0),

$$h(w, x, w_0 , x_0 , \sigma, w, x, t) = [y_0(w_0 + tw, x_0 + tx) - y_0(w_0 , x_0)]/t.$$

Thus, (4.7) and (4.8) give, as $t \to +0$,

$$[y_0(w_0 + tw, x_0 + tx) - y_0(w_0, x_0)]/t \to \gamma(id_W, id_X, w_0, x_0, \sigma)(w, x)$$

uniformly for $(w, x) \in (WX)_1(0)$, for fixed (w_0, x_0, σ). This proves Theorem 3.1.

5. Proof of Theorem 3.2₁

Since the assertions to be proved (that is, the existence and continuity of $D_{(w,x,v)} y_0$ near a point (w^1, x^1, v^1)) are local, we can restrict our attention to v near v^1. It will also be clear that the only $e = (w, x, y)$ which come into consideration are $e \in E_s(0)$, where $s \, (<r)$ is chosen so that $E_s(0)$ contains all of the iterates $e_n = e_n(w, x)$, $n = 1, 2, \dots$, for (w, x, v) near (w^1, x^1, v^1). Thus we can suppose that there is a constant C such that

$$\| D_v A_n \|, \| D_v B_n \|, \| D_v F_n \|, \| D_v G_n \| \leqslant C \tag{5.1}$$

on $E_s \times V^0 \times \Lambda$. By replacing v by a new variable $v^1 + \epsilon v$, we can suppose that C is arbitrarily small, that V^0 is centered at 0. We can also suppose that $V^0 = V_s(0)$.

In what follows, we consider the space W to be replaced by the space VW, and deal with a sequence of maps $T_{11}, T_{21}, \dots$, depending on a parameter λ, from $(VE)_s(0)$ into $VE = VWXY$,

$$T_{n1}(v, e) = (v, K_n w, A_{n1} x + F_{n1}, B_{n1} y + G_{n1}), \tag{5.2}$$

where the linear maps K_n, $A_{n1} = A_n(0, \lambda)$, and $B_{n1} = B_n(0, \lambda)$ depend only on λ, and

$$F_{n1}(v, e, \lambda) = F_n(e, v, \lambda) + [A_n(v, \lambda) - A_n(0, \lambda)]x,$$
$$G_{n1}(v, e, \lambda) = G_n(e, v, \lambda) + [B_n(v, \lambda) - B_n(0, \lambda)]y.$$

Note that the linear map $(v, w) \to (v, K_n w)$ has a norm 1, and that a Lipschitz constant for F_{n1}, G_{n1} with respect to (v, e) is $\delta^2 + 2C + Cs \equiv \delta'^2$. Since C is arbitrarily small, we can suppose that the inequalities (1.4) hold if δ is replaced by $\delta' > 0$.

It follows from Theorem 3.1 that the set $\mathscr{D}(\{T_{n1}\}, (VE)_s(0))$ is a manifold $y = y_0(w, x, v) = y_0(w, x, v, \lambda)$, $(w, x, v) \in (VWX)_s(0)$, and that $D_{(w,x,v)} y_0$ exists and is a continuous function of (w, x, v, λ). By the analogues of (2.1), (2.2), we also have that if $e_0 = (w, x, y_0(w, x, v))$ and $e_n = T_n(e_{n-1})$, so that $(v, e_n) = T_{n1}(v, e_{n-1})$, then $\| D_{(w,x,v)} e_n \| \leqslant 1$ on $(VE)_s(0) \times \Lambda$. This proves Theorem 3.2₁.

6. Proof of Theorem 3.2_k, $k > 1$

Let $k > 1$ and Theorem 3.2_{k-1} hold. For a moment, assume that no parameters v occur in T_n. Consider the map (3.2), (3.3), where $K_{n0} = K_n$ depends only on $\sigma = \lambda$. The map τ_{n0} depends on parameters (w_0, x_0, λ) and, on any given sphere $(\Omega\Xi H)_\rho(0)$, satisfies the conditions of Theorem 3.2_{k-1}, in which (w_0, x_0) plays the role of the parameter v. Thus, by Theorem 3.2_{k-1}, the function (3.4) is of class C^{k-1} with respect to (ω, ξ, w_0, x_0) with derivatives continuous in $(\omega, \xi, w_0, x_0, \lambda)$ and the specified boundedness properties. Theorem 3.2_k then follows, for the case under consideration, from (3.5).

In the case that parameters v do occur in T_n, apply the same arguments to the maps belonging to (5.2) in the same way that (3.2)-(3.3) belongs to (1.2). This completes the proof.

Notes

The idea of reducing the study of the behavior of solutions of ordinary differential equations to the study of maps is due to Poincaré; cf. e.g., VII Appendix. In particular, in connection with periodic solutions of autonomous systems, see Poincaré [3, IV] or [5, III, chap. XXVII]. In the latter context, he introduced the concept of invariant manifolds of a map (actually, "invariant curves" in his case). Poincaré's fundamental idea of studying maps in the theory of differential equations was further exploited by G. D. Birkhoff and has led to a large body of research associated with the name of "dynamical systems" or "topological dynamics."

SECTIONS 5-6. In the case where y_0, z_0 are 1-dimensional and the map T is analytic, Lemma 5.1 is due to Poincaré [3, IV, pp. 202-204] or [5, III, chap. XXVII]. The corresponding result where T is C^1 is due to Hadamard [2] (who, however, did not show that his invariant curve is of class C^1) and when T is C^1 (or as in Exercise 5.1), it is due to Sternberg [1]. D. C. Lewis [1] extended Hadamard's method to the case of arbitrary dimension when A, C have simple elementary divisors, carrying out the details in the analytic case. Lemma 5.1 was stated by Sternberg in [3] where he indicated a rather incomplete sketch of the proof based on the successive approximations (5.13). The proof in the text, using these approximations, is taken from Hartman [20] (cf. [28]). (The proof contained in Exercises 5.3, 5.4 may be new. These exercises give a simple proof of the assertion in Exercise 5.4(b) due to Coffman [2]; a weaker existence assertion and the uniqueness statement in Exercise 5.4(d) are contained in Perron [13]; Exercise 5.3(b) is in Coffman [2].). The analogous result for differential equations (i.e., Theorem 6.1) is due to Lyapunov [2, p. 291], under conditions of analyticity, and to Coddington and Levinson [2, p. 333] in the nonanalytic (and, more general, nonautonomous) case; cf. § X 8 and the reference there to Petrovsky [1].

SECTION 7. Theorem 7.1 with F, R, R^{-1} analytic was proved by Poincaré in 1879 (see [1, pp. xcix–cx]) under the assumptions that the elementary divisors of E are simple and that the eigenvalues $\lambda_1, \ldots, \lambda_d$ of E lie in an open half-plane, Re $e^{i\theta}\lambda > 0$, of the complex λ-plane and (*) $\lambda_j \neq m_1\lambda_1 + \cdots + m_d\lambda_d$ for all sets $(m_1, \ldots, m_d)$ of non-negative integers satisfying $m_1 + \cdots + m_d > 1$. For an analogous result for smooth, but nonanalytic F, R, R^{-1}, when Re $\lambda_j < 0$, see Sternberg [3]; cf. Exercise 8.2(b). For Re $\lambda_j < 0$, $F \in C^2$ and R, $R^{-1} \in C^1$, but without Diophantine conditions of type (*), see Hartman [20]; cf. Exercise 8.1 and 8.2(a). When it is not assumed that $\lambda_1, \ldots, \lambda_d$ lie in an open half-plane Re $e^{i\theta}\lambda > 0$, but Re $\lambda_j \neq 0$, the problem for analytic

F, R, R^{-1} has been considered by C. L. Siegel [2] under conditions stronger than (*). For smooth, but nonanalytic, F, R, R^{-1}, see Sternberg [4], Isé and Nagumo [1], and Chen [1]; cf. Appendix. Theorem 7.1 is due to Grobman [1], [2] and to Hartman [28], [21] with different proofs.

SECTIONS 8–9. The 1-dimensional case of Lemma 8.1 (i.e., y_0 is 1-dimensional and z_0 is absent) with T, R, R^{-1} analytic goes back to Abel [1, II, pp. 36–39]; Schröder [1], [2]; and Koenigs [1], [2]; see Picard [3, chap. IV]. For a similar treatment of the 1-dimensional, smooth (but nonanalytic) case, see Sternberg [2]. The case of a contraction (i.e., y_0 of arbitrary dimension d and z_0 missing) with T, R, R^{-1} analytic was treated by Leau [1] under the assumption that the eigenvalues $\alpha_1, \ldots, \alpha_d$ of A satisfy $\alpha_j \neq \alpha_1^{m_1}\alpha_2^{m_2}\ldots\alpha_d^{m_d}$ for all sets of non-negative integers $(m_1, \ldots, m_d)$ subject to $m_1 + \cdots + m_d > 1$. Lemma 8.1,* as stated, is due to Hartman [21], [28].

The papers of Sternberg [3], [4] and Hartman [20], [21], [28] mentioned in connection with § 7 are relevant here and, in fact, deal principally with maps. The device involving (9.1) which permits the deduction of Theorem 7.1 on differential equations from Lemma 8.1 on maps is due to Sternberg [3] and is used in these related papers. Sternberg [3] also considers the question of normal forms for RTR^{-1} different from the linear ones when there are relations $\alpha_j = \alpha_1^{m_1}\alpha_2^{m_2}\ldots\alpha_d^{m_d}$ and generalizes results of Lattés [1], [2] on 2-dimensional analytic maps. For related papers, see also Sternberg [5], Moser [1], C. L. Siegel [3], and Chen [1]; cf. Appendix. A recent important paper of J. Moser [3] is related to the critical case not considered here when some eigenvalues of the linear part of T have absolute value 1 and concerns the problem of the existence of closed invariant curves.

SECTIONS 10–11. As mentioned earlier, the principal results of these sections are due to Poincaré in the analytic (3-dimensional) case. Their validity in more general cases depends on the extensions of Poincaré's results on maps given in the earlier sections of the chapter.

APPENDIX (SECTIONS 12–14). Theorem 12.1 is due to Sternberg [4] and its generalization, Theorem 12.2, to Chen [1]. (Particular normal forms other than linear ones had also been considered by Lattés [1], [2] and by Sternberg [3].) The proof in the text for Theorem 12.2 is based on Chen [1] which, in turn, is a modification and simplification of Sternberg [4]. Chen's improvement consists essentially of noting that Sternberg's procedure is valid without first using the result in Sternberg [3] to obtain a linearization on the invariant manifolds. An admissible choice of $\lambda = N - n$ in Theorem 12.1 is given by Sternberg [3] for the case of contraction maps; for $n = 1$ and the general case of differential equations, see Isé and Nagumo [1]. Exercise 12.2 is due to Chen [1].

---

*The proof in the text is a simplification of that of Hartman [21], [28] due to Pugh [S1].

Chapter X

Perturbed Linear Systems

This chapter is concerned with methods for the asymptotic integrations of differential equations $\xi' = E\xi + F(t, \xi)$ which can be considered as perturbations of linear systems with constant coefficients $\xi' = E\xi$.

The first section of the chapter concerns the simple but important case $E = 0$. Since a very easy argument, which has wide applications, gives the desired result in this case, it seems worth isolating it.

One of the most important methods to be used for an arbitrary E is based on a simple topological principle, discussed in §§ 2–3. This principle has wide applications beyond the scope of this chapter. A very different method for obtaining results analogous to those of §§ 13 and 16 is discussed in Part III of Chapter XII.

In this chapter, for convenience and generality, we shall allow the components of ξ to be complex-valued, so that a linear change of coordinates permits the assumption that E is in a suitable normal form; cf. § IV 9. Correspondingly, if ξ_1, ξ_2 are two vectors, then $\xi_1 \cdot \xi_2$ denotes the scalar product $\sum_k \xi_1{}^k \bar{\xi}_2{}^k$.

1. The Case $E = 0$

This section concerns the equation

$$(1.1) \qquad \xi' = F(t, \xi),$$

where F is "small" in a suitable sense. The main results are the following:

Theorem 1.1. *Let $F(t, \xi)$ be continuous for $t \geqq 0$, $\|\xi\| < \delta\,(\leqq \infty)$ and satisfy*

$$(1.2) \qquad \|F(t, \xi)\| \leqq \psi(t)\,\|\xi\|,$$

where $\psi(t)$ is a continuous function for $t \geqq 0$ such that

$$(1.3) \qquad \int_0^\infty \psi(t)\,dt < \infty.$$

273

If $\|\xi_0\|$ is sufficiently small, say

(1.4)
$$\|\xi_0\| \exp \int_0^\infty \psi(t)\, dt < \delta,$$

then a solution $\xi(t)$ of (1.1) satisfying $\xi(0) = \xi_0$ exists for $t \geq 0$. Furthermore, if $\xi(t)$ is a solution of (1.1) for large t, say $t \geq t_0$, then

(1.5)
$$\xi_\infty = \lim_{t \to \infty} \xi(t)$$

exists and $\xi_\infty \neq 0$ unless $\xi(t) \equiv 0$.

In other words, the solutions of (1.1) for large t behave like the solutions of $\xi' = 0$, namely, like constants. Theorem 1.1 has the following extension:

Theorem 1.2. *Let $F(t, \xi)$ be as in Theorem 1.1, and let ξ_∞ be an arbitrary vector such that*

(1.6)
$$\|\xi_\infty\| \exp \int_0^\infty \psi(t)\, dt < \delta\ (\leq \infty).$$

Then (1.1) has at least one solution $\xi(t)$ for $t \geq 0$ satisfying (1.5). If, in addition, $F(t, \xi)$ satisfies the following type of Lipschitz condition

(1.7)
$$\|F(t, \xi_1) - F(t, \xi_2)\| \leq \psi(t)\, \|\xi_1 - \xi_2\|,$$

then for a given ξ_∞, there is at most one solution $\xi(t)$ of (1.1) which exists for large t and satisfies (1.5).

The last part of Theorem 1.2 states that condition (1.7) establishes a one-to-one correspondence between solutions $\xi = \xi_\infty = \text{const.}$ of $\xi' = 0$ and solutions of (1.1), with the understanding that $\|\xi_\infty\|$ is sufficiently small when $\delta < \infty$.

Proof of Theorem 1.1. Multiply (1.1) scalarly by ξ, so that (1.1), (1.2) imply that

(1.8)
$$|\xi \cdot \xi'| \leq \psi(t)\, \|\xi\|^2.$$

Since $d\,\|\xi(t)\|^2/dt = 2\, \text{Re}\ \xi \cdot \xi'$, a quadrature gives

(1.9) $$\|\xi(t_0)\| \exp -\left| \int_{t_0}^t \psi(s)\, ds \right| \leq \|\xi(t)\| \leq \|\xi(t_0)\| \exp \left| \int_{t_0}^t \psi(s)\, ds \right|,$$

if $\xi(t)$ exists on a t-interval containing t and t_0, where $t \gtrless t_0$. In particular, if $t_0 = 0$ and $\xi(0) = \xi_0$ satisfies (1.4), then $\xi(t)$ exists for $t \geq 0$.

More generally, if $\xi(t)$ exists for $t \geq t_0$, then it is bounded,

(1.10) $\|\xi(t)\| \leq \|\xi(t_0)\|\, M(t_0),$ where $M(t_0) = \exp \int_{t_0}^\infty \psi(t)\, dt.$

Hence, (1.1), (1.2) show that $\|\xi'(t)\| \leq \psi(t)\, \|\xi(t_0)\|\, M(t_0)$. Consequently,

$\int^{\infty} \xi'(t)\,dt$ is absolutely convergent and so the limit (1.5) exists. In fact

$$(1.11) \qquad \|\xi(t) - \xi_\infty\| \leq \|\xi(t_0)\|\,M(t_0) \int_t^{\infty} \psi(s)\,ds \qquad \text{for} \quad t \geq t_0.$$

Note that the inequality (1.9) shows that $\xi(t) \equiv 0$ if and only if $\xi(t)$ vanishes at some point t_0. When $\xi(t) \not\equiv 0$, the first inequality in (1.9) implies that $\xi_\infty \neq 0$. This proves Theorem 1.1.

Proof of Theorem 1.2. Consider first the existence assertion. For a given $t_0 \geq 0$, let $\xi = \xi(t, t_0)$ be a fixed solution (which is not necessarily unique) of the initial value problem

$$(1.12) \qquad \xi' = F(t, \xi), \qquad \xi(t_0) = \xi_\infty.$$

Since (1.9) holds for any t at which $\xi(t, t_0)$ exists, it follows from (1.6) and $\xi(t_0) = \xi_\infty$, that $\xi(t, t_0)$ exists for $t \geq 0$. Also, $\|\xi(t, t_0)\| \leq \|\xi_\infty\|\,M(0) < \delta$, where $M(0)$ is defined in (1.10). Hence (1.2) shows that $\|\xi'(t, t_0)\| \leq \psi(t)\,\|\xi_\infty\|\,M(0)$ for all $t \geq 0$; thus, for $0 \leq t \leq t_0$,

$$(1.13) \qquad \|\xi(t, t_0) - \xi_\infty\| \leq M(0)\,\|\xi_\infty\| \int_t^{t_0} \psi(s)\,ds.$$

In particular, the family of functions $\xi(t, t_0)$ are uniformly bounded and equicontinuous on every bounded t-interval. Hence there exists a sequence $t_1 < t_2 < \ldots$ of t_0-values such that $t_n \to \infty$ as $n \to \infty$, and

$$\xi(t) = \lim \xi(t, t_n)$$

exists uniformly on every bounded t-interval. Furthermore $\xi = \xi(t)$ is a solution of (1.1). Putting $t_0 = t_n$ in (1.13) and letting $n \to \infty$, with t fixed, gives

$$(1.14) \qquad \|\xi(t) - \xi_\infty\| \leq M(0)\,\|\xi_\infty\| \int_t^{\infty} \psi(s)\,ds.$$

This implies (1.5) and completes the existence proof.

Uniqueness will now be proved under the assumption (1.7). Let $\xi = \xi_1(t), \xi_2(t)$ be two solutions of (1.1) for large t, say $t \geq T$, satisfying (1.5). Let $\xi(t) = \xi_1(t) - \xi_2(t)$. Then (1.1) and (1.7) give (1.8), hence (1.9) for $t_0 \geq t \geq T$. If t is fixed and $t_0 \to \infty$ in (1.9), it follows that $\xi(t) = 0$ since $\xi(t_0) \to 0$ as $t_0 \to \infty$. This completes the proof of Theorem 1.2.

The majorant $\psi(t)\,\|\xi\|$ in (1.2) involving a factor $\|\xi\|$ is convenient in Theorems 1.1 and 1.2 only to assure that certain solutions exist for $t \geq 0$. A simpler result involving existence for "large t" is given in the following exercise.

Exercise 1.1. Let $F(t, \xi)$ be continuous on a product set $\{t \geq 0\} \times D$, where D is a bounded open ξ-set. Let F satisfy $\|F(t, \xi)\| \leq \psi(t)$, $t \geq 0$

and $\xi \in D$, for some continuous function $\psi(t)$ satisfying (1.3). (a) Let $\xi_0 \in D$. Then there exists a number T, depending only on dist $(\xi_0, \partial D)$ and the function $\psi(t)$, such that if $t_0 \geqq T$, then a solution $\xi(t)$ of (1.1) satisfying $\xi(t_0) = \xi_0$ exists for $t \geqq T$. Furthermore, any solution $\xi(t)$ of (1.1) for large t has a limit ξ_∞ as $t \to \infty$. (b) Let $\xi_\infty \in D$. Then there exists a number T, depending only on dist $(\xi_\infty, \partial D)$ and the function $\psi(t)$, such that (1.1) has a solution $\xi(t)$ for $t \geqq T$ satisfying (1.5).

Exercise 1.2. Show that   the solutions of (1.1)   exist for $t \geqq 0$ and are bounded if $F(t, \xi)$ is continuous for $t \geqq 0$ and all ξ and if (1.2) is replaced by

$$(1.2')\qquad\qquad |\xi \cdot F(t, \xi)| \leqq \psi(t)\varphi(\|\xi\|^2),$$

where $\psi(t)$ is as in Theorem 1.1 and $\varphi(r)$ is continuous for $r \geqq 0$ and satisfies

$$(1.3')\qquad\qquad \int^\infty \frac{dr}{\varphi(r)} = \infty.$$

Theorem 1.1 and 1.2 have corollaries for the case that (1.1) is replaced by

$$(1.15)\qquad\qquad \zeta' = A(t)\zeta + G(t, \zeta),$$

where $A(t)$ is a continuous $d \times d$ matrix. Here, solutions of (1.15) should be compared with

$$(1.16)\qquad\qquad \zeta' = A(t)\zeta.$$

Let $Z(t)$ be a fundamental matrix for (1.16), so that the change of variables

$$(1.17)\qquad\qquad \zeta = Z(t)\xi$$

transforms (1.15) into

$$(1.18)\qquad\qquad \xi' = Z^{-1}(t)G(t, Z(t)\xi).$$

Thus an application of Theorems 1.1 and 1.2 to (1.18) gives

Corollary 1.1. *Let $A(t)$ be continuous for $t \geqq 0$ and $Z(t)$ a fundamental matrix for (1.16). Let $G(t, \zeta)$ be continuous for $t \geqq 0$ and all ζ and satisfy*

$$(1.19)\qquad\qquad \|Z^{-1}(t)G(t, Z(t)\xi)\| \leqq \psi(t) \|\xi\|,$$

where $\psi(t)$ is as in Theorem 1.1. *Let $\zeta(t)$ be a solution of* (1.15) *on some t-interval. Then $\zeta(t)$ exists for $t \geqq 0$,*

$$(1.20)\qquad\qquad \xi_\infty = \lim_{t \to \infty} Z^{-1}(t)\zeta(t)$$

exists and $\xi_\infty \neq 0$ unless $\zeta(t) \equiv 0$; conversely, for a given ξ_∞, there is a solution $\zeta(t)$ of (1.15) *satisfying* (1.20).

When $Z(t)$ is bounded for $t \geq 0$, we can formulate a corresponding result when $G(t, \zeta)$ is only defined for $t \geq 0$, $\|\zeta\| < \delta < \infty$. In addition, we can obtain an analogue of the uniqueness assertion of Theorem 1.2. When $A(t) \equiv A$ is a constant matrix, Corollary 1.1 takes the following form:

Corollary 1.2. *Let $G(t, \zeta)$ be continuous for $t \geq 0$ and all ζ and satisfy*

$$(1.21) \qquad \|e^{-At}G(t, e^{At}\xi)\| \leq \psi(t) \|\xi\|,$$

where $\psi(t)$ is as in Theorem 1.1. Let $\zeta(t)$ be a solution of

$$(1.22) \qquad \zeta' = A\zeta + G(t, \zeta)$$

on some t-interval. Then $\zeta(t)$ exists for $t \geq 0$,

$$(1.22^*) \qquad \xi_\infty = \lim e^{-At}\zeta(t)$$

exists and $\xi_\infty \neq 0$ unless $\zeta(t) \equiv 0$; furthermore, if ξ_∞ is given, there is a solution of (1.22) for $t \geq 0$ satisfying (1.22^*).

Exercise 1.3.   Formulate theorems related to Corollaries 1.1 and 1.2 as Exercises 1.1 and 1.2 are related to Theorems 1.1 and 1.2.

Generally, a result of the type given by Corollary 1.2 is only convenient when $e^{\pm At}$ are bounded for $t \geq 0$. For example, suppose that $d = 2$ and $A = \text{diag } [1, -1]$, so that $e^{At} = \text{diag } [e^t, e^{-t}]$. Then, if Corollary 1.2 is applicable, (1.22) has a solution of the form $\zeta = e^t(1 + o(1), o(1))$ as $t \to \infty$, but not necessarily a solution of the form $\zeta = e^{-t}(o(1), 1 + o(1))$ as $t \to \infty$. Furthermore the hypothesis (1.21) can be very severe for the type of conclusion stated in Corollary 1.2. The results obtained in the remainder of this chapter are much better, under less stringent conditions, for the situation just described.

Exercise 1.4.   Suppose that (1.1) is a linear homogeneous system, say

$$(1.23) \qquad \xi' = G(t)\xi,$$

where $G(t)$ is a continuous matrix for $t \geq 0$. The system (1.23) will be said to be of class (*) if (i) every solution $\xi(t)$ of (1.23) has a limit ξ_∞ as $t \to \infty$, and (ii) for every constant vector ξ_∞, there is a solution $\xi(t)$ of (1.23) such that $\xi(t) \to \xi_\infty$ as $t \to \infty$. (*a*) Show that (1.23) is of class (*) if and only if, for one and/or every fundamental matrix $Y(t)$ of (1.23), $Y_\infty = \lim Y(t)$ exists as $t \to \infty$ and is nonsingular (and that this is true if and only if $Y_\infty = \lim Y(t)$ exists as $t \to \infty$ and $\int^\infty \text{tr } G(s) \, ds$ converges, possibly conditionally). (*b*) The system (1.23) is of class (*) if and only if the adjoint system $\xi' = -G^*(t)\xi$ is class (*); cf. § IV 7. (*c*) The system

(1.23) is of class (*) if $\int^{\infty} \|G(t)\| \, dt < \infty$ [or, equivalently, if $G(t) = (g_{jk}(t))$ and $\int^{\infty} |g_{jk}(t)| \, dt < \infty$ for $j, k = 1, \ldots, d$]. This is merely a consequence of Theorems 1.1, 1.2. (d) Show that (c) has the following corollary [which is a refinement of (c)]: The system (1.23) is of class (*) if $G_0(t) = \int_t^{\infty} G(s) \, ds$ converges (possibly just conditionally) and either $\int^{\infty} \|G(t)G_0(t)\| \, dt < \infty$ or $\int^{\infty} \|G_0(t)G(t)\| \, dt < \infty$.

2. A Topological Principle

Let y, f be d-dimensional vectors with real- or complex-valued components and $f(t, y)$ a continuous function defined on an open (t, y)-set Ω. Let Ω^0 be an open subset of Ω, $\partial\Omega^0$ the boundary and $\bar{\Omega}^0$ the closure of Ω^0. Recall, from § III 8, that a point $(t_0, y_0) \in \Omega \cap \partial\Omega^0$ is called an egress point of Ω^0, with respect to the system

$$(2.1) \qquad\qquad y' = f(t, y),$$

if for every solution $y = y(t)$ of (2.1) satisfying the initial condition

$$(2.2) \qquad\qquad y(t_0) = y_0,$$

there is an $\epsilon > 0$ such that $(t, y(t)) \in \Omega^0$ for $t_0 - \epsilon \leq t < t_0$. An egress point (t_0, y_0) of Ω^0 is called a strict egress point of Ω^0 if $(t, y(t)) \notin \bar{\Omega}^0$ for $t_0 < t \leq t_0 + \epsilon$ for a small $\epsilon > 0$. The set of egress points of Ω^0 will be denoted by $\Omega_e^{\,0}$ and the set of strict egress points by Ω_{se}^0.

If U is a topological space and V a subset of U, a continuous mapping $\pi : U \to V$ defined on all of U is called a *retraction* of U onto V if the restriction $\pi \mid V$ of π to V is the identity; i.e., $\pi(u) \in V$ for all $u \in U$ and $\pi(v) = v$ for all $v \in V$. When there exists a retraction of U onto V, V is called a *retract of U*. This notion can be illustrated by the following examples, which will have applications.

Example 1. Let U be a d-dimensional ball $\|y\| \leq r$ in the Euclidean y-space and V its boundary sphere $\|y\| = r$. Then V is not a retract of U. For if there exists a retraction $\pi : U \to V$, then there exists a map of U into itself, $y \to -\pi(y)$, without fixed points, which is impossible by the classical fixed point theorem of Brouwer; for the latter, see Hureiwicz and Wallman [1, pp. 40–41].

Example 2. Let C be the "cylinder" which is the product space of a Euclidean sphere $\|y\| = r$ and a Euclidean u-space, so that $C = \{(y, u): \|y\| = r, u \text{ arbitrary}\}$. Let S be a section of C, say, $S = \{(y, u_0): \|y\| \leq r, u_0 \text{ fixed}\}$; see Figure 1. Then $S \cap C = \{(y, u_0): \|y\| = r,$

u_0 fixed} is a retract of C [as can be seen by choosing the retraction $\pi(y, u) = (y, u_0)$], but $S \cap C$ is not a retract of S by Example 1.

Theorem 2.1. *Let $f(t, y)$ be continuous on an open (t, y)-set Ω with the property that initial values determine unique solutions of (2.1). Let Ω^0 be*

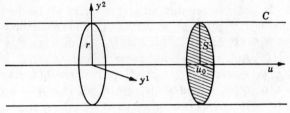

Figure 1.

an open subset of Ω satisfying $\Omega_e{}^0 = \Omega_{se}^0$; i.e., all egress points of Ω^0 are strict egress points. Let S be a nonempty subset of $\Omega^0 \cup \Omega_e{}^0$ such that $S \cap \Omega_e{}^0$ is not a retract of S but is a retract of $\Omega_e{}^0$. Then there exists at least one point $(t_0, y_0) \in S \cap \Omega^0$ such that the solution arc $(t, y(t))$ of (2.1), (2.2) is contained in Ω^0 on its right maximal interval of existence.

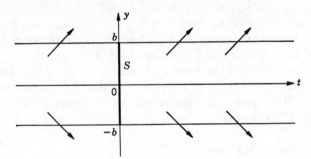

Figure 2.

As an illustration, consider (2.1) where y is a real variable and $f(t, y)$ is continuous on Ω: (t, y) arbitrary. Let Ω^0 be a strip $|y| < b$, $-\infty < t < \infty$; see Figure 2. Thus the part of the boundary of Ω^0 in Ω, i.e., $\partial\Omega^0 \cap \Omega$, consists of the two lines $y = \pm b$. Suppose that $f(t, b) > 0$ and $f(t, -b) < 0$, so that $\Omega_e{}^0 = \Omega_{se}^0 = \partial\Omega^0 \cap \Omega$. Let S be the line segment $S = \{(t, y): t = 0, |y| \leqq b\}$. Then $S \cap \Omega_e{}^0$ is the set of two points $(0, \pm b)$ and is a retract of $\Omega_e{}^0$ but not of S. Thus it follows from Theorem 2.1, that there exists at least one point $(0, y_0)$, $|y_0| < b$, such that a solution of (2.1) determined by $y(0) = y_0$ exists and satisfies $|y(t)| < b$ for $t \geqq 0$.

Proof of Theorem 2.1. Suppose that the theorem is false. Then for $(t_0, y_0) \in S - \Omega_e{}^0$, there exists a $t_1 = t_1(t_0, y_0)$ such that $t_1 > t_0$ and the solution $y(t)$ of (2.1), (2.2) exists on $t_0 \leq t \leq t_1$, $(t, y(t)) \in \Omega^0$ for $t_0 \leq t < t_1$ and $(t_1, y(t_1)) \in \Omega_e^0$ for $t = t_1$. Define a map $\pi_0 \colon S \to \Omega_e^0$ as follows: $\pi_0(t_0, y_0) = (t_1, y(t_1))$ if $(t_0, y_0) \in S - \Omega_e^0$ and $\pi_0(t_0, y_0) = (t_0, y_0)$ if $(t_0, y_0) \in S \cap \Omega_e^0$. Since the solutions of (2.1) depend continuously on initial conditions (Theorem V 2.1) and $\Omega_e^0 = \Omega_{se}^0$, it follows that π_0 is continuous. In order to see this, let $y(t) = \eta(t, t^0, y^0)$ be the solution of (2.1) such that $\eta(t^0, t^0, y^0) = y^0$, so that $\eta(t, t^0, y^0)$ is continuous. Suppose that $(t_0, y_0) \in S \cap \Omega^0$, and (t^0, y^0) is near (t_0, y_0), then $\eta(t, t^0, y^0)$ exists on the interval $[t^0, t_1(t_0, y_0) + \epsilon]$ for some ϵ and $(t, \eta(t, t^0, y^0)) \in \Omega^0$ on $t^0 \leq t \leq t_1(t_0, y_0) - \epsilon$ and $(t, \eta(t, t^0, y^0)) \notin \bar{\Omega}^0$ if $t = t_1(t_0, y_0) + \epsilon$. Thus, $|t_1(t^0, y^0) - t_1(t_0, y_0)| < \epsilon$, and so $(t_1, \eta(t_1(t^0, y^0), t^0, y^0))$ is a continuous function of (t^0, y^0); i.e., π_0 is continuous at (t_0, y_0). A similar argument holds if $(t_0, y_0) \in S \cap \Omega_e^0$.

Let $\pi \colon \Omega_e^0 \to S \cap \Omega_e^0$ be a retraction of Ω_e^0 onto $S \cap \Omega_e^0$. Then the composite map $\pi\pi_0$ is a retraction of S onto $S \cap \Omega_e^0$. The existence of such a retraction gives a contradiction and proves the theorem.

Exercise 2.1. Let U be a topological space; V_1, V_2 subsets of U. The set V_1 is called a *quasi-isotopic deformation retract* of V_2 in U if there exists a continuous map $\pi \colon V_2 \times \{0 \leq s \leq 1\} \to U$ such that (i) $\pi(v_2, 0) = v_2$ for $v_2 \in V_2$; (ii) $\pi(v_1, s) = v_1$ for $v_1 \in V_1$ and $0 \leq s \leq 1$; (iii) $\pi(v_2, 1) \in V_1$ for $v_2 \in V_2$; and (iv) for fixed s on $0 \leq s < 1$, $\pi(v_2, s)$ is a homeomorphism of V_2 onto its image. Let f, Ω, Ω^0 be as in Theorem 2.1; S_1 a subset of Ω_e^0; S a nonempty subset of $\Omega^0 \cup S_1$ such that S_1 is not a quasi-isotopic deformation retract of $S \cup S_1$ in $\Omega^0 \cup S_1$. Then there exists at least one point $(t_0, y_0) \in S \cap \Omega$ such that the solution arc $(t, y(t))$ of (2.1), (2.2) is either in Ω^0 on its right maximal interval of existence or first meets $\partial\Omega^0$ at a point of $S - S_1$.

3. A Theorem of Ważewski

The usefulness of Theorem 2.1 depends on suitable choices of Ω^0. One of the difficulties in the application is the determination of the set of egress points. In some cases to be described, this difficulty can be overcome.

Recall from § III 8 that a real-valued function $u(t, y)$ defined on an open subset of Ω is said to possess a trajectory derivative $\dot{u}(t, y)$ at the point (t_0, y_0) along the solution $y(t)$ of (2.1)–(2.2) if $u(t, y(t))$ has a derivative at $t = t_0$; in this case,

$$(3.1) \qquad \dot{u}(t_0, y_0) = [u(t, y(t))]'_{t=t_0}.$$

If y (hence f) has real-valued components and $u(t, y)$ is of class C^1, this trajectory derivative exists and is

$$(3.2) \qquad \dot{u}(t, y) = \partial u/\partial t + (\mathrm{grad}_y u) \cdot f,$$

where the last term is the scalar product of f and the gradient of u with respect to y.

When y has complex-valued components, a function $u(t, y)$ is said to be of class C^1 if it has continuous partial derivatives with respect to t and the real and imaginary parts of the components of y. Write the kth component y^k of y as $y^k = \sigma^k + i\tau^k$, where σ^k, τ^k are real, so that $\sigma^k = (y^k + \bar{y}^k)/2$, $\tau^k = (y^k - \bar{y}^k)/2i$. This suggests the standard notation, $\partial u/\partial y^k = \frac{1}{2}[\partial u/\partial \sigma^k - i\,\partial u/\partial \tau^k]$ and $\partial u/\partial \bar{y}^k = \frac{1}{2}[\partial u/\partial \sigma^k + i\,\partial u/\partial \tau^k]$. Thus if $\mathrm{grad}_y\, u = (\partial u/\partial y^1, \ldots, \partial u/\partial y^d)$ and $\mathrm{grad}_{\bar{y}}\, u = (\partial u/\partial \bar{y}^1, \ldots, \partial u/\partial \bar{y}^d)$, then (3.2) should be replaced by

$$(3.2^*) \qquad \dot{u}(t, y) = \partial u/\partial t + (\mathrm{grad}_y\, u)\cdot \bar{f} + (\mathrm{grad}_{\bar{y}}\, u)\cdot f,$$

as can be seen by writing (2.1) as a system of $2d$ differential equations for $(\sigma, \tau) = (\sigma^1, \ldots, \sigma^d, \tau^1, \ldots, \tau^d)$.

An open subset Ω^0 of Ω will be called a (u, v)-*subset* of Ω with respect to (2.1) if there exists an (arbitrary) number of real-valued continuous functions, $u_1(t, y), \ldots, u_l(t, y), v_1(t, y), \ldots, v_m(t, y)$, on Ω such that

$$(3.3) \qquad \Omega^0 = \{(t, y): u_j(t, y) < 0 \quad \text{and} \quad v_k(t, y) < 0 \text{ for all } j, k\}$$

and if U_α, V_β are the sets

$$U_\alpha = \{(t, y): u_\alpha(t, y) = 0 \text{ and } u_j(t, y) \leqq 0, \ v_k(t, y) \leqq 0 \text{ for all } j, k\},$$
$$(3.4)$$
$$V_\beta = \{(t, y): v_\beta(t, y) = 0 \text{ and } u_j(t, y) \leqq 0, \ v_k(t, y) \leqq 0 \text{ for all } j, k\},$$

then the trajectory derivatives $\dot{u}_\alpha$, $\dot{v}_\beta$ exist on U_α, V_β and satisfy

$$(3.5) \qquad \dot{u}_\alpha(t, y) > 0 \qquad \text{for} \quad (t, y) \in U_\alpha,$$

$$(3.6) \qquad \dot{v}_\beta(t, y) < 0 \qquad \text{for} \quad (t, y) \in V_\beta,$$

respectively, along all solutions through (t, y). In this definition, either l or m can be zero.

Lemma 3.1. *Let $f(t, y)$ be continuous on an open (t, y)-set Ω and Ω^0 a (u, v)-subset of Ω with respect to (2.1). Then*

$$(3.7) \qquad \Omega_e^{\,0} = \Omega_{se}^{0} = \bigcup_{\alpha=1}^{l} U_\alpha - \bigcup_{\beta=1}^{m} V_\beta.$$

Proof. It is clear that $\partial\Omega^0 \cap \Omega \subset (\bigcup U_\alpha) \cup (\bigcup V_\beta)$. In addition, $\Omega_e^{\,0} \cap V_\beta$ is empty, for if $(t_0, y_0) \in V_\beta$ and $y(t)$ is a solution of (2.1), (2.2), then (3.6) shows that $v_\beta(t, y(t)) > 0$ for $t_0 - \epsilon \leqq t < t_0$ for small $\epsilon > 0$, so that $(t, y(t)) \notin \Omega^0$. Thus

$$(3.8) \qquad \Omega_{se}^{0} \subset \Omega_e^{\,0} \subset (\partial\Omega^0 \cap \Omega) - \bigcup_{\beta=1}^{m} V_\beta \subset \bigcup_{\alpha=1}^{l} U_\alpha - \bigcup_{\beta=1}^{m} V_\beta.$$

Let $(t_0, y_0) \in \bigcup U_\alpha - \bigcup V_\beta$. Then $u_j(t_0, y_0) \leqq 0$ and $v_k(t_0, y_0) < 0$ for all j, k. By (3.5), there is an $\epsilon > 0$ such that $u_\alpha(t, y(t)) < 0$ or > 0 according as $t_0 - \epsilon \leqq t < t_0$ or $t_0 < t \leqq t_0 + \epsilon$ if $(t_0, y_0) \in U_\alpha$, $u_j(t, y(t)) < 0$ for $t_0 - \epsilon \leqq t \leqq t_0 + \epsilon$ if $(t_0, y_0) \notin U_j$, and $v_k(t, y(t)) < 0$ for $t_0 - \epsilon \leqq t \leqq t_0 + \epsilon$ for all k. Hence $(t_0, y_0) \in \Omega_{se}^0$; i.e., $\bigcup U_\alpha - \bigcup V_\beta \subset \Omega_{se}^0$. In view of (3.8), this proves the lemma.

Theorem 3.1. *Let $f(t, y)$ be continuous on an open (t, y)-set Ω with the property that solutions of (2.1) are uniquely determined by initial conditions. Let Ω^0 be a (u, v)-subset of Ω with respect to (2.1). Let S be a nonempty subset of $\Omega^0 \cup \Omega_e^0$ satisfying $S \cap \Omega_e^0$ is not a retract of S but is a retract of Ω_e^0. Then there exists at least one point $(t_0, y_0) \in S \cap \Omega^0$ such that a solution arc $(t, y(t))$ of (2.1), (2.2) is contained in Ω^0 on its right maximal interval of existence.*

This is a corollary of Theorem 2.1 and Lemma 3.1. Sometimes, the requirement of the uniqueness of the solution of (2.1), (2.2) can be omitted:

Corollary 3.1. *Let f, Ω, Ω^0, S be as in Theorem 3.1. except that it is not required that solutions of (2.1) be uniquely determined by initial conditions. But, in addition, let S be compact and let $u_j(t, y), v_k(t, y)$ be of class C^1 (with respect to t and the real and imaginary parts of the components of y). Then the conclusion of Theorem 3.1 is valid.*

Proof. Let $f_1(t, y), f_2(t, y), \ldots$ be a sequence of functions of class C^1 on Ω which tend to $f(t, y)$ uniformly, as $n \to \infty$, on compact subsets of Ω. Let $\Omega_1, \Omega_2, \ldots$ be a sequence of open subsets of Ω, such that $S \subset \Omega_1$, Ω_n has a compact closure $\bar{\Omega}_n \subset \Omega_{n+1}$, and $\Omega = \bigcup \Omega_n$.

By replacing $f_1, f_2, \ldots$ by a subsequence, if necessary, it can be supposed that

$$\partial u_\alpha / \partial t + (\mathrm{grad}_y u_\alpha) \cdot f_n + (\mathrm{grad}_{\bar{y}} u_\alpha) \cdot f_n > 0 \qquad \text{on} \quad U_\alpha \cap \Omega_n,$$
$$\partial v_\beta / \partial t + (\mathrm{grad}_y v_\beta) \cdot f_n + (\mathrm{grad}_{\bar{y}} v_\beta) \cdot f_n < 0 \qquad \text{on} \quad V_\beta \cap \Omega_n.$$

Thus if $\Omega_n^0 = \Omega^0 \cap \Omega_n$, then Ω_n^0 is a (u, v)-subset of Ω_n with respect to the system

$$(3.9) \qquad\qquad y' = f_n(t, y).$$

The set of (strict) egress points Ω_{ne}^0 of Ω_n^0 is $\Omega_e^0 \cap \Omega_n$. Hence $\Omega_{ne}^0 \cap S = \Omega_e^0 \cap S$ is not a retract of S, but $\Omega_{ne}^0 \cap S$ is a retract of $\Omega_{ne}^0 \subset \Omega_e^0$.

Thus by Theorem 2.1 there is a point $(t_n, y_n) \in S$, such that the solution $y = y_n(t)$ of (3.9) satisfying $y_n(t_n) = y_n$ is in Ω_n^0 on its right maximal interval of existence $[t_n, \tau_n)$ relative to Ω_n. If (3.9) is considered on Ω, instead of Ω_n, let the right maximal interval of existence of $y_n(t)$ be $[t_n, \omega_n)$, so that $\tau_n \leqq \omega_n \leqq \infty$ and $\tau_n < \omega_n$ implies that $(\tau_n, y_n(\tau_n)) \subset \partial \Omega_n \cap \Omega$.

Since S is compact, there is a point (t_0, y_0) on S which is a cluster point of the sequence of points (t_1, y_1), (t_2, y_2), ... By Theorem II 3.2, there exists a solution $y(t)$ of (2.1), (2.2) having a right maximal interval of existence $[t_0, \omega)$ and a sequence of integers $n(1) < n(2) < \cdots$ such that $y_{n(k)}(t) \to y(t)$ uniformly as, $k \to \infty$, on any interval $[t_0, t^*] \subset [t_0, \omega)$.

It follows that $(t, y(t)) \subset \bar{\Omega}^0 \cap \Omega$ for $t_0 \leqq t < \omega$. For suppose that there is a t-value t^0, $t_0 < t^0 < \omega$, such that $(t^0, y(t^0)) \notin \bar{\Omega}^0$. Then, for $n = n(k)$ and large k, $(t^0, y_n(t^0)) \notin \bar{\Omega}^0$, so that $(t^0, y_n(t^0)) \notin \bar{\Omega}_n^0$. Hence $\tau_n < t^0 < \omega_n$ for $n = n(k)$ and large k. By choosing a subsequence, if necessary, it can be supposed that $\tau = \lim \tau_{n(k)}$ exists as $k \to \infty$, so that $t_0 \leqq \tau \leqq t^0$ and $(\tau_n, y_n(\tau_n)) \to (\tau, y(\tau))$ as $n = n(k) \to \infty$. But this gives a contradiction for $(\tau_n, y_n(\tau_n)) \in \partial\Omega_n \cap \Omega$, where $n = n(k)$, cannot have a limit point $(\tau, y(\tau)) \in \Omega$.

Since $(t_0, y_0) \in S \subset \Omega^0 \cap \Omega_e^0$ and $\Omega_e^0 = \Omega_{se}^0$, it is seen that $(t_0, y_0) \in \Omega^0$, otherwise $(t, y(t)) \notin \Omega^0$ for $t_0 < t \leqq t_0 + \epsilon$ for some $\epsilon > 0$. By the same argument, $(t, y(t)) \subset \Omega^0$ for $t_0 \leqq t < \omega$. This proves the corollary.

4. Preliminary Lemmas

The theorems of § 3 will be illustrated by using them to obtain results about the asymptotic integrations of

$$(4.1) \qquad \xi' = E\xi + F(t, \xi),$$

where E is a constant matrix and $F(t, \xi)$ is "small", say,

$$(4.2) \qquad \|F(t, \xi)\| \leqq \psi(t) \, \|\xi\|$$

and $\psi(t)$ is "small" for large t. In this section, we state the basic Lemmas 4.1, 4.2, 4.3. Their proofs are given in §§ 5–7 using the results of § 3. Theorems on the asymptotic integration of (4.1) are stated in §§ 8, 11, 13 and 16 and are deduced, respectively, from Lemma 4.1 in §§ 9-10, from Lemmas 4.1–4.2 in § 12, and from Lemmas 4.1–4.3 in §§ 14–15.

If E has at least two eigenvalues with distinct real parts, we can suppose, after a linear change of variables with constant coefficients, that $E = \mathrm{diag}\,[P, Q]$, $\xi = (y, z)$, $E\xi = (Py, Qz)$, where $\dim y + \dim z = \dim \xi$, the real parts of the eigenvalues $p_1, p_2, \ldots, q_1, q_2, \ldots$ of P, Q satisfy

$$(4.3) \qquad \mathrm{Re}\, p_j \leqq \mu, \quad \mathrm{Re}\, q_k > \mu$$

for some number μ. We can also assume that P, Q are in a suitable normal form (cf. § IV 9), so that for an arbitrarily fixed $\epsilon > 0$ and some c,

$$(4.4) \qquad 0 < \epsilon < c,$$

we have the inequalities

(4.5) $\operatorname{Re} y \cdot Py \leqq (\mu + \epsilon) \|y\|^2$, $\operatorname{Re} z \cdot Qz \geqq (\mu + c) \|z\|^2$.

Correspondingly, write (4.1) in the form

(4.6) $y' = Py + F_1(t, y, z)$, $z' = Qz + F_2(t, y, z)$,

where $F = (F_1, F_2)$. The initial conditions will be of the form

(4.7) $y(t_0) = y_0$, $z(t_0) = z_0$.

When (4.2) holds, (4.5) and (4.6) give

(4.8)

$$\operatorname{Re} y \cdot y' \leqq (\mu + \epsilon) \|y\|^2 + \psi(t) \|\xi\| \cdot \|y\|$$

$$\operatorname{Re} z \cdot z' \geqq (\mu + c) \|z\|^2 - \psi(t) \|\xi\| \cdot \|z\|.$$

Sometimes, it will be convenient to suppose that $E = \operatorname{diag}(A_1, A_2, A_3)$, $\xi = (x, y, z)$, $E\xi = (A_1x, A_2y, A_3z)$, where the eigenvalues $\alpha_{j1}, \alpha_{j2}, \ldots$ of A_j satisfy

(4.9) $\operatorname{Re} \alpha_{1k} < \mu - c$, $\operatorname{Re} \alpha_{2k} = \mu$, $\operatorname{Re} \alpha_{3k} > \mu + c$,

where (4.4) holds. Correspondingly, it will be supposed that

(4.10)

$$\operatorname{Re} x \cdot A_1x \leqq (\mu - c) \|x\|^2, \qquad \operatorname{Re} z \cdot A_3z \geqq (\mu + c) \|z\|^2,$$

$$|\operatorname{Re} y \cdot A_2y - \mu \|y\|^2| \leqq \epsilon \|y\|^2.$$

The initial value problem to be considered is

(4.11) $x' = A_1x + F_1$, $y' = A_2y + F_2$, $z' = A_3z + F_3$,

where $F(t, \xi) = (F_1, F_2, F_3)$, and

(4.12) $x(t_0) = x_0$, $y(t_0) = y_0$, $z(t_0) = z_0$.

When (4.2) holds, (4.10) and (4.11) imply that

(4.13)

$$\operatorname{Re} x \cdot x' \leqq (\mu - c) \|x\|^2 + \psi(t) \|\xi\| \cdot \|x\|,$$

$$|\operatorname{Re} y \cdot y' - \mu \|y\|^2| \leqq \epsilon \|y\|^2 + \psi(t) \|\xi\| \cdot \|y\|,$$

$$\operatorname{Re} z \cdot z' \geqq (\mu + c) \|z\|^2 - \psi(t) \|\xi\| \cdot \|z\|.$$

In what follows, x, y, z are (real or complex) Euclidean vectors; $\xi = (y, z)$ or $\xi = (x, y, z)$ and $F = (F_1, F_2)$ or $F = (F_1, F_2, F_3)$ are Euclidean vectors in the corresponding product space. The first lemma refers to (4.6) and (4.7); the last two, to (4.11) and (4.12).

Lemma 4.1. *Let μ, ϵ, c be constants and P, Q constant matrices satisfying (4.4)–(4.5). Let $F(t, \xi) = (F_1, F_2)$ be continuous for $t \geqq 0$ and $\|y\|$,*

$\|z\| < \delta \ (\leqq \infty)$ and satisfy (4.2), where $\psi(t) > 0$ is continuous for $t \geqq 0$, and

$$(4.14) \qquad \tau(t) = \int_t^\infty \psi(s) e^{-(c-\epsilon)(s-t)} \, ds$$

converges, so that there exists a $T \geqq 0$ such that

$$(4.15) \qquad 5\tau(t) \leqq 1 \qquad \text{if} \quad t \geqq T.$$

Let $t_0 > T$ and $0 < \|y_0\| < \delta$. Then there exists at least one z_0, $\|z_0\| < \delta$, such that (4.6)–(4.7) has a solution $y(t)$, $z(t)$ satisfying

$$(4.16) \qquad \|z(t)\| < 5\tau(t) \, \|y(t)\|,$$

$$(4.17) \qquad \|y(t)\| \leqq \|y_0\| \exp \int_{t_0}^t [\mu + \epsilon + 2\psi(s)] \, ds$$

on its right maximal interval $t_0 \leqq t < \omega \ (\leqq \infty)$. In particular, if the right side of (4.17) is less than δ for $t \geqq t_0$, then $\omega = \infty$.

The last assertion is a consequence of Corollary II 3.2. The other parts of Lemma 4.1 will be proved in § 5.

Lemma 4.2. Let μ, ϵ, c be constants and A_1, A_2, A_3 constant matrices satisfying (4.4) and (4.10). Let $F(t, \xi) = (F_1, F_2, F_3)$ be continuous for $t \geqq 0$ and $\|x\|$, $\|y\|$, $\|z\| < \delta \ (\leqq \infty)$ and satisfy (4.2). Let $\psi(t) > 0$ be continuous for $t \geqq 0$,

$$(4.18) \quad \sigma(t) = \int_0^t \psi(s) e^{-(c-\epsilon)(t-s)} \, ds \quad \text{and} \quad \tau(t) = \int_t^\infty \psi(s) e^{-(c-\epsilon)(s-t)} \, ds$$

converges, and let there exist a $T \geqq 0$ such that

$$(4.19) \qquad 7\sigma(t) \leqq 1 \quad \text{and} \quad 7\tau(t) \leqq 1 \qquad \text{if} \quad t \geqq T.$$

Let $t_0 > T$, $\|x_0\| < 7\sigma(t_0) \, \|y_0\|$, $0 < \|y_0\| < \delta$. Then there exists at least one z_0, $\|z_0\| < \delta$, such that (4.11)–(4.12) has a solution $x(t)$, $y(t)$, $z(t)$ satisfying

$$(4.20) \qquad \|x(t)\| < 7\sigma(t) \, \|y(t)\|, \qquad \|z(t)\| < 7\tau(t) \, \|y(t)\|,$$

$$(4.21) \qquad \|y_0\| \exp \int_{t_0}^t (\mu - \epsilon - 3\psi) \, ds \leqq \|y(t)\|$$

$$\leqq \|y_0\| \exp \int_{t_0}^t (\mu + \epsilon + 3\psi) \, ds.$$

on its right maximal interval of existence $t_0 \leqq t < \omega \ (\leqq \infty)$. In particular, if the right side of (4.21) is less than δ for $t \geqq t_0$, then $\omega = \infty$.

In applications of Lemmas 4.1 and 4.2, it is convenient to know when

$$(4.22) \qquad \sigma(t), \quad \tau(t) \to 0 \qquad \text{as} \quad t \to \infty.$$

This is the case if

(4.23) $\psi(t) \to 0$ as $t \to \infty$ or $\int^{\infty} \psi(t)\, dt < \infty$;

or, more generally, if, as $t \to \infty$,

(4.24) $\sup_{s \geq t} (1 + s - t)^{-1} \int_t^s \psi(r)\, dr \to 0$,   i.e., $\int_t^{t+1} \psi(r)\, dr \to 0$.

Hölder's inequality shows that a sufficient condition for (4.24) [hence, for (4.22)] is that

(4.25) $\int^{\infty} |\psi(t)|^p\, dt < \infty$ for some $p \geq 1$.

Actually, the next exercise states that if $\psi \geq 0$ and $c - \epsilon > 0$, then (4.24) is necessary and sufficient for (4.22) to hold.

Exercise 4.1.   Let $\psi(t) \geq 0$ be continuous for $t \geq 0$ and $c - \epsilon > 0$. (a) Show that (4.24) implies (4.22). In fact, if $\delta(t)$ denotes the "sup" in (4.24), then

$$\sigma(t) \leq e^{-(c-\epsilon)(t-T)} \int_0^t \psi(s)\, ds + [1 + (c - \epsilon)^{-1}] \delta(T),$$

$$\tau(t) \leq [1 + (c - \epsilon)^{-1}] \delta(t)   \text{for } 0 \leq T \leq t.$$

(b) Conversely, show that if either $\sigma(t) \to 0$ or $\tau(t) \to 0$ as $t \to \infty$, then (4.24) holds.

Exercise 4.2.   Let ψ, $c - \epsilon$ be as in Exercise 4.1 and let (4.25) hold for some p, $1 \leq p \leq 2$. (a) Show that $\int^{\infty} \sigma^p(t)\, dt < \infty$, $\int^{\infty} \tau^p(t)\, dt < \infty$. (b) Conclude that $\int^{\infty} \psi(t)[\sigma(t) + \tau(t)]\, dt < \infty$.

Exercise 4.3.   Show that Lemma 4.2 remains valid if $\epsilon = \epsilon(t)$, $c = c(t)$, $\mu = \mu(t)$ are continuous functions of t for $t \geq 0$ satisfying (4.4), (4.10) and if (4.18) is replaced by

$$\sigma(t) = \int_0^t \psi(s) \exp \left\{ -\int_s^t [c(r) - \epsilon(r)]\, dr \right\} ds,$$

$$\tau(t) = \int_t^{\infty} \psi(s) \exp \left\{ -\int_t^s [c(r) - \epsilon(r)]\, dr \right\} ds,$$

where it is assumed that the last integral converges and (4.19) holds for some T.

Lemma 4.3.   *In addition to the assumptions of Lemma 4.2, assume that $\psi(t)$ satisfies (4.24) [so that (4.22) holds]; that $\mu = 0$, $\epsilon = 0$ in (4.13); that an equality of the form*

(4.26) $\|y'\| \leq \psi_0(t) \|\xi\|$,   i.e., $A_2 = 0$, $\|F_2\| \leq \psi_0(t) \|\xi\|$, *holds, where $\psi_0(t)$ is a continuous function for $t \geqq 0$ satisfying*

$$(4.27) \qquad \int^{\infty} \psi_0(t)\, dt < \infty;$$

finally, that

$$(4.28) \qquad \|y_0\| \exp 3 \int_{t_0}^{\infty} \psi_0(s)\, ds < \delta.$$

Then $\omega = \infty$ in the assertion of Lemma 4.2,

$$(4.29) \qquad x(t),\ z(t) \to 0 \qquad as \quad t \to \infty \quad and \quad y_\infty = \lim_{t \to \infty} y(t)$$

exists, and $y_\infty \neq 0$. Furthermore there exist $\delta_1 > 0$, $T \geqq 0$ and, for every $t_0 > T$, a positive constant $\delta_2(t_0)$, such that if $y_\infty \neq 0$, x_0 are given vectors and $\|y_\infty\| < \delta_1$, $\|x_0\| < \delta_2 \|y_\infty\|$, then there exist y_0 and z_0 such that (4.11)–(4.12) has a solution for $t \geqq t_0$ satisfying (4.20) and (4.29). [When $\delta = \infty$, δ_1 can be taken to be ∞.]

Remark 1. The proof of this lemma will show that there exists a constant C depending only on the integral of $\psi_0(t)$ over $t_0 \leqq t < \infty$ such that the solutions mentioned satisfy

$$\|y(t) - y_\infty\| \leqq C \|y_*\| \int_t^{\infty} \psi_0(s)\, ds,$$

$$\|x(t)\| \leqq C \|y_*\| \sigma(t), \qquad \|z(t)\| \leqq C \|y_*\| \tau(t),$$

where $t \geqq t_0$ and y_* can be either y_0 or y_∞.

Remark 2. In the proof of the first part of Lemma 4.3, the inequalities (4.13) with $\mu = \epsilon = 0$ and (4.26) need not hold for all $\|\xi\| < \delta$. For in view of (4.21), the proof will involve only y satisfying $c_1 \|y_0\| < \|y\| < c_2 \|y_0\|$, hence

$$(4.30) \qquad c_1 \|y_0\| < \|\xi\| < 3c_2 \|y_0\|,$$

by (4.20), where

$$(4.31) \qquad c_j = \exp 3(-1)^j \int_T^{\infty} \psi_0(s)\, ds \qquad for \quad j = 1, 2.$$

Correspondingly, (4.13) and (4.26) need only be assumed when (4.30) holds. In the second part of Lemma 4.3, the same remains true if (4.30) is replaced by

$$(4.32) \qquad c_1 \|y_\infty\| < \|\xi\| < 3c_2 \|y_\infty\|.$$

These assertions permit the replacement of the assumptions (4.2) and (4.26) in the derivation of (4.13) by another type of hypothesis: For a pair of numbers r, R satisfying $0 < r < R$ $(\leqq \infty)$, let there exist a continuous function $\varphi_{rR}(t) > 0$ for $t \geqq 0$ such that

$$(4.33) \qquad \|F(t, \xi)\| \leqq \varphi_{rR}(t) \qquad \text{if} \quad r < \|\xi\| < R,$$

$$(4.34) \qquad \int^{\infty} \varphi_{rR}(t) \, dt < \infty.$$

Then (4.33) implies that

$$(4.35) \qquad \|F(t, \xi)\| \leqq r^{-1}\varphi_{rR}(t) \, \|\xi\| \qquad \text{for} \quad r < \|\xi\| < R,$$

which is the analogue of (4.2) with

$$(4.36) \qquad \psi(t) = r^{-1}\varphi_{rR}(t).$$

Notice that with this choice of $\psi(t)$ and $\psi_0(t) = \psi(t)$, (4.31) shows that $c_j \to 1$ as $T \to \infty$. Hence if $r < \|y_0\| < R/3$ [or $r < \|y_\infty\| < R/3$], then the first part [or last part] of Lemma 4.3 remains valid.

The case $\mu \neq 0$ can be reduced to $\mu = 0$ by the change of variables $\xi = e^{\mu t}\zeta$ [when $\mu > 0$, it is necessary to assume that $F(t, \xi)$ is defined for $t \geqq 0$ and *all* ξ]:

Corollary 4.1. *In addition to the assumptions of Lemma* 4.2, *assume that $\psi(t)$ satisfies* (4.24) [*so that* (4.22) *holds*] *and that an inequality of the form*

$$(4.37) \quad \|y' - \mu y\| \leqq \psi_0(t) \, \|\xi\|; \qquad i.e., \quad A_2 = \mu I, \quad \|F_2\| \leqq \psi_0(t) \, \|\xi\|,$$

holds, where $\psi_0(t)$ is a continuous function for $t \geqq 0$ satisfying (4.27). *If $\mu > 0$, assume that $\delta = \infty$ (so that F is defined for $t \geqq 0$ and all ξ). Then the assertions of Lemma* 4.3 *remain valid if* (4.29) *is replaced by*

$$(4.38) \quad e^{-\mu t}x(t), \quad e^{-\mu t}z(t) \to 0 \qquad \text{as} \quad t \to \infty \quad \text{and} \quad y_\infty = \lim_{t \to \infty} e^{-\mu t}y(t).$$

Exercise 4.4. Verify Corollary 4.1.

The condition $A_2 = \mu I$ can be replaced by the assumption that A_2 is a diagonal matrix (or has simple elementary divisors) and that all of its eigenvalues have the same real part μ:

Corollary 4.2. *Let the assumptions of Corollary* 4.1 *hold except that $A_2 = \text{diag} [\mu + i\gamma_1, \mu + i\gamma_2, \dots]$, where $\gamma_1, \gamma_2, \dots$ are real numbers, and* (4.37) *is replaced by*

$$(4.39) \qquad \|y' - A_2 y\| \leqq \psi_0(t) \, \|\zeta\|.$$

Then the assertions of Lemma 4.3 *remain valid if* (4.29) *is replaced by*

$$(4.40) \quad e^{-\mu t}x(t), \quad e^{-\mu t}z(t) \to 0 \qquad \text{as} \quad t \to \infty \quad \text{and} \quad y_\infty = \lim_{t \to \infty} e^{-A_2 t}y(t).$$

Note that the last part of (4.40) means that the kth component $y^k(t)$ of $y(t)$ satisfies $e^{-(\mu+i\gamma_k)t}y^k(t) \to y_\infty{}^k$ as $t \to \infty$.

Exercise 4.5.   Reduce Corollary 4.2 to Lemma 4.3 by the change of variables $(x, y, z) \rightarrow (u, v, w)$ given by $x = e^{\mu t}u$, $y = e^{A_2 t}v$, $z = e^{\mu t}w$.

Exercise 4.6.   Let E be a constant matrix with eigenvalues $\lambda_1, \ldots, \lambda_d$ such that $\lambda_1, \ldots, \lambda_k$ are simple eigenvalues with $\operatorname{Re} \lambda_j = 0$ for $j = 1, \ldots, k$ for some k, $1 \leqq k \leqq d$. Of the eigenvalues $\lambda_{k+1}, \ldots, \lambda_d$, let m have positive real parts, n negative real parts, where $0 \leqq m, n \leqq d - k$ and $m + n = d - k$. Let $G(t)$ be a continuous matrix for $t \geqq 0$ such that $G(t) \rightarrow 0$ as $t \rightarrow \infty$ and the elements $g_{ij}(t)$ are of bounded variation for $t \geqq 0$ (i.e., $\int^{\infty} |dg_{ij}(t)| < \infty$). For example, if $G(t)$ is continuously differentiable, let $G(t) \rightarrow 0$ as $t \rightarrow \infty$ and $\int^{\infty} \|G'(t)\| \, dt < \infty$. For large t, the matrix $E + G(t)$ has k simple continuous eigenvalues $\lambda_1(t), \ldots, \lambda_k(t)$ such that $\lambda_j(t) \rightarrow \lambda_j$ as $t \rightarrow \infty$; cf. Exercise IV 9.1. (*a*) Show that the linear system $\xi' = [E + G(t)]\xi$ has n linearly independent exponentially small solutions as $t \rightarrow \infty$. (*b*) If $\operatorname{Re} \lambda_j(t) \leqq 0$ for $j = 1, \ldots, k$, then $\xi' = [E + G(t)]\xi$ has $n + k$ bounded solutions as $t \rightarrow \infty$. (*c*) If $k = 1$, then there exists a vector $c \neq 0$ such that $\xi' = [E + G(t)]\xi$ has a solution of the form $\xi = (c + o(1)) \exp \int^t \lambda_1(s) \, ds$ as $t \rightarrow \infty$. (*d*) If $\int^t \operatorname{Re} \lambda_j(s) \, ds$ is bounded for $j = 1, \ldots, k$, then there exist linearly independent vectors $c_1, \ldots, c_k$ such that $\xi' = [E + G(t)]\xi$ has solutions of the form

$$\xi = [c_j + o(1)] \exp \int^t \lambda_j(s) \, ds \quad \text{as} \quad t \rightarrow \infty \quad \text{for} \quad j = 1, \ldots, k.$$

For applications of the corollaries of Lemma 4.3, see the exercises in § VIII 3. Further applications and extensions of Lemma 4.3 and its corollaries are given in §§ 13–16

Exercise IX 5.4 gives an analogue of Lemma 4.1 for difference equations. Exercises 4.8 and 4.9 to follow give analogues of Lemmas 4.2 and 4.3.

Exercise 4.7.   Let $R = R^d$ be the $\xi = (\xi^1, \ldots, \xi^d)$-space. For $n = 1, 2, \ldots$, let S_n be a map of R into itself and $T_n = S_n \circ S_{n-1} \circ \cdots \circ S_1$. Let S be a compact and $K_0, K_1, K_2 \ldots, K_{00}, K_{10}, K_{20}, \ldots$ closed sets of R such that $S \subset K_0 \cap K_{00}$, $S_n(R - K_{n-1}) \subset R - K_n$, $S_n(K_{n-1} \cap K_{n-1,0}) \subset K_{n0}$, and $K_n \cap T_n(S)$ is not empty for $n = 1, 2, \ldots$. Then there exists a point $\xi_0 \in S$ such that $T_n \xi_0 \subset K_n \cap K_{n0}$ for $n = 1, 2, \ldots$.

Exercise 4.8.   Let A, B, C be square matrices satisfying

$$(4.41) \quad \|Ax\| \leqq (\mu - c) \|x\|, \quad (\mu - \epsilon) \|y\| \leqq \|By\| \leqq (\mu + \epsilon) \|y\|,$$
$$\|Cz\| \geqq (\mu + c) \|z\|,$$

where $\mu > 0$ and $0 < \epsilon < c$. For $n = 1, 2, \ldots$, let X_n, Y_n, Z_n be continuous, vector-valued functions defined for all (x, y, z) which vanish for large $\|x\| + \|y\| + \|z\|$. Let $T_n = S_n \circ S_{n-1} \circ \cdots \circ S_1$, where S_n is the map

$$(4.42) \quad S_n: \quad x_1 = Ax + X_n(x, y, z), \quad y_1 = By + Y_n(x, y, z),$$
$$z_1 = Cz + Z_n(x, y, z).$$

(a) Let $0 < \theta < 1$ and $\|x_0\| \leqq \theta \|y_0\|$. Show that if $0 \leqq \delta \leqq (c - \epsilon)\theta/6$ and

$$(4.43) \qquad \|X_n\|, \|Y_n\|, \|Z_n\| \leqq (\|x\| + \|y\| + \|z\|)\delta,$$

then there exists a z_0 such that $(x_n, y_n, z_n) = T_n(x_0, y_0, z_0)$ satisfies

$$(4.44) \qquad \|x_n\| \leqq \theta \|y_n\|, \|z_n\| \leqq \theta \|y_n\| \quad \text{for} \quad n = 0, 1, \ldots.$$

(b) Show that if $0 < \mu < 1$ and

$$\|X_n\|, \|Y_n\|, \|Z_n\| = o(\|x\| + \|y\| + \|z\|)$$

as $(n, x, y, z) \to (\infty, 0, 0, 0)$, then (4.44) implies that $\|x_n\|/\|y_n\| \to 0$ and $\|z_n\|/\|y_n\| \to 0$ as $n \to \infty$.

Exercise 4.9. Let A, C be matrices satisfying

$$\|Ax\| \leqq a \|x\| \text{ and } \|Cz\| \geqq c \|z\|, \quad \text{where} \quad 0 < a < 1 < c.$$

Let θ, X_n, Y_n, Z_n be as in Exercise 4.8, with $\delta \leqq (c - 1)\theta/6$ in (4.43). Let $T_n = S_n \circ S_{n-1} \circ \cdots \circ S_1$, where S_n is given by (4.42) with $B = I$. Let $\psi_0(1) + \psi_0(2) + \cdots$ be convergent, and

$$\| Y_n(x, y, z)\| \leqq \psi_0(n)(\|x\| + \|y\| + \|z\|) .$$

(a) Let (x_0, y_0, z_0) be such that $(x_n, y_n, z_n) = T_n(x_0, y_0, z_0)$ satisfies (4.44). Show that

$$(4.45) \qquad\qquad y_\infty = \lim_{n \to \infty} y_n$$

exists. (b) In addition, assume that $\delta < (1 - a)\theta/3$ and that $0 \leqq 3\psi_0(n) < 1$. Let x_0, y_∞ be given and satisfy $\|x_0\| \leqq \theta \|y_\infty\|$. Show that there exists a (y_0, z_0) such that $(x_n, y_n, z_n) = T_n(x_0, y_0, z_0)$ satisfies (4.44) and (4.45). (c) Formulate analogues of parts (a) and (b) when the matrix B in (4.42) is μI, $\mu \neq 0$ (instead of I).

5. Proof of Lemma 4.1

In order to apply Corollary 3.1, let

$$\Omega = \{(t, y, z): t > T; \|y\|, \|z\| < \delta; (y, z) \neq 0\},$$
$$\Omega^0 = \{(t, y, z) \in \Omega: \|z\| < 5\tau(t) \|y\|\}.$$

It will be verified that Ω^0 is a (u, v)-subset of Ω determined by the one function $u = \|z\|^2 - 25\tau^2(t) \|y\|^2$. Let

$$U = \{(t, y, z) \in \Omega: u = 0\}.$$

Since $\dot{u} = 2(\mathrm{Re}\, z \cdot z' - 25\tau^2 \,\mathrm{Re}\, y \cdot y' - 25\tau\tau' \|y\|^2)$, it follows from (4.8) that on U, where $5\tau(t) \leqq 1$, $\|z\| = 5\tau(t) \|y\| \leqq \|y\|$, and $\|\xi\| \leqq 2 \|y\|$, we have

$$\dot{u} \geqq 50\tau \|y\|^2 \left[(c - \epsilon)\tau - \tfrac{4}{5}\psi - \tau'\right].$$

The last factor is positive since τ satisfies the differential equation

$$(c - \epsilon)\tau - \psi - \tau' = 0,$$

and $\psi > 0$. Thus Ω^0 is a (u, v)-subset of Ω and $U = \Omega_e^{\,0} = \Omega_{se}^{\,0}$.

Note that, by the definition of Ω, the point $(y, z) = (0, 0)$ is not in Ω; hence $(y, z) \in \Omega_e^{\,0}$ implies that $y \neq 0$.

Let $S = \{(t_0, y_0, z): \|z\| \leqq 5\tau(t_0) \|y_0\|\}$. Thus $S \cap \Omega_e^{\,0} = \{(t_0, y_0, z): \|z\| = 5\tau(t_0) \|y_0\|\}$. S is a ball, $\|z\| \leqq 5\tau(t_0) \|y_0\|$, and $S \cap \Omega_e^{\,0}$ is its boundary and is not a retract of S. Since $U = \Omega_e^{\,0}$, the map $\pi: \Omega_e^{\,0} \to S \cap \Omega_e^{\,0}$ given by $\pi(t, y, z) = (t_0, y_0, z\tau(t_0) \|y_0\|/\tau(t) \|y\|)$ is continuous [since $y \neq 0$ on $\Omega_e^{\,0}$ and $\tau(t) > 0$] and hence is a retraction of $\Omega_e^{\,0}$ onto $S \cap \Omega_e^{\,0}$. The existence of z_0 and a solution $y(t), z(t)$ of (4.6)–(4.7) satisfying (4.16) follows from Corollary 3.1.

Since (4.15), (4.16) imply that $\|z(t)\| \leqq \|y(t)\|$, hence $\|\xi(t)\| \leqq 2 \|y(t)\|$ the inequality (4.17) is a consequence of (4.8). This proves Lemma 4.1.

6. Proof of Lemma 4.2

This proof is similar. It depends on the choices

$$\Omega = \{(t, x, y, z): t > T; \|x\|, \|y\|, \|z\| < \delta; (x, y, z) \neq 0\}$$
$$\Omega^0 = \{(t, x, y, z) \in \Omega: u < 0, v < 0\},$$

where

$$u = \|z\|^2 - 49\tau^2(t) \|y\|^2, \qquad v = \|x\|^2 - 49\sigma^2(t) \|y\|^2.$$

Define the sets

$$U = \{(t, x, y, z) \in \Omega: u = 0, v \leqq 0\},$$
$$V = \{(t, x, y, z) \in \Omega: u \leqq 0, v = 0\}.$$

It is readily verified that (4.13), (4.19), and (4.20) imply that $\|\xi\| \leqq 3 \|y\|$ and

$$\dot{u} > 0 \quad \text{on } U, \qquad \dot{v} < 0 \quad \text{on } V.$$

Hence Ω^0 is a (u, v)-subset of Ω and $\Omega_e^{\,0} = U - V = \{(t, x, y, z) \in \Omega: u = 0, v < 0\}$.

Choose S to be the set $\{(t_0, x_0, y_0, z): \|z\| \leqq 7\tau(t_0) \|y_0\|\}$. As above, it is seen that $S \cap \Omega_e^0$ is not a retract of S but is a retract of Ω_e^0. Thus Lemma 4.2 follows from Corollary 3.1.

7. Proof of Lemma 4.3

Let (x_0, y_0, z_0) and $\xi(t) = (x(t), y(t), z(t))$ be as in Lemma 4.2. By (4.19) and (4.20), $\|\xi(t)\| \leqq 3 \|y(t)\|$. Hence (4.26) gives $\|y'\| \leqq 3\psi_0(t) \|y\|$. It follows from (4.19), (4.20), and (4.28) that $\xi(t)$ exists for $t \geqq t_0$. The first part of (4.29) follows from (4.20) and (4.22). The inequality $\|y'\| \leqq 3\psi_0(t) \|y\|$ implies the existence of the limit y_∞ and $y_\infty \neq 0$ as in the proof of Theorem 1.1.

The last part of Lemma 4.3 will not be deduced from Lemma 4.2 but will be obtained from another application of Corollary 3.1. Let

$$\Omega = \{(t, x, y, z): t > t_0; \|x\|, \|y\|, \|z\| < \delta; (x, y, z) \neq 0\},$$
$$\Omega^0 = \{(t, x, y, z) \in \Omega: u_1 < 0, u_2 < 0, v < 0\},$$

where u_1, u_2, v are defined by

(7.1)
$$u_1 = \|y - y_\infty\|^2 - 49 \|y_\infty\|^2 \left(\int_t^\infty \psi_0 \, ds \right)^2,$$

$$u_2 = \|z\|^2 - 49\tau^2(t) \|y\|^2, \qquad v = \|x\|^2 - 49\sigma^2(t) \|y\|^2,$$

and t_0 is a positive constant to be specified. Let U_α be the subset of Ω where $u_\alpha = 0$ and $u_j \leqq 0$, $v \leqq 0$, and V the subset of Ω where $v = 0$ and $u_1, u_2 \leqq 0$. Then, as in the last section, $\dot{u}_2 > 0$ on U_2, $\dot{v} < 0$ on V. When $u_1, u_2, v \leqq 0$, then $\|\xi\| \leqq 3\|y\|$ and

(7.2)
$$\|y - y_\infty\| \leqq 7 \|y_\infty\| \int_t^\infty \psi_0 \, ds,$$

(7.3)
$$\|y_\infty\| \left(1 - 7\int_t^\infty \psi_0 \, ds \right) \leqq \|y\| \leqq \|y_\infty\| \left(1 + 7\int_t^\infty \psi_0 \, ds \right).$$
Since

$$\dot{u}_1 = 2\left[\mathrm{Re}\,(y - y_\infty) \cdot y' + 49 \|y_\infty\|^2 \psi_0(t) \int_t^\infty \psi_0 \, ds \right],$$

a simple calculation shows that on U_1,

$$\dot{u}_1 \geqq 14 \|y_\infty\|^2 \psi_0(t) \int_t^\infty \psi_0 \, ds \left[-3\left(1 + 7\int_t^\infty \psi_0 \, ds \right) + 7 \right].$$

Let T be so large that

$$8\int_t^\infty \psi_0 \, ds < 1 \qquad \text{for} \quad t \leqq T.$$

Thus, if $t_0 > T$, it follows that $\dot{u}_1 > 0$ on U_1. In this case, Ω^0 is a (u, v)-subset of Ω and $\Omega_e^{\,0} = \Omega_{se}^{0}$ is the subset of Ω where $u_1, u_2 \leqq 0$, $v < 0$ and either $u_1 = 0$ or $u_2 = 0$.

Choose $\delta_2(t_0)$ to be

$$(7.4) \qquad \delta_2(t_0) = \sigma(t_0)\left(1 - 7\int_{t_0}^{\infty} \psi_0 \, ds\right).$$

Thus, by (7.3), $\|x_0\| < \delta_2(t_0) \|y_\infty\|$ implies that $\|x_0\| < \sigma(t_0) \|y\|$ for $y \in \Omega^0$. Let $S = \{(t_0, x_0, y, z) \colon \|z\| \leqq 7\tau(t_0) \|y\|, \|y - y_\infty\| \leqq 7 \|y_\infty\| \int_{t_0}^{\infty} \psi_0 \, ds\}$, so that $S \subset \Omega^0 \cup \Omega_e^{\,0}$. Topologically, S is a ball in the (y, z)-space. (If y, z

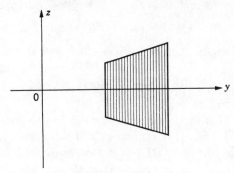

Figure 3.

are 1-dimensional, then S appears as the shaded area in Figure 3.) It is clear that $S \cap \Omega_e^{\,0}$ is the subset of S on which $u_1 = 0$ or $u_2 = 0$, so that, topologically, $S \cap \Omega_e^{\,0}$ is the boundary of S and is not a retract of S. On the other hand, $S \cap \Omega_e^{\,0}$ is a retract of $\Omega_e^{\,0}$ for a retraction $\pi \colon \Omega_e^{\,0} \to S \cap \Omega_e^{\,0}$ is given by $\pi(t, x, y, z) = (t_0, x_0, y^0, z\tau(t_0) \|y^0\|/\tau(t) \|y\|)$, where $y^0 = y^0(t, y)$ is chosen so that $y^0 - y_\infty = \alpha(y - y_\infty)$ and

$$\alpha = \int_{t_0}^{\infty} \psi_0 \, ds \Big/ \int_{t}^{\infty} \psi_0 \, ds.$$

(That $S \cap \Omega_e^{\,0}$ is a retract of $\Omega_e^{\,0}$ is geometrically easy to see, because the projection $(t, x, y, z) \to (t, x_0, y, z)$ of the (t, x, y, z)-space into the (t, y, z)-space carries $\Omega_e^{\,0}$ into a set which is topologically the boundary of a "cylinder" with $S \cap \Omega_e^{\,0}$ corresponding to a section $t = t_0$; cf. Example 2 of § 2.)

Thus by Corollary 3.1, there is a point $(t_0, x_0, y_0, z_0) \in S \cap \Omega^0$ such that a solution arc $(t, x(t), y(t), z(t))$ belonging to (4.11)–(4.12) remains in Ω^0 on its maximal right interval of existence $[t_0, \omega)$. As in the argument at the beginning of the proof of this lemma, $\omega = \infty$ if $\delta = \infty$ or if $\|y_\infty\|$ is sufficiently small; in this case, (4.20), (4.29) hold.

8. Asymptotic Integrations. Logarithmic Scale

Consider again a system of the form

$$(8.1) \qquad \xi' = E\xi + F(t, \xi)$$

in which

$$(8.2) \qquad \|F(t, \xi)\| \leqq \psi(t) \, \|\xi\|$$

holds. In this section it will be supposed that $\xi = (y, z)$, $F = (F_1, F_2)$, and $E = \mathrm{diag}\,[P, Q]$, so that initial value problems associated with (8.1) take the form

$$(8.3) \qquad y' = Py + F_1(t, y, z), \qquad z' = Qz + F_2(t, y, z),$$

$$(8.4) \qquad y(t_0) = y_0, \qquad z(t_0) = z_0.$$

The eigenvalues $p_1, p_2, \ldots$ and $q_1, q_2, \ldots$ of P and Q will be assumed to satisfy

$$(8.5) \qquad \mathrm{Re}\, p_j \leqq \mu, \qquad \mathrm{Re}\, q_k > \mu$$

for some number μ.

Theorem 8.1. *Let* (8.1) *be equivalent to* (8.3) *where the eigenvalues of* P, Q *satisfy* (8.5); $F(t, \xi)$ *is continuous and satisfies* (8.2) *for* $t \geqq 0$ *and* $\|y\|$, $\|z\| < \delta \, (\leqq \infty)$; *and* $\psi(t) > 0$ *is continuous for* $t \geqq 0$ *and satisfies*

$$\sup_{s \geqq t} (1 + s - t)^{-1} \int_t^s \psi(r) \, dr \to 0 \qquad \text{as} \quad t \to \infty.$$

When $\mu \geqq 0$, *assume that* $\delta = \infty$. *Then there exist* $T \geqq 0$ *and* $\delta_1 > 0$ *such that for every* $t_0 \geqq T$ *and* y_0 *satisfying* $\|y_0\| < \delta_1$, *there is a* z_0 *with the property that the initial value problem* (8.3)–(8.4) *has a solution for* $t \geqq t_0$ *satisfying either* $(y(t), z(t)) \equiv 0$ *or* $y(t) \neq 0$ *for* $t \geqq t_0$ *and*

$$(8.6) \qquad \|z(t)\| = o(\|y(t)\|) \qquad \text{as} \quad t \to \infty,$$

$$(8.7) \qquad \limsup_{t \to \infty} t^{-1} \log \|\xi(t)\| \leqq \mu.$$

If μ in (8.7) is replaced by $\mu + \epsilon > \mu$, this follows at once from Lemma 4.1 [with $\delta_1(t_0) = \infty$ if $\delta = \infty$]. Since a linear transformation of the y-variables with constant coefficients does not affect (8.6) but permits an arbitrary choice of $\epsilon > 0$, Theorem 8.1 follows. Assertions (8.6), (8.7) will be improved in § 11 below.

Remark 1. This proof of Theorem 8.1 shows that if the y-variables and z-variables are each subjected to a linear transformation with constant coefficients and $\psi(t)$ is replaced by const. $\psi(t)$ for a suitable constant, then it can be supposed that (4.5) and (4.8) hold. With these choices of coordinates and ψ, the inequalities (4.16)–(4.17) in Lemma 4.1 hold for any solution $(y(t), z(t)) \not\equiv 0$ of (8.3) satisfying (8.6)–(8.7).

Theorem 8.2. *In addition to the conditions of Theorem* 8.1, *assume that F satisfies the Lipschitz condition*

$$(8.8) \qquad \|F(t, \xi_1) - F(t, \xi_2)\| \leqq \psi(t) \|\xi_1 - \xi_2\|,$$

that t_0 is sufficiently large, and that $\|y_0\|$ is sufficiently small. Then z_0 and $(y(t), z(t))$ are unique and $z_0 = g(t_0, y_0)$ is a continuous function (in fact, uniformly Lipschitz continuous on compact subsets of its domain of definition).

If, in addition, F is assumed to be smooth (say, of class C^m, $m \geqq 1$, or analytic), then $z_0 = g(t_0, y_0)$ is of the same smoothness. Here, a function of a vector with complex-valued components is said to be of class C^m if it has continuous, mth order partial derivatives with respect to the real and imaginary parts of its variables. In this terminology, the result for $F \in C^1$ is

Theorem 8.3. *Let the conditions of Theorems* 8.1, 8.2 *hold and let $F(t, \xi)$ have continuous, first order partial derivatives with respect to the real and imaginary parts of the components of ξ. Suppose also that $\mu < 0$. Then $z_0 = g(t_0, y_0)$ is of class C^1. If, in addition, the partial derivatives of F with respect to the real and imaginary parts of the components of ξ vanish at $\xi = 0$ for all t, then the partial derivatives of g with respect to the real and imaginary parts of the components of y_0 vanish at $y_0 = 0$ for all t_0.*

The proofs in §§ 9 and 10 will show that Theorems 8.2 and 8.3 are corollaries of Theorem 8.1, which is, in turn, an immediate consequence of Lemma 4.1. For applications, note that the proofs of Theorems 8.1–8.3 imply the following remark.

Remark 2. Let $\epsilon > 0$ be fixed so small that $\mu + \epsilon < 0$ if $\mu < 0$ and that $\operatorname{Re} q_k > \mu + \epsilon$ in (8.5). Then there exists a number $\rho_\epsilon > 0$ with the property that if the condition on $\psi(t)$ is relaxed to

$$(8.9) \qquad (1 + s - t)^{-1} \int_t^s \psi(r) \, dr \leqq \rho_\epsilon \quad \text{for large } t \text{ and } s \geqq t,$$

then Theorems 8.1–8.3 remain valid if (8.6), (8.7) are replaced by the single condition

$$(8.10) \qquad \limsup_{t \to \infty} t^{-1} \log \|\xi(t)\| \leqq \mu + \epsilon.$$

Notice that the "smallness condition" (8.2) does not seem appropriate if (8.1) is considered only for small ξ, e.g., if $F(t, \xi)$ does not depend on t. In this case, more natural conditions are

$$(8.11) \qquad \frac{\|F(t, \xi)\|}{\|\xi\|} \to 0 \quad \text{as} \quad (t, \xi) \to (\infty, 0)$$

and, of course, $\mu < 0$.

Corollary 8.1 *Let the assumptions of Theorem* 8.1 *hold except that* (8.11) *replaces* (8.2); *also assume that* $\delta < \infty$ *and* $\mu < 0$. *Then the conclusions of Theorem* 8.1 *remain valid. If, in addition,*

$$(8.12) \quad \frac{\|F(t, \xi_1) - F(t, \xi_2)\|}{\|\xi_1 - \xi_2\|} \to 0 \quad as \quad (t, \xi_1, \xi_2) \to (\infty, 0, 0)$$

when $\xi_1 \neq \xi_2$, *then the conclusions of Theorem* 8.2 *hold in the following sense: there exists a small* $\delta_0 > 0$ *with the property that if* t_0 *is sufficiently large and* $\|y_0\|$ *is sufficiently small, then there exists a unique* $z_0 = g(t_0, y_0)$ *such that the solution* $\xi(t) = (y(t), z(t))$ *of* (8.3)–(8.4) *exists and satisfies* $\|\xi(t)\| \leq \delta_0$ *for* $t \geq t_0$ *and the conclusions of Theorem* 8.1; *furthermore,* $g(t_0, y_0)$ *is uniformly Lipschitz continuous. Also, if* F *satisfies the smoothness assumptions of Theorem* 8.3, *then the conclusions of Theorem* 8.3 *are valid.*

This generalizes the last part of Theorem IX 6.1 on the existence of invariant manifolds. The other part will be generalized later in § 11.

Corollary 8.1 follows from the Remark 2 by virtue of the fact that (8.11) implies that, for every $\rho > 0$, there exist $T \geq 0$ and $\delta_0 > 0$ such that

$$(8.13) \quad \|F(t, \xi)\| \leq \rho \|\xi\| \quad for \quad t \geq T \text{ and } \|\xi\| \leq \delta_0$$

and correspondingly, (8.12) gives

$$\|F(t, \xi_1) - F(t, \xi_2)\| \leq \rho \|\xi_1 - \xi_2\| \quad for \quad t \geq T \text{ and } \|\xi_1\|, \|\xi_2\| \leq \delta_0;$$

furthermore, if $\epsilon > 0$ is sufficiently small, then $\mu + \epsilon < 0$ and (8.10), (8.11) imply (8.6), (8.7).

For another deduction of the first part of Corollary 8.1 from Theorem 8.1, make the change of variables

$$(8.14) \quad \xi = e^{-\alpha t}\zeta,$$

where $0 < \alpha < -\mu$. Then (8.1) becomes

$$\zeta' = (E + \alpha I)\zeta + e^{\alpha t}F(t, e^{-\alpha t}\zeta)$$

and μ is replaced by $\mu + \alpha < 0$. For the applicability of Theorem 8.1, it is sufficient to verify the existence of a $\psi(t)$ such that $\psi(t) \to 0$ as $t \to \infty$ and

$$\|e^{\alpha t}F(t, e^{-\alpha t}\zeta)\| \leq \psi(t) \|\zeta\| \quad for \quad \|\zeta\| < \tfrac{1}{2}\delta.$$

Note that $\alpha > 0$ and (8.11) imply that such a $\psi(t)$ is given by

$$\psi(t) = \sup_{\|\zeta\| \leq \frac{1}{2}\delta} \frac{\|F(t, e^{-\alpha t}\zeta)\|}{\|e^{-\alpha t}\zeta\|}.$$

Exercise 8.1. This exercise involves a proof of the conclusions of Theorems 8.1 and 8.2 by the method of successive approximations rather than by the use of Corollary 3.1 (via Lemma 4.1). In view of the change of variables (8.14) with a suitable α, there is no loss of generality in assuming that $\mu < 0$. If $\xi = (y(t), z(t))$ is a solution of (8.3) satisfying (8.7), then it is easy to see that

(8.15)
$$y(t) = e^{P(t-t_0)}y_0 + \int_{t_0}^{t} e^{P(t-s)}F_1(s, y(s), z(s))\, ds,$$

$$z(t) = -\int_{t}^{\infty} e^{Q(t-s)}F_2(s, y(s), z(s))\, ds,$$

where it can be supposed that P, Q are such that

(8.16) $\|e^{Pt}\| \leq e^{(\mu+\epsilon)t}$, $\|e^{-Qt}\| \leq e^{-(\mu+\epsilon)t}$ for $t \geq 0$.

Conversely, if $\xi = (y(t), z(t))$ is a solution of (8.15) satisfying (8.7), then it is a solution of (8.3). Show, by the method of successive approximations, that under the assumptions (8.2), (8.8), where $\psi(t) \to 0$ as $t \to \infty$, (8.15) has a solution (for sufficiently large t_0, small $\|y_0\|$ if $\delta < \infty$) satisfying $y(t_0) = y_0$ and (8.6)–(8.7). Let the 0th approximations be $y_0(t) = e^{P(t-t_0)}y_0$, $z_0(t) = 0$, and the nth approximation be obtained by writing $(y(s), z(s)) = (y_{n-1}(s), z_{n-1}(s))$ on the right of (8.15) and $(y(t), z(t)) = (y_n(t), z_n(t))$ on the left. See Coddington and Levinson [2, Chapter 13]. This gives the existence Theorem 8.1 under the additional condition (8.8) and the condition $\psi(t) \to 0$. Theorems 8.2 and 8.3 can also be proved by the considerations of the successive approximations, but note that Theorems 8.2, 8.3 are deduced in §§ 9 and 10 essentially from Theorem 8.1. (Despite the disadvantages of the method of successive approximations in the present situation, this method has important applications in related problems.)

9. Proof of Theorem 8.2

It can be assumed that (4.4), (4.5), and (4.8) hold; cf. Remark 1 following Theorem 8.1. In terms of the function $\sigma(t)$ in (4.18), define

(9.1) $u(t, y, z) = 25\sigma^2(t)\|z\|^2 - \|y\|^2.$

It is readily verified (cf. § 5) that if (4.8) holds, $T \geq 0$ is sufficiently large, and $t \geq T$, then

(9.2) $\dot{u} > 0$ when $u = 0$.

Uniqueness. Suppose that (8.1) has two solutions $\xi_j(t) = (y_j(t), z_j(t))$, where $j = 1, 2$, satisfying $y_j(t_0) = y_0$ and (8.7), but $z_1(t_0) \neq z_2(t_0)$. Put $\xi(t) = \xi_2(t) - \xi_1(t) = (y(t), z(t))$. Then (8.8) implies (4.8) hence, by (9.2),

$du(t, y(t), z(t))/dt > 0$ if $u(t, y(t), z(t)) = 0$. Since $y(t_0) = 0$, $r(t_0, y(t_0)$, $z(t_0)) > 0$ and, consequently, $u(t, y(t), z(t))$ cannot vanish for $t \geqq t_0$. Thus

$$(9.3) \qquad \|y(t)\| < 5\sigma(t) \|z(t)\| \qquad \text{for} \quad t \geqq t_0.$$

It follows from (4.8) and $\sigma(t) \to 0$ as $t \to \infty$ (cf. Exercise 4.1) that

$$(9.4) \qquad \liminf_{t \to \infty} t^{-1} \log \|z(t)\| \geqq \mu + c.$$

But this contradicts $\|z(t)\| \leqq \|\xi(t)\| \leqq \|\xi_2(t)\| + \|\xi_1(t)\|$, since both $\xi = \xi_1, \xi_2$ satisfy (8.7).

   Continuity of $z_0 = g(t_0, y_0)$. Let $t_0 \geqq T_0$, $\|y_1\| < \delta_1(t_0)$, $z_1 = g(t_0, y_1)$ and $\xi_1(t)$ be the corresponding solution of (8.1). Introduce new variables into (8.1) defined by

$$(9.5) \qquad \zeta = \xi - \xi_1(t),$$

so that (8.1) becomes

$$(9.6) \qquad \zeta' = E\zeta + F(t, \zeta + \xi_1(t)) - F(t, \xi_1(t)),$$

and, by (8.8),

$$(9.7) \qquad \|F(t, \zeta_1 + \xi_1(t)) - F(t, \zeta_2 + \xi_1(t))\| \leqq \psi(t) \|\zeta_1 - \zeta_2\|.$$

It follows from the part of Theorem 8.2 already proved that if $\|y_2 - y_1\|$ is sufficiently small, then (9.6) has a unique solution $\zeta(t)$ which satisfies $\zeta(t) \equiv 0$ or $\limsup t^{-1} \log \|\zeta(t)\| \leqq \mu + \epsilon$ as $t \to \infty$, $\zeta(t_0) = (y_2 - y_1, \ldots)$, and

$$(9.8) \qquad \|z_2(t) - z_1(t)\| \leqq 5\tau(t) \|y_2(t) - y_1(t)\|$$

for $t \geqq t_0$ if $\xi_2(t) = \zeta(t) + \xi_1(t) = (y_2(t), z_2(t))$. (The inequality (9.8) is the analogue of (4.16) in Lemma 4.1.) It follows that $\xi = \xi_2(t)$ is a solution of (8.1) and that $z_2 = g(t_0, y_2)$. Thus $t = t_0$ in (9.8) gives

$$(9.9) \qquad \|g(t_0, y_2) - g(t_0, y_1)\| \leqq 5\tau(t_0) \|y_2 - y_1\|.$$

   Let $\xi = \xi(t, t_0, y_0) = (y(t, t_0, y_0), z(t, t_0, y_0))$ be the unique solution of (8.1) supplied by Theorem 8.1 and the first part of Theorem 8.2. Thus

$$(9.10) \qquad \xi(t_0, t_0, y_0) = (y_0, g(t_0, y_0)).$$

The uniqueness of this solution implies that for $t_1 \geqq t_0$,

$$(9.11) \qquad \xi(t, t_0, y_0) = \xi(t, t_1, y(t_1, t_0, y_0)).$$

In order to examine the continuity of $g(t_0, y_0)$ with respect to t_0, consider $\xi(t, t_0, y_0) - \xi(t, t_1, y_0)$ for $t_1 \geqq t_0$ and small $\|y_0\|$. In view of (9.11), this difference can be written as $\xi(t, t_1, y(t_1, t_0, y_0)) - \xi(t, t_1, y_0)$. The analogue of (9.8) holds and at $t = t_1$ gives

$$\|z(t_1, t_1, y(t_1, t_0, y_0)) - g(t_1, y_0)\| \leqq 5\tau(t_1) \|y(t_1, t_0, y_0) - y_0\|.$$

Since $\xi = \xi(t, t_0, y_0)$ remains in a compact ξ-set for $t_0 \leqq t \leqq t_0 + 1$, and $\|y_0\|$ small, it follows from (8.1) that $\|\xi'(t, t_0, y_0)\| \leqq M$, if M is a bound for F on this set. Hence $\|\xi(t_1, t_0, y_0) - \xi(t_0, t_0, y_0\| \leqq M(t_1 - t_0)$, so that

$$\|z(t_1, t_1, y(t_1, t_0, y_0)) - g(t_0, y_0)\|, \quad \|y(t_1, t_0, y_0) - y_0\| \leqq M(t_1 - t_0).$$

Hence, for $t_0 \leqq t_1 \leqq t_0 + 1$ and $M = M(t_0, y_0)$,

$$(9.12) \qquad \|g(t_1, y_0) - g(t_0, y_0)\| \leqq M(t_1 - t_0)[1 + 5\tau(t_1)].$$

The inequalities (9.9), (9.12) complete the proof of Theorem 8.2.

10. Proof of Theorem 8.3

It will be shown that $\xi(t, t_0, y_0)$ is of class C^1; in particular, $\xi(t_0, t_0, y_0) = (y_0, g(t_0, y_0))$ is of class C^1. The proof will be given as if all variables and functions are real-valued. This is justified since a real system is obtained by separating real and imaginary parts of (8.1); cf. the interpretation $\partial u/\partial y^k = \frac{1}{2}(\partial u/\partial \sigma^k - i\, \partial u/\partial \tau^k)$ if $y^k = \sigma^k + i\tau^k$ mentioned after (3.2).

Let e be a unit vector in the y-space, $h \neq 0$ a small real number. By the Lemma V 3.1, the difference

$$(10.1) \qquad \zeta_h = \frac{\xi(t, t_0, y_0 + he) - \xi(t, t_0, y_0)}{h}$$

satisfies a linear differential equation of the form

$$(10.2) \qquad \zeta' = (E + E_1(t, h, y_0))\zeta,$$

where, in view of the continuity of $\xi(t, t_0, y_0)$,

$$(10.3) \qquad E_1(t, h, y_0) \to \partial_\xi F(t, \xi(t, t_0, y_0)) \qquad \text{as} \quad h \to 0$$

uniformly on bounded t-sets, and $\partial_\xi F$ denotes the Jacobian matrix of F with respect to ξ. By the analogue of (9.8), the function (10.1) is bounded by

$$[1 + 5\tau(t)] \frac{\|y(t, t_0, y_0 + he) - y(t, t_0, y_0)\|}{h}$$

The derivation of (9.8) and the analogue of the inequality (4.17) in Lemma 4.1 show that this is at most

$$[1 + 5\tau(t)] \exp \int_{t_0}^{t} [\mu + \epsilon + 2\psi(s)]\, ds \equiv r(t).$$

Hence, for fixed (t_0, y_0), the family of functions (10.1) is uniformly bounded and equicontinuous in t on bounded t-intervals of $t \geqq t_0$. Thus there exist sequences $h_1, h_2, \ldots$ such that $h_n \to 0$ and the corresponding functions (10.1) tend to a limit $\zeta(t) = \zeta(t, t_0, y_0)$ uniformly for bounded

$t\ (\geqq t_0)$. This limit satisfies the linear system

(10.4) $\zeta' = [E + \partial_\xi F(t, \xi(t, t_0, y_0))]\zeta,$

an initial condition of the form $\zeta(t_0) = (e, z^*)$ for some z^*, and $\|\zeta(t)\| \leqq$ $r(t)$. The last inequality implies that $\zeta(t) = 0$ or

(10.5) $\limsup\limits_{t \to \infty} t^{-1} \log \|\zeta(t)\| \leqq \mu + \epsilon.$

By (8.8), $\|\partial_\xi F(t, \xi)\| \leqq \psi(t)$. Then Theorems 8.1, 8.2 imply that if t_0 is sufficiently large, there is a unique z^* such that (10.4) has a solution satisfying $\zeta(t_0) = (e, z^*)$ and (10.5). Consequently, the selection of the sequence $h_1, h_2, \ldots$ is unnecessary and

(10.6) $\lim\limits_{h \to 0} \dfrac{\xi(t, t_0, y_0 + he) - \xi(t, t_0, y_0)}{h} = \zeta(t, t_0, y_0)$

exists uniformly on bounded t-intervals and is the unique solution of (10.4) satisfying (10.5) and $\zeta(t_0) = (e, z^*)$ for a unique z^*.

Hence $\xi(t, t_0, y_0)$ has partial derivatives with respect to the components of y_0. The continuity of these derivatives as functions of (t, t_0, y_0) follows from (10.4) and arguments similar to those just used to prove (10.6). The existence and continuity of $\partial \xi(t, t_0, y_0)/\partial t_0$ follows by the arguments in the proof of formula (V 3.4) in Theorem V 3.1.

Note that if $\partial_\xi F(t, 0) = 0$, then (10.4) reduces for $y_0 = 0$ to $\zeta' = E\zeta$. The only solutions $\zeta = (y(t), z(t))$ of this linear system satisfying (10.5) have $z(t) = 0$; cf. § IV 5. Thus $\partial_{y_0} z(t, t_0, 0) = 0$ and, at $t = t_0$, this gives $\partial_{y_0} g(t_0, 0) = 0$ and proves Theorem 8.3.

11. Logarithmic Scale (Continued)

The object of this section is to obtain improvements of the assertions of Theorem 8.1 without adding additional assumptions on F. To this end, let $\xi = (x, y, z)$, $E = \operatorname{diag}[A_1, A_2, A_3]$, and $F(t, x, y, z) = (F_1, F_2, F_3)$, so that initial value problems associated with (8.1) take the form

(11.1) $x' = A_1 x + F_1,$ $y' = A_2 y + F_2,$ $z' = A_3 z + F_3,$

(11.2) $x(t_0) = x_0,$ $y(t_0) = y_0,$ $z(t_0) = z_0.$

It will be assumed that the eigenvalues $\alpha_{j1}, \alpha_{j2}, \ldots$ of A_j satisfy

(11.3) $\operatorname{Re} \alpha_{1k} < \mu,$ $\operatorname{Re} \alpha_{2k} = \mu,$ $\operatorname{Re} \alpha_{3k} > \mu$

for some number μ.

Theorem 11.1. *Let* (8.1) *be equivalent to* (11.1), *where* A_1, A_2, A_3 *are matrices satisfying* (11.3), *and let* $F = (F_1, F_2, F_3)$, $\psi(t)$, δ, *and* μ *be as in Theorem* 8.1. *Then there exist* $\delta_1 > 0$, $T \geqq 0$ *and, for every* $t_0 \geqq T$, *a* *constant* $\sigma(t_0) > 0$ *such that if* $\|x_0\| < 7\sigma(t_0) \|y_0\|$ *and* $0 < \|y_0\| < \delta_1$, *then there is a* z_0 *with the properties that* (11.1)–(11.2) *has a solution for* $t \geqq t_0$ *satisfying* $y(t) \not\equiv 0$ *and*

(11.4) $\qquad \|x(t)\|, \quad \|z(t)\| = o(\|y(t)\|) \qquad$ as $\quad t \to \infty$,

(11.5) $\qquad\qquad \lim_{t \to \infty} t^{-1} \log \|\xi(t)\| = \mu.$

This theorem, which concerns *certain* solutions of (8.1), follows at once from Lemma 4.2 (with $\delta_1 = \infty$ when $\delta = \infty$). Note that if μ is the least [or greatest] real part of the eigenvalues of E (so that there are no x [or z] variables), a corresponding statement holds. In fact, this case is contained in Theorem 11.1 since dummy x or z variables can be added to the system (8.1), with suitable choices of A_1 or A_3 and $F_1 \equiv 0$ or $F_3 \equiv 0$. The next theorem concerns *all* solutions of (8.1).

Theorem 11.2. *Assume the hypotheses of Theorem* 11.1 *on* $F(t, \xi)$. *If* $\delta = \infty$, *let* $\xi_0(t) \not\equiv 0$ *be any solution of* (8.1); *and if* $\delta < \infty$, *let* $\xi_0(t) \not\equiv 0$ *be a solution of* (8.1) *for large* t *satisfying*

(11.6) $\qquad\qquad \limsup_{t \to \infty} t^{-1} \log \|\xi_0(t)\| < 0.$

Then the limit (11.5) *exists and is the real part* μ *of an eigenvalue of* E. *If, in addition, coordinates in the* ξ-*space are chosen so that* (8.1) *is of the form* (11.1), *where* (11.3) *holds, then* $\xi(t) = (x(t), y(t), z(t))$ *satisfies* (11.4).

It is clear that the first part of Corollary 8.1 has a similar improvement:

Corollary 11.1. *Let the assumptions of Theorem* 11.1 [*or Theorem* 11.2] *hold except that* (8.2) *is replaced by* (8.11), *and let* $\delta < \infty$, $\mu < 0$. *Then the conclusions of Theorem* 11.1 [*or Theorem* 11.2] *remain valid.*

Exercise 11.1. (a). Consider the case of a linear system of differential equations

(11.7) $\qquad\qquad \xi' = [E + G(t)]\xi,$

where $G(t)$ is a continuous matrix for $t \geqq 0$ such that $\|G(t)\| \leqq \psi(t)$, where $\psi(t)$ is continuous and satisfies (4.24). Let $E = \text{diag}\,[\lambda_1, \ldots, \lambda_d]$, and let the real parts $\mu_1, \ldots, \mu_d$ of $\lambda_1, \ldots, \lambda_d$ be distinct. Then, for any j, $1 \leqq j \leqq d$, (11.7) has a solution $\xi(t) = (\xi^1(t), \ldots, \xi^d(t))$ such that $\xi^j(t) \not\equiv 0$ for large t, $|\xi^k(t)| = o(|\xi^j(t)|)$ as $t \to \infty$ for $k \neq j$, and $t^{-1} \log |\xi^j(t)| \to \mu_j$ as $t \to \infty$. (b) Show that if $\int^\infty \psi(t)\,dt < \infty$, then $\xi^j(t) = [c + o(1)] \exp \lambda_j t$, as $t \to \infty$, for some constant $c \neq 0$.

Exercise 11.2. Let $E = \text{diag}\,[A_1, A_2, A_3]$, where A_j is a square matrix with eigenvalues $\alpha_{j1}, \alpha_{j2}, \ldots$ satisfying (11.3). Let $G(t)$ be a continuous matrix for $t \geqq 0$ and identify (11.1) with (11.7), where $\xi = (x, y, z)$ and $F(t, \xi) = G(t)\xi$. Suppose that $\|G(t)\| \leqq \psi(t)$, where $\psi(t)$ is continuous and satisfies $\int^{\infty} |\psi(t)|^p \, dt < \infty$ for some p, $1 \leqq p \leqq 2$. Let A_2 be a 1×1 matrix, consisting of the constant λ, Re $\lambda = \mu$; thus y is 1-dimensional. Let $\xi(t) = (x(t), y(t), z(t))$ be a solution of (11.7) satisfying (11.4), (11.5). Show that there is a constant $c \neq 0$ such that

$$y(t) = [c + o(1)] \exp \int_0^t [\lambda + g(s)] \, ds,$$

where $g(t)$ is the diagonal element of $G(t)$ which is the coefficient of y in the second equation of (11.1). Note that this equation is the form $y' = \lambda y + \Sigma q_j(t)x^j + g(t)y + \Sigma r_k(t)z^k$, where $(q_1, q_2, \ldots, g, r_1, r_2, \ldots)$ is a row of $G(t)$.

Exercise 11.3.   Let $f(t, y)$ be continuous and have continuous partial derivatives with respect to the components of y on a (t, y)-domain and be periodic of period p in t, $f(t + p, y) = f(t, y)$. Let

(11.8) $$y' = f(t, y)$$

have a periodic solution $y = \gamma(t)$ of period p. Discuss the behavior of solutions of (11.8) and $y(t_0) = y_0$, where (t_0, y_0) is near the curve $(t, \gamma(t))$, $0 \leqq t \leqq p$, on the basis of the following suggestions: Introduce the new variables

(11.9) $$\zeta = y - \gamma(t)$$

Thus (11.8) becomes

$$\zeta' = f(t, \zeta + \gamma(t)) - f(t, \gamma(t)),$$

which can be written as

(11.10) $$\zeta' = P(t)\zeta + H(t, \zeta),$$

where

$$P(t) = \partial_y f(t, y) = \left(\frac{\partial f^j}{\partial y^k}\right) \quad \text{at} \quad y = \gamma(t),$$

(11.11)

$$H(t, \zeta) = f(t, \zeta + \gamma(t)) - f(t, \gamma(t)) - P(t)\zeta.$$

$P(t)$ is a matrix function of period p and $H(t, \zeta)$ is continuous and has continuous partial derivatives with respect to the components of ζ, and $H(t, 0) = 0$, $\partial_\zeta H(t, 0) = 0$. The linear matrix initial value problem

(11.12) $$J' = P(t)J, \quad J(0) = I$$

has a solution which, by the Floquet theory in § IV 6, is of the form

(11.13) $$J(t) = K(t)e^{Et},$$

where $K(t + p) = K(t)$ and E is a constant matrix. The change of variables

(11.14) $$\zeta = K(t)\xi$$

transforms (11.10) into

(11.15) $$\xi' = E\xi + K^{-1}(t)H(t, K(t)\xi).$$

Consider the application of the theorems of § 8 and of this section to (11.15) to obtain generalizations of the results of §§ IX 10, 11. (Note that e^E need not have $\lambda = 1$ as an eigenvalue in the situation here.)

12. Proof of Theorem 11.2

It will be shown that it is sufficient to consider the case of linear equations. Note that (8.2) implies that if the solution $\xi = \xi_0(t)$ of (8.1) vanishes at one t-value, then it vanishes for all t. Hence $\xi_0(t) \neq 0$ for large t, say $t \geqq t_0$. Define a matrix $G(t) = (g_{jk}(t))$ as follows: if $F = (F^1, F^2, \ldots)$, put

(12.1) $$g_{jk}(t) = \frac{F^j(t, \xi_0(t))\bar{\xi}_0{}^k(t)}{\|\xi_0(t)\|^2}$$

for $t \geqq t_0$. Since $\xi = \xi_0(t)$ is a solution of (8.1), it follows that it is a solution of the linear system

(12.2) $$\xi' = (E + G(t))\xi.$$

Note that (8.2) and (12.1) imply that

$$|g_{jk}(t)| \leqq \frac{\psi(t) \|\xi_0(t)\| \cdot |\xi_0{}^k(t)|}{\|\xi_0(t)\|^2} \leqq \psi(t).$$

Hence Theorem 11.2 is contained in the following:

Lemma 12.1 *Let $G(t)$ be a continuous matrix for $t \geqq 0$ such that*

(12.3) $$\|G(t)\| \leqq \psi(t),$$

where $\psi(t) > 0$ is a continuous function satisfying (4.24). Let $\xi = \xi_0(t) \neq 0$ be a solution of (12.2). Then the conclusion of Theorem 11.2 holds.

Proof of Lemma 12.1. Let $\mu_1 < \mu_2 < \cdots < \mu_f$ denote the different real parts of the eigenvalues of E. After a change of coordinates, it can be supposed that $E = \text{diag } [B_1, B_2, \ldots, B_f]$, where the eigenvalues β_{jk} of B_j satisfy $\text{Re } \beta_{jk} = \mu_j$. Correspondingly, let $\xi = (y_1, \ldots, y_f)$, $E\xi = (B_1 y_1, \ldots, B_f y_f)$, and let (12.2) be written as

(12.4) $$y_j' = B_j y_j + \sum_{k=1}^{f} G_{jk}(t)y_k \quad \text{for } j = 1, \ldots, f,$$

where $G_{jk}(t)$ is a rectangular matrix and $\|G_{jk}(t)\| \leqq \psi(t)$.

If $1 \leqq q \leqq f$, t_0 is sufficiently large, and $y_{q0} \neq 0$, then Theorem 11.1 implies that (12.4) has a solution $\xi = (y_1(t), \ldots, y_f(t))$ satisfying

(12.5) $\quad y_k(t_0) = 0 \quad$ if $\ k < q, \quad\ y_q(t_0) = y_{q0},$

(12.6) $\quad \|y_k(t)\| = o(\|y_q(t)\|) \quad$ as $\ t \to \infty \ $ for $\ k \neq q,$

(12.7) $\quad t^{-1} \log \|y_q(t)\| \to \mu_q \quad$ as $\ t \to \infty.$

This solution, say $\xi = \xi_q(t, t_0, y_{q0})$, is unique by Theorem 8.2. In fact, it is unique even if (12.6), (12.7) are replaced by

(12.8) $$\limsup_{t \to \infty} \ t^{-1} \log \|\xi(t)\| \leqq \mu_q$$

(cf. Remark 2 following Theorem 8.3). This uniqueness implies that $\xi_q(t, t_0, y_{q0})$ is linear in y_{q0} (for fixed t, t_0, q).

With the understanding that $\xi_q(t, t_0, 0) \equiv 0$, it follows that there exist unique $y_{10}, \ldots, y_{f0}$ such that the given solution $\xi_0(t)$ is of the form

(12.9) $$\xi_0(t) = \sum_{j=1}^{f} \xi_j(t, t_0, y_{j_0}).$$

In fact, $y_{10}, \ldots, y_{f0}$ are defined recursively as follows: if $\xi_0(t) = (y_1(t), \ldots, y_f(t))$, let $y_{10} = y_1(t_0)$; then let $y_{20} = y_2(t_0) - y_{12}(t_0)$, where $\xi_1(t, t_0, y_{10}) = (y_{11}(t), y_{12}(t), \ldots, y_{1f}(t))$; etc.

Let q be the largest j-value such that $y_{j0} \neq 0$ in (12.9). It is clear that $\xi_0(t) = (y_1(t), \ldots, y_f(t))$ satisfies (12.6), (12.7). This proves the lemma.

13. Asymptotic Integration

The object of this section is to study the asymptotic behavior of solutions $\xi(t)$ of a perturbed linear system

(13.1) $$\xi' = E\xi + F(t, \xi),$$

rather than the behavior of $\|\xi(t)\|$ as in § 11.

Suppose that E is in a Jordan normal form $E = \mathrm{diag}\,[J(1), \ldots, J(g)]$, where $J(j)$ is an $h(j) \times h(j)$ matrix [as in (IV 5.15)–(IV 5.16)]. Thus $J(j) = \lambda(j)I_{h(j)} + K_{h(j)}$, where I_h is the unit $h \times h$ matrix and K_h is 0 if $h = 1$ or is the $h \times h$ matrix with ones on the subdiagonal and other elements zero if $h > 1$. According as $h = 1$ or $h > 1$,

(13.2) $J(j)y_j = \lambda y_j \quad$ or $\quad J(j)y_j = (\lambda y_j{}^1, \lambda y_j{}^2 + y_j{}^1, \ldots, \lambda y_j{}^h + y_j{}^{h-1}),$

where $\lambda = \lambda(j)$, $y_j = (y_j{}^1, \ldots, y_j{}^h)$, $h = h(j)$.

Correspondingly, it is supposed that $\xi = (y_1, \ldots, y_g)$,

$$E\xi = (J(1)y_1, \ldots, J(g)y_g), \quad F = (F_1, \ldots, F_g),$$

and (13.1) is of the form

(13.3) $y_j' = J(j)y_j + F_j(t, \xi)$ for $j = 1, \ldots, g$.

Let μ denote one of the numbers Re $\lambda(1), \ldots,$ Re $\lambda(g)$. An index j will be denoted by $p, q,$ or r according as Re $\lambda(j) < \mu$, Re $\lambda(j) = \mu$ or Re $\lambda(j) > \mu$. Put

(13.4) $h_* = \max_q h(q)$.

Let j_0 be an integer and β a number satisfying

(13.5) $j_0 \leqq h_* - 1$ and $\beta \geqq 1$,

and $l(q), k(q)$ integers, if any, such that

(13.6) $1 \leqq l(q) \leqq k(q) \leqq \min(h(q), \beta)$ and $h(q) - l(q) \leqq j_0$.

The next theorem concerns sufficient conditions for (13.3) to have a solution with the following asymptotic properties as $t \to \infty$,

(13.7)
$$y_q^k = e^{\lambda(q)t}\left\{\sum_I c_q^i \frac{t^{k-i}}{(k-i)!} + o(t^{k-\beta})\right\},$$

$$y_j^k = o(e^{\mu t}t^{1-\beta})   \text{if } j \neq q,$$

where c_q^i are constants,

(13.8) $$\sum_I = \sum_{i=l(q)}^{\min(k,k(q))},$$

and $\sum_I = 0$ if $l(q), k(q)$ do not exist.

Note that if the o-terms are replaced by 0, then, since $1 \leqq i \leqq k$ in $\sum_I$, (13.7) becomes a solution of the linear system

(13.9) $\xi' = E\xi$;   i.e., $y_j' = J(j)y_j$ for $j = 1, \ldots, g$.

The choice of the range of summation $l(q) \leqq i \leqq \min(k, k(q))$ is dictated by several considerations. On the one hand, results permitting $i > k$ can easily (but will not) be obtained as a consequence of Theorem 13.1; also the first term in the first line of (13.7) is not significant unless $i \leqq \beta$, hence the choice $i \leqq \min(k, k(q)) \leqq \min(k, \beta)$ since $k \leqq h(q)$. On the other hand, the condition $i \geqq l(q)$ means that the degree of the polynomial $\sum_I c_q^i t^{k-i}/(k-i)!$ does not exceed the given j_0.

Theorem 13.1. *In the system* (13.3), *let $J(j)$ be a Jordan block; cf.* (13.2). *Let $\mu =$ Re $\lambda(j)$ for some j. Let an index $j = 1, \ldots, g$ be denoted by p, q or r according as Re $\lambda(j) < \mu$,* Re $\lambda(j) = \mu$, *or* Re $\lambda(j) > \mu$, *and*

define h_* by (13.4). *Let j_0 be an integer and β a number satisfying*

(13.10) $0 \leqq j_0 \leqq h_* - 1$, $\beta + j_0 \geqq h_*$;   thus $\beta \geqq 1$.

Let $l(q)$, $k(q)$ be integers (if any) satisfying (13.6). *Let $F(t, \xi) = (F_1, \ldots, F_\sigma)$ be continuous for $t \geqq 0$ and all ξ, and satisfy*

(13.11) $\|F(t, \xi)\| \leqq \psi_1(t) \|\xi\|$,

where $\psi_1(t) > 0$ is a continuous function such that

(13.12) $\displaystyle\int^\infty t^{\beta + j_0 - 1} \psi_1(t) \, dt < \infty.$

Let $m = \sum_p h(p) + \sum_q [h(q) - k(q)]$. For any set of constants $c_q{}^k$, $l(q) \leqq k \leqq k(q)$, not all 0, there exists an m parameter family of solutions $\xi(t)$ of (13.3) *defined for large t and satisfying the asymptotic relations* (13.7) *as $t \to \infty$.*

The part of the assertion concerning "m parameter family of solutions" means essentially that it is possible to specify a partial set of m "initial conditions," as well as the asymptotic behavior (13.7) for $\xi(t)$; cf. the statement following (14.15) in the proof of Theorem 13.1.

Remark 1.   Consider a system of differential equations

(13.13) $\eta' = E^0\eta + F^0(t, \eta),$

where E^0 is a constant matrix and $F^0(t, \eta)$ is continuous for $t \geqq 0$ and all η. Let L be a nonsingular constant matrix such that $L^{-1}E^0L$ is a matrix $E = \mathrm{diag}\,[J(1), \ldots, J(g)]$ in a Jordan normal form. Then the change of variables $\eta = L\xi$ reduces (13.13) to (13.1) [i.e., to (13.3)], where $F(t, \xi) = L^{-1}F^0(t. L\xi)$. The applicability of Theorem 13.1, or at least the condition (13.11), can sometimes be verified without the knowledge of L or the explicit reduction of (13.13) to (13.1). For it is clear that $\|F^0(t, \eta)\| \leqq \psi_1(t) \|\eta\|$ implies that $\|F(t, \xi)\| \leqq c\psi_1(t) \|\xi\|$ if, e.g., $c = \|L^{-1}\| \cdot \|L\|$.

Remark 2.   The derivation of Theorem 13.1 from Lemma 4.3 will show that the theorem remains valid if $F(t, \xi)$ is defined only for $t \geqq 0$, $\|\xi\| < \delta < \infty$ if $\mu < 0$ (or $\mu = 0$, $h_* = 1$, and the constants $|c_q{}^k|$ are sufficiently small).

Theorem 13.1 has a partial "converse" dealing with *all* (rather than *certain*) solutions $\xi(t)$ of (13.1) satisfying

(13.14) $t^{-1} \log \|\xi(t)\| \to \mu$ as $t \to \infty$

(cf. Theorem 11.2):

Theorem 13.2.   *Let $E = \mathrm{diag}\,[J(1), \ldots, J(g)]$ and $F(t, \xi)$ be as in*

Theorem 13.1 *except that* (13.12) *is replaced by*

$$(13.15) \qquad \int^{\infty} t^{h_0-1} \psi_1(t) \, dt < \infty, \qquad \text{where} \quad h_0 \geqq h_*$$

(and h_0 is not necessarily an integer). Let $\xi(t) \neq 0$ be a solution of (13.3) *satisfying* (13.14). *Then there exists constants* c_q^k, $k = 1, \ldots, h(q)$, *not all 0, such that if j_0 is defined by*

$$(13.16) \qquad j_0 = \max \, [k(q) - k] \qquad \text{for} \quad c_q^k \neq 0,$$

$\beta = h_0 - j_0$, *and $l(q)$, $k(q)$ are the least, greatest integers (if any) satisfying* (13.6), *then $\xi(t)$ satisfies the asymptotic relations* (13.7) *as $t \to \infty$.*

Consequences and refinements of Theorem 13.1, 13.2 will be given in § 16; see also § XII 9.

14. Proof of Theorem 13.1

Change of Variables. In order to apply Lemma 4.3, make the linear change of variables

$$(14.1) \qquad \xi = Q(t)\zeta$$

given in terms of $\xi = (y_1, \ldots, y_g)$ and $\zeta = (z_1, \ldots, z_g)$ by the formulae

$$(14.2) \qquad \begin{aligned} y_q^{\ k} &= e^{\lambda(q)t} \left\{ \sum_{\mathrm{I}} \frac{t^{k-i}}{(k-i)!} z_q^{\ i} + t^{k-\beta} \sum_{\mathrm{II}} \frac{1}{(k-i)!} z_q^{\ i} \right\}, \\ y_j^{\ k} &= \epsilon^{-k} e^{\mu t} t^{1-\beta} z_j^{\ k} \qquad \text{if} \quad j \neq q, \end{aligned}$$

where $0 < \epsilon < 1$, $\sum_{\mathrm{I}}$ is the sum over the i-range $l(q) \leqq i \leqq \min \, (k, k(q))$ as in (13.8), and $\sum_{\mathrm{II}}$ is the sum over the other indices i on the range $1 \leqq i \leqq k$, so that

$$\sum_{\mathrm{I}} + \sum_{\mathrm{II}} = \sum_{i=1}^{k}.$$

A solution $\xi(t)$ of (13.3) satisfies (13.7) if the corresponding vector $\zeta(t)$, defined by (14.1), satisfies

$$(14.3) \qquad \begin{aligned} z_q^{\ i} &= c_q^{\ i} + o(t^{i-\beta}) \qquad \text{for} \quad l(q) \leqq i \leqq h(q), \\ z_j^{\ k} &= o(1) \qquad \text{otherwise.} \end{aligned}$$

To clarify the meaning of (14.1) and to calculate the resulting differential equation for ζ, the map (14.1) will be given a decomposition of the form

$$(14.4) \qquad \xi = Q(t)\zeta = Q_0(t) \, D(t)\zeta,$$

to be described. This factorization is suggested by the fact that if $t^{k-\beta}$ in the first formula of (14.2) is replaced by t^{k-i} (and written behind the sign $\sum_{\mathrm{II}}$), then this formula becomes $y_q = e^{J(q)t} z_q$.

The change of variables $\xi = Q_0(t)w$, $w = (w_1, \ldots, w_q)$, is given by

(14.5) $\xi_j = e^{J(j)t}w_j$ if $j = q$, $\xi_j = e^{\mu t}w_j$ if $j \neq q$.

Thus (13.3) becomes

(14.6) $w_q' = e^{-J(q)t}F_q(t, Q_0w)$,

$\quad\quad\quad w_j' = [J(j) - \mu I_h]w_j + e^{-\mu t}F_j(t, Q_0w)$ if $j \neq q$,

where $h = h(j)$. Finally, let $D(t)$ be the diagonal matrix such that $w = D(t)\zeta$ is given by

$$w_q^k = z_q^k \quad \text{if} \quad l(q) \leq k \leq k(q),$$

(14.7) $$w_q^k = t^{k-\beta}z_q^k \quad \text{if} \quad k < l(q) \quad \text{or} \quad k > k(q),$$

$$w_j^k = \epsilon^{1-k}t^{1-\beta}z_j^k \quad \text{if} \quad j \neq q.$$

If the resulting differential equation for ζ is written as

(14.8) $$\zeta' = E_0\zeta + Q^{-1}(t)F(t, Q\zeta),$$

then the linear part $\zeta' = E_0\zeta$ is given by

$$z_q^{k'} = 0 \quad \text{if} \quad l(q) \leq k \leq k(q),$$

(14.9) $$z_q^{k'} = (\beta - k)t^{-1}z_q^k \quad \text{if} \quad k < l(q) \quad \text{or} \quad k > k(q),$$

$$z_j' = [J_\epsilon(j) - \mu I_h + (\beta - 1)t^{-1}I_h]z_j \quad \text{if} \quad j \neq q,$$

where $J_\epsilon(j)$ is the matrix obtained by replacing the ones on the subdiagonal of $J(j)$ by ϵ; cf. § IV 9. The last part of (14.9) is easy to see if the transformation $w_j \to z_j$ is made in two steps $w_j^k \to z_j^k/\epsilon^{k-1} \to t^{1-\beta}z_j^k/\epsilon^{k-1}$.

Finally, replace the independent variable t by s, where

(14.10) $$t = e^s, \quad \text{so that} \quad \frac{d\zeta}{ds} = t\zeta'.$$

Thus (14.8) becomes

(14.11) $$\frac{d\zeta}{ds} = E_0 t\zeta + tQ^{-1}F(t, Q\zeta),$$

where the linear part of this equation is

$$\frac{dz_q^k}{ds} = 0 \quad \text{if} \quad l(q) \leq k \leq k(q),$$

(14.12) $$\frac{dz_q^k}{ds} = (\beta - k)z_q^k \quad \text{if} \quad k < l(q) \quad \text{or} \quad k > k(q),$$

$$\frac{dz_j}{ds} = [(J_\epsilon(j) - \mu I_h)t + (\beta - 1)I_h]z_j \quad \text{if} \quad j \neq q.$$

Preliminary Existence Result. Suppose that there is a continuous function $\psi_0(t)$ for large t such that

(14.13) $\| Q^{-1}(t) F(t, Q\zeta)\| \leqq \psi_0(t) \|\zeta\|$ for $\|\zeta\| < \delta \leqq \infty$,

(14.14) $\displaystyle\int^\infty \psi_0(t)\, dt < \infty.$

The last condition is equivalent to $\displaystyle\int^\infty t\psi_0(t)\, ds < \infty$ since $ds = dt/t$. Then if $\epsilon > 0$ is sufficiently small, Lemma 4.3 is applicable to (14.11) if x is a vector with components $z_p{}^k$, and $z_q{}^k$, $k > k(q)$; y is the vector with components $z_q{}^k$, $l(q) \leqq k \leqq k(q)$; and z is the vector with components $z_r{}^k$ and $z_q{}^k$, $k < l(q)$. Note that (14.12) shows that there is a constant $c > 0$ such that $\mathrm{Re}\,(z_q{}^k \, d\bar{z}_q{}^k/ds) \leqq -c\,|z_q{}^k|^2$ or $\geqq c\,|z_q{}^k|^2$ according as $k > k(q) \geqq \beta$ or $k < l(q) \leqq \beta$; also $\mathrm{Re}\,(z_j \cdot d\bar{z}^j/ds) \leqq -ct\,\|z_j\|^2$ or $\geqq ct\,\|z_j\|^2$ according as $j = p$ (i.e., $\mathrm{Re}\,\lambda(j) < \mu$) or $j = r$ (i.e., $\mathrm{Re}\,\lambda(j) > \mu$) if $\epsilon > 0$ is small and $t > 0$ is large.

Thus, by Lemma 4.3, (14.13)–(14.14) imply that if $c_q{}^k$, $l(q) \leqq k \leqq k(q)$, are given constants, not all 0, then there exists a solution $\zeta(t)$ of (14.11) such that, as $t \to \infty$,

(14.15) $z_q{}^k(t) = c_q{}^k + o(1)$ for $l(q) \leqq k \leqq k(q)$,

$\qquad\qquad\quad z_j{}^k = o(1)$ otherwise.

In fact, we can also specify a set of m initial conditions for ζ: $z_p{}^k(T) = z_{p0}^k$ and $z_q{}^k(T) = z_{q0}^k$ for $k(q) < k \leqq h(q)$ if T is sufficiently large and $|z_{p0}^k|$, $|z_{q0}^k|$ are sufficiently small numbers.

The Norms $\|Q\|$, $\|Q^{-1}\|$. In order to complete the proof, it remains to show that the assumptions (13.11)–(13.12) imply (14.13)–(14.14) and that a solution $\zeta(t)$ of (14.8) satisfying (14.15) also satisfies (14.3). To this end, it will first be verified that there exist positive constants c, c' such that for large t,

(14.16) $c' \leqq \|Q(t)\|\, e^{-\mu t} t^{-j_0} \leqq c$, $c' \leqq \|Q^{-1}(t)\|\, e^{\mu t} t^{1-\beta} \leqq c$.

From (14.2), the norm of $Q(t)$ is easily seen to be $O(e^{\mu t} t^\gamma)$, where $\gamma = \max\,[h_* - \beta, h(q) - l(q)]$ and the max refers to the set q. From (13.6), $h(q) - l(q) \leqq j_0$ and, from (13.10), $h_* - \beta \leqq j_0$; hence $\|Q(t)\| = O(e^{\mu t} t^{j_0})$ as $t \to \infty$. It is similarly seen that $e^{\mu t} t^{j_0} = O(\|Q(t)\|)$ as $t \to \infty$. This gives the first part of (14.16).

The factorization $Q = Q_0 D$ of Q into nonsingular matrices for $t > 0$ shows that Q^{-1} exists and is $Q^{-1} = D^{-1} Q_0^{-1}$. The inverse map

(14.17) $\qquad\qquad\qquad\qquad \zeta = Q^{-1}\xi$

is easily seen, from $\xi = Q_0 w$, $w = D\zeta$ in (14.5), (14.7), to be

$$z_q{}^k = e^{-\lambda(q)t} \sum_{i=1}^{k} \frac{(-1)^{k-i} t^{k-i}}{(k-i)!} y_q{}^i \quad \text{if} \quad l(q) \leqq k \leqq k(q),$$

$$(14.18) \quad z_q{}^k = e^{-\lambda(q)t} \sum_{i=1}^{k} \frac{(-1)^{k-i} t^{\beta-i}}{(k-i)!} y_q{}^i \quad \text{if} \quad k < l(q) \quad \text{or} \quad k > k(q),$$

$$z_j{}^k = \epsilon^{k-1} e^{-\mu t} t^{\beta-1} y_j{}^k \quad \text{if} \quad j \neq q.$$

Thus, for large t, $\|Q^{-1}(t)\|$ is bounded from above and below by a positive constant times $e^{-\mu t} t^\gamma$, where $\gamma = \max [\beta - 1, k(q) - 1]$. Since $k(q) - 1 \leqq \beta - 1$, by (13.6), the last part of (14.16) follows.

Completion of the Proof. In view of (13.11),

$$\|Q^{-1} F(t, Q\zeta)\| \leqq \|Q^{-1}\| \, \psi_1(t) \, \|Q\| \cdot \|\zeta\|.$$

Hence, by (14.16),

$$(14.19) \qquad \|Q^{-1} F(t, Q\zeta)\| \leqq c^2 t^{\beta + j_0 - 1} \psi_1(t).$$

Thus (13.12) implies that (14.13), (14.14) hold if $\psi_0(t) = c^2 t^{\beta + j_0 - 1} \psi_1(t)$, and so (14.8) has a solution $\zeta(t)$ satisfying (14.15).

In view of the first part (14.9), the corresponding equations in (14.8) are

$$z_q^{k\prime} = (qk)\text{th component of } Q^{-1} F(t, Q\zeta),$$

so that, by (14.18),

$$z_q^{k\prime} = e^{-\lambda(q)t} \sum_{i=1}^{k} \frac{(-t)^{k-i}}{(k-i)!} F_q{}^i(t, Q\zeta) \quad \text{if} \quad l(q) \leqq k \leqq k(q),$$

where $F_q{}^i$ is the (qi)th component of F. Hence,

$$z_q^{k\prime} = O(e^{-\mu t} t^{k-1} \psi_1(t) \|Q\zeta\|) \quad \text{as} \quad t \to \infty,$$

by (13.11). In view of (14.16) and the boundedness of $\zeta(t)$ as $t \to \infty$,

$$z_q^{k\prime} = O(t^{k + j_0 - 1} \psi_1(t)) \quad \text{as} \quad t \to \infty.$$

Consequently, $k \leqq \beta$ shows that

$$(14.20) \qquad z_q{}^k(t) - c_q{}^k = t^{k-\beta} O\left(\int_t^\infty s^{\beta + j_0 - 1} \psi_1(s) \, ds \right).$$

This gives the first part of (14.3) and completes the proof of Theorem 13.1.

15. Proof of Theorem 13.2

This theorem can be reduced to the case of linear equations by the device used at the beginning of § 12. Hence we can be suppose that $F(t, \xi) = G(t)\xi$, where $G(t)$ is a matrix satisfying $\|G(t)\| \leqq \psi_1(t)$ and (13.1) is replaced by

$$(15.1) \qquad \xi' = E\xi + G(t)\xi.$$

Let q_0 denote a fixed value of q and k_0 an integer on the range $1 \leqq k_0 \leqq h(q_0)$. Then the equation (15.1) has a solution $\xi_{k_0 q_0}(t)$ satisfying, as $t \to \infty$,

$$y_q^k(t) = \frac{e^{\lambda(q)t} t^{k-k_0}}{(k - k_0)!} + o(e^{\mu t} t^{k-\gamma}) \quad \text{if} \quad q = q_0, \qquad k_0 \leqq k \leqq h(q_0),$$

(15.2)
$$y_q^k(t) = o(e^{\mu t} t^{k-\gamma}) \quad \text{if} \quad q = q_0, \qquad 1 \leqq k < k_0,$$

$$y_q^k(t) = o(e^{\mu t} t^{k-\gamma}) \quad \text{if} \quad q \neq q_0, \qquad 1 \leqq k \leqq h(q),$$

$$y_q^k(t) = o(e^{\mu t} t^{1-\gamma}) \quad \text{if} \quad j \neq q,$$

where $\gamma = h_0 - h(q_0) + k_0 \geqq 1$. This follows from Theorem 13.1 with j_0, β replaced by $h(q_0) - k_0$, $\gamma = h_0 - [h(q_0) - k_0]$ and the choice $c_q^k = 1$ or $c_q^k = 0$ according as $(qk) = (q_0 k_0)$ or $(qk) \neq (q_0 k_0)$.

The set of solutions $\xi_{qk}(t)$ is a set of $\Sigma h(q)$ linearly independent solutions. Also if $n = \Sigma h(p)$, then Theorem 8.2 implies that there are exactly n linearly independent solutions $\xi_1(t), \ldots, \xi_n(t)$ satisfying

$$\limsup t^{-1} \log \|\xi(t)\| < \mu$$

and $n + \Sigma h(q)$ linearly independent solutions satisfying

$$\limsup t^{-1} \log \|\xi(t)\| \leqq \mu.$$

Hence if $\xi(t) \neq 0$ is a solution of (15.1) satisfying (13.14), then there exist constants $c_1, \ldots, c_n$ and c_q^k such that

(15.3)
$$\xi(t) = \sum_{j=1}^{n} c_j \xi_j(t) + \sum_{q} \sum_{k=1}^{k(q)} c_q^k \xi_{qk}(t)$$

and that not all c_q^k are 0. It will be left to the reader to verify that this implies Theorem 13.2.

16. Corollaries and Refinements

When the matrix E in Theorem 13.1 has simple elementary divisors (e.g., when the eigenvalues of E are distinct) or even if $h_* = 1$, then $h_* = 1$, $j_0 = 0$, $\beta \geqq 1$, and condition (13.12) reduces to

$$\int^{\infty} t^\alpha \psi_1(t)\, dt < \infty \quad \text{for} \quad \alpha = \beta - 1 \geqq 0;$$

cf. Corollary 4.2. Here, the asymptotic formulae (13.7) reduce to

$$y_q(t) = e^{\lambda(q)t}[c_q + o(t^{-\alpha})], \qquad y_j(t) = e^{\lambda(q)t} o(t^{-\alpha}) \quad \text{for} \quad j \neq q.$$

For a fixed j_0, the smallest admissible value of β in Theorem 13.1 is $\beta = h_* - j_0$ in which case (13.12) becomes $\int^{\infty} t^{h_* - 1} \psi_1(t)\, dt < \infty$. A larger choice of β has the role of possibly increasing the number of significant terms in the asymptotic formulae (13.7) and of improving the error terms. When (13.12) is strengthened to

$$(16.1) \qquad \int^{\infty} t^{2h_*-1}\psi_1(t)\,dt < \infty,$$

the maximal number of significant terms is possible. In this case, we have

Corollary 16.1. *Let $E = \mathrm{diag}\,[J(1), \ldots, J(g)]$, μ, h_* be as in Theorem 13.1 and let $F(t, \xi)$ be continuous for $t \geq 0$ and all ξ and satisfy* (13.11), *where $\psi_1(t) > 0$ is a continuous function satisfying* (16.1). *Let $\xi = \xi_0(t) \neq 0$ be a solution of the linear system $\xi' = E\xi$ such that $t^{-1}\log \|\xi(t)\| \to \mu$ as $t \to \infty$. Then* (13.1) *has a solution $\xi(t)$ satisfying $\|\xi(t) - \xi_0(t)\|\, e^{-\mu t} \to 0$, $t \to \infty$.*

In this corollary, E is not required to be in a Jordan normal form (cf. Remark 1 following Theorem 13.1). If it is, we can, in addition, assign a partial set of $\Sigma h(p)$ initial conditions, $y_p(t_0) = y_{p0}$ for sufficiently large t_0. Also, $\xi(t)$ satisfies the asymptotic relations (13.7), where $l(q) = 1$, $k(q) = h(q)$, $c_q^{\ k}$ are suitable constants determined by $\xi_0(t)$, j_0 is defined by (13.16), and $\beta = 2h_* - j_0$. This improves the asymptotic relation claimed in the corollary.

The deduction of Theorem 13.1 from Lemma 4.3 shows that assumptions (13.11), (13.12) can be weakened somewhat.

Corollary 16.2. *Let assumptions* (13.11), (13.12) *of Theorem 13.1 be relaxed to*

$$(16.2) \qquad \|Q^{-1}(t)F(t, Q(t)\zeta)\| \leq \psi_0(t)\, \|\zeta\|,$$

or, more generally, to

$$(16.3) \qquad \|Q^{-1}(t)F(t, Q(t)\zeta)\| \leq \psi(t)\, \|\zeta\|$$

$$(16.4) \quad e^{-\mu t}t^{\beta-i}\,|F_q^{\ i}(t, Q(t)\zeta)| \leq \psi_0(t)\, \|\zeta\| \qquad \text{for} \quad i = 1, \ldots, k(q),$$

where $\xi = Q(t)\zeta$ is given by (14.2) *and $\psi(t)$, $\psi_0(t)$ are positive continuous functions for $t > 0$ such that*

$$(16.5) \qquad \sup_{s \geq t} (1 + s - t)^{-1} \int_s^t \psi(r)\,dr \to 0 \qquad \text{as} \quad t \to \infty,$$

$$(16.6) \qquad \int^{\infty} \psi_0(t)\,dt < \infty.$$

Then the conclusions of Theorem 13.1 remain valid.

Exercise 16.1. By referring to Remark 1 following Lemma 4.3 and to the proof of Theorem 13.1, find sharper estimates for the o-terms in (13.7) under the conditions (16.3)–(16.6) of Corollary 16.2.

Remark 2 following Theorem 13.1 and Corollary 16.2 have important consequences. For example, suppose that F in (13.1) does not depend on

t, so that (13.1) can be written as

(16.7) $$\xi' = E\xi + F(\xi),$$

where $F(\xi)$ is defined for $\|\xi\| < \delta < \infty$ and satisfies

(16.8) $$\|F(\xi)\| \leqq C_0 \|\xi\|^{1+\theta}, \qquad \theta > 0, \qquad C_0 = \text{const.},$$

or, more generally,

(16.9) $$\|F(\xi)\| \leqq \frac{C_0 \|\xi\|}{|\log \|\xi\| |^\nu}, \qquad \nu > j_0 + \beta,$$

or even

(16.10) $$\|F(\xi)\| \leqq \varphi(\|\xi\|) \|\xi\|,$$

where $\varphi(\rho)$ is a nondecreasing function for $0 \leqq \rho < \delta$ such that

(16.11) $$\int_{+0} \rho^{-1} |\log \rho|^{j_0+\beta-1} \varphi(\rho) \, d\rho < \infty.$$

Then (14.16) and (16.10) imply that if $\mu < 0$ and $\|\zeta\| \leqq 1$, then

$$\|Q^{-1}\| \cdot \|F(Q\zeta)\| \leqq ce^{-\mu t} t^{\beta-1} \varphi(ce^{\mu t} t^{j_0}) ce^{\mu t} t^{j_0} \|\zeta\|$$

for large t. Thus the analogue of (16.2) holds with

$$\psi_0(t) = c^2 t^{j_0+\beta-1} \varphi(ce^{\mu t} t^{j_0}).$$

If $\rho = ce^{\mu t} t^{j_0}$ is introduced as a new integration variable in the integral in (16.11) and it is noted that $d\rho/\rho \sim \mu \, dt$ and $\log \rho \sim \mu t$ as $t \to \infty$, then it is seen that (16.6) is a consequence of (16.11).

Corollary 16.3. *In* (16.7), *let* $E = \text{diag}\,[J(1), \ldots, J(g)]$ *be as in Theorem* 13.1, *let* $F(\xi)$ *be continuous for* $\|\xi\| < \delta \, (<\infty)$ *and satisfy* (16.10), *where* $\varphi(\rho)$ *is a nondecreasing function of* ρ *satisfying* (16.11). *Let* $\mu < 0$. *Then the conclusions of Theorem* 13.1, *with* (13.3) *replaced by* (16.7), *remain valid.*

Exercise 16.2.   By involving the Remark 2 following Lemma 4.3, show that conditions (16.10), (16.11) in Corollary 16.3 can be replaced by

(1,6.12) $$\|F(t, \xi)\| \leqq \varphi_0(\|\xi\|) \qquad \text{for} \quad \|\xi\| < \delta,$$

where $\varphi_0(\rho)$ is a nondecreasing function of ρ, $0 < \rho < \delta$, such that

(16.13) $$\int^\infty \rho^{-2} |\log \rho|^{\beta+j_0-1} \varphi_0(\rho) \, d\rho < \infty.$$

[This is somewhat more general than Corollary 16.3 for (16.12) implies (16.10) with $\varphi_0(\rho) = \varphi(\rho)\rho$. Although (16.10) is a consequence of (16.12) with $\varphi(\rho) = \varphi_0(\rho)/\rho$, the monotony of φ_0 does not imply that of φ.]

Analogously, we obtain the following consequence of the proofs of Theorems 13.1 and 13.2.

Corollary 16.4.   *Let E, μ, F, φ be as in Corollary 16.3 except that* (16.11) *is replaced by*

(16.14) $$\int^{\infty} \rho^{-1} |\log \rho|^{h_0-1} \varphi(\rho)\, d\rho < \infty, \qquad \text{where} \quad h_0 \geqq h_*$$

(and h_0 need not be an integer).   Then the conclusions of Theorem 13.2, *with* (13.1) *replaced by* (16.7), *are valid.*

17. Linear Higher Order Equations

The results of §§ 4, 11, 13, 16 will be applied in this section to a linear differential equation of order $d > 1$,

(17.1) $u^{(d)} + [a_1 + p_1(t)]u^{(d-1)} + \cdots$
$$+ [a_{d-1} + p_{d-1}(t)]u' + [a_d + p_d(t)]u = 0,$$

for a real- or complex-valued function u. This will be viewed as a perturbation of the equation

(17.2) $$u^{(d)} + a_1 u^{(d-1)} + \cdots + a_{d-1}u' + a_d u = 0$$

with constant coefficients. The characteristic equation for (17.2) is

(17.3) $$\lambda^d + a_1\lambda^{d-1} + \cdots + a_{d-1}\lambda + a_d = 0.$$

Equation (17.1) can be written as a linear system

(17.4) $$\xi' = [R + G(t)]\xi,$$

for the d-dimensional vector $\xi = (u^{(d-1)}, \ldots, u^{(1)}, u^{(0)})$, where $u = u^{(0)}$ and R, $G(t)$ are the matrices

$$R = \begin{pmatrix} -a_1 & -a_2 & -a_3 & \cdots & -a_{d-1} & -a_d \\ 1 & 0 & 0 & \cdots & 0 & 0 \\ 0 & 1 & 0 & \cdots & 0 & 0 \\ \cdot & \cdot & \cdot & \cdots & \cdot & \cdot \\ \cdot & \cdot & \cdot & \cdots & \cdot & \cdot \\ 0 & 0 & 0 & \cdots & 1 & 0 \end{pmatrix},$$

(17.5)

$$G(t) = \begin{pmatrix} -p_1 & -p_2 & -p_3 & \cdots & -p_{d-1} & -p_d \\ 0 & 0 & 0 & \cdots & 0 & 0 \\ 0 & 0 & 0 & \cdots & 0 & 0 \\ \cdot & \cdot & \cdot & \cdots & \cdot & \cdot \\ \cdot & \cdot & \cdot & \cdots & \cdot & \cdot \\ 0 & 0 & 0 & \cdots & 0 & 0 \end{pmatrix}.$$

Note first that if $a_1 = \cdots = a_d = 0$, then R is in the Jordan normal form and consists of one Jordan block with $\lambda = 0$ on its main diagonal. If, the coefficients $p_1(t), \ldots, p_d(t)$ are small, then (17.1) can be considered to be a perturbation of $u^{(d)} = 0$ which has the linearly independent solutions $u = 1, t, \ldots, t^{d-1}$. It will be verified that Corollary 16.2 has the following consequence.

Theorem 17.1 *In* (17.1), *let* $a_1 = \cdots = a_d = 0$ *and let* $p_1(t), \ldots, p_d(t)$ *be continuous complex-valued functions for* $t \geqq 0$ *satisfying*

$$(17.6) \qquad \int^{\infty} t^{k+\alpha-1} |p_k(t)| \, dt < \infty \qquad \text{for some} \quad \alpha \geqq 0$$

and $k = 1, \ldots, d$. *Then, for any* j, $0 \leqq j \leqq d - 1$, (17.1) *has a solution satisfying* $u(t) = (t^j/j!)(1 + o(t^{-\alpha}))$ *as* $t \to \infty$, *and this relation can be "differentiated"* $d - 1$ *times, i.e.,*

$$(17.7) \qquad \begin{aligned} u^{(k)} &= \frac{t^{j-k}}{(j-k)!}\left(1 + o\left(\frac{1}{t^{\alpha}}\right)\right) \qquad \text{for} \quad k = 0, \ldots, j, \\ u^{(k)} &= o(t^{j-k-\alpha}) \qquad \text{for} \quad k = j+1, \ldots, d-1. \end{aligned}$$

It will be clear from the proof that, for a *given* j (rather than for *any* j) on the range $0 \leqq j \leqq d - 1$, a sufficient condition for the existence of a solution satisfying (17.7) is that

$$\int^{\infty} \left(\sum_{k=1}^{d-j-1} |p_k(t)| \, t^{k-1} + \sum_{k=d-j}^{d} |p_k(t)| \, t^{k+\alpha-1} \right) dt < \infty.$$

Proof. Since R in (17.5) is in a Jordan normal form, (17.4) can be identified with (13.3) if $F(t, \xi) = G(t)\xi$, where $\xi = (\xi^1, \ldots, \xi^d)$ and $\xi^k = u^{(d-k)}$. In order to verify the conditions of Corollary 16.2, note that the sets of p and r are vacuous and that there is only one q. Correspondingly, $\lambda(q) = 0$ and $h(q) = d$. Let $j_0 = j$ be the index j in (17.7), $\beta = d - j + \alpha$, and $l(q) = k(q) = d - j$. Thus $\sum_I$ in (13.8) contains no terms if $k < d - j$ or exactly one term $i = d - j$ if $d - j \leqq k \leqq d$. Also, let $c_q{}^i = 1$ or $c_q{}^i = 0$ according as $i = d - j$ or $i \neq d - j$, so that the desired asymptotic relation (17.7) is identical with (the first part of) (13.7).

Consider $F(t, Q(t)\zeta) = G(t)Q(t)\zeta$. Since only the first row of $G(t)$ contains nonzero elements, this can be written as $F(t, Q\zeta) = (F^1, 0, \ldots, 0)$, where, by (14.1)–(14.2) and (17.5),

$$F^1 = -\sum_{k=d-j}^{d} \frac{p_k t^{k-d+j}}{(k-d+j)!} z^{d-j} - \sum_{k=1}^{d} p_k t^{k-\beta} \sum_{II} \frac{1}{(k-i)!} z^i$$

$\zeta = (z^1, \ldots, z^d)$, and $\sum_{II}$ is the sum over the set of i-values, $1 \leqq i \leqq k$ and $i \neq d - j$. Consequently, $\| Q^{-1}(t) \| \leqq ct^{\beta-1}$ implies that

$$\| Q^{-1}(t) \| \cdot \| F(t, Q\zeta) \| \leqq c_0 \left(\sum_{k=1}^{d-j-1} |p_k(t)| \, t^{k-1} + \sum_{k=d-j}^{d} |p_k(t)| \, t^{k+\alpha-1} \right) \| \zeta \|$$

for a suitable constant c_0 and large t. Since the coefficient of $\|\zeta\|$ is a function $\psi_0(r)$ satisfying $\int^\infty \psi_0(t)\, dt < \infty$ by (17.6), Theorem 17.1 follows from Corollary 16.2.

When all the roots of (17.3) are the same, say λ, this can be reduced to the situation of Theorem 17.1 by replacing u by the new dependent variable $v = ue^{-\lambda t}$. In the other extreme case, when λ is a simple root of (17.3), we have

Theorem 17.2 *Let* (17.3) *have a simple root, say λ, and suppose that if λ_0 is any other root, then* $\mathrm{Re}\ \lambda \neq \mathrm{Re}\ \lambda_0$. *Let $p_1(t), \ldots, p_d(t)$ be continuous functions for $t \geqq 0$ satisfying*

$$(17.8) \qquad \int^\infty |p_k(t)|\, t^\alpha\, dt < \infty \qquad \text{for some}\quad \alpha \geqq 0$$

and $k = 1, \ldots, d$. Then (17.1) *has a solution $u(t)$ satisfying*

$$(17.9) \qquad u^{(k)}(t) = e^{\lambda t}\left[\lambda^k + o\!\left(\frac{1}{t^\alpha}\right)\right] \qquad \text{for}\quad k = 0, \ldots, d-1$$

as $t \to \infty$.

Proof. This is the simplest case of Corollary 16.2 when $h(q) = 1$. Let $j_0 = 0$, $\beta = 1 + \alpha$. Let (17.4) be identified with (13.13) in Remark 1 following Theorem 13.1. Then

$$\|F^0(t, \xi)\| = \|G(t)\xi\| \leqq c\,\|\xi\|\sum_{k=1}^{d}|p_k(t)|$$

for some constant c. Thus (13.11) holds with $\psi_1(t) = c\Sigma\,|p_k(t)|$ and the theorem follows from Corollary 16.2.

Consider the general case where (17.3) has a root, say $\lambda = 0$, of multiplicity h, $1 \leqq h \leqq d$.

Theorem 17.3. *Let $\lambda = 0$ be a root of* (17.3) *of multiplicity h, $1 \leqq h \leqq d$; i.e., let $a_{d-h+1} = \cdots = a_d = 0$ and $a_{d-h} \neq 0$; and suppose that if $\lambda_0 \neq 0$ is any other root, then* $\mathrm{Re}\ \lambda_0 \neq 0$. *Let $p_1(t), \ldots, p_d(t)$ be continuous functions for $t \geqq 0$ such that*

$$(17.10) \qquad \int^\infty t^{k-d+h+\alpha-1}|p_k(t)|\, dt < \infty \qquad \text{for}\quad k = d-h+1, \ldots, d.$$

$$(17.11) \qquad \int^\infty |p_k(t)|\, t^\alpha\, dt < \infty \qquad \text{for}\quad k = 1, \ldots, d-h,$$

for some $\alpha \geqq 0$. Then, for any j, $0 \leqq j \leqq h-1$, (17.1) *has a solution $u(t)$ satisfying, as $t \to \infty$,*

$$u^{(k)} = \frac{t^{j-k}}{(j-k)!} + o(t^{j-k-\alpha}) \qquad \text{for}\quad k = 0, \ldots, j,$$

$$(17.12) \qquad u^{(k)} = o(t^{j-k!-\alpha}) \qquad \text{for}\quad k = j+1, \ldots, h-1,$$

$$u^{(k)} = o(t^{j-h+1-\alpha}) \qquad \text{for}\quad k = h, \ldots, d-1.$$

Exercise 17.1.   Prove Theorem 17.3.

Exercise 17.2.   Restate Theorem 17.3 when $\lambda = 0$ is replaced by an arbitrary λ.

Theorems 17.1–17.3 depend on §§ 13, 16; we can also apply the results of § 11: .

Theorem 17.4.   *Let λ be a simple root of* (17.3) *and suppose that if λ_0 is any other root, then* Re $\lambda_0 \neq$ Re λ. *Let $p_1(t), \ldots, p_d(t)$ be continuous functions for $t \geq 0$ satisfying*

$$(17.13) \qquad p_k(t) \to 0 \quad \text{as} \quad t \to \infty \quad \text{for} \quad k = 1, \ldots, d$$

or, more generally,

$$(17.14) \qquad \sup_{s \geq t} (1 + s - t)^{-1} \int_t^s |p_k(r)| \, dr \to 0 \quad \text{as} \quad t \to \infty$$

for $k = 1, \ldots, d$. Then (17.1) *possesses a solution $u(t) \neq 0$ for large t such that*

$$(17.15) \qquad u^{(k)}(t) = u(t)[\lambda^k + o(1)] \text{ for } k = 1, \ldots, d - 1 \text{ as } t \to \infty.$$

Proof.   It is sufficient to prove this theorem in the case that $\lambda = 0$, otherwise $ue^{-\lambda t}$ is introduced as a new dependent variable in (17.1). Thus $a_d = 0$. Write (17.1) as the system (17.4), (17.5). Let Y be a constant nonsingular matrix such that $Y^{-1}RY = E = \text{diag}\,[J(1), \ldots, J(g)]$ is in a Jordan normal form. The first column of Y can be taken to be $(0, \ldots, 0, 1)$, since this is an eigenvector of R belonging to the simple eigenvalue $\lambda = 0$. Thus $J(1)$ is the 1×1 zero matrix and the diagonal elements $\lambda(j)$ of $J(j)$ are such that Re $\lambda(j) \neq 0$ for $j = 2, \ldots, d$. The change of variables $\xi = Y\eta$ reduces (17.4) to

$$(17.16) \qquad\qquad \eta' = E\eta + Y^{-1}G(t)Y\eta.$$

If $\eta = (\eta^1, \ldots, \eta^d)$, it follows from Theorem 11.1 that (17.16) has a solution $\eta(t) \neq 0$ such that $\eta^k(t) = o(|\eta^1(t)|)$ as $t \to \infty$ for $k = 2, \ldots, d$. The corresponding solution $\xi(t) = Y\eta(t)$ of (17.4), where $\xi = (\xi^1, \ldots, \xi^d)$, satisfies $\xi^k(t) = o(|\xi^d(t)|)$ as $t \to \infty$ for $k = 1, \ldots, d - 1$. Since $u(t) = \xi^d(t)$ and $u^{(d-k)} = \xi^k$ for $k = 1, \ldots, d - 1$, the relations (17.15) follow.

It cannot be expected that condition (17.14) in Theorem 17.4 can be improved. This is shown by the following exercise.

Exercise 17.3.   (*a*) In the second order equation,

$$(17.17) \qquad\qquad u'' - [\lambda^2 + q(t)]u = 0,$$

let $q(t)$ be continuous for $t \geq 0$ and Re $\lambda \neq 0$. Show that a necessary condition for (17.17) to possess a solution $u(t)$ which does not vanish for large t and satisfies $u'/u \to \lambda$ as $t \to \infty$ is that

(17.18) $\displaystyle \sup_{s \geq t} (1 + s - t)^{-1} \left| \int_t^s q(r)\, dr \right| \to 0$ as $t \to \infty$.

(b) Prove that the necessary condition (17.18) in (a) is sufficient if λ is a positive number and $q(t)$ is real-valued; see Hartman [5]. For a related result, see Exercise XI 7.5.

Theorem 17.5. *Assume the conditions of Theorem 17.4 with (17.14) strengthened to*

(17.19) $\displaystyle \int^{\infty} |p_k(t)|^p\, dt < \infty$ *for some p, $1 \leq p \leq 2$,*

for $k = 1, \ldots, d$. Then a solution $u(t) \neq 0$ of (17.1) satisfying (17.15) also satisfies

(17.20) $\displaystyle u(t) = [c + o(1)] \exp \int^t [\lambda + g(s)]\, ds$ *as $t \to \infty$,*

where $c \neq 0$ is a constant,

(17.21) $\displaystyle g(t) = - \frac{1}{F'(\lambda)} \sum_{k=1}^{d} \lambda^{d-k} p_k(t),$

$F' = dF/d\lambda$ and F is the polynomial on the left of (17.3) [so that $F'(\lambda) = (\lambda - \lambda_2) \ldots (\lambda - \lambda_d)$ if $\lambda_2, \ldots, \lambda_d$ are the roots of (17.3) distinct from λ].

Proof.   Write (17.1) as the system (17.4), (17.5) and make the change of variables $\xi = Y\eta$, where $Y = Y(0)$ is the constant matrix given in Exercise IV 8.2 and having $(\lambda^{d-1}, \ldots, \lambda, 1)$ as its first column. Then (17.4) becomes (17.16), where $E = \text{diag}\,[J(1), \ldots, J(g)]$ and $J(1)$ is the 1×1 matrix λ. Since Y is a constant matrix, (17.19) implies that the pth power of the absolute values of the elements of $Y^{-1}G(t)Y$ are integrable over $0 \leq t < \infty$. Hence, it follows from Theorem 11.1 that (17.16) has solutions $\eta(t)$ such that if $\eta = (\eta^1, \ldots, \eta^d)$, then $\eta^1(t) \neq 0$ for large t and $\eta^j(t) = o(|\eta^1(t)|)$ as $t \to \infty$ for $j = 2, \ldots, d$. Furthermore, by Exercise 11.2, any such solution satisfies

$$\eta^1(t) = [c + o(1)] \exp \int_0^t [\lambda + g(s)]\, ds \qquad \text{as   } t \to \infty,$$

where $g(t)$ is the element in the first row and first column of $Y^{-1}G(t)Y$.

In order to calculate $g(t)$, note that since the first column of Y is $(\lambda^{d-1}, \ldots, \lambda, 1)$, the element in the first row, first column of $G(t)Y$ is $-\Sigma \lambda^{d-k} p_k(t)$. All elements of $G(t)Y$ not in the first row are 0. Hence, the upper left corner element of $Y^{-1}G(t)Y$ is $-\Sigma \lambda^{d-k} p_d(t)$ times the corresponding element of Y^{-1}. This element of Y^{-1} is the cofactor A of the corresponding element of Y divided by det Y. If the distinct roots of (17.3) and their multiplicities are $\lambda, \lambda(2), \ldots, \lambda(g)$ and 1, $h(2), \ldots, h(g)$, respectively, then

$$\det Y = \prod_{j=2}^{g} [\lambda - \lambda(j)]^{h(j)} \prod_{2 \leq i < j} [\lambda(i) - \lambda(j)]^{h(i)h(j)};$$

see Exercise IV 8.2. The determinant which is the cofactor A has the same form as det Y, except that λ does not occur. It follows that A is the second of the two products above. Hence

$$g = - \frac{A}{\det Y} \sum \lambda^{d-k} p_k = - \prod_{j=2}^{g} [\lambda - \lambda(j)]^{-h(j)} \sum \lambda^{d-k} p_k,$$

i.e., (17.21) holds. The relations $\xi = Y\eta$, $\xi^k = u^{(d-k)}$ and the fact that the first column of Y is $(\lambda^{d-1}, \ldots, \lambda, 1)$ completes the proof of Theorem 17.5.

As an illustration of Theorem 17.5, consider the second order equation (17.17) in which Re $\lambda \neq 0$ and $\int^{\infty} |q(t)|^p \, dt < \infty$ for some p, $1 \leq p \leq 2$. Then (17.17) has a pair of solutions satisfying

$$(17.22) \quad u' \sim \pm \lambda u, \qquad u \sim \exp \pm \int^{t} \left[\lambda + \frac{1}{2\lambda} q(s) \right] ds \qquad \text{as} \quad t \to \infty.$$

Exercise 17.4. Let $q(t)$ be real-valued and continuous for $t \geq 0$, $q(t) \to 0$ as $t \to \infty$, and $q(t)$ of bounded variation for $t \geq 0$ [e.g., let $q(t)$ be monotone or let $q(t)$ have a continuous derivative such that $\int^{\infty} |q'(t)| \, dt < \infty$]. Show that (a) $u'' + [1 + q(t)]u = 0$ has solutions $u(t)$ satisfying

$$u' \sim \pm iu \quad \text{and} \quad u(t) \sim \exp \pm i \int^{t} [1 + q(s)]^{\frac{1}{2}} \, ds \qquad \text{as} \quad t \to \infty,$$

and that (b) $u'' + [-1 + q(t)]u = 0$ has solutions $u(t)$ satisfying

$$u' \sim \pm u \quad \text{and} \quad u(t) \sim \exp \pm \int^{t} [1 - q(s)]^{\frac{1}{2}} \, ds \qquad \text{as} \quad t \to \infty.$$

(c) State an analogous result for (17.17) where $\lambda \neq 0$ and it is not assumed that λ or $q(t)$ are real-valued; cf. Exercises XI 8.4(b).

Exercise 17.5. In the differential equation

$$(17.23) \qquad u'' - f(t)u = 0,$$

let $f(t)$ be a continuously differentiable, complex-valued function for $t \geq 0$ such that

$$(17.24) \qquad \text{Re} f^{\frac{1}{2}}(t) \neq 0 \quad \text{and} \quad \int^{\infty} |\text{Re} f^{\frac{1}{2}}(t)| \, dt = \infty.$$

(a) Show that if $f'/|f| \cdot |\mathrm{Re}\, f^{1/2}| \to 0$ as $t \to \infty$, then (17.23) has solutions satisfying

(17.25) $u' \sim \pm f^{1/2}(t)u$ as $t \to \infty$.

(b) Show that if $\displaystyle\int^\infty |\mathrm{Re}\, f^{1/2}(t)|^{1-p}\, |f'(t)/f(t)|^p\, dt < \infty$ for some p, $1 \leqq$

$p \leqq 2$, then (17.23) has solutions satisfying (17.25) and

(17.26) $u(t) \sim f^{-1/4}(t) \exp \pm \displaystyle\int^t f^{1/2}(r)\, dr$ as $t \to \infty$.

If $f > 0$, condition (17.24) is redundant.

(c) Show that if $f'(t)/f^{3/2}(t)$ is of bounded variation, i.e.,

$$\int^\infty |d(f'/f^{3/2})| < \infty,$$

$f'(t)/f^{3/2}(t) \to 0$ as $t \to \infty$, then (17.23) has a pair of solutions satisfying (17.25) and

(17.27) $u(t) \sim f^{-1/4}(t) \exp \pm \displaystyle\int^t f^{1/2}(1 + f'^2/16 f^3)^{1/2}\, dr$ as $t \to \infty$.

For other results of this type, see § XI 9. For analogous results when $\mathrm{Re}\, f^{1/2} \equiv 0$, see Exercise XI 8.5.

Exercise 17.6. As a simple application of the last exercise, consider Weber's equation

(17.28) $u'' + tu' - 2\lambda u = 0$,

where λ is a constant. (a) By introducing the new independent variable $s = \frac{1}{2}t^2$, deduce from Theorem 17.4 that (17.28) has a pair of solutions $u_0(t)$, $u_1(t)$ which do not vanish for large t and satisfy $u_0' \sim -tu_0$, $u_1' = o(tu_1)$ as $t \to \infty$. (b) Show that (17.28) has a pair of solutions u_0, u_1 satisfying $u_0 \sim t^{-1-2\lambda}e^{-t^2/2}$, $u_1 \sim t^{2\lambda}$ as $t \to \infty$. (c) Find asymptotic relations for derivatives u' of solutions u of (17.28) by differentiating (17.28) and applying (b). (See also Exercise XI 9.7.)

Notes

For references and other treatments of the topics in this chapter, see Cesari [2] and Bellman [4].

SECTION 1. The main results, Theorems 1.1 and 1.2, are due to Wintner [3], [7], [8], who gave the existence assertions essentially in the form stated in Exercise 1.2. Linear cases, where $F(t, \xi) = G(t)\xi$ for a matrix $G(t)$, are much older; see Dunkel [1]. For Exercise 1.1, see Hale and Onuchic [1]; cf. § XII 9. For Exercise 1.4, see Wintner [21].

SECTION 2. Theorem 2.1 was formulated by Ważewski [5] and is a very useful tool in the study of differential equations. Special cases of this theorem and the arguments in its proof had been used earlier; cf. Hartman and Wintner [1] or Nemytzkiĭ and

Stepanov [1, p. 93]. For another type of topological argument, useful for similar purposes, cf. Atkinson [2]. Exercise 2.1 is due to Pliś [1].

SECTION 3. The results of this section are due to Ważewski [5].

SECTIONS 4–7. Lemmas 4.1 and 4.2 are related to results of Ważewski [6], Szmydtówna [1], Lojasiewicz [1], and Hartman and Wintner [17], [19]. The proofs in the text are adapted from those of Ważewski and his students just mentioned; for other proofs, see the papers of Hartman and Wintner. Lemma 4.3 and applications were given in the papers of Hartman and Wintner. Conditions of the type (4.24) were introduced by Hartman [5]. For Exercise 4.6, see Levinson [3] (for the part dealing with boundedness, see Cesari [1]); an analogous result (see Exercise 17.4) on a second order equation was given by Wintner [10]. See Cesari [2, pp. 38–42], for related results and references. For results related to Exercises 4.8, 4.9 and applications, see Coffman [2].

SECTIONS 8–12. Results related to those occurring in § 8 for analytic systems are the oldest in this chapter and go back to Poincaré and to Lyapunov [2]. For particular cases for linear differential equations, see Poincaré [4] and Perron [2]. Cotton [1] and then Perron [9], [10], [12] systematically investigated nonanalytic, nonlinear cases, but under conditions heavier than those in the text. Their results depended on the method of successive approximations. See also Bellman [1], who used fixed point theorems to obtain an analogue of Theorem 8.1, and the references above for §§ 4–7 to Ważewski, Hartman, and Wintner, etc. The relaxation of the condition "$\psi(t) \to 0$ as $t \to \infty$" to (4.24) is due to Hartman [5] and to Hartman and Wintner [19]. A form of Theorem 8.2 involving stronger hypothesis and weaker assertions was given by Petrovsky [1]. The last two parts of Corollary 8.1 are proved in Coddington and Levinson [2, Chap. 13] by the method of successive approximations; cf. Exercise 8.1. For another application of a related method of successive approximations, see Lillo [1]. The comparatively simple proofs in the text for Theorems 8.2, 8.3 and Corollary 8.1 are new. Theorem 11.1 is a slight improvement of a result of Lettenmeyer [2]. Theorem 11.2 is given by Hartman and Wintner [19]. Results of the type in Exercise 11.1 go back to Bôcher [2] and Dunkel [1]; cf. notes on §§ 13–16 below. Exercise 11.2 was first given by Hartman [5] for the case of a second order equation (see Theorem 17.5 with $d = 2$) and generalized to the situation in Exercise 11.2 by Hartman and Wintner [17].

SECTIONS 13–16. Results of the type in Theorem 13.1 were first given by Bôcher [2] for a second order, linear equation. Using successive approximations similar to those of Exercise 8.1, Dunkel [1] generalized Bôcher's result to arbitrary linear systems (13.1), where $F(t, \xi) = G(t)\xi$, but his results are not as sharp as those given here. Theorems 13.1, 13.2 and their corollaries in § 16 are due to Hartman and Wintner [19]. The proofs in the text, which take full advantage of Ważewski's principle of § 2, depend in an essential way on the change of variables (14.1)–(14.2) similar to those introduced by Hartman and Wintner [17] and simplified by Coffman [2]. See also Olech [1].

SECTION 17. When $\alpha = 0$, Theorem 17.1 is due to Bôcher [2] for $d = 2$ and, in a weakened form, it is contained in Dunkel's result [1] for arbitrary d. For $\alpha = 0$, it is given by Faedo [1] and Ghizetti [2]. Theorem 17.2 and a less precise form of Theorem 17.3 with $\alpha = 0$, are also contained in Dunkel [1]; Faedo [1], [2]; and Ghizetti [1]. Theorem 17.4 is a generalization of results of Poincaré [4] and Perron [2] and is contained in Hartman and Wintner [17]. For Exercise 17.3, see Hartman [5]. Theorem 17.5 for $d = 2$ is due to Hartman [5]; the result formulated in the text is new. For a generalization of the case $d = 2$, see Bellman [3]. Exercise 17.4 is due to Wintner [10], [13] and is contained in the more general result of Exercise 4.6; cf. also Exercise XI 8.4(b). Results of the type in Exercise 17.5(a) go back to Wiman [1], [2]; for both parts (a) and (b)* see Hartman and Wintner [17].

*For related results, see Coppel [S1], Chapter IV.

Chapter XI

Linear Second Order Equations

1. Preliminaries

One of the most frequently occurring types of differential equations in mathematics and the physical sciences is the linear second order differential equation of the form

$$(1.1) \qquad u'' + g(t)u' + f(t)u = h(t)$$

or of the form

$$(1.2) \qquad (p(t)u')' + q(t)u = h(t).$$

Unless otherwise specified, it is assumed that the functions $f(t)$, $g(t)$, $h(t)$, and $p(t) \neq 0$, $q(t)$ in these equations are continuous (real- or complex-valued) functions on some t-interval J, which can be bounded or unbounded. The reason for the assumption $p(t) \neq 0$ will soon become clear.

Of the two forms (1.1) and (1.2), the latter is the more general since (1.1) can be written as

$$(1.3) \qquad (p(t)u')' + p(t)f(t)u = p(t)h(t),$$

if $p(t)$ is defined as

$$(1.4) \qquad p(t) = \exp \int_a^t g(s)\, ds$$

for some $a \in J$. As a partial converse, note that if $p(t)$ is continuously differentiable then (1.2) can be written as

$$u'' + \frac{p'(t)}{p(t)}u' + \frac{q(t)}{p(t)}u = \frac{h(t)}{p(t)},$$

which is of the form (1.1).

When the function $p(t)$ is continuous but does not have a continuous derivative, (1.2) cannot be written in the form (1.1). In this case, (1.2) is to be interpreted as the first order, linear system for the binary vector

322

$$x = (x^1, x^2) \equiv (u, p(t)u'),$$

(1.5) $$x^{1\prime} = \frac{x^2}{p(t)}, \qquad x^{2\prime} = -q(t)x^1 + h(t).$$

In other words, a solution $u = u(t)$ of (1.2) is a continuously differentiable function such that $p(t)u'(t)$ has a continuous derivative satisfying (1.2). When $p(t) \neq 0$, $q(t)$, $h(t)$ are continuous, the standard existence and uniqueness theorems for linear systems of § IV 1 are applicable to (1.5), hence (1.2). [We can also deal with more general (i.e., less smooth) types of solutions if it is only assumed, e.g., that $1/p(t)$, $q(t)$, $h(t)$ are locally integrable; cf. Exercise IV 1.2.]

The particular case of (1.2) where $p(t) \equiv 1$ is

(1.6) $$u'' + q(t)u = h(t).$$

When $p(t) \neq 0$ is *real-valued*, (1.2) can be reduced to this form by the change of independent variables

(1.7) $$ds = \frac{dt}{p(t)}, \qquad \text{i.e.,} \qquad s = \int_a^t \frac{dr}{p(r)} + \text{const.}$$

for some $a \in J$. The function $s = s(t)$ has a derivative $ds/dt = 1/p(t) \neq 0$ and is therefore strictly monotone. Hence $s = s(t)$ has an inverse function $t = t(s)$ defined on some s-interval. In terms of the new independent variable s, the equation (1.2) becomes

(1.8) $$\frac{d^2u}{ds^2} + p(t)q(t)u = p(t)h(t),$$

where t in $p(t)q(t)$ and $p(t)h(t)$ is replaced by the function $t = t(s)$. The equation (1.8) is of the type (1.6).

If $g(t)$ has a continuous derivative, then (1.1) can be reduced to an equation of the form (1.6) also by a change of the dependent variable $u \to z$ defined by

(1.9) $$u = z \exp \left(-\tfrac{1}{2} \int_a^t g(s)\, ds \right)$$

for some $a \in J$. In fact, substitution of (1.9) into (1.1) leads to the equation

(1.10) $$z'' + \left[f(t) - \frac{g^2(t)}{4} - \frac{g'(t)}{2} \right]z = h(t) \exp \tfrac{1}{2} \int_a^t g(s)\, ds,$$

which is of the type (1.6).

In view of the preceding discussion, the second order equations to be considered will generally be assumed to be of the form (1.2) or (1.6). The following exercises will often be mentioned.

Exercise 1.1.   (*a*) The simplest equations of the type considered in this chapter are

(1.11) $u'' = 0, \quad u'' - \sigma^2 u = 0, \quad u'' + \sigma^2 u = 0,$

where $\sigma \neq 0$ is a constant. Verify that the general solution of these equations is

(1.12) $u = c_1 + c_2 t, \quad u = c_1 e^{\sigma t} + c_2 e^{-\sigma t}, \quad u = c_1 \cos \sigma t + c_2 \sin \sigma t,$

respectively. (*b*) Let a, b be constants. Show that $u = e^{\lambda t}$ is a solution of

(1.13) $$u'' + bu' + au = 0,$$

if and only if λ satisfies

(1.14) $$\lambda^2 + b\lambda + a = 0.$$

Actually, the substitution $u = z e^{-bt/2}$ [cf. (1.9)] reduces (1.13) to

$$z'' + \sigma^2 z = 0, \qquad \sigma^2 = a - \tfrac{1}{4}b^2.$$

Hence by (*a*) the general solution of (1.13) is

(1.15) $u = e^{-bt/2}(c_1 + c_2 t) \quad$ or $\quad u = c_1 e^{\lambda_1 t} + c_2 e^{\lambda_2 t}$

according as (1.14) has a double root $\lambda = \tfrac{1}{2}b$ or distinct roots $\lambda_1, \lambda_2 = -\tfrac{1}{2}b \pm (\tfrac{1}{4}b^2 - a)^{1/2}$. When a, b are real and $\tfrac{1}{4}b^2 - a < 0$, nonreal exponents in the last part of (1.15) can be avoided by writing

(1.16) $u = e^{-bt/2}[c_1 \cos (a - \tfrac{1}{4}b^2)^{1/2}t + c_2 \sin (a - \tfrac{1}{4}b^2)^{1/2}t].$

(*c*) Let μ be a constant. Show that $u = t^\lambda$ is a solution of

(1.17) $$u'' + \frac{\mu}{t^2}u = 0$$

if and only if λ satisfies

(1.18) $\lambda(\lambda - 1) + \mu = 0, \quad$ i.e., $\quad \lambda = \tfrac{1}{2} \pm (\tfrac{1}{4} - \mu)^{1/2}.$

Thus if $\mu \neq \tfrac{1}{4}$, the general solution of (1.17) is

(1.19) $u = c_1 t^{\lambda_1} + c_2 t^{\lambda_2}, \quad \mu \neq \tfrac{1}{4}, \quad$ and $\quad \lambda_1, \lambda_2 = \tfrac{1}{2} \pm (\tfrac{1}{4} - \mu)^{1/2}.$

If μ is real and $\mu > \tfrac{1}{4}$, the nonreal exponents can be avoided by writing

(1.20) $u = t^{1/2}[c_1 \cos (\mu - \tfrac{1}{4})^{1/2} \log t + c_2 \sin (\mu - \tfrac{1}{4})^{1/2} \log t].$

Actually, the change of variables $u = t^{1/2} z$ and $t = e^s$ transforms (1.17) into

(1.21) $$\frac{d^2 z}{ds^2} + (\mu - \tfrac{1}{4})z = 0.$$

Thus by (a) the general solution of (1.17) is

(1.22) $u = t^{1/2}(c_1 + c_2 \log t)$ or $u = c_1 t^{\lambda_1} + c_2 t^{\lambda_2}$

according as $\mu = \frac{1}{4}$ or $\mu \neq \frac{1}{4}$.

Exercise 1.2.   Consider the differential equation

(1.23) $u'' + q(t)u = 0$.

The change of variables

(1.24) $t = e^s$ and $u = t^{1/2}z$

transforms (1.23) into

(1.25) $\dfrac{d^2z}{ds^2} + t^2\left[q(t) - \dfrac{1}{4t^2}\right]z = 0$,   where $t = e^s$.

For a given constant μ, consider the sequence of functions

$q_0 = \mu - \frac{1}{4}$, $q_1(t) = \mu t^{-2}$, $q_2(t) = t^{-2}(\frac{1}{4} + \mu \log^{-2} t), \ldots,$

defined by $t^2[q_n(t) - 1/4t^2] = q_{n-1}(s)$ if $t = e^s$, so that $q_n(t) = t^{-2}[\frac{1}{4} + q_{n-1}(\log t)]$ or

$$q_n(t) = t^{-2}\left[\frac{1}{4}\sum_{k=0}^{n-2}\left(\prod_{j=1}^{k}\log_j t\right)^{-2} + \mu\left(\prod_{j=1}^{n-1}\log_j t\right)^{-2}\right]$$ for $n \geqq 1$,

$\log_1 t = \log t$, $\log_j t = \log (\log_{j-1} t)$, and the empty product is 1. If $q(t) = q_n(t)$, $n > 0$, in (1.23), then the change of variables (1.24) reduces (1.23) to the case where $t, q_n(t)$ are replaced by $s, q_{n-1}(s)$. In particular, if μ is real and $q = q_n(t)$, $n \geqq 0$, then real-valued solutions of (1.23) have infinitely many zeros for large $t > 0$ if and only if $\mu > \frac{1}{4}$.

2. Basic Facts

Before considering more complicated matters, it is well to point out the consequences of Chapter IV (in particular, § IV 8) for the homogeneous and inhomogeneous equation

(2.1) $(p(t)u')' + q(t)u = 0$,

(2.2) $(p(t)w')' + q(t)w = h(t)$.

To this end, the scalar equations (2.1) or (2.2) can be written as the binary vector equations

(2.3) $x' = A(t)x$,

(2.4) $y' = A(t)y + \begin{pmatrix} 0 \\ h(t) \end{pmatrix}$, where $x = (x^1, x^2), y = (y^1, y^2)$ are the vectors $x = (u, p(t)u'), y = (w, p(t)w')$ and $A(t)$ is the 2×2 matrix

(2.5)
$$A(t) = \begin{pmatrix} 0 & \dfrac{1}{p(t)} \\ -q(t) & 0 \end{pmatrix}.$$

Unless the contrary is stated, it is assumed that $p(t) \neq 0$, $q(t), h(t)$, and other coefficient functions are continuous, complex-valued functions on a t-interval J (which may or may not be closed and/or bounded).

(i) If $t_0 \in J$ and u_0, u_0' are arbitrary complex numbers, then the initial value problem (2.2) and

(2.6)
$$w(t_0) = u_0, \qquad w'(t_0) = u_0'$$

has a unique solution which exists on all of J; Lemma IV 1.1.

(ii) In the particular case (2.1) of (2.2) and $u_0 = u_0' = 0$, the corresponding unique solution is $u(t) \equiv 0$. Hence, if $u(t) \not\equiv 0$ is a solution of (2.1), then the zeros of $u(t)$ cannot have a cluster point in J.

(iii) *Superposition Principles.* If $u(t), v(t)$ are solutions of (2.1) and c_1, c_2 are constants, then $c_1 u(t) + c_2 v(t)$ is a solution of (2.1). If $w_0(t)$ is a solution of (2.2), then $w_1(t)$ is also a solution of (2.2) if and only if $u = w_1(t) - w_0(t)$ is a solution of (2.1).

(iv) If $u(t), v(t)$ are solutions of (2.1), then the corresponding vector solutions $x = (u(t), p(t)u'(t)), (v(t), p(t)v'(t))$ of (2.3) are linearly independent (at every value of t) if and only if $u(t), v(t)$ are linearly independent in the sense that if c_1, c_2 are constants such that $c_1 u(t) + c_2 v(t) \equiv 0$, then $c_1 = c_2 = 0$; cf. § IV 8(iii).

(v) If $u(t), v(t)$ are solutions of (2.1), then there is a constant c, depending on $u(t)$ and $v(t)$, such that their Wronskian $W(t) = W(t; u, v)$ satisfies

(2.7)
$$u(t)v'(t) - u'(t)v(t) = \frac{c}{p(t)}.$$

This follows from Theorem IV 1.2 since a solution matrix for (2.3) is

$$X(t) = \begin{pmatrix} u(t) & v(t) \\ p(t)u'(t) & p(t)v'(t) \end{pmatrix},$$

$\det X(t) = p(t)W(t)$ and $\operatorname{tr} A(t) = 0$; cf. § IV 8(iv). A simple direct proof is contained in the following paragraph.

(vi) *Lagrange Identity.* Consider the pair of relations

(2.8)
$$(pu')' + qu = f, \qquad (pv')' + qv = g,$$

where $f = f(t)$, $g = g(t)$ are continuous functions on J. If the second is multiplied by u, the first by v, and the results subtracted, it follows that

$$(2.9) \qquad [p(uv' - u'v)]' = gu - fv$$

since $[p(uv' - u'v)]' = u(pv')' - v(pu')'$. The relation (2.9) is called the *Lagrange identity*. Its integrated form

$$(2.10) \qquad [p(uv' - u'v)]_a^t = \int_a^t (gu - fv) \, ds,$$

where $[a, t] \subset J$, is called *Green's formula*.

(vii) In particular, (v) shows that $u(t)$ and $v(t)$ are linearly independent solutions of (2.1) if and only if $c \neq 0$ in (2.7). In this case every solution of (2.1) is a linear combination $c_1 u(t) + c_2 v(t)$ of $u(t)$, $v(t)$ with constant coefficients.

(viii) If $p(t) \equiv$ const. [e.g., $p(t) \equiv 1$], the Wronskian of any pair of solutions $u(t)$, $v(t)$ of (2.1) is a constant.

(ix) According to the general theory of § IV 3, if one solution of $u(t) \not\equiv 0$ of (2.1) is known, the determination (at least, locally) of other solutions $v(t)$ of (2.1) are obtained by considering a certain scalar differential equation of first order. If $u(t) \neq 0$ on a subinterval J' of J, the differential equation in question is (2.7), where u is considered known and v unknown. If (2.7) is divided by $u^2(t)$, the equation becomes

$$(2.11) \qquad \left(\frac{v}{u}\right)' = \frac{c}{p(t)u^2(t)}$$

and a quadrature gives

$$(2.12) \qquad v(t) = c_1 u(t) + cu(t) \int_a^t \frac{ds}{p(s)u^2(s)},$$

if a, $t \in J'$; cf. § IV 8(iv). It is readily verified that if c_1, c are arbitrary constants and a, $t \in J'$, then (2.12) is a solution of (2.1) satisfying (2.7) on any interval J' where $u(t) \neq 0$.

(x) Let $u(t)$, $v(t)$ be solutions of (2.1) satisfying (2.7) with $c \neq 0$. For a fixed $s \in J$, the solution of (2.1) satisfying the initial conditions $u(s) = 0$, $p(s)u'(s) = 1$ is $c^{-1}[u(s)v(t) - u(t)v(s)]$. Hence the solution of (2.2) satisfying $w(t_0) = w'(t_0) = 0$ is

$$(2.13) \qquad w(t) = c^{-1} \int_{t_0}^t [u(s)v(t) - u(t)v(s)]h(s) \, ds;$$

cf. § IV 8(v) (or, more simply, verify this directly). The general solution of (2.2) is obtained by adding a general solution $c_1 u(t) + c_2 v(t)$ of (2.1) to (2.13) to give

$$(2.14) \quad w(t) = u(t)\left[c_1 - c^{-1}\int_{t_0}^t v(s)h(s) \, ds\right] + v(t)\left[c_2 + c^{-1}\int_{t_0}^t u(s)h(s) \, ds\right].$$

If the closed bounded interval $[a, b]$ is contained in J, then the choice

$$t_0 = a, \qquad c_1 = c^{-1} \int_a^b v(s) h(s) \, ds \qquad \text{and} \quad c_2 = 0$$

reduces (2.14) to the particular solution

$$(2.15) \qquad w(t) = c^{-1} \left[v(t) \int_a^t u(s) h(s) \, ds + u(t) \int_t^b v(s) h(s) \, ds \right].$$

This can be written in the form

$$(2.16) \qquad\qquad w(t) = \int_a^b G(t, s) h(s) \, ds,$$

where

$$(2.17)
\begin{aligned}
G(t, s) &= c^{-1} v(t) u(s) \qquad \text{if} \quad a \leqq s \leqq t, \\
G(t, s) &= c^{-1} u(t) v(s) \qquad \text{if} \quad t \leqq s \leqq b.
\end{aligned}$$

Remark. If $h(t)$ is (not necessarily continuous but) integrable over $[a, b]$, then $w(t)$ is a "solution" of (2.2) in the sense that $w(t)$ has a continuous derivative w' such that $p(t) w'(t)$ is absolutely continuous and (2.2) holds except on a t-set of measure 0.

Exercise 2.1   Verify that if $\alpha, \beta, \gamma, \delta$ are constants such that

$$\alpha u(a) + \beta p(a) u'(a) = 0, \qquad \gamma v(b) + \delta p(b) v'(b) = 0,$$

then the particular solution (2.15) of (2.2) satisfies

$$\alpha w(a) + \beta p(a) w'(a) = 0, \qquad \gamma w(b) + \delta p(b) w'(b) = 0.$$

An extremely simple but important case occurs if $p \equiv 1$, $q \equiv 0$ so that (2.1) becomes $u'' = 0$. Then $u(t) = t - a$ and $v(t) = b - t$ are the solutions of (2.1) satisfying $u(a) = 0$, $v(b) = 0$, and (2.7) with $c = a - b$. Hence

$$(2.18)$$
$$w(t) = \frac{1}{a - b} \left[(b - t) \int_a^t (s - a) h(s) \, ds + (t - a) \int_t^b (b - s) h(s) \, ds \right]$$

is the solution of $w'' = h(t)$ satisfying $w(a) = w(b) = 0$.

Exercise 2.2.   Let $[a, b] \subset J$. Show that most general function $G(t, s)$ defined for $a \leqq s, t \leqq b$ for which (2.16) is a solution of (2.2) for $a \leqq t \leqq b$ for every continuous function $h(t)$ is given by

$$G(t, s) = c^{-1} \sum_{k=1}^{2} \sum_{j=1}^{2} a_{jk} u_j(t) u_k(s) \qquad \text{if} \quad a \leqq s \leqq t,$$

$$G(t, s) = c^{-1} \sum_{k=1}^{2} \sum_{j=1}^{2} b_{jk} u_j(t) u_k(s) \qquad \text{if} \quad t \leqq s \leqq b,$$

where $A = (a_{jk})$, $B = (b_{jk})$ are constant matrices such that

$$B - A = \begin{pmatrix} 0 & 1 \\ -1 & 0 \end{pmatrix}$$

and $u_1 = u(t)$, $u_2 = v(t)$ are solutions of (2.1) satisfying (2.7) with $c \neq 0$. In this case, $G(t, s)$ is continuous for $a \leq s, t \leq b$.

Exercise 2.3.   Let a (and/or b) be a possibly infinite end point of J which does not belong to J, so that $p(t)$, $q(t)$, $h(t)$ and $u(t)$, $v(t)$ need not have limits as $t \to a + 0$ (and/or $t \to b - 0$). Suppose, however, that h, u, v have the property that the integrals in (2.15) are convergent (possibly, just conditionally). Then (2.15) is a solution of (2.2) on J. [This follows from the derivation of (2.15) or can be verified directly by substituting (2.15) into (2.2).]

(**xi**) *Variation of Constants.*   In addition to (2.1), consider another equation

(2.19) $(p_0(t)w')' + q_0(t)w = 0,$

where $p_0(t) \neq 0$, $q_0(t)$ are also continuous in J. Correspondingly, (2.19) is equivalent to a first order system

(2.20) $y' = A_0(t)y,$

where

(2.21) $y = (u, p_0(t)u')$ and $A_0(t) = \begin{pmatrix} 0 & 1/p_0(t) \\ -q_0(t) & 0 \end{pmatrix}.$

Let $u_0(t)$, $v_0(t)$ be linearly independent solutions of (2.19) such that

(2.22) $Y(t) = \begin{pmatrix} u_0 & v_0 \\ p_0 u_0' & p_0 v_0' \end{pmatrix}$

is a fundamental matrix for (2.20) with det $Y(t) \equiv 1$; i.e.,

$$p_0(u_0 v_0' - u_0' v_0) = 1.$$

Hence

(2.23) $Y^{-1}(t) = \begin{pmatrix} p_0 v_0' & -v_0 \\ -p_0 u_0' & u_0 \end{pmatrix}.$

Consider the linear change of variables

(2.24) $x = Y(t)y = \begin{pmatrix} u_0 y^1 + v_0 y^2 \\ p_0 u_0' y^1 + p_0 v_0' y^2 \end{pmatrix}$

for the system (2.3). The resulting differential equation for the vector y is

(2.25) $y' = C(t)y,$ where $C(t) = Y^{-1}(t)[A(t) - A_0(t)] Y(t);$

cf. Theorem IV 2.1. A direct calculation using (2.5), (2.21), (2.22), and (2.23) shows that

$$(2.26) \quad C(t) = \left(\frac{1}{p} - \frac{1}{p_0}\right) p_0^2 \begin{pmatrix} u_0' v_0' & v_0'^2 \\ -u_0'^2 & -u_0' v_0' \end{pmatrix} + (q - q_0) \begin{pmatrix} u_0 v_0 & v_0^2 \\ -u_0^2 & -u_0 v_0 \end{pmatrix}.$$

In the particular case, $p_0(t) = p(t)$, so that (2.19) reduces to

$$(2.27) \qquad (pw')' + q_0 w = 0,$$

the *matrix* $C(t)$ *depends on* $u_0(t)$, $v_0(t)$ *but not on their derivatives.* Here, (2.1) or equivalently (2.3) is reduced to the binary system

$$(2.28) \qquad y' = (q - q_0) \begin{pmatrix} u_0 v_0 & v_0^2 \\ -u_0^2 & -u_0 v_0 \end{pmatrix} y.$$

Exercise 2.4. In order to interpret the significance of y, i.e., of the components y^1, y^2 of y in (2.28) for a corresponding solution $u(t)$ of (2.1), write (2.1) as $(pw')' + q_0 w = h(t)$, where $w = u(t)$, $h = [q_0(t) - q(t)]u(t)$. Then it is seen that the solution $u(t)$ of (2.1) is of the form (2.14) if $c = 1$ and $u(t)$, $v(t)$ are replaced by $u_0(t)$, $v_0(t)$. Using (2.24), where $p =_. p_0$ and x is the binary vector $(u(t), p(t)u'(t))$, show that the coefficients of $u_0(t)$, $v_0(t)$ in this analogue of formula (2.14) are the component y^1, y^2 of the corresponding solution $y(t)$ of (2.28).

(xii) If we know a particular solution $u_0(t)$ of (2.27) which does not vanish on J, then we can determine linearly independent solutions by a quadrature [cf. (ix)] and hence obtain the matrix in (2.28). Actually this desired result can be obtained much more directly. Let (2.27) have a solution $w(t) \neq 0$ on the interval J. Change the dependent variable from u to z in (2.1), where

$$(2.29) \qquad u = w(t)z.$$

The differential equation satisfied by z is

$$w(pz')' + 2pz'w' + [(pw')' + qw]z = 0.$$

If this is multiplied by w, it follows that

$$(2.30) \qquad (pw^2 z')' + w[(pw')' + qw]z = 0$$

or, by (2.27),

$$(2.31) \qquad (pw^2 z')' + w^2(q - q_0)z = 0;$$

i.e., (2.29) reduces (2.1) to (2.30) or (2.31). Instead of starting with a differential equation (2.27) and a solution $w(t)$, we can start with a function $w(t) \neq 0$ such that $w(t)$ has a continuous derivative $w'(t)$ and $p(t)w'(t)$ has a continuous derivative, in which case $q_0(t)$ is defined by (2.27), so $q_0 = -(pw')'/w$. The substitution (2.29) will also be called a *variation of constants*.

(**xiii**) *Liouville Substitution.* As a particular case, consider (2.1) with $p(t) \equiv 1$,

$$(2.32) \qquad\qquad u'' + q(t)u = 0.$$

Suppose that $q(t)$ has a continuous second derivative, is real-valued, and does not vanish, say

$$(2.33) \qquad\qquad \pm q(t) > 0, \quad \text{where} \quad \pm = \operatorname{sgn} q(t)$$

is independent of t. Consider the variation of constants

$$(2.34) \qquad\qquad u = w(t)z, \quad \text{where} \quad w = |q(t)|^{-1/4} > 0.$$

Then (2.32) is reduced to (2.30), where $p \equiv 1$, i.e., to

$$(2.35) \qquad (|q|^{-1/2} z')' \pm \left(|q|^{1/2} - \frac{q''}{4 |q|^{3/2}} \pm \frac{5q'^2}{16 |q|^{5/2}}\right)z = 0.$$

A change of independent variables $t \to s$ defined by

$$(2.36) \qquad\qquad ds = |q|^{1/2}\, dt$$

transforms (2.35) into

$$(2.37) \qquad\qquad \frac{d^2 z}{ds^2} \pm f(s)z = 0,$$

where

$$(2.38) \qquad\qquad f(s) = 1 - \frac{q''}{4 |q|^2} \pm \frac{5q'^2}{16 |q|^3}$$

and the argument of q and its derivatives in (2.38) is $t = t(s)$, the inverse of the function $s = s(t)$ defined by (2.36) and a quadrature; cf. (1.7). In these formulae, a prime denotes differentiation with respect to t, so that $q' = dq/dt$.

The change of variables (2.34), (2.36) is the Liouville substitution. This substitution, or repeated applications of it, often leads to a differential equation of the type (2.37) in which $f(s)$ is "nearly" constant; cf. Exercise 8.3. For a simple extreme case of this remark, see Exercise 1.1(c).

(**xiv**) *Riccati Equations.* Paragraphs (xi), (xii), and (xiii) concern the transformation of (2.1) into a different second order linear equation or into a suitable binary, first order linear system. (Other such transformations will be utilized later; cf. §§ 8–9.) Frequently, it is useful to transform (2.1) into a suitable *nonlinear* equation or system. In this direction, one of the most widely used devices is the following: Let

$$(2.39) \qquad\qquad r = \frac{p(t)u'}{u},$$

so that $r' = (pu')'/u - p^{-1}(pu'/u)^2$. Thus, if (2.1) is divided by u, the result can be written as

$$(2.40) \qquad r' + \frac{r^2}{p(t)} + q(t) = 0.$$

This is called the Riccati equation of (2.1). (In general, a differential equation of the form $r' = a(t)r^2 + b(t)r + c(t)$, where the right side is a quadratic polynomial in r, is called a *Riccati differential equation*.)

It will be left to the reader to verify that if $u(t)$ is a solution of (2.1) which does not vanish on a t-interval J' ($\subset J$), then (2.39) is a solution of (2.40) on J'; conversely if $r = r(t)$ is a solution of (2.40) on a t-interval J' ($\subset J$), then a quadrature of (2.39) gives

$$(2.41) \qquad u = c \exp \int^t \frac{r(s)\, ds}{p(s)},$$

a nonvanishing solution of (2.1) on J'.

Exercise 2.5. Verify that the substitution $r = u'/u$ transforms

$$u'' + g(t)u' + f(t)u = 0$$

into the Riccati equation

$$r' + r^2 + g(t)r + f(t) = 0.$$

(xv) *Prüfer Transformation.* In the case of an equation (2.1) with real-valued coefficients, the following transformation of (2.1) is often useful (cf. §§ 3, 5): Let $u = u(t) \not\equiv 0$ be a real-valued solution of (2.1) and let

$$(2.42) \qquad \rho = (u^2 + p^2 u'^2)^{\frac{1}{2}} > 0, \qquad \varphi = \arctan \frac{u}{pu'}.$$

Since u and u' cannot vanish simultaneously a suitable choice of φ at some fixed point $t_0 \in J$ and the last part of (2.42) determine a continuously differentiable function $\varphi(t)$. The relations (2.42) transform (2.1) into

$$(2.43) \qquad \varphi' = \frac{1}{p(t)} \cos^2 \varphi + q(t) \sin^2 \varphi,$$

$$(2.44) \qquad \rho' = -\left[q(t) - \frac{1}{p(t)} \right] \rho \sin \varphi \cos \varphi.$$

The equation (2.43) involves only the one unknown function φ. If a solution $\varphi = \varphi(t)$ of (2.43) is known, a corresponding solution of (2.44) is obtained by a quadrature.

An advantage of (2.43) over (2.40) is that any solution of (2.43) exists on the whole interval J where p, q are continuous. This is clear from the relation between solutions of (2.1) and (2.43).

Exercise 2.6.   Verify that if $\tau(t) > 0$ is continuous on J and is locally of bounded variation (i.e., is of bounded variation on all closed, bounded subintervals of J) and if $u = u(t) \not\equiv 0$ is a real-valued solution of (2.1), then

$$(2.45) \qquad \rho = (\tau^2 u^2 + p^2 u'^2)^{\frac{1}{2}} > 0, \qquad \varphi = \arctan \frac{\tau u}{p u'}$$

and a choice of $\varphi(t_0)$ for some $t_0 \in J$ determine continuous functions $\rho(t)$, $\varphi(t)$ which are locally of bounded variation and

$$(2.46) \qquad d\varphi = \left(\frac{\tau}{p} \cos^2 \varphi + \frac{q}{\tau} \sin^2 \varphi \right) dt + (\sin \varphi \cos \varphi) \, d(\log \tau)$$

$$(2.47) \qquad d(\log \rho) = -\left[\left(\frac{q}{\tau} - \frac{\tau}{p} \right) \sin \varphi \cos \varphi \right] dt + (\sin^2 \varphi) \, d(\log \tau),$$

The relations (2.46), (2.47) are understood to mean that Riemann, Stieltjes integrals of both sides of these relations are equal. Conversely (continuous) solutions of (2.46)–(2.47) determine solutions of (2.1), via (2.45). Note that if $q(t) > 0$, $p(t) > 0$, and $q(t)p(t)$ is locally of bounded variation, then the choice $\tau(t) = p^{\frac{1}{2}}(t)q^{\frac{1}{2}}(t) > 0$ gives $q/\tau = \tau/p = p^{\frac{1}{2}}/q^{\frac{1}{2}}$ and reduces (2.45) and (2.46), (2.47) to

$$(2.48) \qquad \rho = (pqu^2 + p^2 u'^2)^{\frac{1}{2}} > 0, \qquad \varphi = \arctan \frac{q^{\frac{1}{2}} u}{p^{\frac{1}{2}} u'},$$

and

$$(2.49) \qquad d\varphi = \frac{q^{\frac{1}{2}}}{p^{\frac{1}{2}}} dt + (\tfrac{1}{2} \sin \varphi \cos \varphi) \, d(\log pq),$$

$$(2.50) \qquad d(\log \rho) = (\tfrac{1}{2} \sin^2 \varphi) \, d(\log pq).$$

3. Theorems of Sturm

In this section, we will consider only differential equations of the type (2.1) having real-valued, continuous coefficient functions $p(t) > 0$, $q(t)$. "Solution" will mean "real-valued, nontrivial ($\not\equiv 0$) solution." The object of interest will be the set of zeros of a solution $u(t)$. For the study of zeros of $u(t)$, the Prüfer transformation (2.42) is particularly useful since $u(t_0) = 0$ if and only if $\varphi(t_0) = 0 \mod \pi$.

Lemma 3.1.   *Let $u(t) \not\equiv 0$ be a real-valued solution of (2.1) on $t_0 \leq t \leq t^0$, where $p(t) > 0$ and $q(t)$ are real-valued and continuous. Let $u(t)$ have exactly $n (\geq 1)$ zeros $t_1 < t_2 < \cdots < t_n$ on $t_0 < t \leq t^0$. Let $\varphi(t)$ be a continuous function defined by (2.42) and $0 \leq \varphi(t_0) < \pi$. Then $\varphi(t_k) = k\pi$ and $\varphi(t) > k\pi$ for $t_k < t \leq t^0$ for $k = 1, \ldots, n$.*

Proof. Note that at a t-value where $u = 0$, i.e., where $\varphi = 0 \bmod \pi$, (2.43) implies that $\varphi' = 1/p > 0$. Consequently $\varphi(t)$ is increasing in the neighborhoods of points where $\varphi(t) = j\pi$ for some integer j. It follows that if $t_0 \leq a \leq t^0$ and $j\pi \leq \varphi(a)$, then $\varphi(t) > j\pi$ for $a < t \leq t^0$; also if $j\pi \geq \varphi(a)$, then $\varphi(t) < j\pi$ for $t_0 \leq t < a$. This implies the assertion.

In the theorems of this section, two equations will be considered

$$(3.1_j) \qquad (p_j(t)u')' + q_j(t)u = 0, \qquad j = 1, 2,$$

where $p_j(t)$, $q_j(t)$ are real-valued continuous functions on an interval J, and

$$(3.2) \qquad p_1(t) \geq p_2(t) > 0 \quad \text{and} \quad q_1(t) \leq q_2(t).$$

In this case, (3.1_2) is called a *Sturm majorant* for (3.1_1) on J and (3.1_1) is a *Sturm minorant* for (3.1_2). If, in addition,

$$(3.3_1) \qquad\qquad\qquad q_1(t) < q_2(t)$$

or

$$(3.3_2) \qquad p_1(t) > p_2(t) > 0 \quad \text{and} \quad q_2(t) \neq 0$$

holds at some point t of J, then (3.1_2) is called a *strict Sturm majorant* for (3.1_1) on J.

Theorem 3.1 (Sturm's First Comparison Theorem). *Let the coefficient functions in (3.1_j) be continuous on an interval $J : t_0 \leq t \leq t^0$ and let (3.1_2) be a Sturm majorant for (3.1_1). Let $u = u_1(t) \not\equiv 0$ be a solution of (3.1_1) and let $u_1(t)$ have exactly $n \ (\geq 1)$ zeros $t = t_1 < t_2 < \cdots < t_n$ on $t_0 < t \leq t^0$. Let $u = u_2(t) \not\equiv 0$ be a solution of (3.1_2) satisfying*

$$(3.4) \qquad \frac{p_1(t)u_1'(t)}{u_1(t)} \geq \frac{p_2(t)u_2'(t)}{u_2(t)}$$

at $t = t_0$. (The expression on the right [or left] of (3.4) at $t = t_0$ is considered to be $+\infty$ if $u_2(t_0) = 0$ [or $u_1(t_0) = 0$]; in particular, (3.4) holds at $t = t_0$ if $u_1(t_0) = 0$.) Then $u_2(t)$ has at least n zeros on $t_0 < t \leq t_n$. Furthermore $u_2(t)$ has at least n zeros on $t_0 < t < t_n$ if either the inequality in (3.4) holds at $t = t_0$ or (3.1_2) is a strict Sturm majorant for (3.1_1) on $t_0 \leq t \leq t_n$.

Proof. In view of (3.4), it is possible to define a pair of continuous functions $\varphi_1(t)$, $\varphi_2(t)$ on $t_0 \leq t \leq t^0$ by

$$(3.5) \quad \varphi_j(t) = \arctan \frac{u_j(t)}{p_j(t)u_j'(t)} \quad \text{and} \quad 0 \leq \varphi_1(t_0) \leq \varphi_2(t_0) < \pi.$$

Then the analogue of (2.43) is

$$(3.6_j) \qquad \varphi_j' = \frac{1}{p_j(t)} \cos^2 \varphi_j + q_j(t) \sin^2 \varphi_j \equiv f_j(t, \varphi_j).$$

Since the continuous functions $f_j(t, \varphi_j)$ are smooth as functions of the variable φ_j, the solutions of (3.6) are uniquely determined by their initial conditions. It follows from (3.2) that $f_1(t, \varphi) \leq f_2(t, \varphi)$ for $t_0 \leq t \leq t^0$ and all φ. Hence the last part of (3.5) and Corollary III 4.2 show that

$$(3.7) \qquad \varphi_1(t) \leq \varphi_2(t) \qquad \text{for} \quad t_0 \leq t \leq t^0.$$

In particular, $\varphi_1(t_n) = n\pi$ implies that $n\pi \leq \varphi_2(t_n)$ and the first part of the theorem follows from Lemma 3.1.

In order to prove the last part of the theorem, suppose first that the sign of inequality holds in (3.4) at $t = t_0$. Then $\varphi_1(t_0) < \varphi_2(t_0)$. Let $\varphi_{20}(t)$ be the solution of (3.6_2) satisfying the initial condition $\varphi_{20}(t_0) = \varphi_1(t_0)$, so that $\varphi_{20}(t_0) < \varphi_2(t_0)$. Since solutions of (3.6_2) are uniquely determined by initial conditions, $\varphi_{20}(t) < \varphi_2(t)$ for $t_0 \leq t \leq t^0$. Thus the analogue of (3.7) gives $\varphi_1(t) \leq \varphi_{20}(t) < \varphi_2(t)$, and so $\varphi_2(t_n) > n\pi$. Hence $u_2(t)$ has n zeros on $t_0 < t < t_n$.

Consider the case that equality holds in (3.4) but either (3.3_1) or (3.3_2) holds at some point of $[t_0, t_n]$. Write (3.6_2) as

$$\varphi_2' = \frac{1}{p_1} \cos^2 \varphi_2 + q_1 \sin^2 \varphi_2 + \epsilon(t),$$

where

$$\epsilon(t) = \left(\frac{1}{p_2} - \frac{1}{p_1} \right) \cos^2 \varphi_2 + (q_2 - q_1) \sin^2 \varphi_2 \geq 0.$$

If the assertion is false, it follows from the case just considered that $\varphi_1(t) = \varphi_2(t)$ for $t_0 \leq t \leq t_n$. Hence, $\varphi_1'(t) = \varphi_2'(t)$ and so $\epsilon(t) = 0$ for $t_0 \leq t \leq t_n$. Since $\sin \varphi_2(t) = 0$ only at the zeros of $u_2(t)$, it follows that $q_2(t) = q_1(t)$ for $t_0 \leq t \leq t_n$ and that $(p_2^{-1} - p_1^{-1}) \cos^2 \varphi_2 = 0$. Hence, $p_2^{-1}(t) - p_1^{-1}(t) > 0$ at some t implies $\cos \varphi_2(t) = 0$; i.e., $u_2' = 0$. If (3.3_1) does not hold at any t on $[t_0, t_n]$, it follows that (3.3_2) holds at some t and hence on some subinterval of $[t_0, t_n]$. But then $u_2' = 0$ on this interval, thus $(p_2 u_2')' = 0$ on this interval. But this contradicts $q_2(t) \neq 0$ on this interval. This completes the proof.

Corollary 3.1 (Sturm's Separation Theorem). *Let (3.1_2) be a Sturm majorant for (3.1_1) on an interval J and let $u = u_j(t) \not\equiv 0$ be a real-valued solution of (3.1_j). Let $u_1(t)$ vanish at a pair of points $t = t_1, t_2 \, (>t_1)$ of J. Then $u_2(t)$ has at least one zero on $[t_1, t_2]$. In particular, if $p_1 \equiv p_2$, $q_1 \equiv q_2$ and u_1, u_2 are real-valued, linearly independent solutions of $(3.1_1) \equiv (3.1_2)$, then the zeros of u_1 separate and are separated by those of u_2.*

Note that the last statement of this theorem is meaningful since the zeros of u_1, u_2 do not have a cluster point on J; see § 2(ii). In addition, $u_1(t), u_2(t)$ cannot have a common zero $t = t_1$; otherwise, the uniqueness of the solutions of (3.1_1) implies that $u_1(t) = cu_2(t)$ with $c = u_1'(t_1)/u_2'(t_1)$ [so that $u_1(t)$, $u_2(t)$ are not linearly independent].

Exercise 3.1. (*a*) [Another proof for Sturm's separation theorem when $p_1(t) \equiv p_2(t) > 0, q_2(t) \geqq q_1(t)$.] Suppose that $u_1(t) > 0$ for $t_1 < t < t_2$ and that the assertion is false, say $u_2(t) > 0$ for $t_1 \leqq t \leqq t_2$. Multiplying (3.1_1) where $u = u_1$ by u_2 and (3.1_2) where $u = u_2$ by u_1, subtracting, and integrating over $[t_1, t]$ gives

$$p(t)(u_1'u_2 - u_1u_2') \geqq 0 \qquad \text{for} \quad t_1 \leqq t \leqq t_2,$$

where $p = p_1 = p_2$; cf. the derivation of (2.9). This implies that $(u_1/u_2)' \geqq 0$; hence $u_1/u_2 > 0$ for $t_1 < t \leqq t_2$ (*b*) Reduce the case $p_1(t) \geqq p_2(t)$ to the case $p_1(t) \equiv p_2(t)$ by the device used below in the proof of Corollary 6.5.

Exercise 3.2. (*a*) In the differential equation

$$(3.8) \qquad\qquad u'' + q(t)u = 0,$$

let $q(t)$ be real-valued, continuous, and satisfy $0 < m \leqq q(t) \leqq M$. If $u = u(t) \not\equiv 0$ is a solution with a pair of zeros $t = t_1$, $t_2(> t_1)$, then $\pi/m^{1/2} \geqq t_2 - t_1 \geqq \pi/M^{1/2}$. (*b*) Let $q(t)$ be continuous for $t \geqq 0$ and $q(t) \to 1$ as $t \to \infty$. Show that if $u = u(t) \not\equiv 0$ is a real-valued solution of (3.8), then the zeros of $u(t)$ form a sequence $(0 \leqq) t_1 < t_2 < \ldots$ such that $t_n - t_{n-1} \to \pi$ as $n \to \infty$. (*c*) Observe that real-valued solutions $u(t) \not\equiv 0$ of (1.17) have at most one zero for $t > 0$ if $\mu \leqq \frac{1}{4}$ and have infinitely many zeros for $t > 0$ if $\mu > \frac{1}{4}$. In the latter case, the zeros cluster at $t = 0$ and $t = \infty$. (*d*) Consider the Bessel equation

$$(3.9) \qquad\qquad v'' + \frac{v'}{t} + \left(1 - \frac{\mu^2}{t^2}\right)v = 0,$$

where μ is a real parameter. The variations of constants $u = t^{1/2}v$ transforms (3.9) into

$$(3.10) \qquad u'' + \left(1 - \frac{\alpha}{t^2}\right)u = 0, \qquad \text{where} \quad \alpha = \mu^2 - \frac{1}{4}.$$

Show that the zeros of a real-valued solution $v(t)$ of (3.9) on $t > 0$ form a sequence $t_1 < t_2 < \ldots$ such that $t_n - t_{n-1} \to \pi$ as $n \to \infty$.

Theorem 3.2 (Sturm's Second Comparison Theorem). *Assume the conditions of the first part of Theorem 3.1 and that $u_2(t)$ also has exactly n zeros on $t_0 < t \leqq t^0$. Then (3.4) holds at $t = t^0$ (where the expression on the right [or left] of (3.4) at $t = t^0$ is taken to be $-\infty$ if $u_2(t^0) = 0$ [or $u_1(t^0) = 0$]). Furthermore the sign of inequality holds at $t = t^0$ in (3.4) if the conditions of the last part of Theorem 3.1 hold.*

Proof. The proof of this assertion is essentially contained in the proof of Theorem 3.1 if it is noted that the assumption on the number of zeros of

$u_2(t)$ implies the last inequality in $n\pi \leqq \varphi_1(t^0) \leqq \varphi_2(t^0) < (n + 1)\pi$. Also, the proof of Theorem 3.1 gives $\varphi_1(t^0) < \varphi_2(t^0)$ under the conditions of the last part of the theorem.

4. Sturm-Liouville Boundary Value Problems

This topic is one of the most important in the theory of second order linear equations. Since a full discussion of it would be very lengthy and since very complete treatments can be found in many books, only a few high points will be discussed here.

In the equation

(4.1λ) $$(p(t)u')' + [q(t) + \lambda]u = 0,$$

let $p(t) > 0$, $q(t)$ be real-valued and continuous for $a \leqq t \leqq b$ and λ a complex number. Let α, β be given real numbers and consider the problem of finding, if possible, a nontrivial ($\not\equiv 0$) solution of (4.1λ) satisfying the boundary conditions

(4.2) $u(a) \cos \alpha - p(a)u'(a) \sin \alpha = 0$, $u(b) \cos \beta - p(b)u'(b) \sin \beta = 0$.

Exercise 4.1.   Show that if λ is not real, then (4.1λ) and (4.2) do not have a nontrivial solution.

Exercise 4.2.   Consider the following special cases of (4.1λ), (4.2):

(4.3) $$u'' + \lambda u = 0,   u(0) = u(\pi) = 0.$$

Show that this has a solution only if $\lambda = (n + 1)^2$ for $n = 0, 1, \ldots$ and that the corresponding solution, up to a multiplicative constant, is $u = \sin (n + 1)t$.

It will be shown that the results of Exercise 4.2 for the special case (4.3) are typical for the general situation (4.1λ), (4.2).

Theorem 4.1.   *Let $p(t) > 0$, $q(t)$ be real-valued and continuous for $a \leqq t \leqq b$. Then there exists an unbounded sequence of real numbers $\lambda_0 < \lambda_1 < \ldots$ such that* (i) *(4.1λ), (4.2) has a nontrivial ($\not\equiv 0$) solution if and only if $\lambda = \lambda_n$ for some n;* (ii) *if $\lambda = \lambda_n$ and $u = u_n(t) \not\equiv 0$ is a solution of (4.1λ_n), (4.2), then $u_n(t)$ is unique up to a multiplicative constant, and $u_n(t)$ has exactly n zeros on $a < t < b$ for $n = 0, 1, \ldots$;* (iii) *if $n \neq m$, then*

(4.4) $$\int_a^b u_n(t)u_m(t) \, dt = 0;$$

(iv) *if λ is a complex number $\lambda \neq \lambda_n$ for $n = 0, 1, \ldots$, then there exists a continuous function $G(t, s; \lambda) = \bar{G}(s, t; \bar{\lambda})$ for $a \leqq s$, $t \leqq b$ with the property that if $h(t)$ is any function integrable on $a \leqq t \leqq b$, then*

(4.5λ) $$(p(t)w')' + [q(t) + \lambda]w = h(t)$$

has a unique solution w =w (t) satisfying

(4.2′) $w(a) \cos \alpha - p(a)w'(a) \sin \alpha = 0$, $w(b) \cos \beta - p(b)w'(b) \sin \beta = 0$

and w(t) is given by

(4.6) $$w(t) = \int_a^b G(t, s; \lambda)h(s) \, ds,$$

also $G(t, s; \lambda)$ is real-valued when λ is real; (v) *if $\lambda = \lambda_n$ and $h(t)$ is a function integrable on $a \leqq t \leqq b$, then $(4.5\lambda_n)$, $(4.2')$ has a solution if and only if*

(4.7) $$\int_a^b u_n(t)h(t) \, dt = 0;$$

in this case, if $w(t)$ is a solution of $(4.5\lambda_n)$, $(4.2')$, then $w(t) + cu_n(t)$ is also a solution and all solutions are of this form; (vi) *if the functions $u_n(t)$ are chosen real-valued (uniquely up to a factor ± 1) so as to satisfy*

(4.8) $$\int_a^b u_n^2(t) \, dt = 1,$$

then $u_0(t)$, $u_1(t)$, ... form a complete orthonormal sequence for $L^2(a, b)$; i.e., if $h(t) \in L^2(a, b)$, then $h(t)$ has the Fourier series

(4.9) $$h(t) \sim \sum_{n=0}^{\infty} c_n u_n(t), \quad \text{where} \quad c_n = \int_a^b h(t)u_n(t) \, dt$$

and

(4.10) $$\int_a^b \left| h(t) - \sum_{k=0}^{n} c_k u_k(t) \right|^2 dt \to 0 \quad \text{as} \quad n \to \infty.$$

If $h(t)$ is not continuous in (iv) or (v), then a solution of (4.5λ) is to be interpreted as in the Remark in § 2(x).

Note the parallel of the assertions concerning the solvability of (4.5λ), $(4.2')$ with the corresponding situation for linear algebraic equations $(\lambda I - L)w = h$, where L is a $d \times d$ Hermitian symmetric matrix, I is the unit matrix and w, h vectors: $(\lambda I - L)u = 0$ has a solution $u \neq 0$ if and only if λ is an eigenvalue $\lambda_1, \ldots, \lambda_d$ of L; $\lambda_1, \ldots, \lambda_d$ are real; if $\lambda \neq \lambda_n$, then $(\lambda I - L)w = h$ has a unique solution w for every h; finally, if $\lambda = \lambda_n$, then $(\lambda I - L)w = h$ has a solution w if and only if h is orthogonal (i.e., $u \cdot h = 0$) to all solutions u of $(\lambda I - L)u = 0$.

Proof. This proof will only be sketched; details will be left to the reader.

On (i) and (ii). In view of Exercise 4.1, it suffices to consider only real λ. Let $u(t, \lambda)$ be the solution of (4.1λ) satisfying the initial condition

(4.11) $$u(a) = \sin \alpha, \quad p(a)u'(a) = \cos \alpha,$$

so $u(t, \lambda)$ satisfies the first of the two conditions (4.2). It is clear that

(4.1λ), (4.2) has a solution ($\not\equiv 0$) if and only if $u(t, \lambda)$ satisfies the second condition in (4.2).

For fixed λ, define a continuous function $\varphi(t, \lambda)$ of t on $[a, b]$ by

$$(4.12) \qquad \varphi(t, \lambda) = \arctan \frac{u(t, \lambda)}{p(t)u'(t, \lambda)}, \qquad \varphi(a, \lambda) = \alpha.$$

Then $\varphi(t, \lambda)$ has a continuous derivative satisfying

$$(4.13) \qquad \varphi' = \frac{1}{p(t)}\cos^2 \varphi + [q(t) + \lambda]\sin^2 \varphi, \qquad \varphi(a) = \alpha;$$

cf. § 2(xv). If follows from Theorem V 2.1 that the solution $\varphi = \varphi(t, \lambda)$ of (4.13) is a continuous function of (t, λ) for $a \leqq t \leqq b$, $-\infty < \lambda < \infty$. The proof of the Sturm Comparison Theorem 3.1 shows that $\varphi(b, \lambda)$ is an increasing function of λ. Without loss of generality, it can be supposed that α satisfies $0 \leqq \alpha < \pi$. Note that

$$(4.14) \qquad \varphi(b, \lambda) \to \infty \qquad \text{as} \quad \lambda \to \infty.$$

In order to see this, introduce the new independent variable defined by $ds = dt/p(t)$ and $s(a) = 0$, so that (4.1λ) becomes

$$(4.15) \qquad \ddot{u} + p(t)[q(t) + \lambda]u = 0, \qquad t = t(s), \qquad \dot{u} = \frac{du}{ds}.$$

If $M > 0$ is any number, $\lambda > 0$ can be chosen so large that $p(t)[q(t) + \lambda] \geqq M^2$ for $a \leqq t \leqq b$. Sturm's Comparison Theorem 3.1 applied to (4.15) and

$$\ddot{u} + M^2 u = 0$$

shows that if n is arbitrary and M is sufficiently large, then a nontrivial real-valued solution of (4.15) has at least n zeros on the s-interval, $0 \leqq s \leqq \int_a^b dt/p(t)$; i.e., $\varphi(b, \lambda) \geqq n$ if $\lambda > 0$ is sufficiently large by Lemma 3.1.

It will be verified that

$$(4.16) \qquad \varphi(b, \lambda) \to 0 \qquad \text{as} \quad \lambda \to -\infty.$$

By Lemma 3.1, $\varphi(b, \lambda) \geqq 0$. Let $-\lambda > 0$ be so large that $p(t)[q(t) + \lambda] \leqq -M^2 < 0$. The solution of

$$\ddot{u} - M^2 u = 0$$

satisfying the analogue of (4.11), where $a = 0$ and $p \equiv 1$, is

$$u(s) = \sin \alpha \cosh Ms + \frac{1}{M}\cos \alpha \sinh Ms.$$

The analogue of $\varphi(t, \lambda)$ is

$$\psi(s, M) = \arctan \frac{u(s)}{\dot{u}(s)}, \qquad \psi(0, M) = \alpha.$$

For any fixed $s > 0$,

$$\frac{u(s)}{\dot{u}(s)} \to 0 \qquad \text{as} \quad M \to \infty;$$

hence $\psi(b_0, M) \to 0$ as $M \to \infty$, where $b_0 = \int_a^b dt/p(t)$. By Sturm's Comparison Theorem 3.1, $\varphi(b, \lambda) \leq \psi(b_0, M)$. This proves (4.16)

The limit relations (4.14), (4.16) and the strict monotony of $\varphi(b, \lambda)$ as a function of λ show that there exist $\lambda_0, \lambda_1, \dots$ such that

$$\varphi(b, \lambda_n) = \beta + n\pi \qquad \text{for} \quad n = 0, 1, \dots,$$

where it is supposed that $0 < \beta \leq \pi$. Furthermore $\varphi(b, \lambda) \neq \beta$ mod π unless $\lambda = \lambda_n$. This implies (i) and (ii).

On (iii). In order to verify (iii), multiply $(4.1\lambda_n)$ by u_m, $(4.1\lambda_m)$ by u_n, subtract and integrate over $a \leq t \leq b$; i.e., apply the Green identity (2.10) to $f = -\lambda_n u_n(t)$, $g = -\lambda_m u_m(t)$.

On (iv). See § 2(x) and Exercise 2.1. Choose $u = u(t, \lambda)$, and $v(t)$ as a solution of (4.1λ) satisfying the second condition in (4.2).

On (v). Suppose first that $(4.5\lambda_n)$, $(4.2')$ has a solution $w = w(t)$. Apply the Green identity (2.10) in the case where q is replaced by $q + \lambda_n$, $f = h$, $w = u$, $v = u_n$, $g = 0$ in (2.8) in order to obtain (4.7).

Conversely, assume that (4.7) holds. Let $u(t) = u_n(t)$ and let $v(t)$ be a solution of $(4.1\lambda_n)$ linearly independent of $u_n(t)$, say $p(t)[uv' - u'v] = c \neq 0$. Then (2.15) is a solution of $(4.5\lambda_n)$. Furthermore $w(t)$ satisfies the first of the boundary conditions in $(4.2')$ since $u = u_n$ does; cf. Exercise 2.1. On the other hand, (4.7) and (2.15) show that $w(b) = w'(b) = 0$. Hence $w(t)$ is a solution (4.5λ) satisfying the boundary conditions $(4.2')$.

On (vi). Although the assertion (vi) is the main part of Theorem 4.1, it is a consequence of elementary theorems on completely continuous, self-adjoint operators on Hilbert space. For the sake of completeness, the proof of the necessary theorems will be sketched and (vi) will be deduced from them. A knowledge of Fourier series (involving, e.g., Bessel's inequality, Parseval's relation, and the theorem of Fischer-Riesz) will be assumed. In order to minimize the required discussion of topics on Hilbert space, some of the definitions or results, as stated, will involve redundant hypotheses.

Introduce the following notation and terminology:

$$(4.17) \qquad (f, g) = \int_a^b f(t)\bar{g}(t)\, dt, \qquad \|f\| = (f, f)^{1/2} \geq 0,$$

where $f, g \in L^2(a, b)$. Thus $|(f, g)| \leq \|f\| \cdot \|g\|$ and $\|f + g\| \leq \|f\| + \|g\|$ by Schwarz's inequality. A sequence of functions $f_1(t), f_2(t), \dots$ in $L^2(a, b)$ will be said to tend to $f(t)$ in $L^2(a, b)$ if $\|f_n - f\| \to 0$ as $n \to \infty$. They will be said to tend to $f(t)$ *weakly* in $L^2(a, b)$ if the sequence $\|f_1\|, \|f_2\|, \ldots$ is bounded and, for every $\varphi(t) \in L^2(a, b)$, $(f_n, \varphi) \to (f, \varphi)$ as $n \to \infty$. (In this last definition, the condition on $\|f_1\|, \|f_2\|, \ldots$ is redundant but this fact will not be needed below.) A subset H of $L^2(a, b)$ is called a linear manifold if f, $g \in H$ implies that $c_1 f + c_2 g \in H$ for all constants c_1, c_2 and it is called *closed* if $f_n \in H$ for $n = 1, 2, \ldots, f \in L^2(a, b)$ and $\|f_n - f\| \to 0$ as $n \to \infty$ imply that $f \in H$. A linear manifold H of $L^2(a, b)$ will be called *weakly closed* if $f_n \in H$ for $n = 1, 2, \ldots, f \in L^2(a, b)$ and $f_n \to f$ weakly as $n \to \infty$ imply that $f \in H$. (The fact that the notions of "closed" and "weakly closed" are equivalent for linear manifolds will not be needed here.)

Lemma 4.1. *Let $f_1, f_2, \ldots$ be a sequence of elements of $L^2(a, b)$ satisfying $\|f_n\| \leq 1$. Then there exists an $f(t) \in L^2(a, b)$ and a subsequence $f_{n(1)}(t)$, $f_{n(2)}(t), \ldots$ of the given sequence such that $\|f\| \leq 1$ and $f_{n(j)} \to f(t)$ weakly as $j \to \infty$.*

Proof. Without loss of generality, it can be supposed that $[a, b] = [0, \pi]$. Thus each $f_n(t)$ has a sine Fourier series

$$f_n(t) \sim \sum_{k=1}^{\infty} c_{nk} \sin kt,$$

where, by Parseval's relation, $\sum_k |c_{nk}|^2 = \|f_n\|^2 \leq 1$. It follows from Cantor's diagonal process (Theorem I 2.1) that there exists a sequence of integers $1 \leq n(1) < n(2) < \ldots$ such that

$$(4.18) \quad c_k = \lim c_{n(j)k} \quad \text{exists as } j \to \infty \quad \text{for } k = 1, 2, \ldots.$$

Note that

$$\sum_{k=1}^{m} |c_k|^2 = \lim_{j \to \infty} \sum_{k=1}^{m} |c_{n(j)k}|^2 \leq 1.$$

Hence $\sum |c_k|^2 \leq 1$ and so, by the theorem of Fischer-Riesz, there exists an $f(t) \in L^2(a, b)$ such that

$$f(t) \sim \sum_{k=1}^{\infty} c_k \sin kt, \qquad \|f\|^2 = \sum |c_k|^2 \leq 1.$$

It follows from (4.18) that $(f_{n(j)}, \varphi) \to (f, \varphi)$ as $j \to \infty$ holds if $\varphi = \sin kt$ for $k = 1, 2, \ldots$. Hence it holds for any sine polynomial $p(t) = a_1 \sin t + \cdots + a_m \sin mt$. For any $\varphi(t) \in L^2(0, \pi)$, there exists a sine polynomial $p(t)$ such that $\|\varphi - p\|$ is arbitrarily small and $|(f_{n(j)} - f, \varphi)| \leq |(f_{n(j)} - f, p)| + |(f_{n(j)} - f, p - \varphi)|$, while $|(f_{n(j)} - f, p - \varphi)| \leq \|f_{n(j)} - f\| \cdot \|p - \varphi\| \leq 2\|p - \varphi\|$. Hence the lemma follows.

Lemma 4.2. *Let G be a self-adjoint, linear operator defined on a weakly closed linear manifold H of $L^2(a, b)$ satisfying $(Gh, h) = 0$ for all $h \in H$. Then $Gh = 0$ for all $h \in H$.*

To say that G is a linear operator on H means that to every $h \in H$ there is associated a unique element $w = Gh \in H$ and that if $w_j = Gh_j$ for $j = 1, 2$, then $c_1 w_1 + c_2 w_2 = G(c_1 h_1 + c_2 h_2)$ for all complex constants c_1, c_2. The assumption that G is *self-adjoint* means that $(Gh, f) = (h, Gf)$ for all $f, h \in H$.

Proof. If $f, h \in H$ and c is a complex number, then $0 = (G(h + cf), h + cf) = 2 \operatorname{Re} \bar{c}(Gh, f)$ since $(Gf, f) = (Gh, h) = 0$. By the choices $c = 1$ and $c = i$, it follows that $(Gh, f) = 0$. On choosing $f = Gh$, it is seen that $Gh = 0$.

Lemma 4.3. *Let G be a completely continuous, self-adjoint linear operator on a weakly closed linear manifold H of $L^2(a, b)$ and let $Gh \neq 0$ for some $h \in H$. Then G has at least one (real) eigenvalue $\mu \neq 0$; i.e., there exists a (real) number $\mu \neq 0$ and an $h_0 \in H$, $h_0 \neq 0$, such that $Gh_0 = \mu h_0$.*

A linear operator G on H is called *completely continuous* if $h_n, h \in H$ and $h_n \to h$ weakly as $n \to \infty$ imply that $\|Gh_n - Gh\| \to 0$ as $n \to \infty$.

Proof. It follows from Lemma 4.1, the complete continuity of G, and the fact that H is weakly closed that G is bounded, i.e., that there exists a constant C such that $\|Gh\| \leq C$ for all $h \in H$ satisfying $\|h\| \leq 1$.

By Schwarz's inequality, $|(Gh, h)| \leq \|Gh\| \cdot \|h\| \leq C$ if $\|h\| \leq 1$. Hence sup (Gh, h) and inf (Gh, h) for all $\|h\| \leq 1$ exist and are finite. Since $Gh \neq 0$ for some $h \in H$, it follows from Lemma 4.2 that at least one of these two numbers is not zero. For the sake of definiteness, let $\mu = $ sup $(Gh, h) \neq 0$. The choice $h = 0$ shows that $\mu \geq 0$, hence $\mu > 0$.

It will be shown that there exists an $h_0 \in H$ such that $(Gh_0, h_0) = \mu$ and $\|h_0\| \leq 1$. For there exist elements $h_1, h_2, \ldots$ in H such that $\|h_n\| \leq 1$ and $(Gh_n, h_n) \to \mu$ as $n \to \infty$. In view of Lemma 4.1, we can suppose that there exists an $h_0 \in L^2(a, b)$ such that $h_n \to h_0$ weakly as $n \to \infty$ and $\|h_0\| \leq 1$. Since H is weakly closed, $h_0 \in H$. The complete continuity of G shows that $\|Gh_n - Gh_0\| \to 0$ as $n \to \infty$. Also $(Gh_0, h_0) = (Gh_n, h_n) + 2 \operatorname{Re} (G(h_0 - h_n), h_n) + (G(h_0 - h_n), h_0 - h_n)$. From the boundedness of G and Schwarz's inequality, we conclude, by letting $n \to \infty$, that $(Gh_0, h_0) = \mu$.

Note that $\mu \neq 0$ implies $h_0 \neq 0$. Also, since $\mu > 0$, it follows that $\|h_0\| = 1$, otherwise $(Gh, h) = \mu/\|h_0\|^2 > \mu$ for $h = h_0/\|h_0\|$ and $\|h\| = 1$.

In order to verify that $Gh_0 = \mu h_0$, let h be any element of H satisfying $\|h\| = 1$ and $(h_0, h) = 0$. Let $h_\epsilon = (h_0 + \epsilon h)/(1 + \epsilon^2)^{1/2}$ for a real ϵ, so that $\|h_\epsilon\|^2 = 1$. Then the function

$$(Gh_\epsilon, h_\epsilon) = (1 + \epsilon^2)^{-1}\{(Gh_0, h_0) + 2\epsilon \operatorname{Re} (Gh_0, h) + \epsilon^2(Gh, h)\}$$

of ϵ has a maximum at $\epsilon = 0$ and hence $\operatorname{Re} (Gh_0, h) = 0$. Since h can be replaced by ih, it follows that $(Gh_0, h) = 0$ for all $h \in H$ satisfying $(h_0, h) = 0$. In particular, $(Gh_0, h) = 0$ if $h = Gh_0 - \mu h_0$. This implies that

$\mu^2 = \|Gh_0\|^2$ and hence $\|Gh_0 - \mu h_0\|^2 = \|Gh_0\|^2 - 2(Gh_0, h_0) + \mu^2 = 0$. This proves $Gh_0 = \mu h_0$ and completes the proof of the lemma.

Completion of Proof of (vi). A standard theorem on Fourier series implies that (vi) is false if and only if there exist functions $h(t) \in L^2(a, b)$, $\|h\| \neq 0$, having zero Fourier coefficients $(h, u_n) = 0$ for $n = 0, 1, \ldots$. Suppose, if possible, that (vi) is false and let H denote the set of all elements $h(t) \in L^2(a, b)$ satisfying $(h, u_n) = 0$ for $n = 0, 1, \ldots$. Then H is a weakly closed, linear manifold in $L^2(a, b)$ and contains elements $h \neq 0$.

Choose a real number $\lambda \neq \lambda_n$ for $n = 0, 1, \ldots$. Then (4.6) defines a linear operator G, $w = Gh$, on $L^2(a, b)$. This operator is self-adjoint since

$$(Gh, f) = \int_a^b \int_a^b G(t, s; \lambda) h(s) f(t) \, ds \, dt = (h, Gf)$$

follows from the fact that $G(t, s; \lambda)$ is real-valued and $G(t, s; \lambda) = G(s, t; \lambda)$.

Also G is completely continuous. In order to verify this, let $h_n \to h$ weakly as $n \to \infty$ and $w_n = Gh_n$, $w = Gh$, then

$$w_n(t) - w(t) = \int_a^b G(t, s; \lambda)[h_n(s) - h(s)] \, ds = (h_n - h, G(t, \cdot \, ; \lambda))$$

tends to 0 as $n \to \infty$ for every fixed t. Furthermore, by Schwarz's inequality

$$|w_n(t) - w(t)|^2 \leq 2C \int_a^b |G(t, s; \lambda)|^2 \, ds \leq \text{const.}$$

if $\|h_n\|^2 \leq C$, $\|h(s)\|^2 \leq C$. Thus $\|w_n - w\|^2 = \int |w_n(t) - w(t)|^2 \, dt \to 0$ as $n \to \infty$ by Lebesgue's theorem on dominated convergence. [Actually, by Theorem I 2.2, $w_n(t) \to w(t)$ as $n \to \infty$ uniformly for $a \leq t \leq b$ since it is easily seen that the sequence $w_1, w_2, \ldots$ is uniformly bounded and equicontinuous.]

Finally, note that if $h \in H$, then $w = Gh$ is in H. In fact, $(h, u_n) = 0$ implies that $(w, u_n) = 0$ as can be seen by applying the Green identity (2.10) to $u = u_n, f = -\lambda_n u_n, v = w, g = -\lambda w + h$. Thus the restriction of G to the weakly closed linear manifold H gives a completely continuous, self-adjoint operator on H.

From (4.5λ) and (4.6), it is seen that $h \not\equiv 0$ implies that $w = Gh \not\equiv 0$. Since H contains elements $h \neq 0$, Lemma 4.3 is applicable. Let $Gh_0 = \mu h_0$, where $h_0 \in H$, $\|h_0\| = 1$, $\mu \neq 0$. Thus, if $w_0 = Gh_0$, it follows from (4.5λ), (4.6) that $u = w_0(t) \not\equiv 0$ is a solution of (4.1 $\lambda - 1/\mu$) satisfying the boundary conditions (4.2). Hence, by part (i), there is a non-negative integer k such that $\lambda - 1/\mu = \lambda_k$ and $w_0 = cu_k$ for some constant $c \neq 0$. But this contradicts $(w_0, u_n) = 0$ for $n = 0, 1, \ldots$ and proves the theorem.

Exercise 4.3.   Let $p_0(t) > 0$, $r_0(t) > 0$ and $q_0(t)$ be real-valued continuous functions on an open bounded interval $a < t < b$. Let $\lambda_0 < \lambda_1 < \dots$. Suppose that

(4.19) $(p_0(t)u')' + [q_0(t) + \lambda_n r_0(t)]u = 0$, $n = 0, 1, \dots$,

has a (real) solution $u_n(t)$ on $a < t < b$ having at most n zeros and
(H.) $\lim u_n(t)/u_0(t) \neq 0$ exist and $\lim p_0(u_0 u'_{n+1} - u'_0 u_{n+1}) = 0$ at a, b.
(a) Show that if $p_1(t) = 1/r_0(t)u_0^2(t) > 0$, $r_1(t) = 1/p_0(t)u_0^2(t) > 0$ and $q_1(t) = -\lambda_0/p_0(t)u_0^2(t)$, then $v_n(t) = p_0(u_0 u'_{n+1} - u'_0 u_{n+1})$ is a solution of

(4.20) $(p_1(t)v')' + [q_1(t) + \lambda_{n+1} r_1(t)]v = 0$, $n = 0, 1, \dots$,

having at most n zeros on $a < t < b$ and such that the analogue of
(H), $t \to a$ and $t \to b$, holds.   (b) Show that there exist
positive continuous functions $a_0(t)$, $a_1(t)$, $\dots$, $a_{k-1}(t)$ on $a < t < b$, such that $u_0(t)$, $\dots$, $u_{k-1}(t)$ are solutions of the kth order linear differential equation

(4.21) $(a_{k-1} \dots \{a_2[a_1(a_0 u)']'\}' \dots)' = 0$.

Exercise 4.4 *(Continuation)*.   (a) Let p_0, r_0, q_0, λ_n, u_n be as in Exercise 4.3. Let $a < t_1 < \dots < t_{k+1} < b$ and $\alpha_1, \dots, \alpha_{k+1}$ be arbitrary numbers. Then there exists a unique set of constants $c_0, \dots, c_k$ such that

(4.22) $c_0 u_0(t_j) + \dots + c_k u_k(t_j) = \alpha_j$ for $j = 1, \dots, k + 1$.

Use induction on k (for *all* systems $u_0, u_1, \dots$) or use Exercise IV 8.3. [This result is, of course, applicable to (real-valued) $u_n(t)$ in Theorem 4.1. If the functions p_0, r_0, q_0 have derivatives of sufficiently high order, then the interpolation property (4.22) can be generalized, as in Exercise IV 8.3(*d*).]
(b) Let $a < t_0 < \dots < t_n < b$. Then $D(t_0, \dots, t_n) \equiv \det (u_j(t_k))$, where $j, k = 0, \dots, n$, is different from 0. (c) Let $c_0, \dots, c_n$ be real numbers and $U_n(t) = c_0 u_0(t) + \dots + c_n u_n(t)$. Then $U_n(t) \equiv 0$ if $U_n(t)$ vanishes at $n + 1$ distinct points of $a < t < b$, and if $U_n(t) \not\equiv 0$ vanishes at n distinct points, then it changes sign at each. (*d*) Every real-valued continuous function $v(t)$ orthogonal to $u_0, \dots, u_n$ on $[a, b]$ (i.e., $\int_a^b v u_j \, dt = 0$ for $j = 0, \dots, n$) changes sign at least $n + 1$ times. (*e*) For any choice of constants $c_m, \dots, c_n$, the function $c_m u_m(t) + \dots + c_n u_n(t)$ changes sign at least m times and at most n times, where $m \leq n$.

5. Number of Zeros

This section will be concerned with zeros of real-valued solutions of an equation of the form

(5.1) $u'' + q(t)u = 0$.

Theorem 5.1. *Let $q(t)$ be real-valued and continuous for $a \leq t \leq b$. Let $m(t) \geq 0$ be a continuous function for $a \leq t \leq b$ and*

$$(5.2) \qquad \gamma_m = \inf \frac{m(t)}{(t-a)(b-t)} \qquad \text{for} \quad a < t < b.$$

If a real-valued solution $u(t) \not\equiv 0$ of (5.1) has two zeros, then

$$(5.3) \qquad \int_a^b m(t)q^+(t)\, dt > \gamma_m(b-a),$$

where $q^+(t) = \max\,(q(t), 0)$; in particular,

$$(5.4) \qquad \int_a^b (t-a)(b-t)q^+(t)\, dt > b - a.$$

Exercise 5.1. Show that the inequality (5.3) is "sharp" in the sense that (5.3) need not hold if γ_m is replaced by $\gamma_m + \epsilon$ for $\epsilon > 0$.

Proof of Theorem 5.1. Assume that (5.1) has a solution ($\not\equiv 0$) with two zeros on $[a, b]$. Since $q^+(t) \geq q(t)$, the equation

$$(5.5) \qquad u'' + q^+(t)u = 0$$

is a Sturm majorant for (5.1) and hence has a solution $u(t) \not\equiv 0$ with two zeros $t = \alpha, \beta$ on $[a, b]$; cf. Theorem 3.1. Since $u'' = -q^+u$, it follows that

$$(\beta - \alpha)u(t) = (\beta - t)\int_\alpha^t (s - \alpha)q^+(s)u(s)\, ds + (t - \alpha)\int_t^\beta (\beta - s)q^+(s)u(s)\, ds;$$

cf. Exercise 2.1, in particular (2.18). Suppose that α, β are successive zeros of u and that $u(t) > 0$ for $\alpha < t < \beta$. Choose $t = t_0$ so that $u(t_0) = \max u(t)$ on (α, β). The right side is increased if $u(s)$ is replaced by $u(t_0)$. Thus dividing by $u(t_0) > 0$ gives

$$\beta - \alpha < (\beta - t)\int_\alpha^t (s - \alpha)q^+(s)\, ds + (t - \alpha)\int_t^\beta (\beta - s)q^+(s)\, ds,$$

where $t = t_0$. Since $\beta - t \leq \beta - s$ for $t \geq s$ and $t - \alpha \leq s - \alpha$ for $s \geq t$,

$$(5.6) \qquad \beta - \alpha < \int_\alpha^\beta (\beta - s)(s - \alpha)q^+(s)\, ds.$$

Finally, note that $(t-a)(b-t)/(b-a) \geq (t-\alpha)(\beta-t)/(\beta-\alpha)$ for $a \leq \alpha \leq t \leq \beta \leq b$; in fact, differentiation with respect to β and α shows that $(t-\alpha)(\beta-t)/(\beta-\alpha)$ increases with β if $t \geq \alpha$ and decreases with α if $t \leq \beta$. Hence (5.4) follows from the last display. The relation (5.3) is a consequence of (5.2) and (5.4). This proves Theorem 5.1.

Since $(t - a)(b - t) \leq (b - a)^2/4$, the choice $m(t) = 1$ in Theorem 5.1 gives the following:

Corollary 5.1 (Lyapunov). *Let $q(t)$ be real-valued and continuous on $a \leq t \leq b$. A necessary condition for (5.1) to have a solution $u(t) \not\equiv 0$ possessing two zeros is that*

$$(5.7) \qquad \int_a^b q^+(t)\, dt > \frac{4}{b - a}.$$

Exercise 5.2.   Let $q(t) \geq 0$ be continuous on $a \leq t \leq b$ and let (5.1) have a solution $u(t)$ vanishing at $t = a, b$ and $u(t) > 0$ in (a, b). (a) Use (5.7) to show that $\int_a^b q(t)\, dt > 2M/A$, where $M = \max u(t)$ and $A = \int_a^b u(t)\, dt$. (b) Show that the factor 2 of M/A cannot be replaced by a larger constant.

Exercise 5.3.   (a) Consider a differential equation $u'' + g(t)u' + f(t)u = 0$ with real-valued continuous coefficients on $0 \leq t \leq b$ having a solution $u(t) \not\equiv 0$ vanishing at $t = 0, b$. Show that

$$b < \int_0^b t(b - t)f^+(t)\, dt + \max\left\{\int_0^b t\, |g|\, dt, \quad \int_0^b (b - t)\, |g|\, dt\right\}.$$

(b) In particular, if $|g| \leq M_1$ and $|f| \leq M_2$, then $1 < M_1 b/2 + M_2 b^2/6$. But this inequality can easily be improved by the use of Wirtinger's inequality $\int_0^b u^2\, dt \leq (b/\pi)^2 \int_0^b u'^2\, dt$ (which can be proved by assuming $b = \pi$, expanding u into a Fourier sine series, and applying Parseval's relation for u, u'). Show that $1 \leq M_1 b/\pi + M_2 b^2/\pi^2$. (c) The result of part (b) can further be improved to $1 \leq 2M_1 b/\pi^2 + M_2 b^2/\pi^2$. See Opial [3]. (d) An analogous result for a dth order equation, $d \geq 2$, is as follows: Let the differential equation $u^{(d)} + p_1(t)u^{(d-1)} + \cdots + p_d(t)u = 0$ have continuous coefficients for $0 \leq t \leq b$ and a solution $u(t) \not\equiv 0$ with d zeros on $[0, b]$. Let $|p_j(t)| \leq M_j$. Then $1 < M_1 b + M_2 b^2/2! + \cdots + M_{d-1} b^{d-1}/(d - 1)! + (M_d b^d/d!)[(d - 1)^{d-1}/d^d]$.

When $q(t) = q^+(t)$ is a positive constant on $[0, T]$, the number N of zeros of a solution ($\not\equiv 0$) of (5.1) on $(0, T]$ obviously satisfies

$$(5.8) \qquad \pi N \leq (q^+)^{\frac{1}{2}} T \equiv \int_0^T (q^+)^{\frac{1}{2}}\, dt \leq \left(T \int_0^T q^+(t)\, dt\right)^{\frac{1}{2}},$$

where the last inequality follows from Schwarz's inequality. It turns out that a similar inequality holds for nonconstant, continuous $q(t)$:

Corollary 5.2. *Let $q(t)$ be real-valued and continuous for $0 \leq t \leq T$.*

*Let $u(t) \not\equiv 0$ be a solution of (5.1) and N the number of its zeros on $0 <$
$t \leqq T$. Then*

(5.9)
$$N < \frac{1}{2}\left(T\int_0^T q^+(t)\,dt\right)^{1/2} + 1.$$

Proof. In order to prove this, let $N \geqq 2$ and let the N zeros of u on
$(0, T]$ be $(0 <) t_1 < t_2 < \cdots < t_N (\leqq T)$. By Corollary 5.1,

(5.10)
$$\int_u^v q^+(t)\,dt > \frac{4}{v - u} \qquad \text{if} \quad u = t_k, \quad v = t_{k+1},$$

for $k = 1, \ldots, N - 1$. Since the harmonic mean of $N - 1$ positive
numbers is majorized by their arithmetic mean,

$$\left[\frac{1}{N - 1}\sum_{k=1}^{N-1}\frac{1}{t_{k+1} - t_k}\right]^{-1} \leqq \frac{1}{N - 1}\sum_{k=1}^{N-1}(t_{k+1} - t_k) = \frac{t_N - t_1}{N - 1}.$$

Thus adding (5.10) for $k = 1, \ldots, N - 1$ gives

$$\int_{t_1}^{t_N} q^+(t)\,dt > \frac{4(N - 1)^2}{t_N - t_1} \geqq \frac{4(N - 1)^2}{T},$$

hence (5.9).

Exercise 5.4. Show that N also satisfies

(5.11)
$$N < \int_0^T t q^+(t)\,dt + 1.$$

To this end, use (5.3) with $m(t) = t - a$ in place of (5.7).

Note that if $q(t)$ is a positive constant, then

$$\left|\pi N - \int_0^T q^{1/2}\,dt\right| \leqq \pi.$$

An analogous inequality holds under mild assumptions on nonconstant q:

Theorem 5.2. *Let $q(t) > 0$ be continuous and of bounded variation on*
$0 \leqq t \leqq T$. *Let $u(t) \not\equiv 0$ be a real-valued solution of (5.1) and N the*
number of its zeros on $0 < t \leqq T$. Then

(5.12)
$$\left|\pi N - \int_0^T q^{1/2}(t)\,dt\right| \leqq \pi + \frac{1}{4}\int_0^T \frac{|dq(t)|}{q(t)}$$

Proof. In terms of $u(t)$ define a continuous function $\varphi(t)$ by

$$\varphi(t) = \arctan\frac{q^{1/2}(t)u}{u'}, \qquad 0 \leqq \varphi(0) < \pi.$$

Then [cf. Exercise 2.6; in particular (2.49) where $p(t) \equiv 1$]

$$\varphi(T) = \varphi(0) + \int_0^T q^{1/2}(t)\,dt + \frac{1}{4}\int_0^T \sin 2\varphi(t)\,d(\log q).$$

By Lemma 3.1, N is the greatest integer not exceeding $\varphi(T)/\pi$, so that $\pi N \leqq \varphi(T) \leqq \pi(N + 1)$. This implies (5.12).

Exercise 5.5.   (a) Let $q(t)$ be continuous on $0 \leqq t \leqq T$. Let $u(t) \not\equiv 0$ be a real-valued solution and N the number of its zeros on $0 < t \leqq T$. Show that

$$|\pi N - T| \leqq \pi + \int_0^T |1 - q(t)|\, dt.$$

(b) If, in addition, $q(t) > 0$ has a continuous second derivative, then

$$\left| \pi N - \int_0^T q^{\frac{1}{2}}(t)\, dt \right| \leqq \pi + \int_0^T \left| \frac{5q'^2}{16q^{\frac{5}{2}}} - \frac{q''}{4q^{\frac{3}{2}}} \right| dt.$$

Corollary 5.3.   *Let $q(t) > 0$ be continuous and of bounded variation on $[0, T]$ for every $T > 0$. Suppose also that*

$$(5.13) \qquad \int_0^T q^{-1} |dq| = o\left(\int_0^T q^{\frac{1}{2}}\, dt \right) \qquad as \quad T \to \infty;$$

e.g., suppose that $q(t)$ has a continuous derivative $q'(t)$ satisfying

$$(5.14) \qquad\qquad q'(t) = o(q^{\frac{3}{4}}(t)) \qquad as \quad t \to \infty.$$

Let $u(t) \not\equiv 0$ be a real-valued solution of (5.1) and $N(T)$ the number of its zeros on $0 < t \leqq T$. Then

$$(5.15) \qquad\qquad \pi N(T) \sim \int_0^T q^{\frac{1}{2}}(t)\, dt \qquad as \quad T \to \infty.$$

This is clear from (5.13) and the formula (5.12) in Theorem 5.2. It should be mentioned that if, e.g., q is monotone and $q(t) \to \infty$ as $t \to \infty$, then (5.14) imposes no restriction on the rapidity of growth of $q(t)$ but is a condition on the regularity of growth. This can be seen from the fact that the integral

$$\int^T \frac{q'\, dt}{q^{\frac{3}{4}}} = \frac{2}{q^{\frac{1}{4}}(T)} + \text{const.}$$

tends to a limit as $T \to \infty$; thus, in general, $q'/q^{\frac{3}{4}}$ is "small" for large t.

The conditions of Corollary 5.3 for the validity of (5.15) can be lightened somewhat, as is shown by the following exercises.

Exercise 5.6.   (a) Let $q(t) > 0$ be continuous for $t \geqq 0$ and satisfy

$$(5.16) \qquad \sup_{s \leqq t < \infty} \frac{|\log q(t)/q(s)|}{1 + \int_s^t q^{\frac{1}{2}}(r)\, dr} \to 0 \qquad as \quad s \to \infty.$$

Let $u(t) \not\equiv 0$ be a solution of (5.1) and $N(T)$ the number of its zeros on $0 < t \leqq T$. Then (5.15) holds. (b) Necessary and sufficient for (5.16) is

the following pair of conditions: $\int^{\infty} q^{\frac{1}{2}} \, dt = \infty$ and $q(t + cq^{-\frac{1}{2}}(t))/$
$q(t) \to 1$ as $t \to \infty$ holds uniformly on every fixed bounded c-interval on
$-\infty < c < \infty$.

Exercise 5.7.   Part (b) of the last exercise can be generalized as follows:
Let $q(t) > 0$ be continuous for $t \geqq 0$. Let $m(t) > 0$ be continuous for
$t > 0$ and satisfy $[m(t)/m(s)]^{\pm 1} \leqq C(t/s)^{\gamma}$ for $0 < s < t < \infty$ and some
pair of non-negative constants C, γ. Necessary and sufficient for

$$\sup_{s \leqq t < \infty} \frac{|\log q(t)/q(s)|}{1 + \int_{s}^{t} m(q(r)) \, dr} \to 0 \qquad \text{as} \quad s \to \infty$$

is that $\int^{\infty} m(q(t)) \, dt = \infty$ and that $q(t + c/m[q(t)])/q(t) \to 1$ as min $[t,$
$t + c/m(q(t))] \to \infty$ holds uniformly on every bounded c-interval on
$-\infty < c < \infty$.

An estimate for N of a type very different from those just given is the
following:

Theorem 5.3.   *Let* $p(t) > 0$, $q(t)$ *be real-valued and continuous for*
$0 \leqq t \leqq T$. *Let* $u(t), v(t)$ *be real-valued solutions of*

$$(5.17) \qquad (pu')' + qu = 0$$

satisfying

$$(5.18) \qquad p(t)[u'(t)v(t) - u(t)v'(t)] = c > 0.$$

Let N *be the number of zeros of* $u(t)$ *on* $0 < t \leqq T$. *Then*

$$(5.19) \qquad \left| \pi N - c \int_{0}^{T} \frac{dt}{p(t)[u^2(t) + v^2(t)]} \right| \leqq \pi.$$

Proof.   Let α be an arbitrary real number.   Consider the solutions
$$u^*(t) = u(t) \cos \alpha + v(t) \sin \alpha, \qquad v^*(t) = -u(t) \sin \alpha + v(t) \cos \alpha$$
of (5.17). They satisfy

$$(5.20) \qquad u^2 + v^2 = u^{*2} + v^{*2}, \qquad p[u^{*\prime}v^* - u^*v^{*\prime}] = c > 0.$$

Choose α so that $u^*(0) = 0$ and let N^* be the number of zeros of $u^*(t)$ on
$0 < t \leqq T$.

Since (5.20) implies that u^*, v^* are linearly independent, they have no
common zeros. Hence it is possible to define a continuous function by

$$(5.21) \qquad \varphi(t) = \arctan \frac{u^*(t)}{v^*(t)} \qquad \text{and} \quad \varphi(0) = 0.$$

This function is continuously differentiable and, by (5.20),

$$(5.22) \qquad \varphi'(t) = \frac{c}{p(t)[u^2(t) + v^2(t)]} > 0.$$

Hence $\varphi(t)$ is increasing; also $\varphi(t) = 0$ mod π if and only if $u^*(t) = 0$. Thus N^* is the greatest integer not exceeding $\varphi(T)/\pi$ and a quadrature of (5.22) gives

$$\pi N^* \leqq c \int_0^T \frac{dt}{p(t)[u^2(t) + v^2(t)]} < \pi(N^* + 1).$$

Sturm's separation theorem implies $N^* \leqq N \leqq N^* + 1$, thus (5.19) follows.

Exercise 5.8. Let $p(t)$, $q(t)$, $u(t)$, $v(t)$, and N be as in Theorem 5.3 and, in addition, let $q(t) \geqq 0$. Show that

$$(5.23) \qquad \left| \pi N - c \int_0^T \frac{q(t)\, dt}{p^2(t)[u'^2(t) + v'^2(t)]} \right| \leqq 2\pi$$

(If $q > 0$, the relations (5.19) and (5.23) are particular cases of "duality" in which (u, u', q, dt) are replaced by $(pu', -u, 1/q, q\, dt)$; cf. Lemma XIV 3.1.)

Exercise 5.9. (a) Let $q(t)$ be continuous for $t \geqq 0$. Using (5.9) and (5.19), show that if all solutions of $u'' + q(t)u = 0$ are bounded, then, for large t,

$$(5.24) \qquad \frac{1}{t} \int_0^t q^+(s)\, ds \geqq \text{const.} > 0.$$

Replacing u, v in (5.19) by u/ϵ, ϵv, show that if, in addition, a nontrivial solution $u(t) \to 0$ as $t \to \infty$, then

$$(5.25) \qquad \frac{1}{t} \int_0^t q^+(s)\, ds \to \infty \qquad \text{as} \quad t \to \infty.$$

(b) Let $q(t) \geqq 0$ for $t \geqq 0$. Using (5.9) and (5.23), show that if the first derivatives of all solutions of $u'' + q(t)u = 0$ are bounded, then, for large t,

$$(5.26) \qquad \frac{1}{t} \int_0^t q^+(s)\, ds \leqq \text{const.}$$

If, in addition, $u'(t) \to 0$ as $t \to \infty$ for some solution $u(t) \not\equiv 0$, then

$$(5.27) \qquad \frac{1}{t} \int_0^t q^+(s)\, ds \to 0 \qquad \text{as} \quad t \to \infty.$$

(c) Generalize (a) [or (b)] for the case when $u'' + qu = 0$ is replaced by $(pu')' + qu = 0$ and the assumption that solutions [or derivatives of solutions] are bounded is replaced by the assumption that all solutions satisfy $u(t) = O(1/\Phi(t))$ [or $u'(t) = O(1/\Phi(t))$], where $\Phi(t) > 0$ is continuous.

6. Nonoscillatory Equations and Principal Solutions

A homogeneous, linear second order equation with real-valued coefficient functions defined on an interval J is said to be *oscillatory on J*

if one (and/or every) real-valued solution ($\not\equiv 0$) has infinitely many zeros on J. Conversely, when every solution ($\not\equiv 0$) has at most a finite number of zeros on J, it is said to be *nonoscillatory on J*. In the latter case, the equation is said to be *disconjugate on J* if every solution ($\not\equiv 0$) has at most one zero on J. If $t = \omega$ is a (possibly infinite) endpoint of J which does not belong to J, then the equation is said to be *oscillatory at $t = \omega$* if one (and/or every) real-valued solution ($\not\equiv 0$) has an infinite sequence of zeros clustering at $t = \omega$. Otherwise it is called *nonoscillatory at $t = \omega$*.

Extensions of many of the results of this section to higher order equations or more general systems will be indicated in §§ 10, 11 of the Appendix.

Theorem 6.1. *Let $p(t) > 0$, $q(t)$ be real-valued, continuous functions on a t-interval J. Then*

$$(6.1) \qquad (p(t)u')' + q(t)u = 0$$

is disconjugate on J if and only if, for every pair of distinct points t_1, $t_2 \in J$ and arbitrary numbers u_1, u_2, there exists a unique solution $u = u^*(t)$ of (6.1) satisfying

$$(6.2) \qquad u^*(t_1) = u_1 \quad \text{and} \quad u^*(t_2) = u_2;$$

or, equivalently, if and only if every pair of linearly independent solutions $u(t)$, $v(t)$ of (6.1) satisfy

$$(6.3) \qquad u(t_1)v(t_2) - u(t_2)v(t_1) \neq 0$$

for distinct points t_1, $t_2 \in J$.

Proof. Let $u(t)$, $v(t)$ be a pair of linearly independent solutions of (6.1). Then any solution $u^*(t)$ is of the form $u^* = c_1 u(t) + c_2 v(t)$. This solution satisfies (6.2) if and only if

$$c_1 u(t_1) + c_2 v(t_1) = u_1, \qquad c_1 u(t_2) + c_2 v(t_2) = u_2.$$

These linear equations for c_1, c_2 have a solution for all u_1, u_2 if and only if (6.3) holds. In addition, they have a solution for all u_1, u_2 if and only if the only solution of

$$c_1 u(t_1) + c_2 v(t_1) = 0, \qquad c_1 u(t_2) + c_2 v(t_2) = 0$$

is $c_1 = c_2 = 0$; i.e., if and only if the only solution $u^*(t)$ of (6.1) with two zeros $t = t_1$, t_2 is $u^*(t) \equiv 0$.

Corollary 6.1. *Let $p(t) > 0$, $q(t)$ be as in Theorem 6.1. If J is open or is closed and bounded, then (6.1) is disconjugate on J if and only if (6.1) has a solution satisfying $u(t) > 0$ on J. If J is a half-closed interval or a closed half-line, then (6.1) is disconjugate on J if and only if there exists a solution $u(t) > 0$ on the interior of J.*

The example $u'' + u = 0$ on J: $0 \leq t < \pi$ shows that, in the last part of the theorem, there need not exist a solution $u(t) > 0$ on J.

Exercise 6.1.   Deduce Corollary 6.1 from Theorem 6.1 (another proof follows from Exercise 6.6).

Exercise 6.2.   Let $p(t) > 0$, $q(t) \geqq 0$ be continuous on an interval $J: a \leqq t < \omega$ ($\leqq \infty$) such that $\int^{\omega} dt/p(t) = \infty$, then (6.1) is disconjugate on J if and only if it has a solution $u(t)$ such that $u(t) > 0$, $u'(t) \geqq 0$ for $a < t < \omega$.

A very useful criterion for (6.1) to be disconjugate is a "variational principle" to be stated as the next theorem. A real-valued function $\eta(t)$ on the subinterval $[a, b]$ of J will be said to be *admissible of class* $A_1(a, b)$ [or $A_2(a, b)$] if (i) $\eta(a) = \eta(b) = 0$, and (ii$_1$) $\eta(t)$ is absolutely continuous and its derivative $\eta'(t)$ is of class L^2 on $a \leqq t \leqq b$ [or (ii$_2$) $\eta(t)$ is continuously differentiable and $p(t)\eta'(t)$ is continuously differentiable on $a \leqq t \leqq b$]. Put

$$(6.4) \qquad I(\eta; a, b) = \int_a^b (p\eta'^2 - q\eta^2)\, dt \qquad \text{for} \quad \eta \in A_1(a, b).$$

If η is admissible $A_2(a, b)$, the first term can be integrated by parts and it is seen that

$$(6.5) \qquad I(\eta; a, b) = -\int_a^b \eta[(p\eta')' + q\eta]\, dt \qquad \text{for} \quad \eta \in A_2(a, b).$$

Theorem 6.2.   *Let $p(t) > 0$, $q(t)$ be real-valued continuous functions on a t-interval J. Then (6.1) is disconjugate on J if and only if, for every closed bounded subinterval $a \leqq t \leqq b$ of J, the functional (6.4) is positive-definite on $A_1(a, b)$ [or $A_2(a, b)$]; i.e., $I(\eta; a, b) \geqq 0$ for $\eta \in A_1(a, b)$ [or $\eta \in A_2(a, b)$] and $I(\eta; a, b) = 0$ if and only if $\eta \equiv 0$.*

The "only if" half of the theorem is stronger for $A_1(a, b)$ and the "if" half is stronger for $A_2(a, b)$.

Proof ("Only if").   Suppose that (6.1) is disconjugate on $a \leqq t \leqq b$. Then, by Corollary 6.1, there is a solution $u(t) > 0$ on $a \leqq t \leqq b$. If $\eta(t) \in A_1(a, b)$, put $\zeta(t) = \eta(t)/u(t)$. Then

$$(6.6) \qquad I(\eta; a, b) = \int_a^b [\zeta^2(pu'^2 - qu^2) + p(u^2\zeta'^2 + 2\zeta\zeta'uu')]\, dt.$$

An integration by parts [integrating u' and differentiating $(pu')\zeta^2$] shows that the first term is

$$\int_a^b \zeta^2 pu'^2\, dt = [\zeta^2 puu']_a^b - \int_a^b [\zeta^2 u(pu')' + 2p\zeta\zeta'uu']\, dt.$$

The integrated terms vanish since $\eta(a) = \eta(b) = 0$ imply that $\zeta(a) = \zeta(b) = 0$. The last two formula lines and $\zeta^2 u[(pu')' + qu] = 0$ give

$$(6.7) \qquad I(\eta; a, b) = \int_a^b pu^2\zeta'^2\, dt \qquad \text{for} \quad \eta = u\zeta \in A_1(a, b).$$

It is clear that $I(\eta; a, b) \geqq 0$ and $I(\eta; a, b) = 0$ if and only if $\zeta(t) \equiv 0$. This proves the "only if" part of the theorem.

Proof ("If"). Suppose that $I(\eta; a, b)$ is positive definite on $A_2(a, b)$ for every $[a, b] \subset J$. Let $\eta(t)$ be a solution of (6.1) having two zeros $t = a, b \in J$. It will be shown that $\eta(t) \equiv 0$. In fact $\eta(t) \in A_2(a, b)$; thus (6.5) holds. Hence, $I(\eta; a, b) = 0$ because η is a solution of (6.1). Since (6.4) is positive definite on $A_2(a, b)$, it follows that $\eta(t) \equiv 0$. This implies that (6.1) is disconjugate on J and completes the proof of the theorem.

Exercise 6.3. Suppose that J is not a closed bounded interval. Show that, in Theorem 6.2, (6.1) is disconjugate on J if $I(\eta; a, b) \geqq 0$ for all $[a, b] \subset J$ and all $\eta \in A_2(a, b)$.

Exercise 6.4. Deduce Sturm's separation theorem (Corollary 3.1) from Theorem 6.2.

If P is a constant positive definite Hermitian matrix, then there exists a positive definite Hermitian matrix P_1 which is the "square root" of P in the sense that $P = P_1{}^2 = P_1{}^* P_1$; cf. Exercise XIV 1.2. An analogue of this algebraic fact will be obtained for the differential operator

$$L[\eta] = -(p(t)\eta')' - q(t)\eta.$$

Note that (6.5) can be written as

$$I(\eta; a, b) = (L[\eta], \eta) \quad \text{for} \quad \eta \in A_2(a, b);$$

cf. (4.17). Also, (6.7) can be written as

$$I(\eta; a, b) = \int_a^b \frac{p(\eta'u - \eta u')^2}{u^2} \, dt.$$

In addition to the quadratic functional (6.4), consider the bilinear form

$$I(\eta_1, \eta_2; a, b) = \int_a^b (p\eta_1'\eta_2' - q\eta_1\eta_2) \, dt$$

for $\eta_1, \eta_2 \in A_1(a, b)$. If $\eta_1 \in A_2(a, b)$, an integration by parts shows that

$$I(\eta_1, \eta_2; a, b) = -\int_a^b \eta_2[(p\eta_1')' + q(\eta_1)] \, dt = (L[\eta_1], \eta_2).$$

If $u(t)$ is a solution of (6.1) and $u(t) > 0$ on $[a, b]$, it is readily verified that, for $\eta_1, \eta_2 \in A_2(a, b)$ and $\zeta_1 = \eta_1/u$, $\zeta_2 = \eta/u$,

$$I(\eta_1, \eta_2; a, b) = \int_a^b pu^2\zeta_1'\zeta_2' \, dt.$$

or

$$I(\eta_1; \eta_2; a, b) = \int_a^b \frac{p(\eta_1'u - \eta_1u')(\eta_2'u - \eta_2u')}{u^2} \, dt$$

Thus if the first order differential operator L_1 is defined by

$$L_1[\eta] = \frac{p^{\frac{1}{2}}(t)[\eta'u(t) - \eta u'(t)]}{u(t)} = p^{\frac{1}{2}}u\left(\frac{\eta}{u}\right)',$$

i.e., by

(6.8) $$L_1[\eta] = p^{\frac{1}{2}}(t)\eta' - \frac{p^{\frac{1}{2}}(t)u'(t)}{u(t)}\eta = p^{\frac{1}{2}}u\left(\frac{\eta}{u}\right)',$$

then it follows that

(6.9) $(L[\eta_1], \eta_2) = (L_1[\eta_1], L_1[\eta_2])$ for $\eta_1, \eta_2 \in A_2(a, b)$.

Consequently, if L [i.e., (6.4)] is positive definite on $A_2(a, b)$, so that there exists a positive solution $u(t) > 0$ of (6.1) on $[a, b]$, then formally

$$L = L_1{}^*L_1.$$

In' fact this relation is not only formally correct but is correct in the following sense:

Corollary 6.2.   *Let $p(t) > 0$, $q(t)$ be continuous on J and let (6.1) have a solution $u(t) > 0$ on J. Let L_1 be defined by (6.8) and $L_1{}^*$ its formal adjoint*

$$L_1{}^*[\eta] = -(p^{\frac{1}{2}}(t)\eta)' - \frac{p^{\frac{1}{2}}(t)u'(t)}{u(t)}\eta = -\frac{(p^{\frac{1}{2}}u\eta)'}{u};$$

cf. § IV 8 (viii). Then

$$L[\eta] = L_1{}^*\{L_1[\eta]\}$$

for all continuously differentiable functions η for which $p(t)\eta'$ is absolutely continuous (i.e., for all η for which $L[\eta]$ is usually defined).

This can be deduced from the identity (6.9) or, more easily, by a straightforward verification. See Appendix for generalizations of this result.

Theorem 5.3 and its proof have the following consequence.

Theorem 6.3.   *Let $p(t) > 0$, $q(t)$ be real-valued and continuous on an interval J. Then (6.1) is nonoscillatory on J if and only if every pair of linearly independent solutions $u(t)$, $v(t)$ of (6.1) satisfy*

$$\int_J \frac{dt}{p(t)(|u|^2 + |v|^2)} < \infty.$$

Furthermore, (6.1) is disconjugate on J if and only if

$$|c|\int_a^b \frac{dt}{p(t)(u^2 + v^2)} < \pi$$

for every pair of real-valued solutions $u(t)$, $v(t)$ satisfying $p(u'v - uv') = c \neq 0$ and every interval $[a, b] \subset J$.

If J is a half-open interval, say $J:a \leqq t < \omega \ (\leqq \infty)$ and (6.1) is non-oscillatory at $t = \omega$, then (6.1) has real-valued solutions $u(t)$ for which $\int^{\omega} dt/pu^2$ is convergent and solutions for which it is divergent. The latter type of solution will be called a *principal solution* of (6.1) at $t = \omega$.

Theorem 6.4. *Let* $p(t) > 0$, $q(t)$ *be real-valued and continuous on* J: $a \leqq t < \omega \ (\leqq \infty)$ *and such that* (6.1) *is nonoscillatory at* $t = \omega$. *Then there exists a real-valued solution* $u = u_0(t)$ *of* (6.1) *which is uniquely determined up to a constant factor by any one of the following conditions in which* $u_1(t)$ *denotes an arbitrary real-valued solution linearly independent of* $u_0(t)$: (i) u_0, u_1 *satisfy*

$$(6.10) \qquad \frac{u_0(t)}{u_1(t)} \to 0 \quad \text{as} \quad t \to \omega;$$

(ii) u_0, u_1 *satisfy*

$$(6.11_0) \quad \int^{\omega} \frac{dt}{p(t)u_0^2(t)} = \infty \quad \text{and} \quad (6.11_1) \quad \int^{\omega} \frac{dt}{p(t)u_1^2(t)} < \infty;$$

(iii) *if* $T \in J$ *exceeds the largest zero, if any, of* $u_0(t)$ *and if* $u_1(T) \neq 0$, *then* $u_1(t)$ *has one or no zero on* $T < t < \omega$ *according as*

$$(6.12_0) \quad \frac{u_1'}{u_1} < \frac{u_0'}{u_0} \quad \text{or} \quad (6.12_1) \quad \frac{u_1'}{u_1} > \frac{u_0'}{u_0}$$

holds at $t = T$; *in particular*, (6.12_1) *holds for all* $t \ (\in J)$ *near* ω.

It is understood that in (6.10) and (6.11) only t-values exceeding the largest zeros, if any, of u_0, u_1 are considered. A solution $u_0(t)$ satisfying one (and/or) all of the conditions (i), (ii), (iii) will be called a *principal solution* of (6.1) (at $t = \omega$). A solution $u(t)$ linearly independent of $u_0(t)$ will be termed a *nonprincipal solution* of (6.1) (at $t = \omega$). In view of (6.10), (6.11), the terms "principal" and "nonprincipal" might well be replaced by "small" and "large." The expressions "small," "large" will not be used in this context because of the relative nature of these terms. Consider, e.g., the equations $u'' - u = 0$, $u'' = 0$ and $u'' + u/4t^2 = 0$ for $t \geqq 1$. Examples of principal and nonprincipal solutions at $t = \infty$ for the first equation are $u = e^{-t}$ and $u = e^t$; for the second, $u = 1$ and $u = t$; for the third, $u = t^{1/2}$ and $u = t^{1/2} \log t$; cf. Exercise 1.1. The proof of (ii) will lead to the following:

Corollary 6.3. *Assume the conditions of Theorem 6.4. Let* $u = u(t) \not\equiv 0$ *be any real-valued solution of* (6.1) *and let* $t = T$ *exceed its last zero. Then*

$$(6.13) \qquad u_1(t) = u(t) \int_T^t \frac{ds}{p(s)u^2(s)}$$

is a nonprincipal solution of (6.1) *on* $T \leqq t < \omega$. *If, in addition, $u(t)$ is a nonprincipal solution of* (6.1), *then*

$$(6.14) \qquad u_0(t) = u(t) \int_t^\omega \frac{ds}{p(s)u^2(s)}$$

is a principal solution on $T \leqq t < \omega$.

Proof of Theorem 6.4 and Corollary 6.3

On (i). Let $u(t)$, $v(t)$ be a pair of real-valued linearly independent solutions of (6.1) such that

$$(6.15) \qquad p(u'v - uv') = c \neq 0.$$

If T exceeds the largest zero, if any, of $v(t)$, then (6.15) is equivalent to

$$(6.16) \qquad \left(\frac{u}{v}\right)' = \frac{c}{pv^2} \neq 0,$$

for $T \leqq t < \omega$. Hence u/v is monotone on this t-range and so

$$(6.17) \qquad C = \lim_{t \to \omega} \frac{u(t)}{v(t)}$$

exists if $C = \pm\infty$ is allowed.

It will be shown that u, v can be chosen so that $C = 0$ in (6.17). If this is granted and if $u(t)$ is called $u_0(t)$, then (i) holds. In fact, a solution $u_1(t)$ is linearly independent of $u_0(t)$ if and only if it is of the form $u_1(t) = c_0 u_0(t) + c_1 v(t)$ and $c_1 \neq 0$; in which case, $C = 0$ implies that $u_1 = [c_1 + o(1)]v(t)$; thus $u_0 = o(u_1)$ as $t \to \omega$.

If $C = \pm\infty$ in (6.17) and if u, v are interchanged, then (6.17) holds with $C = 0$. If $|C| < \infty$ and if $u(t) - Cv(t)$ is renamed $u(t)$, then (6.15) still holds and (6.17) holds with $C = 0$. This proves (i).

On (ii). Note that (6.16), (6.17) give

$$C = \frac{u(T)}{v(T)} + c \int_T^\omega \frac{ds}{p(s)v^2(s)}$$

whether or not $|C| = \infty$ or $|C| < \infty$. If u, v is a pair u_0, u_1, so that $C = 0$, then (6.11_1) holds. If u, v is a pair u_1, u_0, so that $C = \pm\infty$, then (6.11_0) holds.

On Corollary 6.3. Note that if $u(t)$ is a solution of (6.1) and $u(t) \neq 0$ for $T \leqq t < \omega$, then (6.13) defines a solution $u_1(t)$ linearly independent of

$u(t)$ and that the same is true of (6.14) when the integral is convergent; see § 2 (ix). By (i), this implies Corollary 6.3.

On (iii). Since u_0, u_1 can be replaced by $-u_0, -u_1$, respectively, without affecting the zeros of u_1 or the inequalities (6.12), it can be supposed that

$$(6.18) \qquad u_0(t) > 0 \quad \text{for} \quad T \leqq t < \omega \quad \text{and} \quad u_1(T) > 0.$$

Multiplying (6.12) by $u_0(T)u_1(T) > 0$ shows that the case (6.15), where $(u, v) = (u_1, u_0)$ holds with $c < 0$ or $c > 0$ according as (6.12_0) or (6.12_1) holds. Hence $u_1(t)/u_0(t) \to \mp \infty$ as $t \to \omega$ according as (6.12_0) or (6.12_1) holds. Since $u_1(T)/u_0(T) > 0$ and, by the Sturm separation theorem, u_1 has at most one zero on $T < t < \omega$, the statement concerning the zeros of u_1 on $T < t < \omega$ follows.

It remains to show that property (iii) is characteristic of a principal solution; i.e., if $u_0(t)$ has the property (iii) for every solution $u_1(t)$ linearly independent of $u_0(t)$, then $u_0(t)$ is a principal solution. In particular (6.12_1) holds for $t \ (\in J)$ near ω. Consequently $|u_0(t)| \leqq$ const. $|u_1(t)|$ for $t \to \omega$. This is a contradiction if $u_0(t)$ is not a principal solution and $u_1(t)$ is chosen to be a principal solution.

Exercise 6.5. Assume (i) that the conditions of Theorem 6.4 hold; (ii) that (6.1) has a nonvanishing real-valued solution for $(a \leqq) \ T \leqq t < \omega$; and (iii) that $u_{0r}(t)$ is the unique solution of (6.1) satisfying $u_{0r}(T) = 1$, $u_{0r}(r) = 0$, where $T < r < \omega$; cf. Theorem 6.1. (a) Show that $u_0(t) = \lim u_{0r}(t)$ exists as $r \to \omega$ uniformly on compact intervals of J and is the principal solution of (6.1) at $t = \omega$ satisfying $u_0(T) = 1$. (b) Show that (a) is false if condition (ii) is relaxed to the condition that (6.1) is disconjugate on $T \leqq t < \omega$.

Exercise 6.6. Let $p(t) > 0$, $q(t)$ be real-valued and continuous functions such that (6.1) is disconjugate on a t-interval J having $t = \omega \ (\leqq \infty)$ as right endpoint. Let $u_0(t)$ be a principal solution of (6.1) at $t = \omega$. Then $u_0(t) \neq 0$ on the interior of J.

Sturm's comparison theorem implies that "$q(t) \leqq 0$ on J" is sufficient for (6.1) to be disconjugate on J. In this case, we can give some additional information about a principal solution.

Corollary 6.4. *Let $p(t) > 0$, $q(t) \leqq 0$ be continuous on $J : a \leqq t < \omega$. Then (6.1) has a principal solution satisfying*

$$(6.19) \qquad u_0(t) > 0, \qquad u_0'(t) \leqq 0 \quad \text{for} \quad a \leqq t < \omega$$

and a nonprincipal solution $u_1(t)$ such that

$$(6.20) \qquad u_1(t) > 0, \qquad u_1'(t) > 0 \quad \text{for} \quad a \leqq t < \omega.$$

Exercise 6.7.   (*a*) In Corollary 6.4, the conditions (6.19) uniquely determine $u_0(t)$, up to a constant factor, if and only if

(6.21) $$\int^\omega \frac{dt}{p(t)} = \infty \quad \text{or} \quad -\int^\omega q(t)\, dt = \infty.$$

(*b*) Assume the first part of (6.21). Using Corollary 9.1, show that a principal solution in Corollary 6.4 satisfies $u_0(t) \to 0$ as $t \to \omega$ if and only if $-\int^\infty q(t) \left(\int^t dr/p(r) \right) dt = \infty.$

For generalizations, related results, and a different proof of Corollary 6.4, see XIV §§ 1, 2.

Proof.   Assume first that $p(t) \equiv 1$, so that (6.1) is of the form

(6.22) $$u'' + q(t)u = 0,$$

where $q \leq 0$. Hence the graph of a solution $u = u(t)$ of (6.22) in the (t, u)-plane is concave upwards when $u(t) > 0$. Let $u(t)$ be the solution of (6.22) determined by $u(a) = 1$, $u'(a) = 1$. Then $u = u(t)$ has a graph which is concave upward for $a \leq t < \omega$. In particular, $u(t) > u(a) = 1$, $u'(t) \geq u'(a) = 1$; so that $u(t) \geq 1 + t$. Thus $\int^\omega dt/u^2(t)$ is convergent, and so $u(t)$ is a nonprincipal solution of (6.22). By Corollary 6.3,

$$u_0(t) = u(t) \int_t^\omega \frac{ds}{u^2(s)} > 0$$

is a principal solution of (6.22). Differentiating this formula gives

$$u_0'(t) = u'(t) \int_t^\omega \frac{ds}{u^2(s)} - \frac{1}{u(t)}$$

Since $u'(t)$ is nondecreasing,

$$u_0'(t) \leq \int_t^\omega u'(s) \frac{ds}{u^2(s)} - \frac{1}{u(t)} = -\lim_{s \to \omega} \frac{1}{u(s)} \leq 0.$$

This gives (6.19). The case $p(t) > 0$ can be reduced to the case $p(t) \equiv 1$ by the change of independent variables (1.7). This completes the proof.

Exercise 6.8.   Give a proof of the part of Corollary 6.4 concerning $u_0(t)$ along the following lines: Let $a < T < \omega$ and let $u_T(t)$ be the solution of (6.1) satisfying $u_T(a) = 1$, $u_T(T) = 0$; cf. Theorem 6.1. Show that $u_0(t) = \lim u_T(t)$ exists as $T \to \omega$ uniformly on compact intervals of $[a, \omega)$, is a principal solution of (6.1) and satisfies (6.19); cf. Exercise 6.5.

Corollary 6.5.   *In the two differential equations*

(6.23$_j$) $$(p_j(t)u')' + q_j(t)u = 0,$$

where $j = 1, 2$, let $p_j(t) > 0$, $q_j(t)$ be real-valued and continuous on J : $a \leqq t < \omega$; let (6.23_2) be a Sturm majorant for (6.23_1), i.e.,

$$(6.24) \qquad p_1 \geqq p_2 > 0 \quad \text{and} \quad q_1 \leqq q_2;$$

let (6.23_2) be disconjugate [so that (6.23_1) is also]. Let $u_2(t) \not\equiv 0$ be a real-valued solution of (6.23_2). Then (6.23_1) has principal and nonprincipal solutions, $u_{10}(t)$ and $u_{11}(t)$, which satisfy

$$(6.25) \qquad \frac{p_1 u_{10}'}{u_{10}} \leqq \frac{p_2 u_2'}{u_2} \leqq \frac{p_1 u_{11}'}{u_{11}}$$

for all t beyond the last zero, if any, of $u_2(t)$.

The rough content of this corollary is that the principal [nonprincipal] solutions of (6.23_1) are smaller [larger] than the principal [nonprincipal] solutions of (6.23_2). If $p_1 \equiv p_2$ and u_2, u_{10}, u_{11} are normalized by suitable constant factors, (6.25) implies that $u_{10} \leqq u_2 \leqq u_{11}$ for t near ω.

Exercise 6.9. In Corollary 6.5, the principal solutions u_{10} of (6.23_1) satisfy $\int^\omega u_{10}^2(q_2 - q_1)\, dt < \infty$. In particular, if $q \leqq 0$ in (6.1), then a principal solution u_0 of (6.1) satisfies $\int^\omega u_0^2 |q|\, ds < \infty$.

Proof.

Case 1 ($p_1 \equiv p_2$). Suppose that $u_2(t) > 0$ for $T \leqq t < \omega$. Make the variation of constants $u = u_2 z$ in (6.23_1). Then (6.23_1) is transformed [cf. (2.31) of § 2 (xii)] into

$$(6.26) \qquad (p_1 u_2^2 z')' + u_2^2(q_1 - q_2)z = 0,$$

where $q_1 - q_2 \leqq 0$ and

$$(6.27) \qquad \frac{u'}{u} = \frac{u_2'}{u_2} + \frac{z'}{z}.$$

By Corollary 6.4, (6.26) has solutions $z_0(t), z_1(t)$ satisfying $z_0 > 0$, $z_0' \leqq 0$, and $z_1 > 0$, $z_1' > 0$ for $T \leqq t < \omega$. The desired solutions of (6.23_1) are $u_{10} = u_2 z_0$, $u_{11} = u_2 z_1$.

Case 2 ($p_1 \not\equiv p_2$). The function $r = p_2 u_2'/u_2$ satisfies the Riccati equation $r' + r^2/p_2 + q_2 = 0$ belonging to (6.23_2); cf. § 2 (xiv). This equation can be written as

$$(6.28) \qquad r' + \frac{r^2}{p_1} + q_0 = 0,$$

where $q_0 = q_2 + (1/p_2 - 1/p_1)(p_2 u_2'/u_2)^2 \geqq q_2 \geqq q_1$. But (6.28) is the Riccati equation belonging to

$$(6.29) \qquad (p_1 u')' + q_0 u = 0,$$

which is a Sturm majorant for (6.23_1). In addition, (6.29) has the solution [cf. § 2 (xiv)]

$$u = \exp \int_T^t \left(\frac{r}{p_1} \right) ds = \exp \int_T^t \left(\frac{p_2 u_2'}{p_1 u_2} \right) ds$$

satisfying

$$\frac{p_1 u'}{u'} = \frac{p_2 u_2'}{u_2}.$$

Thus application of the Case 1 to (6.23_1), (6.29) gives the desired result.

Exercise 6.10. In the differential equations

$$(6.30_j) \qquad\qquad u'' + g_j(t)u' - f_j(t)u = 0,$$

where $j = 1, 2$, let f_j, g_j be continuous for $0 \leqq t < \omega$ ($\leqq \infty$); let $0 \leqq f_1(t) \leqq f_2(t)$ and $g_1(t) \leqq g_2(t)$; let $u_1(t)$ be a solution of (6.30_1) satisfying $u_1(0) = 1$ and $u_1(t) > 0$, $u_1'(t) \leqq 0$ for $0 \leqq t < \omega$; cf. Corollary 6.4. Then (6.30_2) has a solution $u_2(t)$ satisfying $u_2(0) = 1$, $u_2'(t) \leqq 0$ and $0 < u_2(t) \leqq u_1(t)$ for $0 \leqq t < \omega$ [in fact, satisfying $u_2(0) = 1$ and $0 \leqq u_2/u_1 \leqq 1$, $(u_2/u_1)' \leqq 0$ for $0 \leqq t < \omega$].

The following is a "selection" or "continuity" theorem for principal solutions:

Corollary 6.6. *Let* $p_1(t), p_2(t), \ldots, p_\infty(t)$ *and* $q_1(t), q_2(t), \ldots, q_\infty(t)$ *be continuous functions for* $a \leqq t < \omega$ *satisfying*

$$(6.31_j) \quad p_j(t) > 0, \quad q_j(t) \leqq 0 \quad \text{for} \quad a \leqq t < \omega \quad \text{and} \quad j = 1, 2, \ldots, \infty$$

and

$$(6.32) \qquad p_j(t) \to p_\infty(t), \quad q_j(t) \to q_\infty(t) \qquad \text{as} \quad j \to \infty$$

uniformly on every closed interval of $a \leqq t < \omega$. *For* $1 \leqq j < \infty$, *let* $u_{j0}(t)$ *be a principal solution of*

$$(6.33_j) \qquad\qquad (p_j u')' + q_j(t)u = 0$$

satisfying (6.19) *and*

$$(6.34) \qquad\qquad u(a) = 1.$$

Then there exists a sequence of positive integers $j(1) < j(2) < \cdots$ *such that*

$$(6.35) \qquad u_\infty(t) = \lim_{n \to \infty} u_{j0}(t), \qquad \text{where} \quad j = j(n),$$

exists uniformly on every closed interval of $a \leqq t < \omega$ *and is a solution of* (6.33_∞) *satisfying* (6.19) *and* (6.34).

Of course, a selection is unnecessary (i.e., $j(n) = n$ is permitted) if (6.33_∞) has a unique solution satisfying (6.19) and (6.34); cf. Exercise 6.7. Note that u_∞ need not be a principal solution. E.g., let $a = 1, \omega = \infty$, $p_j = t^2/(1+t/j)$, $p_\infty = t^2$, $q_j = q_\infty = 0$, $u_{j0} = 1$, but $u_{\infty 0} = 1/t$.

Exercise 6.11. This corollary is false if the condition $q_j(t) \leqq 0$ is replaced by the assumption that (6.33_j) is nonoscillatory and (6.19) is deleted from both assumption and assertion.

Proof. Let $u_{j1}(t)$ be the solution of (6.33_j) determined by

$$(6.36) \qquad u_{j1}(a) = 1, \qquad p_j(a)u'_{j1}(a) = 1.$$

Then (6.20) holds and $u_{j1}(t)$ is a nonprincipal solution of (6.33_j); cf. the proof of Corollary 6.4. Hence, by Corollary 6.3, the principal solution $u_{j0}(t)$ of (6.33_j) satisfying (6.34) is given by

$$(6.37) \qquad C_j u_{j0}(t) = u_{j1}(t) \int_t^\omega \frac{ds}{p_j(s)u_{j1}^2(s)} \qquad \text{for} \quad a \leqq t < \omega,$$

where

$$(6.38) \qquad C_j = \int_a^\omega \frac{ds}{p_j(s)u_{j1}^2(s)}.$$

Differentiation of (6.37) gives

$$0 \geqq C_j u'_{j0}(t) = u'_{j1}(t) \int_t^\omega \frac{ds}{p_j u_{j1}^2} - \frac{1}{p_j(t)u_{j1}(t)},$$

so that, if $t = a$,

$$0 \geqq p_j(a)u'_{j0}(a) = 1 - \frac{1}{C_j}.$$

Thus the sequence $p_j(a)u'_{j0}(a)$, $j = 1, 2, \dots$, is bounded if

$$(6.39) \qquad C_j \geqq \text{const.} > 0 \qquad \text{for} \quad j = 1, 2, \dots$$

In order to verify (6.39), note that (6.36) and the assumption on (6.32) imply that $u_{j1}(t) \to u_{\infty 1}(t)$ as $j \to \infty$ uniformly on closed intervals of $a \leqq t < \omega$. Thus, by (6.38),

$$C_j > \int_a^T \frac{ds}{p_j u_{j1}^2} \to \int_a^T \frac{ds}{p_\infty u_{\infty 1}^2} \qquad \text{as} \quad j \to \infty$$

for any fixed T, $a < T < \omega$. This implies (6.39).

Since the sequence of numbers $u_{j0}(a) = 1$ and $u'_{j0}(a)$ for $j = 1, 2, \dots$, are bounded, there exist subsequences which have limits. If $j(1) < j(2) < \cdots$ are the indices of such a subsequence and

$$1 = \lim u_{j0}(a), \qquad u'_{\infty 0} = \lim u'_{j0}(a) \qquad \text{for} \quad j = j(n) \to \infty,$$

then the assumption on (6.32) implies (6.35) uniformly on every interval $[a, T] \subset [a, \omega)$, where $u_\infty(t)$ is the solution of (6.33_∞) satisfying $u_\infty(a) = 1$, $u_\infty'(a) = u'_{\infty 0}$. The solution $u_\infty(t)$ clearly satisfies (6.19) and (6.34). This proves Corollary 6.6.

7. Nonoscillation Theorems

This section will be concerned with conditions, necessary and/or sufficient, for

$$(7.1) \qquad\qquad u'' + q(t)u = 0$$

to be nonoscillatory. In view of the Sturm comparison theorem, the simplest (and one of the most important) sufficient conditions for (7.1) to be nonoscillatory [or oscillatory] is for (7.1) to possess a nonoscillatory [or oscillatory] Sturm majorant [minorant]. For example, if $q(t) \leq 0$ [so that $u'' = 0$ is a Sturm majorant for (7.1)], then (7.1) is nonoscillatory. If $q(t) = \mu t^{-2}$, then (7.1) is nonoscillatory or oscillatory at $t = \infty$ according as $\mu \leq \frac{1}{4}$ or $\mu > \frac{1}{4}$; see Exercise 1.1(c). This gives the following criteria:

Theorem 7.1. *Let $q(t)$ be real-valued and continuous for large $t > 0$. If*

$$(7.2) \qquad -\infty \leq \limsup_{t \to \infty} t^2 q(t) < \frac{1}{4} \qquad \left[or \quad \infty \geq \liminf_{t \to \infty} t^2 q(t) > \frac{1}{4} \right],$$

then (7.1) is nonoscillatory [or oscillatory] at $t = \infty$.

If, e.g., $t^2 q(t) \to \frac{1}{4}$ as $t \to \infty$, then Theorem 7.1 does not apply. In this case, Exercise 1.2 shows that (7.2) can be replaced by

$$-\infty \leq \limsup_{t \to \infty} t^2 \log^2 t \left[q(t) - \frac{1}{4t^2} \right] < \frac{1}{4}$$

or

$$\infty \geq \liminf_{t \to \infty} t^2 \log^2 t \left[q(t) - \frac{1}{4t^2} \right] > \frac{1}{4}$$

In fact, the sequence of functions in Exercise 1.2 gives a scale of tests for (7.1) to be nonoscillatory or oscillatory at $t = \infty$.

The criterion given by Sturm's comparison theorem can be cast in the following convenient form:

Theorem 7.2 *Let $q(t)$ be real-valued and continuous for $J : a \leq t < \omega \, (\leq \infty)$. Then (7.1) is disconjugate on J if and only if there exists a continuously differentiable function $r(t)$ for $a < t < \omega$ such that*

$$(7.3) \qquad\qquad r' + r^2 + q(t) \leq 0.$$

Exercise 7.1. Formulate analogues of Theorem 7.2 when J is open or J is closed and bounded.

Remark. It is clear from § 1 that analogues of Theorem 7.2 remain valid if (7.1) is replaced by an equation of the form $(pu')' + qu = 0$ or $u'' + gu' + fu = 0$ provided that (7.3) is replaced by the corresponding

Riccati differential inequality $r' + r^2/p + q \leqq 0$ or $r' + r^2 + gr + f \leqq 0$, respectively.

Proof. First, if (7.1) is disconjugate on J, then (7.1) has a solution $u = u_0(t) > 0$ for $a < t < \omega$; see Corollary 6.1. In this case, $r = u_0'/u_0$ satisfies the Riccati equation

$$(7.4) \qquad\qquad r' + r^2 + q(t) = 0$$

for $a < t < \omega$. This proves the "only if" part of the theorem.

If there exists a continuously differentiable function $r(t)$ satisfying (7.3), let $q_0(t) \leqq 0$ denote the left side of (7.3) for $a < t < \omega$, so that $r' + r^2 + q - q_0 = 0$. Then

$$u'' + [q(t) - q_0(t)]\, u = 0$$

is a Sturm majorant for (7.1) on $a < t < \omega$ and, by § 2 (xiv), possesses the positive solution $u = \exp \int_c^t r(s)\, ds$, where $a < c < \omega$. This shows that (7.1) is disconjugate on $a < t < \omega$. In order to complete the proof, we must show that if $u_1(t) \not\equiv 0$ is the solution of (7.1) satisfying $u_1(a) = 0$ and $u_1'(a) = 1$, then $u_1(t) \neq 0$ for $a < t < \omega$. Suppose that this is not the case, so that $u_1(t_0) = 0$ for some t_0, $a < t_0 < \omega$. Since u_1 changes sign at $t = t_0$ and solutions of (7.1) depend continuously on initial conditions, it follows that if $\epsilon > 0$ is sufficiently small, then the solution of (7.1) satisfying $u(a + \epsilon) = 0$, $u'(a + \epsilon) = 1$ has a zero near t_0. This contradicts the fact that (7.1) is disconjugate on $a < t < \omega$ and proves the theorem.

Exercise 7.2.   (a) Using the Remark following Theorem 7.2, show that if, in the differential equations

$$(7.5_j) \qquad\qquad u'' + g_j(t)u' + f_j(t)u = 0,$$

where $j = 1, 2$, the coefficient functions are real-valued and continuous on $J : a \leqq t < \omega\ (\leqq \infty)$ such that

$$(7.6) \qquad\qquad g_1(t) \leqq g_2(t), \qquad f_1(t) \leqq f_2(t)$$

and if (7.5_2) has a solution $u(t)$ satisfying $u > 0$, $u' \geqq 0$ for $a < t < \omega$, then (7.5_1) is disconjugate on J. [For an application in Exercise 7.9, note that the conditions on (7.5_2) hold if (7.5_2) is disconjugate on J, $f_2(t) \geqq 0$ and $\int_a^\omega \left[\exp - \int_a^t g_2(s)\, ds \right] dt = \infty$; cf. Exercise 6.2.]   (b) Let $f(t)$ be continuous and $g(t)$ continuously differentiable real-valued functions on $a \leqq t \leqq b$. Then

$$u'' + g(t)u' + f(t)u = 0$$

is disconjugate on $[a, b]$ if there exists a real number c such that

$$f(t) - cg'(t) + c(c - 1)g^2(t) \leqq 0$$

for $a \leqq t \leqq b$.

Corollary 7.1. *Let $q(t)$ be real-valued and continuous on $J : a \leqq t < \omega$, C a constant, and*

$$(7.7) \qquad\qquad Q(t) = C - \int_a^t q(s) \, ds.$$

If the differential equation

$$(7.8) \qquad\qquad u'' + 4Q^2(t)u = 0$$

is disconjugate on J, then (7.1) *is disconjugate on J.*

Exercise 7.3. Show that this corollary is false if the 4 in (7.8) is replaced by a constant $\gamma < 4$.

Proof of Corollary 7.1. In the Riccati equation (7.4) belonging to (7.1), introduce the new variable

$$(7.9) \qquad\qquad \rho = r - Q,$$

so that $\rho' = r' + q$, and (7.4) becomes

$$(7.10) \qquad\qquad \rho' + \rho^2 + 2Q\rho + Q^2 = 0.$$

Since $2Q\rho \leqq \rho^2 + Q^2$, a solution of

$$(7.11) \qquad\qquad \rho' + 2(\rho^2 + Q^2) = 0$$

on some interval satisfies

$$(7.12) \qquad\qquad \rho' + \rho^2 + 2Q\rho + Q^2 \leqq 0.$$

The differential equation (7.11) can be written as

$$(7.13) \qquad\qquad \sigma' + \sigma^2 + 4Q^2 = 0 \qquad \text{if} \quad \sigma = 2\rho.$$

Finally, (7.13) is the Riccati equation for (7.8).

Thus if (7.8) has a solution $u(t) > 0$ on J, then $\sigma = u'/u$ satisfies (7.13). Hence $\rho = \frac{1}{2}\sigma$ satisfies (7.12) and $r = \rho + Q$ is a solution of the differential inequality (7.3) on J. In virtue of Theorem 7.2, this proves the corollary.

Exercise 7.4. A counterpart of Corollary 7.1 can be stated as follows: Let $q(t)$ be real-valued and continuous for $0 \leqq t \leqq b$. Let a be fixed, $0 \leqq a < b$. Suppose that

$$Q(t) = \int_0^t q(s) \, ds$$

has the properties that $Q(t) \geqq 0$ for $a \leqq t \leqq b$ and that if $z(t)$ is a solution of $z'' + Q^2(t)z = 0$, $z'(a) = 0$, then $z(t)$ has a zero on $a < t \leqq b$. Then a solution $u(t)$ of (7.1) satisfying $u'(0) = 0$ has a zero on $0 < t \leqq b$.

One of the main results on equations (7.1) which are nonoscillatory at $t = \infty$ will be based on the following lemma.

Lemma 7.1. *Let $q(t)$ be real-valued and continuous on $0 \leqq t < \infty$ with the property that (7.1) is nonoscillatory at $t = \infty$. Then a necessary and sufficient condition that*

$$(7.14) \qquad \int^{\infty} \left(\frac{u'}{u}\right)^2 dt < \infty$$

holds for one (and/or every) real-valued solution $u(t) \not\equiv 0$ of (7.1) is that

$$(7.15) \qquad \lim_{T \to \infty} \frac{1}{T} \int_0^T \left(\int_0^t q(s)\, ds\right) dt = C \qquad \text{exists}$$

(as a finite number).

Remark. For the application of this lemma, it is important to note that the proof will show that condition (7.15) can be relaxed to

$$(7.16) \qquad \liminf_{T \to \infty} \frac{1}{T} \int_0^T \left(\int_0^t q(s)\, ds\right) dt > -\infty.$$

In other words, when (7.1) is nonoscillatory at $t = \infty$, then (7.16) implies (7.15); in fact, it implies the stronger relation

$$(7.17) \qquad \frac{1}{T} \int_0^T \left| C - \int_0^t q(s)\, ds \right|^2 dt \to 0 \qquad \text{as} \quad T \to \infty.$$

Exercise 7.5. Let $q(t)$ be as in Lemma 7.1. Show that

$$(7.18) \qquad \frac{u'}{u} \to 0 \qquad \text{as} \quad t \to \infty$$

holds for one (and/or every) real-valued relation $u(t) \not\equiv 0$ of (7.1) if and only if

$$(7.19) \qquad \sup_{0 < \sigma < \infty} \frac{1}{1 + \sigma} \left| \int_t^{t+\sigma} q(s)\, ds \right| \to 0 \qquad \text{as} \quad t \to \infty.$$

[Note that (7.19) holds if, e.g., $q(t) \to 0$ as $t \to \infty$ or $\int^{\infty} |q(s)|^{\gamma}\, ds < \infty$ for some $\gamma \geqq 1$.]

Proof. Suppose first that (7.14) holds for a real-valued solution $u(t) \not\equiv 0$ of (7.1). Let $t = a$ exceed the largest zero, if any, of $u(t)$. Put $r = u'/u$ for $t \geqq a$, so that r satisfies the Riccati equation (7.4). A quadrature gives

$$(7.20) \qquad r(t) + \int_a^t r^2(s)\, ds = r(a) - \int_a^t q(s)\, ds$$

for $t \geqq a$. Then (7.14) implies that (7.20) can be written as

$$(7.21) \qquad r(t) - \int_t^\infty r^2(s)\, ds = C - \int_0^t q(s)\, ds,$$

where $C = r(a) - \int_a^\infty r^2(s)\, ds + \int_0^a q(s)\, ds$. By (7.14),

$$\frac{1}{T}\int_a^T r^2(t)\, dt, \qquad \frac{1}{T}\int_a^T \left(\int_t^\infty r^2(s)\, ds\right)^2 dt \to 0 \qquad \text{as} \quad T \to \infty.$$

Hence (7.21) implies (7.17) [by virtue of the inequality $(\alpha + \beta)^2 \leqq 2(\alpha^2 + \beta^2)$ for real numbers α, β]. Since Schwarz's inequality [cf. (7.22)] shows that (7.15) is a consequence of (7.17), it follows that (7.15) is necessary for (7.14).

In order to prove the converse, assume (7.16), that $u(t) \not\equiv 0$ is any real-valued solution of (7.1), and that $u(t) > 0$ for $t \geqq a$. Then (7.20) holds for $r = u'/u$ and a quadrature of (7.20) gives

$$\frac{1}{t}\int_a^t r(s)\, ds + \frac{1}{t}\int_a^t \left(\int_a^s r^2(\sigma)\, d\sigma\right) ds$$
$$= \frac{1}{t} r(a)(t - a) - \frac{1}{t}\int_a^t \left(\int_a^s q(\sigma)\, d\sigma\right) ds.$$

The assumption (7.16) implies that the right side is bounded from above. Suppose, if possible, that (7.14) does not hold, then the second term on the left tends to ∞ as $t \to \infty$, thus

$$-\frac{1}{t}\int_a^t r(s)\, ds \geqq \frac{1}{2t}\int_a^t \left(\int_a^s r^2(\sigma)\, d\sigma\right) ds \qquad \text{for large } t.$$

Schwarz's inequality implies

$$(7.22) \qquad \left|\frac{1}{t}\int_a^t r(s)\, ds\right| \leqq \left(\frac{1}{t}\int_a^t r^2(s)\, ds\right)^{\frac{1}{2}}$$

and, consequently,

$$4t\int_a^t r^2(s)\, ds \geqq \left(\int_a^t \left(\int_a^s r^2(\sigma)\, d\sigma\right) ds\right)^2 \qquad \text{for large } t.$$

This can be written as

$$4tS' \geqq S^2, \qquad \text{where} \quad S(t) = \int_a^t \left(\int_a^s r^2(\sigma)\, d\sigma\right) ds \to \infty$$

as $t \to \infty$. A quadrature gives

$$\text{const.} - \frac{4}{S(t)} \geqq \log t \qquad \text{for large } t.$$

This contradiction shows that the hypotheses that (7.14) fails to hold is untenable and proves the theorem.

Theorem 7.3. *Let $q(t)$ be real-valued and continuous for $0 \leqq t < \infty$. A necessary condition for (7.1) to be nonoscillatory at $t = \infty$ is that either*

$$(7.23) \qquad \liminf_{T \to \infty} \frac{1}{T} \int_0^T \left(\int_0^t q(s)\, ds \right) dt = -\infty$$

or that (7.15) holds [and, in the latter case, (7.17) holds].

It follows, e.g., that if $q(t) \geqq 0$, then, in order for (7.1) to be non-oscillatory at $t = \infty$, it is necessary that $\int^\infty q(t)\, dt < \infty$. In fact, as is seen from Exercise 7.8, it is necessary that $\int^\infty t^\gamma q(t)\, dt < \infty$ for every $\gamma < 1$.

Proof. Suppose that (7.1) is nonoscillatory at $t = \infty$ and that (7.23) fails to hold, so that (7.16) holds. The validity of (7.17) must be verified. But this is clear from the proof of Lemma 7.1 which shows that, on the one hand, (7.16) implies (7.14) for every real-valued solution $u(t) \not\equiv 0$ of (7.1) and, on the other hand, that (7.14) for some solution assures (7.17).

Exercise 7.6. Let $q(t)$ be as in Theorem 7.3 and, in addition, satisfy

$$(7.24) \qquad\qquad q(t) \to 0 \qquad \text{as} \quad t \to \infty$$

or, more generally, (7.19). Then a necessary condition for (7.1) to be nonoscillatory at $t = \infty$ is that either

$$(7.25) \qquad\qquad \int_0^T q(t)\, dt \to -\infty \qquad \text{as} \quad T \to \infty$$

or that

$$(7.26) \qquad \int_0^\infty q(t)\, dt = \lim_{T \to \infty} \int_0^T q(t)\, dt \qquad \text{converges}$$

(possibly conditionally).

Exercise 7.7. (a) Give examples to show that (7.15) in Theorem 7.3 is compatible with each of the possibilities

$$(7.27) \qquad \limsup_{T \to \infty} \frac{1}{T} \int_0^T \left(\int_0^t q(s)\, ds \right) dt \quad \text{is} \quad -\infty, \quad \text{finite,} \quad \text{or} \quad +\infty.$$

(b) Show that if, in Theorem 7.3, $q(t)$ is half-bounded or, more generally, if there exists an $\epsilon > 0$ such that

$$\int_t^{s+t} q(\sigma)\, d\sigma \qquad \text{is half-bounded for} \quad 0 \leqq t < \infty, \quad 0 \leqq s \leqq \epsilon,$$

then a necessary condition that (7.1) be nonoscillatory is that either (7.15) or (7.25) hold. See Hartman [10].

Changes of variables in (7.1) followed by applications of Theorem 7.3 (and its consequences) give new necessary conditions for (7.1) to be non-oscillatory. This is illustrated by the following exercise.

Exercise 7.8. (*a*) Introduce the new independent and dependent variables $s = t^\gamma$, $\gamma > 0$ and $z = t^{(\gamma-1)/2}u$, and state necessary conditions for the resulting equation and/or (7.1) to be nonoscillatory at $t = \infty$. (*b*) In particular, show that if $q(t) \geq 0$ and (7.1) is nonoscillatory at $t = \infty$, then $\int^\infty t^{1-\gamma}q(t) \, dt < \infty$ for all $\gamma > 0$.

The next result gives a conclusion very different from (7.17) in Theorem 7.3 in the case (7.15).

Theorem 7.4. *Let $q(t)$ be as in Theorem 7.3 such that (7.1) is non-oscillatory at $t = \infty$ and (7.23) does not hold [so that (7.15) does]. Then*

$$(7.28) \qquad \int^\infty \exp\left(-4\int_0^t Q^+(s) \, ds\right) dt = \infty \ ,$$

where

$$(7.29) \qquad Q(t) = C - \int_0^t q(s) \, ds, \qquad Q^+(t) = \max\,(Q(t), 0).$$

In applications, interesting cases of this theorem occur if (7.26) holds, so that

$$(7.30) \qquad Q(t) = \int_t^\infty q(s) \, ds.$$

It is readily verified from $q(t) = \mu/t^2$, $t \geq 1$, that the "4" in (7.28) cannot be replaced by a larger constant. It is rather curious that the proof of Corollary 7.1 and Theorem 7.4 depend on the inequality $2Q\rho \leq \rho^2 + Q^2$. In the proof of Corollary 7.1, this inequality is used to deduce (7.12) from (7.11); in the proof of Theorem 7.4, it is used to deduce

$$(7.31) \qquad\qquad\qquad \rho' + 4Q\rho \leq 0$$

from (7.10).

Proof. Let $u = u(t) \not\equiv 0$ be a real-valued solution of (7.1) and suppose T is so large that $u(t) \neq 0$ for $t \geq T$. Since it is assumed that (7.15) holds, the relation (7.14) holds. Thus if $r = u'/u$, a quadrature of the corresponding Riccati equation gives (7.21) as in the proof of Lemma 7.1. Rewrite (7.21) as $r(t) = \rho(t) + Q(t)$, where

$$(7.32) \qquad\qquad\qquad \rho(t) = \int_t^\infty r^2(s) \, ds.$$

Since $\rho' = -r^2 = -(\rho + Q)^2$, the equation (7.10) holds. This gives (7.31). In particular, if $Q(t) \geq 0$, then

$$(7.33) \qquad\qquad\qquad \rho' + 4Q^+\rho \leq 0 \cdot$$

Note that if $Q < 0$, then (7.33) holds since $\rho' \leqq 0$, $\rho \geqq 0$. Hence (7.33) holds for $t \geqq T$. Since the result to be proved is trivial if $q(t) \equiv 0$ for large t, it can be supposed that this is not the case. Hence $r \not\equiv 0$ for large t and so, $\rho(t) > 0$. Consequently, (7.33) gives

$$(7.34) \qquad \rho(t) \leqq \rho(T) \exp\left(-4\int_T^t \overset{+}{Q}(s)\, ds\right) \qquad \text{for} \quad t \geqq T.$$

Suppose, if possible, that (7.28) fails to hold, then (7.32), (7.34) show that

$$\int^\infty \rho(t)\, dt < \infty, \qquad \text{hence} \qquad \int^\infty tr^2(t)\, dt < \infty$$

holds for $r = u'/u$, where $u(t) \not\equiv 0$ is an arbitrary real-valued solution of (7.1). It will be shown that this leads to a contradiction. To this end, note that

$$\log \frac{u(t)}{u(T)} = \int_T^t r(s)\, ds.$$

Thus Schwarz's inequality gives

$$\left[\log \frac{u(t)}{u(T)}\right]^2 \leqq \left(\int_T^t sr^2(s)\, ds\right)\left(\int_T^t \frac{ds}{s}\right).$$

Consequently there exist constants c_0, c such that $|u(t)| \leqq c_0 \exp c(\log t)^{\frac{1}{2}}$ for large t. It follows that

$$\int^\infty \frac{dt}{u^2(t)} \geqq \frac{1}{c_0^2}\int^\infty \exp\left[-2c(\log t)^{\frac{1}{2}}\right] dt = \infty$$

for all real-valued solutions $(\not\equiv 0)$ of (7.1). This contradicts the existence (Theorem 6.4) of nonprincipal solutions and completes the proof.

Exercise 7.9.   In the differential equations

$$(7.35_j) \qquad u'' + q_j(t)u = 0,$$

where $j = 1, 2$, let $q_j(t)$ be real-valued and continuous for large t and such that

$$Q_j(t) = \int_t^\infty q_j(s)\, ds = \lim_{T\to\infty}\int_t^T$$

converges (possibly conditionally) and (7.35_2) is non-oscillatory at $t = \infty$. Show that (7.35_1) is nonoscillatory at $t = \infty$ if $|Q_1(t)| \leqq Q_2(t)$, or if c_1, c_2 are constants, $0 < c_2 \leqq 2$, $c_2^2 Q_2^2 + (1-c_2)q_2 \geqq max(0, c_1^2 Q_1^2 + (1-c_1)q_1)$ and $c_1 Q_1 \leqq c_2 Q_2$.

8. Asymptotic Integrations. Elliptic Cases

In the next two sections, we will consider the problem of the asymptotic integration of equations

$$(8.1) \qquad u'' + q(t)u = 0,$$

where $q(t)$ is continuous for large t. Except for the last part of this section, the main interest will center around the situations where the coefficient $q(t)$ is nearly a constant or (8.1) can be reduced to this case. The last part of this section (see Exercises 8.6, 8.8) deals with bounds for $|u'|$ when $q(t)$ is bounded from above.

When $q(t)$ is a constant, say λ, and λ is real and positive, then the solutions are, roughly speaking, of the same order of magnitude. On the other hand, if λ is not real and positive, then essentially there is one small solution, as $t \to \infty$, and the other solutions are large. These facts indicate that different techniques will be needed when $q(t)$ is nearly a constant λ, and λ is or is not real and positive. In this section, the first case will be considered.

Theorem 8.1. *In the differential equations* (8.1) *and*

$$(8.2) \qquad w'' + q_0(t)w = 0,$$

let $q(t)$, $q_0(t)$ *be continuous, complex-valued functions for* $0 \leqq t < \infty$ *satisfying*

$$(8.3) \qquad \int^{\infty} |w(t)|^2 \, |q_0(t) - q(t)| \, dt < \infty$$

for every solution $w(t)$ *of* (8.2). *Let* $u_0(t)$, $v_0(t)$ *be linearly independent solutions of* (8.2). *Then to every solution* $u(t)$ *of* (8.1), *there corresponds at least one pair of constants* α, β *such that*

$$(8.4) \qquad \begin{aligned} u(t) &= [\alpha + o(1)]u_0(t) + [\beta + o(1)]v_0(t), \\ u'(t) &= [\alpha + o(1)]u_0'(t) + [\beta + o(1)]v_0'(t), \end{aligned}$$

as $t \to \infty$; *conversely, to every pair of constants* α, β, *there exists at least one solution* $u(t)$ *of* (8.1) *satisfying* (8.4).

Note that for a given $u(t)$, (8.4) might hold for more than one pair of constants (α, β). This is true, e.g., if $v_0(t) = o(u_0(t))$ as $t \to \infty$.

An interesting aspect of Theorem 8.1 is the fact that the main condition (8.3) does not involve the derivatives $w'(t)$ of solutions $w(t)$ of (8.2). This advantage is lost if (8.1) or (8.2) is replaced by a more complicated equation as in the Exercise 8.4 below.

Proof. It can be supposed that det $Y(t) = 1$, where

$$Y(t) = \begin{pmatrix} u_0 & v_0 \\ u_0' & v_0' \end{pmatrix}.$$

Write (8.1) as a first order system $x' = A(t)x$ for the binary vectors $x = (u, u')$; cf. (2.5). Then the variations of constants $x = Y(t)y$ reduce the system $x' = A(t)x$, say to $y' = C(t)y$, in (2.28); cf. § 2(xi). Thus

Theorem 8.1 follows from the linear case of Theorem X 1.2, cf. Exercise X 1.4.

Corollary 8.1. *Let $q(t)$ be a continuous complex-valued function on $0 \leqq t < \infty$ satisfying*

$$(8.5) \qquad \int^{\infty} |1 - q(t)|\, dt < \infty.$$

Then if α, β are constants, there exists one and only one solution $u(t)$ of (8.1) satisfying the asymptotic relations

$$(8.6) \qquad \begin{aligned} u &= [\alpha + o(1)] \cos t + [\beta + o(1)] \sin t, \\ u' &= -[\alpha + o(1)] \sin t + [\beta + o(1)] \cos t. \end{aligned}$$

The relations (8.6) can also be written as $u = \delta \cos [t + \gamma + o(1)]$, $u' = -\delta \sin [t + \gamma + o(1)]$ as $t \to \infty$ for some constants γ and δ.

Exercise 8.1. Show that if α, β are constants, there exists a unique solution $v(t)$ of the Bessel equation $t^2 v'' + t v' + (t^2 - \mu^2) v = 0$ for $t > 0$ such that $u(t) = t^{1/2} v(t)$ satisfies (8.6) as $t \to \infty$.

Exercise 8.2. Show that the conclusion of Corollary 8.1 is correct if (8.5) is relaxed to the following conditions in which $f(t) = 1 - q(t)$: the integrals

$$g_0(t) = \int_t^{\infty} f(s)\, ds, \quad g_1(t) = \int_t^{\infty} f(s) \cos 2s\, ds, \quad g_2(t) = \int_t^{\infty} f(s) \sin 2s\, ds$$

exist as (possibly conditional) improper Riemann integrals $\left(\int^{\infty} = \lim \int^T \right.$ as $\left. T \to \infty \right)$ and $\int^{\infty} |g_k(t) f(t)|\, dt < \infty$ for $k = 0, 1, 2$.

Exercise 8.3. (a) Let $q(t)$ be a positive function on $0 \leqq t < \infty$ possessing a continuous second derivative and such that

$$(8.7) \qquad \int^{\infty} q^{1/2}(t)\, dt = \infty \quad \text{and} \quad \int^{\infty} \left| \frac{5 q'^2}{16 q^3} - \frac{q''}{4 q^2} \right| q^{1/2}\, dt < \infty.$$

Then the assertion of Corollary 8.1 remains valid if (8.6) is replaced by

$$q^{1/4} u = [\alpha + o(1)] \cos \int_0^t q^{1/2}(s)\, ds + [\beta + o(1)] \sin \int_0^t q^{1/2}(s)\, ds,$$

$$(q^{1/4} u)' q^{-1/2} = -[\alpha + o(1)] \sin \int_0^t q^{1/2}(s)\, ds + [\beta + o(1)] \cos \int_0^t q^{1/2}(s)\, ds.$$

(b) Show that (8.7) in (a) holds if $0 \neq \alpha > -\frac{1}{2}$, $q(t) \geqq \text{const.} > 0$ for $0 \leqq t < \infty$, and $f(t) = q^{\alpha}(t) \geqq 0$ has a continuous second derivative such that $\int^{\infty} |f''(t)|\, dt < \infty$. [In fact, for the validity of the conclusion of (a), it can be merely supposed that $f(t)$ has a continuous first derivative which is of bounded variation on $0 \leq t < \infty$, i.e., $\int^{\infty} |df'(t)| < \infty$; e.g., $f'(t)$ is monotone and bounded. This last refinement follows from the first part by approximating $q(t)$ by suitable smooth functions.]

Exercise 8.4.   (*a*) In the differential equations

$$(8.8_j) \qquad (p_j u')' + r_j u' + q_j u = 0, \qquad j = 0, 1,$$

let $p_j(t) \neq 0$, $q_j(t)$, $r_j(t)$ be continuous complex-valued functions for $0 \leq t < \infty$ such that

$$(8.9) \quad
\begin{aligned}
&\int^{\infty} |w|^2 \, |q_1 - q_0| \cdot \left| \exp \int^t \frac{r_0 \, ds}{p_0} \right| dt < \infty, \\[2mm]
&\int^{\infty} |p_0 w'|^2 \left| \frac{1}{p_1} - \frac{1}{p_0} \right| \cdot \left| \exp \int^t \frac{r_0 \, ds}{p_0} \right| dt < \infty, \\[2mm]
&\int^{\infty} \left| \frac{r_0}{p_0} - \frac{r_1}{p_1} \right| \left(1 + \left| p_0 w w' \exp \int^t \frac{r_0 \, ds}{p_0} \right| \right) dt < \infty
\end{aligned}$$

hold for all solutions $u = w(t)$ of (8.8_0). Let $u_0(t)$, $v_0(t)$ be linearly independent solutions of (8.8_0). Then to every solution $u(t)$ of (8.8_1), there corresponds at least one pair of constants α, β such that (8.4) holds; conversely, if α, β are constants, then there is at least one solution $u(t)$ of (8.8_1) satisfying (8.4). (*b*) In the differential equation (8.1), let $q(t) \neq 0$ be a continuous, complex-valued function for $0 \leq t < \infty$ such that $q(t)$ is of bounded variation over $0 \leq t < \infty$ $\left(\text{i.e., } \int_0^{\infty} |dq| < \infty \right)$; $c_0 = \lim q(t)$, $t \to \infty$, is a positive constant; and the solutions $u(t)$ of (8.1) are bounded (e.g., if $q(t)$ is real-valued or, more generally, $\int^{\infty} |\mathrm{Im}\, q(t)| \, dt < \infty$, then solutions $u(t)$ and their derivatives $u'(t)$ are bounded). Let α, β be constants. Then (8.1) has a unique solution $u(t)$ satisfying, as $t \to \infty$,

$$u = [\alpha + o(1)] \cos \int_0^t q^{\frac{1}{2}}(s) \, ds + [\beta + o(1)] \sin \int_0^t q^{\frac{1}{2}}(s) \, ds,$$

$$(8.10)$$

$$u' = -[\alpha + o(1)] c_0^{\frac{1}{2}} \sin \int_0^t q^{\frac{1}{2}}(s) \, ds + [\beta + o(1)] c_0^{\frac{1}{2}} \cos \int_0^t q^{\frac{1}{2}}(s) \, ds,$$

where $q^{\frac{1}{2}}(t)$ is any fixed continuous determination of the square root of q.

Exercise 8.5.   Let $f(t)$ be a nonvanishing (possibly complex-valued) function for $t \geq 0$ having a continuous derivative satisfying

$$\int^{\infty} \left| d\left(\frac{f'}{f^{3/2}} \right) \right| < \infty, \qquad \gamma = \lim_{t \to \infty} \frac{f'}{4f^{3/2}}, \qquad \text{and} \quad \gamma^2 \neq 1.$$

Suppose further that

(8.11) $\exp \pm i \int^t f^{\frac{1}{2}} \left(1 - \dfrac{f'^2}{16f^3}\right)^{\frac{1}{2}} dr$ are bounded

as $t \to \infty$. Then the differential equation

(8.12) $$u'' + f(t)u = 0$$

has a pair of solutions satisfying, as $t \to \infty$,

$$u \sim f^{-\frac{1}{4}}(t) \exp \pm i \int^t f^{\frac{1}{2}} \left(1 - \frac{f'^2}{16f^3}\right)^{\frac{1}{2}} dr,$$

$$u' \sim [-\gamma \pm i(1 - \gamma^2)^{\frac{1}{2}}] f^{\frac{1}{2}} u.$$

Here all powers of $f(t)$ that occur can be assumed to be integral (positive or negative) powers of a fixed continuous fourth root $f^{\frac{1}{4}}(t)$ of $f(t)$. Condition (8.11) is trivially satisfied if $f(t)$ is real-valued and positive and $0 < \gamma^2 < 1$.

The object of the next exercise is to obtain bounds for derivatives of solutions of (8.1) or, more generally, the inhomogeneous equation

(8.13) $$u'' + q(t)u = f(t).$$

Exercise 8.6. Let $q(t)$, $f(t)$ be continuous real-valued functions on $0 \leqq t \leqq t_0$. Let the positive constants ϵ, $1/\theta > 1$, C be such that

(8.14) $$0 < \epsilon \leqq \frac{\theta}{C}$$

and

(8.15) $$\int_a^b q(\tau) \, d\tau \leqq C \quad \text{if} \quad b - a \leqq \frac{\theta}{C}$$

and $0 \leqq a < b \leqq t_0$. [The inequality (8.15) holds, e.g., is $q(t) \leqq C^2/\theta$ for $0 \leqq t \leqq t_0$.] Let $u = u(t)$ be a real-valued solution of (8.13). Consider the case

$uu' \geqq 0$ at $t = T$, $0 \leqq T \leqq t_0 - \theta/C$, $S = T + \theta/C$, $U = T + \epsilon$

or the case

$uu' \leqq 0$ at $t = T$, $\theta/C \leqq T \leqq t_0$, $S = T - \theta/C$, $U = T - \epsilon$.

(*a*) Show that, in either case,

(8.16) $|u'(T)| \leqq \dfrac{1}{1 - \theta} \left| \displaystyle\int_T^U |f| \, d\tau \right| + \max \left[\dfrac{C\,|u(T)|}{1 - \theta}, \left(C + \dfrac{1}{\epsilon}\right) |u(U)| \right].$

(b) Show that (8.16) holds if $|u(U)|$ is replaced by $2\epsilon^{-1}\left|\int_T^U |u|\,d\tau\right|$. (c) Part (a) implies that if $\theta/C \leq t \leq t_0 - \theta/C$, then

$$(8.17) \qquad |u'(t)| \leq \frac{1}{1-\theta}\int_{t-\epsilon}^{t+\epsilon} |f|\,d\tau + C_0(\epsilon)\sum_{j=-1}^{1} |u(t+j\epsilon)|,$$

where $C_0(\epsilon) = \max\,(C/(1-\theta),\, C + 1/\epsilon)$. (d) Put

$$(8.18) \qquad r(t) = |u'(t)| + C|u(t)|.$$

Show that there exists a nondecreasing function $K(\Delta)$ for $0 < \Delta < \min\,(\theta, 1-\theta)/C$ such that

$$(8.19) \quad r(t) \leq K(\Delta)\left\{r(a) + r(b) + \int_a^b |f|\,d\tau\right\} \qquad \text{for} \quad a \leq t \leq b$$

if $b - a = \Delta$ and $0 \leq a < b \leq t_0$.

The results of the last exercise can be extended to an equation of the form

$$(8.20) \qquad u'' + p(t)u' + q(t)u = f(t);$$

see Exercise 8.8. In fact, the results for (8.20) can be derived from those on (8.13) by the use of the lemma given in the next exercise which has nothing to do with differential equations.

Exercise 8.7. Let $h(t) \geq 0$ be of bounded variation and $g(t)$ continuous on an interval $a \leq t \leq b$. Then

$$(8.21) \qquad \int_a^b h(t)\,dg(t) \leq (\inf h + \operatorname{var} h)\sup_{a \leq \alpha < \beta \leq b}\int_\alpha^\beta dg(t),$$

where the integrals are Riemann-Stieltjes integrals and var h denotes the total variation of $h(t)$ on $a \leq t \leq b$.

Exercise 8.8. Let $p(t), q(t), f(t)$ be continuous real-valued functions on $0 \leq t \leq t_0$ and $u(t)$ a real-valued solution of (8.20). Let $1/\theta > 1$, C be positive constants such that (8.15) holds. Consider the two cases in Exercise 8.6, with (8.14) replaced by

$$(8.22) \qquad 0 < \epsilon \leq \theta/CE^2, \qquad \text{where} \quad E = \exp\left|\int_T^S |p(\tau)|\,d\tau\right|.$$

(a) Then parts (a), (b) of Exercise 8.6 hold if $|u'(T)|$ in (8.16) is replaced by $|u'(T)|/E$. (b) Part (c) of Exercise 8.6 holds if $|u'(t)|$ in (8.17) is replaced by $|u'(t)|/E$, where

$$(8.23) \qquad 0 < \epsilon \leq \theta/CE^2 \qquad \text{and}$$

$$E = \max \exp\int_t^{t+\theta/C} |p(\tau)|\,d\tau \qquad \text{for} \quad 0 \leq t \leq t_0 - \theta/C.$$

(c) Part (d) of Exercise 8.6 holds if Δ is restricted to $0 < \Delta < \min (\theta, 1 - \theta)/CE^2$, where E is defined in (8.23).

9. Asymptotic Integrations. Nonelliptic Cases

Asymptotic integrations of $u'' + q(t)u = 0$, where $q(t)$ is "nearly" a real, but not positive constant, can be based on Chapter X as were the results of the last section. Instead a different technique will be used in this section; this technique takes greater advantage of the special structure of the second order equation

$$(9.1) \qquad (p(t)u')' + q(t)u = 0.$$

This equation is equivalent to a binary system of the form

$$(9.2) \qquad v' = \beta(t)z, \qquad z' = \gamma(t)v$$

in which the diagonal elements vanish. [This system cannot be reduced to an equation of the form (9.1) unless either $\beta(t)$ or $\gamma(t)$ does not vanish.] The main results on (9.1) will be based on lemmas dealing with (9.2).

A system of the form (9.2) on $0 \leqq t < \omega \ (\leqq \infty)$ will be called of *type Z* at $t = \omega$ if
$$z(\omega) = \lim z(t) \qquad \text{exists as} \quad t \to \omega$$
for every solution $(v(t), z(t))$, and $z(\omega) \neq 0$ for some solution. It is easy to see that (9.2) is of type Z if and only if there exist linearly independent solutions $(v_j(t), z_j(t))$, $j = 0, 1$, such that $\lim z_0(t) = 0$ and $\lim z_1(t) = 1$.

Lemma 9.1. *Let* $\beta(t)$, $\gamma(t)$ *be continuous complex-valued functions for* $0 \leqq t < \omega \ (\leqq \infty)$. *Suppose that*

$$(9.3_1) \quad \int^{\omega} |\gamma(t)| \, dt < \infty; \qquad (9.3_2) \quad \int^{\omega} |\gamma(t)| \left(\int^{t} |\beta(s)| \, ds \right) dt < \infty$$

or, more generally, that

$$(9.4_1) \qquad \int^{\omega} \gamma(t) \, dt = \lim_{T \to \omega} \int^{T} \gamma(t) \, dt \qquad \text{exists}$$

(possibly conditionally) and that

$$(9.4_2) \qquad \int^{\omega} |\beta(s)| \, \Gamma(s) \, ds < \infty, \qquad \text{where} \quad \Gamma(s) = \sup_{s \leqq t < \omega} \left| \int_t^{\omega} \gamma(r) \, dr \right|.$$

Then (9.2) *is of type Z.*

Unless $\beta(t) \equiv 0$, the condition (9.3_2) implies (9.3_1). If the order of integration is reversed, it is seen that (9.3_2) is equivalent to

$$\int^{\omega} |\beta(s)| \left(\int_s^{\omega} |\gamma(t)| \, dt \right) ds < \infty.$$

This shows that (9.3) implies (9.4). Lemma 9.1 has a partial converse.

Lemma 9.2. *If $\beta(t)$ and $\gamma(t)$, where $0 \leq t < \omega$ ($\leq \infty$), are continuous real-valued functions which do not change sign (i.e., $\beta \geq 0$ or $\beta \leq 0$ and $\gamma \geq 0$ or $\gamma \leq 0$) and if (9.2) is of type Z, then (9.3_1)–(9.3_2) hold.*

Exercise 9.1.   Generalize Lemma 9.1 to the case where (9.2) is replaced by a d-dimensional system of the form $v^{j\prime} = \gamma_j(t)v^{j+1}$, where $j = 1, \ldots, d$ and $v^{d+1} = v^1$.

Proof of Lemma 9.1.   Two quadratures of (9.2) give

$$(9.5)\quad v(t) = \int_T^t \beta(s)z(s)\,ds + c_1, \qquad c_1 = v(T),$$

$$(9.6)\quad z(t) = \int_T^t \gamma(s)\int_T^s \beta(r)z(r)\,dr\,ds + c_1\int_T^t \gamma(s)\,ds + c_2, \quad c_2 = z(T).$$

On interchanging the order of integration, the last formula becomes

$$(9.7)\qquad z(t) = \int_T^t \beta(r)z(r)\int_r^t \gamma(s)\,ds\,dr + c_1\int_T^t \gamma(s)\,ds + c_2.$$

If $t \geq T$, then $T \leq r \leq t$ and the definition of Γ in (9.4_2) imply that

$$(9.8)\qquad \left|\int_r^t \gamma(s)\,ds\right| \leq \left|\int_r^\omega \gamma(s)\,ds\right| + \left|\int_t^\omega \gamma(s)\,ds\right| \leq 2\Gamma(r).$$

Consequently

$$|z(t)| \leq 2\int_T^t |\beta(s)|\,\Gamma(s)\,|z(s)|\,ds + C,$$

where

$$(9.9)\qquad\qquad\qquad C = 2\,|c_1|\,\Gamma(T) + |c_2|.$$

By Gronwall's inequality (Theorem III 1.1),

$$(9.10)\quad |z(t)| \leq C \exp 2\int_T^t |\beta(s)|\,\Gamma(s)\,ds \leq C \exp 2\int_T^\omega |\beta(s)|\,\Gamma(s)\,ds$$

for $T \leq t < \omega$. Hence (9.4_2) implies that $z(t)$ is bounded. The relations (9.7) and (9.4_2) then show that $z(\omega) = \lim z(t)$ as $t \to \omega$ exists.

The limit $z(\omega)$ is obtained by writing $t = \omega$ in (9.7). In order to show that $z(\omega) \neq 0$ for some solution of (9.2), choose the initial conditions $c_1 = v(T) = 0$ and $c_2 = z(T) = 1$ in (9.5), (9.6). Thus $C = 1$ in (9.9) and (9.10) and so (9.7), (9.8), and (9.10) give

$$|z(\omega) - 1| \leq 2\left(\int_T^\omega |\beta(r)|\,\Gamma(r)\,dr\right)\exp 2\int_T^\omega |\beta(s)|\,\Gamma(s)\,ds.$$

Since the right side tends to 0 as $T \to \omega$, it follows that if T is sufficiently near to ω, then $z(\omega) \neq 0$. This proves Lemma 9.1.

Proof of Lemma 9.2. Let $(v(t), z(t))$ be a solution of (9.2) such that $z(\omega) = 1$. It can also be supposed that $v(T) = 0$ for some T. Otherwise it is possible to add to $(v(t), z(t))$ a suitable multiple of a solution $(v_0(t), z_0(t)) \not\equiv 0$ for which $z(\omega) = 0$. In fact $v_0(t) \equiv 0$ cannot hold, for then (9.2) shows that $z_0(t) \equiv z_0(\omega) = 0$.

Thus $c_1 = 0$ in (9.6) and $z(\omega) = 1$ shows that (9.3_2) holds (since β, γ do not change sign). If $\beta(t) \not\equiv 0$ for t near ω, (9.3_1) follows. If, however, $\beta(t) \equiv 0$ and (9.2) is of type Z, then (9.3_1) holds when γ does not change signs. This completes the proof.

Let $(v(t), z(t))$, $(v_1(t), z_1(t))$ be solutions of (9.2). Then

(9.11) $$z_1(t)v(t) - v_1(t)z(t) = c_0$$

is a constant. This follows from Theorem IV 1.2 (or can easily be verified by differentiation). If $z_1(t) \neq 0$ and (9.11) is multiplied by $\gamma(t)/z_1{}^2(t)$, it is seen from (9.2) that $(z/z_1)' = c_0\gamma/z_1{}^2$, and hence there is a constant c_1 such that

(9.12) $$z(t) = c_1 z_1(t) + c_0 z_1(t) \int_T^t \frac{\gamma(s)\,ds}{z_1{}^2(s)}$$

if $z_1 \neq 0$ on the interval $[T, t]$. Similarly, if $v_1 \neq 0$ for $[T, t]$ then $(v/v_1)' = -c_0\beta/v_1{}^2$ and

(9.13) $$v(t) = c_1 v_1(t) - c_0 v_1(t) \int_T^t \frac{\beta(s)\,ds}{v_1{}^2(s)}.$$

Conversely, if $z_1 \neq 0$ [or $v_1 \neq 0$] in the t-interval $[T, t]$, then (9.12) [or (9.13)] and (9.11) define a solution $(v(t), z(t))$ of (9.2).

Exercise 9.2. Suppose that (9.2) is of type Z and that $(v_1(t), z_1(t))$ is a solution of (9.2) satisfying $z_1(\omega) = 1$. (a) Show that

$$\int^\omega \frac{\gamma(t)\,dt}{z_1{}^2(t)} = \lim_{T \to \omega} \int^T \frac{\gamma(t)\,dt}{z_1{}^2(t)} \qquad \text{exists}$$

and that (9.2) has a solution $(v_0(t), z_0(t))$ in which

(9.14) $$z_0(t) = z_1(t) \int_t^\omega \frac{\gamma(s)\,ds}{z_1{}^2(s)}$$

for t near ω. (b) If $(v(t), z(t))$ is any solution of (9.2), then

$$v(t) = O\left(1 + \int^t |\beta(s)|\,ds\right) \qquad \text{as} \quad t \to \omega.$$

If, in addition, (9.3) holds, then (9.14) satisfies

(9.15) $$z_0(t) = O\left(\int_t^\omega |\gamma(s)|\,ds\right),$$

$c = \lim v_0(t)$ exists as $t \to \omega$, and

$$(9.16) \qquad v_0(t) = c + O\left(\int_t^\omega |\beta(s)| \int_s^\omega |\gamma(r)|\, dr\, ds\right) \qquad \text{as} \quad t \to \omega.$$

Also, if $\gamma(t)$ is real-valued and does not change signs, then (9.15) can be improved to

$$(9.17) \qquad z_0(t) \sim \int_t^\omega \gamma(s)\, ds.$$

Lemma 9.3. *Let $\beta(t)$, $\gamma(t)$ be as in Lemma 9.1. In addition, suppose that $\beta(t) \geqq 0$ and that*

$$(9.18) \qquad \int^\omega \beta(t)\, dt = \infty.$$

Then (9.2) has a pair of solutions $(v_j(t), z_j(t))$ for $j = 0, 1$, satisfying, as $t \to \omega$,

$$(9.19_0) \qquad v_0 \sim 1, \qquad z_0 = o\left(\frac{1}{\int^t \beta(s)\, ds}\right);$$

$$(9.19_1) \qquad v_1 \sim \int^t \beta(s)\, ds, \qquad z_1 \sim 1.$$

This has a partial converse.

Lemma 9.4. *Let $\beta(t)$, $\gamma(t)$ be continuous real-valued functions such that $\beta(t) \geqq 0$ satisfies (9.18) and $\gamma(t)$ does not change signs. Let (9.2) have a solution satisfying either (9.19_0) or (9.19_1). Then (9.3) holds [so that (9.2) has solutions satisfying (9.19_0) and (9.19_1)].*

Exercise 9.3.   Prove Lemma 9.4.

Proof of Lemma 9.3.   By Lemma 9.1, (9.2) has a solution $(v_1(t), z_1(t))$ such that $z_1(\omega) = 1$. Thus the first part of (9.19_1) follows from the first equation in (9.2). Note that

$$(9.20) \qquad \int^t \frac{\beta(s)\, ds}{\left(\int^s \beta(r)\, dr\right)^2} = \text{const.} - \frac{1}{\int^t \beta(s)\, ds}$$

tends to const. as $t \to \omega$ by (9.18). Consequently, the integral $c_1 = \int_T^\omega \beta(s)\, ds/v_1^2(s)$ is absolutely convergent (for T near ω). It follows from (9.13) with the choice $c_0 = 1$, that (9.2) has a solution $(v, z) = (v_0, z_0)$ satisfying (9.11) with $c_0 = 1$ and

$$v_0(t) = v_1(t) \int_t^\omega \frac{\beta(s)\, ds}{v_1^2(s)}.$$

Then $v_0 \sim 1$ follows from the first part of (9.19_1) and (9.20). Letting

$(v, z, c_0) = (v_0, z_0, 1)$ in (9.11) and solving for z_0 gives the last part of (9.19_0). This completes the proof.

Theorem 9.1. *Let $p(t)$ be a positive and $q_0(t)$ a real-valued continuous function for $0 \leqq t < \omega$ such that*

$$(9.21) \qquad\qquad (p(t)x')' + q_0(t)x = 0$$

is nonoscillatory at $t = \omega$ and let $x_0(t)$, $x_1(t)$ be principal, nonprincipal solutions of (9.21); cf. § 6. Suppose that $q(t)$ is a continuous complex-valued function satisfying

$$(9.22) \qquad\qquad \int^\omega |x_0(t)x_1(t)| \cdot |q(t) - q_0(t)|\, dt < \infty$$

or, more generally,

$$(9.23_1) \qquad \int^\omega (q - q_0)x_0{}^2\, dt = \lim_{T \to \omega} \int^T (q - q_0)x_0{}^2\, dt \qquad \text{exists,}$$

$$(9.23_2) \quad \int^\omega \frac{\Gamma(s)\, ds}{p(s)x_0{}^2(s)} < \infty, \qquad \text{where} \quad \Gamma(s) = \sup_{s \leqq t < \omega} \left| \int_t^\omega (q - q_0)x_0{}^2\, dr \right|.$$

Then (9.1) has a pair of solutions $u_0(t)$, $u_1(t)$ satisfying, as $t \to \omega$,

$$(9.24) \qquad\qquad u_j \sim x_j,$$

$$(9.25) \qquad\qquad \frac{pu_j'}{u_j} = \frac{px_j'}{x_j} + o\left(\frac{1}{|x_0 x_1|}\right),$$

for $j = 0, 1$.

Exercise 9.4. Verify that if $q(t)$ is real-valued, $q(t) - q_0(t)$ does not change signs, and (9.1) has a solution $u_j(t)$ satisfying (9.24)–(9.25) for either $j = 0$ or $j = 1$, then (9.22) holds.

Condition (9.22) in Theorem 9.1 should be compared with (8.3) in Theorem 8.1. The analogue of (8.3) is the stronger condition

$$\int^\omega |x_1|^2 \cdot |q - q_0|\, dt < \infty \qquad \text{since} \quad x_0 = o(x_1) \qquad \text{as} \quad t \to \omega.$$

Remark. It will be clear from the proof of Theorem 9.1 that if $q_0(t)$ is complex-valued but has a pair of solutions asymptotically proportional to real-valued positive functions $x_0(t)$, $x_1(t)$ satisfying (9.22) [or (9.23)] and $\int^\omega ds/px_1{}^2 < \infty, \int^\omega ds/px_0{}^2 = \infty, x_1 \sim x_0 \int^t ds/px_0{}^2$, then Theorem 9.1 remains valid.

Exercise 9.5. Let $p(t) \neq 0$, $q(t)$, $q_0(t)$ be continuous complex-valued functions for $0 \leqq t < \omega \ (\leqq \infty)$ such that (9.21) has a solution $x_1(t)$ which does not vanish for large t and satisfies

$$\int^\omega \frac{dt}{p(t)x_1{}^2(t)} = \lim \int^T \qquad \text{as} \quad T \to \omega \qquad \text{exists}$$

and

$$\int^{\omega} |q - q_0| \cdot |x_1|^2 \Gamma \, dt < \infty, \qquad \text{where} \quad \Gamma(t) = \sup_{t \le s < \omega} \left| \int_s^{\omega} \frac{dr}{px_1^2} \right|.$$

Then (9.1) has a pair of nontrivial solutions $u_0(t)$, $u_1(t)$ such that

$$u_0 = o(|u_1|), \qquad u_1 \sim x_1$$

$$\frac{pu_1'}{u_1} = \frac{px_1'}{x_1} + O\left(\frac{1}{|x_1|^2} \int^t |q - q_0| \cdot |x_1|^2 \, ds \right) \qquad \text{as} \quad t \to \omega.$$

Proof of Theorem 9.1. The variations of constants $u = x_0(t)v$ reduces (9.1) to

(9.26) $$(px_0^2 v')' + x_0^2(q - q_0)v = 0$$

for t near ω; cf. (2.31). Write this as a system (9.2), where

(9.27) $$z = px_0^2 v', \qquad \beta = \frac{1}{px_0^2}, \qquad \gamma = -x_0^2(q - q_0).$$

It will be verified that Lemma 9.3 is applicable. Note that condition (9.18) holds since $x_0(t)$ is a principal solution of (9.21); Theorem 6.4. A nonprincipal solution $x_1(t)$ of (9.21) is given by

(9.28) $$x_1(t) = x_0(t) \int^t \frac{ds}{p(s)x_0^2(s)} = x_0(t) \int^t \beta(s) \, ds$$

and any other nonprincipal solution is a constant times $[1 + o(1)]x_1(t)$ as $t \to \omega$; Corollary 6.3. The condition (9.4) is equivalent to (9.23).

Thus Lemma 9.3 is applicable. Let (v_0, z_0), (v_1, z_1) be the corresponding solutions of (9.2) and $u_0 = x_0 v_0$, $u_1 = x_0 v_1$ the corresponding solutions of (9.1). Then the first part of (9.19_j) for $j = 0, 1$ gives (9.24) for $j = 0, 1$. Note that $u = x_0 v$ implies that $pu'/u = px_0'/x_0 + pv'/v$, so that, by (9.27), $pu'/u = px_0'/x_0 + z/x_0^2 v$. Since $z_0/v_0 = o\left(1 / \int^t \beta(s) \, ds \right)$, the case $j = 0$ of (9.25) follows. Also, $z_1/x_0^2 v_1 = [1 + o(1)]/x_0^2 \int^t \beta(s) \, ds = [1 + o(1)]/x_0 x_1$ and, from (9.28), $px_1'/x_1 = px_0'/x_0 + 1/x_0 x_1$. Consequently, the case $j = 1$ of (9.25) holds. This proves the theorem.

Corollary 9.1. *In the equation*

(9.29) $$u'' - q(t)u = 0,$$

let $q(t)$ be a continuous complex-valued function for large t satisfying

(9.30) $$\int^{\infty} t \, |q(t)| \, dt < \infty$$

or, more generally,

$$(9.31) \quad Q(t) \equiv \int_t^\infty q(s)\,ds = \lim_{T \to \infty} \int_t^T \quad \text{exists and} \quad \int^\infty \sup_{t \le r < \infty} |Q(r)|\,dt < \infty$$

Then (9.29) has a pair of solutions $u_0(t)$, $u_1(t)$ satisfying, as $t \to \infty$,

$$(9.32) \quad\quad u_0(t) \sim 1, \quad u_0'(t) = o\left(\frac{1}{t}\right),$$

$$(9.33) \quad\quad u_1(t) \sim t, \quad u_1'(t) \sim 1.$$

Conversely, if $q(t)$ is real-valued and does not change signs and if (9.29) has a solution satisfying (9.32) or (9.33), then (9.30) holds.

The first part of the corollary follows from Theorem 9.1, where (9.29) and $x'' = 0$ are identified with (9.1) and (9.21), respectively. The latter has the solutions $x_0(t) = 1$, $x_1(t) = t$. [Under the condition (9.30), the existence of u_0, u_1 is also contained in Theorem X 17.1.] The last part of the corollary follows from Lemma 9.4 or Exercise 9.4.

Corollary 9.2. *In the equation*

$$(9.34) \quad\quad u'' - [\lambda^2 + q(t)]u = 0,$$

let $\lambda > 0$ and $q(t)$ be a complex-valued continuous function for large t satisfying

$$(9.35) \quad\quad \int^\infty |q(t)|\,dt < \infty$$

or, more generally,

$$(9.36) \quad \int^\infty q(s)e^{-2\lambda s}\,ds = \lim_{T \to \infty} \int^T \quad \text{exists and}$$

$$\int^\infty e^{2\lambda t} \sup_{t \le s < \infty} \left| \int_s^\infty q(r)e^{-2\lambda r}\,dr \right| dt < \infty.$$

Then (9.34) has solutions $u_0(t)$, $u_1(t)$ satisfying

$$(9.37) \quad\quad u_0 \sim -u_0'/\lambda \sim e^{-\lambda t}, \quad u_1 \sim u_1'/\lambda \sim e^{\lambda t}.$$

Conversely, if $q(t)$ is real-valued and does not change signs and if (9.34) has a solution $u_0(t)$ or $u_1(t)$ satisfying the corresponding conditions in (9.37), then (9.35) holds.

The first part follows from Theorem 9.1 if (9.34) and $x'' - \lambda^2 x = 0$ are identified with (9.1) and (9.21), respectively. The latter has solutions $x_0(t) = e^{-\lambda t}$, $x_1(t) = e^{\lambda t}$. [Under condition (9.35), the existence of u_0, u_1 is also implied by Theorem X 17.2.]

Exercise 9.6.   Let $q(t) > 0$ be a positive function on $0 \leqq t < \infty$ possessing a continuous second derivative and satisfying

$$\int^\infty q^{\frac{1}{2}}(t)\, dt = \infty \quad \text{and} \quad \int^\infty \left| \frac{5q'^2}{16q^3} - \frac{q''}{4q^2} \right| q^{\frac{1}{2}}\, dt < \infty.$$

Then $u'' - q(t)u = 0$ has a pair of solutions satisfying

$$q^{\frac{1}{4}}u \sim \exp \pm \int^t q^{\frac{1}{2}}(s)\, ds, \qquad q^{-\frac{1}{4}}(q^{\frac{1}{4}}u)' \sim \pm q^{\frac{1}{4}}u$$

as $t \to \infty$. (Compare this with Exercise X 17.5.)

Exercise 9.7.   Find asymptotic formulae for the principal and non-principal solutions of Weber's equation

$$u'' + tu' - 2\lambda u = 0$$

(where λ is a real number) by first eliminating the middle term using the analogue of substitution (1.9) and then applying Exercise 9.6 to the resulting equation; cf. Exercise X 17.6.

Corollary 9.3.   *In equation* (9.29), *let* $q(t)$ *be a continuous complex-valued function for large* t *such that* $Q(t)$ *in* (9.31) *satisfies*

$$\int^\infty Q(t)\, dt = \lim \int^T \qquad \text{exists as} \quad T \to \infty.$$

Then a sufficient condition for (9.29) *to have solutions* $u_0(t),\ u_1(t)$ *satisfying*

(9.38) $u_0(t) \sim 1, \qquad u_0'(t) = o(1),$

(9.39) $u_1(t) \sim t, \qquad u_1'(t) \sim 1,$

as $t \to \infty$, *is that*

(9.40) $\displaystyle \int^\infty t\, |Q(t)|^2\, dt < \infty.$

This condition is also necessary if $q(t)$ *is real-valued.*

   Proof.   It is easily verified that

(9.41) $x_0(t) = \exp \left(-\int^t Q(s)\, ds \right)$

is a solution of

(9.42) $x'' - [q(t) + Q^2(t)]x = 0.$

One of the conditions on Q implies that $\lim x_0(t)$ exists as $t \to \infty$ and is not 0. Correspondingly, the solution of (9.42) given by

(9.43) $\displaystyle x_1(t) = x_0(t) \int^t \frac{ds}{x_0^2(s)}$

is asymptotically proportional to t, as $t \to \infty$.

Thus (9.42) has solutions asymptotically proportional to the (positive) functions 1, t. Hence if (9.29) and (9.42) are identified with (9.1), (9.21), respectively, and if (9.40) holds, the Remark following Theorem 9.1 shows that the conclusions of that theorem are valid. Consequently (9.29) has solutions u_0, u_1 satisfying $u_0 \sim x_0$, $u_1 \sim x_1$ as $t \to \infty$. The analogues of (9.25) are

$$\frac{u_0'}{u_0} = -Q + o\left(\frac{1}{t}\right), \qquad \frac{u_1'}{u_1} = -Q + 1 + o\left(\frac{1}{x_0 x_1}\right).$$

Since $Q(t) \to 0$ as $t \to \infty$, it is clear that certain constant multiples of u_0, u_1 satisfy (9.38), (9.39). The last part of the theorem follows from the fact that $q + Q^2 \geqq q$ when q is real-valued; cf. Exercise 9.4.

By the use of a simple change of variables, a theorem about (9.29) for "small" $q(t)$ can be transcribed into a theorem about (9.34) for "small" $q(t)$, and conversely:

Lemma 9.5. *Let $q(t)$ be a continuous complex-valued function for large t. Then the change of variables, where $\lambda > 0$,*

$$(9.44) \qquad u = v e^{-\lambda t}, \qquad s = (2\lambda)^{-1} e^{2\lambda t}$$

transforms (9.34) into

$$(9.45) \qquad \frac{d^2 v}{ds^2} - e^{-4\lambda t} q(t) v = 0;$$

while the change of variables

$$(9.46) \qquad u = t^{\frac{1}{2}} v = e^s v, \qquad s = \tfrac{1}{2} \log t$$

transforms (9.29) into

$$(9.47) \qquad \frac{d^2 v}{ds^2} - [1 + 4t^2 q(t)] v = 0.$$

Exercise 9.8.   Verify this lemma.

Exercise 9.9.   (a) Let $\lambda > 0$ and $q(t)$ be a continuous complex-valued function for large t such that

$$Q_\lambda(t) = \int_t^\infty q(s) e^{-2\lambda s}\, ds = \lim_{T \to \infty} \int_t^T \qquad \text{exists},$$

$$\int^\infty Q_\lambda(t) e^{2\lambda t}\, dt = \lim_{T \to \infty} \int^T \quad \text{exists and} \quad \int^\infty |Q(t)|^2\, e^{4\lambda t}\, dt < \infty.$$

Then $u'' - [\lambda^2 + q(t)] u = 0$ has a pair of solutions satisfying, as $t \to \infty$,

$$u \sim e^{\pm \lambda t}, \qquad \frac{u'}{u} = \pm \lambda + e^{2\lambda t} Q_\lambda(t) + o(1).$$

(b) Let $q(t)$ be a continuous complex-valued function for $t \geqq 0$ such that $\int^{\infty} t^{2p-1} |q(t)|^p \, dt < \infty$ for some p on the range $1 \leqq p \leqq 2$. Then $u'' - q(t)u = 0$ has a solution satisfying

$$u \sim \exp - \int^t sq(s) \, ds \quad \text{and} \quad \frac{u'}{u} = o\left(\frac{1}{t}\right)$$

and a solution satisfying

$$u \sim t \exp \int^t sq(s) \, ds \quad \text{and} \quad \frac{u'}{u} \sim \frac{1}{t},$$

as $t \to \infty$.

APPENDIX: DISCONJUGATE SYSTEMS

10. Disconjugate Systems

This appendix deals with systems of equations of the form

(10.1) $[P(t)x' + R(t)x]' - [R^*(t)x' - Q(t)x] = 0$

or, more generally, systems of the form

(10.2) $x' = A(t)x + B(t)y, \qquad y' = C(t)x - A^*(t)y.$

Here x, y are d-dimensional vectors; $A(t)$, $B(t)$, $C(t)$, $P(t)$, $Q(t)$, $R(t)$ are $d \times d$ matrices (with real- or complex-valued entries) continuous on a t-interval J. The object is to obtain generalizations of some of the results of § 6. The difficulty arises from the fact that the theorems of Sturm in § 3 do not have complete analogues.

In dealing with (10.1), it will usually be assumed that

(10.3) $P = P^* \quad \text{and} \quad Q = Q^*,$

(10.4) $\det P \neq 0.$

If the vector y is defined by

(10.5) $y = P(t)x' + R(t)x,$

then (10.1) is of the form (10.2), where

(10.6) $A = -P^{-1}R, \qquad B = P^{-1}, \qquad C = -Q - R^*P^{-1}R;$

so that

(10.7) $B = B^* \qquad \text{and} \qquad C = C^*,$

(10.8) $\det B \neq 0.$

The motivations for the assumptions (10.3), (10.4) are the following: Condition (10.4) makes the system (10.1) or, equivalently, (10.2) non-singular in the sense that the usual existence theorems for initial value problems are applicable. Condition (10.3) makes (10.1) formally "self-adjoint" in the sense that if $-L[x]$ denotes the vector on the left of (10.1), whether or not $L[x] = 0$, then we have *Green's formula*

$$\int_a^b (L[x] \cdot z - x \cdot L[z]) \, dt = [x \cdot (Pz' + Rz) - (Px' + Rx) \cdot z]_a^b$$

for all suitable vector functions $x(t)$, $z(t)$.

In particular, if $x(t)$, $x_0(t)$ are solutions of (10.1), so that $L[x] \equiv 0$, $L[x_0] \equiv 0$, then

$$x \cdot (Px_0' + Rx_0) - (Px' + Rx) \cdot x_0 = \text{const.}$$

When this constant is 0, the solutions x, x_0 will be called *conjugate solutions* of (10.1).

A system (10.1) is called *disconjugate on J* if every solution $x(t) \not\equiv 0$ vanishes at most once on J. Correspondingly, a system of the form (10.2) will be termed *disconjugate on J* if, for every solution $(x(t), y(t)) \not\equiv 0$, the vector $x(t)$ vanishes at most once on J.

Instead of (10.2), it will be more convenient to deal with the matrix equations

(10.9) $U' = A(t)U + B(t)V, \qquad V' = C(t)U - A^*(t)V,$

where U, V are $d \times d$ matrices. Note that $U(t)$, $V(t)$ is a solution of (10.9) if and only if $x(t) = U(t)c$, $y(t) = V(t)c$ is a solution of (10.2) for every constant vector c. Thus all solutions of (10.2) are determined if we know two solutions $(U(t), V(t))$, $(U_0(t), V_0(t))$ of (10.9) such that

(10.10) $\det Y(t) \neq 0, \qquad \text{where} \quad Y(t) = \begin{pmatrix} U(t) & U_0(t) \\ V(t) & V_0(t) \end{pmatrix}$

is a $2d \times 2d$ matrix. In fact, $Y(t)$ is a fundamental matrix for the system (10.2)

(i) If $U(t)$, $V(t)$ is a solution of (10.9) and if (10.8) holds, then $V(t)$ is determined by $U(t)$; in fact,

(10.11) $BV(t) = U' - AU.$

(ii) If (10.7) holds and $(U(t), V(t))$, $(U_0(t), V_0(t))$ are solutions of (10.9), then

(10.12) $U^*V_0 - V^*U_0 = K_0,$

where K_0 is a constant matrix. This is readily verified by differentiating

(10.12). If $K_0 = 0$, the solutions (U, V), (U_0, V_0) of (10.9) will be called *conjugate solutions.*

(iii) In particular, if $U_0 = U$, $V_0 = V$, then $U^*V - V^*U$ is a constant. When this constant matrix is 0, then

(10.13) $U^*V = V^*U$;   i.e., $(U^*V)^* = U^*V$ is Hermitian.

In this case, the solution (U, V) of (10.9) is called *self-conjugate.*

(iv) Let $(U(t), V(t))$ be a solution of (10.9) such that

(10.14) $\det U(t) \neq 0$

on some t-interval and let

(10.15) $K = U^*V - V^*U.$

Consider the "variation of constants"

(10.16) $U_0 = UZ,$

where (U_0, V_0) is a solution of (10.9) satisfying (10.12). Then

$$V_0 = U^{*-1}(K_0 + V^*UZ)$$

or, by (10.15),

(10.17) $V_0 = VZ + U^{*-1}(K_0 - KZ).$

Since $U_0' = AU_0 + BV_0$ and $U' = AU + BV$, it follows from (10.16), (10.17) that

$$Z' = -(U^{-1}BU^{*-1}K)Z + U^{-1}BU^{*-1}K_0.$$

Thus, if $T(t)$ is the fundamental solution of the homogeneous part of this equation, satisfying $T(s) = I$,

(10.18) $T' = -(U^{-1}BU^{*-1}K)T,$ $T(s) = I,$

then solutions Z are of the form

$$Z = T(t)\left\{K_1 + \left(\int_s^t T^{-1}(r)U^{-1}BU^{*-1}\,dr\right)K_0\right\},$$

where K_1 is a constant matrix; see Corollary IV 2.1. Correspondingly, by (10.16),

(10.19) $U_0(t) = U(t)T(t)\left\{K_1 + \left(\int_s^t T^{-1}(r)U^{-1}BU^{*-1}\,dr\right)K_0\right\}.$

Conversely, it is readily verified that if (10.7)–(10.8) and (10.14) hold, then (10.19) and (10.11) determine a solution of (10.9) satisfying (10.12).

(v) If, in this discussion, (U, V) is a self-conjugate solution of (10.2), so that $K = 0$ in (10.15), then $T(t) \equiv I$ and (10.19) reduces to

$$(10.20) \qquad U_0(t) = U(t)\left\{K_1 + \left(\int_s^t U^{-1}BU^{*-1}\,dt\right)K_0\right\}.$$

The corresponding solution $(U_0(t), V_0(t))$ of (10.2) is self-conjugate if and only if $K_1^*K_0 = K_0^*K_1$; i.e., $(K_1^*K_0)^* = K_1^*K_0$.

Let $(U_0(t), V_0(t))$ be self-conjugate, $\det U_0(t) \neq 0$, and $\det K_1 \neq 0$. Interchanging (U, V), (U_0, V_0) in (10.12) changes the sign of K_0. Thus the argument leading to (10.20) gives

$$(10.21) \qquad U(t) = U_0(t)\left\{K_1^{-1} - \left(\int_s^t U_0^{-1}BU_0^{*-1}\,dt\right)K_0\right\},$$

since $U_0(s) = U(s)K_1$.

Theorem 10.1 *Let $A(t)$, $B(t)$, $C(t)$ be continuous on J. Then (10.2) is disconjugate on J if and only if, for every pair of points $t = t_1$, $t_2 \in J$ and arbitrary vectors x_1, x_2, the equation (10.2) has a (unique) solution $(x(t), y(t))$ satisfying $x(t_1) = x_1$, $x(t_2) = x_2$.*

The proof will be omitted as it is similar to that of Theorem 6.1. It does not depend on the structure of (10.2) and applies equally well to any system $x' = A(t)x + B(t)y$, $y' = C(t)x + D(t)y$, where A, B, C, D are continuous.

In order to avoid an interruption of the proofs to follow, we now prove a lemma (which has nothing to do with differential equations).

Lemma 10.1 (F. Riesz). *Let $A_1, A_2, \ldots$ be Hermitian matrices satisfying*

$$0 \leqq A_1 \leqq A_2 \leqq \cdots \leqq I.$$

Then $A = \lim A_n$ exists as $n \to \infty$.

If A, B are two Hermitian matrices, the inequality $A > B$ [or $A \geqq B$] means that $A - B$ is positive definite [or non-negative definite]. To say that "$A = \lim A_n$ exists" means that "$A_1\eta, A_2\eta, \ldots$ is convergent for every fixed vector η."

Proof. Note that if $A \geqq 0$ and ξ, η are two arbitrary vectors, then the *generalized Schwarz inequality*

$$|A\xi \cdot \eta|^2 \leqq (A\xi \cdot \xi)(A\eta \cdot \eta)$$

holds. In order to see this, let ϵ be real and $\xi_\epsilon = \xi + \epsilon(A\xi \cdot \eta)\eta$, so that

$$0 \leqq A\xi_\epsilon \cdot \xi_\epsilon = A\xi \cdot \xi + 2\epsilon\,|A\xi \cdot \eta|^2 + \epsilon^2\,|A\xi \cdot \eta|^2\,(A\eta \cdot \eta).$$

Since the right side is a quadratic polynomial in ϵ which is non-negative for all real ϵ, its discriminant is non-negative. This fact gives the desired inequality.

Let $A_{mn} = A_n - A_m$ where $n > m$, so that $A_{mn} \geqq 0$. Hence the generalized Schwarz inequality implies that

$$\|A_{mn}\eta\|^4 = |A_{mn}\eta \cdot A_{mn}\eta|^2 \leqq (A_{mn}\eta \cdot \eta)(A^2_{mn}\eta \cdot A_{mn}\eta).$$

Since $0 \leqq A_{mn} \leqq I$, it follows that the norm of A_{mn} is at most 1, and so $(A^2_{mn}\eta \cdot A_{mn}\eta) \leqq \|\eta\|^2$ and

$$\|A_n\eta - A_m\eta\|^4 \leqq (A_n\eta \cdot \eta - A_m\eta \cdot \eta)\|\eta\|^2.$$

The sequence of numbers $A_n\eta \cdot \eta$, for $n = 1, 2, \ldots$ is nondecreasing and bounded, hence convergent, so that $A_n\eta - A_m\eta$ is small for large m and n. Therefore, $\lim A_n\eta$ exists, as was to be proved.

Theorem 10.2. *Let $A(t)$, $B(t) = B^*(t)$, $C(t) = C^*(t)$ be continuous on J and let $B(t)$ be positive definite. (i) If J is a half-open bounded interval or a closed half-line, then (10.2) is disconjugate on J if and only if there exists a self-conjugate solution $(U(t), V(t))$ of (10.9) such that $\det U(t) \neq 0$ on the interior of J. (ii) If J is a closed bounded interval or an open interval, then (10.2) is disconjugate on J if and only if (10.9) has a self-conjugate solution $(U(t), V(t))$ satisfying $\det U(t) \neq 0$ on J.*

Proof of (i). For the sake of definiteness, let $J = [a, \omega)$, $\omega \leqq \infty$. Let $Y(t)$ in (10.10) be the fundamental matrix for (10.2) such that $Y(t)$ is the identity matrix at $t = a$. In particular, $U(a) = I$, $V(a) = 0$, and $U_0(a) = 0$, $V_0(a) = I$; thus both $(U(t), V(t))$, $(U_0(t), V_0(t))$ are self-conjugate solutions of (10.9), since (10.13) and the analogue for U_0, V_0 hold at $t = a$. The general solution of (10.2) vanishing at $t = a$ is given by $x = U_0(t)c$, $y = V_0(t)c$ for an arbitrary vector c. If (10.2) is disconjugate on J, then $\det U_0(t_0) \neq 0$ for $a < t_0 < \omega$. Otherwise there is a $c_0 \neq 0$ such that $x = U_0(t)c_0$ vanishes at $t = a$ and at $t = t_0$. But then $x(t) \equiv 0$. Hence $y = V_0(t)c_0 \equiv 0$ since $\det B \neq 0$.

Conversely, let $(U(t), V(t))$ be a self-conjugate solution of (10.9) such that $\det U(t) \neq 0$ on $a < t < \omega$. Define a solution $U_0(t)$, $V_0(t)$ of (10.9) by (10.20), where $a < s < \omega$, $K_1 = 0$, $K_0 = I$. Thus $U_0(s) = 0$. Since B, hence $U^{-1}BU^{*-1}$, is positive definite, it follows that $\det U_0(t) \neq 0$ for $s < t < \omega$. Clearly, if $(x(t), y(t)) \neq 0$ is a solution of (10.2) such that x vanishes at $t = s$, then it is necessarily of the form $(x, y) = (U_0(t)c, V_0(t)c)$ and x does not vanish for $s < t < \omega$.

It remains to show that if $(x(t), y(t)) \neq 0$ is a solution of (10.2) such that $x(a) = 0$, then $x(t) \neq 0$ for $a < t < \omega$. To this end, it suffices to show that if $a < s < \omega$, then there exists a self-conjugate solution $(U_0(t), V_0(t))$ of (10.9) such that $U_0(a) = 0$ and $\det U_0(t) \neq 0$ for $a < t \leqq s$. Put

$$U_1(t) = U(t)\left\{-I + \int_s^t U^{-1}BU^{*-1}\,dt\right\},$$

so that, by the analogue of (10.20), U_1, $V_1 = B^{-1}(U_1' - AU_1)$ is a self-conjugate solution of (10.9). Since B is positive definite, the factor $\{\ldots\}$ of $U(t)$ in the last formula is negative definite for $a < t \leqq s$. Hence $\det U_1(t) \neq 0$ for $a < t \leqq s$. By the analogue of (10.21),

$$U(t) = U_1(t)\left\{ -I - \int_s^t U_1^{-1}BU_1^{*-1}\,dt \right\}.$$

Consequently,

$$\left(-I + \int_s^t U^{-1}BU^{*-1}\,dt \right)\left(-I - \int_s^t U_1^{-1}BU_1^{*-1}\,dt \right) = I.$$

Since the first factor is negative definite for $a < t \leqq s$, the second factor (which is the inverse of the first) is negative definite. Hence

$$S(t) \equiv -\int_s^t U_1^{-1}BU_1^{*-1}\,dt = \int_t^s U_1^{-1}BU_1^{*-1}\,dt$$

is positive definite for $a < t \leqq s$, increases with decreasing t (in the sense that $S(r) - S(t) > 0$ for $a < r < t \leqq s$), and $S(t) \leqq I$. It follows from Lemma 10.1 that

$$S(a) = \lim_{t \to a} S(t) = \int_a^s U_1^{-1}BU_1^{*-1}\,dt \quad \text{exists.}$$

For $a \leqq t \leqq s$, put

$$U_0(t) = U_1(t)\int_a^t U_1^{-1}BU_1^{*-1}\,dt,$$

which can also be written in the form

$$U_0(t) = U_1(t)\left\{ S(a) - \int_s^t U_1^{-1}BU_1^{*-1}\,dt \right\}.$$

Hence, by (10.20), U_0, $V_0 = B^{-1}(U_0' - AU_0)$ is a self-conjugate solution of (10.9). It is clear that $U_0(a) = 0$ and $\det U_0(t) \neq 0$ for $a < t \leqq s$. This completes the proof for the case of $[a, \omega)$.

Proof of (ii). If J is a closed bounded interval $[a, b]$, we can extend the definitions of the continuous $A(t)$, $B(t)$, $C(t)$ to an interval $J' = [a - \delta, b + \delta] \supset J$, so that $B = B^*$, $C = C^*$, and $B(t)$ is positive definite. It is clear that if $\delta > 0$ is sufficiently small, then (10.2) is disconjugate on $[a - \delta, b + \delta]$ if and only if it is disconjugate on J. Since $[a, b] \subset [a - \delta, b + \delta)$, the case $J = [a, b]$ of the theorem follows from the case $[a, \omega)$ just treated.

Consider J open. The arguments used in the case $[a, \omega)$ show that if (10.9) has a self-conjugate solution $(U_0(t), V_0(t))$ with $\det U_0 \neq 0$ on J, then (10.2) is disconjugate on J. The converse is implied by Exercise 10.6.

Consider the functional

$$(10.22) \quad I(\eta; a, b) = \int_a^b \{[P(t)\eta' + R(t)\eta] \cdot \eta' + [R^*(t)\eta' - Q(t)\eta] \cdot \eta\} \, dt,$$

where the matrices P, Q, R are determined by (10.6). A vector function $\eta(t)$ on a subinterval $[a, b]$ of J will be said to be admissible of class $A_1(a, b)$ [or $A_2(a, b)$] if (i) $\eta(a) = \eta(b) = 0$, and (ii$_1$) $\eta(t)$ is absolutely continuous and its derivative $\eta'(t)$ is of class L^2 on $a \leqq t \leqq b$ [or (ii$_2$) $\eta(t)$ is continuously differentiable and $P\eta' + R\eta$ is continuously differentiable on $a \leqq t \leqq b$]. If $\eta(t) \in A_2(a, b)$, then an integration by parts [integrating η' and differentiating $P(t)\eta' + R(t)\eta$] gives

$$(10.23) \quad I(\eta; a, b) = - \int_a^b [(P\eta' + R\eta)' - (R^*\eta' - Q\eta)] \cdot \eta \, dt.$$

If $-L[x]$ denotes the vector on the right of (10.1), whether or not $L[x] = 0$, then

$$(10.24) \quad I(\eta; a, b) = \int_a^b L[\eta] \cdot \eta \, dt \quad \text{for} \quad \eta \in A_2(a, b).$$

Theorem 10.3. *Let $A(t)$, $B(t) = B^*(t)$, $C(t) = C^*(t)$ be continuous on J, and $B(t)$ positive definite. Then (10.2) is disconjugate on J if and only if for every closed bounded interval $[a, b]$ in J, the functional $I(\eta; a, b)$ is positive definite on $A_1(a, b)$ [or $A_2(a, b)$]; i.e., $I(\eta; a, b) \geqq 0$ for every $\eta \in A_1(a, b)$ [or $\eta \in A_2(a, b)$] and $I(\eta; a, b) = 0$ if and only if $\eta \equiv 0$.*

Proof ("*Only if*"). The proof of "only if" is similar to that of the "only if" portion of Theorem 6.2. Suppose that (10.2) is disconjugate on J. Then, if $[a, b] \subset J$, there is a self-conjugate solution $(U(t), V(t))$ of (10.9) such that $\det U(t) \neq 0$ on $[a, b]$.

For a given $\eta(t) \in A_1(a, b)$, define $\zeta(t)$ by $\eta(t) = U(t)\zeta(t)$. Using the analogue $V = PU' + RU$ of (10.5) and the analogue $V' = R^*U' - QU$ of (10.1), it is seen that the integrand of (10.22) is

$$(PU\zeta' + V\zeta) \cdot (U'\zeta + U\zeta') + (V'\zeta + R^*U\zeta') \cdot U\zeta.$$

Since $PU\zeta' \cdot U'\zeta = U\zeta' \cdot PU'\zeta$ and $R^*U\zeta' \cdot U\zeta = U\zeta' \cdot RU\zeta$, the integrand can be written as

$$PU\zeta' \cdot U\zeta' + U\zeta' \cdot V\zeta + V\zeta \cdot U'\zeta + V\zeta \cdot U\zeta' + V'\zeta \cdot U\zeta.$$

In addition, the second term is $V^*U\zeta' \cdot \zeta = U^*V\zeta' \cdot \zeta = V\zeta' \cdot U\zeta$ since (U, V) is self-conjugate and satisfies (10.13). Hence the integrand in (10.22) is $PU\zeta' \cdot U\zeta' + (V\zeta \cdot U\zeta)'$, and so

$$(10.25) \quad I(\eta; a, b) = \int_a^b (PU\zeta' \cdot U\zeta') \, dt \quad \text{if} \quad \eta = U\zeta \in A_1(a, b)$$

and $\zeta(a) = \zeta(b) = 0$. Since $B(t)$ is positive definite, $P = B^{-1}$ is also positive definite. Hence $I(\eta, a, b) \geqq 0$ and $I(\eta; a, b) = 0$ if and only if $\zeta \equiv 0$.

Proof ("*If*"). This proof is identical to the corresponding part of the proof of Theorem 6.2 and will be omitted.

Exercise 10.1 (*Jacobi*). Let $A_{3/2}(a, b)$ denote the set of vector-valued functions on $[a, b]$ such that (i) $\eta(a) = \eta(b) = 0$; (ii) $\eta(t)$ is continuous on $a \leqq t \leqq b$; and (iii) the interval $[a, b]$ has a decomposition (depending on η) $a = t_0 < t_1 < \cdots < t_m = b$, such that η is continuously differentiable and $P\eta' + R\eta$ is continuously differentiable on each interval $[t_j, t_{j+1}]$ for $j = 0, \ldots, m - 1$. Thus $A_1[a, b] \supset A_{3/2}[a, b] \supset A_2[a, b]$. Let $A(t)$, $B(t)$, $C(t)$ be as in Theorem 10.3 and suppose that J is not a closed bounded interval. Show that (10.2) is disconjugate on J if and only if $I(\eta; a, b) \geqq 0$ for all $\eta \in A_{3/2}(a, b)$ and all $[a, b] \subset J$.

Exercise 10.2. Let $P_j(t)$, $Q_j(t)$ be continuous $d \times d$ matrices for $j = 1, 2$ on $J: a \leqq t \leqq b$ such that (i) $P_2 = P_2{}^*$ and $Q_2 = Q_2{}^*$ are Hermitian; (ii) $0 < P_2 \leqq \operatorname{Re} P_1$ and $\operatorname{Re} Q_1 \leqq Q_2$ if the inequality $P_2 > 0$ means that P_2 is positive definite and, e.g., $P_2 \leqq \operatorname{Re} P_1$ means that $P_2(t)\eta \cdot \eta \leqq \operatorname{Re} [P_1(t)\eta \cdot \eta]$ for all constant vectors η; finally (iii) let $(P_2 x')' + Q_2 x = 0$ be disconjugate on J. Then $(P_1 x')' + Q_1 x = 0$ is disconjugate on J.

If $\xi(t)$, $\eta(t)$ are of class L^2 on $[a, b]$, introduce the "scalar product"

$$(\xi, \eta) = \int_a^b \xi(t) \cdot \eta(t) \, dt.$$

Thus, in this notation, (10.24) can be written as $I(\eta; a, b) = (L[\eta], \eta)$ for $\eta \in A_2(a, b)$. Correspondingly (10.25) and $\eta = U\zeta$ imply that

$$(10.26) \qquad (L[\eta], \eta) = (L_1[\eta], L_1[\eta]) \qquad \text{for} \quad \eta \in A_2(a, b),$$

where

$$(10.27) \qquad L_1[\eta] = P^{1/2}(t) U(t)(U^{-1}(t)\eta)'$$

and where $P^{1/2}(t)$ is the unique positive definite Hermitian matrix such that $(P^{1/2}(t))^2 = P(t)$, so that $P(t)$ is continuous on J; cf. Exercise XIV 1.2. The bilinear relation corresponding to (10.26) is

$$(10.28) \quad (L[\eta_1], \eta_2) = (L_1[\eta_1], L_1[\eta_2]) \qquad \text{for} \quad \eta_1, \eta_2 \in A_2(a, b).$$

Formally, then, $L = L_1{}^*L$. Actually, this is correct in the following sense: (10.27) can be written as

$$(10.29) \qquad L_1[\eta] = P^{1/2}(t)\eta' - P^{1/2}(t)U'(t)U^{-1}(t)\eta,$$

for $(U^{-1})' = -U^{-1}U'U^{-1}$ as can be seen by differentiating the relation $U^{-1}U = I$. The formal adjoint of L_1 is therefore

$$(10.30) \qquad L_1^*[\eta] = -(P^{\frac{1}{2}}(t)\eta)' - U^{*-1}(t)U^{*'}(t)P^{\frac{1}{2}}(t)\eta,$$

cf. IV § 8(viii); i.e.,

$$(10.31) \qquad L_1^*[\eta] = -U^{*-1}(U^*P^{\frac{1}{2}}\eta)'.$$

Corollary 10.1. *Let A, B, C be as in Theorem* 10.3. *Let* (10.9) *have a self-conjugate solution* $(U(t), V(t))$ *such that* $\det U(t) \neq 0$ *on J. Then $L = L_1^*L_1$; i.e.,*

$$(10.32) \qquad L[\eta] = -(P\eta' + R\eta)' + (R^*\eta' - Q\eta)$$

is given by

$$(10.33) \qquad L[\eta] = L_1^*\{L_1[\eta]\}$$

for every continuously differentiable $\eta(t)$ such that $P\eta' + R\eta$ is continuously differentiable on J.

This can be deduced from (10.28) or, more easily, by straightforward differentiation (using the relations $V = PU' + RU$, $V' = R^*U' - QU$).

Theorem 10.4 *Let $A(t)$, $B(t) = B^*(t)$, $C(t) = C^*(t)$ be continuous on $J : a \leqq t < \omega$ $(\leqq \infty)$ and let $B(t)$ be positive definite. Let* (10.2) *be disconjugate on J and $(U(t), V(t))$ a solution of* (10.9) *such that $\det U(t) \neq 0$ on $s \leqq t < \omega$ for some $s \in [a, \omega)$, and let $T(t)$ be defined by* (10.18). *Then*

$$(10.34) \qquad S_s(t) = \int_s^t T^{-1}(r)U^{-1}BU^{*-1}\,dr$$

is nonsingular for $s < t < \omega$ and

$$(10.35) \qquad \lim_{t \to \omega} S_s^{-1}(t) = M \qquad \text{exists,}$$

where M depends on s and the matrix function $U(t)$.

If, in Theorem 10.4, $M = 0$, the solution $(U(t), V(t))$ of (10.9) will be called a *principal solution* of (10.9). It will turn out that a principal solution is necessarily self-conjugate. In this case $T^{-1}(r) = I$, and $M = 0$ can be expressed as

$$(10.36) \qquad \int^t U^{-1}BU^{*-1}\,dt \to \infty \qquad \text{as} \quad t \to \omega,$$

in the sense that

$$\left\| \int^t U^{-1}BU^{*-1}\,dt\,c \right\| \to \infty \qquad \text{as} \quad t \to \omega$$

uniformly for all unit vectors c. [Compare (10.36), where $B = P^{-1}$, and the definition (6.11) for principal solution in § 6.]

Theorem 10.5. *Let A, B, C and J be as in Theorem* 10.4. (i) *Then* (10.9) *possesses a principal solution* $(U_0(t), V_0(t))$. (ii) *Another solution* $(U(t), V(t))$ *is a principal solution if and only if* $(U(t), V(t)) = (U_0(t)K_1, V_0(t)K_1)$, *where K_1 is a constant nonsingular matrix.* (iii) *Let* $(U(t), V(t))$ *be a solution of* (10.9). *Then the constant matrix K_0 in* (10.12) *is nonsingular if and only if* $\det U(t) \neq 0$ *for t near ω and*

$$(10.37) \qquad U^{-1}(t)U_0(t) \to 0 \qquad \text{as} \quad t \to \omega,$$

in which case M in (10.35) *is nonsingular.*

The proofs of Theorems 10.4 and 10.5 will be given together.

Proof of Theorem 10.5 (i). This proof is essentially contained in the proof of Theorem 10.2(i). Since (10.2) is disconjugate on J, there exists a self-conjugate solution $(U_1(t), V_1(t))$ of (10.9) such that $\det U_1(t) \neq 0$ for $a < t < \omega$. Let $a < \alpha < \omega$ and put

$$(10.38) \qquad U_2(t) = U_1(t)\left\{ I + \int_\alpha^t U_1^{-1}BU_1^{*-1}\, dt \right\}.$$

Then U_2, $V_2 = B^{-1}(U_2' - AU_2)$ is a self-conjugate solution of (10.9), $\det U_2(t) \neq 0$ for $\alpha \leqq t < \omega$, and

$$(10.39) \qquad U_1(t) = U_2(t)\left\{ I - \int_\alpha^t U_2^{-1}BU_2^{*-1}\, dt \right\},$$

by (10.21). Thus

$$\left(I + \int_\alpha^t U_1^{-1}BU_1^{*-1}\, dt \right)\left(I - \int_\alpha^t U_2^{-1}BU_2^{*-1}\, dt \right) = I.$$

As in the proof of Theorem 10.2(i), it follows that

$$(10.40) \qquad M_2 = \lim_{t \to \omega} S(t) = \int_\alpha^\omega U_2^{-1}BU_2^{*-1}\, dt \quad \text{exists}$$

if $S(t)$ is defined by

$$(10.41) \qquad S(t) = \int_\alpha^t U_2^{-1}BU_2^{*-1}\, dt.$$

The limit M_2 is nonsingular since $M_2 > S(t)$ and $S(t)$ is positive definite for $\alpha < t < \omega$.

In terms of $U_2(t)$, define

$$(10.42) \qquad U_0(t) = U_2(t)\int_t^\omega U_2^{-1}BU_2^{*-1}\, dt,$$

or, equivalently,

$$(10.43) \qquad U_0(t) = U_2(t)\left\{ M_2 - \int_\alpha^t U_2^{-1}BU_2^{*-1}\, dt \right\};$$

therefore, U_0, $V_0 = B^{-1}(U_0' - AU_0)$ is a self-conjugate solution of (10.9) and det $U_0(t) \neq 0$ for $\alpha \leqq t < \omega$. Thus, by (10.21),

$$U_2(t) = U_0(t)\left\{M_2^{-1} + \int_\alpha^t U_0^{-1}BU_0^{*-1}\,dt\right\}.$$

Consequently,

$$\left(M_2^{-1} + \int_\alpha^t U_0^{-1}BU_0^{*-1}\,dt\right)\left(\int_t^w U_2^{-1}BU_2^{*-1}\,dt\right) = I,$$

and hence

$$\left(\int_\alpha^t U_0^{-1}BU_0^{*-1}\,dt\right)\left(\int_t^w U_2^{-1}BU_2^{*-1}\,dt\right) = I + o(1) \qquad \text{as} \quad t \to \omega,$$

so that

$$\left(\int_\alpha^t U_0^{-1}BU_0^{*-1}\,dt\right)^{-1} \to 0 \qquad \text{as} \quad t \to \omega.$$

Thus $(U_0(t), V_0(t))$ is a principal solution of (10.9); cf. (10.35).

Proof of Theorem 10.4. Let $(U(t), V(t))$ be a solution of (10.9) such that det $U(t) \neq 0$ for $s \leqq t < \omega$. Thus the matrix $S_s(t)$ in (10.34) is defined for $t \geqq s$.

It will first be shown that det $S_s(t) \neq 0$ for $s < t < \omega$. Put

$$U^0(t) = U(t)T(t)S_s(t)$$

and $V^0 = B^{-1}(U^{0'} - AU^0)$. This defines a solution (U^0, V^0) of (10.9); cf. (10.19). Suppose that det $S_s(t_0) = 0$ for some $t_0 > s$, then there is a constant vector $c_0 \neq 0$ such that $x(t) = U^0(t)c_0$ vanishes at $t = s, t_0$. Since (10.2) is disconjugate on J, it follows that $U^0(t)c_0 \equiv 0$. Hence $S_s(t)c_0 \equiv 0$, and so $S_s'(t)c_0 \equiv 0$. Since $S_s' = T^{-1}U^{-1}BU^{*-1}$ is nonsingular, this is a contradiction and shows that det $S_s(t) \neq 0$ for $s < t < \omega$.

It will next be shown that the limit (10.35) exists. Let $(U_0(t), V_0(t))$ be the principal solution of (10.9) just constructed, so that $U_0(t)$ is given by (10.42) in terms of a self-conjugate solution $(U_2(t), V_2(t))$ with det $U_2(t) \neq 0$ for $\alpha \leqq t < \omega$. Let $\alpha \leqq s < r < w$ and consider the function

(10.44)
$$U_{0r}(t) = U_2(t)\int_t^r U_2^{-1}BU_2^{*-1}\,dt.$$

It is clear that (U_{0r}, V_{0r}), where $V_{0r} = B^{-1}(U_{0r}' - AU_{0r})$ is a solution of (10.9), for $U_{0r}(t)$ can be written as

$$U_{0r}(t) = U_2(t)\left\{\int_\alpha^r U_2^{-1}BU_2^{*-1}\,dt - \int_\alpha^t U_2^*BU_2^{*-1}\,dt\right\};$$

cf. (10.19). It follows that if K_r is the constant matrix

(10.45)
$$K_r = U^*V_{0r} - V^*U_{0r},$$

then, by the analogue of (10.19), where $U_0(s) = U(s)K_1$,

$$U_{0r}(t) = U(t)T(t)\{U^{-1}(s)U_{0r}(s) + S_r(t)K_r\}.$$

Since $U_{0r}(t) = 0$ when $t = r$, it is seen that, in the last line, $\{\ldots\} = 0$ when $t = r$. Hence

(10.46) $$S_s^{-1}(r) = -K_r U_{0r}^{-1}(s)U(s).$$

It is clear from (10.42) and (10.44) that $U_{0r}(t) \to U_0(t)$ as $r \to \omega$; hence $K_r \to K_0$ as $r \to \infty$, where

(10.47) $$K_0 = U^*V_0 - V^*U_0.$$

Thus

(10.48) $$M = \lim_{r \to \omega} S_s^{-1}(r) \quad \text{exists and is} \quad -K_0 U_0^{-1}(s)U(s).$$

This proves Theorem 10.4.

Proof of Theorem 10.5 (ii). Let $(U(t), V(t))$ be a principal solution of (10.9). Let s be such that $\det U(t) \neq 0$ for $s \leq t < \omega$ and that the limit M in (10.35) is 0. In view of (10.48), $M = 0$ holds if and only if $K_0 = 0$ in (10.47). Since (U_0, V_0) is self-conjugate, it follows that $V_0 = U_0^{*-1}V_0^*U_0$. Hence $K_0 = 0$ gives

$$V(t) = V_0(t)U_0^{-1}(t)U(t) \quad \text{for} \quad t \geq s.$$

Let $K_1 = U_0^{-1}(s)U(s)$. Then $(U(t), V(t))$ satisfies the initial conditions

(10.49) $$U(t) = U_0(t)K_1, \qquad V(t) = V_0(t)K_1$$

at $t = s$. Hence (10.49) holds for all t. This proves (ii).

Proof of Theorem 10.5 (iii). Let $(U(t), V(t))$ be a solution of (10.9) and let K_0 be given by (10.47). Then

$$U(t) = U_0(t)\left\{U_0^{-1}(s)U(s) - \left(\int_s^t U_0^{-1}BU_0^{*-1}\, dt\right)K_0\right\}.$$

Since (10.36) holds with U replaced by U_0, it is clear from (10.36) that if K_0 is nonsingular, then $U(t)$ is nonsingular for t near ω. In this case,

$$I = U^{-1}(t)U_0(t)\left\{U_0^{-1}(s)U(s) - \left(\int_s^t U_0^{-1}BU_0^{*-1}\, dt\right)K_0\right\}$$

and (10.37) follows from (10.36), where U is replaced by U_0.

Conversely, if $U(t)$ is nonsingular for t near ω and (10.37) holds, then the last formula line shows that K_0 is nonsingular. Thus M is nonsingular by virtue of (10.48). This completes the proof of Theorem 10.5.

Exercise 10.3. State an analogue of Corollary 6.3.

Exercise 10.4 (*Analogue of* (6.11$_1$) *in Theorem 6.4*). Let the conditions of Theorem 10.5 hold. Let $(U(t), V(t))$ be a self-conjugate solution of (10.9) such that $\det U(t) \neq 0$ for t near ω and the limit M in (10.35) is nonsingular. [Note that $T(t) \equiv I$.] Let $c \neq 0$ be a constant vector and $x(t) = U(t)c$ a solution of (10.1). Then $\int^{\omega} [P(t)x(t) \cdot x(t)]^{-1} dt < \infty$.

Exercise 10.5 (*Analogue of Exercise 6.5*). Let the conditions of Theorem 10.5 hold and let $a < T < \omega$. Let $(U_{0r}(t), V_{0r}(t))$ be the solution of (10.9) satisfying $U_{0r}(T) = I$, $U_{0r}(r) = 0$; cf. Theorem 10.1. Then $\lim (U_{0r}(t), V_{0r}(t)) = (U_0(t), V_0(t))$ exists as $r \to \omega$ and is a principal solution.

Exercise 10.6. Let J be an open interval; $A(t), B(t) = B^*(t), C(t) = C^*(t)$ continuous on J; and $B(t)$ positive definite. Let (10.2) be disconjugate on J. Let ω ($\leq \infty$) be the right endpoint of J and $[a, \omega) \subset J$. Let $(U_0(t), V_0(t))$ be a principal solution of (10.9) on $[a, \omega)$. Then $\det U_0(t) \neq 0$ on J.

Exercise 10.7 (*Analogue of (iii) in Theorem 6.4*). Let A, B, C, and J be as in Theorem 10.4. Let $(U_0(t), V_0(t))$ be a principal solution of (10.9) and let $\det U_0(t) \neq 0$ for $a \leq t < \omega$. Let $(U(t), V(t))$ be a self-conjugate solution of (10.9) satisfying $U(a) \neq 0$. Let $\Delta = V_0(a)U_0^{-1}(a) - V(a)U^{-1}(a)$, so that $\Delta = \Delta^*$. Then $\Delta \leq 0$ (i.e., $\Delta x \cdot x \leq 0$ for all vectors x) if and only if $\det U(t) \neq 0$ for $t \geq a$.

11. Generalizations

The methods of the last section are applicable to more general situations which will be indicated here. The material of the last section can be considered from the following point of view: First, what are conditions (necessary and/or sufficient) for the functional $I(\eta; a, b)$ in (10.22) to be positive definite for all $[a, b] \subset J$ on certain classes of functions $A_1(a, b)$ or $A_2(a, b)$? Second, if this is the case, what are some consequences for the solutions of the corresponding Euler-Lagrange equations (10.1) or their Hamiltonian form (10.2)?

In this section, a similar problem is considered, but the assumption that P is positive definite is relaxed and the classes $A_1(a, b)$, $A_2(a, b)$ are replaced by more restricted classes of function $\eta(t)$. In particular, it will be required that the competing functions $\eta(t)$ satisfy certain side conditions, namely, certain linear differential equations (as in Bolza's problem in the calculus of variations).

Let $P(t) = P^*(t)$, $Q(t) = Q^*(t)$, $R(t)$ be continuous $d \times d$ matrices on J and consider the functional

$$(11.1) \quad I(\eta; a, b) = \int_a^b [(P\eta' + R\eta) \cdot \eta' + (R^*\eta' - Q\eta) \cdot \eta] dt.$$

In addition, consider a set of e first order linear differential equations

(11.2) $$M(t)\eta' + N(t)\eta = 0,$$

where $M(t)$, $N(t)$ are $e \times d$, $e < d$, matrices of complex-valued continuous functions on J. It will be assumed that the $(d + e) \times (d + e)$ Hermitian matrix

(11.3) $$P_0(t) = \begin{pmatrix} P(t) & M^*(t) \\ M(t) & 0 \end{pmatrix} \quad \text{is nonsingular.}$$

In particular, $\det P_0(t) \neq 0$ and the rank of $M(t)$ is e.

For the variational problem (11.1) subject to the side conditions (11.2), the Euler-Lagrange equations are

(11.4) $$(Px' + Rx + M^*z)' - (R^*x' - Qx + N^*z) = 0,$$

(11.5) $$Mx' + Nx = 0,$$

where z is an e-dimensional vector. [The derivation of (11.4) and significance of z need not concern us here.]

The matrix inverse to (11.3) is of the form

(11.6) $$P_0^{-1}(t) = \begin{pmatrix} E(t) & F^*(t) \\ F(t) & G(t) \end{pmatrix}, \quad \text{where} \quad E = E^*, \quad G = G^*,$$

E is a $d \times d$, G an $e \times e$, and F an $e \times d$ matrix. Introduce the variables

(11.7) $$y = P(t)x' + R(t)x + M^*z.$$

Then equations (11.5), (11.7) can be written as $P_0(x', z) = (y - Rx, -Nx)$ or $(x', z) = P_0^{-1}(y - Rx, -Nx)$; i.e., $x' = Ey - (ER + F^*N)x$ and $z = Fy - (FR + GN)x$. Hence (11.4), (11.5) become

(11.8) $$x' = A(t)x + B(t)y, \qquad y' = C(t)x - A^*(t)y,$$

where

(11.9) $$A = -(ER + F^*N), \qquad B = E,$$
$$C = -Q - R^*ER - R^*F^*N - N^*FR - N^*GN.$$

In particular, (11.8) implies (11.5).

The assumptions on P, Q, R will be $P = P^*$, $Q = Q^*$,

(11.10) $$P(t)\eta \cdot \eta > 0 \quad \text{whenever} \quad \eta \neq 0 \quad \text{and} \quad M(t)\eta = 0.$$

Correspondingly, $B = B^*$, $C = C^*$, and, by (11.3) and (11.6),

(11.11) $$B(t)M^*(t) \equiv 0; \qquad BP\eta = \eta \quad \text{if} \quad M\eta = 0;$$

$B(t)$ is non-negative definite and of rank $d - e$ (for $BP\eta \cdot P\eta = P\eta \cdot \eta$ if $M\eta = 0$).

There is a complication for the system (11.8) which did not arise in the last section. If $(x(t), y(t))$ is a solution of (11.8), it does not follow that $y(t)$ is determined by $x(t)$. This difficulty will be avoided by an extra assumption:

(11.12)   (11.8) is identically normal,

where (11.12) means that if $(x(t), y(t))$ is a solution of (11.8) such that $x(t) \equiv 0$ on some subinterval of J, then $x(t) \equiv 0$, $y(t) \equiv 0$ on J.

Notions of "disconjugate," "conjugate solutions of (10.9)," "self-conjugate solutions of (10.9)" can be defined as before.

Exercise 11.1. Verify the validity of analogues of Theorem 10.1 and Theorem 10.2. (As in the proof of Theorem 10.2, postpone one half of the proof of the statement concerning open J.)

Let the classes $A_1(a, b)$ [or $A_2(a, b)$] of vector functions $\eta(t)$ be defined as before with the additional condition: (iii) $\eta(t)$ satisfies the differential equations (11.2) except for a t-set of measure 0.

Exercise 11.2. Verify the validity of the analogue of Theorem 10.3. For the proof, note that if $(x(t), y(t))$ is a solution of (11.8), then $\eta = x(t)$ is a solution of (11.2). Hence if $(U(t), V(t))$ is a solution of (10.9), then $MU' + NU = 0$. Thus if $\eta(t)$ is a solution of (11.2) and $\eta = U\zeta$, then $M(U\zeta') = 0$, so that $PU\zeta' \cdot U\zeta' \equiv 0$ only if $\eta \equiv 0$.

Exercise 11.3. If $P(t)$ is of rank $d - e$ [so that $P(t)$ is non-negative definite], let $P^{1/2}(t)$ denote the unique, non-negative Hermitian square root of $P(t)$; thus $P^{1/2}(t)$ is continuous; cf. Exercise XIV 1.2. If $P(t)$ is of rank $\geq d - e$, let $E_0(t)$ be the orthogonal projection of the vector space onto the null manifold of $M(t)$ and let $P^{1/2}(t) = (E_0(t)P(t)E_0(t))^{1/2}$. Actually, $P(t)$ can be replaced by $E_0(t)P(t)E_0(t)$ in (11.1)–(11.12), since $E_0\eta = \eta$ if $M\eta = 0$. Using $P^{1/2}(t)$, state and prove the analogue of Corollary 10.1. [An additional condition for the validity of (10.33) will be that $\eta(t)$ satisfy (11.2).]

Exercise 11.4. Verify the validity of the analogues of Theorem 10.4 and 10.5 with an analogous definition of *principal solution.*

As an example and application, consider a formally self-adjoint differential equation of order $2d$ for a scalar function u,

(11.13) $L\{u\} = 0,$

where

(11.14) $L\{u\} = \sum_{k=0}^{d} (p_{2k}u^{(k)})^{(k)} + i \sum_{k=1}^{d} [(p_{2k-1}u^{(k-1)})^{(k)} + (p_{2k-1}u^{(k)})^{(k-1)}],$

where $p_0(t), \ldots, p_{2d}(t)$ are real-valued functions on an interval J, $i = (-1)^{1/2}$,

(11.15) $(-1)^d p_{2d}(t) > 0,$

and $p_{2k}(t), p_{2k-1}(t)$ have k continuous derivatives. For a given function $u(t)$ of class C^{2d} on J, define a vector $y = (y^1, \ldots, y^d)$ by

$$(11.16) \quad y^k = (-1)^k \sum_{j=k}^{d} [p_{2j}u^{(j)} + i(p_{2j-1}u^{(j-1)} + p_{2j+1}u^{(j+1)})]^{(j-k)}$$

for $k = 1, \ldots, d$, where $p_{2d+1} \equiv 0$. The operator $L\{u\} = 0$ is termed formally self-adjoint because of the validity of the Green relation

$$\int_a^b (L\{u\}\bar{v} - uL\{\bar{v}\}) \, dt = \sum_{k=1}^{d} [\bar{z}^k u^{(k-1)} - \bar{v}^{(k-1)} y^k]_a^b,$$

where $z = (z^1, \ldots, z^d)$ belongs to v as y belongs to u. In particular if u, u_0 are solutions of (11.14), then

$$\sum_{k=1}^{d} [\bar{u}_0^{(k-1)} y^k - \bar{y}_0^k u^{(k-1)}] = \text{const.},$$

where y_0 belongs to u_0. When this constant is 0, the solutions u, u_0 will be called *conjugate*. If $u = u_0$ and u, u are conjugate solutions, then u is called a *self-conjugate solution*. [When $p_1 \equiv p_3 \equiv \cdots \equiv p_{2d-1} \equiv 0$, so that (11.14) is a real equation, then all real-valued solutions are self-conjugate.]

Consider the functional

$$(11.17) \qquad\qquad I\{u; a, b\} = \int_a^b L\{u\}\bar{u} \, dt.$$

A formal integration by parts (ignoring the integrated terms which vanish if $u = u' = \cdots = u^{(d-1)} = 0$ at $t = a, b$) gives

$$(11.18) \quad I\{u; a, b\} = \int_a^b \left\{ \sum_{k=0}^{d} (-1)^k p_{2k} |u^{(k)}|^2 \right.$$
$$\left. - 2 \sum_{k=1}^{d} (-1)^k p_{2k-1} \, \text{Im} \, \bar{u}^{(k)} u^{(k-1)} \right\} dt.$$

Then $I(\eta; a, b) = I\{u; a, b\}$ is of the type (11.1) for $\eta = (u, u', \ldots, u^{(d-1)})$ and the conditions (11.2) become

$$(11.19) \qquad\qquad \eta^{j'} - \eta^{j+1} = 0 \quad \text{for} \quad j = 1, \ldots, d-1.$$

Here

$$P = \text{diag}\,[0, \ldots, 0, (-1)^d p_{2d}], \quad Q = \text{diag}\,[p_0, -p_1, \ldots, (-1)^{d-1} p_{2d-2}]$$
$$R = i \, \text{diag}\,[-p_1, p_3, \ldots, (-1)^d p_{2d-1}];$$

also M, N are the $(d-1) \times d$ matrices

$$M = \begin{pmatrix} 1 & 0 & . & . & . & 0 & 0 \\ 0 & 1 & . & . & . & 0 & 0 \\ . & . & . & & & . & . \\ . & . & . & & & . & . \\ . & . & . & & & . & . \\ 0 & 0 & . & . & . & 1 & 0 \end{pmatrix}, \qquad N = \begin{pmatrix} 0 & 1 & 0 & . & . & . & 0 \\ 0 & 0 & 1 & . & . & . & 0 \\ . & . & . & . & & & . \\ . & . & . & & . & & . \\ . & . & . & & & . & . \\ 0 & 0 & 0 & . & . & . & 1 \end{pmatrix}.$$

Correspondingly, A is the matrix with ones on the superdiagonal, diagonal elements $(0, \ldots, 0, -ip_{2d-1}/p_{2d})$, and other elements zero; $B = \mathrm{diag}\,[0, \ldots, 0, (-1)^d/p_{2d}]$; and C is the Hermitian matrix with superdiagonal elements $(ip_1, -ip_3, \ldots, (-1)^{d-1}ip_{2d-1})$, diagonal elements $(p_0, -p_2, \ldots, (-1)^{d-2}p_{2d-4}, (-1)^{d-1}p_{2d-2} + |p_{2d-1}|^2/p_{2d})$, and the elements not on the main diagonal and superdiagonal are zero.

When $(x(t), y(t))$ is a solution of (11.8), then $x(t) = (u, u', \ldots, u^{(d-1)})$, where u is a solution of (11.13) and the components of y are given by (11.16). Conversely, if u is a solution of (11.13), this choice of x, y gives a solution of (11.8). The condition (11.15) assures (11.3), (11.10), and (11.12).

Note that if $(U(t), V(t))$ is a solution of (10.9) and the kth column of $U(t)$ is $(u_k, u_k', \ldots, u_k^{(d-1)})$, where $u = u_k$ is a solution of (11.13), then $\det U(t)$ is the Wronskian of the solutions $u_1, \ldots, u_d$. The solution $(U(t), V(t))$ is self-conjugate, if and only if u_j, u_k are conjugate solutions of (11.13) for $j, k = 1, \ldots, d$.

Exercise 11.5. State the analogue of Theorem 10.3, specifying the classes $A_1(a, b)$, $A_2(a, b)$ in terms of scalar functions u.

Consider finally the analogue of Corollary 10.1; cf. Exercise 11.3. Since $P(t) = \mathrm{diag}\,[0, \ldots, 0, (-1)^d p_{2d}]$, the matrix $P^{1/2}(t)$ is $P^{1/2} = \mathrm{diag}\,(0, \ldots, 0, |p_{2d}|^{1/2})$. The vector $\eta(t)$ satisfies the analogue of (11.2) if and only if $\eta(t)$ is of the form $\eta = (v(t), v'(t), \ldots, v^{(d-1)}(t))$ for a scalar function $v(t)$. In this case, (10.29) is a vector of the form $(0, \ldots, 0, L_1\{v\})$, where $L_1\{v\} = \alpha_0(t)v^{(d)} + \cdots + \alpha_d(t)v$ is a differential operator of order d and $\alpha_0(t), \ldots, \alpha_d(t)$ are continuous complex-valued functions. In fact, it is clear from (10.29) that $\alpha_0(t) = |p_{2d}(t)|^{1/2} > 0$. It is also clear that $L_1\{v\} = 0$ if $v = u_k$ and $(u_k, u_k', \ldots, u_k^{(d-1)})$ is the kth column of $U(t)$. Since the Wronskian of $u_1, \ldots, u_d$ is $\det U(t) \neq 0$, it follows that $v = u_1, \ldots, u_d$ are d linearly independent solutions of $L_1\{v\} = 0$.

Consequently, if $W(w_1, \ldots, w_j)$ denotes the Wronskian of the j functions $w_1, \ldots, w_j$, then

$$(11.20) \qquad L_1\{v\} = |p_{2d}(t)|^{1/2}\, \frac{W(u_1, \ldots, u_d, v)}{W(u_1, \ldots, u_d)}.$$

In order to see this, note that the expression on the right is a differential operator of order d with leading coefficient $|p_{2d}(t)|^{1/2}$ and solutions $u = u_1, \ldots, u_d$. Thus if either side of (11.19) is written as a linear homogeneous system $\eta' = \Omega(t)\eta$ for $\eta = (v, v', \ldots, v^{d-1})$ in the obvious way, then the system has a fundamental solution $U(t)$ and so $\Omega = U'U^{-1}$. This proves the identity (11.20).

Exercise 11.6. If (11.13) has solutions $u_1(t), \ldots, u_d(t)$ on J, which are pairwise conjugate (and self-conjugate) and which have a nonvanishing Wronskian $W(u_1, \ldots, u_d) \neq 0$ on J, then $L\{u\} \equiv L_1^*\{L_1[u]\}$ for all functions u of class C^{2d} on J. Here, if $L_1\{u\} = \Sigma \, \alpha_k(t)u^{(k)}$, then $L_1^*\{u\} = \Sigma \, (-1)^k\{\bar{\alpha}_k(t)u\}^{(k)}$, where the sum Σ is over $0 \leqq k \leqq d$.

Notes

SECTIONS 1 AND 2. See notes on relevant portions of Chapter IV. For the substitution (2.34) in § 2 (xiii), see Liouville [1, II, pp. 22–23]. The substitution $r = u'/u$, which transforms (2.32) into a Riccati equation, was used in special cases by Euler (circa 1765) and Liouville (1841) for dth order linear differential equations. For the transformation (2.42) in § 2(xv), see Prüfer [1, p. 503].

SECTION 3. The results of this section are due to Sturm [1]; see Bôcher [1], [3]. The proofs in the text are suggested by Prüfer's work [1] and are given in detail by Kamke [3]. The proof for Sturm's separation theorem (Corollary 3.1) given in Exercise 3.1(a) for the case $p_1 = p_2$ goes back to arguments of Sturm; a similar proof for the general case is due to Picone [1].

SECTION 4. Theorem 4.1 goes back to the work of Sturm [1] and Liouville [1]. The proof of (i)–(iii) in the text follows that of Prüfer [1]. The proof of (vi) is based on results of Hilbert and E. Schmidt on integral equations with "Hilbert-Schmidt" kernels; cf. Riesz and Sz.-Nagy [1, pp. 239–242]. For useful and interesting results on the asymptotic behavior of the eigenvalues λ_n, see Borg [1] and references there to Weyl. For a complete characterization of spectra of singular boundary value problem in terms of zeros of solutions, see Hartman [6] and Wolfson [1]. For Exercises 4.3 and 4.4, see Prüfer [1].

SECTION 5. Theorem 5.1 is due to Hartman and Wintner [9] and generalizes Corollary 5.1, which is an interpretation of a result of Lyapunov [1]. The proof of Theorem 5.1 in the text is that of Nehari [1].* That the factor 4 in (5.7) cannot be increased was first proved by van Kampen and Wintner [1]. The proof in Hints for Exercise 5.1 is given in Hartman and Wintner [9] and is adapted from Borg [2]. For Exercise 5.2, see Hartman and Wintner [10]. For Exercise 5.3(a), see Hartman and Wintner [21]; for part (c), see Opial [3]; part (d) is a slight improvement of a result of de la Vallée Poussin [1] (cf. Sansone [1, I, p. 183]; see also Nehari [2]). Corollary 5.2 and Exercise 5.4 are due to Hartman and Wintner [5]; for generalizations, see Hartman [7]. Theorem 5.2 is a result of Hartman and Wintner; see Hartman [7, p. 642]. Corollary 5.3 is a result of Wiman [1]; for the generalization in Exercise 5.6(a), see Hartman and Wintner [2]. For Exercises 5.6(b) and 5.7, see Hartman [18]. For Theorem 5.3, see Milne [1]; cf. Hartman and Wintner [5]. Generalizations of Exercise 5.9(a)*and (b) to binary systems of first order were obtained by Petty [1] by different methods.

SECTION 6. The use of the term "disconjugate" here is suggested by Wintner [20]. Theorem 6.2 is a classical result in the calculus of variations (Jacobi, Weierstrass,

*See footnotes, page 403.

Erdmann); cf. Bolza [1, chap. 2 and 3] or Morse [2, chap. 1]. The proof in the text is based on Clebsch's [1] transformation of the second variation; cf. Bolza [1, p. 632]. Corollary 6.2 is a particular case of a result of Heinz [2]; cf. Exercise 11.6. The notion of a "principal solution" for a disconjugate equation was introduced by Leighton and Morse [1]; cf. Leighton [1] for the use of the term "principal." The proof in the text of Theorem 6.4 is adapted from Hartman and Wintner [18, Appendix]. Corollary 6.4 is a particular case of a theorem of A. Kneser [2] on second order (not necessarily linear equations); cf. Chapter XII, Part I in this book. The proof in the text is that of Hartman [3]; Kneser's proof is similar to the one suggested in Exercise 6.8. For Corollary 6.5, see Hartman and Wintner [6, p. 635]. For Corollary 6.6, see Hartman and Wintner [3]. (For an application of Theorem 6.4 and Corollary 6.5 to differential geometry, see E. Hopf [1].)

SECTION 7. Theorem 7.1 is due to A. Kneser [1]. The remark following Theorem 7.1 on the use of the functions in Exercises 1.2 is due to Hille [1] and to Hartman [4]. Theorem 7.2*was given by Wintner [20]. For Corollary 7.1 and Exercise 7.4, see Hartman [9] and [25], respectively. Exercise 7.2(b) is a result of Hartman and Wintner [20] and generalizes a result of Picard [4, p. 8]. Hartman [10] contains Lemma 7.1, Theorems 7.3–7.4, and Exercises 7.5–7.8. Theorem 7.4 is a generalization of a result of Wintner [20]. Exercise 7.8 is related to results of Wintner [9], [15], [20]; Hille [1]; and Leighton [2]. Exercise 7.9*may be new (it was first given by Hille [1] under the additional assumption that $q_1(t) \geq 0$, $q_2(t) \geq 0$ and then by Wintner [24] under the conditions $0 \leq Q_1(t) \leq Q_2(t)$; the proof suggested in the Hints is much simpler than the proofs of these authors). For some results related to this section, see Wintner [14], Zlámal [1], Olech, Opial and Ważewski [1], and Opial [4]. For a study of zeros of solutions of certain fourth order equations, see Leighton and Nehari [1].

SECTION 8. Theorem 8.1 is a variant of a result of Wintner [10]. Corollary 8.1 is a result of Bôcher [2]. For Exercise 8.2, see Prodi [1]. For Exercise 8.3, see Wintner [12]. Exercise 8.4(b) is due to Wintner [10] and sharpens a result of Cesari [1]. For Exercise 8.5, see Hartman and Wintner [17]; for related results, see Atkinson [1] and references there. For Exercises 8.6 and 8.8, see Hartman [25]. For Exercise 8.7* see Ganelius [1]; this result was first used in connection with linear, second order differential equations by Brinck [1].

SECTION 9. The general procedure in this section is suggested by unpublished notes of Hartman and Wintner. Lemma 9.1 is related to results of Wintner [13], [17] on a second order equation. Analogues of Lemmas 9.1 and 9.2 for (9.1) when $p(t) > 0$ and $q(t) \leq 0$ go back to Weyl [1]. Theorem 9.1 is an unpublished result of Hartman and Wintner. The first part of Corollary 9.1 is a result of Bôcher [2] under the condition (9.30) and of Wintner [17] under condition (9.31). Similarly, the first part of Corollary 9.20 is due to Bôcher [2] under condition (9.35) and to Hartman and Wintner [12] under condition (9.36). For Exercise 9.6, see Wintner [12]. For Corollary 9.3 and Exercises 9.7 and 9.8, see Hartman and Wintner [12], where analogues and generalizations are given. For results related to this section, see Opial [2], [5], [6]; Ráb [1]; Zlámal [1].

SECTION 10. The use of the term "conjugate solutions" is the same as that suggested by von Escherich [1] for the case of real systems (10.1); the analogous Lagrange relation [see displays following (10.8)] on which this definition is based is due to Clebsch [1]. In relation to (10.18)–(10.19), see Kaufman and R. L. Sternberg [1], Barrett [1], and Reid [3]. The remarks above concerning Theorem 6.2 are applicable to Theorem 10.3 in the case of real matrices P, Q, R. For the complex case, see Reid [2], [3], [5], [6]. The proof in the text is based on Clebsch's transformation of the second variation. Exercise 10.1 is a special case of Jacobi's classical theorem on

*See footnotes, page 403.

conjugate points. Exercise 10.2 when $P_1 = P_1{}^*$ and $Q_1 = Q_1{}^*$ is the simplest case of a result of Morse [1]; see Hartman and Wintner [22], where P_j, Q_j are assumed real. Corollary 10.1 is suggested by Heinz [2]; cf. Exercise 11.6. The proof in the text is much simpler than that of Heinz.

The concept of a "principal solution" for systems (10.1) with $R(t) \equiv 0$, was introduced by Hartman [12] who proved an analogue of Theorem 10.5 dealing however, with only self-conjugate solutions $(U(t), V(t))$. The definition of principal solution in the text, and Theorems 10.4, 10.5 are due to Reid [3]. Although a principal solution in Reid's sense turns out to be self-conjugate and hence identical with a principal solution in Hartman's sense, the handling of nonprincipal solutions is more convenient by Reid's definition. The proof in the text for the existence of principal solutions [Theorem 10.5(i)] follows Hartman [12]. Reid's existence proof is outlined in Exercise 10.4. The proofs of Theorem 10.4 and the other parts of Theorem 10.5 are based on Reid [3]. Lemma 10.1 and its proof are due to F. Riesz; cf. Riesz and Sz.-Nagy [1]. For Exercise 10.3, see Hartman [12].

SECTION 11. The results of this section given in Exercises 11.1–11.5 are due to Reid [3]; see also Sandor [1] and Reid [6] for related results and generalizations. For Exercise 11.6, see Heinz [2] (and, for $d = 1$, cf. Brinck [1]).

For a correct proof and generalization of Nehari s result [2] on n-th order equations; see Hartman [S2] ; for related results. see references there to A.Yu. Levin and Hukuhara. For further discussion and bibliography, see Coppel [S2].

The first part of Exercise 5.9(a) is due to Wintner [S1].

Theorem 7.2 goes back to Bocher [3] and de la Vallee Poussin [1].

The case $c_k = 1$ of Exercise 7.9 was given by Taam [S1].

For Exercise 8.7, see also Wüst [S1].

For an extension of some of the results of Sections 4-7 (e.g., Sturm comparison theorems, principal solutions, disconjugacy criteria, etc.) to n-th order equations, see Hartman [S1] and Levin [S1]. Coppel s book [S2] also deals with these subjects among others.

Chapter XII

Use of Implicit Function and
Fixed Point Theorems

Many different problems in the theory of differential equations are solved by the use of implicit function theory—either of the classical type or of a more general type involving fixed point theorems and/or functional analysis. This will be illustrated in this chapter. Part I deals with the existence of periodic solutions of linear and nonlinear differential equations. Part II deals with solutions of certain second order boundary value problems. In Part III, a general abstract theory is formulated. Use of this general theory is illustrated by an application to a problem of asymptotic integration.

Although Parts I and II are applications of the general theory of Part III, there are several reasons for giving them separate treatments. The first reason is the importance and comparative simplicity of the situations involved. The second reason is that Parts I and II serve as motivation for the somewhat abstract theory of Part III. The third and most important reason is the fact that, as usual, a general theory in the theory of differential equations only provides a guide for the procedure to be followed. Its use in a particular situation generally involves important problems of obtaining appropriate estimates in order to establish the applicability of the general theory.

Two general theorems will be used. The first is a very simple fact:

Theorem 0.1. *Let $\mathfrak{D}$ be a Banach space of elements $x, y, \ldots$ with norms $|x|, |y|, \ldots$. Let T_0 be a map of the ball $|x| \leqq \rho$ in $\mathfrak{D}$ into $\mathfrak{D}$ satisfying $|T_0[x] - T_0[y]| \leqq \theta |x - y|$ for some θ, $0 < \theta < 1$. Let $m = |T_0[0]|$ and $m \leqq \rho(1 - \theta)$. Then there exists a unique fixed point x_0 of T_0, i.e., a unique point x_0 satisfying $T_0[x_0] = x_0$. In fact, x_0 can be obtained as the limit of successive approximations $x_1 = T_0[0], x_2 = T_0[x_1], x_3 = T_0[x_2], \ldots$.*

Remark. If T_0 maps the ball $|x| \leqq \rho$ into itself, then the condition $m \leqq \rho(1 - \theta)$ can be omitted.

Exercise 0.1. Verify this theorem and the Remark.

A much more sophisticated fixed point theorem is the following:

Theorem 0.2 (Tychonov).   *Let* $\mathfrak{D}$ *be a linear, locally convex, topological space. Let S be a compact, convex subset of* $\mathfrak{D}$ *and* T_0 *a continuous map of S into itself. Then* T_0 *has a fixed point* $x_0 \in S$, *i.e.,* $T_0[x_0] = x_0$.

The following corollary of this will be used subsequently.

Corollary 0.1.   *Let* $\mathfrak{D}$ *be a linear, locally convex, topological, complete Hausdorff space (e.g., let* $\mathfrak{D}$ *be a Banach or a Fréchet space). Let S be a closed, convex subset of* $\mathfrak{D}$ *and* T_0 *a continuous map of S into itself such that the image* $T_0 S$ *of S has a compact closure. Then* T_0 *has a fixed point* $x_0 \in S$.

Theorem 0.2 was first proved by Schauder under the assumption that $\mathfrak{D}$ is a Banach space and this case of the theorem is usually called "Schauder's fixed point theorem." For a proof of Theorem 0.2, see Tychonov [1].

Parts I and II will use the cases of Corollary 0.1 when $\mathfrak{D}$ is the Banach space C^0, C^1. Part III will use the case when $\mathfrak{D}$ is a simple Fréchet space, namely, the space of continuous functions on $J : 0 \leqq t < \omega (\leqq \infty)$ with the topology of uniform convergence on closed intervals in J.

Corollary 0.1 is obtained from Theorem 0.2 in the following way: Let $\mathfrak{D}$, S, T_0 be as in Corollary 0.1. Let S_1 be the closure of $T_0 S$, so that S_1 is compact. Also $S_1 \subset S$ since S is closed. Under the assumptions on $\mathfrak{D}$, the convex closure of S_1 (i.e., the smallest closed convex set containing S_1) is compact since S_1 is. (This is an immediate consequence of Arzela's theorem in the applications below; cf., e.g., the Remark following the proof of Theorem 2.2.) Let S^0 denote this convex closure of S_1. Since S is convex $S^0 \subset S$. Thus T_0 is a continuous map of the convex compact S^0 into itself (in fact, $T_0 S^0 \subset T_0 S \subset S_1 \subset S^0$) and the corollary follows from Theorem 0.2.

Part III will depend on the "open mapping theorem" in functional analysis. This theorem will be used in the following form:

Theorem 0.3 (Open Mapping Theorem).   *Let* X_1, X_2 *be Banach spaces and* T_0 *a linear operator from* X_1 *onto* X_2 *with a domain* $\mathscr{D}(T_0)$, *which is necessarily a linear manifold in* X_1, *and range* $\mathscr{R}(T_0) = X_2$. *Let* T_0 *be a closed operator, i.e., let the graph of* T_0, $\mathscr{G}(T_0) = \{(x_1, T_0 x_1) : x_1 \in \mathscr{D}(T_0)\}$ *be a closed set in the Banach space* $X_1 \times X_2 = \{(x_1, x_2) : x_1 \in X_1, x_2 \in X_2\}$ *with norm* $|(x_1, x_2)| = \max (|x_1|, |x_2|)$. *Then there exists a constant* K *with the property that, for every* $x_2 \in X_2$, *there is at least one* $x_1 \in \mathscr{D}(T_0)$ *such that* $T_0 x_1 = x_2$ *and* $|x_1| \leqq K |x_2|$. *[In particular, when* T_0 *is one-to-one, so that* x_1 *is unique, then* $|x_1| \leqq K |T_0 x_1|$ *holds for all* $x_1 \in \mathscr{D}(T_0)$.]*

For a proof of the open mapping theorem in the form that "if P is a continuous, linear map from a Banach space X to another Banach space X_2 with domain $\mathscr{D}(P) = X$ and range $\mathscr{R}(P) = X_2$, then P maps open sets into open sets," see Banach [1, pp. 38–40]. Theorem 0.3 results by applying this to the projection map $P : \mathscr{G}(T_0) \to X_2$, where $P(x_1, T_0 x_1) = T_0 x_1$ and noting that a sphere about the origin in $\mathscr{G}(T_0)$ has a P-image which contains a sphere about the origin in X_2.

As a motivation for the procedures to be followed consider the problem of finding a solution of the differential equation

$$(0.1) \qquad\qquad y' = f^0(t, y)$$

in a certain set S of functions $y(t)$. Write this differential equation as

$$y' = A(t)y + f(t, y), \qquad \text{where} \quad f(t, y) = f^0(t, y) - A(t)y,$$

for some choice of $A(t)$. Suppose that for every $x(t) \in S$, the equation

$$(0.2) \qquad\qquad y' = A(t)y + f(t, x(t))$$

has a solution $y(t) \in S$. Define an operator $T_0 : S \to S$ by putting $y(t) = T_0[x(t)]$, where $y(t) \in S$ is a suitably selected solution of (0.2). It is clear that a fixed point $y_0(t)$ of T_0 [i.e., $T_0[y_0(t)] = y_0(t)$] is a solution of (0.1) in S.

For the applicability of the theorems just stated, it will be assumed that S is a subset of a suitable topological vector space $\mathfrak{D}$. It will generally be convenient to introduce another space $\mathfrak{B}$ and two operators L and T_1. The operator L is the linear differential operator $L[y] = y' - A(t)y$, so that $g(t) = L[y(t)]$ if

$$(0.3) \qquad\qquad y' = A(t)y + g(t).$$

It will also be assumed that if $x(t) \in S$, then $g(t) = f(t, x(t))$ is in $\mathfrak{B}$ and $T_1 : S \to \mathfrak{B}$ is defined by $g(t) = T_1[x(t)]$. Investigations of T_0 are then reduced to examinations of the linear differential operator L and of the nonlinear operator T_1.

The applicability of Theorem 0.1 can arise in the following type of situation: Suppose that $\mathfrak{B}$, $\mathfrak{D}$, are Banach spaces and that $|g|_{\mathfrak{B}}$, $|y|_{\mathfrak{D}}$ denote the norms of elements $g \in \mathfrak{B}$, $y \in \mathfrak{D}$, respectively. Assume that for every $g(t) \in \mathfrak{B}$, the equation (0.3) (i.e., $L[y] = g$) has a unique solution $y(t) \in S \subset \mathfrak{D}$, that $y(t)$ depends linearly on $g(t)$, and that there exists a constant K such that $|y|_{\mathfrak{D}} \leqq K|g|_{\mathfrak{B}}$. Suppose that, for the map $T_1 : S \to \mathfrak{B}$ there is a constant θ such that $|T_1[x_1] - T_1[x_2]|_{\mathfrak{B}} \leqq \theta |x_1 - x_2|_{\mathfrak{D}}$ for x_1, $x_2 \in S$. Then T_0 satisfies $|T_0[x_1(t)] - T_0[x_2(t)]|_{\mathfrak{D}} \leqq \theta K |x_1 - x_2|_{\mathfrak{D}}$. According to Theorem 0.1, the sequence of successive approximations

$$x_1, \; x_2 = T_0[x_1], \; x_3 = T_0[x_2], \ldots$$

will converge to a fixed point of T_0 (under suitable conditions on S, x_1, and θK).

In some situations, the equation $L[y] = g$ may have solutions y satisfying $|y|_\mathfrak{D} \leqq K |g|_\mathfrak{B}$ although y is not unique; cf., e.g., Theorem 0.3. In this case, y need not depend linearly on g but it might be possible to form convergent successive approximations in the following way: For a given x_1, let $x_2 = y$ be a solution of $L[y] = T_1[x_1(t)]$. If $x_1, x_2, \ldots, x_{n-1}$ have been defined for $n > 2$, determine an x_n from the equation $L[x_n - x_{n-1}] = T_1[x_{n-1}] - T_1[x_{n-2}]$ and the inequality $|x_n - x_{n-1}|_\mathfrak{D} \leqq K |T_1[x_{n-1}] - T_1[x_{n-2}]|_\mathfrak{B}$. This situation will not arise below.

When the inequality $|T_1[x_1(t)] - T_1[x_2(t)]|_\mathfrak{B} \leqq \theta |x_1 - x_2|_\mathfrak{D}$ is not available, Theorem 0.2 may still be applicable to assure the existence of a fixed point of T_0.

PART I. PERIODIC SOLUTIONS

1. Linear Equations

In this section, unless otherwise specified, the components of the d-dimensional vectors y, z are real- or complex-valued. Let $p > 0$ be fixed. Consider an inhomogeneous system of linear equations

$$(1.1) \qquad y' = A(t)y + g(t)$$

and the corresponding homogeneous system

$$(1.2) \qquad y' = A(t)y,$$

where $A(t)$ is a continuous $d \times d$ matrix and $g(t)$ a continuous vector-valued function for $0 \leqq t \leqq p$. In addition, consider a set of boundary conditions

$$(1.3) \qquad My(0) - Ny(p) = 0,$$

where M, N are constant $d \times d$ matrices. For example, if $M = N = I$ and $A(t), g(t)$ are periodic of period p, then a solution $y(t)$ of (1.1) or (1.2) satisfying (1.3) is of period p.

Lemma 1.1. *Let $A(t)$ be continuous for $0 \leqq t \leqq p$ and M, N constant $d \times d$ matrices. Let $Y(t)$ be a fundamental matrix for (1.2). Then a necessary and sufficient condition for (1.2) to have a nontrivial ($\not\equiv 0$) solution satisfying (1.3) is that $M Y(0) - N Y(p)$ be singular. In fact, the number k, $0 \leqq k \leqq d$, of linearly independent solutions of (1.2), (1.3) is the number of linearly independent vectors c satisfying*

$$(1.4) \qquad [M Y(0) - N Y(p)]c = 0;$$

i.e., $d - k = rank [M Y(0) - N Y(p)]$.
This is clear since the general solution of (1.2) is $y = Y(t)c$.

Exercise 1.1.   Let $A(t)$ be periodic of period p and

(1.5) $Y(t) = Z(t)e^{Rt}$,   where $Z(t + p) = Z(t)$

and R is a constant matrix; cf. the Floquet theory in § IV 6. Then (1.2) has a nontrivial ($\not\equiv 0$) solution of period p if and only if $\lambda = 1$ is a characteristic root of (1.2); i.e., $e^{Rp} - I$ is singular. In fact, the number of linearly independent solutions of period p is the number of linearly independent solutions c of

(1.6) $[Y(0) - Y(p)]c = 0$,   i.e., $(e^{Rp} - I)c = 0$.

For algebraic linear equations, the inhomogeneous system $Cy = g$ has a solution y for every g if and only if the only solution of $Cy = 0$ is $y = 0$. The analogous situation is valid here.

Theorem 1.1.   *Let $A(t)$ be continuous for $0 \leqq t \leqq p$; M, N constant $d \times d$ matrices such that the $d \times 2d$ matrix (M, N) is of rank d. Then (1.1) has a solution $y(t)$ satisfying (1.3) for every continuous $g(t)$ if and only if (1.2), (1.3) has no nontrivial ($\not\equiv 0$) solution; in which case $y(t)$ is unique and there exists a constant K, independent of $g(t)$, such that*

(1.7) $\|y(t)\| \leqq K \displaystyle\int_0^p \|g(s)\|\, ds$ for $0 \leqq t \leqq p$.

Proof.   The general solution of (1.1) is given by

(1.8) $y(t) = Y(t)\left\{ c + \displaystyle\int_0^t Y^{-1}(s)g(s)\, ds \right\}$;

Corollary IV 2.1.   This solution satisfies (1.3) if and only if

(1.9) $[MY(0) - NY(p)]c = NY(p) \displaystyle\int_0^p Y^{-1}(s)g(s)\, ds$.

Assume that (1.2), (1.3) has no nontrivial solution. Then, by Lemma 1.1, the matrix $V = MY(0) - NY(p)$ is nonsingular, thus (1.9) has a unique solution. Substituting this value of c in (1.8) gives the unique solution of (1.1), (1.3):

(1.10) $y(t) = Y(t)\left\{ V^{-1}N \displaystyle\int_0^p Y^{-1}(s)g(s)\, ds + \displaystyle\int_0^t Y^{-1}(s)g(s)\, ds \right\}$.

It is clear that there exists a constant K satisfying (1.7) for $0 \leqq t \leqq p$.

This proves one-half of Theorem 1.1 (and this part did not use the assumption that rank $(M, N) = d$). The converse follows from Theorem 1.2.

Exercise 1.2.   What is the Green's function $G(t, s)$ in the last part of Theorem 1.1, i.e., what is the function $G(t, s)$, $0 \leq s, t \leq p$, such that

$$y(t) = \int_0^p G(t, s)g(s)\, ds$$

is the unique solution (1.10) of (1.1), (1.3)?

Consider the equations adjoint to (1.1), (1.2)

(1.11) $z' + A^*(t)z + h(t) = 0,$

(1.12) $z' + A^*(t)z = 0,$

where A^* is the complex conjugate transpose of A; cf. § IV 7. Consider also a set of boundary conditions

(1.13) $Pz(0) - Qz(p) = 0,$

where P, Q are constant $d \times d$ matrices. If $y(t)$ is a solution of (1.1) and $z(t)$ a solution (1.11), the Green formula (IV 7.3) is

(1.14) $\int_0^p [g(s) \cdot z(s) - y(s) \cdot h(s)]\, ds = [y(t) \cdot z(t)]_0^p.$

When do the boundary conditions (1.3) and (1.13) imply that

(1.15) $y(p) \cdot z(p) - y(0) \cdot z(0) = 0,$

i.e., that the right side of (1.14) is 0? Note that if M, Q are nonsingular, then this is the case if and only if $0 = y(p) \cdot Q^{-1}Pz(0) - M^{-1}Ny(p) \cdot z(0) = (P^*Q^{*-1} - M^{-1}N)y(p) \cdot z(0) = [M^{-1}(MP^* - NQ^*)Q^{*-1}]y(p) \cdot z(0)$. In this case, necessary and sufficient for (1.3), (1.13) to imply (1.15) is that

(1.16) $MP^* - NQ^* = 0.$

Lemma 1.2.   *Let M, N be constant $d \times d$ matrices such that* rank$(M, N) = d$. *Then there exist $d \times d$ matrices P, Q satisfying* rank $(P, Q) = d$, (1.16), *and having the property that the relations* (1.3), (1.13) *imply* (1.15). *The pairs of vectors $z(0)$, $z(p)$ satisfying* (1.13) *are independent of the choice of P, Q.*

Proof.   Since rank $(M, N) = d$, there exist $d \times d$ matrices M_1, N_1 such that the $2d \times 2d$ matrix

(1.17) $W = \begin{pmatrix} M & -N \\ M_1 & N_1 \end{pmatrix}$

is nonsingular. Write the inverse of W as

$$W^{-1} = \begin{pmatrix} P_1^* & P^* \\ Q_1^* & Q^* \end{pmatrix} \quad \text{or} \quad W^{*-1} = \begin{pmatrix} P_1 & Q_1 \\ P & Q \end{pmatrix},$$

so that (1.16) holds and rank $(P, Q) = d$.

Let y_1, y_2, z_1, z_2 be d-dimensional vectors and $\eta = (y_1, y_2)$, $\zeta = (z_1, z_2)$ be corresponding $2d$-dimensional vectors. Then

$$(1.18) \qquad \eta \cdot \zeta = W^{-1}W\eta \cdot \zeta = W\eta \cdot W^{*-1}\zeta;$$

thus

$$(1.19) \quad My_1 - Ny_2 = 0, \qquad Pz_1 + Qz_2 = 0 \quad \text{imply that} \quad \eta \cdot \zeta = 0.$$

The choices $y_1 = y(0)$, $y_2 = y(p)$, $z_1 = z(0)$, $z_2 = -z(p)$ show that (1.3), (1.13) imply (1.16). This completes the existence proof.

The formulation (1.19) of the implication (1.3), (1.13) $\Rightarrow$ (1.16) makes the last part of the lemma clear. For if $\eta = (y_1, y_2) \neq 0$ satisfies $My_1 - Ny_2 = 0$, then $M_1y_1 + N_1y_2 \neq 0$. In fact, since rank $(P, Q) = d$, the set of vectors $\zeta = (z(0), -z(p))$ satisfying $Pz(0) - Qz(p) = 0$ is the set of vectors satisfying $\eta \cdot \zeta = 0$ for all $\eta = (y_1, y_2)$ such that $My_1 - Ny_2 = 0$. Since this set of vectors $\zeta = (z(0), -z(p))$ is determined by M, N, the proof of the theorem is complete.

Boundary conditions (1.13) satisfying the conditions of Lemma 1.2 will be called the *adjoint boundary conditions of* (1.3). Correspondingly, the problems (1.2)–(1.3) and (1.12)–(1.13) will be called "*adjoint problems.*" (Note that the adjoint of the "periodic boundary conditions" $y(p) = y(0)$, i.e., $M = N = I$, are equivalent to the "periodic conditions" $z(p) = z(0)$, i.e., $P = Q = I$.)

There is an analogue of the algebraic fact that if C is a $d \times d$ matrix, then the number of linearly independent solutions of $Cy = 0$ and of the "adjoint" equation $C^*z = 0$ is the same:

Lemma 1.3. *Let* $A(t)$ *be continuous for* $0 \leq t \leq p$; M, N *constant* $d \times d$ *matrices such that rank* $(M, N) = d$; *and* (1.13) *boundary conditions adjoint to* (1.3). *Then* (1.2)–(1.3) *and* (1.12)–(1.13) *have the same number of linearly independent solutions.*

Proof. Since the relationship between (1.2)–(1.3) and (1.12)–(1.13) is symmetric, it suffices to show that if (1.12)–(1.13) has k linearly independent solutions, where $0 \leq k \leq d$, then (1.2)–(1.3) has at least k linearly independent solutions.

Let $Y(t)$ be a fundamental matrix of (1.2), then $Y^{*-1}(t)$ is a fundamental solution of (1.12) by Lemma IV 7.1. In terms of (1.17), define a constant $2d \times 2d$ matrix

$$(1.20) \qquad U = W \operatorname{diag}[Y(0), Y(p)] = \begin{pmatrix} M\,Y(0) & -N\,Y(p) \\ M_1\,Y(0) & N_1\,Y(p) \end{pmatrix};$$

so that U is nonsingular and

$$U^{*-1} = W^{*-1} \operatorname{diag}[Y^{*-1}(0), Y^{*-1}(p)] = \begin{pmatrix} P_1\,Y^{*-1}(0) & Q_1\,Y^{*-1}(p) \\ P\,Y^{*-1}(0) & Q\,Y^{*-1}(p) \end{pmatrix}.$$

Thus, if c_0 is a constant d-dimensional vector such that $z(t) = Y^{*-1}(t)c_0$ is a solution of (1.12)–(1.13), then $U^{*-1}(c_0, -c_0) = (b, 0)$. Here b is a d-dimensional vector, and if c_0 varies over a set of k linearly independent vectors, then b varies over a set of k linearly independent vectors, since U^{*-1} is nonsingular. From (1.20), it is easy to see that the equation $(c_0, -c_0) = U^*(b, 0)$ gives

$$(1.21) \qquad c_0 = Y^*(0)M^*b = Y^*(p)N^*b,$$

so that

$$(1.22) \qquad [Y^*(0)M^* - Y^*(p)N^*]b = 0.$$

Hence the matrix $Y^*(0)M^* - Y^*(p)N^*$ annihilates k linearly independent vectors b; therefore, the same is true of its complex conjugate transpose $MY(0) - NY(p)$. In view of Lemma 1.1, this proves Lemma 1.3.

Remark. For the purpose of the next proof, note that the lemma just proved implies that (1.22) holds if and only if the vector c_0 in (1.21) is such that the solution $z = Y^{*-1}(t)c_0$ of (1.12) satisfies (1.13).

Another algebraic fact is that if C is a singular matrix, then $Cy = g$ has a solution y if and only if g is orthogonal (i.e., $g \cdot z = 0$) to all solutions z of the homogeneous "adjoint" system $C^*z = 0$. Again an analogous situation is valid here:

Theorem 1.2. *Let $A(t)$ be continuous for $0 \leq t \leq p$, M and N constant $d \times d$ matrices such that* rank $(M, N) = d$, *and let* (1.2)–(1.3) *and* (1.12)–(1.13) *be adjoint problems. Suppose that* (1.2)–(1.3) *has exactly k linearly independent solutions $y_1(t), \ldots, y_k(t)$ and let $z_1(t), \ldots, z_k(t)$ be linearly independent solutions of* (1.12)–(1.13). *Let $g(t)$ be continuous for $0 \leq t \leq p$. Then* (1.1) *has a solution $y_0(t)$ satisfying* (1.3) *if and only if*

$$(1.23) \qquad \int_0^p g(s) \cdot z_j(s)\, ds = 0 \qquad \text{for} \quad j = 1, \ldots, k.$$

In this case, the solutions of (1.1), (1.3) *are given by $y_0(t) + \alpha_1 y_1(t) + \cdots + \alpha_k y_k(t)$, where $\alpha_1, \ldots, \alpha_k$ are arbitrary constants.*

Proof. Note that, by the proof of Theorem 1.1, the problem (1.1), (1.3) has a solution if and only if (1.9) has a solution c. This is the case if and only if

$$\left(NY(p) \int_0^p Y^{-1}(s)g(s)\, ds \right) \cdot b = 0$$

for all solutions b of (1.22). In view of (1.21), this is equivalent to the condition that

$$0 = \int_0^p [g(s) \cdot Y^{*-1}(s)Y^*(p)N^*b]\, ds = \int_0^p g(s) \cdot z(s)\, ds$$

for all solutions $z = Y^{*-1}(s)c_0$ of (1.12)–(1.13), i.e., that (1.23) holds. This proves the theorem.

The next theorem is a rather particular result for the case that $A(t)$, $g(t)$ are of period p.

Theorem 1.3. *Let $A(t)$ be continuous and of period p. Then, for a fixed continuous $g(t)$ of period p, (1.1) has a solution of period p if and only if (1.1) has at least one bounded solution for $t \geqq 0$.*

Proof. The necessity of the existence of a bounded solution is clear. In order to prove the converse, assume that (1.1) has a solution $y(t)$ bounded for $t \geqq 0$. Let $Y(t)$ be the fundamental matrix of (1.2) satisfying $Y(0) = I$. Then (1.1) has a solution of period p if and only if the equation $c = Y(p)c + b$, where

$$b = Y(p) \int_0^p Y^{-1}(s)g(s)\,ds,$$

has a solution c; cf. (1.9) in the proof of Theorem 1.1.

If $c = y(0)$ in (1.8), then $y(p) = Y(p)y(0) + b$ holds for every solution $y(t)$ of (1.1). Since $y(t + p)$ is also a solution, $y(2p) = Y(p)y(p) + b = Y^2(p)y(0) + Y(p)b + b$, or more generally,

$$y(np) = Y^n(p)y(0) + \left(\sum_{k=0}^{n-1} Y^k(p) \right) b.$$

Suppose, if possible, that $[I - Y(p)]c = b$ has no solution. Then $[Y(p) - I]^*$ is singular and there exists a vector c_0 such that $[Y(p) - I]^*c_0 = 0$ and $b \cdot c_0 \neq 0$. Thus $c_0 = Y^*(p)c_0$ and $c_0 = (Y^k(p))^*c_0$ for $k = 0, 1, \ldots$. Multiply the equation in the last formula line scalarly by c_0 to obtain

$$y(np) \cdot c_0 = y(0) \cdot c_0 + n(b \cdot c_0),$$

since $Y^k(p)y(0) \cdot c_0 = y(0) \cdot (Y^k(p))^*c_0$. As $b \cdot c_0 \neq 0$ and the sequence $y(p)$, $y(2p)$, ... is bounded, a contradiction results. This proves the theorem.

2. Nonlinear Problems

This section deals with the existence of periodic solutions for nonlinear systems. With very minor changes, the methods and results are applicable to the situation when the requirement of "periodicity" is replaced by boundary conditions of the type (1.3). The results depend on those of the last section for linear equations and, in particular, on the "a priori bound" for certain solutions of (1.1) given by (1.7). The first two theorems concern a nonlinear system of the form

$$(2.1) \qquad\qquad y' = A(t)y + f(t, y)$$

in which y is a vector with real- or complex-valued components.

Theorem 2.1. *Let $A(t)$ be continuous and periodic of period p and such that (1.2) has no nontrivial solution of period p. Let K be as in (1.7) in Theorem 1.1, where $M = N = I$. Let $f(t, y)$ be continuous for all (t, y), of period p in t for fixed y, and satisfy a Lipschitz condition of the form*

$$(2.2) \qquad \|f(t, y_1) - f(t, y_2)\| \leqq \theta \|y_1 - y_2\|$$

for all t, y_1, y_2 with a Lipschitz constant θ so small that $K\theta p < 1$. Then (2.1) has a unique solution of period p.

Actually, it is not necessary that $f(t, y)$ be defined for all y. If $m = \max \|f(t, 0)\|$, it is sufficient to require that $f(t, y)$ be defined for $\|y\| \leqq r$, where

$$(2.3) \qquad \frac{Kpm}{1 - K\theta p} \leqq r.$$

Proof. Introduce the Banach space $\mathfrak{D}$ of continuous periodic functions $g(t)$ of period p with the norm $|g| = \max \|g(t)\|$. Thus convergence of $g_1(t), g_2(t), \ldots$ in $\mathfrak{D}$ is equivalent to the usual uniform convergence over $0 \leqq t \leqq p$.

Let $g(t)$ be a continuous function of period p satisfying $\|g(t)\| \leqq r$. Thus by Theorem 1.1 the equation

$$(2.4) \qquad y' - A(t)y = f(t, g(t))$$

has a unique solution $y(t)$ of period p. Define an operator T_0 on the set of all such $g(t)$ by putting $y(t) = T_0[g]$. Note that (1.7), (2.4) and (2.2) show that if $z(t) = T_0[h]$, then

$$(2.5) \quad |y - z| \leqq Kp\theta \, |g - h|; \qquad \text{i.e.,} \quad |T_0[g] - T_0[h]| \leqq Kp\theta \, |g - h|,$$

where $|y| = \max \|y(t)\|$ for $0 \leqq t \leqq p$. In addition, if $m = \max \|f(t, 0)\|$, then $|T_0[0]| \leqq Kpm$.

Thus Theorem 2.1 follows from Theorem 0.1, for $y_0(t)$ is a fixed point of T_0, $T_0[y_0] = y_0$, if and only if $y_0(t)$ is a solution of (2.1) of period p; cf. (2.4) where $y = T_0[g]$.

In Theorem 2.1, we can omit assumption (2.2) when $\|f(t, y)\|$ is "small," at the cost of losing "uniqueness."

Theorem 2.2. *Let $A(t)$, K be as in Theorem 2.1. Let $f(t, y)$ be continuous for all t and $\|y\| \leqq r$, of period p in t for fixed y, and satisfy*

$$(2.6) \qquad Kp \, \|f(t, y)\| \leqq r \qquad \text{for} \quad 0 \leqq t \leqq p, \quad \|y\| \leqq r.$$

Then (2.1) has at least one periodic solution of period p.

Proof. As in the last proof, define $y(t) = T_0[g]$ as the unique solution of (2.4) of period p, where $g(t)$ is of period p and $|g| \leqq r$. In order to prove the theorem, it suffices to show that T_0 has a fixed point y_0, $T_0[y_0] = y_0$. This will be proved by an appeal to Corollary 0.1 of Tychonov's theorem.

It follows from (1.7) and (2.6) that $y = T_0[g]$ satisfies $|y| \leqq r$. In other words, if $\mathfrak{D}$ is the same Banach space as in the last proof, then T_0 maps the sphere $|g| \leqq r$ of $\mathfrak{D}$ into itself. Also, (1.7) gives

$$|T_0[g] - T_0[h]| \leqq K \int_0^p \|f(t, g(t)) - f(t, h(t))\| \, dt.$$

Since f is continuous, it is clear that if $|g - h| = \max \|g(t) - h(t)\| \to 0$, then $T_0[g] - T_0[h] \to 0$. Thus T_0 is a continuous map.

If $y = T_0[g]$, then $\|y(t)\| \leqq r$ and (2.4) show that there is a constant C, independent of g, such that $\|y'(t)\| \leqq C$. This implies that the set of functions $y(t) = T_0[g]$ in the range of T_0 is bounded and equicontinuous. Hence, by Arzela's theorem, it has a compact closure in $\mathfrak{D}$ (i.e., any sequence $y_1, y_2, \ldots$ has a uniformly convergent subsequence). Consequently, Corollary 0.1 implies that T_0 has a fixed point y_0. Clearly $y = y_0(t)$ is a periodic solution of period p. This proves the theorem.

Remark. In the deduction of Corollary 0.1 from the Tychonov Theorem 0.2, it is necessary to know that the convex closure of the range $\mathcal{R}(T_0)$ of T_0 is compact. This is clear in the proof just completed, for $y(t)$ in the range of T satisfies the conditions: (i) $y(t)$ is continuous of period p; (ii) $\|y(t)\| \leqq r$; and (iii) $\|y(t) - y(s)\| \leqq C |t - s|$. The convex hull of $\mathcal{R}(T_0)$ [i.e., the smallest convex set containing $\mathcal{R}(T_0)$] is the set of functions $y(t)$ representable in the form $\lambda_1 y_1(t) + \cdots + \lambda_n y_n(t)$, where $n = 1, 2, \ldots$; $\lambda_j \geqq 0$ and $\lambda_1 + \cdots + \lambda_n = 1$. It is clear that functions in this set satisfy (i)–(iii). The closure of this set of functions under the norm of $\mathfrak{D}$ (i.e., under uniform convergence over $0 \leqq t \leqq p$) gives a set of functions satisfying (i)–(iii). Thus the compactness of this set in $\mathfrak{D}$ is clear from Arzela's theorem. (A remark similar to this can be made for the other applications of Corollary 0.1 in this chapter; see Theorem 4.2 and Theorem 8.2.)

Consider now a system of nonlinear differential equations depending on a parameter μ,

(2.7) $$x' = F(t, x, \mu),$$

where F is continuous, of period p in t for fixed (x, μ), and x, F are real d-dimensional vectors. Suppose that for $\mu = 0$, (2.7) has a periodic solution $x = g_0(t)$. Write $y = x - g_0(t)$; then (2.7) becomes

$$y' = F(t, y + g_0(t), \mu) - F(t, g_0(t), 0).$$

If F has continuous partial derivatives with respect to x and $A(t) = \partial_x F(t, g_0(t), 0)$, where $\partial_x F$ is the Jacobian matrix of F with respect to x, then the last equation is of the form (2.1), where

$$f(t, y) = F(t, y + g_0(t), \mu) - F(t, g_0(t), 0) - A(t)y$$

and $\|f(t, y)\|/\|y\| \to 0$ as $(y, \mu) \to 0$ uniformly in t for $0 \leqq t \leqq p$. In particular, when $|\mu|$ is small, (2.6) holds for small $r > 0$; in fact, (2.2) holds for small $\|y_1\|$, $\|y_2\|$ with arbitrarily small θ and $f(t, 0) \equiv 0$. It follows from Theorem 2.1 that if (1.2) has no nontrivial periodic solution of period p, then (2.7) has a unique solution $x(t) = x(t, \mu)$ of period p for each small $|\mu|$. The proof of Theorem 2.1 can also be used to show that if F depends smoothly on μ, then $x(t, \mu)$ depends smoothly on μ. All of these assertions can, however, be proved more directly by the use of the classical implicit function theorem.

Theorem 2.3. *Let x, F be real vectors. Let $F(t, x, \mu)$ be continuous for all t, small $|\mu|$, and x on some d-dimensional domain. Let F be of period p in t for fixed (x, μ) and have continuous partial derivatives with respect to the components of x. Let (2.7), where $\mu = 0$, have a solution $x = g_0(t)$ of period p with the property that if $A(t) = \partial_x F(t, g_0(t), 0)$, then (1.2) has no nontrivial solution of period p. Then, for each small $|\mu|$, (2.7) has a unique solution $x = x(t, \mu)$ of period p with initial point $x(0, \mu)$ near $g_0(0)$; $x(t, \mu)$ is a continuous function of (t, μ), and $x(t, 0) = g_0(t)$. If, in addition, F has a continuous partial derivative with respect to μ, then $x(t, \mu)$ is of class C^1.*

It will be clear from the proof that if more smoothness is assumed for F (e.g., $F \in C^k$ or F analytic), then $x(t, \mu)$ is correspondingly smoother (e.g., $x(t, \mu) \in C^k$ or $x(t, \mu)$ analytic).

Proof. Let $x = \xi(t, x_0, \mu)$ be the unique solution of (2.7) satisfying the initial condition $x(0) = x_0$. Then $\xi(t, x_0, \mu)$ is continuous and has continuous partial derivatives with respect to t and the components of x_0; see Corollary V 3.3. Also, if x_0 is near to $g_0(0)$, then $\xi(t, x_0, \mu)$ exists on the interval $0 \leqq t \leqq p$; see Theorem V 2.1. The solution $x = \xi(t, x_0, \mu)$ is periodic of period p if and only if

$$(2.8) \qquad \xi(p, x_0, \mu) - x_0 = 0.$$

Since $\xi(t, g_0(0), 0) = g_0(t)$, the equation (2.8) is satisfied if $(x_0, \mu) = (g_0(0), 0)$. Hence it can be solved for $x_0 = x_0(\mu)$ if the Jacobian matrix of the left side, $\partial_{x_0} \xi(p, x_0, \mu) - I$, is nonsingular at $(x_0, \mu) = (g_0(0), 0)$. The partial derivatives of $\xi(t, x_0, \mu)$ with respect to a component of x_0, when $(x_0, \mu) = (g_0(0), 0)$, is a solution of the equations of variation (1.2); see Theorem V 3.1. In fact, $Y(t) = \partial_{x_0} \xi(t, g_0(0), 0)$ is a fundamental matrix for (1.2) satisfying $Y(0) = I$. Hence the assumption that (1.2) has no periodic solution is equivalent to the assumption that $Y(p) - I$ is nonsingular; cf. Lemma 1.1, where $M = N = I$. Thus the implicit function theorem is applicable to (2.8) and gives a continuous function $x_0 = x_0(\mu)$. Correspondingly, $x = \xi(t, x_0(\mu), \mu)$ is a periodic solution of (2.7) of period p and the only such solution with initial point x_0 near $g_0(0)$. The other assertions of Theorem 2.3 also follow from the implicit function theorem.

The question of the existence of periodic solutions when

$$\det \left[Y(p) - I \right] = 0$$

has a vast literature and will not be pursued here.

Note that if F in (2.7) does not depend on t and $g_0(t) \not\equiv$ const., then the conditions of Theorem 2.3 cannot be satisfied since $x = g_0'(t)$ is a nontrivial periodic solution of the equations of variation (1.2). Here, however, we have the following analogue.

Theorem 2.4. *Let x, F be real vectors. Let $F(x, \mu)$ be continuous for small $|\mu|$ and for x on some d-dimensional domain and have continuous partial derivatives with respect to the components of x. When $\mu = 0$, let*

$$(2.9) \qquad\qquad x' = F(x, \mu)$$

have a solution $x = g_0(t) \not\equiv$ const. of period $p_0 > 0$ such that if $A(t) = \partial_x F(g_0(t), 0)$, then exactly one of the characteristic roots of (1.2) is 1 [i.e., e^{Rp_0} has $\lambda = 1$ as a simple eigenvalue; cf. (1.5) where $p = p_0$]. Then, for small $|\mu|$, (2.9) has a unique periodic solution $x = x(t, \mu)$ with a period $p(\mu)$, depending on μ, such that $x(t, \mu)$ is near $g_0(t)$ and the period $p(\mu)$ is near p_0; furthermore $x(t, \mu)$, $p(\mu)$ are continuous, $x(t, 0) = g_0(t)$, and $p(0) = p_0$.

Remarks similar to those for Theorem 2.3 concerning the smoothness of F and corresponding smoothness of $x(t, \mu)$, $p(\mu)$ hold.

The geometrical considerations in the proof to follow are clarified by reference to Lemma IX 10.1, which shows that we obtain all solutions of (2.9) near $x = g_0(t)$ by considering solutions with initial points $x(0) = x_0$ near to $g_0(0)$ and x_0 restricted to be on the hyperplane π normal to $F(g_0(0), 0)$ and passing through $g_0(0)$.

Proof. Let $x = \xi(t, x_0, \mu)$ be the unique solution of (2.9) satisfying $x(0) = x_0$. This solution is of period p if and only if (2.8) holds. The equation (2.8) is satisfied when $(p, x_0, \mu) = (p_0, g_0(0), 0)$.

Since solutions of (2.9) are uniquely determined by initial conditions and $g_0(t) \not\equiv$ const., it follows that $F(g_0(t), 0) \neq 0$ for all t. Suppose that the coordinates in the x-space are chosen so that $g_0(0) = 0$ and $F(0, 0) = (0, \ldots, 0, \alpha)$, $\alpha \neq 0$, and let π denote the hyperplane $x^d = 0$ through the point $g_0(0) = 0$ normal to $F(0, 0)$. Consider x_0 on this hyperplane, $x_0 = (x_0^1, \ldots, x_0^{d-1}, 0)$. Then for small $|\mu|$, the equation (2.8) has a unique solution for p, x_0, in terms of μ if the Jacobian matrix of $\xi(t, x_0, \mu) - x_0$ with respect to $x_0^1, \ldots, x_0^{d-1}$ and t is nonsingular at $(t, x_0, \mu) = (p_0, 0, 0)$.

The matrix $Y(t)$, in which the columns are the vectors $\partial \xi / \partial x_0^1, \ldots, \partial \xi / \partial x_0^{d-1}$ and ξ' at $(x_0, \mu) = (0, 0)$, is a fundamental matrix for (1.2) and its last column is $F(g_0(t), 0)$. At $t = 0$,

$$(2.10) \qquad\qquad Y(0) = \mathrm{diag}\,[I_{d-1}, \alpha] = Z(0),$$

by (1.5). Since (1.2) has, up to constant factors, only the last column $g_0'(t)$ of $Y(t)$ at $(x_0, \mu) = (0, 0)$ as a periodic solution of period p_0, the matrix $Y(p_0) - Y(0)$ annihilates vectors c of the form $c = (0, \ldots, 0, c^d)$ and no others.

The Jacobian matrix J of $\xi(t, x_0, \mu) - x_0$ with respect to $x_0^1, \ldots, x_0^{d-1}$, and t at $(t, x, \mu) = (p_0, 0, 0)$ is

$$J = Y(p_0) - \text{diag } [I_{d-1}, 0],$$

and the last column of $Y(p_0)$ is $F(g_0(0), 0) = (0, \ldots, 0, \alpha)$, so that $J = [Y(p_0) - Y(0)] + \text{diag } [0, \ldots, 0, \alpha]$. If J is singular, then there exists a vector $c = (c^1, \ldots, c^d) \neq 0$ such that $Jc = 0$; i.e.,

$$[Y(p_0) - Y(0)]c + (0, \ldots, 0, \alpha c^d) = 0.$$

In view of (2.10) and $Z(0) = Z(p_0)$, this is the same as

$$Z(0)\{(e^{Rp_0} - I)c + (0, \ldots, 0, c^d)\} = 0 \quad \text{or}$$
$$(e^{Rp_0} - I)c + (0, \ldots, 0, c^d) = 0.$$

If $c^d = 0$, then $c = 0$ for $e^{Rp_0} - I$ only annihilates vectors of the form $(0, \ldots, 0, c^d)$. If $c^d \neq 0$, then $(e^{Rp_0} - I)^2 c = 0$. But this implies that $\lambda = 1$ is at least a double eigenvalue of e^{Rp_0}. This contradiction shows that J is nonsingular.

Hence the implicit function theorem is applicable to (2.8) and gives the desired functions $x_0^1(\mu), \ldots, x_0^{d-1}(\mu)$, and $p(\mu)$. Correspondingly, if $x_0(\mu) = (x_0^1(\mu), \ldots, x_0^{d-1}(\mu), 0)$, then $x(t, \mu) = \xi(t, x_0(\mu), \mu)$ is a periodic solution of (2.9) and is the only periodic solution having an initial point x_0, with $x_0^d = 0$, near to $g_0(0)$ and a period near to p_0. This proves Theorem 2.4.

Exercise 2.1. Let dim $x = 2$; $F(t, x)$ continuous for all t and x, periodic of period p in t for fixed x. Let the solution $x = x(t, t_0, x_0)$ of

(2.11) $x' = F(t, x)$

satisfying $x(t_0) = x_0$ be unique for all t_0, x_0 and exist for $t \geq t_0$. Finally, for some (t_0, x_0), let $x(t, t_0, x_0)$ be bounded for $t \geq t_0$. Then (2.11) has at least one periodic solution of period p. See Massera [1].

Exercise 2.2. Let $\alpha(t) = (\alpha^1(t), \ldots, \alpha^d(t))$, $\beta(t) = (\beta^1(t), \ldots, \beta^d(t))$ be piecewise continuously differentiable for $0 \leq t \leq p$; $\alpha^j(t) \leq \beta^j(t)$ for $j = 1, \ldots, d$; and $\alpha(0) = \alpha(p)$, $\beta(0) = \beta(p)$. Let

$$f(t, y) = (f^1(t, y), \ldots, f^d(t, y))$$

be continuous on an open set containing $\Omega^0 = \{(t, y): \alpha^j(t) \leq y^j \leq \beta^j(t) \text{ for } 0 \leq t \leq p\}$ and let $f(t, y)$ be uniformly Lipschitz continuous with respect to y. Suppose finally that the functions $u^j(t, y) = \alpha^{j\prime}(t) - f^j(t, y^1, \ldots, y^{j-1}, \alpha^j(t), y^{j+1}, \ldots, y^d)$ and $v^j(t, y) = \beta^{j\prime}(t) - f^j(t, y^1, \ldots, y^{j-1}, \beta^j(t),$

$y^{j+1}, \ldots, y^d)$ do not change signs (e.g., $u^j \geqq 0$ or $u^j \leqq 0$) and that $u^j v^j \leqq 0$ for all $(t, y) \in \Omega^0$. Then $y' = f(t, y)$ has at least one solution $y = y(t)$, $0 \leqq t \leqq p$, such that $(t, y(t)) \in \Omega^0$ and $y(0) = y(p)$. See Knobloch [1].

PART II. SECOND ORDER BOUNDARY VALUE PROBLEMS

3. Linear Problems

This part of the chapter concerns boundary value problems involving a system of second order equations. Consider first a linear inhomogeneous system of the form

$$(3.1) \qquad x'' = B(t)x + F(t)x' + h(t)$$

and the corresponding homogeneous system

$$(3.2) \qquad x'' = B(t)x + F(t)x'$$

for a d-dimensional vector x (with real- or complex-valued components). The problem involves solutions satisfying boundary conditions

$$(3.3) \qquad x(0) = x_0, \qquad x(p) = x_p,$$

when $p > 0$, x_0, x_p are given. For the inhomogeneous equation (3.1), the conditions (3.3) are not more general than

$$(3.4) \qquad x(0) = 0, \qquad x(p) = 0,$$

for if $x - [(x_p - x_0)t/p + x_0]$ is introduced as a new dependent variable, the equation (3.1) goes over into another equation of the same form with $h(t)$ replaced by $h(t) + B(t)(x_p - x_0)t/p + B(t)x_0 + F(t)(x_p - x_0)/p$.

Actually, the theory of the boundary value problem (3.1), (3.4) is contained in § 1. In order to see this, write (3.1) as a first order system

$$(3.5) \qquad y' = A(t)y + g(t),$$

where $y = (x, x')$ is a $2d$-dimensional vector, $g(t) = (0, h(t))$, and $A(t)$ is a $2d \times 2d$ matrix

$$(3.6) \qquad A(t) = \begin{pmatrix} 0 & I \\ B(t) & F(t) \end{pmatrix}.$$

The boundary conditions (3.4) can be written as

$$(3.7) \qquad My(0) - Ny(p) = 0,$$

where M, N are the constant $2d \times 2d$ matrices

$$(3.8) \qquad M = \begin{pmatrix} I & 0 \\ 0 & 0 \end{pmatrix} \quad \text{and} \quad N = \begin{pmatrix} 0 & 0 \\ I & 0 \end{pmatrix}.$$

Note that

$$\text{rank}\,(M, N) = \text{rank} \begin{pmatrix} I & 0 & 0 & 0 \\ 0 & 0 & I & 0 \end{pmatrix} = 2d.$$

Instead of restricting M, N to be of the type (3.8), it is possible to choose more general matrices; in this case, (3.4) is replaced by conditions of the form

$$M_{j1}x(0) + M_{j2}x'(0) - N_{j1}x(p) - N_{j2}x'(p) = 0 \qquad \text{for} \quad j = 1, 2,$$

where M_{jk}, N_{jk} are constant $d \times d$ matrices such that

$$(M, N) = \begin{pmatrix} M_{11} & M_{12} & N_{11} & N_{12} \\ M_{21} & M_{22} & N_{21} & N_{22} \end{pmatrix}$$

is of rank $2d$. For the sake of simplicity, only the choice (3.8), i.e., only the boundary conditions (3.4), will be considered.

Lemma 1.1 implies the following:

Lemma 3.1. *Let $B(t)$, $F(t)$ be continuous $d \times d$ matrices for $0 \leqq t \leqq p$; $U(t)$ the $d \times d$ matrix solution of*

$$(3.9) \qquad U'' = B(t)U + F(t)U', \qquad U(0) = 0, \qquad U'(0) = I.$$

Then (3.2) has a nontrivial solution ($\not\equiv 0$) solution satisfying (3.4) if and only if $U(p)$ is singular. In fact, the number k, $0 \leq k \leq d$, of linearly independent solutions of (3.2), (3.4) is the number of linearly independent vectors c satisfying $U(p)c = 0$.

The corresponding corollary of Theorem 1.1 is

Theorem 3.1. *Let $B(t)$, $F(t)$ be continuous for $0 \leqq t \leqq p$. Then (3.1) has a solution $x(t)$ satisfying (3.4) for every $h(t)$ continuous on $[0, p]$ if and only if (3.2), (3.4) has no nontrivial ($\not\equiv 0$) solution. In this case, $x(t)$ is unique and there exists a constant K such that*

$$(3.10) \qquad \|x(t)\|,\ \|x'(t)\| \leqq K \int_0^p \|h(s)\|\,ds.$$

Exercise 3.1. Verify Theorem 3.1.

The homogeneous adjoint system for (3.5) is $y' = -A^*(t)y$ which is not equivalent to a second order system without additional assumptions on B or F. The simplest assumption of this type is that $F(t)$ is continuously differentiable. In this case, the homogeneous adjoint system $y' = -A^*(t)y$ is equivalent to

$$(3.11) \qquad z'' = [B^*(t) - F^{*\prime}(t)]z - F^*(t)z'$$

and the corresponding inhomogeneous system is

$$(3.12) \qquad z'' = [B^*(t) - F^{*\prime}(t)]z - F^*(t)z' + f(t).$$

[Actually, the differentiability condition can be avoided by writing the terms involving F^* as $(F^*z)'$, and interpreting (3.11), (3.12) as first order systems for the $2d$-dimensional vector $(-z' - F^*z, z)$.]

In order to obtain the corresponding Green's relation, multiply (3.1) scalarly by z, (3.12) by x, subtract and integrate over $[0, p]$ to obtain

(3.13) $$\int_0^p [h(t) \cdot z(t) - x(t) \cdot f(t)] \, dt = [x' \cdot z - x \cdot z' - Fx \cdot z]_0^p.$$

Thus, if x satisfies (3.4) and z satisfies

(3.14) $$z(0) = 0, \qquad z(p) = 0$$

then

(3.15) $$\int_0^p [h(t) \cdot z(t) - x(t) \cdot f(t)] \, dt = 0,$$

so that (3.4) and (3.14) are adjoint boundary conditions.

Exercise 3.2.   Verify that (3.2), (3.4) and (3.11), (3.14) are adjoint boundary problems in the sense of § 1.

Lemma 3.2.   *Let $B(t)$ be continuous and $F(t)$ continuously differentiable for $0 \leqq t \leqq p$. Then (3.2), (3.4) have the same number of linearly independent solutions as the adjoint problem (3.11), (3.14).*

Finally, a corollary of Theorem 1.2 is

Theorem 3.2.   *Let $B(t)$ be continuous and $F(t)$ continuously differentiable on $[0, p]$ and such that (3.2), (3.4) has k, $1 \leqq k \leqq d$, linearly independent solutions. Let $z_1(t), \ldots, z_k(t)$ be k linearly independent solutions of (3.11), (3.14). Let $h(t)$ be continuous on $[0, p]$. Then (3.1), (3.4) has a solution if and only if*

(3.16) $$\int_0^p h(t) \cdot z_j(t) \, dt = 0 \qquad for \quad j = 1, \ldots, k.$$

The next uniqueness theorem has no analogue in § 1.

Theorem 3.3.   *Let $B(t)$, $F(t)$ be continuous $d \times d$ matrices on $0 \leqq t \leqq p$ such that*

(3.17) $$\mathrm{Re} \, [(B(t) - \tfrac{1}{4}F(t)F^*(t))x \cdot x] \geqq 0$$

for all vectors x (i.e., let the Hermitian part of the matrix $B - \tfrac{1}{4}FF^*$ be non-negative definite). Let $g(t)$ be continuous for $0 \leqq t \leqq p$. Then

(3.18) $$x'' = B(t)x + F(t)x' + h(t)$$

has at most one solution satisfying given boundary conditions $x(0) = x_0$, $x(p) = x_p$.

Remark 1.   Actually, Theorem 3.3 remains valid if (3.17) is relaxed to

$$(3.19) \qquad 2 \operatorname{Re} [(B(t) - \tfrac{1}{4}F(t)F^*(t))x \cdot x] > -(\pi/p)^2 \|x\|^2$$

for all vectors $x \neq 0$; cf. Exercise 3.3.

Proof.   Since the difference of two solutions of the given boundary value problem is a solution of

$$(3.20) \qquad x'' = B(t)x + F(t)x', \qquad x(0) = x(p) = 0,$$

it suffices to show that the only solution of (3.20) is $x \equiv 0$.

Let $x(t)$ be a solution of (3.20). Put $r(t) = \|x(t)\|^2$. Then $r' = 2 \operatorname{Re} x \cdot x'$ and $r'' = 2 \operatorname{Re} (x \cdot x'' + \|x'\|^2)$, so that $r'' = 2 \operatorname{Re} [(B(t)x + F(t)x') \cdot x + \|x'\|^2]$. It is readily verified that

$$\operatorname{Re} (B(t)x + F(t)x') \cdot x + \|x'\|^2 = \|x' + \tfrac{1}{2}F^*x\|^2 + \operatorname{Re} (Bx - \tfrac{1}{4}FF^*x) \cdot x.$$

Thus

$$(3.21) \qquad r'' = 2 \|x' + \tfrac{1}{2}F^*x\|^2 + 2 \operatorname{Re} [(B - \tfrac{1}{4}FF^*)x \cdot x].$$

Hence (3.17) implies that $r'' \geq 0$. Since the last part of (3.20) means that $r(0) = r(p) = 0$, it follows that $r(t) \equiv 0$ for $0 \leq t \leq p$. This proves Theorem 3.3.

Exercise 3.3.   (*a*) Show that if there exists a continuous real-valued function $q(t)$, $0 \leq t \leq p$, such that the equation

$$r'' + q(t)r = 0$$

has no solution $r(t) \not\equiv 0$ with two zeros on $0 \leq t \leq p$ [e.g., $q(t) < (\pi/p)^2$] and (3.17) is relaxed to

$$(3.22) \qquad 2 \operatorname{Re} [(B(t) - \tfrac{1}{4}F(t)F^*(t))x \cdot x] \geq -q(t) \|x\|^2$$

for all vectors x, then the conclusion of Theorem 3.3 remains valid. (*b*) Let there exist a continuously differentiable $d \times d$ matrix $K(t)$ on $[0, p]$ such that

$$(3.23) \qquad \operatorname{Re} [B - K' + (\tfrac{1}{2}F - K^H)(\tfrac{1}{2}F^* - K^H)]x \cdot x \geq 0$$

for all vectors x and $0 \leq t \leq p$, where $K^H = \tfrac{1}{2}(K + K^*)$. Then the conclusion of Theorem 3.3 is valid. [Note that (3.23) reduces to (3.17) if $K(t) \equiv 0$, so that (*b*) generalizes Theorem 3.3, but not part (*a*) of this exercise.] The 2 in (3.22), hence in (3.19), is not needed if $F \equiv 0$.

Remark 2.   If $F(t)$ has a continuous derivative, then (3.20) implies that $x \equiv 0$ if and only if $z \equiv 0$ is the only solution of

$$(3.24) \qquad z'' = [B^*(t) - F^{*'}(t)]z - F^*(t)z', \qquad z(0) = z(p) = 0;$$

cf. Lemma 3.2. Hence, the conclusion of Theorem 3.3 is valid if B, F in the criteria (3.17), (3.22), (3.23) are replaced by $B^* - F^{*'}$, $-F^*$, respectively.

4. Nonlinear Problems

Let x and f denote vectors with real-valued components. This section deals with second order equations of the form

$$(4.1) \qquad x'' = f(t, x, x')$$

and the question of the existence of solutions satisfying the boundary conditions

$$(4.2) \qquad x(0) = 0, \qquad x(p) = 0$$

or, for given x_0 and x_p,

$$(4.3) \qquad x(0) = x_0, \qquad x(p) = x_p.$$

The equation (4.1) will be viewed as an "inhomogeneous form" of

$$(4.4) \qquad x'' = 0.$$

The problem (4.2), (4.4) has no nontrivial solution. Thus, by Theorem 3.1, an equation

$$(4.5) \qquad x'' = h(t)$$

has a unique solution satisfying (4.2). In fact, this solution is given by

$$(4.6) \qquad x(t) = -\frac{1}{p}\left[(p - t)\int_0^t sh(s)\,ds + t\int_t^p (p - s)h(s)\,ds\right].$$

This can be verified by differentiating (4.6) twice; cf. (XI 2.18). The relation (4.6) can be abbreviated to

$$(4.7) \qquad x(t) = -\int_0^p G(t, s)h(s)\,ds,$$

where

$$(4.8) \qquad G(t, s) = \frac{1}{p}(p - t)s \quad \text{or} \quad G(t, s) = \frac{1}{p}t(p - s)$$

according as $0 \leqq s \leqq t \leqq p$ or $0 \leqq t \leqq s \leqq p$. Thus

$$(4.9) \qquad 0 \leqq G(t, s) \leqq \frac{p}{4}, \qquad \int_0^p G(t, s)\,ds = \frac{1}{2}t(p - t) \leqq \frac{p^2}{8},$$

$$\int_0^p |G_t(t, s)|\,ds = \frac{1}{2p}[t^2 + (p - t)^2] \leqq \frac{p}{2},$$

where $G_t = \partial G/\partial t$. Thus (4.6) or (4.7) and its differentiated form imply

$$(4.10) \qquad \|x(t)\| \leqq \frac{p^2}{8} \max \|h(s)\|, \qquad \|x'(t)\| \leqq \frac{p}{2} \max \|h(s)\|,$$

where the max refers to $0 \leqq s \leqq p$.

Theorem 4.1. *Let $f(t, x, x')$ be continuous for $0 \leqq t \leqq p$ and all (x, x') and satisfy a Lipschitz condition with respect to x, x' of the form*

$$(4.11) \quad \|f(t, x_1, x_1') - f(t, x_2, x_2')\| \leqq \theta_0 \|x_1 - x_2\| + \theta_1 \|x_1' - x_2'\|$$

with Lipschitz constants θ_0, θ_1 so small that

$$(4.12) \qquad \frac{\theta_0 p^2}{8} + \frac{\theta_1 p}{2} < 1.$$

Then (4.1) has a unique solution satisfying (4.2).

Remark 1.   Instead of requiring f to be defined for $0 \leqq t \leqq p$ and all (x, x'), it is sufficient to have f defined for $0 \leqq t \leqq p$, $\|x\| \leqq R$, $\|x'\| \leqq 4R/p$, where R satisfies either

$$(4.13) \qquad \frac{mp^2}{8} \leqq R \left[1 - \left(\frac{\theta_0 p^2}{8} + \frac{\theta_1 p}{2} \right) \right]$$

if $m = \max \|f(t, 0, 0)\|$ for $0 \leqq t \leqq p$, or merely

$$(4.14) \qquad \frac{Mp^2}{8} \leqq R$$

if $M = \max \|f(t, x, x')\|$ for $\|x\| \leqq R$, $\|x'\| \leqq 4R/p$.

Proof.   Let $\mathfrak{D}$ be the Banach space of functions $h(t)$, $0 \leqq t \leqq p$, having continuous first derivatives and the norm

$$(4.15) \qquad |h| = \max \left(\max_{0 \leqq t \leqq p} \|h(t)\|, \frac{p}{4} \max_{0 \leqq t \leqq p} \|h'(t)\| \right).$$

Consider an $h(t)$ in the sphere $|h| \leqq R$ of $\mathfrak{D}$.   Let $x(t)$ be the unique solution of

$$(4.16) \qquad \qquad x'' = f(t, h(t), h'(t))$$

satisfying $x(0) = x(p) = 0$.   Define an operator T_0 on the sphere $|h| \leqq r$ of $\mathfrak{D}$ by putting $T_0[h(t)] = x(t)$.

If $x_0 = T_0[0]$ and $\|f(t, 0, 0)\| \leqq m$, then

$$(4.17) \qquad \qquad \|x_0(t)\| \leqq \frac{mp^2}{8}, \qquad \frac{p}{4} \|x_0'(t)\| \leqq \frac{mp^2}{8}$$

by the case $h = f(t, 0, 0)$ of (4.10).   Thus the norm $x_0(t) = T_0[0] \in \mathfrak{D}$ satisfies

$$(4.18) \qquad \qquad |T_0[0]| \leqq \frac{mp^2}{8}.$$

Also, if $x_1 = T_0[h_1]$, $x_2 = T_0[h_2]$, then, by (4.10) and (4.11),

$$\|x_1(t) - x_2(t)\| \leq \frac{p^2}{8} (\theta_0 \max \|h_1 - h_2\| + \theta_1 \max \|h_1' - h_2'\|),$$

$$\|x_1'(t) - x_2'(t)\| \leq \frac{p}{2} (\theta_0 \max \|h_1 - h_2\| + \theta_1 \max \|h_1' - h_2'\|).$$

If the last inequality is multiplied by $p/4$ and $\theta_1(p^2/8) \max \|h_1' - h_2'\|$ is written as $(\theta_1 p/2)[(p/4) \max \|h_1' - h_2'\|]$, it follows that

(4.19) $$|T_0[h_1] - T_0[h_2]| \leq \left(\frac{\theta_0 p^2}{8} + \frac{\theta_1 p}{2}\right) |h_1 - h_2|.$$

Thus the inequalities (4.12), (4.13) and (4.18) show that Theorem 0.1 is applicable and give Theorem 4.1.

Similarly, if $\|f(t, x, x')\| \leq M$ for $\|x\| \leq R$, $\|x'\| \leq 4R/p$, then the derivation of (4.17) shows that if $|h| \leq R$, then $x = T_0[h]$ satisfies $|x| \leq Mp^2/8$. Thus if (4.14) holds, T_0 maps the sphere $|h| \leq R$ into itself and the Remark following Theorem 0.1 is applicable in view of (4.12). Hence the proof of Theorem 4.1 and Remark 1 following it is complete.

Corollary 4.1. *Let $f(t, x, x')$ be continuous for $0 \leq t \leq p$, $\|x\| \leq R_0$, $\|x'\| \leq R_1$ and satisfy (4.11), (4.12) and $\|f(t, x, x')\| \leq M$. Let*

(4.20) $$\frac{Mp^2}{8} + \|x_0\| \leq R_0, \qquad \frac{Mp}{2} + \frac{\|x_0\|}{p} \leq R_1.$$

Then (4.1) has a unique solution satisfying

(4.21) $$x(0) = 0 \quad \text{and} \quad x(p) = x_0.$$

Exercise 4.1. (a) Prove Corollary 4.1. (b) In Corollary 4.1, let $\|f(t, x, x')\| \leq M$ be relaxed to $\|f(t, tx_0/p, x_0/p)\| \leq m$ for $0 \leq t \leq p$ and R be defined by replacing "$\leq$" by "$=$" in (4.13). Show that the conclusion of Corollary 4.1 remains valid if $R + \|x_0\| \leq R_0, 4R/p + \|x_0\|/p \leq R_1$ replaces (4.20).

Theorem 4.2. *Let $f(t, x, x')$ be continuous and bounded, say,*

$$\|f(t, x, x')\| \leq m,$$

for $0 \leq t \leq p$ and all (x, x'). Then (4.1) has at least one solution $x(t)$ satisfying $x(0) = x(p) = 0$ and

(4.22) $$\|x(t)\| \leq \frac{mp^2}{8}, \qquad \|x'(t)\| \leq \frac{mp}{2}.$$

It is sufficient to require that $f(t, x, x')$ be defined only for $\|x\| \leq mp^2/8$, $\|x'\| \leq mp/2$.

Proof. Let $\mathfrak{D}$ be the Banach space of continuously differentiable functions $h(t)$, $0 \leq t \leq p$, with norm $|h|$ defined (4.15). Consider $h(t)$ in the sphere $|h| \leqq mp^2/8$ of $\mathfrak{D}$. For such an h, put $x = T_0[h]$, where $x(t)$ is the unique solution of (4.16) satisfying $x(0) = x(p) = 0$. Then $\|x(t)\| \leqq mp^2/8$ and $\|x'(t)\| \leqq mp/2$, so that T_0 maps the sphere $|h| \leqq mp^2/8$ into itself.

If $|h_1|, |h_2| \leqq mp^2/8$ and $x_1 = T_0[h_1]$, $x_2 = T_0[h_2]$, then (4.7) and (4.9) imply that

$$|x_1 - x_2| \leqq \frac{p}{4} \int_0^p \|f(t, h_1(t), h_1'(t)) - f(t, h_2(t), h_2'(t))\| \, dt.$$

Since f is a continuous function, it follows that if $|h_1 - h_2| \to 0$, then $|x_1 - x_2| \to 0$. Thus T_0 is continuous.

For any $x(t)$ in the range of T_0, i.e., $x = T_0[h]$ for some h, (4.16) implies $\|x''(t)\| \leqq m$. It follows that the set of functions $x(t)$ in the range of $T_0[h]$, $|h| \leqq mp^2/8$, are such that $x(t), x'(t)$ are bounded and equicontinuous since

$$\|x(t_1) - x(t_2)\| \leqq \tfrac{1}{2} mp \, |t_1 - t_2|, \quad \|x'(t_1) - x'(t_2)\| \leqq m \, |t_1 - t_2|.$$

Hence Arzela's theorem implies that the range of $T_0[h]$ has a compact closure. Consequently, Tychonov's theorem is applicable and gives Theorem 4.2.

Corollary 4.2. *Let $f(t, x, x')$ be continuous and satisfy $\|f\| \leqq M$ for $0 \leqq t \leqq T$, $\|x\| \leqq R_0$, $\|x'\| \leqq R_1$. Let p and x_0 satisfy $0 < p \leqq T$ and (4.20). Then (4.1) has a solution satisfying (4.21). (In particular, if $0 < T < \min((8R_0/M)^{1/2}, 2R_1/M)$, then there exists a $\delta > 0$ such that if $\|x_0\| \leqq \delta$, then (4.1) has a solution satisfying (4.21) for $p = T$.)*

Exercise 4.2. Prove Corollary 4.2.

Exercise 4.3. Let $f(t, x, x')$ be continuous for $0 \leqq t \leqq p$, $\|x\| \leqq R_0$, and arbitrary x'. Let there exist positive constants a, b such that $\|f(t, x, x')\| \leqq a \|x'\|^2 + b$ for $0 \leqq t \leqq p$, $\|x\| \leqq R_0$. Assume that $a, b, \|x_0\|$ are such that $a(bp^2 + 2 \|x_0\|) < 1$ and $r^* = (ap)^{-1}\{1 - [1 - a(bp^2 + 2 \|x_0\|)]^{1/2}\}$ satisfies $r^*p + 3 \|x_0\| \leqq 4R_0$. Then the boundary value problem (4.1),(4.21) has a solution.

Note that Corollaries 4.1 and 4.2 are similar except that in Corollary 4.1 there is the extra assumption that (4.11) and (4.12) hold; correspondingly, there is the extra assertion that the solution of (4.1), (4.21) is unique. We can prove another type of uniqueness theorem.

Theorem 4.3. *Let $f(t, x, x')$ be continuous for $0 \leqq t \leqq p$ and for (x, x') on some $2d$-dimensional convex set. Let $f(t, x, x')$ have continuous partial derivatives with respect to the components of x and x'. Let the Jacobian matrices of f with respect to x, x'*

$$(4.23) \quad B(t, x, x') = \partial_x f(t, x, x'), \quad F(t, x, x') = \partial_{x'} f(t, x, x')$$

satisfy

(4.24) $$2(B - \tfrac{1}{4}FF^*)z \cdot z > -\frac{\pi^2}{p^2}\|z\|^2$$

for all (constant) vectors $z \neq 0$. *Then* (4.1) *has at most one solution satisfying given boundary conditions* $x(0) = x_0$, $x(p) = x_p$.

By the use of Exercise 3.3(a), condition (4.24) can be relaxed to

$$2(B - \tfrac{1}{4}F^*F)z \cdot z \geqq -q(t)\|z\|^2,$$

where $q(t)$ satisfies the conditions of Exercise 3.3(a). Here and in (4.24), "2" is not needed if f is independent of x'.

Proof. Suppose that there exist two solutions $x_1(t)$, $x_2(t)$. Put $x(t) = x_2(t) - x_1(t)$, so that

$$x'' = f(t, x_2(t), x_2'(t)) - f(t, x_1(t), x_1'(t)), \qquad x(0) = x(p) = 0.$$

This can be written as

$$x'' = B_1(t)x + F_1(t)x', \qquad x(0) = x(p) = 0,$$

where

(4.25) $$B_1(t) = \int_0^1 B \, ds, \qquad F_1(t) = \int_0^1 F \, ds,$$

and the argument of B, F in (4.25) is

(4.26) $$(t, (1 - s)x_1(t) + sx_2(t), (1 - s)x_1'(t) + sx_2'(t)).$$

This is a consequence of Lemma V 3.1.

For any constant vector z, an application of Schwarz's inequality to the formula in (4.25) for each component of $F_1^*(t)z$ gives

$$\|F_1^*(t)z\|^2 \leqq \int_0^1 \|F^*z\|^2 \, ds,$$

where the argument of F^* is (4.26). Hence,

$$[B_1(t) - \tfrac{1}{4}F_1(t)F_1^*(t)]z \cdot z \geqq \int_0^1 [B - \tfrac{1}{4}FF^*]z \cdot z \, ds.$$

Thus by (4.24)

$$2[B_1(t) \dot{-} \tfrac{1}{4}F_1(t)F_1^*(t)]z \cdot z > -\frac{\pi^2}{p^2}\|z\|^2$$

for all vectors $z \neq 0$. Consequently, Theorem 3.3 and Remark 1 following it imply that $x(t) \equiv 0$. This proves the theorem.

Exercise 4.4. Let $f(t, x, x')$ be continuous for $0 \leqq t \leqq p$ and (x, x') on some $2d$-dimensional domain and satisfy a Lipschitz condition of the form (4.11), where

(4.27) $$2\theta_0 + \tfrac{1}{2}\theta_1^2 < \frac{\pi^2}{p^2}.$$

Then (4.1) has at most one solution satisfying given boundary conditions $x(0) = x_0$, $x(p) = x_p$.

Exercise 4.5.   Let $f(t, x, x')$ be continuous for $0 \leq t \leq p$ and (x, x') on a $2d$-dimensional domain. Let $\Delta x = x_2 - x_1$, $\Delta x' = x_2' - x_1'$, $\Delta f = f(t, x_2, x_2') - f(t, x_1, x_1')$, where x_1, x_2, x_1', x_2' are independent variables and assume that

$$\Delta x \cdot \Delta f + |\Delta x'|^2 > 0 \qquad \text{if} \quad \Delta x \neq 0, \quad \Delta x \cdot \Delta x' = 0.$$

Then the boundary value problem $x'' = f(t, x, x')$, $x(0) = x_0$, $x(p) = x_p$ has at most one solution.

Exercise 4.6.   (a) Let x be a real variable. Let $f(t, x, x')$ be continuous and strictly increasing in x for fixed (t, x'). Then (4.1) can have at most one solution satisfying given boundary conditions $x(0) = x_0$, $x(p) = x_p$. (b) Show that (a) is false if "strictly increasing" is replaced by "nondecreasing." (c) Show that if, in part (a), "strictly increasing" is replaced by "nondecreasing" and, in addition, f satisfies a uniform Lipschitz condition with respect to x', then the conclusion in (a) is valid. [For an existence theorem under the conditions of part (c), see Exercise 5.4.]

Exercise 4.7 (*Continuity Method*).   Let x be a real variable. Let $\alpha(t, x')$, $\beta(t, x')$ be real-valued, continuous functions for $-\infty < t$, $x' < \infty$ with the properties that (i) α, β are periodic of period $p > 0$ in t for fixed x'; (ii) $\alpha > 0$; (iii) $|\beta(t, x')| \to \infty$ and $|\alpha(t, x')/\beta(t, x')| \to 0$ as $|x'| \to \infty$ uniformly in t. (a) Show that

(4.28) $x'' = x\alpha(t, x') + \beta(t, x')$

has at most one solution of period p,

(4.29) $x(0) - x(p) = 0$, $x'(0) - x'(p) = 0$.

(b) Show that if $C = \max |\beta(t, 0)|/\alpha(t, 0)$ and K is so large that $C\alpha(t, x') \leq \frac{1}{2} |\beta(t, x')|$ and $|\beta(t, 0)| \leq |\beta(t, x')|/4$ when $|x'| \geq K$, then any periodic solution $x(t)$ of (4.28) satisfies $|x(t)| \leq C$, $|x'(t)| \leq K$. (c) Assume that α, β are of class C^1. By showing that the set of λ-values on $0 \leq \lambda \leq 1$ for which

$$x'' = x\alpha(t, x') + \beta(t, x') - \beta(t, 0) + \lambda\beta(t, 0)$$

has a periodic solution is open and closed on $0 \leq \lambda \leq 1$, prove that (4.28) has a unique periodic solution. (d) Show that the assumption in (c) that α, β are of class C^1 can be omitted.

Exercise 4.8 (*Continuation*).   Let $\alpha(t, x, x')$, $\beta(t, x, x')$ be continuous for $-\infty < t, x, x' < \infty$ with the properties that (i) α, β are periodic of period $p > 0$ in t for fixed (x, x'); (ii) $\alpha > 0$; (iii) there is a constant C such that $|\beta(t, x, 0)| \leq C\alpha(t, x, 0)$ for $-\infty < t, x < \infty$; (iv) $|\beta(t, x, x')| \to \infty$ and $|\alpha(t, x, x')/\beta(t, x, x')| \to 0$ as $|x'| \to \infty$ uniformly on bounded

(t, x)-sets. Show that

$$x'' = x\alpha(t, x, x') + \beta(t, x, x')$$

has at least one periodic solution.

5. A Priori Bounds

The proofs for the existence theorems for solutions of boundary value problems in the last section depended on finding bounds for the solution and its derivative. This section deals with more a priori bounds and their applications. The main problem to be considered is of the following type: Given a d-dimensional vector function $x(t)$ of class C^2 on some interval $0 \leq t \leq p$, a bound for $\|x(t)\|$, and some majorants for $\|x''\|$, find a bound for $\|x'\|$. The following result holds for the case when x is a real-valued function:

Lemma 5.1. *Let $\varphi(s)$, where $0 \leq s < \infty$, be a positive continuous function satisfying*

$$(5.1) \qquad \int^{\infty} \frac{s \, ds}{\varphi(s)} = \infty.$$

Let $R \geq 0$ and $\tau > 0$. Then there exists a number M [depending only on $\varphi(s)$, R, τ] with the following property: If $x(t)$ is a real-valued function of class C^2 for $0 \leq t \leq p$, where $p \geq \tau$, satisfying

$$(5.2) \qquad |x| \leq R, \qquad |x''| \leq \varphi(|x'|),$$

then $|x'| \leq M$ for $0 \leq t \leq p$.

Proof.   In view of (5.1), there exists a number M such that

$$(5.3) \qquad \int_{2R/\tau}^{M} \frac{s \, ds}{\varphi(s)} = 2R.$$

It will be shown that M has the desired property. [Instead of assumption (5.1), it would be sufficient to assume the existence of an M satisfying (5.3).]

Let $|x'(t)|$ assume its maximum value at a point $t = a, 0 \leq a \leq p$. We can suppose that $x'(a) > 0$, otherwise x is replaced by $-x$. If $x'(a) > 2R/\tau$, then there exists a point t on $0 \leq t \leq p$ where $x'(t) \leq 2R/p \leq 2R/\tau$. Otherwise $x(p) - x(0) > 2R$ which contradicts $|x| \leq R$. Assume $x'(a) > 2R/\tau$ and let $t = b$ be a point nearest $t = a$ where $x'(t) = 2R/\tau$. For sake of definiteness, let $b > a$. Thus $0 \leq 2R/\tau = x'(b) \leq x'(t) \leq x'(a)$ for $a \leq t \leq b$.

If the second inequality in (5.2) is multiplied by $x'(t) > 0$, a quadrature over $a \leq t \leq b$ gives

$$\left| \int_a^b \frac{x'(t)x''(t) \, dt}{\varphi(x'(t))} \right| \leq \int_a^b x'(t) \, dt \leq 2R.$$

Even though it is not assumed that $x'' \neq 0$, the formal change of variables $s = x'(t)$ is permitted on the left and gives

$$\int_{2R/\tau}^{x'(a)} \frac{s\,ds}{\varphi(s)} \leq 2R;$$

cf. Lemma I 4.1. From (5.3), it is seen that $x'(a) \leq M$. Thus it follows that either $x'(a) \leq 2R/\tau$ or $x'(a) \leq M$. In either case $x'(a) \leq M$. Since $x'(a) = \max |x'(t)|$ for $0 \leq t \leq p$, the lemma follows.

Lemma 5.1 is false if x is a d-dimensional vector, $d \geq 2$, and absolute values are replaced by norms in (5.2). In order to see this, note that $\varphi(s) = \gamma s^2 + C > 0$, where γ and C are constants, satisfies the condition of Lemma 5.1. Let $x(t)$ denote the binary vector $x(t) = (\cos nt, \sin nt)$. Thus $\|x\| = 1$, $\|x'(t)\| = |n|$, $\|x''(t)\| = n^2 = \|x'\|^2$. Thus the inequalities analogous to (5.2),

(5.4) $$\|x\| \leq R, \qquad \|x''\| \leq \varphi(\|x'\|),$$

hold for $R = 1$, $\varphi(s) = s^2 + 1$. But there does not exist a number M such that $\|x'(t)\| \leq M$ for all choices of n. The main result for vector-valued functions will be the next lemma.

Lemma 5.2. *Let $\varphi(s)$, where $0 \leq s < \infty$, be a positive continuous function satisfying (5.1). Let α, K, R, τ be non-negative constants. Then there exists a constant M [depending only on $\varphi(s)$, α, R, τ, K] with the following property: If $x(t)$ is a vector-valued function of class C^2 on $0 \leq t \leq p$, where $p \geq \tau$, satisfying (5.4) and*

(5.5) $$\|x\| \leq R, \qquad \|x''\| \leq \alpha r'' + K, \qquad \text{where} \quad r = \|x\|^2,$$

then $\|x'\| \leq M$ on $0 \leq t \leq p$.

Proof. The first step of the proof is to show that (5.5) alone implies the existence of a bound for $\|x'(t)\|$ on any interval $[\mu, p - \mu]$, $0 < \mu \leq \frac{1}{2}p$. Let $0 < \mu < p$ and $0 \leq t \leq p - \mu$, then

(5.6) $$x(t + \mu) - x(t) - \mu x'(t) = \int_{t}^{t+\mu} (t + \mu - s)x''(s)\,ds,$$

$t + \mu - s \geq 0$, and (5.5) imply that

$$\mu \|x'(t)\| \leq 2R + \int_{t}^{t+\mu} (t + \mu - s)(\alpha r''(s) + K)\,ds.$$

This inequality and the analogue of (5.6) in which x is replaced by r give

$$\mu \|x'(t)\| \leq 2R + \alpha[r(t + \mu) - r(t) - \mu r'(t)] + \tfrac{1}{2}K\mu^2;$$

hence

(5.7) $$\mu \|x'(t)\| \leq 2R(1 + \alpha R) + \tfrac{1}{2}K\mu^2 - \alpha\mu r'(t) \qquad \text{for} \quad 0 \leq t \leq p - \mu$$

Similarly, for $\mu \leq t \leq p$, the relation

$$x(t) - x(t - \mu) - \mu x'(t) = -\int_{t-\mu}^{t} (t - \mu - s)x''(s)\, ds$$

implies that

(5.8) $\quad \mu \, \|x'(t)\| \leq 2R(1 + \alpha R) + \tfrac{1}{2}K\mu^2 + \alpha\mu r'(t) \qquad$ for $\quad \mu \leq t \leq p.$

Let

(5.9) $$M_1(\tfrac{1}{2}p) = \frac{4R(1 + \alpha R)}{p} + 4Kp.$$

The choice $\mu = \tfrac{1}{2}p$ in (5.7) and (5.8) gives

(5.10) $\qquad \|x'(t)\| \leq M_1(\tfrac{1}{2}p) - \alpha r'(t) \qquad$ for $\quad 0 \leq t \leq \tfrac{1}{2}p,$

(5.11) $\qquad \|x'(t)\| \leq M_1(\tfrac{1}{2}p) + \alpha r'(t) \qquad$ for $\quad \tfrac{1}{2}p \leq t \leq p.$

Adding (5.10), (5.11) for $t = p/2$ shows that

(5.12) $\qquad\qquad\qquad \|x'(\tfrac{1}{2}p)\| \leq M_1(\tfrac{1}{2}p).$

The assumption (5.4) and (5.10)–(5.11) imply that

(5.13) $\qquad \dfrac{|x' \cdot x''|}{\varphi(\|x'\|)} \leq \|x'\| \leq M_1(\tfrac{1}{2}p) \pm \alpha r',$

where $\pm$ is required according as $t \geq \tfrac{1}{2}p$ or $t \leq \tfrac{1}{2}p$. Let $\Phi(s)$ be defined by

(5.14) $$\Phi(s) = \int_0^s \frac{u\, du}{\varphi(u)}.$$

Then, by Lemma I 4.1,

(5.15) $\qquad |\Phi(\|x'(t)\|) - \Phi(\|x'(\tfrac{1}{2}p)\|)| = \left| \int x' \cdot x'' \, \dfrac{dt}{\varphi(\|x'\|)} \right|,$

where the integral is taken over the t-interval with endpoints t and $p/2$. In view of (5.13), the integral is majorized by

$$\tfrac{1}{2}pM_1(\tfrac{1}{2}p) + \alpha\, |r(t) - r(\tfrac{1}{2}p)| \leq \tfrac{1}{2}pM_1(\tfrac{1}{2}p) + 2\alpha R^2.$$

Hence

$$\Phi(\|x'(t)\|) \leq \Phi(\|x'(\tfrac{1}{2}p)\|) + \tfrac{1}{2}pM_1(\tfrac{1}{2}p) + 2\alpha R^2.$$

In view of (5.12) and the fact that Φ is an increasing function, $\|x'(t)\| \leqq M(p)$, where

$$M(p) = \Phi^{-1}[\Phi(M_1(\tfrac{1}{2}p)) + \tfrac{1}{2}pM_1(\tfrac{1}{2}p) + 2\alpha R^2]$$

and Φ^{-1} is the function inverse to Φ. If $p \geqq \tau$, then $t \in [0, p]$ is contained in an interval of length τ in $[0, p]$. Thus the considerations just completed show that p can be replaced by τ, and the lemma is proved with $M(\tau)$ as an admissible choice of M.

Exercise 5.1.   Show that an analogue of Lemma 5.2 remains valid if (5.5) is replaced by

$$\|x\| \leqq R, \qquad \|x''\| \leqq \rho'',$$

where $\rho(t)$ is real-valued function of class C^2 on $0 \leqq t \leqq p$ such that $|\rho(t)| \leqq K_1$. In this case, M depends only on $\varphi(s)$, α, R, τ, and K_1.

The choice $\varphi(s) = \gamma s^2 + C$ in Lemma 5.1 gives the following:

Corollary 5.1. *Let* γ, C, α, K, R, τ *be non-negative constants. Then there exists a constant M [depending only on γ, C, α, R, τ, K] such that if $x(t)$ is of class C^2 on $0 \leqq t \leqq p$, where $p \geqq \tau$, satisfying (5.5) and*

$$(5.16) \qquad \|x\| \leqq R, \qquad \|x''\| \leqq \gamma \|x'\|^2 + C,$$

then $\|x'\| \leqq M$ *for* $0 \leqq t \leqq p$.

Remark 1.   If γ in (5.16) satisfies $\gamma R < 1$, then (5.5) holds with

$$(5.17) \qquad \alpha = \frac{\gamma}{2(1 - \gamma R)}, \qquad K = \frac{C}{1 - \gamma R}.$$

Thus assumption (5.5) is redundant in Corollary 5.1 when $\gamma R < 1$ (but the example preceding Lemma 5.2 shows that (5.5) cannot be omitted if $\gamma R = 1$). Also if α in (5.5) satisfies $2\alpha R < 1$, then (5.16) holds with

$$(5.18) \qquad \gamma = \frac{2\alpha}{1 - 2\alpha R} \quad \text{and} \quad C = \frac{K}{1 - 2\alpha R},$$

so that (5.16) is redundant in this case. Even if $d = 1$ (so that $x(t)$ is real-valued), condition (5.16) cannot be omitted if $2\alpha R > 1$).

In order to verify the first part of Remark 1, note that

$$(5.19) \qquad r'' = 2(x \cdot x'' + \|x'\|^2).$$

Hence (5.16) shows that $r'' \geqq 2[(1 - \gamma R) \|x'\|^2 - CR]$. Another application of (5.16) gives $\gamma r'' \geqq 2[(1 - \gamma R)(\|x''\| - C) - CR\gamma] = 2[(1 - \gamma R) \|x''\| - C]$. This is the same as (5.5) with the choices (5.17). The proof of the remark concerning (5.18) is similar.

Exercise 5.2.   Show that if $2\alpha R > 1$, then assumption (5.5) cannot be dropped in Corollary 5.1.

The following simple fact will be needed subsequently.

Lemma 5.3. *Let $f(t, x, x')$ be a continuous function on a set*

$$(5.20) \quad E(p, R) = \{(t, x, x'): 0 \leq t \leq p, \|x\| \leq R, x' \text{ arbitrary}\},$$

and let f have one or more of the following properties:

$$(5.21) \quad x \cdot f + \|x'\|^2 > 0 \quad \text{when} \quad x \cdot x' = 0 \quad \text{and} \quad \|x\| > 0,$$

$$(5.22) \quad x \cdot f + \|x'\|^2 > 0 \quad \text{when} \quad x \cdot x' = 0 \quad \text{and} \quad \|x\| = R,$$

$$(5.23) \quad \quad \quad \|f\| \leq \varphi(\|x'\|),$$

$$(5.24) \quad \quad \quad \|f\| \leq 2\alpha(x \cdot f + \|x'\|^2) + K.$$

Let $M > 0$. Then there exists a continuous bounded function $g(t, x, x')$ defined for $0 \leq t \leq p$ and arbitrary (x, x') satisfying

$$(5.25) \quad g(t, x, x') \equiv f(t, x, x') \quad \text{for} \quad 0 \leq t \leq p, \quad \|x\| \leq R, \quad \|x'\| \leq M$$

and having the corresponding set of properties among the following:

$$(5.21') \ x \cdot g + \|x'\|^2 > 0 \quad \text{when} \quad x \cdot x' = 0 \quad \text{and} \quad \|x\| \geq 0,$$

$$(5.22') \ x \cdot g + \|x'\|^2 > 0 \quad \text{when} \quad x \cdot x' = 0 \quad \text{and} \quad \|x\| \geq R,$$

$$(5.23') \quad \quad \quad \|g\| \leq \varphi(\|x'\|),$$

$$(5.24') \quad \quad \quad \|g\| \leq 2\alpha(x \cdot g + \|x'\|^2) + K.$$

Proof. We can obtain such a function g as follows: Let $\delta(s)$, where $0 \leq s < \infty$, be a real-valued continuous function satisfying $\delta = 1$, $0 < \delta < 1$, $\delta = 0$ according as $\delta \leq M$, $M < s \leq 2M$, $s > 2M$. Put

$$g(t, x, x') = \delta(\|x'\|)f(t, x, x') \quad \text{on} \quad E(p, R),$$

$$g(t, x, x') = \frac{R}{\|x\|} g\left(t, \frac{Rx}{\|x\|}, x'\right) \quad \text{for} \quad \|x\| > R.$$

On $E(p, R)$, the identity

$$x \cdot g + \|x'\|^2 = \delta(\|x'\|)(x \cdot f + \|x'\|^2) + [1 - \delta(\|x'\|)] \|x'\|^2$$

makes it clear that g has the desired properties on $E(p, R)$. Furthermore the validity of any of the relations $(5.21')$–$(5.24')$ for $\|x\| = R$ implies its validity for $\|x\| > R$. This proves the lemma.

Note that inequalities of the type (5.23), (5.24) imply that solutions of

$$(5.26) \quad \quad \quad x'' = f(t, x, x')$$

satisfy (5.4), (5.5), respectively; cf. (5.19).

Theorem 5.1. *Let $f(t, x, x')$ be a continuous function on the set $E(p, R)$ in (5.20) satisfying*

$$(5.27) \qquad x \cdot f + \|x'\|^2 \geqq 0 \qquad if \quad x \cdot x' = 0 \quad and \quad \|x\| = R,$$

(5.24) and (5.23), where $\varphi(s)$, $0 \leqq s < \infty$, is a positive continuous function satisfying (5.1). Let $\|x_0\|$, $\|x_p\| \leqq R$. Then (5.26) has at least one solution satisfying $x(0) = x_0$, $x(p) = x_p$.

It will be clear from the proof that assumption (5.23) can be omitted if $2\alpha R < 1$. Furthermore, if f satisfies

$$(5.28) \qquad\qquad \|f\| \leqq \gamma \|x'\|^2 + C,$$

where γ, C are non-negative constants and $\gamma R < 1$, then both assumptions (5.23) and (5.24) can be omitted.

If the vector x is 1-dimensional, Lemma 5.1 can be used in the proof instead of Lemma 5.2. This gives the following:

Corollary 5.2. *Let x be a real variable and $f(t, x, x')$ be a real-valued function in Theorem 5.1. Then the conclusion of Theorem 5.1 remains valid if condition (5.24) is omitted.*

Note that, in this case, condition (5.27) becomes simply $f(t, +R, 0) \geqq 0$ and $f(t, -R, 0) \leqq 0$ for $0 \leqq t \leqq p$.

Proof of Theorem 5.1. The proof will be given first for the case that f satisfies (5.22) instead of (5.27). Let $M > 0$ be a constant (with $p = \tau$) supplied by Lemma 5.2. Let $g(t, x, x')$ be a continuous bounded function for $0 \leqq t \leqq p$ and arbitrary (x, x') satisfying (5.25), (5.22'), (5.23'), and (5.24'). By Theorem 4.2, the boundary value problem

$$x'' = g(t, x, x'), \qquad x(0) = x_0, \qquad and \quad x(p) = x_p$$

has a solution $x(t)$. Condition (5.22') means that $r = \|x(t)\|^2$ satisfies $r'' > 0$ if $r' = 0$ and $r \geqq R^2$; cf. (5.19). Hence $r(t)$ does not have a maximum at any point t, $0 < t < p$, where $r(t) \geqq R^2$. Since $r(0) = \|x_0\|^2$, $r(p) = \|x_p\|^2$ satisfy $r(0)$, $r(p) \leqq R^2$, it follows that $r(t) \leqq R^2$ (i.e., $\|x(t)\| \leqq R$) for $0 \leqq t \leqq p$. By virtue of $x'' = g$ and (5.23'), (5.24'), Lemma 5.2 is applicable to $x(t)$ and implies that $\|x'(t)\| \leqq M$ for $0 \leqq t \leqq p$.

Consequently, (5.25) shows that $x(t)$ is a solution of (5.26). This proves Theorem 5.1 provided that (5.27) is strengthened to (5.22). In order to remove this proviso, note that if $\epsilon > 0$, the function $f(t, x, x') + \epsilon x$ satisfies the conditions of Theorem 5.1 as well as (5.22) if φ, K in (5.23), (5.24) are replaced by $\varphi + \epsilon R$, $K + \epsilon R$, respectively. Hence

$$x'' = f(t, x, x') + \epsilon x$$

has a solution $x = x_\epsilon(t)$ satisfying the boundary conditions. It is clear that $\|x_\epsilon(t)\| \leqq R$ and that there exists a constant M (independent of ϵ, $0 < \epsilon \leqq 1$) such that $\|x_\epsilon'(t)\| \leqq M$. Consequently, if $N = \max \|f(t, x, x')\| + 1$

for $0 \leqq t \leqq p$, $\|x\| \leqq R$, $\|x'\| \leqq M$, then $\|x_\epsilon''(t)\| \leqq N$. Thus the family of functions $x_\epsilon(t)$, $x_\epsilon'(t)$ for $0 \leqq t \leqq p$ are uniformly bounded and equicontinuous. By Arzela's theorem, there is a sequence $1 > \epsilon_1 > \epsilon_2 > \cdots$ such that $\epsilon_n \to 0$ as $n \to \infty$, and $x(t) = \lim x_\epsilon(t)$ exists as $\epsilon = \epsilon_n \to 0$ and is a solution of (5.26) satisfying $x(0) = x_0$, $x(p) = x_p$. This completes the proof of Theorem 5.1.

Exercise 5.3. Show that if (5.27) in Theorem 5.1 is strengthened to

$$(5.29) \qquad x \cdot f + \|x'\|^2 \geqq 0 \qquad \text{when} \quad x \cdot x' = 0,$$

then (5.26) has a solution $x(t)$ satisfying $x(0) = x_0$, $x(p) = 0$, and

$$(5.30) \qquad r \geqq 0, \qquad r' \leqq 0 \qquad \text{if} \quad r = \|x\|^2.$$

Exercise 5.4. Let u be a real variable. Let $h(t, u, u')$ be real-valued and continuous for $0 \leqq t \leqq p$ and all (u, u'), and satisfy the following conditions: (i) h is a nondecreasing function of u for fixed (t, u'); (ii) $|h| \leqq \varphi(|u'|)$ where $\varphi(s)$ is a positive, continuous, nondecreasing function for $s \geqq 0$ satisfying (5.1); (iii) $u'' = h(t, u, u')$ has at least one solution $u_0(t)$ which exists on $0 \leqq t \leqq p$ [e.g., (ii) and (iii) hold if $|h| \leqq \alpha |u'| + K$ for constants α, K]. Let u_0, u_p be arbitrary numbers. Then $u'' = h(t, u, u')$ has at least one solution $u(t)$ satisfying $u(0) = u_0$, $u(p) = u_p$. [For a related uniqueness assertion, see Exercise 4.6(c).]

Theorem 5.2. *Let $f(t, x, x')$ be continuous in*

$$(5.31) \qquad E(R) = \{(t, x, x'): \ 0 \leqq t < \infty, \ \|x\| \leqq R, \ x' \text{ arbitrary}\}.$$

For every $p > 0$, let f satisfy the conditions of Theorem 5.1 on $E(p, R)$ in (5.20), where $\varphi(s)$ and the constants α, K in (5.23), (5.24) can depend on p. Let $\|x_0\| \leqq R$. Then (5.26) has a solution $x(t)$ which satisfies $x(0) = x_0$ and exists for $t \geqq 0$.

Exercise 5.5. (a) Prove Theorem 5.2. (b) Show that if, in addition, (5.27) is strengthened to (5.29) in Theorem 5.2, then the solution $x(t)$ can be chosen so that (5.30) holds. (c) Furthermore, if (5.29) is strengthened to $x \cdot f + \|x'\|^2 \geqq 0$, then $r \geqq 0$, $r' \leqq 0$, $r'' \geqq 0$ for $t \geqq 0$. (d) If x is 1-dimensional, show that condition (5.24) can be omitted from Theorem 5.2 and parts (b) and (c) of this exercise.

Exercise 5.6. Let $f(t, x, x')$ be continuous on the set $E(R)$ in (5.31). For every m, $0 < m < R$, let there exist a continuous function $h(t) = h(t, m)$ for large t such that $\int^\infty th(t) \, dt = \infty$ and $x \cdot f(t, x, x') \geqq h(t) \geqq 0$ for large t, $0 < m \leqq \|x\| \leqq R$, x' arbitrary. Let $x(t)$ be a solution of (5.26) for large t. Then $x(t) \to 0$ as $t \to \infty$.

Exercise 5.7. Let $f(t, x, x')$ be continuous on $E(R)$ in (5.31) and have continuous partial derivatives with respect to the components of x, x'; let the Jacobian matrices (4.23) satisfy $\frac{1}{2}(B + B^*) - \frac{1}{4}FF^* \geqq 0$; cf. (3.17).

Let $\|x_0\| \leqq R$. Then (5.26) has at most one solution satisfying $x(0) = x_0$ and $\|x(t)\| \leqq R$ for $t \geqq 0$.

Remark 2. The main role of the assumptions involving (5.23) and/or (5.24) in Theorems 5.1, 5.2 is to assure that the following holds:

Assumption (A_p). There exists a constant $M = M(p)$ with the property that if $x(t)$ is a solution of $x'' = f(t, x, x')$ for $0 \leqq t \leqq p$ satisfying $\|x(t)\| \leqq R$, then $\|x'(t)\| \leqq M$ for $0 \leqq t \leqq p$.

Exercise 5.8. Let $f(t, x, x')$ be continuous on $E(R)$ in (5.31) and satisfy assumption (A_p) for all $p \geqq p_0 > 0$. Suppose that, for each x_0 in $\|x_0\| \leqq R$, (5.26) has exactly one solution $x(t) = x(t, x_0)$ satisfying $x(0) = x_0$ and existing for $t \geqq 0$ (cf., e.g., Theorem 5.2 and Exercise 5.7.) (*a*) Show that $x(t, x_0)$ is a continuous function of (t, x_0) for $t \geqq 0$, $\|x_0\| \leqq R$. (*b*) Suppose, in addition, that $f(t, x, x')$ is periodic of period p_0 in t for fixed (x, x'). Then (5.26) has at least one solution $x(t)$ of period p_0.

PART III. GENERAL THEORY

6. Basic Facts

The main objects of study in this part of the chapter will be a linear inhomogeneous system of differential equations

$$(6.1) \qquad y' = A(t)y + g(t),$$

the corresponding homogeneous system

$$(6.2) \qquad y' = A(t)y,$$

and a related nonlinear system

$$(6.3) \qquad y' = A(t)y + f(t, y).$$

Let J denote a fixed t-interval $J: 0 \leqq t < \omega (\leqq \infty)$. The symbols $x, y, f, g, \dots$ denote elements of a d-dimensional Banach space Y over the real or complex number field with norms $\|x\|$, $\|y\|$, $\|f\|$, $\|g\|, \dots$. (Here $\|x\|$ is not necessarily the Euclidean norm.) In (6.1), $g = g(t)$ is a locally integrable function on J (i.e., integrable on every closed, bounded subinterval of J). $A(t)$ is an endomorphism of Y for (almost all) fixed t and is locally integrable on J. Thus if a fixed coordinate system is chosen on Y, $A(t)$ is a locally integrable $d \times d$ matrix function on J.

When $y(t)$ is a solution of (6.1) on the interval $[0, a] \subset J$, the fundamental inequality

(6.4) $\|y(t)\| \leqq \left\{ \|y(t')\| + \int_0^a \|g(s)\| \, ds \right\} \exp \int_0^a \|A(s)\| \, ds$ for $0 \leqq t, t' \leqq a$

follows from Lemma IV 4.1. If this relation is integrated with respect to t' over $[0, a]$, we obtain

(6.5) $\|y(t)\| \leqq \left\{ \frac{1}{a} \int_0^a \|y(s)\| \, ds + \int_0^a \|g(s)\| \, ds \right\} \exp \int_0^a \|A(s)\| \, ds$

for $0 \leqq t \leqq a$.

Let $L = L_J$ denote the space of real-valued functions $\varphi(t)$ on J with the topology of convergence in the mean L^1 on compact intervals of J. Thus L is a Fréchet (= complete, linear metric) space. For example, the following metric, which will not be used below, can be introduced on L: let $0 = t_0 < t_1 < t_2 < \ldots, t_n \to \omega$ as $n \to \infty$, and let the distance between $\varphi, \psi \in L$ be

$$d(\varphi, \psi) = \sum_{n=1}^{\infty} \frac{2^n}{2^n[1 + I(n)]}, \qquad \text{where} \quad I(n) = \int_0^{t_n} |\varphi - \psi| \, dt.$$

Correspondingly, let $C = C_J$ denote the space of continuous, real-valued functions $\varphi(t)$ on J with the topology of uniform convergence on compact interval of J. Thus C is also a Fréchet space. A metric on C, e.g., is

$$d(\varphi, \psi) = \sum_{n=1}^{\infty} \frac{m(n)}{2^n[1 + m(n)]}, \qquad \text{where} \quad m(n) = \max_{[0, t_n]} |\varphi(t) - \psi(t)|.$$

The symbols $L^p = L_J^p$, $1 \leqq p \leqq \infty$, denote the usual Banach spaces of real-valued functions $\varphi(t)$ on $J: 0 \leqq t < \omega \ (\leqq \infty)$ with the norm

$$|\varphi|_p = \left(\int_J |\varphi(t)|^p \, dt \right)^{1/p} \qquad \text{if} \quad 1 \leqq p < \infty,$$

$$|\varphi|_\infty = \operatorname*{ess\,sup}_J |\varphi(t)| \qquad \text{if} \quad p = \infty.$$

L_0^∞ is the subspace of L^∞ consisting of functions $\varphi(t)$ satisfying $\varphi(t) \to 0$ as $t \to \omega$. For other Banach spaces B of real-valued, measurable functions $\varphi(t)$ in J, the notation $|\varphi|_B$ will be used for the norm of $\varphi(t)$ in B.

Remark. Strictly speaking, the spaces L, L^∞, $L_0^\infty, \ldots$ are not spaces of "real-valued functions" but rather spaces of "equivalence classes of real-valued functions," where two functions are in the same equivalence class if they are equal except on a set of Lebesgue measure zero. Since no confusion will arise, however, over this "abuse of language," the abbreviated terminology will be used. In this terminology, the meaning of a "continuous function in L" or the "intersection $L \cap C$" is clear.

$L(Y)$, $L^p(Y)$, $B(Y)$, ... will represent the space of measurable vector-valued functions $y(t)$ on $J: 0 \leq t < \omega$ ($\leq \infty$) with values in Y such that $\varphi(t) = \|y(t)\|$ is in L, L^p, B, ... With L^p or B, the norm $|\varphi|_p$ or $|\varphi|_B$ will be abbreviated to $|y|_p$ or $|y|_B$.

A Banach space $\mathfrak{D}$ will be said to be *stronger than* $L(Y)$ when (i) $\mathfrak{D}$ is contained in $L(Y)$ algebraically and (ii) for every a, $0 < a < \omega$, there is a number $\alpha = \alpha_{\mathfrak{D}}(a)$ such that $y(t) \in \mathfrak{D}$ implies

$$(6.6) \qquad \int_0^a \|y(t)\| \, dt \leq \alpha |y|_{\mathfrak{D}}, \qquad \text{where} \quad \alpha = \alpha_{\mathfrak{D}}(a).$$

[It is easily seen from the Open Mapping Theorem 0.3 that condition (ii) is equivalent to: "convergence in $\mathfrak{D}$ implies convergence in $L(Y)$."]

If $\mathfrak{D}$ is a Banach space stronger than $L(Y)$, a $\mathfrak{D}$-solution $y(t)$ of (6.1) or (6.2) means a solution $y(t) \in \mathfrak{D}$. Let $Y_{\mathfrak{D}}$ denote the set of initial points $y(0) \in Y$ of $\mathfrak{D}$-solutions $y(t)$ of (6.2). Then $Y_{\mathfrak{D}}$ is a subspace of Y. Let Y_1 be a subspace of Y complementary to $Y_{\mathfrak{D}}$; i.e., Y_1 is a subspace of Y such that $Y = Y_{\mathfrak{D}} \oplus Y_1$ is the direct sum of $Y_{\mathfrak{D}}$ and Y_1, so that every element $y \in Y$ has a unique representation $y = y_0 + y_1$ with $y_0 \in Y_{\mathfrak{D}}, y_1 \in Y_1$ (e.g., if Y is a Euclidean space, Y_1 can be, but need not be, the subspace of Y orthogonal to $Y_{\mathfrak{D}}$). Let P_0 be the projection of Y onto $Y_{\mathfrak{D}}$ annihilating Y_1; thus if $y = y_0 + y_1$ with $y_0 \in Y_{\mathfrak{D}}, y_1 \in Y_1$, then $P_0 y = y_0$.

Lemma 6.1. *Let $A(t)$ be locally integrable on J and let $\mathfrak{D}$ be a Banach space stronger than $L(Y)$. Then there exist constants C_0, C_1 such that if $y(t)$ is a $\mathfrak{D}$-solution of (6.2), then*

$$(6.7) \qquad |y|_{\mathfrak{D}} \leq C_0 \|y(0)\| \quad \text{and} \quad \|y(0)\| \leq C_1 |y|_{\mathfrak{D}}.$$

Proof. $Y_{\mathfrak{D}}$ is a subspace of the finite dimensional space Y. In addition, there is a one-to-one, linear correspondence between solutions $y(t)$ of (6.2) and their initial points $y(0)$. Thus the set of $\mathfrak{D}$-solutions of (6.2) is a finite dimensional subspace of $\mathfrak{D}$ which is in one-to-one, linear correspondence with $Y_{\mathfrak{D}}$. It is a well known and easily verified fact that if two finite-dimensional, normed linear spaces can be put into one-to-one correspondence, then the norm of an element of one space is majorized by a constant times the norm of the corresponding element of the other space. [For example, an admissible choice of C_1 is a

$$a^{-1}\alpha_{\mathfrak{D}}(a) \exp \int_0^a \|A(s)\| \, ds$$

for any a, $0 < a < \omega$. This follows from (6.6) and the choices $t = 0$, $g(s) = 0$ in (6.5).]

Let $\mathfrak{B}$, $\mathfrak{D}$ be Banach spaces stronger than $L(X)$. Define an operator $T = T_{\mathfrak{B}\mathfrak{D}}$ from $\mathfrak{D}$ to $\mathfrak{B}$ as follows: The domain $\mathscr{D}(T) \subset \mathfrak{D}$ of T is the set of functions $y(t)$, $t \in J$, which are absolutely continuous (on compact subintervals of J), $y(t) \in \mathfrak{D}$, and $y'(t) - A(t)\,y(t) \in \mathfrak{B}$. For such a function $y(t)$, Ty is defined to be $y'(t) - A(t)\,y(t)$. In other words, $Ty = g$, where $g(t) \in \mathfrak{B}$ is given by (6.1).

Lemma 6.2. *Let $A(t)$ be locally integrable on J and let $\mathfrak{B}$, $\mathfrak{D}$ be Banach spaces stronger than $L(Y)$. Then $T = T_{\mathfrak{B}\mathfrak{D}}$ is a closed operator; that is the graph of T, $\mathscr{G}(T) = \{(y(t), g(t)) : y(t) \in \mathscr{D}(T), g = Ty\}$, is a closed set of the Banach space $\mathfrak{D} \times \mathfrak{B}$.*

Proof. In order to prove this, it must be shown that if $y_1(t), y_2(t), \dots$ are elements of $\mathscr{D}(T)$, $g_n = Ty_n$, $y(t) = \lim y_n(t)$ exists in $\mathfrak{D}$ and $g(t) = \lim g_n(t)$ exists in $\mathfrak{B}$, then $y(t) \in \mathscr{D}(T)$ and $g(t) = Ty$.

The basic inequality (6.5) combined with (6.6) and the analogue of (6.6) for the space $\mathfrak{B}$ give

$$\|y_n(t) - y_m(t)\| \leqq \left\{\frac{1}{a}\,\alpha_{\mathfrak{D}}\,|y_n - y_m|_{\mathfrak{D}} + \alpha_{\mathfrak{B}}\,|g_n - g_m|_{\mathfrak{B}}\right\} \exp\int_0^a \|A(s)\|\,ds.$$

Hence $y(t)$ is the uniform limit of $y_1(t), y_2(t), \dots$ on any interval $[0, a] \subset J$.

The differential equation (6.1) is equivalent to the integral equation

$$y(t) = y(a) + \int_a^t A(s)\,y(s)\,ds + \int_a^t g(s)\,ds.$$

Since the convergence of $g_1, g_2, \dots$ in $\mathfrak{B}$ implies its convergence in $L(Y)$, it follows that (6.1) holds where $y = \lim y_n(t)$ in $\mathfrak{D}$, $g = \lim g_n(t)$ in $\mathfrak{B}$. Finally, $y \in \mathfrak{D}$, $g \in \mathfrak{B}$ show that $y \in \mathscr{D}(T)$. This proves Lemma 6.2.

The pair of Banach spaces $(\mathfrak{B}, \mathfrak{D})$ is said to be *admissible* for (6.1) or for $A(t)$ if each is stronger than $L(Y)$ and, for every $g(t) \in \mathfrak{B}$, the differential equation (6.1) has a $\mathfrak{D}$-solution. In other words, the map $T = T_{\mathfrak{B}\mathfrak{D}} : \mathscr{D}(T) \to \mathfrak{B}$ is onto, i.e., the range of T is $\mathfrak{B}$. (For example, if $J : 0 \leqq t < \infty$, $A(t)$ is continuous of period p, and $\mathfrak{B} = \mathfrak{D}$ is the Banach space of continuous functions $y(t)$ of period p with norm $|y|_{\mathfrak{D}} = \sup\|y(t)\|$, then $(\mathfrak{B}, \mathfrak{D})$ is admissible for (6.1) if and only if (6.2) has no nontrivial solution of period p; see Theorem 1.1.)

Lemma 6.3. *Let $A(t)$ be locally integrable on J, let $(\mathfrak{B}, \mathfrak{D})$ be admissible for (6.1), and let $y_0 \in Y_{\mathfrak{D}}$. Then, if $g(t) \in \mathfrak{B}$, (6.1) has a unique $\mathfrak{D}$-solution $y(t)$ such that $P_0y(0) = y_0$. Furthermore, there exist positive constants C_0 and K, independent of $g(t)$, satisfying*

$$(6.8) \qquad\qquad |y|_{\mathfrak{D}} \leqq C_0\|y_0\| + K\,|g|_{\mathfrak{B}}.$$

Proof. Consider first the case that $y_0 = 0$, so that we seek $\mathfrak{D}$-solutions $y(t)$ with $y(0) \in Y_1$. For any $g \in \mathfrak{B}$, (6.1) has a solution $y(t) \in \mathfrak{D}$, by assumption. Let $y(0) = y_0 + y_1$, where $y_0 = P_0y(0) \in Y_{\mathfrak{D}}$, $y_1 \in Y_1$. Let $y_0(t)$ be the solution of the homogeneous equation (6.2) such that $y_0(0) = y_0$, so that $y_0(t) \in \mathfrak{D}$. Then $y_1(t) = y(t) - y_0(t) \in \mathfrak{D}$ is a solution of (6.1) and $y_1(0) = y_1 \in Y_1$.

It is clear that $y_1(t)$ is a unique $\mathfrak{D}$-solution of (6.2) with initial point in Y_1. Thus there is a one-to-one linear correspondence between $g \in \mathfrak{B}$ and $\mathfrak{D}$-solutions $y_1(t)$ of (6.2) with $y_1(0) \in Y_1$. The proof of Lemma 6.2 shows that if T_1 is the restriction of $T = T_{\mathfrak{B}\mathfrak{D}}$ with domain consisting of elements $y(t) \in \mathcal{D}(T)$ satisfying $y(0) \in Y_1$, then T_1 is closed. Thus T_1 is a closed, linear, one-to-one operator which maps its domain in $\mathfrak{D}$ onto $\mathfrak{B}$. By the Open Mapping Theorem 0.3, there is a constant K such that if $T_1 y = g$, then $|y|_{\mathfrak{D}} \leqq K |g|_{\mathfrak{B}}$. This proves the theorem for $y_0 = 0$.

If $y_0 \neq 0$, let $y_1(t)$ be the unique $\mathfrak{D}$-solution of (6.2) satisfying $y_1(0) \in Y_1$. Let $y_0(t)$ be the unique $\mathfrak{D}$-solution of the homogeneous equation (6.2) satisfying $y_0(0) = y_0$. Then $y(t) = y_0(t) + y_1(t)$ is a $\mathfrak{D}$-solution of (6.1), $P_0 y(0) = y_0$, and $|y|_{\mathfrak{D}} \leqq |y_0(t)|_{\mathfrak{D}} + |y_1(t)|_{\mathfrak{D}}$. By the part of the lemma already proved, $|y_1(t)|_{\mathfrak{D}} \leqq K |g|_{\mathfrak{B}}$ and, by Lemma 6.1, $|y_0(t)|_{\mathfrak{D}} \leqq C_0 \|y_0\|$. This completes the proof of Lemma 6.3.

7. Green's Functions

Let $h_{0a}(t)$ be the characteristic function of the interval $0 \leqq t \leqq a$, so that $h_{0a}(t) = 1$ or 0 according as $0 \leqq t \leqq a$ does or does not hold. Similarly, let $h_a(t)$ be the characteristic function of the half-line $t \geqq a$, so that $h_a(t) = 1$ or 0 according as $t \geqq a$ or $t < a$.

A Banach space $\mathfrak{B}$ of functions on J: $0 \leqq t < \omega$ ($\leqq \infty$) will be called *lean at* $t = \omega$ if $\psi(t) \in \mathfrak{B}$ and $0 < a < \omega$ imply that $h_{0a}(t)\psi(t)$, $h_a(t)\psi(t) \in \mathfrak{B}$; $|h_{0a}\psi|_{\mathfrak{B}}$, $|h_a\psi|_{\mathfrak{B}} \leqq |\psi|_{\mathfrak{B}}$; and $|h_a\psi|_{\mathfrak{B}} \to 0$ as $a \to \omega$. Since $h_a(t)\psi(t) = \psi(t) - h_{0a}(t)\psi(t)$ on J, the property "lean at $t = \omega$" implies that the set of functions $h_{0a}(t)\psi(t)$ of $\mathfrak{B}$ vanishing outside of compact intervals $[0, a] \subset J$ is dense in $\mathfrak{B}$.

Let $\mathfrak{D}$ be a Banach space stronger than $L(Y)$. As above, let $Y_1 = Y_{1\mathfrak{D}}$ be a subspace of Y complementary to $Y_{\mathfrak{D}}$. Let $P_0 = P_{0\mathfrak{D}}$ be the projection of Y onto $Y_{\mathfrak{D}}$ annihilating Y_1, and $P_1 = I - P_0$ the projection of Y onto Y_1 annihilating $Y_{\mathfrak{D}}$. In terms of a fixed basis on Y, P_0 and P_1 are representable as matrices.

Let $U(t)$ be the fundamental matrix for (6.1) on $0 \leqq t < \omega$ satisfying $U(0) = I$. For $0 \leqq s, t < \omega$, define a (matrix) function $G(t, s)$ by

(7.1)
$$G(t, s) = U(t)P_0 U^{-1}(s) \quad \text{for} \quad 0 \leqq s \leqq t,$$

$$G(t, s) = -U(t)P_1 U^{-1}(s) \quad \text{for} \quad 0 \leqq t < s.$$

For a fixed t, $G(t, s)$ is continuous on $0 \leqq s < \omega$, except at $s = t$, where it has left and right limits, $U(t)P_0 U^{-1}(t)$ and $-U(t)P_1 U^{-1}(t)$.

Theorem 7.1. *Let $A(t)$ be locally integrable on J. Suppose that $\mathfrak{B}$, $\mathfrak{D}$ are Banach spaces stronger than $L(Y)$; that $\mathfrak{B}$ is lean at ω; and that*

$\mathfrak{D}$ *has the property that if* $y(t)$, $y_1(t)$ *are continuous functions from* J *to* Y *and* $y(t) - y_1(t) \equiv 0$ *near* $t = \omega$ *(i.e.,* $y_1 - y_2 = 0$ *except on an interval* $[0, a] \subset J$), *then* $y(t) \in \mathfrak{D}$ *implies that* $y_1(t) \in \mathfrak{D}$. *Then* $(\mathfrak{B}, \mathfrak{D})$ *is admissible for* (6.1) *if and only if, for every* $g(t) \in \mathfrak{B}$,

$$(7.2) \qquad y(t) = \int_0^\omega G(t, s)g(s)\, ds = \lim_{a \to \omega} \int_0^a G(t, s)g(s)\, ds$$

exists in $\mathfrak{D}$. *In this case, the limit is uniform on compact intervals of* J *and is the unique* $\mathfrak{D}$-*solution of* (6.1) *with* $y(0) \in Y_1$.

Proof. *"Only if"*. Let $g(t) \in \mathfrak{B}$, $g_a(t) = h_{0a}(t)g(t)$. Then (7.2) becomes

$$(7.3) \qquad y_a(t) = \int_0^\omega G(t, s)g_a(s)\, ds = \int_0^a G(t, s)g(s)\, ds,$$

where the integral exists as a Lebesgue integral for every fixed t, since $G(t, s)$ is bounded for $0 \leq s \leq a$ and $g_a(s)$ is integrable over J. In view of the first part of (7.1), the contribution of $0 \leq s \leq t$ to (7.3) is

$$U(t)P_0 \int_0^t U^{-1}(s)g_a(s)\, ds = U(t) \int_0^t U^{-1}(s)g_a(s)\, ds - U(t)P_1 \int_0^t U^{-1}(s)g_a(s)\, ds.$$

Hence, by the second part of (7.1), (7.3) is

$$(7.4) \qquad y_a(t) = U(t) \int_0^t U^{-1}(s)g_a(s)\, ds + U(t)y_a(0),$$

where

$$(7.5) \qquad y_a(0) = -P_1 \int_0^\omega U^{-1}(s)g_a(s)\, ds.$$

It follows from (7.4) and Corollary IV 2.1 that $y_a(t)$ is a solution of (6.1) when $g(t)$ is replaced by $g_a(t)$.

An analogue of the derivation of (7.4) gives

$$y_a(t) = -U(t) \int_t^a U^{-1}(s)g_a(s)\, ds + U(t)P_0 \int_0^a U^{-1}(s)g_a(s)\, ds.$$

Hence

$$U^{-1}(a)y_a(a) = P_0 \int_0^a U^{-1}(s)g_a(s)\, ds \in Y_{\mathfrak{D}}.$$

Thus for $a \leq t < \omega$, $y_a(t)$ is identical with the solution $U(t)\, U^{-1}(a)y_a(a)$ of the homogeneous equation (6.2). Since the initial point of the latter solution is in $Y_{\mathfrak{D}}$, the property assumed for $\mathfrak{D}$ implies that $y_a(t) \in \mathfrak{D}$.

Since $y_a(0) \in Y_1$ by (7.5), it follows that $y_a(t)$ is the unique solution of (6.1), where $g = g_a(t)$, satisfying $y_a(0) \in Y_1$. Hence, by Lemma 6.3, $|y_a|_{\mathfrak{D}} \leq K |g_a|_{\mathfrak{B}}$.

Let $0 < a < b < \omega$. Then, since $\mathfrak{B}$ is lean at $t = \omega$,

$$|y_a - y_b|_{\mathfrak{D}} \leqq K |g_a - g_b|_{\mathfrak{B}} \leqq 2K |h_a g|_{\mathfrak{B}} \to 0 \qquad \text{as} \quad a \to \omega$$

Thus $y = \lim y_a(t)$ exists in $\mathfrak{D}$ as $a \to \omega$. Also $g = \lim g_a(t)$ in $\mathfrak{B}$. Since $T = T_{\mathfrak{B}\mathfrak{D}}$ in Lemma 6.2 is closed, $y(t)$ is a $\mathfrak{D}$-solution of (6.1). The proof of Lemma 6.3 shows that $y = \lim y_a(t)$ uniformly on compact intervals of J. Hence, $y(0) = \lim y_a(0) \in Y_1$. This proves "only if" in Theorem 7.1. The "if" part is easy.

Corollary 7.1. *Let $\omega = \infty$; B and D be Banach spaces of class $\mathcal{T}^{\#}$; B' be the space associate to B; cf. § XIII 9. For the admissibility of $(B(Y), D(Y))$,(i) it is necessary that $\|G(t, \cdot)\| \in B'$ for fixed t—thus the integrals in (7.2) are Lebesgue integrals; (ii) when B is lean at ∞, it is necessary and sufficient that (7.2) define a bounded operator $g \to y$ from $B(Y)$ to $D(Y)$; (iii) it is sufficient that $r(t) \in D$ where $r(t) = |\, \|G(t, \cdot)\| \,|_{B'}$; (iv) when $D = L^\infty$, it is necessary and sufficient that $r(t) \in L^\infty$.*

Exercise 7.1. Verify this corollary.

8. Nonlinear Equations

Lemmas 6.1–6.3 will be used to study the nonlinear equation

$$(8.1) \qquad y' = A(t)y + f(t, y).$$

Let $\mathfrak{B}$, $\mathfrak{D}$ be Banach spaces stronger than $L(Y)$ and Σ_ρ the closed ball

$$\Sigma_\rho = \{y(t): y(t) \in \mathfrak{D}, \quad |y|_{\mathfrak{D}} \leqq \rho\} \qquad \text{in } \mathfrak{D}.$$

Theorem 8.1. *Let J; $0 \leqq t < \omega$ $(\leqq \infty)$; $A(t)$ a locally, integrable $d \times d$ matrix function on J, and $(\mathfrak{B}, \mathfrak{D})$ admissible for (6.1). Let $f(t, y(t))$ be an element of $\mathfrak{B}$ for every $y(t) \in \Sigma_\rho$ and satisfy*

$$(8.2) \qquad |f(t, y_1(t)) - f(t, y_2(t))|_{\mathfrak{B}} \leqq \theta |y_1(t) - y_2(t)|_{\mathfrak{D}}$$

for all $y_1(t), y_2(t) \in \Sigma_\rho$ and some constant θ; $r = |f(t, 0)|_{\mathfrak{B}}$; $y_0 \in Y_{\mathfrak{D}}$. Suppose that if C_0, K are the constants in Lemma 6.3, then $\theta, r, \|y_0\|$ are so small that

$$(8.3) \qquad C_0\|y_0\| + Kr \leqq \rho(1 - \theta K) \qquad \text{and} \quad \theta K < 1.$$

Then (8.1) has a unique solution $y(t) \in \Sigma_\rho$ satisfying

$$(8.4) \qquad P_0 y(0) = y_0.$$

It will be clear from the proof that the first part of (8.3) can be replaced by the assumption

$$(8.5) \qquad C_0 \|y_0\| + K |\, f(t, y(t)|_{\mathfrak{B}} \leqq \rho \qquad \text{for all} \quad y(t) \in \Sigma_\rho.$$

In fact, the role of the assumption in (8.3) is to assure (8.5). In (8.4), P_0 is the projection of Y onto $Y_{\mathfrak{D}}$ annihilating a fixed subspace Y_1, where $Y = Y_{\mathfrak{D}} \oplus Y_1$.

Proof. Theorem 8.1 is an immediate consequence of Theorem 0.1 and Lemma 6.3. Since $f(t, x(t)) \in \mathfrak{B}$ for any $x(t) \in \Sigma_\rho$, Lemma 6.3 and the assumption that $(\mathfrak{B}, \mathfrak{D})$ is admissible imply that

$$(8.6) \qquad\qquad y' = A(t)y + f(t, x(t))$$

has a unique $\mathfrak{D}$-solution $y(t)$ satisfying (8.4) and (6.8), where $g(t) = f(t, x(t))$. Define the operator T_0 from Σ_ρ into $\mathfrak{D}$ by $y(t) = T_0[x(t)]$. In particular, if $m = |T_0[0]|_{\mathfrak{D}}$, then

$$(8.7) \qquad m \leqq C_0 \|y_0\| + Kr, \qquad \text{where} \quad r = |f(t, 0)|_{\mathfrak{B}}.$$

If $x_1(t), x_2(t) \in \Sigma_\rho$ and $y_1 = T_0[x_1]$, $y_2 = T_0[x_2]$, it follows that $y_1(t) - y_2(t)$ is the unique $\mathfrak{D}$-solution of

$$y' = A(t)y + f(t, x_1(t)) - f(t, x_2(t))$$

satisfying $P_0 y(0) = 0$. Hence, by Lemma 6.3 and by (8.2),

$$(8.8) \qquad\qquad |y_1 - y_1|_{\mathfrak{D}} \leqq \theta K |x_1 - x_2|_{\mathfrak{D}}.$$

Consequently, Theorem 0.1 is applicable, and so T_0 has a unique fixed point $y(t) \in \Sigma_\rho$. This proves Theorem 8.1.

The statement of the next theorem involves the space $C(Y)$ of continuous functions $y(t)$ from J to Y with the topology of uniform convergence on compact intervals in J. The theorem will also involve an assumption concerning the continuity of the map $T_1[y(t)] = f(t, y(t))$ from the closure of the subset $\Sigma_\rho \cap C(Y)$ of $C(Y)$ into $\mathfrak{B}$. This condition is rather natural in dealing with Banach spaces $\mathfrak{B}, \mathfrak{D}$ of continuous functions on J with norms which imply uniform convergence on J. This is the case in Parts I and II, where J is replaced by a closed bounded interval $0 \leqq t \leqq p$. This continuity condition will also be satisfied under different circumstances in Corollary 8.1.

Theorem 8.2. *Let $A(t)$ be locally integrable on J; $\mathfrak{B}, \mathfrak{D}$ Banach spaces stronger that $L(Y)$; Σ_ρ the closed ball of radius ρ in $\mathfrak{D}$; and S the closure of $\Sigma_\rho \cap C(Y)$ in $C(Y)$. Let $A(t)$ and $f(t, y)$ satisfy* (i) *$(\mathfrak{B}, \mathfrak{D})$ is admissible for* (6.1); (ii) *$y(t) \to f(t, y(t))$ is a continuous map of the subset S of the space $C(Y)$ into $\mathfrak{B}$;* (iii) *there exists an $r > 0$ such that*

$$(8.9) \qquad |f(t, y(t))|_{\mathfrak{B}} \leqq r \qquad \text{for} \quad y(t) \in S;$$

and (iv) *there exists a function $\lambda(t) \in L$ such that*

$$(8.10) \qquad \|f(t, y(t))\| \leqq \lambda(t) \qquad \text{for} \quad t \in J, \quad y(t) \in S.$$

Let C_0, K be the constants of Lemma 6.3 and let $y_0 \in Y_\mathfrak{D}$. Let r, $\|y_0\|$ be so small that

$$(8.11) \qquad\qquad C_0\|y_0\| + Kr \leqq \rho.$$

Then (8.1) has at least one solution $y(t) \in \Sigma_\rho$ satisfying $P_0 y(0) = y_0$.

Proof. As in the last proof, define an operator T_0 of S into $\mathfrak{D}$ by putting $y = T_0[x]$, where $x(t) \in S$ and $y(t)$ is the unique $\mathfrak{D}$-solution of (8.6) satisfying (8.4). Thus, by Lemma 6.3,

$$|y|_\mathfrak{D} \leqq C_0 \|y_0\| + K |f(t, x(t))|_\mathfrak{B} \leqq C_0 \|y_0\| + Kr.$$

Hence assumption (8.11) implies that T_0 maps S into itself, in fact, into $\Sigma_\rho \cap C(Y) \subset S$.

Note that the basic inequality (6.5) implies that

$$\|y(t)\| \leqq \left\{ \frac{1}{a} \int_0^a \|y(s)\| \, ds + \int_0^a \|g(s)\| \, ds \right\} \exp \int_0^a \|A(s)\| \, ds$$

for $0 \leqq t \leqq a$ if $g(t) = f(t, x(t))$. Since $\mathfrak{D}$ is stronger than $L(Y)$, (6.6) holds. Also there is a similar inequality for elements $g \in \mathfrak{B}$ with a suitable constant $\alpha_\mathfrak{B} (a)$. Hence, for $0 \leqq t \leqq a$,

$$(8.12) \qquad \|y(t)\| \leqq \left\{ \frac{1}{a} \alpha_\mathfrak{D}(a) |y|_\mathfrak{D} + \alpha_\mathfrak{B}(a) |g|_\mathfrak{B} \right\} \exp \int_0^a \|A(s)\| \, ds.$$

It will first be verified that $T_0 \colon S \to S$ is continuous where S is considered to be a subset of $C(Y)$. Let $x_j(t) \in S$, $g_j(t) = f(t, x_j(t))$, $y_j(t) = T_0[x_j(t)]$ for $j = 1, 2$, then $y_1(t) - y_2(t)$ is the unique $\mathfrak{D}$-solution of (6.1), where $g = g_1 - g_2$, satisfying $P_0[y_1(0) - y_2(0)] = 0$. Hence Lemma 6.3 implies that

$$|y_1 - y_2|_\mathfrak{D} \leqq K |g_1 - g_2|_\mathfrak{B}.$$

Also, (8.12) holds if $y = y_1 - y_2$ and $g = g_1 - g_2$. Thus, for $0 \leqq t \leqq a$

$$\|y_1(t) - y_2(t)\| \leqq \left\{ \frac{1}{a} \alpha_\mathfrak{D}(a)K + \alpha_\mathfrak{B}(a) \right\} |g_1 - g_2|_\mathfrak{B} \exp \int_0^a \|A(s)\| \, ds.$$

Since, by assumption (ii), $x_1(t) \to x_2(t)$ in $C(Y)$ implies $g_1 \to g_2$ in $\mathfrak{B}$, it follows that $y_1(t) \to y_2(t)$ uniformly on intervals $[0, a]$ of J; i.e., $y_1(t) \to y_2(t)$ in $C(Y)$. This proves the continuity of $T_0 \colon S \to S$.

It will now be shown that the image $T_0 S$ of S has a compact closure in $C(Y)$. It follows from (8.12), where $g(t) = f(t, x(t))$ and $y(t) = T_0[x(t)]$ that, for $0 \leqq t \leqq a$,

$$\|y(t)\| \leqq \left\{ \frac{1}{a} \alpha_\mathfrak{D}(a)\rho + \alpha_\mathfrak{B}(a)r \right\} \exp \int_0^a \|A(s)\| \, ds.$$

Thus the set of functions $y(t) \in T_0 S$ are uniformly bounded on every interval $[0, a]$ of J. If $c(a)$ is the number on the right of the last inequality, then (8.6) and (8.10) show that

$$\|y(t) - y(s)\| \leqq c(a) \int_s^t \|A(u)\| \, du + \int_s^t \lambda(u) \, du \qquad \text{for} \quad 0 \leqq s \leqq t \leqq a.$$

Therefore, the functions $y(t)$ in the image $T_0 S$ of S are equicontinuous on every interval $[0, a] \subset J$. Consequently, Arzela's theorem shows that $T_0 S$ has a compact closure in $C(Y)$. Since S is convex and closed in $C(Y)$, it follows from Corollary 0.1 that T_0 has a fixed point $y(t) \in S$. Thus Theorem 8.2 is a consequence of the fact that $y(t) = T_0[y(t)] \in \Sigma_p \cap C(Y)$.

It is convenient to have conditions on $\mathfrak{B}$, $\mathfrak{D}$, $f(t, y)$, $\lambda(t)$ which imply (ii), (iii), (iv) in Theorem 8.2.

Assumption (H_0) on $\mathfrak{B} = \mathbf{B(X)}$: *Let $\mathfrak{B} = B(X)$ (cf. § 6), where X is a subspace of Y and B is a Banach space of real-valued functions on J such that (i) B is stronger than L; (ii) B is lean at $t = \omega$ (cf. § 7); (iii) B contains the characteristic function $h_{0a}(t)$ of the intervals $[0, a] \subset J$; and (iv) if $\varphi_1(t) \in B$ and $\varphi_2(t)$ is a measurable function on J such that $|\varphi_2(t)| \leqq |\varphi_1(t)|$, then $\varphi_2(t) \in B$ and $|\varphi_2|_\mathfrak{B} \leqq |\varphi_1|_\mathfrak{B}$.*

It is important to have $\mathfrak{B} = B(X)$ rather than $\mathfrak{B} = B(Y)$ for applications to higher order equations. If such equations are written as systems of differential equations of the first order, the "inhomogeneous term $f(t, y)$" will generally belong to a subspace X of Y; e.g., $f(t, y)$ might be of the form $(h, 0, \ldots, 0)$.

Examples of spaces B satisfying the conditions in (H_0) are $B = L^p$, $1 \leqq p < \infty$, and $B = L_0^\infty$ (but not $B = L^\infty$). Other such spaces B can be obtained as follows: Let $\psi(t) > 0$ be a measurable function such that $\psi(t)$ and $1/\psi(t)$ are bounded on every interval $0 \leqq t \leqq a$ ($< \omega$). Denote by $B = L_{\psi 0}^\infty$ the space of functions $\varphi(t)$ on J such that $\varphi(t)/\psi(t) \in L_0^\infty$ with the norm $|\varphi|_\mathfrak{B} = |\varphi/\psi|_\infty$. The space $B = L_{\psi 0}^\infty$ satisfies conditions (i)–(iv). For this space, $\lambda(t) \in B$ holds if

$$(8.13) \qquad 0 \leqq \lambda(t) \leqq \psi(t) \quad \text{and} \quad \frac{\lambda(t)}{\psi(t)} \to 0 \quad \text{as} \quad t \to \omega.$$

Assumption (H_1) on $f(t, y)$: *Let $f(t, y)$ be continuous on the product set of J and the ball $\|y\| \leqq \rho$ in Y, let f have values in X, and let there exist a function $\lambda(t) \in L$ such that*

$$(8.14) \qquad \qquad \|f(t, y)\| \leqq \lambda(t) \quad \text{for} \quad t \in J, \|y\| \leqq \rho.$$

Corollary 8.1. *Let $A(t)$ be locally integrable on J, $(\mathfrak{B}, \mathfrak{D})$ admissible for (6.1), $\mathfrak{B}$ satisfies (H_0), $\mathfrak{D} = L^\infty(Y)$ [or $\mathfrak{D} = L_0^\infty(Y)$], $f(t, y)$ satisfies (H_1) and $\lambda(t) \in B$ with $r = |\lambda|_B$. Let $y_0 \in Y_\mathfrak{D}$. Then, if (8.11) holds, (8.1) has* *at least one solution $y(t)$ on $0 \leq t < \omega$ satisfying $P_0 y(0) = y_0$, $\|y(t)\| \leq \rho$ [and $y(t) \to 0$ as $t \to \omega$].*

Exercise 8.1.   Verify Corollary 8.1.

Exercise 8.2.   Let Y be expressed as a direct sum $Y_\mathfrak{D} \oplus Y_1$; let P_0 be the projection of Y onto $Y_\mathfrak{D}$ annihilating Y_1, and $P_1 = I - P_0$ the projection of Y onto Y_1 annihilating $Y_\mathfrak{D}$. Let $A(t)$ be locally integrable on J: $0 \leq t < \infty$. Define $G(t, s)$ by (7.1) and suppose that there exist constants N, $\nu > 0$ such that $\|G(t, s)\| \leq Ne^{-\nu|t-s|}$ for s, $t \geq 0$. Let $f(t, y)$ be continuous for $0 \leq t < \infty$, $\|y\| \leq \rho$, and let $\|f(t, y)\| \leq r$. Let $y_0 \in Y_\mathfrak{D}$. Show that if $\|y_0\|$ and $r > 0$ are sufficiently small, then (8.1) has a solution $y(t)$ for $0 \leq t < \infty$ satisfying $\|y(t)\| \leq \rho$ and $P_0 y(0) = y_0$. (For necessary and sufficient conditions assuring these assumptions on G, see Theorems XIII 2.1 and XIII 6.4.)

9. Asymptotic Integration

In this section, let J be the half-line J: $0 \leq t < \infty$ (so that $\omega = \infty$). As a corollary of Theorem 8.2, we have:

Theorem 9.1. *Let $A(t)$ be continuous on J: $0 \leq t < \infty$. Let $f(t, y)$ be continuous for $t \geq 0$, $\|y\| \leq \rho$, satisfy*

$$(9.1) \qquad \|f(t, y)\| \leq \lambda(t) \quad \text{for} \quad t \geq 0, \quad \|y\| \leq \rho,$$

and have values in a subspace X of Y. Assume either (i) that $\lambda(t) \in L^1$ and that $(L^1(X), \mathfrak{D})$, where $\mathfrak{D} = L^\infty(Y)$ [or $\mathfrak{D} = L_0^\infty(Y)$], is admissible for

$$(9.2) \qquad y' = A(t)y + g(t);$$

or (ii) that there exists a measurable function $\psi(t) > 0$ on J such that $\psi(t)$ and $1/\psi(t)$ are locally bounded, that

$$(9.3) \qquad 0 \leq \lambda(t) \leq \psi(t) \quad \text{and} \quad \frac{\lambda(t)}{\psi(t)} \to 0 \quad \text{as} \quad t \to \infty,$$

and that for every $g(t) \in L(X)$, for which

$$(9.4) \qquad \frac{g(t)}{\psi(t)} \to 0 \quad \text{as} \quad t \to \infty,$$

(9.2) has a $\mathfrak{D}$-solution. Then if t_0 is sufficiently large, the system

$$(9.5) \qquad y' = A(t)y + f(t, y)$$

has a solution for $t \geq t_0$ such that $\|y(t)\| \leq \rho$ [and $y(t) \to 0$ as $t \to \infty$].

Remark 1.   Assumption (ii) merely means that $(L_{\psi 0}^\infty(X), \mathfrak{D})$ is admissible for (9.2). Actually, assumption (i) is a special case of (ii) but is isolated for convenience. For a discussion of conditions necessary and sufficient for $(L^1(X), L^\infty(X))$ or $(L^1(X), L_0^\infty(X))$ to be admissible for (9.2), where $X = Y$, see Theorem XIII 6.3.

Remark 2. Let $U(t)$ be the fundamental solution for

$$(9.6) \qquad\qquad y' = A(t)y$$

satisfying $U(0) = I$. Let $y_0 \in Y_{\mathfrak{D}}$. Then if $\|y_0\|$ is sufficiently small and $t_0 \geqq 0$ is sufficiently large, the solution $y(t)$ in Theorem 9.1 can be chosen so as to satisfy

$$U^{-1}(t_0)y(t_0) = y_0.$$

Let C_0, K be the constants of Lemma 6.3 associated with the admissibility of the appropriate pairs of spaces $(L^1(X), \mathfrak{D})$ or $(L_{\psi 0}^\infty(X), \mathfrak{D})$. According as (i) or (ii) is assumed, the conditions of smallness on $\|y_0\|$ and largeness of t_0 are

$$C_0 \|y_0\| + K \int_{t_0}^{\infty} \lambda(t)\, dt \leqq \rho \quad \text{or} \quad C_0 \|y_0\| + \frac{K\lambda(t)}{\psi(t)} \leqq \rho \quad \text{for } t \geqq t_0.$$

Proof. Let $\mathfrak{B} = L^1(X)$ or $\mathfrak{B} = L_{\psi 0}^\infty(X)$ according as (i) or (ii) is assumed. Then Theorem 9.1 is a consequence of Corollary 8.1 obtained by replacing $f(t, y)$, $\lambda(t)$ by the functions $h_a(t)f(t, y)$, $h_a(t)\lambda(t)$, where $a = t_0$ and $h_a(t)$ is 1 or 0 according as $t \geqq a$ or $t < a$.

Exercise 9.1. The following type of question often arises: Let $y_1(t)$ be a solution of the homogeneous linear system (9.6). When does (9.5) have a solution $y(t)$ for large t such that $y - y_1 \to 0$ as $t \to \infty$? Deduce sufficient conditions from Theorem 9.1.

As an application of Theorem 9.1, consider a second order equation

$$(9.7) \qquad\qquad u'' = h(t, u, u')$$

for a real-valued function u. Assume that $h(t, u, u')$ is continuous for $t \geqq 0$ and arbitrary (u, u'). Let α, β be constants and consider the question whether (9.7) has a solution for large t satisfying

$$(9.8) \qquad u(t) - \alpha t - \beta \to 0 \quad \text{and} \quad u'(t) - \alpha \to 0 \quad \text{as} \quad t \to \infty$$

Introduce the change of variables $u \to v$, where

$$(9.9) \qquad\qquad u = \alpha t + \beta + v,$$

then (9.7) becomes

$$(9.10) \qquad\qquad v'' = h(t, \alpha t + \beta + v, \alpha + v')$$

and (9.8) is $v, v' \to 0$ as $t \to \infty$. Theorem 9.1 implies the following:

Corollary 9.1 *Let $h(t, u, u')$ be continuous for $t \geqq 0$ and arbitrary (u, u') such that*

$$|h(t, \alpha t + \beta + u, \alpha + u')| \leqq \lambda(t) \qquad \text{for} \quad |u|, |u'| \leqq \rho,$$

where $\lambda(t)$ is a function satisfying

$$\int^{\infty} t\lambda(t)\,dt < \infty.$$

Then (9.7) *has a solution $u(t)$ for large t satisfying* (9.8).

Exercise 9.2. (*a*) Verify Corollary 9.1. (*b*) Apply it to the case that $h = f(t)g(u)$, where $\alpha \neq 0$ or $\alpha = 0$. (*c*) Generalize it by replacing (9.7) by $u^{(d)} = h(t, u, u', \ldots, u^{(d-1)})$.

Actually Corollary 9.1 is a special case of Theorem X 13.1, but Theorem X 13.1 can itself be deduced from Theorem 9.1; cf. Exercise 9.3 below.

Many problems involving asymptotic integrations can be solved by the use of Theorem 9.1. Often these problems can be put into the following form: Let $Q(t)$ be a continuously differentiable matrix for $t \geqq 0$. Does the nonlinear system (9.5) have a solution $y(t)$ such that if

$$(9.11) \qquad\qquad y = Q(t)x,$$

then $c = \lim x(t)$ exists as $r \to \infty$? The differential equation for $x(t)$ is

$$(9.12) \qquad x' = Q^{-1}(t)[A(t)Q(t) - Q'(t)]x + Q^{-1}(t)f(t, Q(t)x).$$

The change of variables
$$(9.13) \qquad\qquad z = x - c$$
transforms (9.12) into

$$(9.14) \qquad\qquad z' = Q^{-1}(AQ - Q')z + g(t, z, c),$$

where

$$(9.15) \qquad g(t, z, c) = Q^{-1}(AQ - Q')c + Q^{-1}f(t, Qz + Qc).$$

The problem is thus reduced to the question: Does (9.14) have a solution $z(t)$ for large t such that $z(t) \to 0$ as $t \to \infty$? Clearly, Theorem 9.1 is adapted to answer such questions.

We should point out that if the answer is affirmative, then (9.11) and the conclusion $x(t) - c \to 0$ as $t \to \infty$ need not be very informative unless estimates for $\|x(t) - c\|$ are obtained [e.g., if $Q(t)$ is the 2×2 matrix $Q(t) = (q_{jk}(t))$, where $q_{k1} = (-1)^k e^{-t}, q_{k2} = e^t$ for $k = 1, 2$ and $c = (1, 0)$, then we can only deduce $y(t) = o(e^t)$, but not an asymptotic formula of the type $y(t) = (-1 + o(1), 1 + o(1))e^{-t}$ as $t \to \infty$.]

Exercise 9.3. Follow the procedure just mentioned and deduce Theorem X 13.1 by using Theorem 9.1 (instead of Lemma X 4.3).

Notes

INTRODUCTION. The use of fixed point theorems in function spaces was initiated by Birkhoff and Kellogg [1]. For Theorem 0.2, see Tychonov [1]. For Schauder's fixed point theorem, see Schauder [1]. For the remark at the end of the Introduction, see Graves [1]. As mentioned in the text, Theorem 0.3 is a result of Banach [1].

SECTION 1. Results analogous to those of this section but dealing with one equation of the second order, e.g., go back to Sturm. Boundary value problems for systems of second order equations were considered by Mason [1]. The results of this section (except for Theorem 1.3) are due to Bounitzky [1]; the treatment in the text follows Bliss [1]. These results are merely the introduction to the subject which is usually concerned with eigenfunction expansions; see Bliss [1] for older references to Hildebrandt, Birkhoff, Langer, and others. For an excellent recent treatment for the singular, self-adjoint problem; see Brauer [2]. Theorem 1.3 is given by Massera [1], who attributes the proof in the text to Bohnenblust.

SECTION 2. Theorems 2.1 and 2.2 are similar to Theorems 4.1 and 4.2, respectively. Exercise 2.1 is a result of Massera [1] and generalizes a theorem of Levinson [2]; its proof depends on a (2-dimensional) fixed point theorem of Brouwer. Exercise 2.2 is a result of Knobloch [1], who uses a variant of Brouwer's fixed point theorem due to Miranda [1]; cf. Conti and Sansone [1, pp. 438–444].

Theorems 2.3 and 2.4 are due to Poincaré [5, I, chap. 3 and 4]; see Picard [2, III, chap. 8]. Problems concerning "degenerate" cases of Theorems 2.3 and 2.4 when the Jacobians in the proofs vanish were also treated by Poincaré and since then by many others, including Lyapunov. For some more recent work and older references, see E. Hölder [1], Friedrichs [1], and J. Hale [1]; for the problem in a very general setting, see D. C. Lewis [4].

SECTION 3. The scalar case of Theorem 3.3 is a result of Picard [4]; the extension to systems is in Hartman and Wintner [22]. In the scalar case, (3.17) can be relaxed to the condition Re $B(t)x \cdot x \geqq 0$, Rosenblatt [2]; see also Exercise 4.5(c). The uniqueness criterion in Exercise 3.3(b), among others, is given by Hartman and Wintner [22]. Sturm types of comparison theorems for self-adjoint systems have been given by Morse [1].

SECTION 4. Theorem 4.1 and its proof are due to Picard [4, pp. 2–7]. For related results in the scalar case, see Nagumo [2], [4], references in Hartman and Wintner [8] and Lees [1] to Rosenblatt, Cinquini, Zwirner, and others. Theorem 4.2 is a result of Scroza-Dragoni [1]. The uniqueness Theorem 4.3 is due to Hartman [19]. For Exercise 4.6(b), see Hartman and Wintner [8]; for part (c), with the additional condition that f has a continuous partial derivative $\partial f/\partial x \geqq 0$, see Rosenblatt [2]. For Exercises 4.7 and 4.8, see Nirenberg [1].

SECTION 5. Lemma 5.1 and Corollary 5.2 are results of Nagumo [2]. The example following Lemma 5.1 is due to Heinz [1]. The other theorems of this section are contained in Hartman [19]. Exercise 5.4 is a generalization of a result of Lees [1] who gives a very different proof from that in the *Hints*. For the scalar case in Exercise 5.5(d), see Hartman and Wintner [8]; this result was first proved by A. Kneser [2] (see Mambriani [1]) for the case when f does not depend on x'. For related results, see Exercises XIV 2.8 and 2.9. A generalization of Exercise 5.9 involving almost periodic functions is given in Hartman [19] and is based on a paper of Amerio [1].

SECTION 6. Part III is an outgrowth of a paper of Perron [12], whose results were carried farther by Persidskiĭ [1], Malkin [1], Krein [1], Bellman [2], Kučer [1], and Maizel' [1]. Except for Kučer, these authors deal, for the most part, with the case $\mathfrak{B} = L^\infty(Y)$, $\mathfrak{D} = L^\infty(Y)$. (For a statement concerning the results of these earlier papers, see Massera and Schäffer [1, I].) The results of this section are due to Massera and Schäffer [1] who deal with the more general situation when the space Y need not be finite-dimensional.

SECTION 7. For the notion of "lean at ω," see Schäffer [2, VI]. The Green's functions G of this section occur in Massera and Schäffer [1, I and IV]. Theorem 7.1 and Corollary 7.1 may be new.

SECTION 8. Theorem 8.1 is a result of Corduneanu [1]. Theorem 8.2 is a corrected version of a similar result of Corduneanu[1] (see Hartman and Onuchic [1]); also Massera [8]. For Corollary 8.1, see Hartman and Onuchic [9]. For Exercise 8.2, see Massera and Schäffer [1, I or IV].

SECTION 9. This application of the results of § 8 is given by Hartman and Onuchic [1]. For Corollary 9.1, see Hale and Onuchic [1].

Chapter XIII

Dichotomies for Solutions of Linear Equations

For $t \geqq 0$, consider an inhomogeneous linear system of differential equations

(0.1) $$y' - A(t)y = g(t)$$

and the corresponding homogeneous system

(0.2) $$y' - A(t)y = 0,$$

or, more generally, an inhomogeneous, linear system of equations of $(m + 1)$st order

(0.3) $$u^{(m+1)} + \sum_{k=0}^{m} P_k(t)u^{(k)} = f(t)$$

and the corresponding homogeneous system

(0.4) $$u^{(m+1)} + \sum_{k=0}^{m} P_k(t)u^{(k)} = 0.$$

Suppose that $\mathfrak{B}$, $\mathfrak{D}$ are Banach spaces of vector-valued functions and that (0.1) [or (0.3)] has a solution $y(t) \in \mathfrak{D}$ [or $u(t) \in \mathfrak{D}$] for every $g(t) \in \mathfrak{B}$ [or $f(t) \in \mathfrak{B}$]; i.e., that $(\mathfrak{B}, \mathfrak{D})$ is admissible in the sense of § XII 6. Then, under suitable conditions on the coefficients and the spaces $\mathfrak{B}$ and $\mathfrak{D}$, this implies a [an exponential] dichotomy for the solutions of the homogeneous equations, roughly in the sense that some of the solutions are small [or exponentially small] and that others are large [or exponentially large] as $t \to \infty$. This type of assertion and its converse will be the subject of this chapter.

In particular, for (0.2), we shall obtain conditions necessary and/or sufficient in order that there exist Green's matrices $G(t, s)$ defined as in § XII 7, satisfying

$$\|G(t, s)\| \leqq K \qquad \text{or} \qquad \|G(t, s)\| \leqq Ke^{-v|t-s|}$$

for $s, t \geqq 0$, where K, $v > 0$ are constants. The main results for (0.2) are given in § 6; corresponding results for (0.4) are given in § 7. The main

450

tools will be the Open Mapping Theorem XII 0.3 (in fact, analogues of the lemmas of § XII 6) and the following basic inequality for solutions of (0.1): Let y be a point of the (real or complex) vector space Y with norm $\|y\|$ and let $\|A(t)\| = \sup \|A(t)y\|$ for $\|y\| = 1$, then solutions of (0.1) satisfy

$$(0.5) \qquad \|y(t)\| \leqq \left\{ \|y(s)\| + \left| \int_s^t \|g(r)\| \, dr \right| \right\} \exp \left| \int_s^t \|A(r)\| \, dr \right|$$

for arbitrary s, t. When Y is the Euclidean space, this can be strengthened to

$$(0.6) \qquad \|y(t)\| \leqq \left\{ \|y(s)\| + \left| \int_s^t \|g(r)\| \, dr \right| \right\} \exp \left| \int_s^t \mu(r) \, dr \right|,$$

where $\mu(t) = \sup |\mathrm{Re}\, A(t)y \cdot y|$ for $\|y\| = 1$; see § IV 4.

It will be convenient to write (0.1), (0.2) as equations $Ty = g$, $Ty = 0$, where T is the operator $Ty = y' - A(t)y$. In order to avoid a special treatment of (0.3), this equation will be written as (0.1), where $y = (u, u', \dots, u^{(m)})$. It will be advantageous, however, to introduce a "projection" operator P, which in the general case of (0.1) is the identity $Py = y$ but in the case (0.3) of (0.1) is $Py = u$.

This chapter will be divided into two parts: Part I deals with analogues of (0.1), (0.3) and Part II with the analogues of the adjoint equations.

PART I. GENERAL THEORY

1. Notations and Definitions

(i) Below $y, z, \dots$ [or $u, v, \dots$] are elements of a finite dimensional real or complex Banach space Y [or U] with given norms $\|y\|, \|z\|, \dots$ [or $\|u\|, \|v\|, \dots$]. It will not be assumed that these spaces are Euclidean. For example, in dealing with a product space $X \times Y$, it is often more convenient to use the norm $\|(x, y)\| = \max (\|x\|, \|y\|)$. It is also more convenient to work with the angular distance between two nonzero elements $y, z \in Y$ defined by

$$(1.1) \qquad \gamma[y, z] = \left\| \frac{y}{\|y\|} - \frac{z}{\|z\|} \right\|$$

than to assume that Y is Euclidean and to deal, e.g., with the Euclidean angle between y, z or with $|\sin (y, z)|$. (If Y is Euclidean, then γ in (1.1) is $2 |\sin \frac{1}{2}(y, z)|$.)

Note that $\|y\| \cdot \|z\| \, \gamma[y, z]$ is the norm of $y \|z\| - \|y\| z$ and hence is $\|(y - z) \|z\| - (\|y\| - \|z\|)z\| \leqq 2 \|z\| \cdot \|y - z\|$. Interchanging y and z shows that

$$(1.2) \qquad \gamma[y, z] \max (\|y\|, \|z\|) \leqq 2 \|y - z\|.$$

If X is a linear manifold in Y and $y \in Y$, let $d(X, y) = \text{dist}(X, y) = \inf \|x - y\|$ for $x \in X$. In particular,

(1.3) $\|y\| \geqq d(x, y)$.

The condition

(1.4) $\|y\| \leqq \lambda \, d(X, y)$ and $\lambda \geqq 1$

will frequently occur, where X is a subspace (i.e., a closed linear manifold). Note that the inequality (1.4) can hold for $y \in X$ only if $y = 0$. If $y \neq 0$ and λ is an admissible number in (1.4), then $1/\lambda$ can be interpreted as a "rough measure of the angle between y and the linear manifold X." This is clear if (1.4) is written as $1/\lambda \leqq d(X, y \|y\|^{-1}) = \inf \|x - y \|y\|^{-1}\|$ for $x \in X$.

Y^* denotes the space dual to Y and $\langle y, y^* \rangle$ is the corresponding pairing (i.e., "scalar product") of $y \in Y$, $y^* \in Y^*$.

(ii) It will be supposed that a coordinate system in Y [or U] is fixed. Thus an element $y \in Y$ can be represented as $y = (y^1, \ldots, y^d)$, where $d = \dim Y$, and a linear operator from Y to Y is a $d \times d$ matrix A with the norm $\|A\| = \sup \|Ay\|$ for $\|y\| = 1$. (This is only for the purpose of making the theorems of Chapter IV, as stated, available here.)

(iii) Let J denote the closed half-line $0 \leqq t < \infty$ and J' a bounded subinterval of J. The characteristic function of J' will be denoted by $h_{J'}(t)$, so that $h_{J'}(t) = 0$ or 1 according as $t \notin J'$ or $t \in J'$. Correspondingly, $h_s(t)$ is the characteristic function of the half-line $s \leqq t < \infty$ and $h_{s\epsilon}(t)$ the characteristic function of $J' = [s, s + \epsilon]$.

$\varphi_{s\epsilon}(t)$ will always denote a non-negative, integrable function on J with support on $[s, s + \epsilon]$ [i.e., vanishing for $t < s$ or $t > s + \epsilon$, so that $\varphi_{s\epsilon}(t) = \varphi_{s\epsilon}(t)h_{s\epsilon}(t)$].

(iv) Let $\mathscr{T}$ denote the set of normed spaces Φ whose elements are (equivalence classes modulo null sets of) real-valued measurable functions $\varphi(t)$ on J satisfying the following conditions: (a) $\Phi \neq \{0\}$; (b) the elements $\varphi(t)$ of Φ are locally integrable and for every bounded J' there exists a number $\alpha = \alpha(J', \Phi)$ such that

$$\int_{J'} |\varphi(t)| \, dt \leqq \alpha \, |\varphi|_\Phi \qquad \text{for all} \quad \varphi \in \Phi;$$

the least number α satisfying this relation will be denoted by $|h_{J'}|_{\Phi'}$, so that

(1.5) $\int_{J'} |\varphi(t)| \, dt \leqq |\varphi|_\Phi \, |h_{J'}|_{\Phi'}$ for all $\varphi \in \Phi$

cf. § 9. (c) if $\varphi \in \Phi$ and ψ is a real-valued measurable function on J such that $|\psi(t)| \leqq |\varphi(t)|$, then $\psi \in \Phi$ and $|\psi|_\Phi \leqq |\varphi|_\Phi$; (d) if $\varphi \in \Phi$, $s > 0$, and $\psi(t) = 0$ or $\psi(t) = \varphi(t - s)$ according as $0 \leqq t < s$ or $t \geqq s$, then

$\psi \in \Phi$ and $|\psi|_\Phi = |\varphi|_\Phi$; *(e)* the characteristic functions $h_{J'}(t)$ of bounded intervals J' are elements of Φ.

Unless the contrary is implied, B and D below denote Banach spaces in $\mathscr{T}$. It is clear that all of the spaces L^p on J, $1 \leq p \leq \infty$ are in $\mathscr{T}$. Also, the subspace L_0^∞ of L^∞, consisting of functions $\varphi(t) \in L^\infty$ satisfying $\varphi(t) \to 0$ as $t \to \infty$, is in $\mathscr{T}$.

When $\Phi = L^p$, $1 \leq p \leq \infty$, the norm $|\varphi|_{L^p}$ will be abbreviated to $|\varphi|_p$.

(v) M will denote the Banach space of (equivalence classes modulo null sets of) locally integrable functions $\varphi(t)$ on J with the norm

$$(1.6) \qquad |\varphi|_M = \sup_{t \geq 0} \int_t^{t+1} |\varphi(s)| \, ds.$$

Clearly, $M \in \mathscr{T}$.

(vi) If $\Phi \in \mathscr{T}$, Φ_∞ denotes the linear manifold of functions $\varphi(t) \in \Phi$ with compact support, i.e., functions $\varphi(t) \in \Phi$ vanishing for large t. If, in addition, Φ is a Banach space (i.e., complete), then $\overline{\Phi}_\infty$ is the completion (closure) of Φ_∞ in Φ.

(vii) If $\Phi \in \mathscr{T}$ and Y is a finite-dimensional Banach space (over the real or complex numbers), $\Phi(Y)$ will denote the normed vector space of (equivalence classes modulo null sets of) measurable functions $y(t)$ from J to Y [i.e., functions $y(t)$ with components which are measurable functions] such that $\varphi(t) = \|y(t)\|$ is in Φ with the norm $|y(t)|_{\Phi(Y)}$ defined to be $|\varphi|_\Phi$. For brevity, the norm of $y(t) \in \Phi(Y)$ will be denoted by $|y|_\Phi$. It is easy to see that if Φ is Banach space, then so is $\Phi(Y)$.

(viii) Let L denote the space of (equivalence classes modulo null sets of) real-valued measurable functions $\varphi(t)$ on J with the topology of convergence in the mean L^1 on bounded intervals. Correspondingly, $L(Y)$ is the space of locally integrable functions $y(t)$ from J to Y with the topology of convergence in the mean L^1 on bounded intervals.

Condition *(b)* in (iv), cf. (1.5), on spaces $\Phi \in \mathscr{T}$ means that Φ is stronger than L, so that convergence in Φ implies convergence in L; see § XII 6.

(ix) A space $\Phi \in \mathscr{T}$ is called *quasi-full* if it has the property that $\varphi(t) \in L$, $\varphi(t) \notin \Phi$ implies that either $h_{0\Delta}(t)\varphi(t) \notin \Phi$ for some $\Delta > 0$ or that $|h_{0\Delta}\varphi|_\Phi \to \infty$ as $\Delta \to \infty$. Clearly, the spaces $\Phi = L^p$ for $1 \leq p \leq \infty$ are quasi-full.

(x) *Dichotomies.* Let Y, W be Banach spaces and $\mathscr{N}$ a linear manifold of functions $y = y(t)$ from J to Y. With each $y(t) \in \mathscr{N}$, let there be associated a non-negative function $\rho_y(t)$ on J and an element $y[0]$ of W. We shall assume that the map Q from $\mathscr{N}$ to W given by $Qy(t) = y[0]$ is linear and one-to-one; $y[0]$ will be called the "initial value" of $y(t)$. Let W_0 be a linear manifold in the range of Q.

When W_0 is a subspace (i.e., closed linear manifold), it is said to induce a *partial dichotomy* for $(\mathcal{N}, \rho_y, y[0])$ if there exist a positive constant M_0 and a non-negative number θ^0 such that

(a) if $y(t) \in \mathcal{N}$ with $y[0] \in W_0$, then

$$(1.7) \qquad \rho_y(t) \leq M_0 \rho_y(s) \qquad \text{if} \quad \theta^0 \leq s \leq t;$$

(b) if $z(t) \in \mathcal{N}$ with $\|z[0]\| \leq \lambda \, d(W_0, z[0])$ and $\lambda > 1$, then

$$(1.8) \qquad \rho_z(t) \leq \lambda M_0 \rho_z(s) \qquad \text{if} \quad s \geq \theta^0, \quad 0 \leq t \leq s.$$

A subspace W_0 is said to induce a *total dichotomy* for $(\mathcal{N}, y[0])$ if it induces a partial dichotomy for $(\mathcal{N}, \rho_y, y[0])$, where $\rho_y(t) = \|y(t)\|$, and in addition the following holds:

(c) there exists a constant $\gamma_0 > 0$ such that if $y(t), z(t)$ are as in (a) and (b), respectively, then

$$(1.9) \qquad \lambda\gamma[y(t), z(t)] \geq \gamma_0 > 0 \qquad \text{when} \quad t \geq \theta^0$$

and $y(t) \neq 0$, $z(t) \neq 0$.

A subspace W_0 is said to induce an *exponential dichotomy* for $(\mathcal{N}, \rho_y, y[0])$ if there exist a non-negative number θ^0, positive numbers M_1, ν, ν' and, for every $\lambda > 1$, a positive number $M_1' = M_1'(\lambda)$ such that

(a) if $y(t) \in \mathcal{N}$ with $y[0] \in W_0$, then

$$(1.10) \qquad \rho_y(t) \leq M_1 e^{-\nu(t-s)} \rho_y(s) \qquad \text{for} \quad \theta^0 \leq s \leq t;$$

(b) if $z(t) \in \mathcal{N}$ with $\|z[0]\| \leq \lambda \, d(W_0, z[0])$ and $\lambda > 1$, then

$$(1.11) \qquad \rho_z(t) \geq M_1' e^{\nu'(t-s)} \rho_z(s) \qquad \text{for} \quad t \geq \theta^0, \quad 0 \leq s \leq t.$$

A subspace W_0 is said to induce a *total exponential dichotomy* for $(\mathcal{N}, y[0])$ if it induces an exponential dichotomy for $(\mathcal{N}, \rho_y, y[0])$, where $\rho_y(t) = \|y(t)\|$, and condition (c) of a total dichotomy holds (i.e., W_0 induces a total dichotomy for $(\mathcal{N}, y[0])$ and an exponential dichotomy for $(\mathcal{N}, \rho_y, y[0])$ with $\rho_y(t) = \|y(t)\|$).

A manifold W_0 (not necessarily closed) is said to induce an *individual partial* [or *exponential*] *dichotomy* for $(\mathcal{N}, \rho_y, y[0])$ if

(a) for every $y(t) \in \mathcal{N}$ with $y[0] \in W_0$, there exist constants $\theta^0 \geq 0$, $M_0 > 0$ [or $M_1, \nu > 0$] depending on $y(t)$ such that (1.7) [or (1.10)] holds;

(b) for every $z(t) \in \mathcal{N}$ with $z(0) \notin W_0$, there exist constants $\theta^0 \geq 0$, $M_0' > 0$ [or $M_1', \nu' > 0$] depending on $z(t)$ such that

$$(1.12) \qquad \rho_z(t) \leq M_0' \rho_z(s) \qquad \text{if} \quad s \geq \theta^0 \quad \text{and} \quad 0 \leq t \leq s$$

[or (1.11)] holds.

If no confusion results $(\mathcal{N}, \rho_y, y[0])$ will be shortened to $(\mathcal{N}, \rho_y)$; also, if $\rho_y(t) = \|y(t)\|$, $\mathcal{N}$ will be written in place of $(\mathcal{N}, \rho_y)$.

2. Preliminary Lemmas

The notation $Y, \mathcal{N}, W, W_0, Q, \rho_y(t)$ is the same in this section as in paragraph (x) of the last section. It will sometimes be assumed that $\rho_y(t)$, for fixed t, is a "semi-norm,"

$$(2.1) \quad \rho_{cy}(t) \leqq |c| \, \rho_y(t), \qquad \rho_{y-z}(t) \leqq \rho_y(t) + \rho_z(t) \qquad \text{for} \quad y(t), z(t) \in \mathcal{N},$$

c an arbitrary constant; and/or that there exist $\theta^1 \geqq 0$ and $K_0 > 0$ such that

$$(2.2) \qquad\qquad \rho_y(t) \leqq K_0 \, \|y[0]\| \qquad \text{for} \quad t \geqq \theta^1,$$

$$(2.3) \qquad\qquad \|z[0]\| \leqq \lambda K_0 \rho_z(t) \qquad \text{for} \quad t \geqq \theta^1,$$

whenever

$$(2.4) \qquad\qquad y(t) \in \mathcal{N} \qquad \text{with} \quad y[0] \in W_0,$$

$$(2.5) \qquad z(t) \in \mathcal{N} \qquad \text{with} \quad \|z[0]\| \leqq \lambda \, d(W_0, z[0]), \qquad \lambda > 1,$$

respectively.

Note that if W_0 is a subspace, a sufficient condition for W_0 to induce a partial dichotomy for $(\mathcal{N}, \rho_y(t))$ is that there exist $\theta^0 \geqq 0$, $M_0 > 0$ such that (2.4), (2.5) imply that

$$(2.6) \quad \max\,(\rho_z(r), \rho_y(t)) \leqq \lambda M_0 \rho_{y-z}(s) \qquad \text{for} \quad 0 \leqq r \leqq s \leqq t, \qquad s \geqq \theta^0.$$

In fact, conditions (a), (b) follow from the cases $z \equiv 0$, arbitrary $\lambda > 1$, and the case $y \equiv 0$ of (2.6).

The first lemma will be useful and will illustrate the meaning of "total dichotomy."

Lemma 2.1. *Let W_0 be a subspace in the range of Q. Let $y(t)$, $z(t)$ denote arbitrary elements of $\mathcal{N}$ satisfying (2.4), (2.5), respectively. If there exist $\theta^0 \geqq 0$, $M_0 > 0$ such that*

$$(2.7) \quad \max\,(\|z(r)\|, \|y(t)\|) \leqq \lambda M_0 \, \|z(s) - y(s)\| \qquad \text{for} \quad 0 \leqq r \leqq s \leqq t$$

and $s \geqq \theta^0$, then W_0 induces a total dichotomy for $\mathcal{N}$ (with the corresponding θ^0, M_0, $\gamma_0 = 1/M_0$). Conversely, if W_0 induces a total dichotomy for $\mathcal{N}$, then

$$(2.8) \quad \max\left(\frac{1}{\lambda}\,\|z(r)\|, \|y(t)\|\right) \leqq 2\lambda M_0 \gamma_0^{-1} \, \|z(s) - y(s)\| \quad \text{for } 0 \leqq r \leqq s \leqq t$$

and $s \geqq \theta^0$.

Remark 1.   The factor $1/\lambda$ of $\|z(t)\|$ on the left of (2.8) can be removed under some additional conditions: If W_0 induces a total dichotomy for $\mathcal{N}$, and there exist $\theta^1 \geqq 0$, $K_0 > 0$ such that $\rho_y(t) = \|y(t)\|$ satisfies (2.2),

(2.3) whenever (2.4), (2.5) hold, then there exists a constant $M_0' > 0$ such that

$$\|z(r)\| \leqq \lambda M_0' \|z(s) - y(s)\| \qquad \text{for} \quad \max(\theta^0, \theta^1) \leqq r \leqq s.$$

Proof of Lemma 2.1.   Assume (2.7). The cases $z \equiv 0$, $\lambda > 1$ arbitrary, and $y \equiv 0$ give conditions (a), (b) of a partial dichotomy. Replacing $y(t), z(t) \in \mathcal{N}$ by the elements $y(t)/\|y(s)\|$, $z(t)/\|z(s)\| \in \mathcal{N}$ for a fixed $s \geqq \theta^0$ for which $y(s) \neq 0$, $z(s) \neq 0$ gives (1.9) when $r = s$ or $t = s$ with $\gamma_0 = 1/M_0$.

Conversely, assume that W_0 induces a total dichotomy for $\mathcal{N}$. Then (1.9) and (1.2) show that

$$\gamma_0 \max\left(\|z(s)\|, \|y(s)\|\right) \leqq 2\lambda \|y(s) - z(s)\|$$

if $s \geqq \theta^0$ and $z(s) \neq 0$, $y(s) \neq 0$. Conditions (a), (b) of a total dichotomy give (2.8) if $s \geqq \theta^0$ and $z(s) \neq 0$, $y(s) \neq 0$.

The proof of the last part of this lemma can obviously be modified to give

Corollary 2.1.   *Let W_0 be a subspace of Q inducing a total exponential dichotomy for $\mathcal{N}$. Let $y(t), z(t)$ be as in Lemma 2.1. Then, for $0 \leqq r \leqq s \leqq t$ and $s \geqq \theta^0$,*

$$(2.9) \quad \gamma_0 \max\left(M_1' e^{\nu'(s-r)} \|z(r)\|, M_1^{-1} e^{\nu(t-s)} \|y(t)\|\right) \leqq 2\lambda \|z(s) - y(s)\|,$$

where $\nu, \nu', M_1' = M_1'(\lambda) > 0, M_1$, and γ_0 occur in the definition of total exponential dichotomy.

Instead of proving Remark 1 following Lemma 2.1, the following more general assertion will be proved.

Lemma 2.2.   *Let W_0 be a subspace of W with the property that there exist $\theta^0 \geqq 0$, $M_0 > 0$, $\lambda_0 > 1$ such that (2.4), (2.5) with $\lambda = \lambda_0$ imply (2.6) with $\lambda = \lambda_0$. Assume that $\rho_y(t)$ satisfies (2.1) and that there exist $\theta^1 \geqq 0$, $K_0 > 0$ such that (2.4), (2.5) imply (2.2), (2.3). Then there exists an $M_0' > 0$ such that (2.4), (2.5) imply*

$$(2.10) \qquad \rho_z(r) \leqq \lambda M_0' \rho_{y-z}(s) \qquad \text{for} \quad \theta_0 \leqq r \leqq s$$

where $\theta_0 = \max(\theta^0, \theta^1)$.

Proof.   The definition of $d(W_0, z[0])$ shows that there exist elements $y^0 \in W^0$ such that $\|y^0 - z[0]\|$ is arbitrarily near to $d(W_0, z[0])$. Since $\lambda_0 > 1$, y^0 can be chosen so that $z^0 = y^0 - z[0]$ satisfies $\|z^0\| \leqq \lambda_0 d(W_0, z[0]) = \lambda_0 d(W_0, z^0)$. As W_0 is in the range of Q, there is a $y^0(t) \in \mathcal{N}$ such that $y^0[0] = y^0$ and $z^0(t) = y^0(t) - z(t) \in \mathcal{N}$ with $z^0[0] = z^0$. Since $\|z[0]\| \leqq \lambda d(W_0, z[0]) = \lambda d(W_0, z^0) \leqq \lambda \|z^0[0]\|$, it follows that

$$(2.11) \quad \|y^0[0]\| \leqq \|z^0[0]\| + \|z[0]\| \leqq (\lambda + 1)\|z^0[0]\| \leqq 2\lambda \|z^0[0]\|.$$

From $\rho_z(r) \leqq \rho_{z^0}(r) + \rho_{y^0}(r)$ and the case $z \equiv 0$ and $\lambda = \lambda_0$ of (2.6), $\rho_{y^0}(r) \leqq \lambda_0 M_0 \rho_{y^0}(\theta_0)$. The inequality (2.2) implies that $\rho_{y^0}(\theta_0) \leqq K_0 \|y^0[0]\|$, so that $\rho_z(r) \leqq \rho_{z^0}(r) + M_0 K_0 \lambda_0 \|y^0[0]\|$. By (2.3) applied to $z = z^0(t)$ with $\lambda = \lambda_0$,

$$\rho_z(r) \leqq \left\{ 1 + M_0 K_0^2 \lambda_0^2 \frac{\|y^0[0]\|}{\|z^0[0]\|} \right\} \rho_{z^0}(r).$$

Hence the last two displays and $\lambda > 1$ give

$$(2.12) \qquad \rho_z(r) \leqq \lambda(1 + 2M_0 K_0^2 \lambda_0^2)\rho_{z^0}(r).$$

Thus (2.10) with $M_0' = (1 + 2M_0 K_0^2 \lambda_0^2)\lambda_0 M_0$ follows from (2.6), where (λ, z, y) are replaced by $(\lambda_0, z^0, y^0 + y)$.

The following lemma, which is of interest in itself, will be used several times. (It is false if the assumption that W is finite dimensional is omitted.)

Lemma 2.3. *Let W be finite dimensional, W_0 a subspace of W in the range of Q with the property that there exist $\theta^0 \geqq 0$, $M_0 > 0$ such that (2.4), (2.5) imply (2.6). Assume also that $\rho_y(t)$ is a continuous function of t for each $y(t) \in \mathcal{N}$, that (2.1) holds and that there exist $\theta^1 \geqq 0$, $K_0 > 0$ such that (2.4), (2.5) imply (2.2), (2.3). Let X be the set of initial values $y[0]$ of elements $y(t) \in \mathcal{N}$ satisfying $\rho_y(t) \to 0$ as $t \to \infty$. Then $X \subset W_0$ is a subspace of W and there exists a constant $M_0' > 0$ such that the conditions*

$$(2.13) \qquad \begin{aligned} y(t) &\in \mathcal{N} \quad \text{with} \quad y[0] \in X, \\ z(t) &\in \mathcal{N} \quad \text{with} \quad \|z[0]\| \leqq \lambda\, d(X, z[0]), \quad \lambda > 1, \end{aligned}$$

imply that

$$(2.14) \qquad \max (\rho_z(r), \rho_y(t)) \leqq \lambda M_0' \rho_{y-z}(s) \quad \text{for} \quad \theta_0 \leqq r \leqq s \leqq t,$$

where $\theta_0 = \max (\theta^0, \theta^1)$.

This clearly has the following corollary:

Corollary 2.2. *Let W be finite dimensional and W_0 a subspace of W which induces a total dichotomy for $\mathcal{N}$. Assume that $\|y(t)\|$ is a continuous function of t for $y(t) \in \mathcal{N}$ and that (2.4), (2.5) imply (2.2), (2.3) for $\rho_y(t) = \|y(t)\|$ and $\theta^1 = 0$ (e.g., if $y[0] = y(0)$). Let X be the set of initial values $y[0]$ of elements $y(t) \in \mathcal{N}$ satisfying $\|y(t)\| \to 0$ as $t \to \infty$. Then $X \subset W_0$ induces a total dichotomy for $\mathcal{N}$.*

Proof of Lemma 2.3. If the norm $\|w\|$ in W is replaced by an equivalent norm $\|w\|_0$ (i.e., if $c_1 \|w\| \leqq \|w\|_0 \leqq c_2 \|w\|$ for constants $c_1, c_2 > 0$), then assumption and assertion of this lemma remain unchanged. Thus, without loss of generality, we can suppose that W is a Euclidean space.

Let W_1 be the subspace of W orthogonal to W_0. Then, if $y(t) \in \mathcal{N}$ with $y[0] \in W_0$ and $z(t) \in \mathcal{N}$ with $z[0] \in W_1$, (2.6) implies that

$$(2.15) \quad \max(\rho_z(r), \rho_y(t)) \leqq M_0 \rho_{y-z}(s) \quad \text{for} \quad 0 \leqq r \leqq s \leqq t, s \geqq \theta^0$$

since (2.6) holds for *all* $\lambda > 1$.

It is clear from (2.6) that $X \subset W_0$. Let X^1 be the subspace of W_0 orthogonal to X. First we will show that there exists a constant $\alpha > 0$ such that if $x^1(t) \in \mathcal{N}$ with $x^1[0] \in X^1$, then

$$(2.16) \quad \rho_{x^1}(t) \leqq \alpha \rho_{x^1}(s) \quad \text{if} \quad \theta_0 \leqq s, t < \infty.$$

To this end, it is sufficient to show the existence of a constant $\alpha_1 > 0$ satisfying

$$(2.17) \quad \|x^1[0]\| \leqq \alpha_1 \rho_{x^1}(s) \quad \text{for} \quad s \geqq \theta_0.$$

For then (2.16) follows from (2.2) and (2.17) with $\alpha = \alpha_1 K_0$. In order to verify the existence of an α_1, note that, by (2.2), $y[0] = 0$ gives $\rho_y(t) = 0$ for $t \geqq \theta_0$. In particular $\rho_{x^1}(t)$ is uniquely determined by $x^1[0]$. Let $\beta(x^1[0]) = \inf \rho_{x^1}\|(t)\|$ for $t \geqq \theta_0$. It is clear that $\beta(x^1[0]) > 0$ unless $x^1[0] = 0$ (otherwise $\rho_{x^1}(t) \to 0$, $t \to \infty$, but $0 \neq x^1[0] \in X^1$). Since (2.2) shows that convergence of $x^1[0]$ in W implies the convergence of $\rho_{x^1}(t)$ in the norm "sup $\rho_{x^1}(t)$ for $t \geqq \theta_0$," it follows that $\beta(x^1[0])$ is a continuous function of $x^1[0]$. Thus if $X^1 \neq \{0\}$, $\beta(x^1[0])$ has a positive minimum $1/\alpha_1$ on the sphere $\|x^1[0]\| = 1$. This gives (2.17) if $X^1 \neq \{0\}$ while (2.17) is trivial if $X^1 = \{0\}$.

It will next be verified that if $x(t) \in \mathcal{N}$ with $x[0] \in X$, $x^1(t) \in \mathcal{N}$ with $x^1[0] \in X^1$, then

$$(2.18) \quad \max(\rho_{x^1}(s), \rho_x(s)) \leqq 3\alpha M_0 \rho_{x^1-x}(s) \quad \text{for} \quad s \geqq \theta_0.$$

Suppose that (2.18) is false. Then there exist $x(t)$, $x^1(t)$ as specified and an $s \geqq \theta_0$ such that

$$(2.19) \quad \max(\rho_{x^1}(s), \rho_x(s)) > 3\alpha M_0 \rho_{x^1-x}(s).$$

Since $y = x^1(t) - x(t) \in \mathcal{N}$ with $y[0] \in W_0$, it follows that $M_0 \rho_{x^1-x}(s) \geqq \rho_{x^1-x}(t)$ for $t \geqq s \geqq \theta_0$. By (2.1), (2.16), and the fact that $\rho_x(t) \to 0$ as $t \to \infty$, it follows that, for large t,

$$\max(\rho_{x^1}(s), \rho_x(s)) > 3\alpha M_0(\rho_{x^1}(t) - \rho_x(t)) \geqq 3\rho_{x^1}(s),$$

where the last inequality is a consequence of (2.16). Since the last two formula lines hold, $\max(\rho_{x^1}(s), \rho_x(s)) = \rho_x(s)$, and so $\rho_x(s) \geqq 3\rho_{x^1}(s)$. Thus by (2.1) $\rho_{x^1-x}(s) \geqq 2\rho_x(s)/3$. Hence (2.19) implies $\rho_x(s) > 2\alpha M_0 \rho_x(s)$, which is impossible since the constants α, M_0 in (2.15), (2.16) must satisfy $\alpha \geqq 1$, $M_0 \geqq 1$. Hence (2.18) holds. An immediate consequence of (2.18) is that

(2.20) $\max\,(\rho_{x^1}(r),\,\rho_x(t)) \leqq 3\alpha^2 M_0{}^2 \rho_{x^1-x}(s)$ if $\theta_0 \leqq r \leqq s \leqq t.$

The subspace X^0 of W orthogonal to X is the direct sum of X^1 and W_1. It will now be shown that if $x(t) \in \mathcal{N}$ with $x[0] \in X$, $x^0(t) \in \mathcal{N}$ with $x^0[0] \in X^0$, then

(2.21) $\max\,(\rho_{x^0}(r),\rho_x(t)) \leqq M_1 \rho_{x^0-x}(s)$ for $\theta_0 \leqq r \leqq s \leqq t$

if $M_1 = 6\alpha^2 M_0{}^3$. To this end, let $x^0[0] = x_1[0] + x^1[0]$ be the decomposition of $x^0[0]$ into orthogonal components $x_1[0] \in W_1$, $x^1[0] \in X^1$. Since $x^1[0] \in X^1 \subset W_0$ is in the range of Q, there is an $x^1(t) \in \mathcal{N}$ with $Qx^1(t) = x^1[0]$. Let $x_1(t) = x^0(t) - x^1(t) \in \mathcal{N}$, so that $Qx_1(t) = x_1[0]$. By (2.15) applied to $z = -x_1$, $y = x^1(t) - x(t)$,

$$\max\,(\rho_{x_1}(r),\,\rho_{x^1-x}(s)) \leqq M_0 \rho_{x^0-x}(s)   \text{for}   \theta_0 \leqq r \leqq s.$$

Hence (2.20) gives

$$\max\,(\rho_{x_1}(r),\,\rho_{x^1}(r),\,\rho_x(t)) \leqq 3\alpha^2 M_0{}^3 \rho_{x^0-x}(s)   \text{for}   \theta_0 \leqq r \leqq s \leqq t$$

and (2.21) follows with $M_1 = 6\alpha^2 M_0{}^3$ since $\rho_{x^0} \leqq \rho_{x^1} + \rho_{x_1}$ by (2.1).

Lemma 2.3 now follows from Lemma 2.2 (or its proof), where $\lambda_0 = 1$ is permitted here since orthogonal decomposition can be used in the Euclidean space W.

The proof of the existence of exponential dichotomies below will generally depend on proving first the existence of a dichotomy and then the applicability of the following:

Lemma 2.4. *Let $\sigma(t)$ be non-negative for $a \leqq t < \infty$ with the properties that there exist positive constants $\theta < 1$, M_0, δ such that $\sigma(t) \leqq M_0\sigma(s)$ for $a \leqq s \leqq t \leqq s + \delta$ and $\sigma(t + \delta) \leqq \theta\sigma(t)$ for $t \geqq a$. Then $\sigma(t) \leqq M_0\theta^{-1}e^{-\nu(t-s)}\sigma(s)$ for $a \leqq s \leqq t < \infty$, where $\nu = -\delta^{-1}\log\theta > 0$.*

If, in this lemma, the assumption $\sigma(t + \delta) \leqq \theta\sigma(t)$ holds only for $t \geqq b$ $(\geqq a)$, then the main inequality in this assertion is valid for $b \leqq s \leqq t < \infty$. It can, however, be replaced by $\sigma(t) \leqq K'\theta^{-1}e^{-\nu(t-s)}\sigma(s)$ valid for $a \leqq s \leqq t < \infty$ if $K' = M_0{}^m e^{\nu(b-a)}$ and $m = 1 + b - a$.

Proof of Lemma 2.4. Clearly, $\sigma(s + n\delta) \leqq \theta^n\sigma(s)$ for $s \geqq a$ and $n = 0, 1, \ldots$. Hence,

$$s + n\delta \leqq t < s + (n + 1)\delta   \text{implies}   \sigma(t) \leqq (M_0\theta^{-1})\theta^{n+1}\sigma(s).$$

Since $e^{-\nu(t-s)} \geqq e^{-\nu(n+1)\delta} = \theta^{n+1}$, the assertion follows.

Applications of Lemma 2.4, in proving the existence of exponential dichotomies, generally lead to an exponent ν' in (1.11) which depends on $\lambda > 1$. In order to get a ν' independent of λ, the following will be used. It is derived by the arguments used in the proof of Lemma 2.2.

Lemma 2.5. *Let W_0 be a subspace of W in the range of Q. Let there exist $\theta^0 \geqq 0$, $M_0{}' > 0$, $\nu' > 0$ such that if $z(t) \in \mathcal{N}$ with $\|z[0]\| \leqq \lambda_0 \, d(W_0, z[0])$ for a fixed $\lambda_0 > 1$, then*

$$(2.22) \qquad \rho_z(t) \geqq M_0{}' e^{\nu'(t-s)} \rho_z(s) \qquad \text{for} \quad \theta^0 \leqq s \leqq t.$$

Assume that $\rho_y(t)$ satisfies (2.1) and that there exist $\theta^1 \geqq 0$, $K_0 > 0$, $\Delta \geqq 0$ such that (2.4), (2.5) imply (2.2), (2.3), and

$$(2.23) \qquad \rho_z(t) \leqq \lambda K_0 \rho_{z-y}(t + \Delta) \qquad \text{for} \quad t \geqq \theta^1.$$

Finally, suppose that condition (b) for a partial dichotomy holds; cf. (1.8). Then condition (b) of an exponential dichotomy holds with the given ν' (for all $\lambda > 1$); cf. (1.11).

Proof. Let (2.5) hold and let $z(t) = y^0(t) - z^0(t)$ be the decomposition used in the proof of Lemma 2.2, so that (2.11) holds. Then (2.2) and (2.3) with $z = z^0$, $\lambda = \lambda_0$ give

$$\rho_{y^0}(t) \leqq 2K_0{}^2 \lambda_0 \lambda \rho_{z^0}(t) \qquad \text{for} \quad t \geqq \theta^1.$$

Since $z(t) = y^0(t) - z^0(t)$, (2.1) implies that

$$(2.24) \qquad \rho_z(t) \leqq (1 + 2K_0{}^2 \lambda_0 \lambda) \rho_{z^0}(t) \qquad \text{for} \quad t \geqq \theta^1.$$

Applying (2.22) for $z = z^0$, with s and t interchanged, gives

$$M_0{}' \rho_{z^0}(t) \leqq e^{-\nu'(s-t)} \rho_{z^0}(s) \qquad \text{for} \quad \theta^0 \leqq t \leqq s,$$

so that if s is replaced by $s - \Delta$,

$$M_0{}' \rho_{z^0}(t) \leqq e^{\nu'\Delta} e^{-\nu'(s-t)} \rho_{z^0}(s - \Delta) \qquad \text{for} \quad \theta^0 \leqq t \leqq s - \Delta.$$

By (2.23), with z, t, y, λ replaced by z^0, $s - \Delta$, y^0, λ_0,

$$\rho_{z^0}(s - \Delta) \leqq \lambda_0 K_0 \rho_z(s) \qquad \text{for} \quad s - \Delta \geqq \theta^1.$$

Thus (2.24) and the last two inequalities give

$$\rho_z(t) \leqq K' e^{-\nu'(s-t)} \rho_z(s) \qquad \text{for} \quad \theta^1 \leqq t \leqq s - \Delta$$

if $K' = (1 + 2K_0{}^2 \lambda_0 \lambda) \lambda_0 K_0 e^{\nu'\Delta} / M_0{}' > 0$. Finally, (1.8) in condition (b) for a partial dichotomy shows that

$$\rho_z(t) \leqq \lambda K' M_0 e^{\nu'\Delta} e^{-\nu'(s-t)} \rho_z(s) \qquad \text{if} \quad \theta^1 \leqq t \leqq s,$$

and $s \geqq \theta^0 + \Delta$. Another use of condition (b) of a partial dichotomy allows the removal of the restriction $t \geqq \theta^1$ by a suitable alteration of the factor $\lambda K' M_0 e^{\nu'\Delta}$. This proves Lemma 2.5.

The preceding lemmas and their proofs can be used to obtain a characterization of "total dichotomy" or "total exponential dichotomy" for the linear manifold $\mathcal{N}$ of solutions of the homogeneous equation (0.2), where Y is a finite dimensional (Banach) space. Let $U(t)$ be the fundamental matrix of (0.2) satisfying $U(0) = I$. Let W_0, W_1 be complementary subspaces of Y (i.e., $Y = W_0 \oplus W_1$), and let P_0 [or P_1] be the projection of Y onto W_0 [or W_1] annihilating W_1 [or W_0]. Define a Green's matrix $G(t, s)$ by

$$(2.25) \quad G(t, s) = U(t)P_0U^{-1}(s) \quad \text{or} \quad G(t, s) = -U(t)P_1U^{-1}(s)$$

according as $0 \leqq s \leqq t < \infty$ or $0 \leqq t < s < \infty$; cf. (XII 7.1).

Theorem 2.1 *Let $A(t)$ be a matrix of locally integrable (real- or complex-valued) functions on $t \geqq 0$, $\mathcal{N}$ the set of solutions of (0.2), $W = Y$, and $y[0] = y(0) \in Y = W$. A subspace W_0 of Y induces a total dichotomy [or total exponential dichotomy] for $(\mathcal{N}, \|y(t)\|, y(0))$ if and only if, for one and/or every subspace W_1 of Y complementary to W_0, the norm of the Green's matrix (2.25) satisfies*

$$(2.26) \quad\quad\quad \|G(t, s)\| \leqq K \quad \text{for} \quad s, t \geqq 0$$

for some constant $K = K(W_1)$ [or

$$(2.27) \quad\quad\quad \|G(t, s)\| \leqq Ke^{-v|t-s|} \quad \text{for} \quad s, t \geqq 0$$

for some constants $K = K(W_1)$, $v = v(W_1) > 0$].

Proof (*"Only if"*). Let W_0 induce a total dichotomy for $(\mathcal{N}, \|y(t)\|, y(0))$ and W_1 be a subspace of Y complementary to W_0. There exists a $\lambda_0 > 1$ such that if $z(0) \in W_1$, then $\|z(0)\| \leqq \lambda_0 d(W_0, z(0))$. Hence Lemma 2.1 implies that if $y(t)$, $z(t)$ are solutions of (0.2) and $y(0) \in W_0$, $z(0) \in W_1$, then (2.7) holds with $\lambda = \lambda_0$.

Let $c \in Y$ be arbitrary. Then $y(t) = U(t)P_0U^{-1}(s)c$, for a fixed s, is a solution of (0.2) and $y(0) = P_0U^{-1}(s)c \in W_0$, $y(s) = U(s)P_0U^{-1}(s)c$. Also $z(t) = -U(t)P_1U^{-1}(s)c$, for a fixed s, is a solution of (0.2) and $z(0) = -P_1U^{-1}(s)c \in W_1$, $z(s) = -U(s)P_1U^{-1}(s)c$. Thus (2.7) and $P_0 + P_1 = I$ give

$$(2.28) \quad \max{(\|U(r)P_1U^{-1}(s)c\|, \quad \|U(t)P_0U^{-1}(s)c\|)} \leqq K\|c\|$$

for $0 \leqq r \leqq s \leqq t$ if $K = \lambda_0M_0$. In view of (2.25), this is equivalent to (2.26).

Similarly, if W_0 induces a total exponential dichotomy for $\mathcal{N}$, then a use of (2.9) instead of (2.7) leads to (2.27) with K, v in (2.27) given, e.g., by $2\lambda_0\gamma_0^{-1} \max{(M_1, 1/M_1'(\lambda_0))}$, $\min{(v, v'(\lambda_0))}$, respectively, in terms of the constants in (2.9).

Exercise 2.1. Prove the "if" portion of Theorem 2.1.

3. The Operator T

The general theory will be presented in a somewhat abstract form which can then be applied to (0.1) or (0.3) or other situations. In what follows,

B and D are Banach spaces in $\mathcal{T}$. The results of this section are analogues of the lemmas of § XII 6.

Let Y, F be finite dimensional Banach spaces. The objects of study will be a linear operator T from $L(Y)$ to $L(F)$,

$$(3.1) \qquad\qquad Ty(t) = f(t),$$

and the elements $y = y(t)$ of its null space $\mathcal{N}(T)$

$$(3.2) \qquad\qquad Ty(t) = 0.$$

The domain of definition of T will be denoted by $\mathscr{D}(T)$.

Let U be a finite dimensional Banach space and W another Banach space. (It will not be assumed that W is finite dimensional although in the applications below this will be the case. In applications, e.g., to difference-differential equations, this need not be the case.) P will denote an operator from $L(Y)$ to $L(U)$ and Q an operator from $L(Y)$ to W with the same domain of definition as T, $\mathscr{D}(P) = \mathscr{D}(Q) = \mathscr{D}(T)$. The element $w = Qy(t) \in W$ will be called the "initial value" of $y(t)$ and denoted by $y[0]$. There will be no confusion even though $\| \ldots \|$ denotes norm in either Y, F, U, or W.

Remark. It will be convenient to illustrate various statements in the general theory from time to time by references to (0.1). In such references, it is always assumed that $A(t), g(t)$ are defined on J and are integrable over all finite intervals $J' \subset J$ and that $y(t)$ is an absolutely continuous solution (0.1). In this case, the spaces Y, F, U, W are taken to be identical; $Ty(t)$ is defined by $Ty = y'(t) - A(t)y(t)$ and $\mathscr{D}(T)$ is the set of functions $y(t)$ from J to Y which are absolutely continuous (on every J'); $Py(t) = y(t)$ is the identity operator; and $Qy(t) = y(0)$. In applying the general theory to (0.3), where u can be a vector, it is supposed that (0.3) is written as a system (0.1) for $y = (u, u', \ldots, u^{(m)})$, but $Py(t) = u(t)$, $Qy(t) = y(0)$.

Definition. *PD-Solutions and W_D.* Let $D \in \mathcal{T}$ be a Banach space. $y(t)$ is called a PD-solution of (3.1) for a given $f(t) \in L(F)$ if (3.1) holds and $Py(t) \in D(U)$. $W_D = W_D(P)$ denotes the linear manifold in W consisting of initial values $w = y[0]$ of PD-solutions of (3.2).

Definition. *P-Admissibility.* The pair (B, D) of Banach spaces in $\mathcal{T}$ is called P-admissible for (3.1) if, for every $f(t) \in B(F)$, (3.1) has at least one PD-solution $y(t)$.

Various assumptions (A_0), (A_1), ... or (B_1), (B_2), ... concerning T will be made from time to time. These and some of their consequences will now be discussed.

(A_0) If $y(t) \in \mathscr{D}(T)$, then $u(t) = Py(t)$ is (essentially) bounded on every bounded interval J' of J.

This assumption implies that the answer to the question whether or not $y(t)$ is a PD-solution of (3.1) depends only on the behavior of $u(t) = Py(t)$ for large t; cf. conditions (c), (e) for $\Phi = D \in \mathcal{T}$ in § 1 (iv).

(A_1) *Uniqueness for Q.* (a) If (3.2) holds and $y[0] = 0$, then $y(t) \equiv 0$. (b) There exist positive constants a, C_1 such that (3.1) implies that

$$(3.3) \qquad \|y[0]\| \leqq C_1 \left\{ a^{-1} \int_0^a \|y(t)\| \, dt + \int_0^a \|f(t)\| \, dt \right\}.$$

This assumption is, of course, suggested by the inequality (0.5).

(A_1') Same as (A_1) except that (3.3) is replaced by

$$(3.3') \qquad \|y[0]\| \leqq C_1 \left\{ a^{-1} \int_0^a \|Py(t)\| \, dt + \int_0^a \|f(t)\| \, dt \right\}.$$

(A_2) *Normality for P.* If (3.1) holds, then $y(t)$ is uniquely determined by $Py(t)$ and $f(t)$; furthermore, the linear map from $L(U) \times L(F)$ to $L(Y)$ defined by $(Py(t), Ty(t)) \to y(t)$ is continuous in the following sense: if $y_n(t) \in \mathcal{D}(T)$ and the two limits $u(t) = \lim Py_n(t)$ in $L(U)$, $f(t) = \lim Ty_n(t)$ in $L(F)$ exist, then $y(t) = \lim y_n(t)$ in $L(Y)$ exists and $y(t) \in \mathcal{D}(T)$, $Py(t) = u(t)$, $Ty(t) = f(t)$.

The main role of (A_2) is the following (cf. Lemma XII 6.2):

Lemma 3.1. *Assume* (A_2). *The operator T_0 from $D(U)$ to $B(F)$ defined by $T_0[Py(t)] = Ty(t)$, with a domain $\mathcal{D}(T_0)$ consisting of those elements $u = Py(t)$ in the range of P for which $u = Py(t) \in D(U)$ and $Ty(t) \in B(F)$, is closed. Also, for every $t > 0$, there exists a constant $C_2 = C_2(t)$ such that*

$$(3.4) \qquad \int_0^t \|y(s)\| \, ds \leqq C_2(t)[|Py|_D + |f|_B].$$

Proof. In order to prove that T_0 is closed, let $(u_1, f_1), (u_2, f_2), \ldots$ be a convergent sequence of elements in the graph of T_0, where $u_n = Py_n(t)$, $f_n = Ty_n(t)$ and $y_n(t) \in \mathcal{D}(T)$. Thus $u = \lim u_n(t)$ in $D(U)$ and $f = \lim f_n(t)$ in $B(F)$ exist. Since convergence in B or D implies convergence in L [cf. condition (1.5) on $\Phi = B$, $D \in \mathcal{T}$], (A_2) implies that $y(t) = \lim y_n(t)$ exists in $L(Y)$ and $y(t) \in \mathcal{D}(T)$, $Py = u$, $Ty = f$. Hence $u \in \mathcal{D}(T_0)$ and $T_0 u = f$; thus T_0 is closed.

Let T_1 be the map from $D(U) \times B(F)$ to the space $L^1_{[0,t]}(Y)$ of Y-valued functions which are integrable over the interval $[0, t]$, where the domain $\mathcal{D}(T_1)$ is the graph of T_0, and T_1 is defined by $T_1(Py, Ty) = y$. Thus T_1 is defined on a subspace of $D(U) \times B(F)$. (A_2) implies that T_1 is continuous, hence bounded. The inequality (3.4) is equivalent to the boundedness of T_1.

Although the trivial space $B = \{0\}$ is not in $\mathcal{T}$, this choice of B is permitted in Lemma 3.1. The fact that T_0 is closed gives

Corollary 3.1. *Assume* (A_2). *The set of* $u = Py(t)$, *where* $y(t)$ *varies over the PD-solutions of* (3.2), *is a subspace* (*i.e., closed linear manifold*) *of* $D(U)$.

(A_3) (B, D) *is* P-*admissible.*

Lemma 3.2. (A_1) [*or* (A_1')], (A_2), *and* (A_3) *imply that there exist constants* C_3, C_{30} *such that if* $f(t) \in B(F)$, *then* (3.1) *has a PD-solution* $y(t)$ *satisfying*

$$(3.5_1) \qquad |Py|_D \leq C_3 |f|_B; \qquad (3.5_2) \quad \|y[0]\| \leq C_{30} |f|_B.$$

Proof. Let T_0 be the operator from $D(U)$ to $B(F)$ occurring in Lemma 3.1. Thus T_0 is closed by Lemma 3.1 and onto by (A_3). Hence the existence of C_3 follows from the Open Mapping Theorem XII 0.3.

If (A_1) is assumed, then (3.3), (3.4) with $t = a$, and (3.5_1) give

$$\|y[0]\| \leq C_1 \left\{ a^{-1} C_2(a)(C_3 + 1) |f|_B + \int_0^a \|f\| \, dt \right\}.$$

Thus, by (1.5), (3.5_2) holds with $C_{30} = C_1\{a^{-1}C_2(a)(C_3 + 1) + |h_{0a}|_B\}$. If (A_1') holds, it is similarly seen that (1.5), (3.3'), and (3.5_1) imply (3.5_2) with $C_{30} = C_1\{a^{-1}|h_{0a}|_D \cdot C_3 + |h_{0a}|_B\}$.

(A_4) W_D *is closed.*

This assumption is of course trivial if W, hence W_D, is finite dimensional.

Lemma 3.3. *Assume* (A_1) [*or* (A_1')] *and* (A_2). *Let* W_{D0} *be a subspace of* W_D [*e.g., if* (A_4) *holds, let* $W_{D0} = W_D$]. *Then there exists a constant* $C_4 = C_4(W_{D0})$ *such that if* $y(t) \in \mathcal{N}(T)$ *and* $y[0] \in W_{D0}$, *then*

$$(3.6) \qquad |Py|_D \leq C_4 \|y[0]\|.$$

Proof. Let Q_0 be the operator from $D(U)$ to W_{D0} defined by $Q_0[Py(t)] = y[0]$, where $\mathscr{D}(Q_0)$ is the set of $u = Py(t)$ such that $y(t)$ is a PD-solution of (3.2) with $y[0] \in W_{D0}$. Q_0 is closed by (3.3) [or (3.3')] and (A_2), one-to-one by $(A_1 a)$, and onto since $W_{D0} \subset W_D$. Thus the open mapping theorem is applicable to Q_0 and implies Lemma 3.3.

Lemma 3.4. *Assume* (A_1) [*or* (A_1')], (A_2), (A_3), *and* (A_4). *Then there exists a constant* C_5 *with the property that if* $\lambda \geq 1$, $Ty = f$, $Py \in D(U)$, $f \in B(F)$, *and* $\|y[0]\| \leq \lambda d(W_D, y[0])$, *then*

$$(3.7) \qquad |Py|_D \leq \lambda C_5 |f|_B, \qquad \|y[0]\| \leq \lambda C_{30} |f|_B,$$

where C_{30} *is the same as in* (3.5). (*In particular,* $\lambda = 1$ *is permitted in* (3.7) *if* $y[0] = 0$.)

Proof. Let $y = y_0(t)$ be a PD-solution of (3.1) supplied by Lemma 3.3, so that $|Py_0|_D \leq C_3 |f|_B$, $\|y_0[0]\| \leq C_{30} |f|_B$. Since $y(t) - y_0(t)$ is a PD-solution of (3.2), (3.6) gives

$$|Py - Py_0|_D \leq C_4 \|y[0] - y_0[0]\| \leq C_4(\|y[0]\| + \|y_0[0]\|).$$

As $w = y[0] - y_0[0] \in W_D$, we have

$$\|y_0[0]\| = \|y[0] - w\| \geqq d(W_D, y[0]) \geqq \lambda^{-1} \|y[0]\|.$$

Thus the second inequality in (3.7) holds. In addition,

$$|Py|_D \leqq |Py_0|_D + C_4(1 + \lambda)\|y_0[0]\|.$$

Thus $\lambda \geqq 1$ shows that the first inequality in (3.7) holds with $C_5 = C_3 + 2C_4C_{30}$.

4. Slices of $\|Py(t)\|$

Recall that $\varphi_{s\epsilon}(t)$ always denotes a non-negative, integrable function on J with support on $[s, s + \epsilon]$ and that $|\ldots|_1$ refers to the L^1 norm on J.

($B_1\epsilon$) Let $\theta_0 > 0$, $\epsilon > 0$ and $K > 0$ be fixed. For every pair of solutions $y(t)$, $z(t)$ of (3.2), with $y(t) - z(t) \not\equiv 0$, and for any given function $\varphi_{s\epsilon}(t)$ as in § 1(iii), with $s \geqq \theta_0$, let there exist a function $y_1(t) \in \mathscr{D}(T)$ with the properties that (i) $y_1[0] = \text{(const.)} z[0]$; (ii) $\|Pz(t)\| \leqq K \|Py_1(t)\|/|\varphi_{s\epsilon}|_1$ for $0 \leqq t \leqq s$; (iii) $\|Py(t)\| \leqq K \|Py_1(t)\|/|\varphi_{s\epsilon}|_1$ for $t \geqq s + \epsilon$; (iv) there exists a constant K' (depending on y, z, $\varphi_{s\epsilon}$) such that $\|Py_1(t)\| \leqq K'\|Py(t)\|$ for $t \geqq s + \epsilon$; finally, (v) $\|Ty_1(t)\| \leqq K\varphi_{s\epsilon}(t) \|Py(t) - Pz(t)\|$ for $t \geqq 0$. Also, if $y(t) \not\equiv 0$ is a solution of (3.2), then $1/\|Py(t)\|$ is integrable over any closed interval J' of J.

Remark 1.   This assumption will be used only in the particular case

$$(4.1) \qquad \varphi_{s\epsilon}(t) = \frac{h_{s\epsilon}(t)}{\|Py(t) - Pz(t)\|}.$$

For (4.1), condition (v) becomes

$$(4.2) \qquad \|Ty_1(t)\| \leqq Kh_{s\epsilon}(t)$$

and, by Hölder's inequality,

$$(4.3) \qquad \frac{1}{|\varphi_{s\epsilon}|_1} \leqq \epsilon^{-2} \int_s^{s+\epsilon} \|Py(t) - Pz(t)\| \, dt.$$

Remark 2.   Note that assumption ($B_1\epsilon$) holds for all $\theta_0 > 0$, $\epsilon > 0$ with $K = 1$ if T is the operator associated with (0.1) as in the Remark in § 3; thus $Ty(t) = y'(t) - A(t)y(t)$, $Py(t) = y(t)$ and $y[0] = y(0)$. In fact, let

$$y_1(t) = y(t) \int_0^t \varphi_{s\epsilon}(r) \, dr + z(t) \int_t^\infty \varphi_{s\epsilon}(r) \, dr.$$

Then $Ty_1 = y_1' - A(t)y_1 = \varphi_{s\epsilon}(t)[y(t) - z(t)]$ and $y_1(t) = |\varphi_{s\epsilon}(t)|_1 y(t)$ for $t \geqq s + \epsilon$, $y_1(t) = |\varphi_{s\epsilon}|_1 z(t)$ for $0 \leqq t \leqq s$.

Theorem 4.1. *Assume* (A_0), (A_1) *[or* (A_1')*]*, (A_2), (A_3), (A_4), *and* $B_1(\epsilon)$ *for a fixed* ϵ. *Let* $y(t) \in \mathcal{N}(T)$, $y[0] \in W_D$, *and* $z(t) \in \mathcal{N}(T)$, $\|z[0]\| \leq \lambda\, d(W_D, \|z[0]\|)$ *for a* $\lambda > 1$, *and* $s \geq \theta_0$. *Then*

$$(4.4) \quad \max\left(|h_{0s}Pz|_D, |h_{s+\epsilon}Py|_D\right) \leq \lambda K^2 C_5 \epsilon^{-2}\, |h_{0\epsilon}|_B \int_s^{s+\epsilon} \|P(y-z)\|\, dt,$$

where K *and* C_5 *are the constants in* $(B_1\epsilon)$ *and Lemma 3.4, respectively. In particular, if* $\Delta \geq \epsilon$ *is fixed and either*

$$(4.5) \quad \rho_y(t) = |h_{t\Delta}Py|_D \quad \text{or} \quad \rho_y(t) = \int_t^{t+\Delta} \|Py(\tau)\|\, d\tau,$$

then W_D *induces a partial dichotomy for* $(\mathcal{N}(T), \rho_y(t))$ *with* $\theta^0 = \theta_0$ *and*

$$(4.6) \quad M_0 = 1 + 2K^2 C_5 \Gamma(\epsilon), \quad \text{where} \quad \Gamma(\epsilon) = \epsilon^{-2}\, |h_{0\epsilon}|_B\, |h_{0\epsilon}|_{D'}.$$

In the applications of Theorem 4.1, it will be important that M_0 depends on ϵ, but not on Δ. The inequalities (4.10), (4.12) in the following proof will be used in the proofs of Theorems 4.2 and 4.3.

Proof. Apply $(B_1\epsilon)$ with the choice (4.1) of $\varphi_{s\epsilon}(t)$, so that (v) implies (4.2). Assumption (A_0) and (iv) in $(B_1\epsilon)$ show that $Py_1 \in D(U)$. Since $\|y_1[0]\| \leq \lambda\, d(W_D, \|y_1[0]\|)$ by (i) and the condition on $z[0]$, Lemma 3.4 and (4.2) give $|Py_1|_D \leq \lambda K C_5\, |h_{0\epsilon}|_B$. It follows from (ii) and (iii) in $(B_1\epsilon)$ that

$$(4.7) \quad \max\left(|h_{0s}Pz|_D, |h_{s+\epsilon}Py|_D\right) \leq \frac{\lambda K^2 C_5\, |h_{0\epsilon}|_B}{|\varphi_{s\epsilon}|_1}.$$

Hence (4.4) is a consequence of (4.3).

In order to prove the assertions concerning (4.5), note that $|h_{t\Delta}Py|_D \leq |h_{s+\epsilon}Py|_D$ for $t \geq s + \epsilon$, $\Delta > 0$, and $|h_{r\Delta}Pz|_D \leq |h_{0s}Pz|_D$ if $r + \Delta \leq s$. Thus, for any $\Delta > 0$ and $s \geq \theta_0$, the inequality (4.4) implies that, for $r + \Delta \leq s$, $s + \epsilon \leq t$,

$$(4.8) \quad \max\left(|h_{r\Delta}Pz|_D, |h_{t\Delta}Py|_D\right) \leq \lambda K^2 C_5 \epsilon^{-2}\, |h_{0\epsilon}|_B \int_s^{s+\epsilon} \|P(y-z)\|\, d\tau.$$

The relations (1.5) and $\Delta \geq \epsilon$ show that

$$(4.9) \quad \int_s^{s+\epsilon} \|P(y-z)\|\, dt \leq |h_{0\epsilon}|_{D'}\, |h_{s\epsilon}P(y-z)|_D \leq |h_{0\epsilon}|_{D'}\, |h_{s\Delta}P(y-z)|_D.$$

Thus, for $\Delta \geq \epsilon$, $s \geq \theta_0$, $r \geq 0$, $r + \Delta \leq s$, $s + \epsilon \leq t$,

$$(4.10) \quad \max\left(|h_{r\Delta}Pz|_D, |h_{t\Delta}Py|_D\right) \leq \lambda K^2 C_5 \Gamma(\epsilon)\, |h_{s\Delta}P(y-z)|_D,$$

where $\Gamma(\epsilon)$ is defined in (4.6).

The case $z \equiv 0$ and $\lambda = 1$ of (4.10) combined with

$$(4.11) \quad |h_{t\Delta}Py|_D \leq |h_{s\Delta}Py|_D + |h_{s+\Delta,\Delta}Py|_D \quad \text{for} \quad s \leq t \leq s + \Delta$$

gives condition *(a)* [i.e., (1.7)] for a partial dichotomy for $\rho_y(t) = |h_{t\Delta}Py|_D$

with $\theta^0 = \theta_0$ and M_0 given by (4.6), even with the factor 2 omitted from the second term. Similarly, the choice $y \equiv 0$ in (4.10) combined with an analogue of (4.11) gives condition (b) [i.e., (1.8)] with $\theta^0 = \theta_0$ and the same M_0.

In order to deal with the second function in (4.5), apply an analogue of (4.9) to the left side of (4.8) with $\Delta = \epsilon$ to obtain

$$(4.12) \quad \max \left(\int_r^{r+\epsilon} \|Pz\| \, d\tau, \int_t^{t+\epsilon} \|Py\| \, d\tau \right) \leq \lambda K^2 C_5 \Gamma(\epsilon) \int_s^{s+\epsilon} \|P(y-z)\| \, d\tau$$

if $r + \epsilon \leqq s$, $t \geqq s + \epsilon$, $s \geqq \theta_0$. For $\Delta \geqq \epsilon$, let $k \geqq 1$ be an integer such that $k\epsilon \leqq \Delta < (k+1)\epsilon$. Then it follows, by replacing t by $t + j\epsilon$ and s by $s + j\epsilon$ for $j = 0, \dots, k-1$ and adding, that

$$\int_t^{t+k\epsilon} \|Py\| \, d\tau \leqq \lambda K^2 C_5 \Gamma(\epsilon) \int_s^{s+k\epsilon} \|P(y-z)\| \, d\tau.$$

In addition,

$$\int_{t+k\epsilon}^{t+\Delta} \|Py\| \, d\tau \leqq \lambda K^2 C_5 \Gamma(\epsilon) \int_s^{s+\epsilon} \|P(y-z)\| \, d\tau.$$

If the upper limits of integration on the right of the last two inequalities are replaced by $s + \Delta$, we obtain the second of the inequalities contained in

$$(4.13) \quad \max \left(\int_r^{r+\Delta} \|Pz\| \, d\tau, \int_t^{t+\Delta} \|Py\| \, d\tau \right) \leqq 2\lambda K^2 C_5 \Gamma(\epsilon) \int_s^{s+\Delta} \|P(y-z)\| \, d\tau$$

for $r + \Delta \leqq s$, $t \geqq s + \epsilon$, $s \geqq \theta_0$. The other inequality, involving the first integral, is obtained similarly. Combining (4.13) with inequalities of the type

$$\int_t^{t+\Delta} \|Py\| \, d\tau \leqq \int_s^{s+\Delta} \|Py\| \, d\tau + \int_{s+\Delta}^{s+2\Delta} \|Py\| \, d\tau \quad \text{for} \quad s \leqq t \leqq s + \Delta$$

leads to a partial dichotomy for the second function in (4.5) with $\theta^0 = \theta_0$ and M_0 given by (4.6). This proves Theorem 4.1.

Corollary 4.1. *In addition to the assumptions of Theorem* 4.1, *assume that D is quasi-full (cf. (ix) in § 1) and that z(t) is a non-PD-solution of (3.2), then*

$$\int_t^{t+\epsilon} \|Pz(r)\| \, dr \to \infty \quad \text{as} \quad t \to \infty.$$

This follows from (4.4) with $y(t) \equiv 0$ and the definition of a quasi-full space.

Corollary 4.2. *In addition to the assumptions of Theorem* 4.1, *assume that W is finite dimensional, that $\rho_y(t)$ is defined in (4.5) and, in the case of the first choice, that $\rho_y(t)$ is a continuous function of t. Let W_0 be the set of* *initial values* $y[0]$ *of* $y(t) \in \mathcal{N}(T)$ *satisfying* $\rho_y(t) \to 0$ *as* $t \to \infty$. *Then* $W_0 \subset W_D$ *and* W_0 *induces a partial dichotomy for* $(\mathcal{N}(T), \rho_y)$.

Exercise 4.1. Verify this corollary which follows from Lemma 2.3 combined with Theorem 4.1 and its proof.

Theorem 4.2. *Let the assumptions of Theorem* 4.1 *hold with* $B = L^1$ *and* $D = L^\infty$ *or* $D = L_0^\infty$; *in addition, let* $(B_1 \epsilon)$ *hold for all* ϵ, $0 < \epsilon \leq \epsilon_0$, *with* θ_0, K *independent of* ϵ. *Let* $P\mathcal{N}$ *be the set of functions* $Py(t)$ *with* $y(t) \in \mathcal{N}$ *and let the "initial value"* $y[0]$ *be assigned to* $Py(t)$. *Assume that* $\|Py(t)\|$ *is a continuous function of* t *for* $y(t) \in \mathcal{N}$. *Then* W_D *induces a total dichotomy for* $P\mathcal{N}$.

Proof. By the proof of Theorem 4.1, (4.12) holds for $r + \epsilon \leq s$, $t \geq s + \epsilon$, $s \geq \theta_0$. Since $B = L^1$ and $D = L^\infty$ (or L_0^∞), $\Gamma(\epsilon)$ in (4.6) is $\epsilon^{-2} \cdot \epsilon \cdot \epsilon = 1$; cf. (1.5). Hence

$$\max \left(\epsilon^{-1} \int_r^{r+\epsilon} \|Pz\| \, d\tau, \, \epsilon^{-1} \int_t^{t+\epsilon} \|Py\| \, d\tau \right) \leq \lambda K^2 C_5 \epsilon^{-1} \int_s^{s+\epsilon} \|P(y-z)\| \, d\tau.$$

Letting $\epsilon \to 0$ gives

$$\max \left(\|Pz(r)\|, \|Py(t)\| \right) \leq \lambda K^2 C_5 \|P(y-z)(s)\|$$

for $0 \leq r \leq s \leq t$, $s \geq \theta^0$. In view of Lemma 2.1, this proves Theorem 4.2.

Theorem 4.3. *Assume* (A_0), (A_1'), (A_2), (A_3), (A_4), *and* $B_1(\epsilon)$ *for all* $\epsilon \geq \epsilon_0 > 0$ *and with* θ_0, K *independent of* ϵ. *Assume also that either*

$$(4.14) \qquad \Delta^{-1} \, |h_{0\Delta}|_B \to 0 \qquad \text{or} \qquad \Delta^{-1} \, h|_{0\Delta}|_{D'} \to 0 \qquad \text{as} \quad \Delta \to \infty.$$

Then W_D *induces an exponential dichotomy for the functions* (4.5) *for every fixed* $\Delta \geq \epsilon_0$, *with* $\theta^0 = \theta_0$ *and constant* $M_1'(\lambda, \Delta)$ *depending on* Δ, *but* M_1, ν, $\nu' > 0$ *independent of* Δ.

Note that if $(B, D) = (L^p, L^q)$, where $1 \leq p, q \leq \infty$, then (4.14) holds when $(p, q) \neq (1, \infty)$.

Proof. Theorem 4.3 will be deduced from Lemmas 2.4, 2.5, and Theorem 4.1. In the proof consider only the first function $\rho_y(t) = |h_{t\Delta} Py|_D$ in (4.5); the proof for the other function is similar. The condition (4.14) is equivalent to

$$(4.15) \qquad \Gamma(\epsilon) \equiv \epsilon^{-2} |h_{0\epsilon}|_B \, |h_{0\epsilon}|_{D'} \to 0 \qquad \text{as} \quad \epsilon \to \infty.$$

In order to see this, note that if $k \geq 1$ is an integer such that $0 < k\epsilon \leq \eta < (k+1)\epsilon$, then $|h_{0\eta}|_{D'} \leq (k+1) |h_{0\epsilon}|_{D'}$. This is a consequence of the fact that $|h_{J'}|_{D'}$ is the "best" constant in (1.5). Hence $\eta^{-1} |h_{0\eta}|_{D'} \leq (k+1)k^{-1}\epsilon^{-1} |h_{0\epsilon}|_{D'} \leq 2\epsilon^{-1} |h_{0\epsilon}|_{D'}$ for all $\eta \geq \epsilon$ and so the second function of Δ in (4.14) is bounded for $\epsilon \geq \epsilon_0$. Also $\liminf \epsilon^{-1} |h_{0\epsilon}|_{D'} = 0$ as $\epsilon \to \infty$ implies that $\epsilon^{-1} |h_{0\epsilon}|_{D'} \to 0$ as $\epsilon \to \infty$. Similar remarks apply to the first function in (4.14). This makes it clear that (4.14) and (4.15) are equivalent.

Let ϵ $(\geqq \epsilon_0)$ be fixed so large that

$$(4.16) \qquad\qquad \theta \equiv K^2 C_5 \Gamma(\epsilon) < 1.$$

Let $y(t) \in \mathcal{N}(T)$, $y[0] \in W_D$, so that (1.7) in condition (a) of a partial dichotomy is applicable for $\Delta \geqq \epsilon_0$. By the case $z \equiv 0$, $\lambda = 1$ of (4.10), it follows that

$$(4.17) \qquad |h_{t\Delta} P y|_D \leqq \theta\, |h_{s\Delta} P y|_D \qquad \text{if } t \geqq s + \epsilon$$

and $\Delta \geqq \epsilon$. Thus, for fixed $\Delta \geqq \epsilon$, condition (a) [cf. (1.10)] for an exponential dichotomy with $\theta^0 = \theta_0$, $M_1 = M_0 \theta^{-1}$, and $\nu = -\epsilon^{-1} \log \theta$ follows from Lemma 2.4 applied to $\sigma(t) = |h_{t\Delta} P y|_D$. (Hence M_1 and ν are independent of $\Delta \geqq \epsilon$.) If $\epsilon_0 \leqq \Delta < \epsilon$, let $k \geqq 1$ be an integer such that $(k-1)\Delta < \epsilon \leqq k\Delta$; thus $k < 1 + \epsilon/\epsilon_0$. Then, by what has been proved,

$$|h_{t\Delta} P y|_D \leqq |h_{t,k\Delta} P y|_D \leqq M_1 e^{-\nu(t-s)} |h_{s,k\Delta} P y|_D,$$

for $\theta_0 \leqq s \leqq t$. By (1.7) in a partial dichotomy,

$$|h_{s,k\Delta} P y|_D \leqq k M_0 |h_{s\Delta} P y|_D.$$

These two inequalities give condition (a) of an exponential dichotomy with $\theta^0 = \theta_0$, $M_1 = k M_0 \theta^{-1}$, and $\nu = -\epsilon^{-1} \log \theta$.

In order to obtain condition (b) [cf. (1.11)], consider first a fixed $\lambda_0 > 1$. Let $\epsilon = \epsilon(\lambda_0)$ be so large that $\theta \equiv \lambda_0 K^2 C_5 \Gamma(\epsilon) < 1$. Let $z(t) \in \mathcal{N}(T)$, $\|z[0]\| \leqq \lambda_0\, d(W_D, z[0])$. Then (1.8) in condition (b) of a partial dichotomy is applicable for $\Delta \geqq \epsilon$. The case $y = 0$, $\lambda = \lambda_0$, $\epsilon = \epsilon(\lambda_0)$ of (4.10) gives

$$|h_{r\Delta} P z|_D \leqq \theta\, |h_{s\Delta} P z|_D \qquad \text{if } r + \Delta \leqq s \text{ and } \Delta \geqq \epsilon.$$

An application of Lemma 2.4 to $\sigma(t) = 1/|h_{t\Delta} P z|_D$ gives condition (b) for an exponential dichotomy for $\lambda = \lambda_0$ with $\theta^0 = \theta_0$, $M_1' = \lambda_0 M_0 \theta^{-1}$, and $\nu' = -\Delta \log \theta$.

When $\Delta \geqq \epsilon$, the corresponding condition (b) for all $\lambda > 1$ with the same ν' will be deduced from Lemma 2.5. In fact, condition (2.22) has just been verified. (2.1) is clear and (2.2) follows from (3.6) with $K_0 = C_4$, $\theta^1 = 0$. Condition (2.3) follows from (3.3') applied to $y = z$, $f = 0$ which, together with (1.5), gives

$$\|z[0]\| \leqq C_1 a^{-1} \int_0^a \|Pz(t)\|\, dt \leqq C_1 a^{-1} |h_{0a} P z|_D\, |h_{0a}|_{D'}.$$

In fact, since (1.8) implies that $|h_{0a} P z|_D \leqq \lambda k M_0 |h_{t\Delta} P z|_D$ for $t \geqq \max(a, \theta_0)$ if $a \leqq k\Delta$, (2.3) follows with $K_0 = C_1 a^{-1} |h_{0a}|_{D'}\, k M_0$, $\theta^1 = \max(a, \theta_0)$. Finally, (2.23) follows from (4.10) with $K_0 = K^2 C_5 \Gamma(\epsilon)$. Consequently, condition (b) of an exponential dichotomy holds for $\Delta \geqq \epsilon$ with $M_1' = M_1'(\lambda, \Delta)$, $\nu' = \nu'(\Delta)$.

As in case (a), it can be shown that (b) holds for $\epsilon_0 \leqq \Delta < \epsilon$ with a suitable $M_1'(\lambda, \Delta)$ and $\nu' = \nu'(\epsilon)$ independent of Δ. Finally, an analogous argument shows that it is possible to choose $\nu' = \nu'(\epsilon)$ for all $\Delta \geqq \epsilon_0$. This proves Theorem 4.3.

Theorem 4.4. *Let the conditions of Theorem 4.3 hold; in addition, let there exist a subspace W_0 of W which induces a partial dichotomy for $(\mathcal{N}, \|Py(t)\|)$. Then $W_0 = W_D$ and W_D induces an exponential dichotomy for $(\mathcal{N}, \|Py(t)\|)$.*

Proof. In view of Theorem 4.3, W_D induces an exponential dichotomy for $(\mathcal{N}, \rho_y(t))$, where $\rho_y(t) = \displaystyle\int_t^{t+\Delta} \|Py(\tau)\| \, d\tau$ for a suitable $\Delta > 0$. This makes it clear that if $Tz(t) = 0$ and $z[0] \notin W_D$, then $\rho_z(t)$ is not bounded as $t \to \infty$, and so $z[0] \notin W_0$. Thus $W_0 \subset W_D$. Also, if $z[0] \notin W_0$, then "$\rho_y(t) \to 0$ as $t \to \infty$" does not hold, and so $z[0] \notin W_D$. Thus $W_D \subset W_0$ and, consequently, $W_0 = W_D$.

Using the fact that $W_D = W_0$ induces a partial dichotomy for $(\mathcal{N}, \|Py(t)\|)$ and an exponential dichotomy for $(\mathcal{N}, \rho_y(t))$, it is easy to see that it induces an exponential dichotomy for $(\mathcal{N}, \|Py(t)\|)$. Details will be left to the reader.

Theorem 4.1 and 4.4 are immediately applicable to the operator T associated with (0.1); the results will be given in § 6. These theorems are not applicable to operators associated with (0.3) without some boundedness conditions on the coefficients $P_k(t)$. The difficulty arises from the fact that, in general, condition $(B_1\epsilon)$ does not hold. The next section leads to theorems applicable to (0.3) as well as (0.1).

5. Estimates for $\|y(t)\|$

In this section, the role of $(B_1\epsilon)$ will be played by the following condition:

$(B_2\epsilon)$ Let $\epsilon > 0$, $\theta_0 > 0$, $K_2 > 0$ be fixed. For every $s \geqq \theta_0$ and every pair $y(t), z(t) \in \mathcal{N}(T)$, where $y[0] \in W_D$, there exists a $\varphi_{s\epsilon}(t)$ as in § 1(iii) and a $y_1(t) \in \mathscr{D}(T)$ such that (i) $y_1[0] = $ (const.) $z[0]$; (ii) $\|Pz(t)\| \leqq K_2 \|Py(t)\|$ for $0 \leqq t \leqq s$; (iii) $\|Py(t)\| \leqq K_2 \|Py_1(t)\|$ for $s + \epsilon \leqq t < \infty$; (iv) $Py_1(t) \in D(U)$; (v) $Ty_1(t)$ satisfies

$$(5.1) \qquad \|Ty_1(t)\| \leqq \varphi_{s\epsilon}(t) \|y(t) - z(t)\|;$$

(vi) $\varphi_{s\epsilon}(t) \in B$ and there exists a number $b(\epsilon)$ satisfying

$$(5.2) \qquad |\varphi_{s\epsilon}|_B \leqq b(\epsilon) \qquad \text{for } s \geqq \theta_0.$$

Note that (5.1) implies that $Ty_1(t) = 0$ unless $s \leqq t \leqq s + \epsilon$.

$(B_3\delta)$ Let $\delta \geqq 0$ be fixed. There exists a constant $K_3(\delta)$ such that

(i) if $y(t) \in \mathcal{N}(T)$ and $y[0] \in W_D$, then $y(t) \in D(Y)$ and $|h_{s+\delta}y|_D \leqq K_3(\delta) |h_s Py|_D$ for $s \geqq 0$; (ii) if $y(t) \in \mathcal{N}(T)$, then $h_{0s}(t)Py(t) \in D(U)$, $h_{0s}(t)y(t) \in D(Y)$ and $|h_{0s-\delta}y|_D \leqq K_3(\delta) |h_{0s}Py|_D$ for $s \geqq \delta$.

It is clear that if $(B_3\delta)$ holds for $\delta = \delta_0$, then it holds for all $\delta \geqq \delta_0$. For convenient reference, the following variant of $(B_3\delta)$, which will be needed in Part II of this chapter, is stated here.

$(B_{30}\delta)$ Let $\delta \geqq 0$ be fixed. There exists a constant $K_{30}(\delta)$ such that if $f(t) \in B_\infty(F)$ [cf. (vi) in § 1], $f(t) = 0$ on an interval $[s - \delta, s + \Delta + \delta]$ for some $s \geqq \delta$, and $y(t)$ is a PD-solution of $Ty = f$, then $y(t) \in D(Y)$ and $|h_{s\Delta}y|_D \leqq K_{30}(\delta) |Py|_D$.

$(B_4\Delta)$ Let $\Delta > 0$ be fixed. The solutions $y(t)$ of $Ty = 0$ are continuous functions of t (from J to Y) and there exists a constant $K_4(\Delta)$ such that if $y(t) \in \mathcal{N}(T)$, then $\|y(t)\| \leqq K_4(\Delta) \|y(s)\|$ if $|s - t| \leqq \Delta$.

It is clear that if $(B_4\Delta)$ holds for some $\Delta > 0$, then it holds for all $\Delta > 0$.

Remark. If $A(t)$ satisfies

$$(5.3) \qquad \int_t^{t+1} \|A(s)\| \, ds \leqq \text{const.} \qquad \text{for } t \geqq 0$$

or if Y is Euclidean and

$$(5.4) \qquad \int_t^{t+1} \sup_{\|y\|=1} |\text{Re } A(s)y \cdot y| \, ds \leqq \text{const.} \qquad \text{for } t \geqq 0,$$

then the operator T belonging to (0.1), as described in the Remark of § 3, satisfies $(B_4\Delta)$. This is clear from (0.5) or (0.6).

Theorem 5.1. *Let $\epsilon, \delta > 0$ be fixed. Assume (A_1)–(A_4); $(B_2\epsilon)$ with $b(\epsilon)$ independent of y, z; $(B_3\delta)$; and $(B_4\Delta)$. Then W_D induces a total dichotomy for $\mathcal{N} = \mathcal{N}(T)$ with $\theta^0 = 0$. If, in addition, D is quasi-full and $z(t)$ is a non-PD-solution of $Tz = 0$, then $\|z(t)\| \to \infty$ as $t \to \infty$.*

If the assumption that $b(\epsilon)$ can be taken independent of y, z is omitted, then we obtain an individual partial dichotomy instead of a total dichotomy.

Proof. In view of (5.1) and $(B_4\epsilon)$,

$$\|Ty_1(t)\| \leqq K_4(\epsilon)\varphi_{s\epsilon}(t) \|y(t_1) - z(t_1)\| \qquad \text{for } s \leqq t_1 \leqq s + \epsilon$$

and $s \geqq \theta_0$. Hence $|Ty_1|_B \leqq K_4(\epsilon) |\varphi_{s\epsilon}|_B \|y(t_1) - z(t_1)\|$ for $s \leqq t_1 \leqq s + \epsilon$. If $\|z[0]\| \leqq \lambda \, d(W_D, z[0])$, then, since $Py_1 \in D(U)$, Lemma 3.4 gives

$$|Py_1|_D \leqq \lambda C_5 K_4(\epsilon) |\varphi_{s\epsilon}|_B \|y(t_1) - z(t_1)\|,$$

thus (ii) and (iii) in $(B_2\epsilon)$ imply that

$$\max (|h_{0s}Pz|_D, |h_{s+\epsilon}Py|_D) \leqq \lambda K_2 C_5 K_3(\epsilon) |\varphi_{s\epsilon}|_B \|y(t_1) - z(t_1)\|.$$

By $(B_3\delta)$,

$$\max (|h_{0,s-\delta}z|_D, |h_{s+\epsilon+\delta}y|_D) \leqq \lambda K_3(\delta)C_5K_4(\epsilon) |\varphi_{s\epsilon}|_B \|y(t_1) - z(t_1)\| .$$

The inequality in $(B_4\epsilon)$ shows that, for $0 \leqq \tau \leqq t_0 \leqq \tau + \epsilon$,

$$\|y(t_0)\| \cdot |h_{0\epsilon}|_D \leqq K_4(\epsilon) |h_{\tau\epsilon}y|_D$$

and that a similar inequality holds for z. Consequently,

$$\max (\|z(r)\|, \|y(t)\|) \leqq \frac{\lambda K(\epsilon, \delta) |\varphi_{s\epsilon}|_B}{|h_{0\epsilon}|_D} \|y(t_1) - z(t_1)\|$$

where $0 \leqq r \leqq s - \epsilon - \delta$, $t \geqq s + \epsilon + \delta$, $s \geqq \theta_0$, and $K(\epsilon, \delta) = K_3(\delta)C_5K_4^2(\epsilon)$. Finally, applications of $(B_4\epsilon + \delta)$ and $(B_4\theta_0)$ give

$$(5.5) \qquad \max (\|z(r)\|, \|y(t)\|) \leqq \lambda M_0 \|y(s) - z(s)\|$$

for $0 \leqq r \leqq s \leqq t$ if $M_0 = \lambda K(\epsilon, \delta)b(\epsilon)K_4(\epsilon + \delta)K_4(\theta_0)/|h_{0\epsilon}|_D$. Hence Theorem 5.1 follows from Lemma 2.1.

Theorem 5.2. *Assume* (A_1)–(A_4); $(B_2\epsilon)$ *for all* $\epsilon \geqq \epsilon_0 > 0$ *with* θ_0, K_2 *independent of* ϵ, $|\varphi_{s\epsilon}|_B$ *independent of* y, z, *and*

$$(5.6) \qquad \frac{|\varphi_{s\Delta B}|}{|h_{0\Delta}|_D} \to 0 \qquad as \quad \Delta \to \infty$$

uniformly for large s; $(B_3\delta)$; *and* $(B_4\Delta)$. *Then* W_D *induces a total exponential dichotomy for* $\mathcal{N} = \mathcal{N}(T)$ *with* $\theta^0 = 0$.

Proof. Let $z \equiv 0$ in $(B_1\epsilon)$. Then (5.1), $t \geqq s$, Theorem 5.1 imply $\|T_1y(t)\| \leqq M_0\varphi_{s\epsilon}(t) \|y(s)\|$. Arguing as in the last proof, it follows that $|h_{s+\epsilon+\delta}y|_D \leqq C_5K_3(\delta)M_0 |\varphi_{s\epsilon}|_B \|y(s)\|$. Also, by condition (1.7) of a total dichotomy, $M_0 |h_{\tau\epsilon}y|_D \geqq \|y(\tau + \epsilon)\| \cdot |h_{0\epsilon}|_D$. Hence if $t \geqq s + 2\epsilon + \delta$,

$$\|y(t)\| \leqq \frac{C_5K_3(\delta)M_0^2 |\varphi_{s\epsilon}|_B}{|h_{0\epsilon}|_D} \|y(s)\| .$$

In view of (5.6), ϵ and s_0 can be chosen so large that

$$\frac{C_5K_3(\delta)M_0^2 |\varphi_{s\epsilon}|_B}{|h_{0\epsilon}|_D} \leqq \theta < 1 \qquad for \quad s \geqq s_0,$$

so that $\|y(t)\| \leqq \theta \|y(s)\|$ for $s \geqq s_0$, $t \geqq s + 2\epsilon + \delta$. An application of Lemma 2.4 and the remark following it to $\sigma(t) = \|y(t)\|$ give condition (a) for an exponential dichotomy with $\theta^1 = 0$; cf. (1.10).

Let $\lambda_0 > 1$ and $z(t) \in \mathcal{N}(T)$, $\|z[0]\| \leqq \lambda_0 d(W_D, z[0])$. Choosing $y \equiv 0$ in $(B_2\epsilon)$ and arguing as before shows that if $\epsilon = \epsilon(\lambda_0)$, $s = s(\lambda_0)$ are so large that

$$\frac{\lambda_0^2 K_3(\delta)M_0^2 |\varphi_{s\epsilon}|_B}{|h_{0\epsilon}|_D} \leqq \theta < 1 \qquad for \quad s \geqq s_0,$$

then $\|z(r)\| \leqq \theta \|z(s)\|$ if $r \leqq s - \epsilon - \delta$, $s \geqq s_0$. Applying Lemma 2.4 to $\sigma(t) = 1/\|z(t)\|$ gives condition (b) of an exponential dichotomy for $\lambda = \lambda_0$ with $\theta^0 = 0$; cf. (1.11).

Condition (b) for all $\lambda > 1$, with ν' independent of λ, can now be deduced from Lemma 2.5, where $\rho_y(t) = \|y(t)\|$. In fact, (2.22) has just been verified and (2.1) is clear. In order to deduce (2.2), note that (3.6) and $(B_3\delta)$ imply that

$$|h_{t\Delta} y|_D \leqq K_3(\delta)\, |Py|_D \leqq K_3(\delta) C_4\, \|y[0]\|$$

for any $t \geqq \delta$, $\Delta > 0$. By condition (a) of a partial dichotomy, $M_0\, |h_{t\Delta} y|_D \geqq \|y(t)\|\, |h_{0\Delta}|_D$. This gives (2.2) with $K_0 = K_3(\delta) C_4 M_0 / |h_{0\Delta}|_D$ and $\theta^1 = \delta$. To obtain (2.3), begin with (3.3) applied to $y = z$, $f = 0$. Then, by condition (b) of a partial dichotomy,

$$\|z[0]\| \leqq \lambda C_1 M_0\, \|z(t)\| \qquad \text{for} \quad t \geqq a,$$

which is (2.3) with $K_0 = C_1 M_0$ and $\theta^1 = a$. Finally, (2.23) is implied by (5.5). Hence Lemma 2.5 implies condition (1.11) of an exponential dichotomy for $\mathcal{N}(T)$.

We can obtain results analogous to Theorems 5.1–5.2 under a condition somewhat weaker than $(B_4\Delta)$:

$(B_5\Delta)$ Let $\Delta > 0$ be fixed. The solutions $y(t)$ of $Ty = 0$ are continuous functions of t (from J to Y) and there exists a number $K_5(\Delta)$ such that if $y(t) \in \mathcal{N}(T)$, then

$$(5.7) \qquad \|y(t)\| \leqq K_5(\Delta)(\|y(s)\| + \|y(s + \delta)\|) \quad \text{if } s \leqq t \leqq s + \delta, \ \delta \leqq \Delta.$$

Condition $(B_5\Delta)$ is useful for applications to second order equations and is suggested by Exercises XI 8.6 and 8.8. For applications, see Exercises 7.1, 13.1, and 13.2, below.

Choosing $s = t - \frac{1}{2}\delta$ in (5.7) and integrating with respect to δ over an interval $[0, \delta]$ gives

$$(5.8) \qquad \|y(t)\| \leqq 2K_4(\Delta)\delta^{-1} \int_{t-\delta/2}^{t+\delta/2} \|y(r)\|\, dr \qquad \text{if } t \geqq \tfrac{1}{2}\delta > 0, \ \delta \leqq \Delta.$$

Introduce the Banach space $Y^{(2)} = Y \times Y$ with the norm of $\eta = (y_1, y_2) \in Y^{(2)}$ defined by $\|\eta\| = \max (\|y_1\|, \|y_2\|)$.

Theorem 5.3. Let $\epsilon, \Delta > 0$ be fixed. Assume (A_1)–(A_4); $(B_2\epsilon)$ with $b(\epsilon)$ independent of y, z; $(B_3\frac{1}{2}\epsilon)$; and $(B_5\Delta)$ with $\Delta \geqq 2\epsilon$. For $\delta > 0$, let $\mathcal{N}_\delta$ be the manifold of functions $\eta = \eta^\delta(t) = (y(t), y(t + \delta))$ from J to $Y^{(2)}$, where $y(t) \in \mathcal{N}(T)$, and let $\eta^\delta[0] = y[0]$. If $\epsilon \leqq \delta \leqq \Delta$, then W_D induces a partial dichotomy for $(\mathcal{N}_\delta, \|\eta^\delta(t)\|, \eta^\delta[0])$ with $\theta^0 = \max (\theta_0, \epsilon)$ and $M_0 = M_0(\epsilon, \delta)$. If $3\epsilon \leqq \delta \leqq \Delta$, then W_D induces a total dichotomy for $\mathcal{N}_\delta$ with $\theta^0 = \max\,(\theta_0 - \epsilon, 0)$ and $\gamma_0 = \gamma_0(\epsilon, \delta)$. If, in addition, D is *quasi-full,* $\epsilon \leq \delta \leq \Delta$, *and* $y(t)$ *is a non-PD-solution of* $Ty = 0$, *then* $\|\eta^\delta(t)\| \to \infty$ *as* $t \to \infty$.

Exercise 5.1.   Prove Theorem 5.3.

Theorem 5.4.   *Let* $\epsilon_0 > 0$, $\Delta_0 \geq 2\epsilon_0$ *and* $\epsilon_0 \leq \delta \leq \Delta_0$. *Assume* (A_1)–(A_4); $(B_2\epsilon)$ *for* $\epsilon \geq \epsilon_0$ *with* θ_0, K_2 *independent of* ϵ, $|\varphi_{s\epsilon}|_B$ *independent of* $y, z,$ *and* (5.6) *holds uniformly for large* s; $(B_3\frac{1}{2}\epsilon_0)$; *and* $(B_5\Delta_0)$. *Then* W_D *induces a total exponential dichotomy for the manifold* $\mathcal{N}_\delta$, *defined in Theorem 5.3, with* $\theta^0 = \max(\theta_0, \epsilon_0)$.

Exercise 5.2.   Prove Theorem 5.4.

6. Applications to First Order Systems

As was pointed out in §§ 3–4, the assumptions (A_0)–(A_4) and $(B_1\epsilon)$ are satisfied by the operator T associated with (0.1) in the Remark of § 3, with $Py = y$, $Qy = y(0)$. Hence Theorems 4.1–4.3 imply

Theorem 6.1.   *Let* $A(t)$ *be a matrix of locally integrable (real- or complex-valued) functions for* $t \geq 0$. *Let* B, D *be Banach spaces in* Φ *and let* (B, D) *be admissible for* (0.1) *in the sense that for every* $g(t) \in B(Y)$, (0.1) *has a solution* $y(t) \in D(Y)$. *Let* $\mathcal{N}$ *denote the set of solutions* $y(t)$ *of* (0.2) *and* Y_D *be the set of initial conditions* $y(0)$ *belonging to solutions* $y(t) \in \mathcal{N} \cap D(Y)$. (i) *Then* Y_D *induces a partial dichotomy for* $(\mathcal{N}, \rho_y(t), y(0))$, *where*

$$(6.1) \qquad \rho_y(t) = |h_{t\Delta}y|_D \quad \text{or} \quad \rho_y(t) = \int_t^{t+\Delta} \|y(\tau)\| \, d\tau$$

and $\Delta > 0$ *fixed arbitrarily.* (ii) *If, in addition,* D *is quasi-full and* $z(t) \in \mathcal{N}$, *but* $z(t) \notin D(Y)$, *then, for every* $\Delta > 0$,

$$\int_t^{t+\Delta} \|z(\tau)\| \, d\tau \to \infty \qquad \text{as} \quad t \to \infty.$$

(iii) *If* (4.14) *holds, then* Y_D *induces an exponential dichotomy for* $(\mathcal{N}, \rho_y(t), y(0))$, *where the exponents* ν, ν' *can be chosen independent of* $\Delta > 0$. (iv) *If* $B = L^1$ *and* $D = L^\infty$ *or* $D = L_0^\infty$, *then* Y_D *induces a total dichotomy for* $\mathcal{N}$.

Exercise 6.1.   State the consequences of Corollary 4.2 for parts (i) and (iv) of Theorem 6.1.

The next theorem, except for condition (c) of a total dichotomy [cf. (1.9)], is easily deduced from Theorem 6.1 by virtue of the Remark concerning (5.3), (5.4). The entire theorem, however, will be deduced from Theorems 5.1 and 5.2.

Theorem 6.2.   *Let* $A(t)$ *be a matrix of locally integrable, real- or complex-valued functions satisfying* (5.3) *or* (5.4). *Let* B, D *be Banach spaces in* Φ *and* (B, D) *admissible for* (0.1), *and* $\mathcal{N}$, Y_D *as in Theorem 6.1. Then* Y_D

induces a total dichotomy for $\mathcal{N}$. If, in addition, D is quasi-full and $z(t) \in \mathcal{N}$, but $z(t) \notin D(Y)$, then

$$(6.2) \qquad\qquad \|z(t)\| \to \infty \quad as \quad t \to \infty.$$

Finally, if

$$(6.3) \qquad \Delta^{-1} |h_{0\Delta}|_B \to 0 \quad or \quad |h_{0\Delta}|_D \to \infty \quad as \quad \Delta \to \infty,$$

then Y_D induces a total exponential dichotomy for $\mathcal{N}$.

Proof. It suffices to verify the conditions of Theorem 5.1 and/or 5.2 for the operator T, with $Py = y$ and $y[0] = y(0)$, associated with (0.1).

Conditions (A_1)–(A_4) have already been verified. As pointed out before, $(B_1\epsilon)$ holds for all $\theta_0 > 0$, $\epsilon > 0$ with $K = 1$ and arbitrary $\varphi_{s\epsilon}(t)$. Choosing $\varphi_{s\epsilon}(t) = \epsilon^{-1} h_{s\epsilon}(t)$ in Remark 2 preceding Theorem 4.1 shows that $(B_2\epsilon)$ holds for all $\theta_0 > 0$, $\epsilon > 0$ with $K_2 = 1$ (since $K = 1$ and $|\varphi_{s\epsilon}|_1 = 1$) and

$$|\varphi_{s\epsilon}|_B = \epsilon^{-1} |h_{0\epsilon}|_B$$

independent of s and $y(t)$, $z(t)$. Condition $(B_3\delta)$ is trivial since $y = Py$. The Remark preceding Theorem 5.1 shows that (5.3) or (5.4) implies condition $(B_4\Delta)$. Thus Theorem 5.1 is applicable and gives the statements concerning a total dichotomy and (6.2).

In order to apply Theorem 5.2, the condition (5.6), which is

$$\frac{|h_{0\Delta}|_B}{\Delta\,|h_{0\Delta}|_D} \to 0 \quad as \quad \Delta \to \infty,$$

must be verified. But is readily seen that this is equivalent to (6.3); cf., e.g., the beginning of the proof of Theorem 4.2. Hence the proof of Theorem 6.2 is complete.

The next theorem is the main result on total dichotomies for solutions of (0.2) and admissibility for (0.1). For the sake of brevity, let Y_p and $Y_{\infty 0}$ denote the subspace Y_D of Y, when $D = L^p$ and $D = L_0^\infty$, respectively; i.e., Y_p and $Y_{\infty 0}$ denote the set of initial points $y(0) \in Y$ of solutions $y(t)$ of (0.2) of class L^p, $1 \leq p \leq \infty$, and L_0^∞, respectively.

Theorem 6.3. *Let $A(t)$ be a matrix of locally integrable (real- or complex-valued) functions of $t \geq 0$ and $\mathcal{N}$ the set of solutions of (0.2). Then there exists a subspace W_0 of $Y = W$ which induces a total dichotomy for $(\mathcal{N}, \|y(t)\|, y(0))$ if and only if (L^1, L^∞) and/or (L^1, L_0^∞) is admissible for (0.1). In this case, $Y_{\infty 0} \subset W_0$ and both Y_∞ and $Y_{\infty 0}$ induce total dichotomies for $(\mathcal{N}, \|y(t)\|, y(0))$.*

Exercise 6.2. The proof of Theorem 6.3 will depend on Lemma 2.3. Without using this lemma, prove the parts of Theorem 6.3 which do not involve L_0^∞ (an analogous result is applicable even if $Y = W$ is not finite dimensional and Lemma 2.3 is not applicable).

Proof. If (L^1, L_0^∞), hence (L^1, L^∞), is admissible for (0.1), then Y_∞ and $Y_{\infty 0}$ induce total dichotomies for $\mathscr{N}$ by (iv) in Theorem 6.1.

Conversely, if there exists a subspace W_0 of Y which induces a total dichotomy for $\mathscr{N}$, then $Y_{\infty 0} \subset W_0$ and $Y_{\infty 0}$ induces a total dichotomy for $\mathscr{N}$ by Corollary 2.2. Thus, in order to complete the proof, it is sufficient to show that if $Y_{\infty 0}$ induces a total dichotomy for $\mathscr{N}$, then (L^1, L^∞) or (L^1, L_0^∞) is admissible for (0.1).

The proof of this part will use the easier ("only if") portion of Theorem 2.1. Let Z be a subspace of Y complementary to $Y_{\infty 0}$, so that Y is the direct sum $Y_{\infty 0} \oplus Z$ (e.g., if Y is Euclidean, let Z be the orthogonal complement of $Y_{\infty 0}$ in Y). Let P_0 be the projection of Y on $Y_{\infty 0}$ annihilating Z, and $P_1 = 1 - P_0$ the projection of Y on Z annihilating $Y_{\infty 0}$. Let $U(t)$ be the fundamental matrix of (0.2) satisfying $U(0) = I$, and $G(t, s)$ the matrix function

$$G(t, s) = U(t)P_0 U^{-1}(s) \quad \text{for} \quad 0 \leqq s \leqq t,$$
(6.4)
$$G(t, s) = -U(t)P_1 U^{-1}(s) \quad \text{for} \quad 0 \leqq t < s.$$

Then, by Theorem 2.1, there exists a constant K such that

(6.5) $$\|G(t, s)\| \leqq K \quad \text{for} \quad s, t \geqq 0.$$

Let $g(t)$ be an arbitrary element of $L^1(Y)$. In order to show that (0.1) has an L_0^∞-solution, put

(6.6) $$y(t) = \int_0^\infty G(t, s)g(s)\, ds.$$

The integral is absolutely convergent [in view of (6.5) and $g(t) \in L^1(Y)$] and defines a solution of (0.1). Also $\|y(t)\| \leqq K |g|_1$; in particular, $y(t) \in L^\infty(Y)$.

It will be verified that $y(t)$ in (6.6) is in $L_0^\infty(Y)$. By (6.4) and (6.5),

$$\|y(t)\| \leqq \int_0^\infty \|h_{0t}(s)U(t)P_0 U^{-1}(s)g(s)\|\, ds + K \int_t^\infty \|g(s)\|\, ds.$$

The last integral tends to 0 as $t \to \infty$. For fixed s, the solution $U(t)P_0 U^{-1}(s)g(s)$ of (0.2) tends to 0 as $t \to \infty$, since its initial value $P_0 U^{-1}(s)g(s) \in Y_{\infty 0}$. Furthermore,

$$\|h_{0t}(s)U(t)P_0 U^{-1}(s)g(s)\| \leqq K \|g(s)\| \quad \text{for} \quad s \geqq 0 \quad \text{and} \quad t \geqq 0.$$

Thus Lebesgue's term-by-term integration theorem (with majorized convergence) shows that the first integral tends to 0 as $t \to \infty$. Consequently, $y(t) \in L_0^\infty(Y)$. This proves the theorem.

The main result on total exponential dichotomies is the following analogue of Theorem 6.3.

Theorem 6.4. *Let $A(t)$ be a matrix of locally integrable (real- or complex-valued) functions of $t \geqq 0$ and $\mathcal{N}$ the set of solutions of (0.2). Then there exists a subspace W_0 of Y which induces a total exponential dichotomy for $(\mathcal{N}, \|y(t)\|, y(0))$ if and only if (L^p, L^q) for $(p, q) = (1, \infty)$ and for some $(p, q) \neq (1, \infty)$, $1 \leqq p \leqq q \leqq \infty$, are admissible for (0.1). In this case, $W_0 = Y_p = Y_{\infty 0}$, and (L^1, L_0^∞) and (L^p, L^q) are admissible for all (p, q), where $1 \leqq p \leqq q \leqq \infty$.*

Proof. If (L^1, L^∞) and (L^p, L^q) for some $(p, q) \neq (1, \infty)$ and $1 \leqq p$, $q \leqq \infty$ are admissible for (0.1), then $Y_\infty = Y_q = Y_{\infty 0}$ induces a total exponential dichotomy for $\mathcal{N}$ by (iv) in Theorem 6.1 and by Theorem 4.4.

Conversely, let a subspace W_0 of Y induce a total exponential dichotomy for $\mathcal{N}$. Let W_1 be a subspace of Y complementary to W_0 and let $G(t, s)$ be the Green's matrix (6.4) defined in terms of projections P_0, P_1 of Y onto W_0, W_1 annihilating W_1, W_0, respectively. Then, by Theorem 2.1,

$$(6.7) \qquad \|G(t, s)\| \leqq K e^{-v|t-s|} \qquad \text{for} \quad s, t \geqq 0$$

and some constants $K, v > 0$. Let $g(t) \in L^p(Y)$, where $1 \leqq p \leqq \infty$. Then the integral in (6.6) is absolutely convergent and defines a solution of (0.1). Thus it remains to show that $y(t) \in L^q(Y)$ for $p \leqq q \leqq \infty$.

Consider first the case that $p = 1$. As in the last proof, it is easy to see that $y(t) \in L_0^\infty(Y)$. Also $y(t) \in L^1(Y)$ for

$$(6.8) \qquad \int_0^\infty dt \int_0^\infty e^{-v|t-s|} \|g(s)\|\, ds \leqq \int_0^\infty \|g(s)\|\, ds \int_{-\infty}^\infty e^{-v|t|}\, dt < \infty.$$

Hence $y(t) \in L^q(Y)$ for $1 \leqq q \leqq \infty$. Also, if $p = \infty$, then $y(t) \in L^\infty(Y)$ for

$$\int_0^\infty e^{-v|t-s|} \|g(s)\|\, ds \leqq |g|_\infty \int_{-\infty}^{+\infty} e^{-v|s|}\, ds.$$

Consider $1 < p < q < \infty$. Let $\alpha, \beta > 0$, $\alpha + \beta = 1$, and $\varphi(t) = \|g(t)\|$. Then repeated applications of Hölder's inequality show that $\|y(t)\|^q$ is at most

$$K^q \left| \int_0^{+\infty} (e^{-v|t-s|\alpha} \varphi^{1-p/q})(e^{-v|t-s|\beta} \varphi^{p/q})\, ds \right|^q$$

$$\leqq K^q \left(\int_0^{+\infty} e^{-v|t-s|\alpha q/(q-1)} \varphi^{(q-p)/(q-1)}\, ds \right)^{q-1} \left(\int_0^{+\infty} e^{-v|t-s|\beta q} \varphi^p\, ds \right)$$

$$\leqq K^q \left(\int_{-\infty}^{+\infty} e^{-v|s|\alpha p/(q-1)}\, ds \right)^{q(p-1)/p} \left(\int_0^\infty \varphi^p\, ds \right)^{(q-p)/p} \int_0^{+\infty} e^{-v|t-s|\beta q} \varphi^p\, ds.$$

Hence $(\int \|y(t)\|^q\, dt)^{1/q}$ is finite and is at most

$$K \left(\int_{-\infty}^{+\infty} e^{-v|s|\alpha p/(q-1)}\, ds \right)^{(p-1)/p} \left(\int_0^\infty \varphi^p\, ds \right)^{1/p} \left(\int_{-\infty}^{+\infty} e^{-v|t|\beta q}\, dt \right)^{1/q};$$

cf. (6.8). Consequently, $y(t) \in L^q(Y)$ for $p < q < \infty$. Since the majorant remains finite as $q \to p$ or $q \to \infty$, it follows that $y(t) \in L^q(Y)$ for $p \leqq q \leqq \infty$. This proves the theorem.

Since (6.7) is a necessary condition for W_0 to induce a total exponential dichotomy, it is possible to conclude in this case that many pairs (B, D) are admissible. Such a result is given by the following:

Exercise 6.3.   Let $A(t)$ be a matrix of locally integrable functions for $t \geqq 0$ and $\mathcal{N}$ the set of solutions of (0.2). Let B, D be Banach spaces in $\mathcal{T}$ such that

(6.9)
$$\psi(t) \in D \Rightarrow \psi(t + s) \in D \qquad \text{for every fixed} \quad s \geqq 0,$$
$$\varphi(t) \in B \Rightarrow \int_t^{t+1} \varphi(s)\,ds \in D.$$

Let there exist a subspace W_0 of Y which induces a total exponential dichotomy for $\mathcal{N}$. Then (B, D) is admissible for (0.1).

7. Applications to Higher Order Systems

In dealing with (0.3), it will be assumed that U is a finite dimensional vector space with elements u and norm $\|u\|$. $\tilde{U}$ is the corresponding Banach space of linear operators P_0 of U into itself with the norm $\|P_0\| = \sup \|P_0 u\|$ for $\|u\| = 1$. P_0 can, of course, be considered to be a matrix if a coordinate system is fixed on U.

A function $u(t)$ from a t-interval to U will be called an $(m + 1)st$ *integral*, $m \geqq 1$, if $u(t)$ has continuous derivatives, $u = u^{(0)}, u'(t), \ldots, u^{(m)}(t)$ such that $u^{(m)}(t)$ is absolutely continuous [with a derivative $u^{(m+1)}(t)$ almost everywhere].

In this section, let $Y = U^{(m+1)} = U \times \cdots \times U$ and if $y = (u^{(0)}, \ldots, u^{(m)}) \in Y$, put

(7.1)
$$\|y\| = \max \left(\|u^{(0)}\|, \ldots, \|u^{(m)}\| \right).$$

If $f(t) \in L(U)$ and $P_k(t) \in L(\tilde{U})$ for $k = 0, \ldots, m$, then $u(t)$ is a solution of (0.3) if it is an $(m + 1)st$ integral and $y(t) = (u^{(0)}(t), \ldots, u^{(m)}(t))$ satisfies the first order system (0.1) corresponding to $g(t) = (0, \ldots, 0, f(t))$, and the $m + 1$ equations: $u^{(j)\prime} = u^{(j+1)}$ for $j = 0, \ldots, m - 1$ and (0.3). It is clear that in this identification of (0.3) with (0.1), (7.1) implies that the norm of $A(t) \in \tilde{Y}$ satisfies

(7.2)
$$\|A(t)\| \leqq \max \left(1, \sum_{k=0}^{m} \|P_k(t)\| \right)$$

Correspondingly, the inequality (0.5) is applicable.

In dealing with (0.3), T is an operator from $L(Y)$ to $L(U)$. The domain $\mathfrak{D}(T)$ of T is the set of $y = (u(t), u'(t), \ldots, u^{(m)}(t))$, where $u(t)$ is an $(m + 1)$st integral for $t \geqq 0$, and $u', \ldots, u^{(m)}$ are its derivatives. Also, $f = Ty$ is defined by (0.3), $Py(t) = u(t)$, and $y[0] = y(0)$; thus $Y = W$, $F = U$; cf. § 3. Finally, $\mathcal{N} = \mathcal{N}(T)$ is the set of $y(t)$ where $u(t) = Py(t)$ is a solution of the homogeneous equation (0.4); i.e., $\mathcal{N}$ is the set of solutions $y(t)$ of (0.2).

The following preliminary lemma has nothing to do with differential equations.

Lemma 7.1. *Let m be a positive integer. Then there exists a constant $c_m > 0$ with the property that if $u(t) \in U$ is an $(m + 1)$st integral on an interval $\tau \leqq t \leqq \tau + \Delta, \Delta > 0$, then its derivatives satisfy*

$$(7.3) \quad \|u^{(k)}(t)\| \leqq c_m \Delta^{-k} \sum_{j=0}^{m} \|u(\tau + j\Delta/m)\| + \Delta^{m-k} \int_{\tau}^{\tau+\Delta} \|u^{(m+1)}(r)\| \, dr$$

for $k = 0, 1, \ldots, m$.

Proof. It is sufficient to verify the corresponding inequality

$$(7.4) \quad |\varphi^{(k)}(t)| \leqq c_m \Delta^{-k} \sum_{j=0}^{m} |\varphi(\tau + j\Delta/m)| + \Delta^{m-k} \int_{\tau}^{\tau+\Delta} |\varphi^{(m+1)}(r)| \, dr,$$

when $\varphi(t)$ is a real-valued function on $[\tau, \tau + \Delta]$ which is an $(m + 1)$st integral. If $\tau \leqq t_1 \leqq \tau + \Delta$, let $u^* \in U^*$ be chosen so that $\|u^*\| = 1$ and the "scalar" product $\langle u^{(k)}(t_1), u^* \rangle = \|u^{(k)}(t_1)\|$. If $\varphi(t) = \text{Re} \langle u(t), u^* \rangle$, then (7.4) at $t = t_1$ implies (7.3) at $t = t_1$.

In order to prove (7.4), let $\varphi(t) = p(t) + \psi(t)$, where $p(t)$ is the polynomial of degree m satisfying $p(t) = \varphi(t)$ for $t = \tau + j\Delta/m, j = 0, \ldots, m$; i.e.,

$$p(t) = \sum_{j=0}^{m} \varphi(\tau + j\Delta/m) \prod_{k \neq j} \frac{(m/\Delta)(t - \tau) - k}{j - k}.$$

It is clear that there is a constant $c_m > 0$ such that

$$(7.5) \quad |p^{(k)}(t)| \leqq c_m \Delta^{-k} \sum_{j=0}^{m} |\varphi(\tau + j\Delta/m)| \quad \text{for} \quad \tau \leqq t \leqq \tau + \Delta$$

and $k = 0, \ldots, m$. Since $\psi(t) = \varphi(t) - p(t)$ vanishes at the $m + 1$ points $t = \tau + j\Delta/m, j = 0, \ldots, m$, it follows that there is a t-value $t' = t_k'$ such that $\psi^{(k)}(t') = 0$. Hence, for $\tau \leqq t \leqq \tau + \Delta$,

$$(7.6) \qquad |\psi^{(k)}(t)| \leqq \left| \int_{t'}^{t} \psi^{(k+1)} \, dr \right| \leqq \int_{\tau}^{\tau+\Delta} |\psi^{(k+1)}| \, dr.$$

Integrating over $[\tau, \tau + \Delta]$ gives

$$\int_{\tau}^{\tau+\Delta} |\psi^{(k)}| \, dr \leqq \Delta \int_{\tau}^{\tau+\Delta} |\psi^{(k+1)}| \, dr,$$

so that, by induction, (7.6) implies that

$$(7.7) \qquad |\psi^{(k)}(t)| \leqq \Delta^{m-k} \int_\tau^{\tau+\Delta} |\psi^{(m+1)}| \, dr.$$

Since $p(t)$ is a polynomial of degree m, $\psi^{(m+1)}(t) = \varphi^{(m+1)}(t)$. Hence $\varphi = p + \psi$ and (7.5) and (7.7) give (7.4).

Lemma 7.2. *Let $P_k(t) \in L(\tilde{U})$, $k = 0, \ldots, m$, and let $\alpha = \alpha(t_0)$ be a positive integer satisfying*

$$(7.8) \qquad \sum_{k=0}^m \int_s^{s+1} \|P_k(r)\| \, dr \leqq \tfrac{1}{2}\alpha \qquad \text{for} \quad 0 \leqq s \leqq t_0.$$

Let $f(t) \in L(U)$ and $u = u(t)$ a solution of (0.3). Then, for $0 \leqq s \leqq t_0$, $s \leqq t \leqq s + 1$, and $k = 0, \ldots, m$,

$$(7.9) \qquad \|u^{(k)}(t)\| \leqq C^1 \sum_{j=0}^{m\alpha} \|u(s + j/m\alpha)\| + C^2 \int_s^{s+1} \|f(r)\| \, dr,$$

where $C^1 = 2e^\alpha c_m \alpha^m$ and $C^2 = 3e^\alpha$.

Remark. Note that $t_0 > 0$ can be chosen arbitrarily and α (hence C^1, C^2) independent of t_0 if $P_k(t) \in M(\tilde{U})$, $k = 0, \ldots, m$ and

$$(7.10) \qquad 2 \sum_{k=0}^m |P_k|_M \leqq \alpha;$$

cf. (1.6) and (7.8).

Proof. Let $s \geqq 0$. Then (7.8) implies that there exists an integer $i = i(s)$, $0 \leqq i < \alpha$, such that

$$(7.11) \qquad \sum_{k=0}^m \int_\tau^{\tau+\Delta} \|P_k(r)\| \, dr \leqq \tfrac{1}{2} \qquad \text{for} \quad \tau = s + i/\alpha, \ \Delta = 1/\alpha.$$

By (0.3), $\|u^{(m+1)}(t)\| \leqq \Sigma \|P_k(t)\| \cdot \|u^{(k)}(t)\| + \|f(t)\|$. If the relations (7.3) are inserted into this inequality, an integration of the resulting inequality over $[\tau, \tau + \Delta]$ gives, by (7.11),

$$(7.12) \qquad \int_\tau^{\tau+\Delta} \|u^{(m+1)}\| \, dt \leqq c_m \alpha^m \sum_{j=0}^m \|u(\tau + j/m\alpha)\| + 2 \int_\tau^{\tau+\Delta} \|f\| \, dt.$$

Thus, if $y = (u, u^{(1)}, \ldots, u^{(m)})$, (7.3) implies that

$$\|y(\tau)\| \leqq 2c_m \alpha^m \sum_{j=0}^{m\alpha} \|u(s + j/m\alpha)\| + 2 \int_s^{s+1} \|f\| \, dr.$$

Therefore, (7.9) follows from (7.2), (7.8), and the corresponding inequality (0.5) with s replaced by τ.

Corollary 7.1. *Under the conditions of Lemma 7.2,*

$$(7.13) \qquad \|u^{(k)}(t)\| \leqq C^1(1 + m\alpha) \int_{t-1}^{t+1} \|u(s)\| \, ds + C^2 \int_{t-1}^{t+1} \|f(s)\| \, ds$$

for $1 \leqq t \leqq t_0 - 1$ and $k = 0, \ldots, m$.

This follows by integrating (7.9) with respect to s for $t - 1 \leqq s \leqq t$.

Corollary 7.2. *Assume the conditions of Lemma 7.2. Let $f_1(t), f_2(t), \ldots$ be functions in $L(U)$ and $u = u_n(t)$ a solution of (0.3) when $f = f_n$. Suppose that $f = \lim f_n$, $u = \lim u$ exist in $L(U)$. Then the function $u = u(t)$ (up to an equivalence modulo a null set) is a solution of (0.3) and $u_n^{(k)}(t) \to u^{(k)}(t)$, $n \to \infty$, uniformly on bounded subsets of J for $k = 0, \ldots, m$.*

This follows by replacing u, f in (7.13) by $u_p - u_q, f_p - f_q$ for $p, q = 1, 2, \ldots$.

Theorem 7.1. *Let $P_k(t) \in M(\tilde{U})$ for $k = 0, \ldots, m$ and*

$$(7.14) \quad h_{s1}(t)P_k(t) \in B(\tilde{U}), \quad \sup_{s \geq 0} |h_{s1}P_k|_B < \infty \quad \text{for} \quad k = 1, \ldots, m.$$

Suppose that (B, D) is admissible for (0.3); i.e., that for every $f(t) \in B(U)$, (0.3) has a solution $u(t) \in D(U)$. Let Y_D be the set of initial conditions $y(0) = (u(0), u'(0), \ldots, u^{(m)}(0))$ of solutions $u(t) \in D(U)$ of (0.4). Then Y_D induces a total dichotomy for $\mathcal{N}$ (with $\theta^0 = 0$). If, in addition, D is quasi-full and $u(t) \notin D(U)$ is a solution of (0.4), then $\|y(t)\| \to \infty$ as $t \to \infty$.

The conditions (7.14) on $P_k(t)$ are satisfied if, e.g., $P_k(t) \in L^\infty(\tilde{U})$. Note that (7.14) is not required for $k = 0$. The condition that (B, D) is admissible for (0.3) is the condition that (B, D) be P-admissible for the operator T described at the beginning of this section. It does not correspond to the admissibility of (B, D) for the corresponding linear system (0.1) for we do not consider arbitrary $g \in B(Y)$ but only g of the form $g = (0, \ldots, 0, f)$, $f \in B(U)$.

Proof. This theorem will be proved by verifying the applicability of Theorem 5.1. (A_1) follows from (0.5) in view of the identification of (0.1) and (0.3); (A_2) follows from Corollary 7.2; (A_3) is an explicit assumption of the theorem; (A_4) is trivial since $W = Y$ is finite dimensional.

In order to verify $(B_2\epsilon)$, let $\epsilon \geq 1$ and $u(t), v(t)$ solutions of (0.4). Let $\psi(t)$ be a non-negative function on $-\infty < t < \infty$ of class C^m which vanishes except on $[0, 1]$ and satisfies

$$(7.15) \qquad \int_0^1 \psi \, dt = 1.$$

Put $\psi_{s\epsilon}(t) = \epsilon^{-1}\psi(\epsilon^{-1}(t - s))$ and

$$(7.16) \qquad u_1(t) = u(t) \int_0^t \psi_{s\epsilon}(r) \, dr + v(t) \int_t^\infty \psi_{s\epsilon}(r) \, dr.$$

Then for $k = 1, \ldots, m + 1$,

$$(7.17) \quad u_1^{(k)}(t) = u^{(k)}(t) \int_0^t \psi_{s\epsilon} \, dr + v^{(k)}(t) \int_t^\infty \psi_{s\epsilon} \, dr$$
$$+ \sum_{j=0}^{k-1} C_{kj}(u - v)^{(j)} \psi_{s\epsilon}^{(k-j-1)}(t),$$

where C_{kj} is the binomial coefficient $k!\,/j!\,(k-j)!$ It follows that $u = u_1$ is a solution of (0.3), where f is

$$(7.18) \qquad f_1(t) = \sum_{k=1}^{m+1} \sum_{j=0}^{k-1} C_{kj} P_k(t)(u-v)^{(j)} \psi_{s\epsilon}^{(k-j-1)}(t)$$

and $P_{m+1}(t) = I$ is the identity operator.

Since $\psi_{s\epsilon}(t) = 0$ for $t \leq s$ or $t \geq s + \epsilon$, it follows that $y_1(t) = z(t)$ for $0 \leq t \leq s$ and $y_1(t) = y(t)$ for $t \geq s + \epsilon$ if $y = (u, \ldots, u^{(m)})$, $z = (v, \ldots, v^{(m)})$, $y_1 = (u_1, \ldots, u_1^{(m)})$. Thus, if $s > 0$, $y_1(0) = z(0)$ and $P y_1 = u_1 \in D(U)$. There is a constant $c > 0$ such that $|\psi_{s\epsilon}^{(k)}(t)| \leq c h_{s\epsilon}(t)/\epsilon$ for $k = 0, \ldots, m$ and $s, t \geq 0$ and $\epsilon \geq 1$. Thus if $T y_1 = f_1$,

$$(7.19) \qquad \|T y_1\| \leq 2^{m+1} c \epsilon^{-1} h_{s\epsilon}(t) \sum_{k=1}^{m+1} \|P_k(t)\| \cdot \|y(t) - z(t)\|.$$

Thus $(B_2\epsilon)$ with $\epsilon \geq 1$ holds for arbitrary $\theta_0 > 0$ and $K_2 = 1$ when

$$(7.20) \qquad \varphi_{s\epsilon}(t) = 2^{m+1} c \epsilon^{-1} h_{s\epsilon}(t) \left(1 + \sum_{k=1}^{m} \|P_k(t)\| \right).$$

By virtue of (7.14), $\varphi_{s\epsilon}(t) \in B$ and

$$(7.21) \qquad |\varphi_{s\epsilon}|_B \leq 2^{m+1} c \epsilon^{-1} \left(|h_{0\epsilon}|_B + \sum_{k=1}^{m} |h_{s\epsilon} P_k|_B \right).$$

Lemma 7.2 with $f = 0$, and the Remark following it show that $(B_3\delta)$ holds for $\delta \geq 1$ if $K_3(\delta) = C^1(1 + m\alpha)$. Finally, $(B_4\Delta)$ is a consequence of (0.5) and (7.2), since $P_0(t), \ldots, P_m(t) \in M(\tilde{U})$. Thus Theorem 7.1 follows from Theorem 5.1.

Theorems 6.3 and 7.1 have a curious consequence. Let (0.2) be the first order system obtained in the standard way [as before (7.2)] from (0.4). Then we can consider (0.1) without the restriction that g is of the form $(0, \ldots, 0, f)$ and we obtain from Theorem 6.3:

Corollary 7.3. *Let the conditions of Theorem 7.1 hold. Then (L^1, L^∞) and/or (L^1, L_0^∞) is admissible for (0.3) if and only if it is admissible for (0.1).*

For example, if $P_k(t) \in L^\infty(\tilde{U})$, $k = 0, 1, \ldots, m$, and if some pair (B, D) is admissible for (0.3), then (L^1, L_0^∞) is admissible for (0.1) [and for (0.3)].

Theorem 7.2. *Let the assumptions of Theorem 7.1 hold. In addition, assume either the condition*

$$(7.22) \qquad |h_{0\Delta}|_D \to \infty \qquad \text{as} \quad \Delta \to \infty$$

or the condition

$$(7.23) \qquad \Delta^{-1} \left(|h_{0\Delta}|_B + \sum_{k=1}^{m} |h_{s\Delta} P_k|_B \right) \to 0 \qquad \text{uniformly for large} \quad s$$

as $\Delta \to \infty$. Then Y_D induces a total exponential dichotomy for $\mathcal{N}$ (with $\theta^0 = 0$).

When $|h_{0\Delta}|_B/\Delta \to 0$ as $\Delta \to \infty$ and $P_k(t) \in L^\infty(U)$ for $k = 1, \ldots, m$, then (7.23) holds, since $|h_{0\Delta}|_B \leq (1 + \Delta) |h_{01}|_B$.

Proof. This is a consequence of Theorem 5.2. It suffices to verify the analogue of condition (5.6). By (7.21), (5.6) follows either from (7.14), (7.22) or from (7.23).

Exercise 7.1. This exercise concerns real second order equations

$$(7.24) \qquad u'' + p(t)u' + q(t)u = f(t),$$

$$(7.25) \qquad u'' + p(t)u' + q(t)u = 0,$$

for $t \geq 0$. Let U be the (Banach) space of real numbers; W the product space $U^{(2)} = U \times U$ with norm $\|w\| = \max(|w^1|, |w^2|)$ if $w = (w^1, w^2)$; and Y the product space $U^{(4)}$ with norm $\|y\| = \max(|y^1|, |y^2|, |y^3|, |y^4|)$ if $y = (y^1, y^2, y^3, y^4)$. For $\delta > 0$, let $\mathcal{N}_\delta$ denote the linear manifold of functions $y^\delta(t) = (u(t), u'(t), u(t + \delta), u'(t + \delta)) \in Y$ for $t \geq 0$ where $u(t)$ is a solution of (7.25). Let the "initial value" of $y^\delta(t)$ be $y^\delta[0] = (u(0), u'(0)) \in W$. Assume that $p(t)$ is a real-valued function in J: $t \geq 0$ which satisfies

$$(7.26) \qquad h_{s1}(t)p(t) \in B, \qquad \sup_{s \geq 0} |h_{s1}(t)p(t)|_B < \infty.$$

Let $q(t)$ be a real-valued element of L for which there exists a constant C satisfying the one-sided inequality

$$(7.27) \qquad \int_s^t q(r)\, dr \leq C \qquad \text{for} \quad 0 \leq s \leq t \leq s + 1.$$

Finally, suppose that (B, D) is admissible for (7.24); i.e., if $f(t) \in B$, then (7.24) has a solution $u(t) \in D$. Let W_D be the 0-, 1-, or 2-dimensional manifold of initial conditions $(u(0), u'(0))$ of D- solutions of (7.25). (a) Then, for sufficiently small $\delta > 0$, W_D induces a total dichotomy for $\mathcal{N}_\delta$. (b) If, in addition, D is a quasi-full and $u(t) \notin D$ is a solution of (7.25), then $\|y^\delta(t)\| \to \infty$ as $t \to \infty$. (c) If, in addition to the conditions of (a), either (7.22) or

$$(7.28) \qquad \Delta^{-1}(|h_{0\Delta}|_B + |h_{s\Delta}(t)p(t)|_B) \to 0 \qquad \text{uniformly for large } s$$

as $\Delta \to \infty$, then W_D induces a total exponential dichotomy for $\mathcal{N}_\delta$ if $\delta > 0$ is sufficiently small.

8. $P(B, D)$-Manifolds

The main role of assumptions (A_3), (A_4) is in the proof of Lemma 3.2. An analogous lemma follows from a similar pair of assumptions (A_5), (A_7). The notation in this section is the same as in §§ 3–4.

(A_5) If $Ty(t) = f(t)$ and $f(t) = 0$ for large t, then there exists a unique solution $y_\infty(t)$ of $Ty = 0$ such that $y_\infty(t) = y(t)$ for large t.

Definition. P(B, D)-Manifold. Assume (A_5). Let B, D be Banach spaces in $\mathcal{T}$ and B_∞ be defined as in (vi) of § 1. A submanifold X of W_D is called a $P(B, D)$-manifold if there exists a constant C_3 with the property that, for $f(t) \in B_\infty(F)$, there exists a PD-solution $y(t)$ of $Ty = f$ such that $|Py|_D \leqq C_3 |f|_B$, and $y_\infty(t)$ [defined by (A_5)] satisfies $y_\infty[0] \in X$. [When (A_0) is assumed, then even without the assumption $y_\infty[0] \in X$, it follows that $y_\infty[0] \in W_D$.]

Exercise 8.1.   Assume (A_5). Show that there exists a $P(B, D)$-manifold if and only if $(\bar{B}_\infty, D)$ is P-admissible.

(A_6) X is a $P(B, D)$-manifold.

Lemma 8.1.   *Assume (A_1) [or (A_1')], (A_2), (A_5), and (A_6). Let f, y, y_∞ be as in the definition of a $P(B, D)$-manifold. Then there exists a constant C_{30} such that $\|y[0]\| \leqq C_{30} |f|_B$.*

This is a consequence of (3.3) and (3.4) [or (3.3')], (3.5_1) and (1.5).

(A_7) X is a $P(B, D)$-subspace, i.e., a closed $P(B, D)$-manifold.

Lemma 8.2.   *Assume (A_1) [or (A_1')], (A_2), (A_5) and (A_7). Let $\lambda > 1$, $f \in B_\infty(F)$, $y(t)$ any PD-solution of $Ty = f$ with $y_\infty[0] \in X$, $\|y[0]\| \leqq \lambda d(X, y[0])$. Then there exist constants C_5, C_{30} such that (3.7) holds (with $\lambda = 1$ permitted if $y[0] = 0$).*

The proof is identical to that of Lemma 3.4. If, in addition, to (A_1) [or (A_1')]–(A_4), (A_0) and (A_5) are assumed, then Lemma 3.4 is contained in Lemma 8.2, for $X = W_D$ is a $P(B, D)$-subspace in this case.

Remark.   In view of the uses of Lemmas 3.3 and 3.4, it follows that (A_3), (A_4) can be replaced by assumptions (A_5), (A_6), (A_7) in the theorems of §§ 4–5 (and their applications in §§ 6–7) if conclusions of the type "W_D induces a . . . dichotomy" are replaced by "X induces a . . . dichotomy."

When (A_4) holds (e.g., if W is finite dimensional), then (A_5) and (A_6) imply, by Exercise 8.1, that (A_4) holds with B replaced by $\bar{B}_\infty$. The point in the Remark above, however, is that a subspace X [which can be smaller than W_D] can replace W_D in the conclusions of the theorems of §§ 4–5.

PART II.   ADJOINT EQUATIONS

9. Associate Spaces

In this part of the chapter, the concept of an *associate space* will be needed. Let $\mathcal{T}^{\#}$ denote the set of normed spaces $\Phi \in \mathcal{T}$ satisfying the additional condition: $(d^{\#})$ *if* $\varphi(t) \in \Phi$, $s \geqq 0$, *and* $\psi(t) = \varphi(t + s)$ *for* $t \geqq 0$, *then* $\psi(t) \in \Phi$.

Hypothesis.   *In the remainder of this chapter, it is assumed that B, D are Banach spaces in $\mathcal{T}^{\#}$.*

Let Φ be a Banach space in $\mathcal{T}^{\#}$. Let Φ' be the set of (equivalence classes modulo null sets of) real-valued measurable functions $\psi(t)$, $t \geqq 0$, such that for all $\varphi(t) \in \Phi$, $\varphi(t)\psi(t) \in L^1$. It is easy to see that there exists a constant $\alpha = \alpha(\psi)$ satisfying

$$(9.0) \qquad \left| \int_0^\infty \varphi(t)\psi(t)\, dt \right| \le \alpha |\varphi|_\Phi \qquad \text{for all} \quad \varphi \in \Phi.$$

Otherwise, there are elements $\varphi_n \in \Phi$ such that $|\varphi_n|_\Phi \le 2^{-n}$, $\varphi_n(t)\psi(t) \ge 0$, and $\int_0^\infty \varphi_n(t)\psi(t)\, dt \ge 1$. Then $\varphi(t) = \Sigma\, \varphi_n(t) \in \Phi$, but, by Lebesgue's theorem on monotone convergence,

$$\int_0^\infty \varphi(t)\psi(t)\, dt = \Sigma \int_0^\infty \varphi_n(t)\psi(t) \ge \Sigma\, 1 = \infty.$$

Let the least constant α satisfying (9.0) be denoted by $|\psi|_{\Phi'}$, so that

$$\left| \int_0^\infty \varphi(t)\psi(t)\, dt \right| \le |\varphi|_\Phi\, |\psi|_{\Phi'} \qquad \text{for all} \quad \varphi \in \Phi.$$

This is clearly equivalent to

$$(9.1) \qquad \int_0^\infty |\varphi(t)\psi(t)|\, dt \le |\varphi|_\Phi\, |\psi|_{\Phi'} \qquad \text{for all} \quad \varphi \in \Phi.$$

Lemma 9.1. *Let $\Phi \in \mathscr{T}^\#$ be a Banach space. Then Φ', with the norm $|\psi|_{\Phi'}$ for $\psi \in \Phi'$, is a Banach space in $\mathscr{T}$ (in fact, in $\mathscr{T}^\#$) and is quasi-full.*

Exercise 9.1. Prove this lemma. The assumption $\Phi \in \mathscr{T}^\#$, rather than $\Phi \in \mathscr{T}$, assures that $\Phi' \in \mathscr{T}$.

It is clear that Φ' is isomorphic and isometric to a subspace of the dual space Φ^* of Φ.

Exercise 9.2. Let $1 \le p \le \infty$, $1/p + 1/q = 1$. Show that the associate space of L^p is L^q.

Lemma 9.2. *Let $\Phi \in \mathscr{T}^\#$ be a Banach space and Y a (finite-dimensional) Banach space with Y^* its dual space. Let $y^*(t) \in L(Y^*)$. Then $y^*(t) \in \Phi'(Y^*)$ if and only if there exists a constant $\alpha = \alpha(y^*)$ such that for every $y(t) \in \Phi_\infty(Y)$, $\langle y(t), y^*(t) \rangle \in L^1$, and*

$$(9.2) \qquad \left| \int_0^\infty \langle y(t), y^*(t) \rangle\, dt \right| \le \alpha\, |y|_\Phi.$$

In this case, the least constant α satisfying (9.2) is $|y^*|_{\Phi'(Y^*)} = |y^*|_{\Phi'}$.

Recall that $\langle y, y^* \rangle$ denotes the pairing ("scalar product") of elements $y \in Y$, $y^* \in Y^*$. In the case under consideration (when Y is finite dimensional), the nontrivial part of this lemma can be reduced to the 1-dimensional case by the introduction of suitable bases in Y, Y^* and examining the components of $y^*(t)$.

Exercise 9.3. Prove Lemma 9.2.

10. The Operator T'

This part of the chapter concerns an operator T as in § 3, an *associate operator*

(10.1) $$T'y^*(t) = f^*(t),$$

and the corresponding null space $\mathcal{N}(T')$.

(10.2) $$T'y^*(t) = 0.$$

In dealing with the pair T and T', the following will be assumed:

(C_0) T is a linear operator from $L(Y)$ to $L(U)$, i.e., $F = U$; the elements $y(t) \in \mathcal{D}(T)$ are continuous (so as to avoid ambiguities on null sets); and $y[0] = y(0)$, so $W = Y$. Correspondingly, T' is a linear operator from $L(Y^*)$ to $L(U^*)$; the elements $y^*(t) \in \mathcal{D}(T')$ are continuous with $y^*[0] = y^*(0)$; and $u^*(t) = P'y^*(t)$ is a linear operator from $L(Y^*)$ to $L(U^*)$ with $\mathcal{D}(P') = \mathcal{D}(T')$.

(C_1) For each $t \geq 0$, there exists a bounded bilinear form $V_t(y, y^*)$ on $Y \times Y^*$ such that for $y(t) \in \mathcal{D}(T)$, $y^*(t) \in \mathcal{D}(T')$, $u(t) = Py(t)$, $u^*(t) = P'y^*(t)$, and for $0 \leq \sigma < \tau < \infty$, the following "Green's relation" holds:

(10.3) $$\int_\sigma^\tau \langle Ty, u^* \rangle \, dt - \int_\sigma^\tau \langle u, T'y^* \rangle \, dt = [V_t(y(t), y^*(t))]_\sigma^\tau.$$

For $t \geq 0$, $\beta(t)$ denotes a number satisfying

(10.4) $$|V_t(y, y^*)| \leq \beta(t) \|y\| \cdot \|y^*\| \qquad \text{for all} \quad y \in Y, y^* \in Y^*.$$

Note that (10.3) implies that $V_t(y(t), y^*(t))$ is constant on any interval where $Ty = 0$, $T'y^* = 0$. In particular, $V_t(y(t), y^*(t))$ is constant on J for $y(t) \in \mathcal{N}(T)$, $y^*(t) \in \mathcal{N}(T')$.

Definition. If X is a manifold in Y, let X^V be the subspace of Y^* defined by $X^V = \{y^* : V_0(y, y^*) = 0 \text{ for all } y \in X\}$. Correspondingly, if X^* is a manifold in Y^*, let X^{*V} denote the subspace of Y defined by $X^{*V} = \{y : V_0(y, y^*) = 0 \text{ for all } y^* \in X^*\}$.

Assumption $(A_n)^*$ or $(B_n\epsilon)^*$ will mean the analogue of (A_n) or $(B_n\epsilon)$ with T, Y, $F = U$, P, B, D, $W_D = Y_D$, constants $C_j, \dots$ replaced by T', Y^*, U^*, P', D', B', $Y_{B'}^*$, constants $C_j^*, \dots$, respectively, where B', D' are the associate spaces of B, D. Note the replacement of the ordered pair (B, D) by (D', B').

11. Individual Dichotomies

The next theorem involves the following notion: A set Σ of real-valued functions $\varphi(t) \in L$ will be called *small at* ∞ if every $\varphi(t) \in \Sigma$ is small at ∞ in the sense that

$$\liminf_{t \to \infty} \int_t^{t+1} |\varphi(r)| \, dr = 0.$$

Theorem 11.1. *Assume* (A_1) *[or* (A_1')*]–*(A_4)*;* $(B_{30}\delta)$*;* $(B_3\delta)^*$*,* $(B_4\Delta)^*$*; and* (C_0)*,* (C_1) *with a* $\beta(t) \in M$ *in* (10.4). *Suppose also that* $\beta(t)$ *or* B' *or* D *is small at* ∞*. Then* $Y_{B'}^*$*, induces an individual partial dichotomy for* $(\mathcal{N}(T'), \|y^*(t)\|)$*.*

It will be clear from the proof that θ_0, M_0 in (1.7) of condition (*a*) of an individual partial dichotomy can be chosen independent of $y^*(t)$.

The assumption that $\beta(t)$ or B' or D is small at ∞ will be used only in the derivation of the condition (*a*) of an individual partial dichotomy involving $y^*(t) \in \mathcal{N}(T')$ with $P'y^* \in B'$. At the cost of allowing M_0 to depend on y^* in condition (*a*) and of making the additional hypothesis (A_5) of § 8, the condition of "smallness at ∞" will be eliminated in Theorem 11.3.

In view of Theorem 12.3, it follows that the main point of the theorems of this section is that no assumption of the type $(B_2\epsilon)^*$ occurs. The importance of this can be seen by noting that assumption (7.14) occurs in Theorem 7.1 only to insure $(B_2\epsilon)$ for the operator T there.

Proof of Theorem 11.1.

Condition (a).   The condition $\beta(t) \in M$ implies that, for $\delta > 0$,

$$(11.1) \qquad \beta_0(\delta) = \sup \int_t^{t+\delta} \beta(r)\, dr < \infty \qquad \text{for } t \geqq 0.$$

The condition that $\beta(t)$ or B' or D is small at ∞ will be used as follows: if three functions $\beta(t)$, $\|y(t)\|$, $\|y^*(t)\|$ are in M[e.g., $\beta(t) \in M$, $y(t) \in D(Y)$, $y^*(t) \in B'(Y^*)$; cf. (9.1)] and at least one is small at ∞, then

$$(11.2) \qquad \liminf_{\tau \to \infty} \beta(\tau) \|y(\tau)\| \cdot \|y^*(\tau)\| = 0.$$

In order to see this, note that if $\varphi(t) \geqq 0$ is integrable on an interval $[t, t + 1]$, the measure of the set of s-values where $\varphi(s) > 3 \int_t^{t+1} \varphi(r)\, dr$ is less than $1/3$. If this remark is applied to $\varphi = \beta$, $\|y\|$, $\|y^*\|$, it follows that for any $t \geqq 0$, there is a common t-value $\tau \in [t, t + 1]$ satisfying

$$(11.3) \qquad \varphi(\tau) \leqq 3 \int_t^{t+1} \varphi(r)\, dr \qquad \text{for } \varphi = \beta,\ \|y\|,\ \|y^*\|.$$

By the assumption $\varphi \in M$, the integral on the right is bounded for $t \geqq 0$ and, for at least one of the functions, $\varphi(t) \to 0$ for suitable choices of $t = t_n \to \infty$ as $n \to \infty$. Hence (11.2) follows.

Let $y^*(t) \in \mathcal{N}(T')$ with $P'y^* \in B'(U^*)$. Let $s \geqq 0$ and $f(t)$ any element of $B(U)$ with compact support on $t \geqq s + 3\delta$. By Lemma 3.2, $Ty = f$ has a *PD*-solution $y(t)$ satisfying (3.5), hence $|Py|_D \leqq C_3 |f|_B$. By

$(B_{30}\delta)$ and $(B_{3}2\delta)^*$, it is seen that $y(t) \in D(Y)$, $y^*(t) \in D'(Y^*)$. Since $T'y^* = 0$ and $Ty = f$, the Green relation (10.3) gives

$$\int_\sigma^\tau \langle f, u^* \rangle \, dr = [V_t(y(t), y^*(t))]_\sigma^\tau,$$

where $u^*(t) = P'y^*(t)$. The left side is independent of $\sigma \leq s + 3\delta$ and of large τ [since $f(t) = 0$ for $t \leq s + 3\delta$ and large t]. Thus (10.4) and (11.2) give for $0 \leq \sigma \leq s + 3\delta$,

$$(11.4) \qquad \left| \int_{s+3\delta}^\infty \langle f, u^* \rangle \, dr \right| \leq \beta(\sigma) \, \|y(\sigma)\| \cdot \|y^*(\sigma)\|.$$

The argument leading to (11.2) shows that for a suitable choice of σ, $s \leq \sigma \leq s + 3\delta$,

$$\beta(\sigma) \, \|y(\sigma)\| \leq \left(\frac{1}{3\delta} \int_s^{s+3\delta} \beta \, dr \right) \left(\frac{1}{3\delta} \int_s^{s+3\delta} \|y\| \, dr \right).$$

The first factor on the right is at most $(3\delta)^{-1}\beta_0(3\delta)$ by (11.1); the second fact is at most $(3\delta)^{-1} |h_{0,3\delta}|_{D'} |h_{s,3\delta}y|_D$ by (1.5), hence, at most

$$(3\delta)^{-1} |h_{0\delta}|_{D'} K_{30}(\delta) \, |Py|_D$$

by $(B_{30}\delta)$. Since $|Py|_D \leq C_3 |f|_B$, the right side of (11.4) can be replaced by $K' |f|_B \, \|y^*(\sigma)\|$, where $K' = K'(\delta)$ is a constant and σ is some point of $s \leq \sigma \leq s + 3\delta$.

Since $f(t)$ is an arbitrary element of $B(U)$ with compact support on $t \geq s + 3\delta$, Lemma 9.2 implies that

$$(11.5) \quad |h_{s+3\delta}u^*|_{B'} \leq K' \sup \|y^*(\sigma)\| \qquad \text{for} \quad s \leq \sigma \leq s + 3\delta;$$

or by $(B_3\delta)^*$,

$$(11.6) \quad |h_{s+4\delta}y^*|_{B'} \leq K'K_3^*(\delta) \sup \|y^*(\sigma)\| \qquad \text{for} \quad s \leq \sigma \leq s + 3\delta.$$

Arguments involving $(B_4\Delta)^*$ similar to those used in the proof of Theorem 5.1 give condition (a) of a partial dichotomy for $(\mathcal{N}(T'), \|y^*(t)\|)$.

Condition (b). Let $z^*(t) \in \mathcal{N}(T')$, $P'z^* = v^*$, and $v^* \notin B'(Y^*)$. Let $f(t)$ be any element of $B(U)$ with support on $[0, s]$ and $y(t)$ a solution of $Ty = f$ supplied by Lemma 3.2. Then $\tau \geq s$, Green's relation, and (10.4) give

$$\left| \int_0^s \langle f, v^* \rangle \, dr \right| \leq \beta(0) \, \|y(0)\| \cdot \|z^*(0)\| + \beta(\tau) \, \|y(\tau)\| \cdot \|z^*(\tau)\|.$$

The use of (3.5) and the arguments in the derivation of (11.5) show that

$$(11.7) \quad |h_{0s}v^*|_{B'} \leq C_{30}\beta(0) \, \|z^*(0)\| + K' \sup \|z^*(\tau)\|$$

$$\text{for} \quad s + \delta \leq \tau \leq s + 3\delta,$$

where $K' = K'(\delta)$ is the same as in (11.5).

Since $v^* \notin B'(Y^*)$, it follows from the fact that B' is quasi-full (Lemma 9.1) that $|h_{0s}v^*|_{B'} \to \infty$ as $s \to \infty$. Thus there exists an s_0 depending on the solution $z^*(t)$ such that $|h_{0s}v^*|_{B'} \geq 2C_{30}\beta(0)\,\|z^*(0)\|$ for $s \geq s_0$. Then by (11.7)

$$|h_{0s}v^*|_{B'} \leq 2K' \sup \|z^*(\tau)\| \qquad \text{for} \quad s + \delta \leq \tau \leq s + 3\delta, \quad s \geq s_0.$$

Using $(B_3\delta)^*$,

(11.8) $|h_{0s}z^*|_{B'} \leq 2KK_3^*(\delta) \sup \|z^*(\tau)\|$

$$\text{for} \quad s + 2\delta \leq \tau \leq s + 5\delta, \quad s \geq s_0 + \delta.$$

The proof can be completed by the arguments in the proof of Theorem 5.1.

Theorem 11.2. *Let the conditions of Theorem 6.1 hold. In addition, assume either*

(11.9) $\Delta^{-1}|h_{0\Delta}|_{D'} \to 0$ or $|h_{0\Delta}|_{B'} \to \infty$ as $\Delta \to \infty.$

Then $Y_{B'}^*$ induces an individual exponential dichotomy for $(\mathcal{N}(T')$, $\|y^*(t)\|)$, where θ^0, M_1, ν do not depend on $y^*(t)$ in condition (a).

Exercise 11.1.   Prove this theorem.

The elimination of the "smallness at ∞" condition will depend on the following lemma, which involves assumption (A_5) of §8. In this lemma, the notation of (A_5) is used; i.e., if $Ty = f$ and $Ty_1 = f_1$, where f, f_1 vanish for large t, then y_∞, $y_{1\infty}$ are the corresponding solutions suppplied by (A_5).

Lemma 11.1.   *Assume (A_1)–(A_5) and (C_0)–(C_1). Let $y^*(t)$ be a $P'B'$- solution of $T'y^* = 0$. Then there exist constants C_3 and C_{30}, depending on $y^*(0)$, such that for any $f(t) \in B_\infty(U)$, $Ty = f$ has a PD-solution $y(t)$ satisfying (3.5) and $V_0(y_\infty(0), y^*(0)) = 0$; i.e., if $\{y^*(0)\}$ is the 1-dimensional manifold in Y^* spanned by $y^*(0)$, then $Y_D \cap \{y^*(0)\}^V$ is a PD-manifold for (3.1).*

Proof.   We can suppose that there is a PD-solution $y_0(t)$ of $Ty = 0$ satisfying $\alpha \equiv V_0(y_0(0), y^*(0)) \neq 0$. For if not, Lemma 11.1 follows from Lemma 3.2. Let $y(t)$ be the solution of $Ty = f$ supplied by Lemma 3.2 and put

(11.10) $$y_1(t) = y(t) - V_0(y_\infty(0), y^*(0))\alpha^{-1}y_0(t)$$

Then $y_1(t)$ is a PD-solution of $Ty_1 = f$ satisfying $V_0(y_{1\infty}(0), y^*(0)) = 0$.

By Green's relation and $T'y^* = 0$,

(11.11) $$\int_0^\tau \langle f, u^* \rangle \, dt = [V_t(y_1(t), y^*(t))]_0^\tau$$

if $u^* = P'y^*$. Since f vanishes for large τ, (A_5) implies that $y_1(t) = y_{1\infty}(t)$ for large t, so that $V_t(y_1(t), y^*(t)) = V_t(y_{1\infty}(t), y^*(t))$ for large t. But the latter expression does not depend on t (since $Ty_{1\infty} = 0$, $T'y^* = 0$) and is therefore $V_0(y_{1\infty}(0), y^*(0)) = 0$. Thus (11.11) shows that

$$(11.12) \qquad \int_0^\infty \langle f, u^* \rangle \, dt = -V_0(y_1(0), y^*(0)).$$

Thus, by (11.10),

$$|V_0(y_\infty(0), y^*(0))| \leqq \left| \int_0^\infty \langle f, u^* \rangle \, dt \right| + |V_0(y(0), y^*(0))|.$$

The right side is at most $|f|_B(|P'y^*|_{B'} + \beta(0)C_{30} \|y^*(0)\|)$, by (10.4) and Lemma 3.2. Hence (11.10) shows that

$$|Py_1|_D \leqq |Py|_D + |f|_B(|P'y^*|_{B'} + \beta(0)C_{30} \|y^*(0)\|)\alpha^{-1} |Py_0|_D.$$

By (3.5_1) in Lemma 3.2, $|Py|_D \leqq C_3 |f|_B$. Thus the analogue of (3.5_1) holds with y replaced by y_1 and C_3 replaced by the constant $C_3 + (|P'y^*|_{B'} + \beta(0)C_{30} \|y^*(0)\|)\alpha^{-1} |Py_0|_D$. The analogue of (3.5_2) follows as in Lemma 8.1.

Theorem 11.3. *In Theorems 11.1 or 11.2, replace the assumption that "$\beta(t)$ or B' or D is small at ∞" by the assumption* (A_5). *Then the conclusions of these theorems remain valid.*

Exercise 11.2. Prove this theorem. In view of the remarks following Theorem 11.1, only condition (*a*) need be considered. The proof of condition (*a*) here is similar to (but simpler than) the proof of the corresponding condition in Theorem 11.1 or 11.2. The solution of $Ty = f$ supplied by Lemma 11.1 is used in place of that given by Lemma 3.2.

12. P′-Admissible Spaces for T'

The object of this section is to show that, under suitable conditions, if (B, D) is P-admissible for T, then $\overline{(D_\infty{}', B')}$ or (D', B') is P'-admissible for T'.

Lemma 12.1 *Assume* (A_1)–(A_2), (A_5)–(A_6); $(A_5)^*$; *and* (C_0)–(C_1). [*In particular, X is the $P(B, D)$-manifold in* (A_6).] *If* $f^*(t) \in D_\infty{}'(U^*)$ *and* $y^*(t)$ *is a solution of* $T'y^* = f^*$ *with* $y_\infty{}^*(0) \in X^V$, *then* $y^*(t)$ *is a* $P'B'$-*solution and*

$$(12.1) \qquad |P'y^*|_{B'} \leqq C_3 |f^*|_{D'} + \beta(0)C_{30} \|y^*(0)\|,$$

where C_3, C_{30} *are the constants in the definition of a $P(B, D)$-manifold X and Lemma 8.1. In particular* $X^V \subset Y_{B'}^*$.

Proof. Let $f \in B_\infty(U)$ and $y(t)$ the solution of $Ty = f$ supplied by (A_6), so that $y_\infty(0) \in X$ and (3.5) holds. For large t, $V_t(y(t), y^*(t)) = V_t(y_\infty(t), y_\infty{}^*(t)) = V_0(y_\infty(0), y_\infty{}^*(0)) = 0$. Thus Green's formula gives

$$\left| \int_0^\infty \langle f, P'y^* \rangle \, dr \right| \leqq \left| \int_0^\infty \langle Py, f^* \rangle \, dr \right| + |V_0(y(0), y^*(0))|,$$

where the right side is at most $|Py|_D |f^*|_{D'} + \beta(0) \|y(0)\| \cdot \|y^*(0)\| \leq |f|_B(C_3 |f^*|_{D'} + \beta(0)C_{30} \|y^*(0)\|)$ by (3.5). Hence the assertion follows from Lemma 9.2.

Let $y^* \in Y^*$. Then (C_1) implies the existence of a unique $x^* \in Y^*$ such that the linear functional $V_0(y, y^*)$ on y is representable in the usual pairing of Y, Y^* as $V_0(y, y^*) = \langle y, x^* \rangle$ for all $y \in Y$ and $\|x^*\| \leq \beta(0) \|y^*\|$; i.e., there is a unique linear map $x^* = Sy^*$ of Y^* into itself satisfying

$$(12.2) \qquad V_0(y, y^*) = \langle y, Sy^* \rangle \qquad \text{for} \quad y \in Y, \quad y^* \in Y^*.$$

In this section, the following will be assumed:

(C_2) The (unique) bounded linear map $S: Y^* \to Y^*$ defined by (12.2) is onto (and hence has a unique inverse S^{-1} defined on all of Y^*).

It is clear that for a manifold $X \subset Y$, we have $X^V = S^{-1}X^0$, where X^0 is the usual annihilator of X; i.e., $X^0 = \{y^* \in Y^* : \langle y, y^* \rangle = 0$ for all $y \in X\}$.

Lemma 12.2. *Assume* (A_1)–(A_5); $(A_5)^*$; *and* (C_0)–(C_2). *Let* $f^*(t) \in D_\infty'(U^*)$ *and* $y^*(t)$ *a solution of* $T'y = f^*$ *with* $y_\infty^*(0) \in Y_D^V$. *Then* $d(Y_D^V, y^*(0)) \leq C_4 \|S^{-1}\| \cdot |f^*|_{D'}$, *where* C_4 *is the constant in* (3.6) *of Lemma 3.3 and* $\|S^{-1}\|$ *is the norm of the operator* S^{-1} *from* Y^* *to* Y^*.

Proof. Let $y(t)$ be a PD-solution of $Ty = 0$. Then $V_\tau(y(\tau), y^*(\tau)) = V_\tau(y(\tau), y_\infty^*(\tau))$ for large τ. By Green's relation, the last expression is the constant $V_0(y(0), y_\infty^*(0)) = 0$. By Green's formula,

$$|V_0(y(0), y^*(0))| = \left| \int_0^\infty \langle Py, f^* \rangle \, dt \right| \leq |Py|_D |f^*|_{D'}.$$

From the inequality $|Py|_D \leq C_4 \|y(0)\|$ in Lemma 3.3 and from (12.2), it follows that $|\langle y(0), Sy^*(0) \rangle| \leq C_4 \|y(0)\| \cdot |f^*|_{D'}$ for all $y(0) \in Y_D$. Thus $Sy^*(0)$ considered as a bounded linear functional on Y_D (i.e., as an element of the dual space Y_D^*) has a norm not exceeding $C_4 |f^*|_{D'}$. Since Y_D^* is the quotient space Y^*/Y_D^0 in which the norm of an "element" y^* is $d(Y_D^0, y^*)$ it follows that $d(Y_D^0, Sy^*(0)) \leq C_4 |f^*|_{D'}$. Hence $Y_D^V = S^{-1}Y_D^0$ implies that $d(Y_D^V, y^*(0)) \leq \|S^{-1}\| d(Y_D^0, Sy^*(0))$, and so Lemma 12.2 follows.

Theorem 12.1. *Assume* (A_1)–(A_5); (C_0)–(C_2); $(A_5)^*$; *and that* $T'y = f^*$ *has a solution* $y^*(t)$ *for every* $f^*(t) \in D_\infty'(U^*)$. *Then* $(\overline{D_\infty'}, B')$ *is P-admissible for* T'; *in fact,* Y_D^V *is a* $P'(D', B')$-subspace for T' (with permissible constants analogous to C_3, C_{30} being $C_3^* = C_3 + \lambda C_4 C_{30} \|S^{-1}\|$ and $C_{30}^* = \lambda C_4 \|S^{-1}\|$ for any fixed $\lambda > 1$).

Proof. Let $f^*(t) \in D_\infty'(U^*)$. Then $T'z^* = f^*$ has a solution $z^*(t)$ which vanishes for large t. [For if $z^*(t)$ is any solution of $T'z^* = f^*$, then, since $z_\infty^*(t)$ exists by $(A_5)^*$, the desired solution is $z^*(t) - z_\infty^*(t)$.]

Hence $z_\infty^*(t) \equiv 0$. In particular, $0 = z_\infty^*(0) \in Y_D^V$, so that, by Lemma 12.2, $d(Y_D^V, z^*(0)) \leqq C_4 \|S^{-1}\| \cdot |f^*|_{D'}$.

Let $\lambda > 1$ be fixed and decompose the element $z^*(0) \in Y^*$ into $z^*(0) = x_1^* + y_0^*$, where $x_1^* \in Y_D^V$ and $\|y_0^*\| \leqq \lambda\, d(Y_D^V, z^*(0))$. Consequently

$$\|y_0^*\| \leqq C_{30}^* |f^*|_{D'}, \quad \text{where} \quad C_{30}^* = \lambda C_4 \|S^{-1}\|.$$

Since $x_1^* \in Y_D^V \subset Y_{B'}^*$ by Lemma 12.1, $Tx_1^*(t) = 0$ has a solution $x_1^*(t)$ beginning at x_1^* for $t = 0$. Let $y^*(t) = z^*(t) - x_1^*(t)$, so that $Ty^* = f^*$ and $y_\infty^*(t) = -x_1^*(t)$. Hence $y_\infty^*(0) = -x_1^* \in Y_D^V$. In addition $y^*(0) = z^*(0) - x_1^* = y_0^*$. By Lemma 12.1,

$$|P'y^*|_{B'} \leqq C_3 |f^*|_{D'} + C_{30} \|y_0^*\|.$$

Consequently, $|P'y^*|_{B'} \leqq C_3^* |f^*|_{D'}$, where C_3^* is the specified constant. This proves the theorem.

Theorem 12.2. *Assume* (A_1)–(A_5); (C_0)–(C_2); *and that for every* $y_0^* \in Y^*$ *and* $f^* \in D'(U^*)$, $T'y^* = f^*$ *has a solution* $y^*(t)$ *satisfying* $y^*(0) = y_0^*$. *Then* (D', B') *is* P'-*admissible for* T'. (*Furthermore, permissible constants analogous to* C_3, C_{30} *are* $C_3^* = C_3 + \beta(0)C_{30}C_4\|S^{-1}\|$ *and* $C_{30}^* = C_4\|S^{-1}\|$.)

Proof. Let $f^* \in D'(U^*)$. It must be shown that $Ty^* = f^*$ has a $P'B'$-solution $y^*(t)$. Let $y(t) \in \mathcal{N}(T)$ with $y(0) \in Y_D$. Put

$$(12.3) \qquad \varphi(y(0)) = \int_0^\infty \langle Py, f^* \rangle \, dt.$$

Then $|\varphi(y(0))| \leqq |Py|_D |f^*|_{D'} \leqq C_4 \|y(0)\| \cdot |f^*|_{D'}$ by Lemma 3.3. In other words $\varphi(y)$ is a bounded linear functional on Y_D with norm $\leqq C_4 |f^*|_{D'}$ which, therefore, has an extension to Y with the same norm. Consequently, there is an element $x_0^* \in Y^*$ such that $\|x_0^*\| \leqq C_4 |f^*|_{D'}$ and $\varphi(y) = \langle y, x_0^* \rangle$ for all $y \in Y$. By (C_2), $y_0^* = S^{-1}x_0^*$ exists, so that

$$(12.4) \qquad \varphi(y) = V_0(y, y_0^*)$$

for all $y \in Y$ and

$$(12.5) \qquad \|y_0^*\| \leqq C_4 \|S^{-1}\| \cdot |f^*|_{D'}.$$

By assumption, $Ty^* = f^*$ has a solution $y^*(t)$ satisfying $y^*(0) = y_0^*$.

Let $f \in B_\infty(U)$, $y(t)$ a PD-solution of $Ty = f$ supplied by Lemma 3.2. By Green's formula applied to $y_\infty(t)$ and $y^*(t)$,

$$V_t(y_\infty(t), y^*(t)) = V_0(y_\infty(0), y^*(0)) - \int_0^t \langle Py_\infty, f^* \rangle \, dr$$

for large t. Since $y(t)' = y_\infty(t)$ for large t, it follows from (12.3), (12.4) that, for large t,

$$V_t(y(t),\, y^*(t)) = \int_t^\infty \langle Py_\infty, f^* \rangle \, dr \to 0 \qquad \text{as} \quad t \to \infty.$$

Thus Green's formula applied to $y(t)$ and $y^*(t)$ gives

$$\int_0^\infty \langle f, P'y^* \rangle \, dr = \int_0^\infty \langle Py, f^* \rangle \, dr - V_0(y(0), y^*(0)).$$

Consequently, (3.5) and (12.5) imply that

$$\left| \int_0^\infty \langle f, P'y^* \rangle \, dr \right| \leqq |f|_B (C_3 + \beta(0) C_{30} C_4 \, \|S^{-1}\|) \, |f^*|_{D'}.$$

By Lemma 9.2, $P'y^* \in B'$ and $|P'y^*|_{B'} \leqq (C_3 + \beta(0) C_{30} C_4 \, \|S^{-1}\|) \, |f^*|_{D'}$. In view of (12.5), this proves the theorem.

The usefulness of theorems like Theorems 12.1, 12.2 will be illustrated by an application of Theorem 12.1 and of Theorem 11.3 with T, T' interchanged. Note that $D' = (\overline{D_\infty})'$, so that the second associate space $D'' = (D')'$ of D is the same as $(\overline{D_\infty}')'$.

Theorem 12.3. *Assume the conditions of Theorem* 12.1; $(B_{30}\delta)^*$; $(B_3\delta)$ *for a fixed* $\delta > 0$ *with* D *replaced by* D''; $(B_4\Delta)$; $\beta(t) \in M$ *in* (10.4). *Then* $Y_{D'}$ *induces an individual partial dichotomy for* $\mathcal{N}(T)$, *and* $\|y(t)\| \to \infty$ *as* $t \to \infty$ *if* $y(t)$ *is a non-PD''-solution of* $Ty = 0$. *If, in addition,*

$$(12.6) \qquad \Delta^{-1} |h_{0\Delta}|_B \to 0 \qquad \text{or} \qquad |h_{0\Delta}|_{D'} \to \infty \qquad \text{as} \quad \Delta \to \infty,$$

then $Y_{D'}$ *induces an individual exponential dichotomy for* $\mathcal{N}(T)$.

If the condition "$\beta(t)$ or B'' or D' is small at ∞" is assumed, then the constants in condition (a) of the dichotomies do not depend on the solution $y(t)$ involved.

Exercise 12.1. Verify Theorem 12.3.

13. Applications to Differential Equations

The systems formally adjoint to (0.1), (0.2) are

$$(13.1) \qquad\qquad y^{*\prime} + A^*(t)y^* = -g^*(t),$$

$$(13.2) \qquad\qquad y^{*\prime} + A^*(t)y^* = 0,$$

where $A^*(t)$ is the complex conjugate transpose of $A(t)$; cf. § IV 7. Let T be the operator associated with (0.1) with $Py(t) = y(t)$ and $y[0] = y(0)$ and T' is the negative of the corresponding operator associated with (13.1) [i.e., $T'y^* = -(y^{*\prime} + A^*(t)y^*) = g^*$], then T, T' are associate operators in the sense of § 10. The corresponding Green identity is

$$(13.3) \qquad \int_\sigma^\tau \langle Ty, y^* \rangle \, dr - \int_\sigma^\tau \langle y, T'y^* \rangle \, dr = [\langle y(t), y^*(t) \rangle]_\sigma^\tau,$$

as can be seen by differentiating with respect to τ. Thus $V_t(y, y^*) = \langle y, y^* \rangle$. Clearly, (C_0)–(C_2) hold [with $\beta(t) \equiv 1$ and $Sy^* = y^*$] and Theorem 12.2 implies

Theorem 13.1 *Let $A(t)$ be a matrix of locally integrable functions for $t \geqq 0$. Suppose that (B, D) is admissible for (0.1). Then (D', B') is admissible for (13.1).*

Thus the theorems of § 6 become applicable to (13.1).

In order to consider the systems adjoint to (0.3) and (0.4), suppose that $P_k(t)$ is a kth integral (i.e., has $k - 1$ absolutely continuous derivatives). The equations formally adjoint to (0.3), (0.4) are

$$(13.4) \qquad (-1)^{m+1} u^{*(m+1)} + \sum_{k=0}^m Q_k^*(t) u^{*(k)} = f^*(t),$$

$$(13.5) \qquad (-1)^{m+1} u^{*(m+1)} + \sum_{k=0}^m Q_k^*(t) u^{*(k)} = 0,$$

where

$$(13.6) \qquad Q_k(t) = \sum_{j=k}^m (-1)^j C_{jk} P_j^{(j-k)}(t);$$

an asterisk on a matrix denotes complex conjugate transposition, and indices in parentheses denote differentiation; cf. § IV 8(viii). For $y = (u^{(0)}, u^{(1)}, \ldots, u^{(m)}) \in Y = U^{(m+1)}$ and $y^* = (u^{*(0)}, \ldots, u^{*(m)}) \in Y^*$, put

$$(13.7) \qquad V_t(y, y^*) = \sum_{k=0}^m \sum_{j=k}^m (-1)^j C_{j-k, i} \langle P_{j+1}^{(i)}(t) u^{(k)}, u^{*(j-k-1)} \rangle,$$

where $P_{m+1}(t) \equiv I$ is the identity operator. This is a bilinear form in y, y^*. The following Green's formula is readily verified

$$(13.8) \qquad \int_\sigma^\tau \langle f, u^* \rangle \, dr - \int_\sigma^\tau \langle u, f^* \rangle \, dr = [V_t(y(t), y^*(t))]_\sigma^\tau,$$

if $u(t)$, $u^*(t)$ are solutions of (0.3), (13.4), respectively.

A rearrangement of the sums in (13.7) shows that

$$(13.9) \qquad V_t(y, y^*) = \sum_{k=0}^m \left\langle u^{(k)}, \sum_{i=0}^{m-k} \sum_{j=i}^{m-k} (-1)^{k+i} C_{j, j-i} P_{j+k+1}^{*(j-i)} u^{*(i)} \right\rangle.$$

Thus, if $y^* = (v^{*(0)}, \ldots, v^{*(m)}) \in Y^*$ is arbitrary, the $m + 1$ equations

$$\sum_{i=0}^{m-k} \sum_{j=i}^{m-k} (-1)^{k+i} C_{j, j-1} P_{j+k+1}^{*(j-i)}(t) u^{*(i)} = v^{*(k)}$$

can be solved recursively for $k = m$, $m - 1, \ldots, 0$ since $P_{m+1} = I$. For $t = 0$, this implies the analogue of condition (C_2) in § 12.

Let T', P' be the operators associated with (13.4) in the same way that T, P are associated with (0.3) in § 7. The analogue of (0.5) shows that $(B_4\Delta)^*$ holds for T' if $Q_k{}^* \in M(\widetilde{U^*})$ [i.e., $Q_k \in M(\tilde{U})$]; also, Lemma 7.2 shows that the same condition on Q_k implies $(B_3\delta)^*$. Thus Theorem 11.3 gives

Theorem 13.2. *Let $P_k(t) \in L(\tilde{U})$, $k = 0, \ldots, m$ be a kth integral; $Q_k(t) \in M(\tilde{U})$ for $k = 0, \ldots, m$; and let there exist a $\beta(t) \in M$ such that (13.7) satisfies (10.4). Let (B, D) be admissible for (0.3). Then $Y^*_{B'}$ induces an individual partial dichotomy for $\mathcal{N}(T')$; and if $u^*(t)$ is a non-B'-solution of (13.5), then $\|y^*(t)\| \to \infty$ as $t \to \infty$. If, in addition, (11.9) holds, then $Y^*_{D'}$ induces an individual exponential dichotomy for $\mathcal{N}(T')$.*

An immediate corollary of Theorem 12.2 is the following:

Theorem 13.3. *Let $P_k(t) \in L(\tilde{U})$, $k = 0, \ldots, m$, be a kth integral and let (B, D) be admissible for (0.3). Then (D', B') is admissible for (13.4).*

Thus, under the appropriate conditions on the coefficients $Q_k{}^*(t)$ of (13.4), the theorems of § 7 become applicable. An application of Theorem 12.3 gives individual dichotomies for $\mathcal{N}(T)$ without a condition of the type (7.14), but with a condition on $Q_k{}^*(t)$.

Theorem 13.4. *Let the conditions of Theorem 13.3 hold. Let $P_k(t) \in M(\tilde{U})$, $Q_k \in M(\tilde{U})$ for $k = 0, 1, \ldots, m$ and let there exist a $\beta(t) \in M$ such that (13.7) satisfies (10.4). Then $Y_{D'}$ induces an individual partial dichotomy for $\mathcal{N}(T)$; and if $u(t)$ is a non-D''-solution of (0.4), then $\|y(t)\| \to \infty$ as $t \to \infty$. If, in addition, (12.6) holds, then $Y_{D''}$ induces an individual exponential dichotomy for $\mathcal{N}(T)$.*

It is clear from (13.7) that a function satisfying (10.4) is

$$\beta(t) = c^m \left(1 + \sum_{k=1}^{m} \sum_{j=0}^{k-1} \|P_k^{(j)}(t)\| \right)$$

for a suitable constant c^m depending only on m. Thus if

(13.10) $P_k^{(j)}(t) \in M(\tilde{U})$ for $0 \leq j \leq k - 1$, $1 \leq k \leq m$

holds, then $\beta(t) \in M$, also $Q_k \in M(\tilde{U})$ for $k = 1, \ldots, m$. Thus (13.10) and the condition P_0, $Q_0 \in M(\tilde{U})$ imply that the conditions of the second sentence of Theorem 13.4 hold.

Note that if $P_0 \in M(\tilde{U})$ and (13.10) hold with $j = k$ also permitted, then $Q_0 \in M(\tilde{U})$. But in this case, P_k, $Q_k \in L^\infty(\tilde{U})$ for $k = 1, \ldots, m$, and Theorem 13.4 is contained in the theorems of § 7. (This statement concerning L^∞ follows from the fact that if $\varphi(t)$ is absolutely continuous for $t \geq 0$ and φ, $\varphi' \in M$, then $\varphi \in L^\infty$.)

Exercise 13.1.   If $p(t)$ is absolutely continuous, the equations formally adjoint to the real equations (7.24), (7.25) are

(13.11) $u^{*''} - p(t)u^{*'} + [p'(t) + q(t)]u^* = f^*(t),$

(13.12) $u^{*''} - p(t)u^{*'} + [p'(t) + q(t)]u^* = 0.$

The corresponding Green's relation is

(13.13) $\int_\sigma^\tau fu^* \, dr - \int_\sigma^\tau f^*u \, dr = V_t(w(t), w^*(t))|_\sigma^\tau,$

where $w(t) = (u, u')$, $w^*(t) = (u^*, u^{*'})$, and

(13.14) $V_t(w, w^*) = [p(t)u^* - u^{*'}]u + u^*u'.$

Using the notation of Exercise 7.1, let $\mathcal{N}_\delta^*$ denote the linear set of functions

$$y^{*\delta}(t) = (u^*(t), u^{*'}(t), u^*(t + \delta), u^{*'}(t + \delta)) \in Y^*,$$

where $u^*(t)$ is a solution of (13.12). Let the "initial value" of $y^{*\delta}(t)$ be $y^{*\delta}[0] = (u^*(0), u^{*'}(0)) \in W^*$. Let $p(t)$ be a real-valued, absolutely continuous function and $q(t) \in L$ such that

(13.15) $h_{s1}(t)p(t) \in D'$ and $\sup_{s \geq 0} |h_{s1}(t)p(t)|_{D'} < \infty,$

and, for some constant C',

(13.16) $p(t) - p(s) + \int_s^t q(r) \, dr \leq C'$ for $0 \leq s \leq t \leq s + 1.$

Suppose, finally, that (B, D) is admissible for (7.24). (*a*) Then, for sufficiently small $\delta > 0$, $W_{B'}^*$ induces a total dichotomy for $\mathcal{N}_\delta^*$ with $\theta^0 = 0$. (*b*) If, in addition, either $|h_{0\Delta}|_{B'} \to \infty$ or $\Delta^{-1}(|h_{0\Delta}|_{D'} + |h_{s\Delta}p|_{D'}) \to 0$ as $\Delta \to \infty$ uniformly for large s, then $W_{B'}^*$ induces an exponential dichotomy for $\mathcal{N}_\delta^*$ for sufficiently small $\delta > 0$.

Exercise 13.2.   Let $p(t) \in M$ be a real-valued, absolutely continuous function on J and $q(t) \in L$ such that there exist constants C, C' satisfying (7.27) and (13.16). Let (B, D) be admissible for (7.24). Then, for small $\delta > 0$, (*a**)$W_{B'}^*$ induces an individual partial dichotomy for $\mathcal{N}_\delta^*$; and if $u^*(t) \notin B'$ is a solution of (13.12), then $\|y^{*\delta}(t)\| \to \infty$ as $t \to \infty$; (*a*) $W_{D'}$ induces an individual partial dichotomy for $\mathcal{N}_\delta$; and if $u(t) \notin D''$ is a solution of (7.25), then $\|y^\delta(t)\| \to \infty$ as $t \to \infty$; (*b**) if (11.9) holds, then $W_{B'}^*$ induces an individual exponential dichotomy for $\mathcal{N}_\delta^*$; (*b*) if (6.3) holds, then $W_{D'}$ induces an individual exponential dichotomy for $\mathcal{N}_\delta$.

14. Existence of *PD*-Solutions

Lemma 12.1 has the following consequence:

Theorem 14.1. *Let $A(t) \in L(\tilde{U})$ and let (B, D) be admissible for* (0.1). *If* (0.2) *has no solution $y(t) \not\equiv 0$ in $D(Y)$, then every solution $y^*(t)$ of* (13.2) *is in $B'(Y^*)$.*

For $X = Y_D$ is $\{0\}$, hence $Y_D{}^V = Y^* \subset Y_B^*$; i.e., $Y^* = Y_B^*$.

In some situations, it is easy to deduce for (0.2) the existence of solutions $y(t) \notin D(Y)$. Suppose that there is a dichotomy [or exponential dichotomy] for the solutions (e.g., let the theorems of § 6 be applicable). If all solutions $y(t)$ of (0.2) are in $D(Y)$, then all are bounded [or exponentially small] as $t \to \infty$. The same is then true of det $U(t)$ if $U(t)$ is a fundamental matrix of (0.2). Since

$$\det U(t) = [\det U(0)] \exp \int_0^t \operatorname{tr} A(s)\, ds,$$

the integral must be bounded from above [or bounded from above by a negative constant times t]. Thus if $\int^t \operatorname{tr} A(s)\, ds$ does not satisfy this condition, then not all solutions $y(t)$ of (0.2) are in $D(Y)$.

Analogues of Theorem 14.1 and the remarks following it hold if (0.1) is replaced by (0.3). This will be illustrated for scalar second order equations, first in the formally self-adjoint form:

$$(14.1) \qquad (p(t)u')' + q(t)u = f(t),$$

$$(14.2) \qquad (p(t)u')' + q(t)u = 0.$$

Let $\mathcal{C}$ denote the Banach space of complex numbers.

Theorem 14.2. *Let $p(t)$, $1/p(t)$, $q(t)$ be locally integrable complex-valued functions on $t \geq 0$. Suppose that (B, D) is admissible for* (14.1). *Then either* (14.2) *has a solution $u(t) \not\equiv 0$ in $D(\mathcal{C})$ or every solution $u(t)$ of* (14.2) *is in $B'(\mathcal{C})$.*

This is a consequence of Lemma 12.1 for if $u(t)$ is a solution of (14.1) and $v(t)$ a solution of (14.1) with $f(t)$ replaced by $g(t)$, then the corresponding Green's relation is

$$\int_\sigma^\tau (fv - gu)\, dt = [p(t)(u'(t)v(t) - u(t)v'(t))]_\sigma^\tau;$$

cf. the proof of Theorem 14.3.

A variation on the proof of Lemma 12.1 gives a result on non-self-adjoint equations

$$(14.3) \qquad u'' + p(t)u' + q(t)u = f(t),$$

$$(14.4) \qquad u'' + p(t)u' + q(t)u = 0.$$

Theorem 14.3. *Let* $p(t)$, $q(t)$ *be locally integrable complex-valued functions on* $t \geqq 0$. *Suppose that* (B, D) *is admissible for* (14.3). *Then either* (14.4) *has a solution* $0 \not\equiv u(t) \in D(\mathbb{C})$ *or every solution* $u(t)$ *of* (14.4) *satisfies* $u(t) \exp \int_0^t p(r)\, dr \in B'(\mathbb{C})$.

Proof. Suppose no solution $u(t) \not\equiv 0$ of (14.4) is in $D(\mathbb{C})$. Let $f(t) \in B_\infty(\mathbb{C})$. Then, by assumption, (14.3) has a solution $u = v(t) \in D(C)$, which is necessarily unique and vanishes for large t. Let (14.3) for $u = v$ be written as

$$(E(t)v')' + E(t)q(t)v = E(t)f(t), \qquad \text{where} \quad E(t) = \exp \int_0^t p(r)\, dr.$$

If $u(t)$ is any solution of (14.4), a corresponding relation holds with $E(t)f(t)$ replaced by 0. Thus Green's relation gives

$$\int_0^\infty E(r)f(r)u(r)\, dr = E(0)(u'(0)v(0) - u(0)v'(0)).$$

The analogue of inequalities (3.5_2) for $y(t) = (v(t), v'(t))$ gives an inequality of the type

$$\left| \int_0^\infty E(r)f(r)u(r)\, dr \right| \leqq C \, |f|_B,$$

where C depends only on C_{30}, $E(0)$, $(u(0), u'(0))$ and choice of norm on $\mathbb{C} \times \mathbb{C}$. In view of Lemma 9.2, the assertion follows.

Notes

PART I. The idea that "admissibility" of some pair (B, D) leads to some sort of a "dichotomy" for solutions of the homogeneous equation occurs in a paper of Wintner [18] on a self-adjoint equation of the second order with $B = D = L^2$ (see also Putnam [1] and Hartman [8]) and in a paper by Maizel' [1] dealing with first order systems and $B = D = L^\infty$. The main results (§ 6) on the system (0.2) are due to Massera and Schäffer [1, particularly, IV] and Schäffer [2, VI]. For the first order differential operator $Ty = y' - A(t)y$, these authors have written a series of papers treating systematically many of the questions considered in this chapter. An attempt to obtain a unified treatment for (0.1), (0.3), and for other problems (such as those involving difference-differential equations) led to the introduction of the more general operators T of § 3 in Hartman [25]. The results (§ 7) on solutions of the higher order system (0.3) are due to Hartman [25]. The procedures and arguments of Part I are based on those of Massera and Schäffer. (The papers of Massera, Schäffer, and Hartman just mentioned deal with the case when dim $Y \leqq \infty$.) The classes $\mathscr{T}$, $\mathscr{T}^{\ddagger}$ of linear spaces of §§ 1, 9 are discussed by Schäffer [1]. The definitions of dichotomies in § 1 vary somewhat from those of Massera and Schäffer. The results of § 2 are adapted from discussions in Massera and Schäffer [1, IV] and Schäffer [2, VI]. The notion of a (B, D)-manifold in connection with linear systems (0.1) (with $P = I$) is used in Schäffer [2, VI].

PART II. Most of this part of the chapter is an adaptation of results and methods of Schäffer [2, VI] dealing with first order systems (on arbitrary paired Banach spaces Y, Y'). The treatment follows that of Hartman [25]. The idea of obtaining an individual dichotomy for the "adjoint" equation as in Theorem 11.1 without using "test" functions [say as supplied by assumptions of the type $(B_1\epsilon)^*$ or $(B_2\epsilon)^*$] is due to Hartman. (It should be mentioned that "associate spaces" have been discussed by Luxemburg and Zaanen; see Schäffer [1] for references and pertinent results.)

Chapter XIV

Miscellany on Monotony

This chapter contains miscellaneous results related only by the fact that one of the main features of either the assumptions, conclusions, or proofs depends on the notion of "monotony."

Part I deals principally with linear systems of differential equations. Most of the conclusions of the theorems are to the effect that some functions of particular solutions are monotone. Some of these results, in conjunction with the theorem of Hausdorff-Bernstein, imply that certain solutions can be represented as Laplace-Stieltjes transforms of monotone functions.

Part II deals with a very special problem. It is concerned with a singular, boundary value problem related to a particular third order, nonlinear differential equation. This problem had its origins in boundary layer theory in fluid mechanics.

Part III is a discussion of the stability in the large for a trivial or periodic solution of a nonlinear autonomous system. An interesting feature of the proof of Theorem 14.2 is that it essentially reduces a d-dimensional problem to 2-dimensional considerations by dealing only with 1-parameter families of solutions at any one time.

PART I. MONOTONE SOLUTIONS

1. Small and Large Solutions

Consider a system of linear differential equations

$$(1.1) \qquad y' = A(t)y$$

for a vector $y = (y^1, \ldots, y^d)$ with real- or complex-valued components on $0 \leqq t < \omega \, (\leqq \infty)$. Let $\|y\|$ denote the Euclidean norm

$$(1.2) \qquad \|y\| = (|y^1|^2 + \cdots + |y^d|^2)^{\frac{1}{2}} \, (\geqq 0).$$

This section concerns systems (1.1) with the property that, for every solution $y(t)$, either

$$(1.3_0) \qquad \lim_{t \to \omega} \|y(t)\| < \infty \qquad \text{exists (and is finite)}$$

or

(1.3_∞) $$\lim_{t \to \omega} \|y(t)\| \leqq \infty \qquad \text{exists.}$$

(For example, a sufficient condition for (1.3_0) or (1.3_∞), respectively, is that the Hermitian part $A^H(t) = \frac{1}{2}[A(t) + A^*(t)]$ of $A(t)$ be nonpositive definite or nonnegative definite for $0 \leqq t < \omega$; so that $\|y(t)\|$ is nonincreasing or nondecreasing.)

When (1.3_0) [or (1.3_∞)] holds for all solutions, it is natural to ask whether there is a solution $y_0(t)$ satisfying

$(1.4_{0[\infty]})$ $$\lim_{t \to \omega} \|y_0(t)\| = 0 \qquad [\text{or } \lim_{t \to \omega} \|y_0(t)\| = \infty].$$

The next theorem gives an answer to this question.

Theorem 1.1$_{0[\infty]}$. *Let $A(t)$ be a $d \times d$ matrix with (complex-valued) continuous entries for $0 \leqq t < \omega \ (\leqq \infty)$ such that (1.3_0) [or (1.3_∞)] holds for every solution of (1.1). Then a necessary and sufficient [or sufficient] condition for (1.1) to have a solution $y_0(t) \not\equiv 0$ satisfying (1.4_0) [or (1.4_∞)] is that*

$(1.5_{0[\infty]})$ $$\int^t \mathrm{Re} \ \mathrm{tr} \ A(s) \ ds \to -\infty \quad [\text{or } \infty] \qquad \text{as} \quad t \to \omega.$$

Although (1.5_0) is necessary and sufficient for the existence of a solution $y_0(t)$ satisfying (1.4_0), (1.5_∞) is not necessary for the existence of a solution satisfying (1.4_∞). The second order equation $u'' + 3u/16t^2 = 0$ for $t > 0$ has the linearly independent solutions $u = t^{3/4}$ and $u = t^{1/4}$; cf. Exercise XI 1.1(c). Thus if this equation is written as a system (1.1) for the binary vector $y = (u, u')$, then every solution satisfies $\|y(t)\| \to \infty$ as $t \to \infty$. But for this system $\mathrm{tr} \ A(t) = 0$, so that (1.5_∞) does not hold.

Proof of Theorem 1.1$_0$. In this proof, "limit" means "finite limit." Since it is assumed that (1.3_0) holds for every solution, it follows that if $y_1(t)$, $y_2(t)$ is any pair of solutions of (1.1), then the limit of the scalar product $y_1(t) \cdot y_2(t)$ exists as $t \to \omega$. This follows from the relations

$$\|y_1 + y_2\|^2 = \|y_1\|^2 + 2 \ \mathrm{Re} \ y_1 \cdot y_2 + \|y_2\|^2,$$
$$\|y_1 + iy_2\|^2 = \|y_1\|^2 + \mathrm{Im} \ y_1 \cdot y_2 + \|y_2\|^2,$$

where $i^2 = -1$.

Let $Y(t)$ be a fundamental matrix of (1.1). Since the elements of the matrix product $Y^*(t) Y(t)$ are complex conjugates of the scalar products of pairs of solutions of (1.1), it follows that $C = \lim Y^*(t) Y(t)$ exists as $t \to \omega$. In particular, $\det |Y^*Y| = |\det Y|^2$ tends to a limit as $t \to \omega$.

If c is an arbitrary constant vector, the general solution of (1.1) is $y = Y(t)c$ and

$$\|y(t)\|^2 = Y(t)c \cdot Y(t)c = Y^*(t) Y(t)c \cdot c.$$

Hence

$$\lim_{t \to \omega} \|y(t)\|^2 = Cc \cdot c \geqq 0.$$

Since C is Hermitian and non-negative definite, $Cc_0 \cdot c_0 = 0$ can hold for $c_0 \neq 0$ if and only if $Cc_0 = 0$. (Note that for a Hermitian matrix C, the minimum of $Cc \cdot c$, when $\|c\| = 1$, is the least eigenvalue of C.) The equation $Cc_0 = 0$ has a solution $c_0 \neq 0$ if and only if $\det C = 0$. Thus (1.1) has a solution $y_0(t) \not\equiv 0$ satisfying (1.4_0) if and only if

$$\det C = \lim_{t \to \omega} |\det Y(t)|^2 \quad \text{is} \quad 0.$$

In view of Theorem IV 1.2,

(1.6) $$\det Y(t) = [\det Y(0)] \exp \int^t \operatorname{tr} A(s)\, ds.$$

Thus Theorem 1.1_0 follows.

Exercise 1.1.   Prove Theorem 1.1_∞.

The next theorem concerns a linear system of second order equations

(1.7) $$y'' + A(t)y = 0$$

for a vector $y = (y^1, \ldots, y^d)$.

Theorem 1.2$_{0[\infty]}$.   *Let $A(t)$ be a $d \times d$ matrix of continuous, complex-valued functions for $0 \leqq t < \omega$ $(\leqq \infty)$ with the properties that $A(t)$ is Hermitian, positive definite, and monotone (i.e.,*

$(1.8_{0[\infty]})$ $A(t) \leqq A(s)$ [or $A(t) \geqq A(s)$]   *for* $t \geqq s$

in the sense that $A(t) - A(s)$ is non-positive [or non-negative] definite). Then, if $y(t)$ is a solution of (1.7),

$(1.9_{0[\infty]})$ $d\{y \cdot y + A^{-1}y' \cdot y'\} \geqq 0$ [or $\leqq 0$]

and

$(1.10_{0[\infty]})$ $d\{Ay \cdot y + y' \cdot y'\} \leqq 0$ [or $\geqq 0$].

If, in addition,

$(1.11_{0[\infty]})$ $\det A(t) \to 0$ [or ∞]   *as* $t \to \omega$,

then (1.7) possesses a pair of ($\not\equiv 0$) solutions $y_0(t), y_1(t)$ satisfying

$(1.12_{0[\infty]})$ $y_0 \cdot y_0 + A^{-1}y_0' \cdot y_0' \to \infty$ [or 0],

$(1.13_{0[\infty]})$ $Ay_1 \cdot y_1 + y_1' \cdot y_1' \to 0$ [or ∞].

When (1.7) represents Euler-Lagrange equations in mechanics, then the expression $\{\ldots\}$ in (1.10) is essentially "energy." The first part of the theorem implies that if $A(t)$ is positive definite and monotone, then the "energy" is monotone along every solution. It will be undecided if

$y_0(t)$, $y_1(t)$ are linearly independent solutions. For the 1-dimensional case of these theorems, see § 3. The proofs will depend on some facts about Hermitian, positive definite matrices given by the following exercise.

Exercise 1.2. Let A be an Hermitian positive definite matrix. (*a*) Verify that A has an Hermitian, positive definite square root $A^{1/4}$, i.e., $(A^{1/2})^2 = A$. (*b*) Show that $A^{1/2}$ is unique. (*c*) Show that if $A = A(t)$ is a continuously differentiable function of t, then $A^{1/2}(t)$ is continuously differentiable.

Proof of Theorem 1.2$_{0[\infty]}$. Suppose first that $A(t)$ is continuously differentiable. Write (1.7) as a system of first order differential equations for a 2d-dimensional vector (y, z), where

$$(1.14) \qquad\qquad z = A^{-1/2}(t)y'$$

and $A^{-1/2} = (A^{1/2})^{-1} = (A^{-1})^{1/2}$. The resulting system is

$$(1.15) \qquad\qquad y' = A^{1/2}z, \qquad z' = -A^{1/2}y - A^{-1/2}(A^{1/2})'z.$$

It is clear that (1.7) and (1.15) are equivalent by virtue of (1.14).

A 2d-dimensional vector solution $(y(t), z(t))$ of (1.15) has the Euclidean squared length

$$(1.16) \qquad\qquad F \equiv y \cdot y + z \cdot z = y \cdot y + A^{-1}y' \cdot y'.$$

Differentiation of F with respect to t gives $F' = y' \cdot y + y \cdot y' + z' \cdot z + z \cdot z'$; or, by (1.15),

$$F' = -[A^{-1/2}(A^{1/2})' + (A^{1/2})'(A^{-1/2})]z \cdot z$$

since $A^{1/2}$ and its derivative are Hermitian. Differentiation of $(A^{1/2})^2 = A$ shows that $A^{1/2}(A^{1/2})' + (A^{1/2})'(A^{1/2}) = A'$, or

$$(1.17) \qquad\qquad A^{-1/2}(A^{1/2})' + (A^{1/2})'A^{-1/2} = A^{-1/2}A'A^{-1/2}.$$

Hence

$$F' = -A^{-1/2}A'A^{-1/2}z \cdot z = -A'(A^{-1/2}z) \cdot (A^{-1/2}z).$$

Since $A' \leq 0$ or $A' \geq 0$ according as A is nonincreasing or nondecreasing, (1.9$_{0[\infty]}$) follows from (1.16) and (1.8$_{0[\infty]}$).

In order to verify (1.10$_{0[\infty]}$), write (1.7) as a system of first order for (x, y'), where

$$(1.18) \qquad\qquad x = A^{1/2}(t)y.$$

This system is

$$(1.19) \qquad\qquad x' = (A^{1/2})'A^{-1/2}x + A^{1/2}y', \qquad y'' = -A^{1/2}x.$$

The squared (Euclidean) length of (x, y') is

$$(1.20) \qquad\qquad E \equiv x \cdot x + y' \cdot y' = Ay \cdot y + y' \cdot y'.$$

Along a solution $(x(t), y'(t))$ of (1.19), the derivative of E is

$$E' = [(A^{1/2})'A^{-1/2} + (A^{-1/2})(A^{1/2})']x \cdot x,$$

so that, by (1.17),

$$E' = A^{-1/2}A'A^{-1/2}x \cdot x = A'(A^{-1/2}x) \cdot (A^{-1/2}x).$$

Thus $(1.10_{0[\infty]})$ follows from (1.20) and $(1.8_{0[\infty]})$. This proves the first part of the theorem.

It follows that the squared Euclidean lengths $\|(y, z)\|$, $\|(x, y')\|$ of solutions of (1.15), (1.19) tend to limits ($\leq \infty$) as $t \to \infty$. The existence of the appropriate solutions $y_0(t), y_1(t)$ of (1.7) will be obtained by applying Theorem $1.1_{0[\infty]}$ to the systems (1.15), (1.19).

Let $T(t)$ be the trace of the matrix of coefficients in (1.15). Then $T(t)$ is the trace of $-A^{-1/2}(A^{1/2})'$. Hence

(1.21) $2 \operatorname{Re} T(t) = -\operatorname{tr} [A^{-1/2}(A^{1/2})' + (A^{1/2})'A^{-1/2}] = -\operatorname{tr} (A^{-1/2}A'A^{-1/2}).$

It will be shown that

(1.22) $[\log \det A(t)]' = -2 \operatorname{Re} T(t)$

for $0 \leq t < \omega$.

In order to prove (1.22), it is sufficient to consider t-values near a fixed $t = t_0$. If $A(t)$ is multiplied by a positive constant, neither side of (1.22) is affected. Hence it can be supposed that there are constants ϵ, θ such that $0 < \epsilon < \theta < 1$ and $\epsilon \|y\|^2 \leq A(t)y \cdot y \leq \theta \|y\|^2$ for all vectors y and t near t_0. In particular, $\|I - A(t)\| \leq 1 - \epsilon < 1$. Define a matrix, called $\log A(t)$, by the convergent series

(1.23) $\log A(t) = -\sum_{n=1}^{\infty} \frac{[I - A(t)]^n}{n}$

[in analogy with $\log (1 - r) = -\Sigma r^n/n$]; cf. § IV 6. This series can be differentiated term-by-term. Since $[(I - A(t))^n]' = -[A'(I - A)^{n-1} + (I - A)A'(1 - A)^{n-2} + \cdots]$ and $\operatorname{tr} CD = \operatorname{tr} DC$ for any pair of matrices,

$$[\operatorname{tr} \log A(t)]' = \operatorname{tr} \left[\sum_{n=0}^{\infty} [I - A(t)]^n A' \right].$$

This can be written as

$$[\operatorname{tr} \log A(t)]' = \operatorname{tr} A^{-1}A' = \operatorname{tr} (A^{-1/2}A'A^{-1/2}),$$

since

$$A^{-1} = \sum_{n=0}^{\infty} (I - A)^n.$$

Hence, by (1.21),

(1.24) $[\operatorname{tr} \log A(t)]' = -2 \operatorname{Re} T(t).$

It is readily verified from (1.23) that if, for a fixed t, $\lambda = \lambda(t)$ is an eigenvalue of $A(t)$ and if y is a corresponding eigenvector of $A(t)$, then

$$[\log A(t)]y = \left[-\sum_{n=1}^{\infty} \frac{(1-\lambda)^n}{n} \right] y = (\log \lambda)y.$$

Thus $\log \lambda$ is an eigenvalue of $\log A$ and y is a corresponding eigenvector. Hence if $\lambda_1, \ldots, \lambda_d$ are the eigenvalues of the (Hermitian) matrix $A(t)$, then $\log \lambda_1, \ldots, \log \lambda_d$ are those of the (Hermitian) matrix $\log A(t)$. Hence $\operatorname{tr} \log A(t) = \log \lambda_1 + \cdots + \log \lambda_d = \log(\lambda_1 \lambda_2 \ldots \lambda_d) = \log \det A(t)$ and (1.22) follows from (1.24).

Consequently,

$$(1.25) \qquad 2 \int^t \operatorname{Re} T(s)\, ds = -\log \det A(t) + \text{const.}$$

Thus the existence of $y_0(t)$ in Theorem $1.2_{0[\infty]}$ follows from Theorem $1.1_{\infty[0]}$.

If $S(t)$ is the trace of the matrix of coefficients in (1.19), then $S(t)$ is $\operatorname{tr}(A^{\frac{1}{4}})'(A^{-\frac{1}{4}})$. Thus $\operatorname{Re} S(t) = -\operatorname{Re} T(t)$ and the existence of the solution $y_1(t)$ in Theorem $1.2_{0[\infty]}$ follows from Theorem $1.1_{0[\infty]}$.

This proves Theorem $1.2_{0[\infty]}$ under the extra assumption that $A(t)$ has a continuous derivative. If this is not the case, $A(t)$ can be suitably approximated by a sequence of smooth matrix functions $A_1(t), A_2(t), \ldots$ each of which satisfies the assumptions of Theorem $1.2_{0[\infty]}$. The approximations can be made so that $A(t) - A_n(t)$ are so "small" that the solutions of (1.7) and $x'' + A_n x = 0$ are "close"; cf. § X 1. Theorem $1.2_{0[\infty]}$ then follows from a limit process.

Exercise 1.3. Let $A(t)$ satisfy the assumptions of Theorem $1.2_{0[\infty]}$. Let $B(t)$ be a continuous matrix on $0 \leq t < \omega$ ($\leq \infty$) and consider

$$y'' + B(t)y' + A(t)y = 0$$

in place of (1.7). (*a*) The assertion $(1.9_{0[\infty]})$ remains valid if, in addition to $(1.8_{0[\infty]})$, it is assumed that $B(t)A(t) + A(t)B^*(t) \leq 0$ [or ≥ 0]. (When $A(t)$ is continuously differentiable, $(1.8_{0[\infty]})$ and this condition on $BA + AB^*$ can be replaced by the single condition that $A' + BA + AB^* \leq 0$ [or ≥ 0].) Also, the assertion concerning the existence of $y_0(t)$ is valid if $(1.11_{0[\infty]})$ is replaced by

$$[\det A(t)] \exp \left[2 \int^t \operatorname{Re} \operatorname{tr} B(s)\, ds \right] \to 0 \qquad [\text{or } \infty] \qquad \text{as} \quad t \to \omega.$$

(*b*) The assertion $(1.10_{0[\infty]})$ remains valid if, in addition to $(1.8_{0[\infty]})$, it is assumed that $B + B^* \leq 0$ [or ≥ 0]. Also, the assertion concerning the existence of $y_1(t)$ is valid if $(1.11_{0[\infty]})$ is replaced by

$$[\det A(t)] \exp \left[-2 \int^t \mathrm{Re\ tr}\ B(s)\ ds \right] \to 0 \qquad [\text{or } \infty] \qquad \text{as } t \to \omega.$$

2. Monotone Solutions

In contrast to the last section, the notation $A \geq 0$ or $A > 0$ for an arbitrary (not necessarily Hermitian) matrix $A = (a_{jk})$ will mean that $a_{jk} \geq 0$ or $a_{jk} > 0$ hold for $j, k = 1, \ldots, d$. Similarly, $y \geq 0$ or $y > 0$ for a vector $y = (y^1, \ldots, y^d)$ will mean that $y^j \geq 0$ or $y^j > 0$ for $j = 1, \ldots, d$.

Theorem 2.1. *Let $A(t)$ be continuous for $0 \leq t < \omega$ ($\leq \infty$) and satisfy $A(t) \geq 0$. Then the system*

$$(2.1) \qquad\qquad y' = -A(t)y$$

has at least one solution $y(t) \not\equiv 0$ satisfying

$$(2.2) \qquad y(t) \geq 0, \qquad y'(t) \leq 0 \qquad \text{for } 0 \leq t < \omega.$$

Remark. If the interval $0 \leq t < \omega$ ($\leq \infty$) is replaced by $0 < t < \omega$ ($\leq \infty$) in both assumption and assertion, this theorem (and its corollaries) remain valid. For, if $0 < a < \omega$, then Theorem 2.1 implies the existence of a solution satisfying $0 \not\equiv y(t) \geq 0$, $y'(t) \leq 0$ for $a \leq t < \omega$. But then these inequalities also hold for $0 < t < a$; cf. the proof of the theorem.

Proof. Since $A(t) \geq 0$, it follows from (2.1) that a solution of (2.1) satisfies $y'(t) \leq 0$ on any interval on which $y(t) \geq 0$. In particular, if $0 < a < \omega$ and $y(a) > 0$, then $y(t) \geq y(a) > 0$ on $0 \leq t \leq a$.

Let $y_0 > 0$ be fixed; e.g., $y_0 = (1, \ldots, 1)$. Let $y_{a0}(t)$ be the solution of (2.1) satisfying the initial condition $y_{a0}(a) = y_0$, where $0 < a < \omega$. Thus $y_{a0}(t) \geq y_0 > 0$ on $0 \leq t \leq a$. Let $c(a) = \|y_{a0}(0)\|$, so that $c(a) \geq \|y_0\| > 0$.

Let $y_a(t) = y_{a0}(t)/c(a)$. Hence $y_a(t)$ is a solution of (2.1), $y_a(t) \geq y_0/c(a) > 0$ for $0 \leq t \leq a$, and $\|y_a(0)\| = 1$. Choose $0 < a_1 < a_2 < \cdots$ satisfying $a_n \to \omega$ as $n \to \infty$ and

$$y^0 = \lim_{n \to \infty} y_a(0), \qquad \text{where } a = a_n, \quad \text{exists.}$$

Then $\|y^0\| = 1$. In addition,

$$(2.3) \qquad y^0(t) = \lim_{n \to \infty} y_a(t), \qquad \text{where } a = a_n,$$

exists uniformly on closed intervals of $[0, \omega)$ and is the solution of (2.1) satisfying $y^0(0) = y^0$; see Corollary IV 4.1. In view of $y_a(t) \geq 0$ on $0 \leq t \leq a$, (2.3) implies that $y^0(t) \geq 0$ on $0 \leq t < \omega$. Also, $y^0(0) = y^0 \neq 0$. This proves the theorem.

Exercise 2.1.   (*a*) Let $y(t)$ be a solution supplied by Theorem 2.1. Then $y(\omega) = \lim y(t)$ as $t \to \omega$ exists and $\int^{\omega} \|y'(t)\| \, dt < \infty$. (*b*) If $y^m(\omega) > 0$ for some m, $1 \leqq m \leqq d$, and $A(t) = (a_{jk}(t))$ where $j, k = 1, \ldots, d$, then

$$(2.4) \qquad \int^{\omega} a_{jm}(t) \, dt < \infty \qquad \text{for} \quad j = 1, \ldots, d.$$

(*c*) Show that if (2.4) holds for some fixed m, it does not follow that $y^m(\omega) > 0$. (*d*) The condition (2.4) for $m = 1, \ldots, d$ is necessary and sufficient that $y(t)$ in Theorem 2.1 can be chosen so that $y(\omega) > 0$ (i.e., $y^m(\omega) > 0$ for $m = 1, \ldots, d$).

Exercise 2.2.   The following is a theorem of Perron-Frobenius: Let R be a constant $d \times d$ matrix satisfying $R \geqq 0$. Then R has at least one real, non-negative eigenvalue $\lambda \geqq 0$ and a corresponding eigenvector $y \geqq 0$, $y \neq 0$. Furthermore, if $R > 0$, then $\lambda > 0$ and $y > 0$. Deduce this from Theorem 2.1.

Corollary 2.1.   *Let $A(t)$ be completely monotone on $0 \leqq t < \infty$ [i.e., let $A(t) \in C^{\infty}$ for $t \geqq 0$ and $(-1)^n A^{(n)}(t) \geqq 0$ for $n = 0, 1, \ldots$; in other words, $A(t) \geqq 0$, $A'(t) \leqq 0$, $A''(t) \geqq 0, \ldots$]. Then (2.1) has a solution $y(t) \not\equiv 0$ which is completely monotone [i.e., $(-1)^n y^{(n)}(t) \geqq 0$ for $n = 0, 1, \ldots$] on $0 \leqq t < \infty$.*

It follows, therefore, from the theorem of Hausdorff-Bernstein that there exist monotone nondecreasing functions $\sigma^j(t)$ on $t \geqq 0$ for $j = 1, \ldots, d$, such that the components $y^j(t)$ of $y(t)$ have representations of the form

$$y^j(t) = \int_0^{\infty} e^{-ts} \, d\sigma^j(s) \qquad \text{for} \quad t \geqq 0, \quad \text{where} \quad d\sigma^j \geqq 0,$$

$j = 1, \ldots, d$. (For the theorem of Hausdorff-Bernstein, see, e.g., Widder [1].)

Exercise 2.3.   (*a*) Prove Corollary 2.1. (*b*) If $p = (p^1, \ldots, p^d)$ is a vector use the notation $py = (p^1 y^1, \ldots, p^d y^d)$. Show that the conclusion of Corollary 2.1 is true if (2.1) is replaced by

$$(2.5) \qquad p(t)y'(t) = -A(t)y,$$

where $A(t)$ satisfies the conditions of Corollary 2.1 and $p(t) \in C^{\infty}$ for $0 \leqq t < \infty$ satisfies

$$(2.6) \quad p(t) > 0 \qquad \text{and} \quad (-1)^n p^{(n+1)}(t) \geqq 0 \qquad \text{for} \quad n = 0, 1, \ldots$$

[i.e., $p(t) > 0$ and $p(t)$ has a completely monotone derivative $p'(t)$ for $t \geqq 0$].

Corollary 2.2. *In the linear differential equation,*

$$(2.7) \qquad p_0(t)u^{(d)} + \sum_{k=1}^{d} (-1)^{k+1} p_k(t) u^{(d-k)} = 0,$$

let the coefficients $p_0(t), \ldots, p_d(t)$ be continuous (real-valued) functions on $0 \leq t < \omega \ (\leq \infty)$ satisfying

$$(2.8) \quad p_0(t) > 0 \quad \text{and} \quad p_k(t) \geq 0 \quad \text{for} \quad k = 2, \ldots, d \quad \text{and} \quad 0 \leq t < \omega$$

(while $p_1(t)$ is arbitrary). Then (2.7) has at least one solution $u(t)$ satisfying

$$(2.9) \qquad u(t) > 0 \quad \text{and} \quad (-1)^n u^{(n)}(t) \geq 0$$

for $n = 0, \ldots, d - 1$. If, in addition, $p_1(t) \geq 0$, then (2.9) holds also for $n = d$.

Exercise 2.4.   Deduce Corollary 2.2 from Theorem 2.1.

For a different proof in the case $d = 2$, see Corollary XI 6.4.

Corollary 2.3. *In (2.7), let $d \geq 2$ and let the coefficient functions be of class C^∞ for $t \geq 0$, $p_0(t) > 0$ and $-p_0''(t)$, $-p_1'(t)$, $p_2(t), \ldots, p_d(t)$ completely monotone for $t \geq 0$ [so that*

$$(2.10) \quad p_0 > 0, \qquad (-1)^{n+1} p_0^{(n+2)} \geq 0,$$

$$(-1)^{n+1} p_1^{(n+1)} \geq 0 \quad \text{and} \quad (-1)^n p_k^{(n)} \geq 0$$

for $k = 2, \ldots, d$ and $n = 0, 1, \ldots, 0 \leq t < \infty$]. Then (2.7) has a solution $u(t)$ satisfying (2.9) for $n = 1, 2, \ldots$ on $0 \leq t < \infty$.

There is no condition on p_0' or p_1. In view of the theorem of Hausdorff-Bernstein, Corollary 2.3 implies that if $p_0 > 0$ and $-p_0''$, $-p_1'$, $p_2, \ldots, p_d$ have representations of the form

$$\int_0^\infty e^{-ts} \, d\sigma(s) \qquad \text{for} \quad 0 \leq t < \infty,$$

where $\sigma(s)$ is nondecreasing for $s \geq 0$, then (2.7) has a solution $u(t) > 0$ representable in this form.

Exercise 2.5.   Prove Corollary 2.3.

Exercise 2.6.   (*a*) The differential equation for the associated Legendre functions

$$(x^2 - 1)\frac{d^2 u}{dx^2} + 2x \frac{du}{dx} - \left[\nu(\nu + 1) + \frac{\mu^2}{x^2 - 1} \right] u = 0$$

is transformed into the differential equation for the toroidal functions

$$u'' + \left(\frac{\cosh t}{\sinh t} \right) u' - \left[n^2 - \frac{1}{4} + \frac{m^2}{\sinh^2 t} \right] u = 0$$

by the substitution $\nu = n - \frac{1}{2}$, $\mu = m$, $x = \cosh t$. If $n^2 \geq \frac{1}{4}$, show that this last equation has a solution $u(t) > 0$ completely monotone for $t > 0$

(and that this solution is unique up to constant factors if and only if $n^2 > \frac{1}{4}$). (b) In the hypergeometric equation

$$x(1 - x)\frac{du^2}{dx^2} + [c - (a + b + 1)]\frac{du}{dx} - abu = 0,$$

make the change of independent variables $2x - 1 = \cosh t$, where $1 < x < \infty$, or $2x - 1 = -\cosh t$, where $-\infty < x < 0$, so that $0 < t < \infty$. The resulting equation is of the form

$$u'' + p_1(t)u' - p_2(t)u = 0.$$

Show that if $ab \leq 0$ and $a + b \geq \max(2c - 1, 0)$, then there exists a completely monotone solution $u > 0$ on $t > 0$ (and that this solution is unique up to constant factors if and only if either $ab < 0$ or $a = b = 0$). (c) Kummer's form of the confluent hypergeometric equation

$$tu'' + (c - t)u' - au = 0,$$

has a completely monotone solution $u(t) > 0$ for $t > 0$ if $a \geq 0$, c arbitrary; this solution is unique up to constant factors. Also, if t is replaced by $-t$, the new equation has a completely monotone solution for $t > 0$ if $a \leq 0$ and $c \geq 0$ (and this solution is unique up to constant factors if and only if $a < 0$). (d) Whittaker's normal form of the confluent hypergeometric equation

$$u'' + \left\{-\frac{1}{4} + \frac{k}{t} - \frac{4m^2 - 1}{4t^2}\right\}u = 0$$

has a completely monotone solution $u = W_{km}(t) > 0$ for $t > 0$ if $k \leq 0$ and $m \geq \frac{1}{2}$. This solution is unique up to constant factors.

Corollary 2.2 has the following generalization in which

$$W_k(t; u_1, \ldots, u_k) = \det(u_i^{(j-1)}),$$

where $i, j = 1, \ldots, k$, denotes the Wronskian determinant of the functions $u_1, \ldots, u_k$.

Corollary 2.4. *Let m be fixed, $0 < m \leq d$. In (2.7), let the coefficients be continuous for $0 \leq t < \omega\ (\leq \infty)$ and have the properties that*

$$(2.11) \qquad p_k(t) \geq 0 \qquad \text{for} \quad k = m + 1, \ldots, d,$$

and that the mth order differential equation

$$(2.12) \qquad p_0(t)u^{(m)} + \sum_{k=1}^{m}(-1)^{k-1}p_k(t)u^{(m-k)} = 0$$

has a set of solutions $u_1(t), \ldots, u_m(t)$ such that

$$(2.13) \quad W_k(t; u_1, \ldots, u_k) > 0 \qquad \text{for} \quad k = 1, \ldots, m \quad \text{on} \quad 0 \leq t < \omega.$$

Then (2.7) *has a solution satisfying* (2.9) *for* $n = 0, 1, \ldots, d - m$ *on* $0 \leqq t < \omega$.

Exercise 2.7. Prove Corollary 2.4.

Exercise 2.8. Let $f = (f^1, \ldots, f^d)$ and $y = (y^1, \ldots, y^d)$. Assume that $f(t, y)$ is continuous for $t \geqq 0, y \geqq 0$; that $f(t, 0) \equiv 0$; and that $f(t, y) \geqq 0$. Let c be any nonnegative number. Show that $y' = -f(t, y)$ has at least one solution $y(t)$ for $t \geqq 0$ satisfying $\|y(0)\| = c$ and $y(t) \geqq 0$, $y'(t) \leqq 0$ for $t \geqq 0$.

Exercise 2.9. Let y, f be real-valued. (*a*) Assume that $f(t, y, y')$ is continuous for $t \geqq 0, y \geqq 0, y' \leqq 0$; that $f(t, 0, 0) \equiv 0$; that $f(t, y, y') \geqq 0$; and that solutions of $y'' = f(t, y, y')$ are uniquely determined by initial conditions. Show that there exists a $c_0, 0 < c_0 \leqq \infty$, such that if $0 < c < c_0$, then $y'' = f(t, y, y')$ has at least one solution $y(t)$ for $t \geqq 0$ satisfying $y(0) = c, y(t) \geqq 0$ and $y'(t) \leqq 0$ for $t \geqq 0$. This is not contained in Exercise 2.8, where the corresponding initial condition is

$$(|y(0)|^2 + |y'(0)|^2)^{1/2} = c.$$

(*b*) Show that it is not always possible to take $c_0 = \infty$ in (*a*). (*c*) Let $f(t, y, y')$ be continuous for $t \geqq 0, y \geqq 0, y' \leqq 0$; $f(t, 0, 0) = 0$; $f(t, y, 0) \geqq 0$ for $t \geqq 0, y \geqq 0$; for every $R > 0$, let there exist a positive continuous function $\varphi(z) = \varphi_R(z)$ for $z \leqq 0$ such that $\int^{\infty} u \, du / \varphi(-u) = \infty$ and $|f(t, y, z)| \leqq \varphi(z)$ for $0 \leqq t \leqq R, 0 \leqq y \leqq R, z \leqq 0$. Let $c > 0$. Show that $y'' = f(t, y, y')$ has a solution on $t \geqq 0$ satisfying $y(0) = c$ and $y(t) \geqq 0, y'(t) \leqq 0$. (This is a special case of Theorem XII 5.2 and Exercise XII 5.3.)

3. Second Order Linear Equations

This and the next section will be concerned principally with solutions of oscillatory equations (cf. § XI 6) of the form

$$(3.1) \qquad\qquad u'' + q(t)u = 0,$$

where $q(t)$ is a monotone function of t.

Theorem $3.1_{0[\infty]}$. *Let* $q(t) > 0$ *be continuous for* $0 \leqq t < \omega \ (\leqq \infty)$ *and monotone; i.e.,*

$$(3.2_{0[\infty]}) \qquad\qquad dq \leqq 0 \qquad [\text{or} \geqq 0].$$

Then, for any solution $u(t)$, *the functions* $u^2 + u'^2/q$ *and* $qu^2 + u'^2$ *are monotone, in fact,*

$$(3.3_{0[\infty]}) \qquad d\left\{u^2 + \frac{1}{q} u'^2\right\} = \frac{-u'^2 \, dq}{q^2} \geqq 0 \qquad [\text{or} \leqq 0],$$

$$(3.4_{0[\infty]}) \qquad d\{qu^2 + u'^2\} = u^2 \, dq \leqq 0 \qquad [\text{or} \geqq 0].$$

If $u(t) \not\equiv 0$ *has a* (*finite or infinite*) *set of zeros* $(0 \leqq) t_0 < t_1 < \ldots$, *then*

$(3.5_{0[\infty]})$ $t_n - t_{n-1}$ is nondecreasing [or nonincreasing] with n.

Furthermore, if

$(3.6_{0[\infty]})$ $$q(t) \to 0 \quad [\text{or } q(t) \to \infty]$$

as $t \to \omega$, *then* (3.1) *possesses linearly independent solutions* $u_0(t)$, $u_1(t)$ *satisfying, as* $t \to \omega$,

$(3.7_{0[\infty]})$ $$u_0^2 + \frac{1}{q} u_0'^2 \to \infty \quad [\text{or } 0],$$

$(3.8_{0[\infty]})$ $$q u_1^2 + u_1'^2 \to 0 \quad [\text{or } \infty].$$

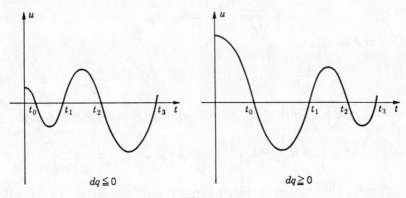

Figure 1

If (3.1) is oscillatory at $t = \omega$, then $q > 0$ implies that the graph of $u = |u(t)|$ in the (t, u)-plane consists of a sequence of convex arches. The assertion (3.3) implies that the successive "amplitudes" (i.e., maxima of $|u|$), which occur at the points where $u' = 0$, are monotone. Correspondingly, the successive maxima of $|u'|$, which occur at the points where $u'' = 0$ or, equivalently, where $u = 0$, are monotone by (3.4). See Figure 1. The Sturm comparison theorems imply (3.5) and even more:

Exercise 3.1. Let $q(t) \geqq 0$ be continuous and nonincreasing for $\tau_1 \leqq t \leqq \tau_5$ and let (3.1) have a solution $u(t)$ with exactly three zeros $t = \tau_1, \tau_3, \tau_5$, where $\tau_1 < \tau_3 < \tau_5$. (*a*) Show that $u'(t)$ has exactly two zeros $t = \tau_2, \tau_4$ satisfying $\tau_1 < \tau_2 < \tau_3 < \tau_4 < \tau_5$ and that $\tau_{j+1} - \tau_j \leqq \tau_{j+2} - \tau_{j+1}$ for $j = 1, 2, 3$. (*b*) After a reflection across a vertical line $t = \tau_{j+1}$ for $j = 1, 2$ or 3, the graph of $u = |u(t)|$ for a quarter-wave $\tau_{j+1} \leqq t \leqq \tau_{j+2}$ lies over the graph of $u = |u(t)|$ for the preceding quarter-wave $\tau_j \leqq t \leqq \tau_{j+1}$; i.e., $|u(\tau_{j+1} - t)| \leqq |u(\tau_{j+1} + t)|$ for $0 \leqq t \leqq \tau_{j+1} - \tau_j$.

The assertions (3.3), (3.4), (3.7), and (3.8) are consequences of Theorem 1.2, but slightly different arguments for their proofs will be indicated.

Except for the assertion (3.5), either of the two cases of the theorem, corresponding to (3.2) and (3.2_∞), is a consequence of the other. This can be seen from the following lemma which can be interpreted as a "duality principle" between equations (3.1) and (3.11) in which u, u', q, dt are replaced by $u', -(\text{sgn } q)u, 1/q, |q(t)| \, dt$, respectively:

Lemma 3.1. *Let $q(t) \neq 0$ be continuous for $0 \leqq t < \omega$. Introduce the new dependent variable v and independent variable s defined by*

$$(3.9) \qquad v = u', \quad ds = |q(t)| \, dt, \quad \text{and} \quad s(0) = 0.$$

Then

$$(3.10) \qquad \frac{dv}{ds} = \frac{u''}{|q|} = -(\text{sgn } q)u \, ;$$

(3.1) and the equation

$$(3.11) \qquad \frac{d^2v}{ds^2} + Q(s)v = 0, \quad \text{where} \quad Q(s) = \frac{1}{q(t)}$$

and $0 \leqq s < \int_0^\omega dt/|q(t)| \leqq \infty$, are equivalent by virtue of (3.9); finally,

$$(3.12) \quad u^2 + \frac{1}{q} u'^2 = Qv^2 + \left(\frac{dv}{ds}\right)^2, \quad qu^2 + u'^2 = v^2 + \frac{1}{Q}\left(\frac{dv}{ds}\right)^2.$$

Proof. This lemma is trivial for, by (3.10), $dv/ds + (\text{sgn } q)u = 0$. Differentiating this relation with respect to t and dividing by $|q(t)|$ gives (3.11). The relations (3.12) follow from (3.9) and (3.10).

Exercise 3.2. Find the analogue of Lemma 3.1 if (3.1) is replaced by $(p(t)u')' + q(t)u = 0$, where $p(t) > 0$, $q(t) \neq 0$.

Proof of Theorem 3.1. Note that if q is monotone, the functions $u^2 + u'^2/q$, $qu^2 + u'^2$ are clearly of bounded variation on any interval $[0, a] \subset [0, \omega)$. The relations (3.3), (3.4) follow from (3.1).

In the case $(3.2_{0[\infty]})$, the existence of a solution u_1 [or u_0] in (3.8_0) [or (3.7_∞)] implies the existence of a solution u_0 [or u_1] in (3.7_0) [or (3.8_∞)]. This can be seen as follows: Let u_0, u_1 be linearly independent solutions of (3.1). Then their Wronskian is a nonzero constant

$$u_0'u_1 - u_0u_1' = c \neq 0.$$

Since this Wronskian is

$$\left(\frac{1}{q^{1/2}} u_0'\right)(q^{1/2}u_1) - u_0u_1' = c,$$

it follows from Schwarz's inequality that

$$(3.13) \qquad 0 < c^2 \leqq \left(u_0^2 + \frac{1}{q} u_0'^2 \right) (q u_1^2 + u_1'^2).$$

Hence (3.7_∞) implies (3.8_∞) [and interchanging u_0 and u_1, it is seen that (3.8_0) implies (3.7_0)].

Thus in view of Lemma 3.1, it only remains to verify $(3.5_{0[\infty]})$ and the existence of a solution u_0 satisfying (3.7_∞) in the case (3.2_∞):

Exercise 3.3.   Assuming (3.2_∞), verify the existence of a solution $u_0 \not\equiv 0$ satisfying (3.7_∞). Apply either Theorem 1.2_∞ or apply arguments similar to those used in the proof of Theorem 1.1 directly to the quantity $u^2 + u'^2/q$ (instead of $\|y\|^2$).

Exercise 3.4.   State and prove the analogue of Theorem $3.1_{0[\infty]}$ when (3.1) is replaced by $(p(t)u')' + q(t)u = 0$.

Corollary 3.1.   *Let $q(t) > 0$ be continuous and nondecreasing for $0 \leqq t < \omega$ ($\leqq \infty$) and (3.1) oscillatory at $t = \omega$. Let $u_0(t)$ be a solution of (3.1) satisfying $u_0(t) \to 0$ as $t \to \omega$. Then*

$$(3.14) \qquad \int_0^\omega u_0(s)\, ds = \lim_{t \to \omega} \int_0^t u_0(s)\, ds \qquad converges$$

(*possibly conditionally*).

Exercise 3.5.   Verify Corollary 3.1.

Exercise 3.6.   Let $J_\mu(t)$ be the Bessel function of order μ. There exists a constant c such that

$$0 \leqq \int_0^t J_\mu(s)\, ds \leqq c \qquad \text{for all} \quad t \geqq 0 \quad \text{and} \quad \mu \geqq \tfrac{1}{2}.$$

See Lorch and Szego [S1] (or Hartman and Wilcox [1, p.239]).

Note that if the conditions $(3.2_{0[\infty]})$ of Theorem 3.1 hold for a continuous $q(t) > 0$ and $(3.6_{0[\infty]})$ does not hold, then $q(\omega) = \lim q(t)$ as $t \to \omega$ satisfies $0 < q(\omega) < \infty$. If $\omega = \infty$ and $0 < q(\infty) < \infty$, then (6.1) has a pair of solutions $u_0,\ u_1$ satisfying

$$u_0 = \cos \int_0^t q^{1/2}(s)\, ds + o(1), \qquad u_0' = -q^{1/2}(\infty) \sin \int_0^t q^{1/2}(s)\, ds + o(1),$$

$$(3.15)$$

$$u_1 = \sin \int_0^t q^{1/2}(s)\, ds + o(1), \qquad u_1' = q^{1/2}(\infty) \cos \int_0^t q^{1/2}(s)\, ds + o(1),$$

as $t \to \infty$; cf. Exercise X 17.4(a) or XI 8.4(b). In particular.

$$(3.16) \qquad u_0^2 + u_1^2 \to 1 \qquad \text{as} \quad t \to \infty.$$

This will be used in the next section.

When $q(t)$ tends monotonously to ∞ [or 0], then there exists at least one solution $u_0(t) \not\equiv 0$ which tends to 0 [or is unbounded]. When $\omega = \infty$ and $q(t)$ is of sufficiently smooth growth, then all solutions tend to 0 [or are unbounded] as can be seen from the asymptotic formula supplied, e.g., by Exercise XI 8.3 or XI 8.5.

A similar statement, without involving asymptotic integration, is given by the following exercise:

Exercise 3.7. A monotone function $H(t)$ on $a \leq t < \infty$ satisfying $H(t) \to \infty$ as $t \to \infty$ will be said to be of "irregular growth" if, for every $\epsilon > 0$, there is an unbounded sequence of t-values $a = t_0 < t_1 < \dots$ such that the open sets $B = (t_0, t_1) \cup (t_2, t_3) \cup \dots$ and $C(n) = (t_1, t_2) \cup (t_3, t_4) \cup \dots \cup (t_{2n-1}, t_{2n})$ have the properties

$$\int_B dH(t) < \infty \quad \text{and} \quad \limsup_{n \to \infty} \frac{1}{t_{2n}} \int_{C(n)} dH(t) < \epsilon.$$

If $H(t)$ is not of "irregular growth," then it will be said to be of "regular growth." Show that if $q(t)$ is continuous and satisfies $(3.2_{0[\infty]})$ and $(3.6_{0[\infty]})$, and if $|\log q(t)|$ is of "regular growth" on $a \leq t < \infty$ for large a, then all nontrivial solutions of (3.1) satisfy $(3.7_{0[\infty]})$ and $(3.8_{0[\infty]})$. See Hartman [23].

Exercise 3.8. Let $n \geq 0$. Let $q(t)$ possess $n + 1$ continuous derivatives for $t \geq 0$ satisfying $(-1)^j q^{(j+1)}(t) \geq 0$ for $j = 0, \dots, n$ and $0 < q(\infty) \leq \infty$. Let $f(t)$ have $n + 1$ continuous derivatives for $t \geq 0$ and $(-1)^j f^{(j)}(t) \geq 0$ for $j = 0, \dots, n + 1$ and $f(\infty) = 0$. Then $v'' + q(t)v = f(t)$ has a unique solution $v(t)$ such that $(-1)^j v^{(j)}(t) \geq 0$ for $j = 0, \dots, n$ and $v^{(j)}(t) \to 0$ as $t \to \infty$ for $j = 0, \dots, n + 2$. Prove this for $n = 0, 1, 2$. (The cases $n \geq 2$ are more complicated; see Hartman [22].)

Exercise 3.9. Let $0 \leq t < \infty$ and assume that (3.1) is *nonoscillatory* at $t = \infty$. For a solution $u \not\equiv 0$, put $r = u'/u$ and $E = qu^2 + u'^2$; let u_0 denote a principal solution (Theorem XI 6.4), $r_0 = u_0'/u_0$ and $E_0 = qu_0^2 + u_0'^2$. (a) Let $q(t) \geq 0$. Show that $r' \leq 0$ and $\int_t^\infty q(s)\, ds \leq r(t) \leq 1/(t - t_0)$ for large t. (b) Under the additional assumptions of Theorem 3.1_0 with $\omega = \infty$, $E \to 0$ as $t \to \infty$ for every solution $u(t)$ of (3.1) if and only if $\int^\infty tq(t)\, dt = \infty$. (c) Let $q(t) < 0$, $dq \geq 0$ [or $dq \leq 0$] and $q(\infty) = 0$ [or $q(\infty) = -\infty$]. Then, for the principal solution, $r_0 < 0$, $r_0' > 0$, $E_0 < 0$, $dE_0 \geq 0$ [or $r_0 < 0$, $r_0' < 0$, $E_0 > 0$, $dE_0 \leq 0$] for $0 \leq t < \infty$ and $r_0(\infty) = 0$, $E_0(\infty) = 0$ [or $r_0(\infty) = -\infty$, $E_0(\infty) = 0$]; and, for a nonprincipal solution, $r > 0$, $r' < 0$, $E > 0$, $dE \geq 0$ [or $r > 0$, $r' > 0$, $E < 0$, $dE \leq 0$] for large t and $r(\infty) = 0$ [or $r(\infty) = \infty$, $E(\infty) = -\infty$]. In the case $q(\infty) = 0$, $E(\infty) = \infty$ for all solutions if and only if

$-\int^{\infty} tq(t)\,dt = \infty$. (d) Let $q(t) < 0$, $dq \geqq 0$ [or $dq \leqq 0$] and $q(\infty) = -\lambda^2$, where $\lambda > 0$. Then, for the principal solutions $r_0 < -\lambda$, $r_0' > 0$, $E_0 < 0$, $dE_0 \geqq 0$ [or $r_0 > -\lambda$, $r_0' < 0$, $E_0 > 0$, $dE_0 \leqq 0$] for $0 \leqq t < \infty$ and $r_0(\infty) = -\lambda$, $E_0(\infty) = 0$ in both cases. On the other hand, $\pm \int^{\infty} \{\exp 2 \int^t [-q(s)]^{1/2}\,ds\}\,dq(t) = \infty$ is necessary and sufficient in order that the nonprincipal solutions satisfy $E(\infty) = \infty$ [or $E(\infty) = -\infty$]; in this case, $r > \lambda$, $r' < 0$, $E > 0$, $dE \geqq 0$ [or $r < \lambda$, $r' > 0$, $E < 0$, $dE \leqq 0$] for large t.

4. Second Order Linear Equations (Continuation)

This section concerns the function $r = {}'[u_0{}^2(t) + u_1{}^2(t)]^{1/2} \geqq 0$, where $u_0(t)$ and $u_1(t)$ are certain solutions of

$$(4.1) \qquad u'' + q(t)u = 0$$

when $q(t)$ is monotone. The desired result will be deduced from results analogous to those concerning (3.15), (3.16) and the following simple lemma.

Lemma 4.1. *Suppose that $q(t) > 0$ is continuous on $0 < t < \omega$ $(\leqq \infty)$, satisfies*

$$(4.2) \qquad q(t) \to 1 \qquad \text{as} \quad t \to \omega,$$

and is monotone,

$$(4.3-) \quad dq \leqq 0 \qquad \text{or} \qquad (4.3+) \quad dq \geqq 0.$$

Then there exist real-valued solutions $u_0(t), u_1(t)$ of (4.1) such that the complex-valued solution $z(t) = u_0(t) + iu_1(t)$ satisfies

$$(4.4) \qquad z \sim \exp i \int^t q^{1/2}(t)\,dt, \qquad z' \sim iz$$

as $t \to \omega$; in this case,

$$(4.5-) \qquad |z|^2 q \geqq 1 \geqq |z|^2 \quad \text{and} \quad |z'|^2 \geqq 1 \geqq \frac{1}{q}|z'|^2,$$

or

$$(4.5+) \qquad |z|^2 q \leqq 1 \leqq |z|^2 \quad \text{and} \quad |z'|^2 \leqq 1 \leqq \frac{1}{q}|z'|^2.$$

Remark 1. It will be convenient to note that in (4.5+), the inequality $|z|^2 \geqq 1$ holds for $0 < t < \omega$ if the conditions $q > 0$, $dq \geqq 0$ are relaxed to $q \leqq 0$ for $0 < t \leqq t_0$ and $q > 0$, $dq \geqq 0$ for $t_0 < t < \omega$ for some fixed t_0, $0 < t_0 < \omega$; cf. the end of the proof of the lemma.

Proof. If $\omega = \infty$, the existence of solutions u_0, u_1 satisfying (4.4) follows from Exercise X 17.4(*a*) or XI 8.4(*b*); cf. (3.15)–(3.16). If $\omega < \infty$, then $q(t)$ can be defined at $t = \omega$ so as to be continuous there and the existence of u_0, u_1 is then trivial.

Lemma 3.1 shows that the assertions (4.5±) follow when (4.5+) is proved in the case (4.3+). For a fixed φ, consider the solution of (4.1) given by

(4.6) $$u = u_0(t) \cos \varphi + u_1(t) \sin \varphi = \text{Re} \{e^{-i\varphi} z(t)\}.$$

Note that $u' = \text{Re} \{e^{-i\varphi} z'(t)\} = -\text{Im} \{e^{-i\varphi} z(t)\} + o(1)$ as $t \to \omega$, by (4.4). Hence $u^2 + u'^2 = |z|^2 + o(1) = 1 + o(1)$ as $t \to \omega$.

By Theorem 3.1_∞, assumption (4.3+) implies that (3.3_∞), (3.4_∞) hold. Since $qu^2 + u'^2 \to 1$ and $u^2 + u'^2/q \to 1$ as $t \to \omega$ by (4.2), it is seen that

(4.7) $$qu^2 + u'^2 \leq 1 \leq u^2 + \frac{1}{q} u'^2 \quad \text{for} \quad 0 < t < \omega.$$

In particular $qu^2 \leq 1$. For fixed t, choose φ so that $u(t) = |z(t)|$. This gives the first of the four inequalities in (4.5+). Also, by (4.7), $u^2(t) \geq 1$ if $u'(t) = 0$. For t fixed, choose φ so that $u'(t) = 0$. By Schwarz's inequality, $|u(t)| \leq |z(t)|$ and so $|z(t)| \geq 1$. This is the second of the inequalities in (4.5+). The last two are obtained similarly. This completes the proof of Lemma 4.1.

As to Remark 1 following Lemma 4.1, note that the Wronskian of u_0, u_1 is 1,

(4.8) $$u_0 u_1' - u_0' u_1 \equiv 1,$$

by (4.4). By the lemma, (4.5+) holds for $t \geq t_0$, so that $|z(t_0)| \geq 1$, $|z'(t_0)| \leq 1$. Choose φ so that $\cos \varphi = u_1'(t_0)/|z'(t_0)|$ and $\sin \varphi = -u_0'(t_0)/|z'(t_0)|$. Then u satisfies the initial conditions

$$u(t_0) = \frac{1}{|z'(t_0)|} \geq 1, \quad u'(t_0) = 0.$$

If $q \leq 0$ for $0 < t \leq t_0$, then an argument involving convexity shows that $u(t) > 0$ and $u''(t) \geq 0$ for $0 < t \leq t_0$. Hence $u'(t) \leq u'(t_0) = 0$ and so $u(t) \geq u(t_0) \geq 1$ for $0 < t \leq t_0$. As before, $|z(t)| \geq u(t) \geq 1$. This proves Remark 1.

Exercise 4.1. In Lemma 4.1, show that $|z|^2 q + |z'|^2 \leq 1 + q$ in both cases (4.3±).

Lemma 4.2. *Let* $q(t) > 0$ *be of class* C^2 *for* $t > t^0$ *with the properties that*

(4.9) $$Q(t) \equiv 1 + \frac{5q'^2}{16q^3} - \frac{q''}{4q^2} = 1 - \frac{1}{4q^{3/4}} \left(\frac{q'}{q^{5/4}}\right)'$$

satisfies

(4.10) $$Q(t) \to 1 \quad \text{as} \quad t \to \infty$$

and that either

(4.11−) $$dQ \leqq 0 \quad \text{for} \quad t > t^0$$

or, for some $t_0 \geqq t^0$,

(4.11+) $\quad Q \leqq 0 \quad$ for $\quad t^0 < t \leqq t_0 \quad$ and
$$Q > 0, \quad dQ \geqq 0 \quad \text{for} \quad t > t_0.$$

Then (4.1) possesses a pair of real-valued solutions $u_0(t)$, $u_1(t)$ such that $z(t) = u_0(t) + iu_1(t)$ satisfies, as $t \to \infty$,

(4.12) $\quad q^{\frac{1}{4}} z \sim \exp i \int^t Q^{\frac{1}{2}}(s) q^{\frac{1}{2}}(s)\, ds, \quad (q^{\frac{1}{4}} z)' \sim i q^{\frac{1}{2}} (q^{\frac{1}{4}} z)$

and, for $t > t^0$, either

(4.13−) $\quad q \cdot |z|^4 \leqq 1 \quad$ or $\quad$ (4.13+) $\quad q\,|z|^4 \geqq 1,$

according as (4.11−) or (4.11+) holds.

Remark 2. Note that if q is of class C^3, then Q has a continuous derivative given by

$$Q' = \frac{18qq'q'' - 15q'^3 - 4q^2q'''}{16q^4}.$$

Hence $dQ \leqq 0$ is implied by

(4.14) $$q > 0, \quad q' \geqq 0, \quad q'' \leqq 0, \quad q''' \geqq 0.$$

Remark 3. If $q(t)$ satisfies the conditions of Lemma 4.2 and if $q(t)$ is monotone for large t with $0 < q(\infty) \leqq \infty$, then (4.10) is redundant. For the monotony of Q implies that $1 - Q(t)$ does not change signs for large t, so that $q'/q^{\frac{5}{4}}$ is monotone by (4.9). Hence

$$4 \int^t |1 - Q|\, q^{\frac{3}{4}}\, dt = \pm \int^t \frac{q'\, dt}{q^{\frac{5}{4}}} = \pm \left(\text{const.} - \frac{4}{q^{\frac{1}{4}}(t)} \right)$$

tends to a limit as $t \to \infty$. Since $0 < q(\infty) \leqq \infty$ and Q is monotone for large t, the relation (4.10) holds.

Proof. By the Liouville change of variables

(4.15) $$U = uq^{\frac{1}{4}}, \quad ds = q^{\frac{1}{2}}(t)\, dt,$$

(4.1) is transformed into

(4.16) $$\frac{d^2U}{ds^2} + Q(t)U = 0, \quad \text{where} \quad t = t(s)$$

and Q is defined by (4.9); cf. § XI 2(xiii). The t-interval $t^0 < t < \infty$ is transformed into some s-interval $(-\infty \leqq) \, a < s < \omega \, (\leqq \infty)$.

By Lemma 4.1 and the Remark following it, the differential equation (4.16) has a pair of real-valued solutions $U_0(s)$, $U_1(s)$ such that $Z(s) = U_0 + iU_1$ satisfies, as $s \to \omega$,

$$Z \sim \exp i \int^s Q^{\frac{1}{2}}(s) \, ds, \qquad \frac{dZ}{ds} \sim iZ$$

and the analogues of $(4.5\pm)$ if z, z', q are replaced by Z, dZ/ds, Q. In particular $|Z|^2 \leqq 1$ or $|Z|^2 \geqq 1$ according as $(4.11-)$ or $(4.11+)$ holds. By (4.15), the equation (4.1) has the solutions $u_0 = U_0/q^{\frac{1}{4}}$, $u_1 = U_1/q^{\frac{1}{4}}$, and $z = u_0 + iu_1$; thus $q^{\frac{1}{4}}z = Z$ at $t = t(s)$. This gives Lemma 4.2.

Theorem 4.1. *Let $q(t)$ satisfy the conditions and u_0, u_1 the assertions of Lemma 4.1 with $\omega = \infty$.*

(i) *Let $(4.11-)$ hold. Then*

$$(4.17) \qquad |z| > 0, \qquad |z|'' \geqq 0 \qquad \text{for} \quad t > t^0.$$

If, in addition, $q(t)$ is continuous for $t > 0$, $q(t) \leqq 0$ for $0 \leqq t < t^0$, and $q(t) \geqq$ const. > 0 for large t, then

$$(4.18) \quad |z| > 0, \qquad |z|' \leqq 0, \qquad |z|'' \geqq 0 \qquad \text{for} \quad 0 < t < \infty.$$

(ii) *Let $(4.11+)$ hold. Then*

$$(4.19) \qquad |z| > 0, \qquad |z|' \geqq 0, \qquad |z|'' \leqq 0 \qquad \text{for} \quad t > t^0.$$

Proof. Let $r = |z| = (u_0{}^2 + u_1{}^2)^{\frac{1}{2}} \geqq 0$. Then two differentiations of r show that, by virtue of (4.1),

$$(4.20) \qquad r'' = -qr + r^{-3} = r^{-3}(1 - qr^4).$$

Since $(4.13\pm)$ mean that $1 - qr^4 \geqq 0$ or $\leqq 0$, the first and last inequalities in (4.17), (4.19) follow. Also, if $q \leqq 0$, then $r'' \geqq 0$, so that the third inequality in (4.18) holds.

From $(4.13-)$, it is seen that $r(t)$ is bounded as $t \to \infty$ if $q(t) \geqq$ const. > 0 for large t. Thus $r' \leqq 0$ follows from $r > 0$, $r'' \leqq 0$ in (4.18). Also $r' \geqq 0$ follows from $r > 0$, $r'' \leqq 0$ in (4.19). This proves the theorem.

Corollary 4.1. *Let $q(t)$ be continuous for $t > 0$ and of class C^3 for $t > t^0 \geqq 0$. Let $q \leqq 0$ for $0 < t \leqq t^0$ and (4.14) hold for $t > t^0$. Then (4.18) holds.*

Exercise 4.2. Consider Bessel's equation

$$(4.21) \qquad v'' + \frac{1}{t}v' + \left(1 - \frac{\mu^2}{t^2}\right)v = 0.$$

The variation of constants $u = t^{1/2}v$ transforms it into

$$(4.22) \qquad u'' + \left(1 - \frac{\alpha}{t^2}\right)u = 0, \qquad \text{where} \quad \alpha = \mu^2 - \tfrac{1}{4},$$

so that $\alpha \gtrless 0$ according as $(0 \leqq) \; \mu \gtrless \tfrac{1}{2}$. The real-valued solutions $v = J_\mu(t)$, $Y_\mu(t)$ of (4.21) are such that, for some real number θ,

$$(4.23) \qquad z = (\tfrac{1}{2}\pi t)^{1/2} e^{i\theta}(J_\mu - iY_\mu)$$

satisfies $z \sim e^{it}$, $z' \sim ie^{it}$ as $t \to \infty$. Use these facts in the following:

(a) Show that

$$(4.24) \quad t(J_\mu^2 + Y_\mu^2)\left(1 - \frac{\alpha}{t^2}\right) < \frac{2}{\pi} < t(J_\mu^2 + Y_\mu^2) \qquad \text{if} \quad \mu > \tfrac{1}{2},$$

$$(4.25) \quad t(J_\mu^2 + Y_\mu^2)\left(1 - \frac{\alpha}{t^2}\right) > \frac{2}{\pi} > t(J_\mu^2 + Y_\mu^2) \qquad \text{if} \quad 0 \leqq \mu < \tfrac{1}{2}.$$

(b) Furthermore,

$$(4.26) \quad (t^2 - \mu^2 + \tfrac{1}{4})^{1/2}(J_\mu^2 + Y_\mu^2) < \frac{2}{\pi} \quad \text{if} \quad t > (\mu^2 - \tfrac{1}{4})^{1/2} \quad \text{and} \quad \mu > \tfrac{1}{2},$$

$$(4.27) \quad (t^2 - \mu^2 + \tfrac{1}{4})^{1/2}(J_\mu^2 + Y_\mu^2) > \frac{2}{\pi} \quad \text{if} \quad t > 0 \quad \text{and} \quad \mu < \tfrac{1}{2}.$$

(c) The function $r = t^{1/2}(J_\mu^2 + Y_\mu^2)^{1/2} > 0$ satisfies (4.18) or (4.19) for $t^0 = 0$ according as $\mu > \tfrac{1}{2}$ or $0 \leqq \mu < \tfrac{1}{2}$. (d) Show that

$$(4.28) \qquad (t^2 - \mu^2)^{1/2}(J_\mu^2 + Y_\mu^2) < \frac{2}{\pi} \qquad \text{if} \quad t \geqq \mu \geqq 0.$$

Exercise 4.3. Let $n \geqq 0$. Let $q(t)$ be continuous for $t \geqq 0$ and possess $n + 2$ continuous derivatives satisfying $(-1)^j q^{(j+1)} \geqq 0$ for $j = 0, \ldots,$ $n + 1$ and $0 < q(\infty) \leqq \infty$. (a) Show that (4.1) has a pair of solutions $u_0(t)$, $u_1(t)$ such that $u_0'u_1 - u_0u_1' = 1$ and $w = u_0^2 + u_1^2$ satisfies $(-1)^j w^{(j)} \geqq 0$ for $j = 0, \ldots, n + 1$ and $w^{(j)} \to 0$ as $t \to \infty$ for $j = 1,$ $\ldots, n + 3$, while $w \to 1$ or $w \to 0$ as $t \to \infty$ according as $q(\infty) < \infty$ or $q(\infty) = \infty$. See Hartman [22]. (b) Let $u(t) \not\equiv 0$ be a real-valued solution of (4.1) and let its zeros be $(0 \leqq) \; t_0 < t_1 < \ldots$. Let $\Delta^{j+1}t_k = \Delta(\Delta^j t_k)$; thus $\Delta^1 t_k = \Delta t_k = t_{k+1} - t_k$, $\Delta^2 t_k = t_{k+2} - 2t_{k+1} + t_k, \ldots$. Show that $(-1)^{j+1}\Delta^j t_k \geqq 0$ for $j = 1, \ldots, n + 1$.

PART II. A PROBLEM IN BOUNDARY LAYER THEORY

5. The Problem

This part deals with a generalization of a problem in boundary layer theory in fluid mechanics. The problem concerns existence and uniqueness questions for a singular boundary value problem involving the autonomous, third order, nonlinear differential equation

$$(5.1) \qquad u''' + uu'' + \lambda(1 - u'^2) = 0,$$

and solutions on $0 \leqq t < \infty$ satisfying the boundary conditions

$$(5.2) \qquad u(0) = \alpha, \qquad u'(0) = \beta, \qquad \text{and} \quad u'(\infty) = 1,$$

where λ, α, β are constants. The problem will further be restricted to the consideration of solutions of (5.1), (5.2) satisfying

$$(5.3) \qquad 0 < u'(t) < 1 \qquad \text{for} \quad 0 < t < \infty;$$

in particular, it will be assumed that $0 \leqq \beta < 1$. [Questions of existence and uniqueness without the restriction (5.3) are not yet completely settled.]

The cases $\lambda = 0$ and $\lambda = \frac{1}{2}$ of (5.1) are often called the Blasius and Homann differential equations, respectively. As far as questions of uniqueness are concerned, the cases $\lambda = 0$, $\lambda > 0$, $\lambda < 0$ are quite different. Although, all cases can be treated in a similar manner, a simple different existence proof for the case $\lambda > 0$ will be given. The existence and uniqueness problems for the case $\lambda > 0$ will be given in § 6, for $\lambda < 0$ in § 7, and for $\lambda = 0$ in § 8. The asymptotic properties of the solutions for all cases will be given in § 9.

6. The Case $\lambda > 0$

The existence theorem in the case $\lambda > 0$ will be based on the following simple topological argument:

Lemma 6.1. *Let y, f be d-dimensional vectors and $f(t, y)$ continuous on an open (t, y)-set Ω such that solutions of initial value problems associated with*

$$(6.1) \qquad y' = f(t, y)$$

are unique. Let Ω^0 be an open subset of Ω with the properties that all egress points from Ω^0 are strict egress points and that the set Ω_e of egress points is not connected. Let Ω_i denote the set of ingress points of Ω^0 and S a connected subset of $\Omega^0 \cup \Omega_e \cup \Omega_i$ such that $S \cap (\Omega^0 \cup \Omega_i)$ contains two points (t_1, y_1), (t_2, y_2) for which the solutions $y_j(t)$ of (6.1) through (t_j, y_j) for $j = 1, 2$ leave Ω_0 with increasing t at points of different (connected) components of Ω_e. Then there exists at least one point $(t_0, y_0) \in S \cap (\Omega^0 \cup \Omega_i)$ such that the solution $y_0(t)$ of (6.1) determined by $y_0(t_0) = y_0$ remains in Ω^0 on its (open) right maximal interval of existence.

For the definition of egress, ingress, and strict egress point, see § III 8.

Proof. If the lemma is false, there exists a continuous map $\pi\colon S \to \Omega_e$

where, for $(t_0, y_0) \in S$, $\pi(t_0, y_0)$ is the first point (t, y), $t \geqq t_0$, where the solution through (t_0, y_0) meets Ω_e. The map π is continuous since every egress point of Ω^0 is a strict egress point and solutions of (6.1) depend continuously on initial conditions (Theorem V 2.1). Consequently, the connectedness of S implies that the image $\pi(S) \subset \Omega_e$ of S is connected. But this contradicts the assumption concerning the existence of (t_1, y_1), (t_2, y_2).

Theorem 6.1. *Let* $\lambda > 0$, $-\infty < \alpha < \infty$, $0 \leqq \beta < 1$. *Then there exists one and only one solution* $u(t)$ *of* (5.1), (5.2), (5.3). *This solution also satisfies*

$$(6.2) \qquad u''(t) > 0 \qquad \text{for} \quad 0 \leqq t < \infty.$$

In the application of Lemma 6.1, the following fact will be needed: the case $\beta = 1$ of (5.1)–(5.2) has the trivial solution

$$(6.3) \qquad u = \alpha + t, \qquad u' = 1, \qquad u'' = 0.$$

This will imply that the set Ω_e below is not connected.

The uniqueness proof will be given in this section for both $\lambda > 0$ and $\lambda = 0$. It will be derived from Exercise III 4.1 and uses (6.2); cf. (8.4).

Proof. *Existence for* $\lambda > 0$. Rewrite the differential equation (5.1) as a system of first order for a 3-dimensional vector $y = (y^1, y^2, y^3)$, where $y^1 = u$, $y^2 = u'$, $y^3 = u''$,

$$(6.4) \qquad y^{1\prime} = y^2, \qquad y^{2\prime} = y^3, \qquad y^{3\prime} = -y^1 y^3 - \lambda[1 - (y^2)^2].$$

Consider this equation on Ω, the entire (t, y)-space. Introduce the open (t, y)-set

$$\Omega^0 = \{(t, y): t, y^1 \text{ arbitrary}, 0 < y^2 < 1, y^3 > 0\},$$

and the boundary sets

$$\Omega^1 = \{(t, y): t, y^1 \text{ arbitrary}, 0 < y^2 < 1, y^3 = 0\},$$

$$\Omega^2 = \{(t, y): t, y^1 \text{ arbitrary}, y^2 = 1, y^3 > 0\},$$

$$\Omega^3 = \{(t, y): t, y^1 \text{ arbitrary}, y^2 = 1, y^3 = 0\},$$

$$\Omega_i = \{(t, y): t, y^1 \text{ arbitrary}, y^2 = 0, y^3 > 0\};$$

see Figure 2. It is readily verified that the set of egress points for Ω^0 is $\Omega^1 \cup \Omega^2$ and that all egress points are strict egress points. The set of ingress points is Ω_i. A solution $y(t)$ through a point (t_0, y_0), where $y_0^2 = y_0^3 = 0$ is not in Ω^0 for small $|t - t_0|$, since $y^{2\prime} = y^3 = 0$ and $y^{2\prime\prime} = y^{3\prime} = -\lambda < 0$ at $t = 0$ imply that $y^2(t) < 0$ for small $|t - t_0| \neq 0$. Note that the points of Ω^3 are neither ingress nor egress points since they are points on the trivial solutions (6.3).

Thus $\Omega_e = \Omega^1 \cup \Omega^2$ is not connected. Let $S = \{(t, y): t = 0, y = (\alpha, \beta, \gamma)$ and $\gamma > 0$ arbitrary$\}$, where α, β are fixed, $0 \leqq \beta < 1$. Thus $S \subset \Omega^0 \cup \Omega_i$ is connected. Let $y_\gamma(t)$ be the solution of (6.4) through the point $(t, y) = (0, \alpha, \beta, \gamma)$.

If $0 < \beta < 1$, the point $(t, y) = (0, \alpha, \beta, 0) \in \Omega^1$ is a strict egress point of Ω^0. This makes it clear that if $\gamma > 0$ is small, the arc $(t, y_\gamma(t))$ leaves Ω^0

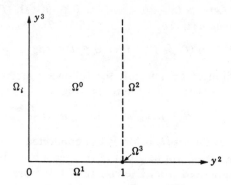

Figure 2.   Projections of Ω^j, Ω_i on (y^2, y^3)-plane.

for some $t > 0$ at a point of Ω^1. The same argument is valid for $\beta = 0$ and small $\gamma > 0$.

It will be shown that if $\gamma > 0$ is large, the solution arc $(t, y_\gamma(t))$ leaves Ω^0 through a point of Ω^2, where $y^2 = 1$. Write the third equation of (6.4) as

(6.5) $$y^{3\prime} = -(y^1 y^2)' + (y^2)^2 - \lambda[1 - (y^2)^2].$$

Along the arc $y = y_\gamma(t)$, with $(t, y_\gamma(t)) \in \bar{\Omega}^0$, the component y^2 is non-decreasing (for $y^{2\prime} = y^3 \geqq 0$). Hence

(6.6) $$y^{3\prime} \geqq -(y^1 y^2)' + \beta^2 - \lambda(1 - \beta^2).$$

A quadrature gives, for $t \geqq 0$,

$$y^3(t) \geqq \gamma + \alpha\beta - y^1(t)y^2(t) + [\beta^2 - \lambda(1 - \beta^2)]t.$$

Since $0 \leqq y^2(t) \leqq 1$ so that $\alpha \leqq y^1(t) \leqq \alpha + t$,

$$y^3(t) \geqq \gamma - |\alpha| \beta - |\alpha| - (\lambda + 1)(1 - \beta^2)t.$$

Consequently, if γ is sufficiently large, then $y^3(t)$ exceeds a given positive constant on a large t-interval $[0, t]$ as long as $(t, y_\gamma(t)) \in \bar{\Omega}^0$. Thus $y^{2\prime} = y^3$ implies that $(t, y_\gamma(t))$ leaves Ω^0 at a point where $y^2 = 1$.

By Lemma 6.1, there is a $\gamma = \gamma_0 > 0$ such that $(t, y_\gamma(t)) \in \Omega^0 \cup \Omega_i$ on its right maximal interval of existence, which is necessarily $0 \leqq t < \infty$. On this solution, $y^{2\prime} = y^3 > 0$ for $t > 0$, so that $y^2 > 0$ for large t. Also, $y^{3\prime} \leqq -\lambda[1 - (y^2)^2] < 0$ for large t shows that $\lim y^3(t)$ exists as $t \to \infty$. This limit is 0 since $y^2(t) < 1$ for all $t \geqq 0$. Consequently, $y^2 \to 1$ as $t \to \infty$ (otherwise $y^{3\prime} < -$const. < 0). This completes the existence proof.

Uniqueness for $\lambda \geqq 0$. The proof depends on the introduction of the following new variables along a solution $u = u(t)$ of (5.1) satisfying $u'(t) > 0$: Let u be the new independent variable and $z = u'^2 > 0$ the new dependent variable; so that $d/dt = u' \, d/du = z^{1/2} \, d/du$ or $d/du = z^{-1/2} \, d/dt$. Thus if a dot denotes differentiation with respect to u, then

$$(6.7) \qquad u' = z^{1/2} \geqq 0, \qquad u'' = \tfrac{1}{2}\dot{z}, \qquad u''' = \tfrac{1}{2}z^{1/2}\ddot{z}.$$

The equation (5.1) is transformed into

$$(6.8) \qquad z^{1/2}\ddot{z} + u\dot{z} + 2\lambda(1 - z) = 0, \qquad \text{where} \quad \dot{z} = dz/du;$$

the boundary conditions (5.2) into

$$(6.9) \qquad\qquad\qquad z(\alpha) = \beta^2,$$

$$(6.10) \qquad\qquad\qquad z(\infty) = 1;$$

and (5.3) into

$$(6.11) \qquad\qquad 0 < z < 1 \qquad \text{for} \quad \alpha < u < \infty.$$

Let $z(u)$ be a solution of (6.8), (6.9) with $0 < z(u) < 1$, $\dot{z}(u) > 0$ on some interval $(\alpha, u_0]$. Let $u = U(z)$ be the function inverse to $z = z(u)$. Put

$$(6.12) \qquad\qquad\qquad V(z) = \dot{z}(U(z)).$$

Then

$$(6.13) \qquad\qquad\qquad \frac{d(-U)}{dz} = -\frac{1}{V};$$

while $dV/dz = \ddot{z} \, dU/dz = \ddot{z}/V$, so that (6.8) gives

$$(6.14) \qquad\qquad \frac{dV}{dz} = -\frac{U}{z^{1/2}} - \frac{2\lambda(1 - z)}{z^{1/2}V}.$$

Then (6.13), (6.14) constitute a system of differential equations for $(-U, V)$, in which the function on the right of (6.13) is an increasing function of $V \, (> 0)$ and the function on the right of (6.14) is an increasing function of $-U$ and a nondecreasing function of V for $V > 0$ and $\lambda \geqq 0$. Hence if $(U_1(z), V_1(z))$, $(U_2(z), V_2(z))$ are any two solutions of (6.13)–(6.14) such that $U_1(\beta)^2 = U_2(\beta^2) = \alpha$, $V_1(\beta^2) > V_2(\beta^2) > 0$. Then $U_2(z) > U_1(z)$, $V_1(z) > V_2(z)$, and $U_2(z) - U_1(z)$, $V_1(z) - V_2(z)$ are increasing functions of z on any interval $\beta^2 < z < z_0$ ($\leqq 1$) on which the solutions exist; cf. Exercise III 4.1(a)–(c).

Now suppose that (5.1)–(5.3) has a pair of distinct solutions $u_1(t)$, $u_2(t)$ on $t \geqq 0$ and suppose that $u_1''(0) > u_2''(0)$. By (6.2) or (8.4), $u_j'' > 0$ for $0 \leqq t < \infty$ and $u_j'' \to 0$ as $t \to \infty$. Let $z_1(u)$, $z_2(u)$ be the corresponding solutions of (6.8) defined by (6.7); $U_1(z)$, $U_2(z)$ the functions inverse to $z_1(u)$, $z_2(u)$ and $V_1(z)$, $V_2(z)$ defined by (6.12). Thus $U_j(z)$, $V_j(z) > 0$ are defined for $\beta^2 \leqq z < 1$. Also $U_1(\beta^2) = U_2(\beta^2) = \alpha$ and $V_1(\beta^2) > V_2(\beta^2)$. Then $u_j'' \to 0$ as $t \to \infty$ implies that $V_1 - V_2 \to 0$ as $z \to 1$, but $V_1 - V_2 > 0$ is increasing with z. This contradiction establishes uniqueness.

Exercise 6.1. Let $u(t, \lambda)$ be the solution supplied by Theorem 6.1. Modify the uniqueness proof, to show that if $0 < \lambda < \mu$, then $u''(0, \lambda) < u''(0, \mu)$ and $u(t, \lambda) < u(t, \mu)$, $u'(t, \lambda) < u'(t, \mu)$ for $0 < t < \infty$.

Exercise 6.2. If $\alpha \geqq 0$, the uniqueness assertion of Theorem 6.1 follows by a variant of Exercise XII 4.6(a) applied to (6.8) where $u > \alpha \geqq 0$ for $t > 0$ [and the interval $0 \leqq t \leqq p$ of Exercise XII 4.6(a) is replaced by $0 \leqq t < \infty$].

Exercise 6.3. Let $\lambda > 0$. Show that if $-\infty < \alpha < \infty$ and $\beta > 1$, then (5.1), (5.2) has one and only one solution $u(t)$ satisfying $u'(t) > 1$ for $0 < t < \infty$. This solution also satisfies $u'' < 0$ for $0 \leqq t < \infty$ and $u'' \to 0$ as $t \to \infty$.

Exercise 6.4. Give another proof of existence in Theorem 6.1 based on the following: Put $z = 1 - u'$. Then (5.1) becomes

$$(6.15) \qquad z'' + uz' - \lambda(1 + u')z = 0, \qquad u = \alpha + t - \int_0^t z(s)\, ds.$$

Define a sequence of successive approximations by letting $z^0(t) = 1$, $u^0(t) \equiv 0$, and if $z^0, \ldots, z^{n-1}, u^0, \ldots, u^{n-1}$ have been defined, let $z^n(t)$ be a solution of

$$(6.16) \qquad z'' + u^{n-1}(t)z' - \lambda[1 + u^{n-1\,\prime}(t)]z = 0$$

satisfying $z(0) = 1 - \beta$, $z > 0$, $z' \leqq 0$ for $t \geqq 0$; cf. Corollary XI 6.4. The function $z^n(t)$ is unique and satisfies $z^n(t) \to 0$ as $t \to \infty$; see Exercise XI 6.7. Let

$$u^n(t) = \alpha + t - \int_0^t z^n(s)\, ds.$$

Show that $1 \equiv z^0 \geqq z^1 \geqq \ldots$ for $t \geqq 0$ and $z^{1\prime} \geqq z^{2\prime} \geqq \ldots$ at $t = 0$;

cf. Corollary XI 6.5. In a similar way, define a sequence of successive approximations $z_0, z_1, \ldots$, with

$$u_n = \alpha + t - \int_0^t z_n(s)$$

starting with $z_0(t) \equiv 0$. Show that $0 \equiv z_0 \leq z_1 \leq z_2 \leq \ldots$ for $t \geq 0$ and $z_1'(0) \leq z_2'(0) \leq \ldots$. Also $z_j(t) \leq z^k(t)$ for $t \geq 0$ and $j, k = 0, 1, \ldots$ and $z_j'(0) \leq z^{k'}(0)$ for $j, k = 1, 2, \ldots$. Show that the limits $z(t) = \lim z^n(t)$ and $z(t) = \lim z_n(t)$ and the last part of (6.15) define solutions of (5.1)–(5.3). (These are the same solution by uniqueness in Theorem 6.1.)

7. The Case $\lambda < 0$

In case $\lambda < 0$, the analogue of Theorem 6.1 becomes

Theorem 7.1.* *Let λ, β be fixed, $\lambda < 0$ and $0 \leq \beta < 1$. Then there exists a number $A = A(\lambda, \beta)$ and a continuous increasing function $\gamma(\alpha)$ defined for $\alpha \geq A$ with the properties that $\gamma(A) = 0$ and that if $u(t)$ is a solution of*

$$(7.1) \qquad u''' + uu'' + \lambda(1 - u'^2) = 0,$$

$$(7.2) \qquad u(0) = \alpha, \qquad u'(0) = \beta,$$

then $u'(\infty) = 1$ and (5.3) hold if and only if $\alpha \geq A$ and $0 \leq u''(0) \leq \gamma(\alpha)$; in this case,

$$(7.3) \qquad u''(t) > 0 \qquad \text{for} \quad 0 < t < \infty.$$

Thus for given $\lambda < 0$ and β on $0 \leq \beta < 1$, the problem (5.1)–(5.3) has one and only one solution if $\alpha = A$. When $\alpha < A$, there is no solution and when $\alpha > A$, there is a family of solutions. In the case $\lambda < 0$, the uniqueness proof of the last section breaks down, for the function on the right of (6.14) is not a nondecreasing function of V.

At one point, the proof will use the simple estimates for solutions of the Weber equation supplied by Exercise X 17.6; cf. Exercise XI 9.7. In the proof, the singular boundary value problem (6.8)–(6.11) will be considered. If $z = z(u)$ is a solution of this problem, a solution of problem (5.1)–(5.3) is obtained by inverting the quadrature

$$t = \int_\alpha^u z^{-\frac{1}{2}}(s) \, ds.$$

Note that the usual existence theorems apply to the differential equation (6.8) only if $z > 0$. Nevertheless, solutions of

$$(7.4) \qquad z^{\frac{1}{2}}\ddot{z} + u\dot{z} + 2\lambda(1 - z) = 0, \qquad \text{where} \quad \dot{z} = dz/du,$$

"determined" by initial conditions

$$(7.5) \qquad z(\alpha) = \beta^2, \qquad \dot{z}(\alpha) = \gamma,$$

*A(λ, β) is continuous; Hartman [S4]. For applications, Hastings [S1], Hastings and Siegel [S1], Hartman [S4].

$0 \leqq \beta < 1, \gamma \geqq 0$, will be considered (at least for small $u - \alpha > 0$) even if $\beta = 0$. This is to be understood in the sense that $u = u(t)$ is the solution of (5.1) satisfying $u(0) = \alpha$, $u'(0) = \beta \geqq 0$, $u''(0) = \frac{1}{2}\gamma \geqq 0$ and $z = z(u)$ is determined by (6.7). In the critical case $z(\alpha) = 0$, $\dot{z}(\alpha) = 0$ (i.e., $\beta = \gamma = 0$), (5.1) implies that $u''' = -\lambda > 0$, hence $u'' > 0$ for small $t > 0$, and so $u' > 0$ for small $t > 0$ and $z > 0$ for small $u - \alpha > 0$.

The proof of Theorem 7.1 will be divided into steps (a)–(k).

Proof. (a) A solution $z(u)$ of (7.4), (7.5) with $0 \leqq \beta^2 < 1$, $\gamma \geqq 0$ satisfies $\dot{z}(u) > 0$ for $u > \alpha$ as long as $0 < z(u) < 1$.

For if there is a point $u_1 > \alpha$, where $\dot{z}(u_1) = 0$, $\dot{z}(u) > 0$ for $\alpha < u < u_1$ and $0 < z(u_1) < 1$, then $\ddot{z}(u_1) \leqq 0$. But this is impossible, for by (7.4), $\ddot{z}(u_1) > 0$.

(b) Let $\beta^2 > 0$ and $\gamma > 0$. Then there exists an $\alpha^0 = \alpha^0(\lambda, \beta, \gamma)$ such that if $\alpha \geqq \alpha^0$, then the solution of (7.4), (7.5) exists on a u-interval $[\alpha^*, \alpha]$, $\dot{z}(u) > 0$ on $[\alpha^*, \alpha]$, and $z(u) = 0$ if $u = \alpha^*$.

Choose α^0 so large that $\alpha \geqq \alpha^0$ implies that

$$\ddot{z}(\alpha) = -\beta^{-1}[\alpha\gamma + 2\lambda(1 - \beta^2)] \qquad \text{is negative.}$$

Thus, by convexity, the statement concerning $z(u)$ is correct unless, in decreasing u from α, we encounter a first point u_1 where $\ddot{z}(u_1) = 0$ before $z(u)$ vanishes. It will be shown that such a point u_1 cannot exist if α^0, hence α, is sufficiently large. If u_1 does exist, then $\dot{z}(u_1) > \dot{z}(u) > \dot{z}(\alpha) = \gamma$ for $u_1 < u < \alpha$. Hence

$$\beta^2 - z(u_1) = \int_{u_1}^{\alpha} \dot{z}(u)\, du > \gamma(\alpha - u_1);$$

thus

$$z(u_1) < \beta^2 + u_1\gamma - \alpha\gamma.$$

From (7.4) and $\ddot{z}(u_1) = 0$, $u_1\dot{z}(u_1) + 2\lambda[1 - z(u_1)] = 0$, so that $u_1 > 0$ and

$$u_1\gamma < u_1\dot{z}(u_1) = 2|\lambda|\,[1 - z(u_1)] < 2|\lambda|.$$

Consequently, the last two formula lines give

$$z(u_1) < \beta^2 + 2|\lambda| - \alpha\gamma,$$

which is negative if α^0 (hence $\alpha \geqq \alpha^0$) is sufficiently large. This contradiction proves the statement (b).

(c) Let $z(u)$ be a solution of (7.4) for large u satisfying $\dot{z}(u) > 0$ and $0 < z(u) < 1$. Then $z(u) \to 1$ as $u \to \infty$.

For the proof of this, use (5.1) rather than (7.4). A differentiation of (5.1) gives

$$u'''' + uu''' + (1 - 2\lambda)u'u'' = 0.$$

By assumption, $0 < u' < 1$ and $u'' > 0$ for large t. At points where $u''' = 0$, it follows that $u'''' < 0$. Hence $u'''(t)$ can have at most one zero. Also, $u''' < 0$ for large t (for if $u''' > 0$, then $u'' > 0$ implies that u' is unbounded).

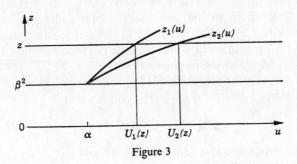

Figure 3

Suppose, if possible, that $\lim u'(t) < 1$. Then (5.1) shows that $uu'' \geqq c > 0$ for large t and some constant c. Hence $u'u'' \geqq cu'/u$, and so it follows that $\tfrac{1}{2}u'^2 > c \log u + \text{const.} \to \infty$ as $t \to \infty$. This contradiction proves (c).

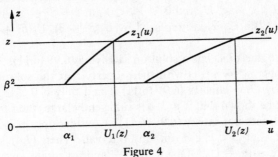

Figure 4

(d) Let $z_1(u)$, $z_2(u)$ be two solutions of (7.4) satisfying either

(7.6) $z_2(\alpha) = z_1(\alpha) = \beta^2,$ $0 \leqq \beta < 1,$ and $0 \leqq \dot{z}_2(\alpha) < \dot{z}_1(\alpha),$

with $\alpha = \alpha_1 = \alpha_2$; or

(7.7) $z_1(\alpha_1) = z_2(\alpha_2) = \beta^2,$ $0 \leqq \beta < 1$ and $\alpha_2 > \alpha_1,$

(7.8) $0 \leqq \dot{z}_2(\alpha_2) \leqq \dot{z}_1(\alpha_1);$

cf. Figures 3 and 4. Let $u = U_j(z)$ be the function inverse to $z = z_j(u)$ and $V_j = \dot{z}_j(U_j(z))$ for $j = 1, 2$. Then, as long as both solutions $z = z_j(u)$ satisfy $\beta^2 < z < 1$,

(7.9) $V_1(z) - V_2(z) > 0,$

(7.10) $U_2(z) - U_1(z) > 0$ is increasing in z.

In particular, the arcs $z = z_1(u)$ and $z = z_2(u)$ in the (u, z)-plane do not intersect for $u > \alpha_2$ as long as $\beta^2 < z_j^2 < 1$.

For a given j, (U_j, V_j) is a solution of (6.13)–(6.14). Since the right sides of these equations are increasing functions in V, $-U$, respectively, the assertions (7.9), (7.10) follow from assertion (a) and from Exercise III 4.1(a)–(c) if $V_1(\alpha_1) > 0$; i.e., $\dot{z}_1(\alpha_1) > 0$ in (7.8). The case $\dot{z}_1(\alpha_1) = \dot{z}_2(\alpha_2) = 0$ follows from continuity considerations (in which we first obtain $U_2 - U_1 > 0$ is nondecreasing in place of (7.10)).

(e) Let $z_1(u)$, $z_2(u)$ be solutions of (7.4) satisfying (7.7) and $\dot{z}_1(\alpha_1) \geqq 0$. Then there exists a positive $\epsilon = \epsilon(\alpha_1, \alpha_2, \beta)$ such that if

$$(7.11) \qquad 0 \leqq \dot{z}_2(\alpha_2) \leqq \dot{z}_1(\alpha_1) + \epsilon,$$

then the arcs $z = z_1(u)$, $z = z_2(u)$ cannot intersect for $u > \alpha_2$ as long as $\beta^2 \leqq z_1, z_2 < 1$; cf. Figure 4.

Let $z_3(u)$ be the solution of (7.4) with $z_3(\alpha_2) = \beta^2$, $\dot{z}_3(\alpha_2) = \dot{z}_1(\alpha_1)$. Let $u = U_3(z)$ be the inverse of $z = z_3(u)$ and $V_3 = \dot{z}_3(U_3(z))$. Then $U_1 < \alpha_2 < U_3$, $0 < V_3 < V_1$ on some small z-interval $\beta^2 \leqq z \leqq \beta^2 + \delta$, in particular, at $z = \beta^2 + \delta$. By continuity, if $\epsilon > 0$ is sufficiently small in (7.11), then $U_1 < \alpha_2 < U_2$ and $0 < V_2 < V_1$ at $z = \beta^2 + \delta$. Assertion (e) follows from (d) if α_1, α_2 are replaced by $U_1(\beta^2 + \delta)$, $U_2(\beta^2 + \delta)$, respectively.

(f) Suppose that (7.4) has a solution satisfying (6.9)–(6.11). Then there exists a number $\gamma^* = \gamma^*(\alpha)$ with the property that the solution of (7.4) determined by (7.5) satisfies (6.9)–(6.11) if and only if $0 \leqq \gamma \leqq \gamma^*$

It will first be shown that if $\gamma > 0$ is sufficiently large, then the solution of (7.4), (7.5) does not satisfy (6.9)–(6.11). Consider only $\alpha < u \leqq \alpha + 1$ and suppose that $0 < z < 1$ on this interval. Then (7.4) shows that $z^{1/2}\ddot{z} > -u\dot{z} \geqq -(\alpha + 1)\dot{z}$. Divide by $z^{1/2}$ and integrate over $[\alpha, u]$ to obtain $\dot{z} \geqq \gamma - 2(\alpha + 1)z^{1/2} \geqq \gamma - 2(1 + |\alpha|)$. Integrating over $\alpha \leqq u \leqq \alpha + 1$ gives the contradiction $z > 1$ at $u = \alpha + 1$ if $\gamma > 2(1 + |\alpha|) + 1$.

By assumption, there is a $\gamma \geqq 0$ such that the solution of (7.4), (7.5) satisfies (6.11). Let $\gamma^* = \sup \gamma$ taken over all such γ. It follows from (d) that solutions of (7.4), (7.5) with $0 \leqq \gamma < \gamma^*$ satisfy (6.11) and solutions of (7.4), (7.5) with $\gamma > \gamma^*$ do not satisfy (6.11). The case $\gamma = \gamma^*$ follows from the hypothesis if $\gamma^* = 0$ or from continuity considerations if $\gamma^* > 0$.

(g) There exist α_0, β_0 (with $0 < \beta_0 < 1$), $\gamma_0 > 0$ such that the solution of (7.4) determined by

$$(7.12) \qquad z(\alpha_0) = \beta_0^2, \qquad \dot{z}(\alpha_0) = \gamma_0$$

exists for $u \geqq \alpha_0$ and

$$(7.13) \qquad \beta_0^2 < z(u) < 1 \qquad \text{for} \quad \alpha_0 < u < \infty.$$

Consider a Riccati equation associated with (7.4) as follows: introduce successively

(7.14) $$w = 1 - z, \qquad r = \dot{w}/w,$$

so that (7.4) becomes

(7.15) $$\dot{r} = -r^2 + \frac{2\lambda - ur}{(1 - w)^{\frac{1}{2}}}.$$

Consider the Weber differential equation

(7.16) $$\ddot{v} + u\dot{v} - 2\lambda v = 0, \qquad \text{where} \quad \dot{v} = dv/du.$$

If s denotes the logarithmic derivative of a nontrivial solution

(7.17) $$s = \dot{v}/v,$$

the corresponding Riccati equation is

(7.18) $$\dot{s} = -s^2 + (2\lambda - us).$$

By Exercise X 17.6(a), (7.16) has a solution $v = v(u)$ such that $s = \dot{v}/v \sim -u$ as $u \to \infty$. Let $v(u) > 0$ for large u and let α_0 be so large that

(7.19) $$0 < v(u) < 1 \quad \text{and} \quad 2\lambda - us > 0 \qquad \text{for} \quad u \geqq \alpha_0 > 0.$$

Define $\beta_0 > 0$, $\gamma_0 > 0$ by

(7.20) $$\beta_0{}^2 = 1 - v(\alpha_0), \qquad \gamma_0 < -\dot{v}(\alpha_0),$$

and let $z(u)$ be the solution of (7.4) and (7.12). Thus

(7.21) $$w(\alpha_0) = 1 - z(\alpha_0) = v(\alpha_0) > 0, \qquad 0 > \dot{w}(\alpha_0) = -\dot{z}(\alpha_0) > \dot{v}(\alpha_0);$$

in particular $r(\alpha_0) > s(\alpha_0)$.

It will be verified that

(7.22) $$r(u) > s(u)$$

for all $u \geqq \alpha_0$ for which $r(u)$ exists. On any interval $\alpha_0 \leqq u \leqq \alpha_1$, where $r(u) \geqq s(u)$ holds, a quadrature shows that $w(u) \geqq v(u) > 0$ by virtue of $r = \dot{w}/w$, $s = \dot{v}/v$, and (7.21). In this case, $0 < z(u) < 1$ for $\alpha_0 \leqq u \leqq \alpha_1$ since $z(u) = 1 - w(u) < 1$.

By (7.21), $r > s$ holds at $u = \alpha_0$. Suppose, if possible, that there exists a first $u = \alpha_1 > \alpha_0$, where (7.22) fails to hold, then $\dot{s}(\alpha_1) \geqq \dot{r}(\alpha_1)$. But, by (7.18) and the last part of (7.19),

$$\dot{s}(\alpha_1) < -s^2(\alpha_1) + \frac{2\lambda - \alpha_1 s(\alpha_1)}{[1 - w(\alpha_1)]^{\frac{1}{2}}} = \dot{r}(\alpha_1).$$

This contradiction proves that (7.22) and $0 < z(u) < 1$ hold for all $u \geqq \alpha_0$.

(h) There exists a number $A_0 = A_0(\lambda, \beta)$ with the property that if $\alpha \geqq A_0$, then (7.4) has a solution satisfying (6.9)–(6.11).

Let $0 < \gamma \leqq \gamma_0$ and $\alpha_1 > \max (\alpha^0, \alpha_0)$, where $(\alpha_0, \beta_0, \gamma_0)$ is given by (g) and $\alpha^0 = \alpha^0(\lambda, \beta_0, \gamma)$ is given by (b). Consider the solution $z_1(u)$ of (7.4) determined by

$$(7.23) \qquad\qquad z(\alpha_1) = \beta_0{}^2, \qquad \dot{z}(\alpha_1) = \gamma.$$

Then, by (d), the solution $z_1(u)$ and the solution $z_0(u)$, determined by (7.12), cannot intersect. Since $z_1(u)$ increases as long as $0 < z_1(u) < 1$ by assertion (a), it follows that $z_1(u)$ exists for $u \geqq \alpha_1$ and $\beta_0{}^2 < z_1(u) < z_0(u) < 1$ for $u > \alpha_1$.

Applying (b), the solution $z_1(u)$ can be extended over an interval $[\alpha^*, \alpha_1]$ such that $z_1(u) \to 0$ as $u \to \alpha^*$. For a given $\beta, 0 \leqq \beta < 1$, there exists a unique u-value $A_0 = A_0(\lambda, \beta)$, $\alpha^* \leqq A_0 < \infty$, satisfying $z_1(A_0) = \beta^2$. Put $\gamma_1 = \dot{z}_1(A_0)$, so that $\gamma_1 > 0$ by (a) or (b) according as $\beta > 0$ or $\beta = 0$.

As before, (d) implies that the solution $z(u)$ of (7.4) satisfying $z(\alpha) = \beta^2$ and $\dot{z}(\alpha) = \gamma$, where $\alpha \geqq A_0$ and $0 \leqq \gamma \leqq \gamma_1$, exists and $0 < z(u) < z_1(u) < 1$ for $u > \alpha$. By (c), this proves (h).

(i) There exists a number $A(\lambda, \beta)$ such that the solutions of (7.4) determined by $z(\alpha) = \beta^2$, $\dot{z}(\alpha) = 0$ satisfy (6.9)–(6.11) if $\alpha \geqq A$; but if $\alpha < A$, then no solution of (7.4) satisfies (6.9)–(6.11).

Let $\alpha_* = -|2^3\lambda|^{-\frac{1}{4}}\pi$. It will first be shown that if $z(u)$ is a solution of (7.4) and (7.5), where $\alpha < \alpha_*$, then $z(u)$ assumes the value 1 for some $u \leqq \alpha - \alpha_* < 0$. In view of (d), it is sufficient to consider the case $\gamma = 0$.

Suppose, if possible, that $0 < z(u) < 1$ for $\alpha < u \leqq \alpha - \alpha_*$. Then $w = 1 - z$ satisfies $0 < w < 1$. Let $r = \dot{w}/w$ as in (7.14), so that (7.15) holds. Note that $w = 1 - \beta^2$ and $\dot{w} = -\gamma = 0$ at $u = \alpha$, so that $r = 0$ and, by (7.15), $\dot{r} = 2\lambda/\beta < 0$. As long as $\alpha \leqq u \leqq 0$, $r \leqq 0$ and $0 < w < 1$, it is seen that $\dot{r} \leqq -r^2 + 2\lambda$. Under these circumstances, $r(u) \leqq R(u)$, where $R(u) = -|2\lambda|^{\frac{1}{4}} \tan |2\lambda|^{\frac{1}{4}}(u - \alpha)$ is the solution of $\dot{R} = -R^2 + 2\lambda$, $R(\alpha) = 0$ [for $\dot{R} = -R^2 + 2\lambda$ is the Riccati equation belonging to $\ddot{v} - 2\lambda v = 0$; cf. § XI 2(xiv)]. Hence

$$\dot{w}/w \leqq -|2\lambda|^{\frac{1}{4}} \tan |2\lambda|^{\frac{1}{4}}(u - \alpha),$$

provided that $\alpha < u \leqq \alpha - \alpha_* < 0$ and $r = \dot{w}/w \leqq 0$. Clearly, the proviso $r \leqq 0$ is not needed, for $0 < |2\lambda|^{\frac{1}{4}}(u - \alpha) < \pi/2$ on this u-range. Since $\tan t$ is not integrable over $0 \leqq t < \frac{1}{2}\pi$, it follows that $w \to 0$ as $u \to u_0$ for some $u_0 \leqq \alpha - \alpha_*$. This contradiction proves the assertion.

If $\alpha = \alpha_1$ has the property that the solution $z(u)$ of (7.4) satisfying $z(\alpha) = \beta^2$, $\dot{z}(\alpha) = 0$ assumes the value 1 for some $u > \alpha_1$, then, by (d), the same is true for all $\alpha < \alpha_1$. Let $A = \sup \alpha_1$. Then $A < \infty$; in fact, $\alpha_* < A \leqq A_0$, where A_0 is given by (h).

If $\alpha > A$, then solutions of (7.4) and $z(\alpha) = \beta^2$, $\dot{z}(\alpha) = 0$ satisfy (6.9)–(6.11). By continuity, the same holds for $\alpha = A$. It is clear from (d) that if $\alpha < A$, then no solution of (7.4), (7.5) satisfies (6.9)–(6.11).

(j) The number $A(\lambda, \beta)$ in (i) is also given by $A = \inf A_0(\lambda, \beta)$ taken over the set of numbers $A_0(\lambda, \beta)$ with the property specified in (h).

In fact, $A \leqq \inf A_0(\lambda, \beta)$ is obvious from the inequality $A \leqq A_0(\lambda, \beta)$ in the last proof. Also if $\alpha_1 < \inf A_0(\lambda, \beta)$, then α_1 has the property specified in the last proof, so that the inequality $\alpha_1 < \inf A_0(\lambda, \beta)$ implies that $\alpha_1 < A$; hence $A \geqq \inf A_0(\lambda, \beta)$.

(k) **Proof of Theorem 7.1.** By the characterization of $A(\lambda, \beta)$ in (i) and (j), equation (7.4) has a solution satisfying (6.9)–(6.11) if and only if $\alpha \geqq A(\lambda, \beta)$. By ($f$), for $\alpha \geqq A(\lambda, \beta)$, there is a $\gamma^* = \gamma^*(\alpha)$ such that the solutions of (7.4), (7.5) satisfy (6.9)–(6.11) if and only if $0 \leqq \gamma \leqq \gamma^*$.

It is clear that (e) implies that $\gamma^*(\alpha)$ is an increasing function of α. In particular, $\gamma^*(\alpha) > 0$ for $\alpha > A$ [since $\gamma^*(A) \geqq 0$].

It will now be verified that $\gamma^*(A) = 0$. For suppose, if possible, that $\gamma^*(A) > 0$. Consider solutions $z_1(u)$ and $z_2(u)$ of (7.4) determined by $z_1(A) = \beta^2$, $\dot{z}_1(A) = \gamma^*(A)$, and $z_2(A) = \beta^2$, $\dot{z}_2(A) = \frac{1}{2}\gamma^*(A)$, respectively. Let $u = U_j(z)$ be the functions inverse to $z = z_j(u)$ and $V_j(z) = \dot{z}(U_j(z))$. Then $U_2(z) > U_1(z)$ and $V_1(z) > V_2(z)$ for $\beta^2 < z < 1$. Let $\alpha < A$ and $z(u)$ be the solution of (7.4) such that $z(\alpha) = \beta^2$, $\dot{z}(\alpha) = \frac{1}{2}\gamma^*(A)$ and let $U(z)$ be the inverse of $z(u)$ and $V(z) = \dot{z}(U(z))$. Let $\delta > 0$ be fixed. Then, by continuity, $U(\beta^2 + \delta) > U_1(\beta^2 + \delta)$, $V_1(\beta^2 + \delta) > V(\beta^2 + \delta)$ for small $A - \alpha > 0$. Then, by (e), $z(u)$ exists and $z(u) < z_1(u) < 1$ for large u. Consequently, $z(u)$ satisfies (6.9)–(6.11). Since $\alpha < A$, this gives a contradiction and proves $\gamma^*(A) = 0$.

The argument just completed can be used to show that $\gamma^*(\alpha - 0) = \gamma^*(\alpha)$ for $\alpha > A$. By considering solutions of (7.4) satisfying (7.5) and $z(\alpha) = \beta^2$, $\dot{z}(\alpha) = \gamma^*(\alpha)$ and applying a continuity argument, it is seen that the solution of (7.4) determined by $z(\alpha) = \beta^2$, $\dot{z}(\alpha) = \gamma^*(\alpha + 0)$ satisfies (6.9)–(6.11). Hence $\gamma^*(\alpha + 0) \leqq \gamma^*(\alpha)$. This proves the continuity of $\gamma^*(\alpha)$ for $\alpha \geqq A$. Thus Theorem 7.1 follows from the choice $\gamma(\alpha) = 2\gamma^*(\alpha)$ since $u'' = \frac{1}{2}\dot{z}$.

8. The Case $\lambda = 0$

When $\lambda = 0$, the differential equation (5.1) reduces to

$$(8.1) \qquad u''' + uu'' = 0;$$

the boundary and side conditions are the same:

$$(8.2) \qquad u(0) = \alpha, \qquad u'(0) = \beta, \qquad \text{and} \quad u'(\infty) = 1,$$

$$(8.3) \qquad 0 < u'(t) < 1 \qquad \text{for} \quad 0 < t < \infty.$$

Theorem 8.1. *If $0 < \beta < 1$, then (8.1)–(8.3) has one and only one solution for every α, $-\infty < \alpha < \infty$. If $\beta = 0$, there exists a number*

$A \leqq 0$ such that (8.1)–(8.3) has a solution if and only if $\alpha \geqq A$; in this case, the solution is unique. In either case $0 < \beta < 1$ or $\beta = 0$, the solution satisfies

$$(8.4) \qquad u''(t) > 0 \qquad for \quad 0 \leqq t < \infty.$$

Notice that the uniqueness has been proved in § 6 in the course of the proof of Theorem 6.1.

Proof. The proof will be given in steps (a)–(i), but the proofs of some steps will only be sketched because of their similarity to some of the arguments in the proofs of Theorem 6.1 and 7.1.

(a) If $u(t)$ is a solution of (8.1), then either $u''(t) \equiv 0$ or $u''(t) > 0$ or $u''(t) < 0$ for all t for which $u(t)$ exists. For (8.1) is a first order linear equation for u''; thus either $u'' \equiv 0$ or $u'' \not\equiv 0$.

(b) Let $u_\gamma(t)$ denote the solution of (8.1) satisfying the initial conditions

$$(8.5) \qquad u(0) = \alpha, \qquad u'(0) = \beta \geqq 0, \qquad and \quad u''(0) = \gamma > 0.$$

Then $u_\gamma(t)$ exists for $t \geqq 0$, and

$$(8.6) \qquad u_\gamma'(\infty) = \lim_{t \to \infty} u_\gamma'(t) \qquad exists \ and \quad u_\gamma'(\infty) > 0.$$

Since $u_\gamma'' > 0$ for all t where $u_\gamma(t)$ exists and $u_\gamma'(0) \geqq 0$, it is clear that there exists a t_0 such that $u_\gamma(t)$ exists on $0 \leqq t \leqq t_0$ and $u_\gamma(t_0) = 1$. In addition, $u_\gamma(t) \geqq 1$ for all $t \geqq t_0$ for which $u_\gamma(t)$ exists. Thus, for $t \geqq t_1 \geqq t_0$,

$$(8.7) \quad 0 \leqq u_\gamma''(t) \leqq u_\gamma''(t_0) \exp(t_0 - t),$$

$$(8.8) \quad 0 < u_\gamma'(t) - u_\gamma'(t_1) \leqq u_\gamma''(t_0)[\exp(t_0 - t_1) - \exp(t_0 - t)].$$

This makes the assertion clear.

(c) The limit $u_\gamma'(\infty)$ is a continuous function of $\gamma > 0$. It is clear that $t_0 = t_0(\gamma), {}^\alpha u_\gamma''(t_0)$ are continuous functions of γ. Thus, by (8.8), $u_\gamma'(t) \to u_\gamma'(\infty)$ uniformly as $t \to \infty$ on closed bounded intervals of $0 < \gamma < \infty$. Hence $u_\gamma'(\infty)$ is a continuous function.

(d) The limit $u_\gamma'(\infty)$ is an increasing function of $\gamma > 0$. This follows by the arguments used in the proof of uniqueness in Theorem 6.1.

(e) The problem (8.1)–(8.3) has a (unique) solution for $\alpha = 0$, $\beta = 0$. For if $u(t)$ is a solution of (8.1) and $c > 0$, then $cu(ct)$ is also a solution of (8.1). (This follows from a direct verification.) Hence, for $\alpha = \beta = 0$, $cu_1(ct) = u_\gamma(t)$ for $\gamma = c^3$, thus $u_\gamma'(\infty) = \gamma^{2/3} u_1'(\infty)$. Since $u_1'(\infty) \neq 0$, there is a unique $\gamma > 0$ such that $u_\gamma'(\infty) = 1$ and $u_\gamma(t)$ is the desired solution of (8.1)–(8.3) when $\alpha = \beta = 0$.

(f) The limit $u_\gamma'(\infty)$ tends to ∞ as $\gamma \to \infty$. For a moment, denote $u_\gamma(t)$ by $u_{\alpha\beta\gamma}(t)$ to show the dependence on α and β, as well as γ. It is clear from the arguments in step (d) in the proof of Theorem 7.1 that $u'_{\alpha\beta\gamma}(\infty)$ is a nonincreasing function of α and a nondecreasing function of β and of

γ. As in the proof of (e), $u_{\alpha 0 \alpha^3}(t) = \alpha u_{101}(\alpha t)$ for $\alpha > 0$. Hence $u'_{\alpha 0 \alpha^3}(\infty) = \alpha^2 u'_{101}(\infty) \to \infty$ as $\alpha \to \infty$. This implies (f).

(g) If, for a fixed β on $0 \leqq \beta < 1$, the problem (8.1)–(8.3) has a solution $u^0(t)$ when $\alpha = \alpha_0$, then it has a solution for $\alpha \geqq \alpha_0$.

The argument in the step (d) of the proof of Theorem 7.1 shows that the existence of a solution $\alpha = \alpha_0$ implies the existence of a solution $u_\gamma(t)$ of (8.1) with $u_\gamma'(\infty) < 1$, where $\gamma = u^{0''}(0)$. Thus steps (c), (d), (f) of this proof imply (g).

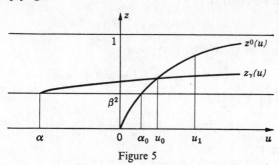

Figure 5

(h) If $\beta = 0$ and $\alpha \leqq -2$, then (8.1)–(8.3) has no solution. Consider the reduction of the problem (8.1)–(8.3) to (6.8)–(6.11) with $\lambda = 0$. The differential equation (6.8) with $\lambda = 0$ is

$$(8.9) \qquad z^{1/2}\ddot{z} + u\dot{z} = 0, \qquad \text{where } \dot{z} = dz/du,$$

and u is on the range $\alpha \leqq u < \infty$. Let $\alpha \leqq -2$ and $\alpha \leqq u \leqq -1$, so that $\ddot{z} = -u\dot{z}/z^{1/2} \geqq \dot{z}/z^{1/2}$. Hence $\dot{z} \geqq \gamma + 2z^{1/2}$ since $\beta = 0$. In particular, $\dot{z} \geqq 2z^{1/2}$ and so $z^{1/2} \geqq u - \alpha \geqq u + 2$. Consequently, z attains the value 1 on $\alpha < u \leqq -1$ for all choices of $\gamma \geqq 0$.

(i) If $0 < \beta < 1$, then (8.1)–(8.3) has a solution.

Consider the differential equation (8.9) and let $z_\gamma(t)$ be the solution of the equation corresponding to the solution $u_\gamma(t)$ of (8.1). Also, let $z^0(t)$ be the solution of (8.9) corresponding to the solution $u^0(t)$ of the problem (8.1)–(8.3) with $\alpha = \beta = 0$; cf. (e).

Let $t = t_\beta$ be the unique t-value where $u^{0'}(t) = \beta$ for $0 < \beta < 1$. Put $\alpha_0 = u^0(t_\beta)$ and let $\alpha < 0$, $u_1 > \alpha_0$; see Figure 5. Note that if $\gamma = 0$, then $u_\gamma''(t) \equiv 0$, $u_\gamma'(t) \equiv \beta > 0$, and so $z_\gamma(u) \equiv \beta^2$, $\dot{z}_\gamma(u) \equiv 0$. In particular, for small $\gamma > 0$, $z_\gamma(u_1) < z^0(u_1)$ and $\dot{z}_\gamma(u) < \dot{z}^0(u)$ for $\alpha_0 \leqq u \leqq u_1$. For such a small $\gamma > 0$, there is a u_0, $\alpha_0 < u_0 < u_1$, where $z_\gamma(u_0) = z^0(u_0)$, $\dot{z}_\gamma(u_0) < \dot{z}^0(u_0)$. Thus, by the arguments of (d) in the proof of Theorem 7.1, $z_\gamma(u) < z^0(u)$ for $u > u_0$. Consequently, $z_\gamma(\infty) \leqq 1$.

The existence assertion (i) for the fixed $\alpha < 0$ follows from (c), (d), and (f). The assertion (i) for all α (for fixed β, $0 < \beta < 1$) is a consequence of (g). This proves (i) and completes the proof of Theorem 8.1.

9. Asymptotic Behavior

In this section, the asymptotic behavior, as $t \to \infty$, of solutions of (5.1)–(5.3) will be discussed. The results will be based on the asymptotic integrations of second order, linear differential equations.

If $u(t)$ is a solution of (5.1), put

$$(9.1) \qquad h(t) = 1 - u'(t).$$

Then $h(t)$ satisfies the differential equation

$$(9.2) \qquad h'' + u(t)h' - \lambda[1 + u'(t)]h = 0.$$

Differentiating (9.2) gives

$$(9.3) \qquad h''' + u(t)h'' + (1 - 2\lambda)u'(t)h' = 0$$

since $h' = -u''$.

In order to eliminate the middle term in (9.2), put

$$(9.4) \qquad h = x \exp -\tfrac{1}{2}\int_0^t u(s)\,ds,$$

so that x satisfies

$$(9.5) \qquad x'' - q(t)x = 0,$$

where

$$(9.6) \quad q(t) = \tfrac{1}{4}u^2 + (\lambda + \tfrac{1}{2})u' + \lambda = \tfrac{1}{4}u^2\left\{1 + \frac{2[(2\lambda + 1)u' + 2\lambda]}{u^2}\right\};$$

cf. (XI 1.9)–(XI 1.10). Thus

$$q' = \tfrac{1}{2}uu' + (\lambda + \tfrac{1}{2})u''$$

and, by (5.1),

$$q'' = -\lambda uu'' + u'^2(\tfrac{1}{2} + \tfrac{1}{2}\lambda + \lambda^2) - \lambda(\lambda + \tfrac{1}{2}).$$

Since $0 < u' < 1$, $u'' > 0$ and $u' \sim 1$, $u \sim t$ as $t \to \infty$, there is a constant C such that for large t

$$\frac{q'^2}{q^{5/2}} \leq C\left(\frac{1}{t^3} + \frac{u''^2}{t^5}\right), \qquad \frac{|q''|}{q^{3/2}} \leq C\left(\frac{u''}{t^2} + \frac{1}{t^3}\right).$$

In addition, $\int^\infty u''\,dt$ is (absolutely) convergent (since $u'(t) \to 1$ as $t \to \infty$), so that

$$(9.7) \qquad \int^\infty \frac{q'^2\,dt}{q^{5/2}} < \infty \qquad \text{and} \qquad \int^\infty \frac{|q''|\,dt}{q^{3/2}} < \infty$$

provided that

(9.8)
$$\int^{\infty} \frac{u''^2\, dt}{t^5} < \infty.$$

It is easy to check (9.8), for an integration by parts (integrating u'' and differentiating u''/t^6) gives

$$\int \frac{u''^2\, dt}{t^5} = \frac{u'u''}{t^5} + \int \frac{u'}{t^5}\left[uu'' + \lambda(1 - u'^2) + \frac{5u''}{t}\right] dt,$$

by (5.1). The last integral is absolutely convergent and $\liminf u''(t) = 0$ as $t \to \infty$. Thus (9.8) holds.

Consequently, (9.7) holds, and thus (9.5) has a principal solution $x(t)$ satisfying, as $t \to \infty$,

(9.9)
$$x \sim cq^{-1/4}(t) \exp\left(-\int^t q^{1/2}(s)\, ds\right),$$

where $c \neq 0$ is a constant, while linearly independent solutions satisfy

(9.10)
$$x \sim cq^{-1/4}(t) \exp\int^t q^{1/2}(s)\, ds;$$

cf. Exercise XI 9.6.

From the last part of (9.6) and $u \sim t$,

$$q^{1/2}(t) = \tfrac{1}{2}u + \tfrac{1}{2}(1 + 2\lambda)\left(\frac{u'}{u}\right) + \frac{\lambda}{u} + O\left(\frac{1}{t^3}\right), \qquad q^{1/4}(t) \sim (\tfrac{1}{2}t)^{1/2};$$

hence

$$\int^t q^{1/2}(s)\, ds = \tfrac{1}{2}\int^t u\, dt + (\tfrac{1}{4} + \lambda)\log u + \lambda\int^t \frac{dt}{u} + c^0 + o(1),$$

where c^0 is a constant. Thus (9.9), (9.10) become

$$x \sim ct^{-1-\lambda}\exp\left(-\int^t \left(\tfrac{1}{2}u + \frac{\lambda}{u}\right) dt\right),$$

$$x \sim ct^\lambda \exp\left(\int^t \left(\tfrac{1}{2}u + \frac{\lambda}{u}\right) dt\right).$$

In view of (9.4), the equation (9.2) has a principal solution satisfying

(9.11)
$$h \sim ct^{-1-\lambda}\exp\left(-\int^t \left(u + \frac{\lambda}{u}\right) ds\right), \qquad c \neq 0,$$

while linearly independent solutions satisfy

(9.12)
$$h \sim ct^\lambda \exp\int^t \frac{\lambda\, ds}{u}, \qquad c \neq 0.$$

By treating (9.3) as a second order equation for h' in the same way that (9.2) was handled, it is seen that (9.3) has principal solutions satisfying

$$(9.11')   \qquad h' = c't^{-2\lambda} \exp\left(-\int^t u\, ds\right), \qquad c' \neq 0,$$

and that the linearly independent solutions satisfy

$$(9.12')   \qquad\qquad\qquad h' = c't^{-1+2\lambda}, \qquad c' \neq 0,$$

as $t \to \infty$.

If (9.1) satisfies (9.11), then, since $u \sim t$, it follows that $\int^\infty th\, dt < \infty$; thus

$$u = t + c_1 + o(1), \qquad \int^t u\, ds = \tfrac{1}{2}t^2 + c_1t + c_2 + o(1)$$

as $t \to \infty$. Substituting this into (9.11), (9.11') gives

$$(9.13)   \qquad 1 - u' \sim c_0 t^{-1-2\lambda} \exp\left(-\tfrac{1}{2}t^2 - c_1t\right), \qquad u'' \sim t(1 - u')$$

as $t \to \infty$, where $c_0 > 0$, c_1 are constants.

If (9.1) satisfies (9.12), then $u \sim t$ implies that $h \equiv 1 - u' \sim ct^{2\lambda+o(1)}$ as $t \to \infty$. Hence $u(t) = t + O(t^{2\lambda+1+\epsilon})$ as $t \to \infty$ for all $\epsilon > 0$. If this is substituted into (9.12), (9.12') and if it is supposed that $\lambda < 0$ (and $2\lambda + \epsilon < 0$), then

$$(9.14)   \qquad\qquad 1 - u' \sim c_0 t^{2\lambda}, \qquad u'' \sim -2\lambda c_0 t^{-1+2\lambda}$$

as $t \to \infty$, where $c_0 > 0$ is a constant.

Theorem 9.1. *Let* $\lambda \geqq 0$ *and let* $u(t)$ *be a solution of* (5.1)–(5.3). *Then there exist constants* $c_0 > 0$, c_1 *such that* (9.13) *holds as* $t \to \infty$.

Proof. For a given $u(t)$, it has to be decided whether $h = 1 - u'$ satisfies (9.11), (9.11') or (9.12), (9.12'). If $\lambda \geqq 0$, (9.12) cannot hold, for otherwise $h = 1 - u' \to 0$, $t \to \infty$ fails to hold. Thus (9.11), (9.11') are valid and, as was seen, this gives (9.13).

Theorem 9.2. *Let* $\lambda < 0$, $0 \leqq \beta < 1$, $\alpha \geqq A(\lambda, \beta)$, *where* $A(\lambda, \beta)$, $\gamma(\alpha)$ *are given by Theorem* 7.1. *Let* $u(t)$ *be a solution of* (5.1)–(5.3). *Then there exist constants* $c_0 > 0$, c_1 *such that* (9.13) *holds if and only if* $u''(0) = \gamma(\alpha)$; *for other solutions* $u(t)$ *of* (5.1)–(5.3), *with* $\alpha > A(\lambda, \beta)$ *and* $0 \leqq u''(0) < \gamma(\alpha)$, *the asymptotic relations* (9.14) *hold (with a suitable constant* $c_0 > 0$).

Proof. (a) If $u^*(t)$ is the solution of (7.1), (7.2) and $u^{*\prime\prime}(0) = \gamma(\alpha)$, then (9.13) holds.

Using the notation of the proof of (g) in § 7, let $z^*(u)$ be related to $u^*(t)$ by (6.7) and let $v(u)$ be a solution of Weber's equation (7.16) satisfying $\dot{v}/v \sim -u$ as $u \to \infty$ and $v(u) > 0$ for large u. Let $r^*(u) = -\dot{z}^*/(1 - z^*)$ and $s(u) = \dot{v}/v$; cf. (7.14), (7.17).

Then, for large u, $r^*(u) \leqq s(u)$. For suppose that $r^*(u) > s(u)$ for some large $u = u_0$. In this case, $r(u) > s(u)$ for $u = u_0$ if $z(u) = -\dot{z}/(1 - z)$

belong to a solution of (7.1), (7.2) with $u''(0) = \gamma(\alpha) + \epsilon$ for small $|\epsilon|$. But then, as in the proof of (g) in § 7, it follows that $r(u) > s(u)$ for all $u \geqq u_0$ and that $u(t)$ satisfies (5.1)–(5.3). This contradicts the main property of $\gamma(\alpha)$.

Hence $r^*(u) \leqq s(u)$ for large u, and so $1 - z^*(u) \leqq c^*v(u)$ for large u and some constant $c^* > 0$. Since $\log v(u) \sim -\frac{1}{2}u^2$ as $u \to \infty$, it follows that $h = 1 - u^{*\prime}$ cannot satisfy (9.14) and therefore satisfies (9.11). This gives (9.13).

(b) The problem (5.1)–(5.3) cannot have two distinct solutions satisfying (9.13).

Suppose, if possible, that there exist two solutions $u_1(t)$, $u_2(t)$ of (5.1)–(5.3) satisfying (9.13) and, say $u_1''(0) > u_2''(0)$. Let $z_j(u)$ be the solution of (7.4) corresponding to $u_j(t)$ by virtue of (6.7) for $j = 1, 2$. Let $U_j(z)$ be the function inverse to $z = z_j(u)$ and $V_j(z) = \dot{z}_j(U_j(z))$.

Then $z_1(u)$, $z_2(u)$ satisfy (7.6) and, by (d) in § 7, the assertions (7.9), (7.10) hold. By (9.13), $u_j''(t) \to 0$ and $u_j''(t) \sim t(1 - u_j')$ as $t \to \infty$. By virtue of (6.7) and $u_j(t) \sim t$ as $t \to \infty$, the latter relation implies that $\dot{z}_j \sim 2u(1 - z_j^{1/2})$ as $u \to \infty$. Or, since $1 - z_j^{1/2} = (1 - z_j)/(1 + z_j^{1/2}) \sim \frac{1}{2}(1 - z_j)$ as $u \to \infty$, we have $\dot{z}_j \sim u(1 - z_j)$ as $u \to \infty$. Thus $V_j \sim U_j(1 - z)$, also $V_j \to 0$ as $z \to 1$ [since $u_j''(t) \to 0$ as $t \to \infty$].

The functions $U = U_j$, $V = V_j$ satisfy the differential equation (6.14). Hence

$$\frac{dV_j}{dz} = -\frac{U_j}{z^{1/2}} + O\left(\frac{1}{U_j}\right) \qquad \text{as} \quad z \to 1.$$

Consequently, as $z \to 1$,

$$\frac{d(V_1 - V_2)}{dz} = \frac{(U_2 - U_1)}{z^{1/2}} + O\left(\frac{1}{U_1}\right) + O\left(\frac{1}{U_2}\right).$$

By (7.10), $U_2(z) - U_1(z) > 0$ is increasing, and so there exists a constant $c > 0$ such that $U_2(z) - U_1(z) \geqq c > 0$ for z near 1. Also $U_j(z) \to \infty$ as $z \to 1$. Therefore, $d(V_1 - V_2)/dz \geqq \frac{1}{2}c > 0$ for z near 1, so that $V_1(z) - V_2(z)$ is increasing for z near 1. Since $V_1(z) - V_2(z) > 0$ by (7.9), this contradicts the fact that $V_j(z) \to 0$ as $z \to 1$ and proves (b) and Theorem 9.2.

PART III. GLOBAL ASYMPTOTIC STABILITY

10. Global Asymptotic Stability

Consider a real autonomous system of differential equations

$$(10.1) \qquad y' = f(y)$$

in which solutions are uniquely determined by initial conditions. Let

$y_0(t)$ be a solution for $t \geqq 0$. This solution is said to be globally asymptotically stable when the system (10.1) has the property that if $y(t)$ is a solution for small $t \geqq 0$, then $y(t)$ exists for all $t \geqq 0$ and $y(t) - y_0(t) \to 0$ as $t \to \infty$.

This contrasts with the notion of asymptotic stability of § III 8 in that it is not assumed here that the initial point $y(0)$ is near the initial point $y_0(0)$ of $y_0(t)$. It will often be assumed that

$$(10.2) \qquad\qquad f(0) = 0$$

and that $y_0(t)$ is the solution $y_0(t) \equiv 0$, as in § III 8.

Let the function $f(y)$ have continuous first order partial derivatives and let $J(y)$ denote the Jacobian matrix $(\partial f/\partial y) = (\partial f^j/\partial y^k)$, where $j, k = 1, \dots, d$. The criteria for global asymptotic stability to be obtained below reduce in simple cases to conditions involving one of the two inequalities

$$(10.3) \qquad\qquad J(y)f(y) \cdot f(y) \leqq 0$$

or

$$(10.4) \qquad\qquad J(y)x \cdot x \leqq 0 \qquad \text{if} \quad x \cdot f(y) = 0,$$

where a dot denotes scalar multiplication. It is very curious that both conditions (10.3) and (10.4), which in a certain sense are complementary, lead to stability.

The condition (10.3) which states that $J(y)x \cdot x \leqq 0$ whenever the vector x is in the direction of $\pm f(y)$ can be replaced by the condition that $J(y)f(y) \cdot x \leqq 0$ whenever $x = Gf(y)$, and G is a constant $d \times d$, positive definite Hermitian matrix; i.e., by

$$(10.5) \qquad\qquad GJ(y)f(y) \cdot f(y) \leqq 0.$$

Correspondingly, (10.4) can be replaced by

$$(10.6) \qquad\qquad GJ(y)x \cdot x \leqq 0 \qquad \text{if} \quad Gx \cdot f(y) = 0,$$

where G is the same as in (10.5). Actually, the conditions (10.5), (10.6) are not more general than (10.3), (10.4) in the following sense:

Exercise 10.1. In (10.1), let $f(y)$ be a function of class C^1 satisfying (10.5) [or (10.6)], where $G = G^*$ is positive definite. Let $G^{1/2}$ be the self-adjoint, square root of G; cf. Exercise 1.2. Introduce the new dependent variable $z = G^{1/2}y$ in (10.1) and show that the resulting system for z satisfies the analogue of (10.3) [or (10.4)].

The general criteria to be obtained will actually be generalizations of (10.3) or (10.4) involving nonconstant, positive definite Hermitian matrices $G(y)$.

11. Lyapunov Functions

Recall that if $V(y)$ is a real-valued function having continuous partial derivatives, then its trajectory derivative $\dot{V}(y)$ with respect to the system

$$(11.1) \qquad y' = f(y)$$

is given by the scalar product

$$(11.2) \qquad \dot{V}(y) = f(y) \cdot \operatorname{grad} V(y).$$

Lemma 11.1. *Let $f(y)$ be continuous on an open set E and such that solutions of (11.1) are uniquely determined by initial conditions. Let $V(y)$ be a real-valued function on E with the following properties:* (i) $V \in C^1$ *on E;* (ii) $V(y)$ *and its trajectory derivative $\dot{V}(y)$ satisfy*

$$(11.3) \qquad \dot{V}(y) \leqq 0$$

on E. Let $y(t)$ be a solution of (11.1) for $t \geqq 0$. Then the ω-limit points of $y(t)$, $t \geqq 0$, in E, if any, are contained in the set $E_0 = \{y : \dot{V}(y) = 0\}$.

Proof. Let $t_n < t_{n+1} \to \infty$, $y(t_n) \to y_0$ as $n \to \infty$ and $y_0 \in E$. Then $V(y(t_n)) \to V(y_0)$ as $n \to \infty$ and $V(y(t)) \geqq V(y_0)$ for $t \geqq 0$. Suppose, if possible, that $y_0 \notin E_0$, so that $\dot{V}(y_0) < 0$. Let $y_0(t)$ be the solution of (11.1) satisfying $y_0(0) = y_0$. Consider $y_0(t)$ for $0 \leqq t \leqq \epsilon$, where $\epsilon > 0$ is small. Then $V(y_0(t)) < V(y_0)$ for $0 < t \leqq \epsilon$.

The continuous dependence of solutions on initial values (Theorem V 2.1) implies that $\|y(t + t_n) - y_0(t)\|$ is small for $0 \leqq t \leqq \epsilon$ and large n. Hence $|V(y(t + t_n)) - V(y_0(t))|$ is small for $0 \leqq t \leqq \epsilon$ and large n. In particular $V(y(t_n + \epsilon)) < V(y_0)$ for large n. But this contradicts $V(y(t)) \geqq V(y_0)$ for $t \geqq 0$ and shows that $y_0 \in E_0$.

Corollary 11.1. *Let f, V be as in Lemma 11.1, where E is the y-space, and let $V(y) \to \infty$ as $\|y\| \to \infty$. Then all solutions $y = y(t)$ of (11.1) starting at $t = 0$ exist for $t \geqq 0$ and are bounded [in fact, $y = y(t)$ is in the set $\{y : V(y) \leqq V(y(0))\}$ for $t \geqq 0$]. If, in addition, there exists a unique point y_0, where $\dot{V}(y_0) = 0$ (i.e., if E_0 reduces to the point y_0), then $\|f(y)\| \gtreqqless 0$ according as $\|y - y_0\| \gtreqqless 0$, and the solution $y_0(t) \equiv y_0$ of (11.1) is globally asymptotically stable.*

Exercise 11.1. Verify Corollary 11.1.

Corollary 11.2. *Let $f(y) \in C^1$ for all y and let $f(y_0) = 0$. Let $G = G^*$ be a real, constant, positive definite, Hermitian matrix and let the Jacobian matrix $J(y) = (\partial f / \partial y)$ satisfy $GJ(y)x \cdot x < 0$ for all $y \neq y_0$ and all vectors $x \neq 0$. Then the solution $y_0(t) \equiv y_0$ of (11.1) is globally asymptotically stable [and, in particular, $f(y) \neq 0$ for $y \neq y_0$].*

Proof. Put $V(y) = G(y - y_0) \cdot (y - y_0)$, so that $V(y) \geqq 0$ according as $\|y - y_0\| \geqq 0$ and $V(y) \to \infty$ as $\|y\| \to \infty$. Also $\dot{V}(y) = 2G(y - y_0) \cdot f(y)$.

It will be verified that $\dot{V}(y) \leqq 0$ according as $\|y - y_0\| \geqq 0$. To this end, we have

$$f(y) \doteq \int_0^1 J(y_0 + s(y - y_0))(y - y_0) \, ds$$

as can be seen by noting that $f(y_0) = 0$ and that the derivative of

$$f(y_0 + s(y - y_0))$$

with respect to s is $J(y_0 + s(y - y_0))(y - y_0)$. Hence

$$G(y - y_0) \cdot f(y) = \int_p^1 GJ(y - y_0) \cdot (y - y_0) \, ds,$$

where the argument of J is the same as before. This shows that $\dot{V}(y) = 2G(y - y_0) \cdot f(y) \leqq 0$ according as $\|y - y_0\| \geqq 0$ and proves the corollary.

Exercise 11.2.   Let $f(y) \in C^1$ for all y and let $GJ(y)x \cdot x < 0$ for all y and all vectors x, where $G = G^*$ is real, positive definite. Let $y_1(t)$, $y_2(t)$ be two distinct solutions of (11.1) starting at $t = 0$. Then $y_1(t)$, $y_2(t)$ exist and $G(y_2(t) - y_1(t)) \cdot (y_2(t) - y_1(t))$ decreases for $t \geqq 0$.

Corollary 11.3.   *Let* $f(y) \in C^1$ *for all* y *and* $\|f(y)\| \to \infty$ *as* $\|y\| \to \infty$. *Let* $G = G^*$ *be a real, positive definite matrix and let* $J(y) = (\partial f / \partial y)$ *satisfy* $GJ(y)f(y) \cdot f(y) \leqq 0$ *according as* $\|y - y_0\| \geqq 0$. *Then* $f(y_0) = 0$ *and the solution* $y_0(t) \equiv y_0$ *of* (11.1) *is globally asymptotically stable.*

Exercise 11.3.   Prove Corollary 11.3 by choosing $V(y) = Gf(y) \cdot f(y)$.

The condition $\|f(y)\| \to \infty$ as $\|y\| \to \infty$ in Corollary 11.3 can be considerably weakened. Also, the constant matrix G can be replaced by suitable matrix functions $G(y)$ in Corollaries 11.2 and 11.3. This type of result will be considered in the §§ 12–13; cf. Corollary 12.1 and Theorem 13.1.

12. Nonconstant G

Let E be a connected open y-set. Let $G(y) = G^*(y)$ be a (real) positive definite matrix and let $G(y)$ be continuous on E. We can associate with the matrix $G(y)$, the Riemann element of arclength

$$(12.1) \qquad ds^2 = G(y) \, dy \cdot dy = \sum_{j=1}^d \sum_{k=1}^d g_{jk}(y) \, dy^j \, dy^k$$

if $G = (g_{jk}(y))$. By this is meant that if $C : y = y(t)$, $a \leqq t \leqq b$, is an arc of class C^1 in E, its Riemann arclength $L(C)$ with respect to (12.1) is defined to be

$$(12.2) \qquad L(C) = \int_a^b [G(y(t))y'(t) \cdot y'(t)]^{1/2} \, dt.$$

This is readily seen to be independent of any C^1 parametrization of C.

We can also introduce a new metric $r(y_1, y_2)$ on E by putting

$$(12.3) \qquad\qquad r(y_1, y_2) = \inf_C L(C)$$

taken over all arcs $C : y = y(t), a \leqq t \leqq b$, in E, of class C^1 joining $y_1 = y(a)$ and $y_2 = y(b)$. The function $r(y_1, y_2)$ satisfies the usual conditions for a metric: $r(y_1, y_2) = r(y_2, y_1)$; $r(y_1, y_2) \geqq 0$ according as $\|y_1 - y_2\| \geqq 0$; and the triangular inequality

$$(12.4) \qquad\qquad r(y_1, y_2) \leqq r(y_1, y^0) + r(y^0, y_2).$$

Remark. Since $G(y)$ is continuous and positive definite, it follows that if E_1 is a compact subset of E, then there exist positive constants c_1, c_2 such that $c_1 \, dy \cdot dy \leqq ds^2 \leqq c_2 \, dy \cdot dy$. Thus if C is an arc of class C^1 in E_1, then $c_1 L_e(C) \leqq L(C) \leqq c_2 L_e(C)$, where $L(C)$ is the Riemann and $L_e(C)$ the Euclidean arclength of C. In particular, if y^0 is an arbitrary point of E and $\epsilon > 0$, there exists a $\delta = \delta(y^0, \epsilon) > 0$ with the property that if $\|y_1 - y_0\| \leqq \delta$, $\|y_2 - y_0\| \leqq \delta$, then, in determining $r(y_1, y_2)$ in (12.3), it suffices to consider arcs C in $\|y_0 - y_0\| \leqq \epsilon$. Hence if y^0 is an arbitrary point of E, then there exists a small $\delta = \delta(y^0) > 0$ and a pair of positive constants c_{10} and c_{20}, depending on y^0, such that $c_{10} \|y_1 - y_2\| \leqq r(y_1, y_2) \leqq c_{20} \|y_1 - y_2\|$ if $\|y_j - y_0\| \leqq \delta$ (or if $r(y_j, y_0) \leqq \delta$) for $j = 1, 2$.

The Riemann element of arclength ds will be called *complete* on the y-set E is it has the property that the convergence of the integral in (12.2) for a half-open arc $C : y = y(t)$ of class C^1 in E defined on a half-open interval $a \leqq t < b (\leqq \infty)$, implies that $y(b) = \lim y(t)$ as $t \to b$ exists and is in E; i.e., ds is complete if half-open arcs C of finite length (12.2) have an endpoint in E.

This concept of "complete" is equivalent to the usual notion that the set E considered as a metric space with the metric (12.3) be complete. But the fact will only be used in § 13. The following simple lemma will be used subsequently.

Lemma 12.1 *Let E be the y-space or the part of y-space $\|y\| \geqq \alpha > 0$ exterior to a ball. Let $G(y)$ be of class C^1 on E and $G(y) = G^*(y)$ positive definite. Then ds in (12.1) is complete on E if and only if every unbounded arc $C : y(t)$, $a \leqq t < b (\leqq \infty)$ of class C^1 in E has an infinite Riemann arclength $L(C)$.*

Exercise 12.1. (*a*) Verify Lemma 12.1. (*b*) If, in Lemma 12.1, $G(y) = p^2(y)I$, where $p(y) > 0$ is a function of class C^1, then a sufficient condition for (12.1) to be complete on E is that $p(y) \geqq c > 0$, or that $\|y\| p(y) \geqq c > 0$, or, more generally, that

$$(12.5) \qquad\qquad \int^\infty \left[\min_{\|y\|=u} p(y) \right] du = \infty.$$

Let $f(y)$ be of class C^1 on E and let $y(t)$ be a solution of

(12.6) $$y' = f(y)$$

on some t-interval. Let $x(t)$ be a solution of the equations of variation of (12.7) along (12.3), i.e., a solution of the linear system

(12.7) $$x' = J(y(t))x,$$

where $J(y)$ is the Jacobian matrix $J(y) = (\partial f/\partial y)$. Let $G(y) \in C^1$ on E and consider the function

(12.8) $$v(t) = G(y(t))x(t) \cdot x(t).$$

Its derivative with respect to t is easily seen to be given by

(12.9) $$v'(t) = 2B(y(t))x(t) \cdot x(t),$$

where $B(y) = (b_{jk}(y))$ is the $d \times d$ matrix with elements

(12.10) $$b_{jk} = \sum_{m=1}^{d} g_{jm} \frac{\partial f^m}{\partial y^k} + \frac{1}{2} \sum_{m=1}^{d} f^m \frac{\partial g_{jk}}{\partial y^m}; \quad \text{i.e., } B = GJ + \frac{1}{2} \sum_{m=1}^{d} f^m \frac{\partial G}{\partial y^m}.$$

In particular,

(12.11) $\quad V(y) = G(y)f(y) \cdot f(y) \quad$ implies that $\quad \dot{V}(y) = 2B(y)f(y) \cdot f(y)$

since $y'(t) = f(y(t))$ is a solution of (12.7).

The matrix B has occurred in (V 7.11) and Lemma V 9.1 for a similar purpose, where $G = A^*A$. (For readers familiar with Riemann geometry, it can be mentioned that if $f(y)$ is considered as a contravariant vector field; $f_{,j}^k$ the components of its covariant derivative; and $B^0(y) = (b_{jk}^0(y))$ is defined by $b_{jk}^0 = \Sigma g_{jm} f_{,k}^m$, then $B - B^0$ is a skew-symmetric matrix. Thus (12.9) is not affected if B is replaced by B^0.)

Note that if $G(y) \equiv G$ is a constant matrix, then $B(y) = GJ(y)$.

Theorem 12.1. *Let $f(y)$ be of class C^1 on an open connected set E containing $y = 0$. Let $G(y) = G^*(y)$ be of class C^1 on E, positive definite for each y, and such that ds in (12.1) is complete on E. Let $\varphi(r) > 0$ be nonincreasing for $r \geqq 0$ and satisfy*

(12.12) $$\int_0^\infty \varphi(r)\, dr = \infty.$$

Finally, let $r(y) = r(y, 0)$ [cf. (12.3)] and

(12.13) $$B(y)f(y) \cdot f(y) \leqq -\varphi(r(y))G(y)f(y) \cdot f(y).$$

Then (i) *every solution $y(t)$ of $y' = f(y)$ starting at $t = 0$ exists for $t \geqq 0$;* (ii) *$y(\infty) = \lim y(t)$ exists as $t \to \infty$ and is a stationary point, $f(y) = 0$ at $y = y(\infty)$;* (iii) *if $y(t) \not\equiv y(\infty)$, then*

(12.14) $$v(t) = G(y(t))f(y(t)) \cdot f(y(t))$$

is a decreasing function for $t \geqq 0$ *and tends to* 0 *as* $t \to \infty$; (iv) *the set of stationary points* [*i.e., zeros of* $f(y)$] *is connected*; *hence* (v) *if the stationary points of* $f(y)$ *are isolated* (*e.g., if* $\det B(y) \neq 0$ *whenever* $f(y) = 0$), *then* $f(y)$ *has a unique stationary point* y_0 *and the solution* $y_0(t) \equiv y_0$ *is globally asymptotically stable.*

The proof will give a priori bounds for solutions $y(t)$. Let

$$(12.15) \qquad \Phi(r) = \int_0^r \varphi(\sigma)\, d\sigma$$

and let $\Psi(r)$ be the function inverse to $\Phi(r)$, then it will be seen that

$$(12.16) \qquad r(y(t)) \leqq c, \qquad \text{where} \quad c = \Psi(\Phi[r(y(0))] + v^{1/2}(0)).$$

In addition, $r(y(t)) \leqq c$ implies that

$$(12.17) \qquad 0 \leqq v(t) \leqq v(0)e^{-2\varphi(c)t} \qquad \text{for} \quad t \geqq 0;$$

and since

$$r(y(t), y(\infty)) \leqq \int_t^\infty v^{1/2}(\sigma)\, d\sigma,$$

we have

$$(12.18) \qquad r(y(t), y(\infty)) \leqq \frac{v^{1/2}(0)e^{-\varphi(c)t}}{\varphi(c)} \qquad \text{for} \quad t \geqq 0.$$

If (12.12) does not hold but the initial point $y(0)$ of a particular solution $y(t)$ is such that the definition of c in (12.16) is meaningful, then assertions (i)–(iii) are valid for this $y(t)$.

Exercise 12.2.   Using the example of the binary system where $f(y) = (-y^1, 0)$, $G = I$, and E is the y-space, show that the additional assumption in (v) concerning isolated stationary points cannot be omitted.

Proof.   (i)–(iii).   Let $y(t)$ be a solution of $y' = f$ starting at $t = 0$. Then the Riemannian length of the solution arc $y = y(t)$ over $[0, t]$ is the integral of $v^{1/2}(t)$, where $v(t)$ is given by (12.14). Put $r(t) = r(y(t)) = r(y(t), 0)$ and

$$(12.19) \qquad u(t) = r(0) + \int_0^t v^{1/2}(\sigma)\, d\sigma.$$

By the triangular inequality (12.4) and by (12.3), it is clear that $r(t) \leqq u(t)$.

Since $y(t)$ is a solution of $y' = f$, its derivative $x = y'(t) = f(y(t))$ is a solution of the equations of variation (12.7). Thus (12.8)–(12.9) hold and so, by (12.13), $v'(t) \leqq -2\varphi(r(t))v(t)$. Consequently,

$$(12.20) \qquad [v^{1/2}(t)]' \leqq -\varphi(r(t))[v^{1/2}(t)].$$

By (12.19), $u' = v^{1/2}$ and $u'' = [v^{1/2}]' \leqq -\varphi(r(t))u'$. Since $\varphi(r)$ is nonincreasing, $r(t) \leqq u(t)$ implies that

$$u''(t) \leqq -\varphi(u(t))u'(t).$$

Integrating over $[0, t]$ gives

$$u'(t) \leq u'(0) - \int_{u(0)}^{u(t)} \varphi(w) \, dw.$$

Since $u(0) = r(0)$ and $u' = v^{1/2}$, this can be written as

$$v^{1/2}(t) \leq v^{1/2}(0) + \Phi(r(0)) - \Phi(u(t)),$$

by (12.15). This inequality, $v^{1/2} \geq 0$, and $r(t) \leq u(t)$ show that

$$\Phi(r(t)) \leq \Phi(u(t)) \leq v^{1/2}(0) + \Phi(r(0)).$$

Consequently, (12.16) holds on any interval $[0, t]$ on which $y(t)$ exists.
Thus the monotony of φ and (12.20) give

$$[v^{1/2}(t)]' \leq -\varphi(c)[v^{1/2}(t)],$$

and so the inequalities in (12.17) hold on any interval $[0, t]$ on which
$y(t)$ exists. Consequently,

$$(12.21) \qquad \int_0^t v^{1/2}(\sigma) \, d\sigma \leq \frac{v^{1/2}(0)(1 - e^{-\varphi(c)t})}{\varphi(c)} \leq \frac{v^{1/2}(0)}{\varphi(c)},$$

and if $0 \leq t < \omega \ (\leq \infty)$ is the right maximal interval of existence of
$y(t)$, then the last integral converges as $t \to \omega$. Since this integral is the
Riemann arclength of the arc $y = y(t)$, $0 \leq t < \omega$, the completeness of
ds implies that $y(\omega) = \lim y(t)$ exists as $t \to \omega$ and is in E. But then
$\omega = \infty$ by Lemma II 3.1. This proves (i), the existence of $y(\infty)$ in (ii),
and (iii). The fact that $f(y) = 0$ at $y = y(\infty)$ follows from (iii) since the
integral in (12.21) is convergent as $t \to \infty$.

Proof. (iv)–(v). Let E_0 be the set of zeros of $f(y)$. In order to show
that E_0 is connected, define a map $P : E \to E_0$ of E onto E_0 as follows:
if $y(t)$ is an arbitrary solution of $y' = f$ for $t \geq 0$, put $Py(0) = y(\infty)$. It
is clear that the range of P is E_0. Since continuous maps send connected
sets into connected sets, it will follow that E_0 is connected if it is verified
that P is continuous.

Let y^0 be an arbitrary point. It will be shown that P is continuous at
$y = y^0$. If $\|y - y^0\|$ is small, there exist positive constants c_{10}, c_{20} such
that $c_{10} \|y - y^0\| \leq r(y, y^0) \leq c_{20} \|y - y^0\|$; cf. the Remark following
(12.4). Thus, in proving the continuity of $P : E \to E_0$ at $y = y^0$, it can be
supposed that E carries the metric defined by $r(y_1, y_2)$ in (12.3).

Let $y^0(t)$ be the solution of $y(t)$ satisfying $y^0(0) = y^0$ and M_δ the Riemann
sphere $r(y^0, y) \leq \delta$. Since c in (12.16) depends only on the initial point
$y(0)$ of the solution $y(t)$, it follows that the inequalities (12.17)–(12.18)
hold with a constant $c > 0$ which can be chosen independent of $y(0) \in M_\delta$.
Hence if $\epsilon > 0$ is fixed, there exists a number t_ϵ independent of $y(0) \in M_\delta$
such that $r(y(t), y(\infty)) < \epsilon$ if $t \geq t_\epsilon$. Let $\delta = \delta(\epsilon) > 0$ be so small that

$r(y(t), y^0(t)) < \epsilon$ for $0 \leqq t \leqq t_\epsilon$ if $r(y^0, y(0)) < \delta(\epsilon)$. Consequently, $r(y^0(\infty), y(\infty)) < 3\epsilon$ if $r(y^0, y(0)) < \delta(\epsilon)$. This proves the continuity of P at $y = y^0$ and completes the proof of (iv).

The main part of (v) follows from (iv). As to the parenthetical part of (v), note that if $f(y_0) = 0$, then $B(y_0) = G(y_0)J(y_0)$ by (12.10), where $J(y) = (\partial f/\partial y)$. This completes the proof of Theorem 12.1.

Corollary 12.1. *Consider a map T of the y-space into itself given by $T : y_1 = f(y)$, where $f(y)$ is of class C^1 for all y. Let the Jacobian matrix $J(y) = (\partial f/\partial y)$ satisfy $\det J(y) \neq 0$ and $J(y)x \cdot x \leqq -\varphi(\|y\|)\|x\|^2$ for all x and y, where $\varphi(r) > 0$ is nonincreasing for $r \geqq 0$ and satisfies (12.12). Then T is one-to-one and onto [i.e., T has a unique inverse $T^{-1} : y = f_1(y_1)$ defined for all y_1]. In particular, there is a unique point y_0 where $f(y) = 0$; furthermore, the solution $y_0(t) \equiv y_0$ of $y' = f(y)$ is globally asymptotically stable.*

Proof. Let E be the y-space and $G = I$ in Theorem 12.1, and replace $f(y)$ by $f(y) - y^0$ for a fixed y^0.

If x_0 is fixed and the condition on $J(y)$ is relaxed to $J(y)(f(y) - x_0) \cdot (f(y) - x_0) \leqq -\varphi(\|y\|)\|f(y) - x_0\|^2$, then it follows from Theorem 12.1 that the equation $f(y) = x_0$ has at least one solution y.

Exercise 12.3. In Corollary 12.1, show that T is one-to-one and onto if "the assumption that $f(y)$ is of class C^1 and the condition on $J(y)$" is relaxed to the following: "$f(y)$ is continuous and satisfies

$$[f(y_1) - f(y_2)] \cdot (y_1 - y_2) \leqq -\varphi(r) \|y_1 - y_2\|^2$$

for all y_1, y_2 in the sphere $\|y\| \leqq r$." (This generalizes the first part of Corollary 12.1; cf. the proof of Corollary 11.2.)

Exercise 12.4. (a) Let $f(y) \in C^1$ for all y; $\varphi(r)$ as in Theorem 12.1, $p(y) > 0$ of class C^1 for all y and satisfying (12.5). If $\dot{p}(y) = f(y) \cdot \operatorname{grad} p(y)$ and $J(y) = (\partial f/\partial y)$, assume that

$$(12.22) \quad (Jf \cdot f)p + \dot{p} \|f\|^2 \leqq -\varphi\left(\int_0^{\|y\|} \left[\min_{\|x\|=u} p(x)\right] du\right) p \|f\|^2.$$

Show that assertions (i)–(v) of Theorem 12.1 are valid with $G(y) = p^2(y)I$. (b) Verify that if $\|f(y)\| \geqq 0$ according as $\|y\| \geqq 0$ and $p(y)$ satisfies all conditions of (a) except that $p(0) = 0$, that $p(y)$ is merely continuous at $y = 0$, and that (12.22) holds only for $y \neq 0$, then the conclusions of (a) still hold. (c) What are the conditions on $f(y)$ in order that part (b) be applicable with $p(y) = \|f(y)\|$?

13. On Corollary 11.2

In order to obtain an analogue of Corollary 11.2 in which G is replaced by a matrix function $G(y)$, a property of complete Riemann elements of arclength will be needed.

A set of points y on E will be said to be bounded with respect to the metric $r(y_1, y_2)$ if for some (and/or every) fixed point $y^0 \in E$, there is a constant such that $r(y, y^0) \leqq c$ for all y in the set.

Lemma 13.1. *Let $G(y) = G^*(y)$ be continuous on a connected open y-set E and positive definite for each $y \in E$. If ds in (12.1) is complete on E, then every subset of E which is bounded with respect to the metric $r(y_1, y_2)$ has at least one cluster point in E, hence a compact closure in E. In particular, every such subset of E is bounded (with respect to the Euclidean metric on E).*

The converse of this assertion is clear. Lemma 13.1 will be used only for the proof of Theorem 13.1. Its use can be avoided, of course, by making the redundant assumption in Theorem 13.1 that ds has the property specified in Lemma 13.1, as well as being complete. In most applications, this fact will be clear. For a proof of Lemma 13.1, see Hopf and Rinow [1].

Theorem 13.1 *Let $f(y) \in C^1$ on an open connected y-set E. Let $G(y) = G^*(y)$ be of class C^1 on E, positive definite for fixed y, and such that ds in (12.1) is complete on E. Let the matrix $B(y)$ defined by (12.10) satisfy*

$$(13.1) \qquad B(y)x \cdot x < 0 \qquad \text{for all vectors } x \neq 0.$$

Then every solution $y(t)$ of $y' = f(y)$ starting at $t = 0$ exists for $t \geqq 0$; furthermore, if $y_1(t)$, $y_2(t)$ are two distinct solutions for $t \geqq 0$, then

$$(13.2) \qquad r(y_1(t), y_2(t)) \text{ is decreasing.}$$

In particular, if there exists a stationary point y_0, $f(y_0) = 0$, then every solution $y(t) \not\equiv y_0$ satisfies

$$(13.3) \quad r(y(t), y_0) \text{ decreases and } r(y(t), y_0) \to 0 \quad \text{as } t \to \infty$$

(and $f(y) \neq 0$ for $y \neq y_0$).

The following proof could be simplified by using the known fact that if y_1, y_2 are two points of E, then there exists a geodesic arc C of class C^2 joining them such that $r(y_1, y_2)$ is the Riemann arclength $L(C)$ of C.

Proof. Let $y_1(t)$ be a solution of $y' = f$ for $0 \leqq t \leqq T$. Let $y_1 = y_1(0)$, $y_2 \neq y_1$, and $r_0 = r(y_1, y_2)$. The set of points $E^T : \{y : r(y, y_1(t)) < r_0 + 1$ for some t, $0 \leqq t \leqq T\}$ has a compact closure in E, by Lemma 13.1. Thus, by the Remark following (12.4) and a similar remark applied to the form $B(y)x \cdot x$ in (13.1), it follows that there is a constant $c > 0$ such that

$$(13.4) \quad B(y)x \cdot x \leqq -cG(y)x \cdot x \qquad \text{for } y \in E^T \text{ and all } x.$$

Let $C_0 : y = z(u)$, $0 \leqq u \leqq 1$, be an arc of class C^1 satisfying $z(0) = y_1$, $z(1) = y_2$ and $L(C_0)$ is so near to $r_0 = r(y_1, y_2)$ that

$$(13.5) \qquad L(C_0) < r_0 + 1 \quad \text{and} \quad e^{-cT}L(C_0) < r_0.$$

Thus it follows that $C_0 \subset E^T$.

Let $y(t, u)$ be the solution of $y' = f$ satisfying $y(0, u) = z(u)$, so that $y(t, 0) = y_1(t)$, and $y_2(t) = y(t, 1)$ is the solution starting at y_2 for $t = 0$. Then $y(t, u)$ is of class C^1 on its domain of existence; Theorem V 3.1.

By Peano's existence theorem, there is an $S > 0$ independent of u such that the solution $y = y(t, u)$ exists for $0 \leq t \leq S$ for every fixed u, $0 \leq u \leq 1$. It will be shown that $y(t, u)$ exists for $0 \leq t \leq T$. This is clear for small $u \geq 0$ by Theorem V 2.1. Suppose, if possible, there is a least u-value ϵ, $0 < \epsilon \leq 1$, such that if the right maximal interval of $y(t, \epsilon)$ is $0 \leq t < \omega$, then $\omega \leq T$.

For fixed t, $0 \leq t < \omega$, let $L(t)$ be the length $L(C(t))$ of the arc $C(t)$: $y = y(t, u)$, $0 \leq u \leq \epsilon$; i.e.,

$$(13.6) \qquad L(t) = \int_0^\epsilon [G(y(t, u))y_u(t, u) \cdot y_u(t, u)]^{\frac{1}{2}} \, du,$$

where $y_u = \partial y/\partial u$. Note that $x = y_u(t, u)$ is a solution of the equations of variation (12.7) with $y(t) = y(t, u)$. Hence the integrand in (13.6) is $v^{\frac{1}{2}}(t, u)$ where, for fixed u, $v(t, u)$ is given by (12.8) with $y(t) = y(t, u)$, $x(t) = y_u(t, u)$. By (12.9),

$$L'(t) = \int_0^\epsilon v^{-\frac{1}{2}}(t, u)B(y)y_u \cdot y_u \, du.$$

For small fixed $t > 0$, the arc $y(t, u)$ is in E^T since $y(0, u) = z(u)$ is. In this case, (13.4) implies that

$$L'(t) \leq -c \int_0^\epsilon v^{\frac{1}{2}}(t, u) \, du = -cL(t).$$

Consequently, by (13.5),

$$(13.7) \qquad L(t) \leq e^{-ct}L(0) < e^{-ct}(r_0 + 1) < r_0 + 1$$

as long as $y(t, u) \in E^T$. The inequality (13.7) shows that as t increases from 0 to ω $(\leq T)$, $y(t, u)$ cannot leave E^T. Thus (13.7) is valid for $0 \leq t < \omega$.

It follows that the integral (13.6) with $t = \omega$ is convergent. Thus, by the completeness of ds in (12.1), $\lim y(t, u)$ exists as $u \to \epsilon$ and is in E for $t = \omega$ (as well as for $0 \leq t < \omega$). This limit is $y(t, \epsilon)$, so that this solution exists for $0 \leq t \leq \omega$. This contradicts the fact that $0 \leq t < \omega$ is the right maximal interval of existence of $y(t, \epsilon)$. Consequently, the solution $y(t, u)$ exists for $0 \leq t \leq T$ for every u, $0 \leq u \leq 1$.

In particular, $y_2(t) = y(t, 1)$ exists for $0 \leq t \leq T$. Also, if $\epsilon = 1$ and $t = T$ in (13.6), and (13.7), then it follows from (12.3) that $r(y_1(T), y_2(T)) \leq e^{-cT}L(C_0)$. Hence, by (13.5), $r(y_1(T), y_2(T)) < r_0 = r(y_1(0), y_2(0))$. Since T can be replaced in this argument by any t-value $0 < t \leq T$, it follows that (13.2) holds on any interval on which $y_1(t)$ exists.

It will now be shown that $y_1(t)$ exists for $t \geqq 0$. If $y_1(T) = y_1(0)$, then $y_1(t)$ is periodic and exists for all T. If $y_1(T) \neq y_1(0)$, apply the argument just completed with $y_2 = y_1(T)$. Then $y_2(t) = y_1(t + T)$ exists for $0 \leqq t \leqq T$; i.e., $y_1(t)$ exists for $0 \leqq t \leqq 2T$. Repetitions of this argument show that $y_1(t)$ exists for $t \geqq 0$.

If y_0 is a stationary point and $y(t) \not\equiv y_0$ is a solution of $y' = f$, then $r(y(t), y_0)$ is decreasing. In particular, $f(y) \neq 0$ for $y \neq y_0$ and $r(y(t), y_0)$ is bounded for $t \geqq 0$. Hence $C^+: y = y(t), t \geqq 0$, has a compact closure in E by Lemma 13.1.

Note that $V(y) = G(y)f(y) \cdot f(y) \geqq 0$ satisfies $\dot{V}(y) = 2B(y)f(y) \cdot f(y) \leqq 0$ by (12.11) and (13.1). Consequently, by Lemma 11.1, the ω-limit points of C^+ are zeros of $\dot{V}(y)$. But (13.1) shows that $\dot{V}(y) = 0$ if and only if $f(y) = 0$. Hence $y(t) \to y_0$ as $t \to \infty$. This proves Theorem 13.1.

Exercise 13.1.   In Theorem 13.1, let assumption (13.1) be relaxed to $B(y)x \cdot x \leqq 0$ for $y \in E$ and for all x. Show that the following analogue of the first part of Theorem 13.1 is valid: (a) $r(y_1(t), y_2(t))$ is nonincreasing and (b) if $y(t)$ is a solution of $y' = f(y)$ starting at $t = 0$, then $y(t)$ exists for $t \geqq 0$.

14. On "$J(y)x \cdot x \leqq 0$ if $x \cdot f(y) = 0$"

The last three sections have been concerned with the condition $J(y)f(y) \cdot f(y) \leqq 0$ and its generalizations. In this section, the condition "$J(y)x \cdot x \leqq 0$ if $x \cdot f(y) = 0$" and generalizations will be considered. If $f(y)$ in

$$(14.1) \qquad\qquad y' = f(y)$$

is of class C^1 on a set E, and $ds^2 = G(y) \, dy \cdot dy$ a positive definite Riemann element of arclength with $G(y) = G^*(y)$ of class C^1, the generalized condition is

$$(14.2) \qquad B(y)x \cdot x \leqq 0 \qquad \text{if} \quad G(y)f(y) \cdot x = 0,$$

where $B(y)$ is defined in (12.10).

In the first theorem, it will be supposed that

$$(14.3) \qquad f(0) = 0 \quad \text{and} \quad f(y) \neq 0 \qquad \text{for} \quad y \neq 0$$

and that

$$(14.4) \qquad y \equiv 0 \text{ is a locally asymptotically stable solution of (14.1);}$$

cf. § III 8. By the *domain of attraction* of $y = 0$ is meant the set of points y_0 of E such that solutions $y(t)$ of (14.1) starting at y_0 for $t = 0$ exist for $t \geqq 0$ and $y(t) \to 0$ as $t \to \infty$. If E is open, the domain of attraction is open.

Theorem 14.1. (i) *Let $f(y)$ be of class C^1 on a connected open y-set E containing $y = 0$ such that (14.3) and (14.4) hold.* (ii) *Let $\epsilon > 0$ be so small that $\|y\| \leq \epsilon$ is in the domain of attraction of $y = 0$ and let E_ϵ be the set of points $y \in E$ satisfying $\|y\| \geq \epsilon$.* (iii) *On E_ϵ, let $G(y) = G^*(y)$ be of class C^1, positive definite for each y and such that $ds^2 = G(y)\,dy \cdot dy$ is complete on E_ϵ.* (iv) *Finally, let (14.2) hold for $y \in E_\epsilon$ and all x. Then $y = 0$ is globally asymptotically stable.*

Before proceeding to the proof, it will be of interest to formulate some corollaries.

Corollary 14.1. *Let* (i), (ii) *of Theorem* 14.1 *hold with E the y-space. Let there exist a function $p(y) > 0$ of class C^1 on $E_\epsilon: \|y\| \geq \epsilon$ satisfying*

$$(14.5) \qquad \int^\infty \left[\min_{\|y\|=u} p(y) \right] du = \infty$$

and, if $\dot{p}(y) = f(y) \cdot \operatorname{grad} p(y)$ and $J(y) = (\partial f/\partial y)$, then

$$(14.6) \qquad \dot{p}(y)\,\|x\|^2 + p(y)J(y)x \cdot x \leq 0 \qquad \text{when} \quad x \cdot f(y) = 0.$$

Then $y = 0$ is globally asymptotically stable.

Note that if $p \equiv 1$, (14.6) reduces to $J(y)x \cdot x \leq 0$ when $x \cdot f(y) = 0$.

Exercise 14.1. Verify Corollary 14.1 by choosing $G(y) = p^2(y)I$; cf. Exercise 12.4.

Exercise 14.2. Verify that Corollary 14.1 is applicable with $p(y) = \|f(y)\|$ if

$$(14.7) \qquad \int^\infty \left[\min_{\|y\|=u} \|f(y)\| \right] du = \infty$$

$$(14.8) \quad [J(y)f(y) \cdot f(y)]\,\|x\|^2 + \|f\|^2 J(y)x \cdot x \leq 0 \qquad \text{when} \quad x \cdot f(y) = 0.$$

Corollary 14.2. *Let* (i) *in Theorem* 14.1 *hold with E the y-space and let* (14.7) *hold. Let the eigenvalues of the Hermitian part $J^H(y) = \frac{1}{2}(J + J^*)$ of $J(y)$ be $\lambda_1(y) \geq \cdots \geq \lambda_d(y)$ and let*

$$(14.9) \qquad \lambda_1(y) + \lambda_2(y) \leq 0.$$

Then $y = 0$ is globally asymptotically stable. If $d = 2$, then (14.9) is equivalent to

$$(14.10) \qquad \qquad \operatorname{tr} J(y) \leq 0.$$

Exercise 14.3. Verify this corollary by showing that (14.9) implies (14.8).

Theorem 14.1 will be deduced from the following result dealing with a solution $y_0(t)$, not necessarily $y_0(t) \equiv 0$.

Theorem 14.2. *Let $f(y) \neq 0$ be of class C^1 on an open y-set E. Let $G(y) = G^*(y)$ be of class C^1 on E and positive definite for $y \in E$ and let*

(14.2) hold. *Let $y_0(t)$ be a solution of* (14.1) *on the right maximal interval* $0 \leq t < \omega \ (\leq \infty)$ *with the property that there exists a number* $\alpha > 0$ *such that* $r(y_0(t), \partial E) > \alpha > 0$. *Then there are positive constants* δ, *K such that, for any solution* $y(t)$ *of* (4.1) *with* $r(y_0(0), y(0)) < \delta$, *there exists an increasing function* $s = s(t)$, $0 \leq t < \omega$, *such that* $s(0) = 0$, $0 \leq t < s(\omega) \ (\leq \infty)$ *is the right maximal interval of existence of* $y(t)$, *and* $r(y(s(t)), y_0(t)) \leq Kr(y(0), y_0(0))$ *for* $0 \leq t < \omega$.

In this theorem $r(y_1, y_2)$ is the metric associated with $ds^2 = G(y) \, dy \cdot dy$. It is not assumed that ds is complete on E and the assumption $r(y_0(t), \partial E) > \alpha$ means that if $C : y = x(u)$, $0 \leq u < 1$, is a half-open arc of class C^1 starting at the point $y_0(t)$, i.e., $x(0) = y_0(t)$, and the Riemann length $L(C)$ is finite and $L(C) \leq \alpha$, then $x(1) = \lim x(u)$ exists as $u \to 1$ and $x(u) \in E$. Roughly speaking, $r(y_0(t), \partial E) > \alpha$ means that $y_0(t)$ is at least a distance α (in the r-metric) from the boundary ∂E of E [i.e., the set $\{y : r(y, y(t)) \leq \alpha$ for some t, $0 \leq t < \omega\}$ is in E]. Theorem 14.2 will be proved in the next section and Theorem 14.1 in § 16.

Exercise 14.4 Let $f(y) \neq 0$ be of class C^1 on a bounded, connected open set E. Let $y(t)$, $t \geq 0$, be a solution of (14.1) such that dist $(y(t), \partial E) > \alpha > 0$, where "dist" refers to the Euclidean metric. (*a*) Let

$$\gamma(y) = \max J(y)x \cdot x \qquad \text{for} \quad \|x\| = 1, \quad x \cdot f(y) = 0$$

satisfy $\gamma(y) \leq -c < 0$ for some constant $c > 0$. Then the set of ω-limit points of $y(t)$, $t \geq 0$, is a periodic solution $y_0(t)$ of (14.1) which has $d - 1$ characteristic exponents with negative real parts (and so is asymptotically stable, in fact, Theorem IX 11.1 is applicable). See Borg [3]. (*b*) Show that the condition $\gamma(y) \leq -c < 0$ for $y \in E$ can be relaxed to the condition $\Gamma(t) - \Gamma(s) \leq C - c(t - s)$ for $0 \leq s < t < \infty$ and a pair of constants $C, c > 0$, where $\Gamma(t) = \int_0^t \gamma(y(u)) \, du$. See Hartman and Olech [1].

15. Proof of Theorem 14.2

In this proof, notions of the length of a vector x at a point y or orthogonality of vectors x_1, x_2 at y refer to the Riemann geometry; i.e., $(G(y)x \cdot x)^{1/2}$ or $G(y)x_1 \cdot x_2 = 0$. Similarly, arclength of an arc C refers to its Riemann arclength $L(C)$; cf. (12.2).

Let $y_0 = y_0(0)$ and π the piece of the hyperplane

$$\pi : G(y_0)f(y_0) \cdot (y - y_0) = 0$$

through y_0 orthogonal to $f(y_0)$ with a parametrization $\pi : y = z(p, u)$ for $0 \leq p \leq p_1$, where u is any unit vector orthogonal to $f(y_0)$ at y_0, $z(p, u) = y_0 + pu$. It is clear that all solutions of (14.1) with initial points near y_0 cross π.

For a fixed u, let $y = y(t, p)$ be the solution of (14.1) determined by the initial condition $y(0, p) = z(p, u)$. Let $0 \leqq t < \omega(p) \leqq \infty$ be the right maximal interval of existence of $y(t, p)$. Thus $y(t, 0) = y_0(t)$ and $\omega(0) = \omega$.

For fixed u, consider the 2-dimensional surface S: $y = y(t, p)$ defined on a (t, p)-set containing $0 \leqq t < \omega(p)$, $0 \leqq p \leqq p_1$. On S, consider the differential equation for the orthogonal trajectories to the parameter arcs $p = $ const. [i.e., to the solution paths of (14.1) on S] determined by the

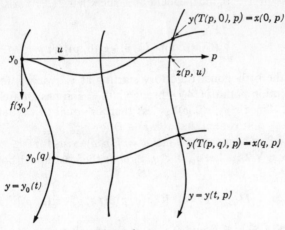

Figure 6

relations $G(y)f(y) \cdot dy/dp = 0$, where $y = y(t, p)$ and $t = t(p)$. Let $t = T(p, q)$ be the solution of this differential equation,

(15.1) $$\frac{dt}{dp} = -\frac{G(y)f(y) \cdot y_p}{G(y)f(y) \cdot f(y)}, \qquad \text{where} \quad y = y(t, p),$$

with initial condition

(15.2) $$T(0, q) = q$$

(so that the corresponding orthogonal trajectory starts at the point $y = y_0(q)$); see Figure 6. In (15.1) and later, subscripts p, q denote partial differentiation.

Since the right side of (15.1) has a continuous partial derivative with respect to the dependent variable t, the solution $t = T(p, q)$ of (15.1), (15.2) is of class C^1 and has a continuous second mixed derivative $T_{pq} = T_{qp}$; see Corollary V 3.1. Furthermore, as a function of p, $T_q(p, q)$ satisfies a homogeneous linear differential equation by Theorem V 3.1, hence

(15.3) $$T_q(p, q) > 0$$

since (15.2) implies $T_q(0, q) = 1 > 0$. The reparametrization of S given by

(15.4) $$S: y = x(q, p) \equiv y(T(p, q), p)$$

will be used.

Let D be the open subset of E which is the union of the "spheres" $r(y, y_0(t)) < \alpha/2$ for $0 \leq t < \omega$. Thus $r(y, \partial E) \geq \alpha/2$ if $y \in D$. There is a constant $\beta > 0$, independent of u, such that $T(p, 0)$ exists for $0 \leq p \leq \beta$ for every u and

(15.5) $$\int_0^\beta [G(x(0, p))x_p(0, p) \cdot x_p(0, p)]^{1/2} \, dp < \alpha/2.$$

Thus the orthogonal trajectory starting at y_0 reaches, for every fixed u, the solution path of (14.1) through $x(0, \beta)$ and has an arclength satisfying (15.5). Since $r(y_0, x(0, p))$ is less than or equal to the integral (15.5) for $0 \leq p \leq \beta$, it is seen that $x(0, p) \in D$ for $0 \leq p \leq \beta$.

The set of q-values for which $t = T(p, q)$ exists for $0 \leq p \leq \beta$ is open by Theorem V 2.1. Let q_0, $0 < q_0 \leq \omega$ be the least upper bound for this set. Put

(15.6) $$L(q, \sigma, \tau) = \int_\sigma^\tau [G(x(q, p))x_p(q, p) \cdot x_p(q, p)]^{1/2} \, dp$$

for $0 \leq \sigma \leq \tau \leq \beta$ and $0 \leq q < q_0$. It will be shown that $L(q, \sigma, \tau)$ is nonincreasing with respect to q for fixed σ, τ. Let $v(q, p)$ denote the square of the integrand. It will suffice to show that $\partial v/\partial q \leq 0$. To this end, note that (15.4) implies that $x_q = T_q f(x)$ and, hence that $x_{qp} = T_q J(x)x_p + T_{qp}f(x)$. In addition, $G(x)f(x) \cdot x_p = 0$ by the definition of $T(p, q)$. Using these facts, (12.10), and

$$\frac{\partial v}{\partial q} = \sum_{k=1}^d x_q^k \left(\frac{\partial G}{\partial y^k}\right)_{y=x} x_p \cdot x_p + 2G(x)x_{pq} \cdot x_p$$

give $\partial v/\partial q = 2T_q B(x)x_p \cdot x_p$. Consequently, $\partial v/\partial q \leq 0$ by (14.2), (15.3).

Thus $L(q, 0, \beta) \leq L(0, 0, \beta)$ if $0 \leq q < q_0$. Since the integral in (15.5) is $L(0, 0, \beta)$, it follows that $L(q, 0, \beta) < \alpha/2$ and so, $x(q, p) \in D$ for $0 \leq q < q_0$, $0 \leq p \leq \beta$.

It will now be shown that

(15.7) $$T(q, p) \to \omega(p) \quad \text{as} \quad q \to q_0$$

for $0 \leq p \leq \beta$, where $0 \leq t < \omega(p)$ is the right maximal interval of existence of $y(t, p)$. Suppose, if possible, that (15.7) fails to hold for some $p = p^0$, $0 \leq p^0 \leq \beta$. Since the arguments to follow do not depend on the position of p^0 on $[0, \beta]$, assume that $p^0 = \beta$. Thus (15.7) fails to hold for $p = \beta$. In particular, $y^0 = \lim y(q, \beta)$ exists as $q \to q_0$, and y^0 is in the closure $\bar{D}$ of D. There exists an orthogonal trajectory $y = x(p)$ on S such that $x(\beta) = y^0$, and $x(p)$ is defined on some interval $(0 \leqq) \sigma < p \leqq \beta$. In particular, the solution $y = y(t, p)$ of (14.1) for $\sigma < p \leqq \beta$ crosses $y = x(p)$ with increasing t near $T(q_0, \beta)$.

From the continuous dependence of solutions on initial conditions, it follows that $x(p)$ [and hence $x_p(p)$] is the uniform limit of $x(q, p)$ [and $x_p(q, p)$, respectively] as $q \to q_0$ on every closed interval $(\sigma <) \tau \leqq p \leqq \beta$. Thus $L(q, \tau, \beta)$ is continuous at $q = q_0$ if $L(q_0, \tau, \beta)$ is defined by

$$L(q_0, \tau, \beta) = \int_\tau^\beta [G(x(p))x_p(p) \cdot x_p(p)]^{1/2}\, dp$$

and $\sigma < \tau < \beta$. By the monotone property of L, $L(q_0, \tau, \beta) \leqq L(0, 0, \beta) < \alpha/2$. Thus the arc $y = x(p)$, $\sigma < p \leqq \beta$, has a finite arclength $< \alpha/2$ and $x(p) \in \bar{D}$, so that $r(x(p), \partial E) > \alpha/2$. Consequently, $x(\sigma) = \lim x(p)$ exists as $p \to \sigma + 0$ and $x(\sigma) \in \bar{D}$.

The limit relation $x(q, p) \to x(p)$ as $q \to q_0$ holds uniformly on the closed interval $\sigma \leqq p \leqq \beta$ if $x(q, p)$ is equicontinuous with respect to p on $\sigma \leqq p \leqq \beta$, for $0 \leqq q \leqq q_0$; see Theorem I 2.2. In order to verify the equicontinuity, note that

(15.8) $\qquad r(x(q, p_1), x(q, p_2)) \leqq L(q, p_1, p_2) \leqq L(0, p_1, p_2)$

and $L(0, p_1, p_2) \to 0$ as $p_2 - p_1 \to 0$.

It is easy to see that $x = x(p)$ can be continued over the interval $0 \leqq p \leqq \beta$. For if $\sigma > 0$, the arguments above can be applied to $p = \sigma$, instead of $p = \beta$, to obtain an extension to an interval $\sigma_1 \leqq p \leqq \beta$, where $0 \leqq \sigma_1 < \sigma$. Furthermore, the set of $p = \sigma_1$ which can be so reached is open and closed relative to $0 \leqq p < \beta$, so that $p = 0$ can be reached.

This means that $y = x(q, p)$ can be defined for $0 \leqq q \leqq q_0$, $0 \leqq p \leqq \beta$ and hence for $0 \leqq q \leqq q_0 + \epsilon$, $0 \leqq p \leqq \beta$ for some $\epsilon > 0$. But this contradicts the definition of q_0. Thus the assumption that (15.7) fails to hold for some $p = p^0$ is untenable. In particular, $q_0 = \omega$.

By (15.8) with $p_1 = 0$, $p_2 = p$, and the definition of $x(q, p)$ in (15.4),

(15.9) $\quad r(y(T(p, q), p), y_0(q)) \leqq L(0, 0, p) \qquad$ for $\quad 0 \leqq q < \omega$.

By the continuity of $G(y)$ at $y = y_0$, there exist constants $\beta > 0$, $c_1 > 0$, c_2 such that $L(0, 0, p) \leqq c_2 p$ if $0 \leqq p \leqq \beta$ and $r(z(p, u), y_0) \geqq c_1 p$; cf. the Remark following (12.4). Furthermore, β, c_1, c_2 can be chosen independent of u.

Thus, if $K = c_2/c_1$, then

(15.10) $\qquad r(y(T(p, t), p), y_0(t)) \leqq Kr(y(0, p), y_0)$

for $0 \leqq t < \omega$ and $0 \leqq p \leqq \beta$. Thus if $y(t)$ is a solution of (14.1) with initial point $y(0) = y(0, p)$ for some p (and u), $0 \leqq p \leqq \beta$, then the assertions of Theorem 14.2, except for $s(0) = 0$, follows with $s(t) = T(p, t)$. On the other hand, if $y(0)$ is near y_0, then there exists a small $|t_1|$ such that $y(t)$ crosses π at $t = t_1$ near y_0; i.e., $y(t_1) = y(0, p)$ for some small $p \geqq 0$ and some u. Also, it is clear that $r(y(0, p), y_0)$ is majorized by a constant times $r(y(0), y_0)$. Thus, if K is suitably altered,

$$r(y(t_1 + T(p, t)), y_0(t)) \leqq Kr(y(0), y_0)$$

for $0 \leqq t < \omega$. Thus the assertions of Theorem 14.2, except for $s(0) = 0$, hold with $s(t) = t_1 + T(p, t)$. In either of the two cases just considered, the modification of $s(t)$ so as to satisfy $s(0) = 0$ is trivial. This proves Theorem 14.2.

16. Proof of Theorem 14.1

Since the domain of attraction of $y = 0$ is an open set containing the sphere $\Sigma(\epsilon): \|y\| \leqq \epsilon$, there exists an $\alpha > 0$ such that $\Sigma (\epsilon + \alpha): \|y\| \leqq \epsilon + \alpha$ is also in the domain of attraction. Let E^* denote the open set $E - \Sigma(\epsilon)$ obtained by deleting $\Sigma(\epsilon)$ from E. Then $f(y) \neq 0$ on E^* and the metric $ds^2 = G(y) \, dy \cdot dy$ satisfies (14.2) on E^*.

Suppose, if possible, that Theorem 14.1 is false, then there exists a point $y_0 \in E^*$ on the boundary of the domain of attraction of $y = 0$. Let $y = y_0(t)$ be the solution of (14.1) satisfying $y_0(0) = y_0$, hence

$$r(y_0(t), \partial E^*) > \alpha$$

on the right maximal interval $0 \leqq t < \omega$.

Thus Theorem 14.2 is applicable with E replaced by E^*. Hence all solutions $y(t)$ starting at points $y(0)$ near y_0 remain close to $y = y_0(t)$ in the sense of Theorem 14.2. In particular $y(t) \in E^*$ on its right maximal interval of existence. But this contradicts the fact that y_0 is on the boundary of the domain of attraction of $y = 0$ and proves the theorem.

Notes

SECTION 1. Theorem 1.1 is due to Hartman [2], [23] and generalizes a result of Milloux [1] on an equation of the second order; see last part of Theorem 3.1. Results of the type (1.9), (1.10) in the first part of Theorem 1.2 were given for the Bessel equations by Watson [1] (cf. [3, pp. 488–489]) and for general scalar second order equations by Milne [1] but, as the proof shows, these are consequences of older theorems on first order systems. (Szego [1] attributes the result to Sonine.) The last part of Theorem 1.2 concerning (1.11)–(1.13) and Exercise 1.3 are in Hartman [23].

SECTION 2. Most of the results of this section are due to Hartman and Wintner. For Theorem 2.1, Corollaries 2.1 and 2.2, see [13]; for Corollaries 2.3 and 2.4, see [14]; for Exercise 2.6 [except part (d)], see [14], [7]; and for Exercise 2.9, see [8]. For Exercise 2.6(d), see Wintner [19]. For a generalization of Theorem 2.1 to Banach spaces, see Coffman [1]. Exercise 2.8 is a modification of a result of Hartman and Wintner [16] and was suggested by Coffman [3]. For related results on a third order, linear equation, see Greguš [1].

SECTION 3. The part of Theorem 3.1 concerning existence in (3.7_∞) is due to Milloux [1]. The reduction of the proof of the theorem via Lemma 3.1 is in Hartman [23]; cf. Wintner [23]. For Corollary 3.1 and Exercise 3.5, see Hartman and Wintner [4]. The result of Exercise 3.7 was stated by Armellini; it was proved independently by Sansone and Tonelli, see Sansone [1, pp. 61–67]; a simple proof is given by Hartman [23]. For Exercise 3.8, see Hartman [22]. For Exercise 3.9, see Hartman and Wintner [15].

SECTION 4. The main results of this section are in Hartman [22]. Parts of Exercise 4.2 are given by Schafheitlin [1]. Exercise 4.3(a) is a particular case of a result on nonlinear differential equations of arbitrary order; see Hartman [22]. For Exercise 4.3(b), see Lorch and Szego [1].

SECTION 5. Cf. Weyl [3] who proved an existence theorem for (5.1)–(5.3) for $\lambda \geqq 0$ and $\alpha = \beta = 0$.

SECTION 6. Theorem 6.1 is due to Iglisch [1], [2]. The existence proof in the text and Exercises 6.1–6.3 are due to Coppel [1]; the uniqueness proof is that of Iglisch [2].

SECTION 7. Theorem 7.1 is a result of Iglisch and Kemnitz [1].

SECTION 8. Cf. Grohne and Iglisch [1]. The arguments in the text are adapted from Iglisch and Kemnitz [1].

SECTION 9. For Theorem 9.1, see Coppel [1]; for Theorem 9.2, see Hartman [29].

SECTION 11. For Lemma 11.1, see LaSalle [3]. Corollary 11.2 is in Hartman [24] and is a generalization of a result of Krasovskiĭ [1], [2], [3] (cf. [4] and Hahn [1, pp. 31–32]). For related results on nonautonomous systems, see Wintner [8, pp. 557–559], Zubov [1], and Hartman [24, pp. 486–492].

SECTIONS 12–13. On the notion of completeness of ds, see Hopf and Rinow [1]. Theorems 12.1 and 13.1 are due to Hartman [24]; see Markus and Yamabe [1] for weaker results. Corollary 12.1 is contained in Hadamard [4] (the proof in the text is in Hartman [24]); the generalization in Exercise 12.3 is due to F. E. Browder [1] with a proof valid for Hilbert space. Theorem 13.1 is related to inequalities of Lewis [2], [3]; cf. Opial [8].

SECTION 14. A condition of the type "$J(y)x \cdot x < 0$ if $x \cdot f(y) = 0$" was introduced by Borg [3]; cf. Exercise 14.4(a). The main results of this section are in Hartman and Olech [1]. They were suggested in part by the 2-dimensional case of Corollary 14.2 in Olech [3].

Hints for Exercises

Chapter II

2.3. Consider the initial value problem $y' = f_y^{-1}(y)\xi, y(0) = 0$. Since $|f_y^{-1}(y)\xi| \leqq M|\xi|$, Theorem 2.1 implies that this initial value problem has a solution $y = Y(t, \xi)$ for $|t| \leqq b/M|\xi|$. By the implicit function theorem, this solution is unique and satisfies $f(Y(t, \xi)) = t\xi$. (Why?) In particular, if $|\xi| = 1$, then $y = Y(t, \xi)$ exists for $|t| \leqq b/M$. Replace ξ by $c\xi$, where $c > 0$, so that $y = Y(t, c\xi)$ satisfies $y' = f_y^{-1}(y)c\xi, y(0) = 0$ for $|t| \leqq b/Mc|\xi|$. By uniqueness, $Y(t, c\xi) = Y(ct, \xi)$ for $|t| \leqq b/Mc|\xi|$. Let $|\xi| = 1$ and $t = 1$ and rename c to t obtaining $Y(1, t\xi) = Y(t, \xi)$ for $|\xi| = 1, |t| \leqq b/M$. Put $y = g(x)$, where $g(x) = Y(1, x)$ for $|x| \leqq b/M$. Thus $f(g(x)) \equiv x$ and, by the implicit function theorem, $g(x)$ is of class C^1 for $|x| \leqq b/M$. The Jacobian matrix $(\partial g/\partial x)$ is $f_y^{-1}(y)$ at $y = g(x)$ and hence is nonsingular. Let D_0 denote the open y-set which is the image of $|x| < b/M$ under the map $y = g(x)$, so that $g(f(y)) = y$ for y on the closure $\bar{D}_0$ of D_0. Then $x = f(y)$ gives a one-to-one map of $\bar{D}_0$ onto $|x| \leqq b/M$ (for if there is an $x_0, |x_0| \leqq b/M$, such that $f(y) = x_0$ has two solutions $y = y_1, y_2 \in \bar{D}_0$, then $g(x)$ is not single-valued; i.e., $g(x_0) = y_1$ and $g(x_0) = y_2$). In order to show that D_0 contains $D_1: |y| \leqq b/MM_1$, apply the result just proved for $x = f(y)$ on $D: |y| \leqq b$ to the map $y = g(x)$ on $|x| \leqq b/M$; this shows that there exists a domain $\bar{D}_1$ in $|x| < b/M$ such that $y = g(x)$ is a continuous one-to-one map of the closure of $\bar{D}_1$ onto $|y| \leqq b/MM_1$.

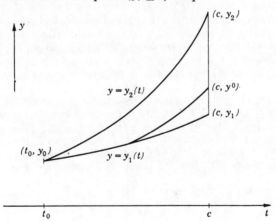

4.1. Consider the continuation to the left of a solution $y' = f(t, y), y(c) = y^0$; see diagram.

4.2. Let $\varphi(r)$ be a continuous function which is 0 for $0 \leqq r \leqq \frac{1}{4}$, 1 for $\frac{3}{4} \leqq r \leqq \frac{5}{4}$, 0 for $r \geqq \frac{3}{2}$; in particular $\varphi(1) = 1$ and $\varphi(r)$ is bounded. Let

557

$U(\theta)$ be periodic of period 2π, odd, and $U(\theta) = 2\theta^{1/2}(\pi - \theta)^{1/2} \geq 0$ for $0 \leq \theta \leq \pi$. Consider the differential equation for $y = (y^1, y^2)$ given by $y' = f(y^1, y^2)$, where $f = (-y^2\varphi(r)U(\theta), y^1\varphi(r)U(\theta))$, $y^1 = r\cos\theta$, $y^2 = r\sin\theta$, and consider the initial point $(y^1, y^2) = (1, 0)$. Find the differential equations for r, θ. Note that $\theta' = U(\theta)$, $\theta(0) = 0$ has the maximal solution $\theta_1(t) = \pi\sin^2 t$ for $0 \leq t \leq \frac{1}{2}\pi$, $\theta_1(t) = \pi$ for $t \geq \frac{1}{2}\pi$ and the minimal solution $\theta = -\theta_1(t)$. Thus S_c, for $c \geq \frac{1}{2}\pi$, is the circle $(y^1)^2 + (y^2)^2 = 1$.

4.3. (a) Show that the set of solutions $y(t)$ of (4.1) is uniformly bounded for $t_0 \leq t \leq c$ in which case the proof of Theorem 4.1 is applicable.

Chapter III

4.1. (a) Consider $z^k(a) \leq y_0^k$ and $D_R z^k(t) \leq f^k(t, z(t))$, $k = 1, \ldots, d$. Let $y_\epsilon(t)$ be a solution of the system $y^{k\prime} = f^k(t, y) + \epsilon$, $y(a) = y_0$ for small $\epsilon > 0$. Suppose that c is such that $a < c \leq b$ and that $y_\epsilon(t)$ exists for $a \leq t \leq c$ for all small ϵ. Suppose, if possible, that $z^k(t) \leq y_\epsilon^k(t)$ does not hold for $a \leq t \leq c, k = 1, \ldots, d$. Then there is a largest $t_0, a < t_0 < c$ such that $z^k(t) \leq y_\epsilon^k(t)$ for $a \leq t \leq t_0, k = 1, \ldots, d$ but, for some j, $z^j(t) > y_\epsilon^j(t)$ holds for some t-values, $t > t_0$ and t arbitrarily near to t_0. In particular, $z^j(t_0) = y_\epsilon^j(t_0)$. But then $D_R z^j(t_0) \leq f^j(t_0, z(t_0)) \leq f^j(t_0, y_\epsilon(t_0)) < y_\epsilon^{j\prime}(t_0)$, so that $z^j(t) < y_\epsilon^j(t)$ for small $t - t_0 > 0$. Contradiction. Thus $z^j(t) \leq y_\epsilon^j(t)$ for $a \leq t \leq c$. Let ϵ tend to 0.

4.1. (b) Let $y_1(t)$ be the solution of $y' = f(t, y)$, $y_1(a) = z(a)$. It suffices to show that $y_1^k(t) < y_0^k(t)$ for $0 < t \leq b, k = 1, \ldots, d$, since $z^k \leq y_1^k \leq y_0^k$. It is clear that $y_1^k < y_0^k$ or that $(y_0^k - y_1^k)' > 0$ at $t = a$ for every $k \neq j$; hence $y_1^k < y_0^k$ for small $t - a > 0$ for $k \neq j$ and for $k = j$. If there is a first t-value $t_0 > a$ at which $y_1^k(t_0) = y_0^k(t_0)$ for some k, then $(y_0^k - y_1^k)' \leq 0$ at $t = t_0$, but $y_1(t_0) \neq y_0(t_0)$ by uniqueness, and the monotony of f^k implies that $(y_0^k - y_1^k)' > 0$ at $t = t_0$.

4.5. Let $U = 1 + r'(u)$, $v(t) = t$, $t_0 = u_0 = v_0 = 0$, and $r = u^4\sin^2 1/u$.

6.1. Let $\varphi(t)$ be defined for $0 < t \leq 1$ and satisfy $t^2 < \varphi(t) < 4t^2/3$, $\varphi'(t)$ is continuous, $\varphi'(t) \geq 2t$ and $\lim \varphi'(t)$ does not exist as $t \to 0$. Put $\omega(t, u) = \varphi'(t)u/\varphi(t)$, so that $\omega(t, u) \geq 3u/2t$. Let y, f be scalars and put $f(t, y) = 0, 4y/3t$, $4t^{1/3}/3$ according as $y < 0, 0 \leq y \leq t^{1/3}$ or $y > t^{1/3}$ and $t \geq 0$.

6.2. Let $f(t, y)$ be 0, $(1 + \epsilon)y/t$ or $(1 + \epsilon)t^\epsilon$ according as $y \leq 0, 0 < y < t^{1+\epsilon}$, or $y \geq t^{1+\epsilon}$.

6.5. Put $\omega_0(t, u) = \sup|f(t, y_1) - f(t, y_2)|$ for $|y_1 - y_2| = u[\text{or} \leq u]$. $\omega_0(t, u)$ is continuous for $t_0 \leq t \leq t_0 + a$ and $0 \leq u \leq 2b$ (why?); [nondecreasing in u (for fixed t);] $\omega_0(t, 0) \equiv 0$; $|f(t, y_1) - f(t, y_2)| \leq \omega_0(t, |y_1 - y_2|)$; $\omega_0(t, u) \leq \omega(t, u)$ for $t_0 < t \leq t_0 + a$, and $0 \leq u \leq 2b$. Suppose that $u' = \omega_0(t, u)$, $u(t_0) = 0$ has a solution $u^0(t) \not\equiv 0$ on some interval $[t_0, t_0 + \epsilon]$. Then $u^{0\prime} = \omega_0(t, u^0) \leq \omega(t, u^0)$. But the proof of Theorem 6.1 shows that this is impossible.

6.6. (a) Suppose that there are two solutions on $0 \leq t \leq \epsilon$ and let $u(t)$ denote their difference. Then $u = u' = \cdots = u^{(d)} = 0$ at $t = 0$ and $|u^{(d)}(t)| \leq \lambda(t)$, where

$$\lambda(t) = \sum_{k=0}^{d-1} \epsilon_k(d - k)! \, t^{-d+k} \, |u^{(k)}(t)|.$$

Successive integrations of $u^{(d)}(t)$ give

$$(d - k - 1)! \, u^{(k)}(t) = \int_0^t (t - s)^{d-k-1} u^{(d)}(s) \, ds \quad \text{for} \quad k = 0, \ldots, d - 1;$$

so that

$$(d - k)! \, t^{-d+k} \, |u^{(k)}(t)| \leqq (d - k)t^{-1} \int_0^t (1 - s/t)^{d-k-1} \lambda(s) \, ds.$$

Multiplying this relation by ϵ_k and adding gives

$$\lambda(t) \leqq \sum_{k=0}^{d-1} \epsilon_k (d - k)t^{-1} \int_0^t (1 - s/t)^{d-k-1} \lambda(s) \, ds.$$

Note that $\lambda(t) \geqq 0$ is continuous for $0 \leqq t \leqq \epsilon$ and $\lambda(0) = 0$. Suppose that $u(t) \not\equiv 0$, then $\lambda(t) \not\equiv 0$. Choose t in the last relation so that $\lambda(t) > 0$ is the maximum value of λ on $[0, \epsilon]$. Then replacing $\lambda(s)$ by $\lambda(t)$ on the right gives the inequality

$$\lambda(t) < \lambda(t) \sum \epsilon_k (d - k)t^{-1} \int_0^t (1 - s/t)^{d-k-1} \, ds.$$

The factor of $\lambda(t)$ on the right is $\sum \epsilon_k = 1$. Contradiction.

6.7. (a) Suppose that there are two solutions $y = y_1(t), y_2(t)$ on $0 \leqq t \leqq \epsilon$. Put $m(t) = |y_1(t) - y_2(t)|$. Then $|D_L m(t)| \leqq \omega_j(t, m(t))$. Suppose that there exists a $t = s, 0 < s \leqq \epsilon$, such that $m(s) > \alpha(s)$. Then the minimal (unique, by Corollary 6.3) solution of $u' = \omega_1(t, u), u(s) = m(s)$ satisfies $u(t) \leqq m(t)$ on its left maximal interval. Then $u(t)$ can be extended over $(0, s]$ and $u(+0) = 0$. This contradicts $u(s) = m(s) > \alpha(s)$. Consequently, $m(t) \leqq \alpha(t)$ for $0 \leqq t \leqq \epsilon$. Similarly, if $m(\epsilon) > 0$, then the minimal (= unique) solution of $v' = \omega_2(t, v)$, $v(\epsilon) = m(\epsilon)$ exists on $0 < t \leqq \epsilon$ and satisfies $0 \leqq v(t) \leqq m(t)$. Thus $0 \leqq \lim v(t)/\beta(t) \leqq \lim m(t)/\beta(t) \leqq \lim \alpha(t)/\beta(t) = 0$ as $t \to 0$. Thus $v(t) \equiv 0$, but $v(\epsilon) = m(\epsilon)$.

6.8. Let $\delta(t) = \|y_2(t) - y_1(t)\|$ and follow proof of Theorem 6.2.

7.1. Consider the scalar initial value problem $u' = 3u^{2/3}, u(t_1) = u_1$.

7.2. Consider the differential equation $u' = g(u)$, where $g(u) > 0$ is continuous for all u. The solutions are the level curves $U(t, u) = \text{const.}$ of $U(t, u) = t - \int_0^u ds/g(s)$.

8.1. It can be supposed that $V(0) = 0$ and that V has a strict minimum at $y = 0$ for otherwise $V(y)$ can be replaced by $\pm[V(y) - V(0)]$. Although it is not assumed that V is of class C^1, V has the trajectory derivative $\dot{V} = 0$. Hence the proof of Theorem 8.1 is applicable.

Chapter IV

2.2. (a) By the standard orthogonalization process, there exists a nonsingular, triangular $Q(t)$ such that $Z(t) = Y(t)Q(t)$ is unitary. This can be verified by first showing that if $\gamma_1, \ldots, \gamma_d$ are d linearly independent (Euclidean) vectors, then there exists a nonsingular, triangular $d \times d$ matrix $R = (r_{jk})$ with $r_{jj} \neq 0$, $r_{jk} = 0$ if $k > j$, such that the (column) vectors $\delta_1, \ldots, \delta_d$ are orthonormal, where $\delta_j = r_{j1}\gamma_1 + \cdots + r_{jj}\gamma_j$. In fact, let $\delta_1 = \gamma_1/\|\gamma_1\|$, so that $r_{11} = 1/\|\gamma_1\|$. In order to obtain δ_2, subtract the component of γ_2 along δ_1 from γ_2 and normalize the result to be a vector of length 1; i.e., $\delta_2 = [\gamma_2 - (\gamma_2 \cdot \delta_1)\delta_1]/\|\gamma_2 - (\gamma_2 \cdot \delta_1)\delta_1\|$, where a dot denotes scalar multiplication. Thus δ_2 is a linear combination $r_{21}\gamma_1 + r_{22}\gamma_2$ of γ_1, γ_2 with $r_{22} \neq 0$. This process continues. If Γ is the matrix with the columns $\gamma_1, \ldots, \gamma_d$ and Δ the matrix with columns $\delta_1, \ldots, \delta_d$, then

$\Delta = R\Gamma$ is unitary. Hence $\Delta^* = \Gamma^* R^*$ is unitary. This can be applied to $\Gamma = Y^*(t)$ to give $Q(t) = R^*$ and $Z(t) = \Delta^*$. The construction shows that $Q(t)$ is continuously differentiable. The desired result follows from Exercise 2.1 for if $C(t) = (c_{jk})$ and $Z(t)$ has the vectors $z_1(t), \ldots, z_d(t)$ as columns, then since $C^H = Z^* A^H Z$ and $c_{jk} = 0$ if $j > k$, it follows that $c_{jj} = \frac{1}{2} z_j \cdot A^H z_j$ and $z_{jk} = z_j \cdot A^H z_k$ if $j < k$. [The construction of $Q(t)$ shows incidentally that the diagonal elements of $Q(t)$ are real-valued; also that $Q(t)$ has only real entries if $Y(t)$ has only real entries.]

2.2. (b) If $y_1(t), \ldots, y_d(t)$ are the columns of $Y(t)$, let

$$Q(t) = \text{diag} \, [1/\|y_1(t)\|, \ldots, 1/\|y_d(t)\|]$$

in Exercise 2.1.

8.2. (c) Induction on $d - g$. If $d - g = 0$, then $h(j) = 1$, the roots $\lambda(1), \ldots, \lambda(d)$ are distinct, and $Y(0)$ is the Vandermonde matrix $(\lambda(j)^{d-k-1})$ for $j, k = 1, \ldots, d$, so that $\det Y(0) = \prod_{i<j} [\lambda(i) - \lambda(j)]$ and (8.11) holds. Assume (8.11) for values of $d - g$ less than a given $d - g > 1$. Assume that $h(1) > 1$. Let $\lambda \neq \lambda(1), \ldots, \lambda(g)$ and replace the first column of $Y(0)$ by $(\lambda^{d-1}, \lambda^{d-2}, \ldots, 1)$. The resulting matrix, say $K(\lambda)$, corresponds to the case where the distinct roots of (8.9) are $\lambda, \lambda(1), \ldots, \lambda(g)$ with corresponding multiplicities $1, h(1) - 1, h(2), \ldots, h(g)$. Thus by the induction hypothesis, $\det K(\lambda)$ is obtained from the right side of (8.11) by replacing $h(1)$ by $h(1) - 1$ and multiplying by $[\lambda - \lambda(1)]^{h(1)-1} \prod_{j>1} [\lambda - \lambda(j)]^{h(j)}$. If the first column of $K(\lambda)$ is differentiated $h(1) - 1$ times with respect to λ at $\lambda = \lambda(1)$, the resulting matrix is $[h(1) - 1]! \, Y(0)$. Hence $[h(1) - 1]! \det Y(0) = \partial^{h(1)-1} \det K(\lambda)/\partial \lambda^{h(1)-1}$ at $\lambda = \lambda(1)$.

8.3. (a) Determine $u_j(t)$ as a solution of (8.1) satisfying $u_j = u_j' = \cdots = u_j^{(d-j-1)} = 0$, $u_j^{(d-j)} = 1$ at $t = a$. Then $W_1(t; u_1) = u_1(t) \neq 0$ on $[a, b]$. Assume that $1 \leq k < d$ and that $W_j(t) = W_j(t; u_1, \ldots, u_j) \neq 0$ on (a, b) for $j = 1, \ldots, k$. Consider the kth order linear differential equation with solutions $u = u_1, \ldots, u_k$ given by

(*) $W(t; u, u_1, \ldots, u_k)/W_k(t) = 0$.

This is an equation of the form $u^{(k)} + \cdots = 0$. Let $a < t_0 < b$; it will be shown that $W_{k+1}(t_0) \neq 0$. Since $W_k(t_0) \neq 0$, it is possible to find constants $c_1, \ldots, c_k$ such that $u_0 = c_1 u_1(t) + \cdots + c_k u_k(t) + u_{k+1}(t)$ has a zero of multiplicity of at least k at $t = t_0$ and, of course, has a zero of multiplicity $d - k - 1$ at $t = a$. Hence, by assumption, $u_0^{(k)}(t_0) \neq 0$. It is clear that $W(t; u_{k+1}, u_1, \ldots, u_k)/W_k(t) = W(t; u_0, u_1, \ldots, u_k)/W_k(t) = u_0^{(k)}(t) \neq 0$ at $t = t_0$. Hence $W_{k+1}(t_0) \neq 0$ for $a < t_0 < b$.

8.3. (b) Define $h(t)$ by $h(t) = L[v]$. Then (8.19) holds if v is written in place of u. Since $a_0 v$ has $d + 1$ zeros on (a, b), its derivative $(a_0 v)'$, hence $a_1(a_0 v)'$, has d zeros on (a, b). Thus $[a_1(a_0 v)']'$, hence $a_2[a_1(a_0 v)']'$, has $d - 1$ zeros on (a, b). Continuing this argument, we find a zero $t = \theta$ for h.

8.3. (c) Let $u(t) \not\equiv 0$ be a solution of (8.1) having d zeros on (a, b). Then $u(t) = c_1 u_1(t) + \cdots + c_d u_d(t)$ for some constants $c_1, \ldots, c_d$. Suppose that $c_{k+1} \neq 0$ but $c_{k+2} = \cdots = c_d = 0$ for some $k, 0 \leq k \leq d - 1$. Let $L_k[u]$ denote the function on the left of (*) in the *Hint* to part (a). Then, by part (b), $L_k[u] = 0$ at some point $t = \theta, a < \theta < b$. This contradicts $W_{k+1}(\theta) \neq 0$.

8.3. (d) Constants $c_1, \ldots, c_d$ such that $u = c_1 u_1(t) + \cdots + c_d u_d(t)$ has the desired property satisfy a set of d linear, inhomogeneous equations. These equations have a unique solution since the only solution for the corresponding homogeneous system is $c_1 = \ldots = c_d = 0$ by part (c).

8.3. (e) Let $u_0(t)$ be any solution of $L[u] \equiv 1$. By (d), there is a solution $u^0(t)$ of (8.1) such that $u(t) = u_0(t) + u^0(t)$ has the desired properties.

8.3. (f) Let $a < t_0 < b$ and suppose that $t_0 \neq t_1, \ldots, t_k$. Then $u_0(t_0) \neq 0$ by (b) since $L[u_0] \equiv 1 \neq 0$. Choose the number α such that $v(t_0) = u(t_0) + \alpha u_0(t_0)$. Then $w(t) \equiv v(t) - u(t) - \alpha u_0(t)$ has at least $d + 1$ zeros on $[\gamma, \delta]$ if $[\gamma, \delta]$ contains $t_0, t_1, \ldots, t_k$. Hence $L[w](\theta) = 0$ for some θ on (γ, δ) by (b). But $L[w](\theta) = L[v](\theta) - \alpha$.

9.1. (a) Let $F(\lambda, t) = \det [\lambda I - R - G(t)]$, $F(\lambda) = \det (\lambda I - R)$ be the characteristic polynomials of $R + G(t)$, R, respectively; so that $F(\lambda, t) \to F(\lambda)$ as $t \to \infty$ uniformly on any bounded λ-set. In the complex λ-plane, let the disc $|\lambda - \lambda_j| \leqq \epsilon$ contain only one zero, $\lambda = \lambda_j$, of $F(\lambda)$. It follows from the theorem of Rouché (in the theory of functions of a complex variable) that, for large t, $F(\lambda, t)$ has exactly one zero $\lambda_j(t)$ in $|\lambda - \lambda_j| \leqq \epsilon$ and $\lambda_j(t) \to \lambda_j$ as $t \to \infty$. By residue calculus,

$$\lambda_j(t) = (2\pi i)^{-1} \oint \lambda F_\lambda(\lambda, t) \, d\lambda / F(\lambda, t),$$

where $F_\lambda = \partial F/\partial \lambda$ and the line integral is taken over the circle $|\lambda - \lambda_j| = \epsilon$.

9.1. (b) Let $R_0 = Q_0^{-1} R Q_0$, $G_0(t) = Q_0^{-1} G(t) Q_0$, and consider the equation for an eigenvector of $R_0 + G_0(t)$ belonging, say, to $\lambda_1(t)$ and having its first component $z^1(t) = 1$. Let R_1 be the $(d - 1) \times (d - 1)$ matrix diag $[\lambda_2, \ldots, \lambda_d, E_0]$ and similarly $G_1(t)$ the matrix obtained from $G_0(t)$ by deleting the first row and column. Then $\det [R_0 + G_0(t) - \lambda_1(t)I] = 0$, but $\det [R + G_1(t) - \lambda_1(t)I_{d-1}] \neq 0$ for large t. Hence, if z_1 is a $(d - 1)$-dimensional vector, then $z = (1, z_1)$ satisfies $[R_0 + G_0(t) - \lambda_1(t)]z = 0$ if and only if $[R_1 + G_1(t) - \lambda_1(t)I_{d-1}]z_1 = -g_1(t)$, where $g_1(t) = (g_{210}(t), \ldots, g_{d10}(t))$ and $(g_{110}, g_{210}, \ldots, g_{d10})$ is the first column of $G_0(t)$. Thus $y_1(t) = Q_0 z(t)$, where $z(t) = (1, z_1(t))$ and $z_1(t) = -[R_1 + G_1(t) - \lambda_1(t)I_{d-1}]^{-1} g_1(t)$.

9.1. (c) Let $Q_1(t)$ be the nonsingular matrix with columns $y_1(t), \ldots, y_k(t)$, $Q_0 e_{k+1}, \ldots, Q_0 e_d$; so that $Q_1^{-1}[R + G(t)]Q_1$ has the form

$$
Q_1^{-1}(t)[R + G(t)]Q_1(t) = \begin{pmatrix}
\lambda_1(t) & 0 & . & . & . & 0 & \\
0 & \lambda_2(t) & . & . & . & 0 & R_{12}(t) \\
. & . & . & . & . & . & \\
0 & 0 & . & . & . & \lambda_k(t) & \\
0 & & & & & & R_{22}(t)
\end{pmatrix},
$$

where $R_{22}(t)$ is a $(d - k) \times (d - k)$ matrix. Let $A(t) = Q_1^{-1}(R + G)Q_1 - \lambda_1(t)I$ and let $v_j = A(t)e_j$. Thus $v_1 = 0$, $v_2 = [\lambda_2(t) - \lambda_1(t)]e_2, \ldots, v_k = [\lambda_k(t) - \lambda_1(t)]e_k$. Note that $A^2(t)y = 0$ implies $A(t)y = 0$, hence $y = (\text{const.}) e_1$, for $\lambda = 0$ is a simple eigenvalue of $A(t)$. This implies that $e_1, \ldots, e_k, v_{k+1}, \ldots, v_d$ are linearly independent, for a linear combination of these vectors is of the form $c_1 e_1 + A(c_2 e_2 + \cdots + c_d e_d)$. Let $w_r = A(t)v_r(t)$ for $r = 2, \ldots, d$ and write $w_r(t)$ as the linear combination $w_r(t) = h_{r1}e_1 + \cdots + h_{rk}e_k + h_{rk+1}v_{k+1} + \cdots + h_{rd}v_d$. Then $h_{r1} = 0$ (otherwise, there is a vector y such that $Ay = e_1$). Thus, with respect to the basis $e_1, e_2, \ldots, e_k, v_{k+1}, \ldots, v_d$, the linear transformation

$A(t)$ corresponds to the matrix

$$
B(t) = \begin{pmatrix}
0 & 0 & 0 & \cdots & 0 & 0 \ldots 0 \\
0 & \lambda_2(t) - \lambda_1(t) & 0 & \cdots & 0 & \\
0 & 0 & \lambda_3(t) - \lambda_1(t) & \cdots & 0 & \\
\cdot & \cdot & \cdot & \cdots & \cdot & H_{12}(t) \\
0 & 0 & 0 & \cdots & \lambda_k(t) - \lambda_1(t) & \\
& & 0 & & & H_{22}(t)
\end{pmatrix},
$$

where $H_{22}(t)$ is a $(d - k) \times (d - k)$ matrix. This means that if $Q_2(t)$ is the matrix with the columns $(e_1, e_2, \ldots, e_k, v_{k+1}(t), \ldots, v_d(t))$ then $Q_2^{-1} A(t) Q_2 = B(t)$, i.e., $(Q_1 Q_2)^{-1}[R + G(t)](Q_1 Q_2) = B(t) + \lambda_1(t)I$. The desired conclusion follows from this in the case $k = 1$. If $k > 1$, repeat these operations on the $(d - 1) \times (d - 1)$ matrix in the lower right corner of $B(t) + \lambda_1(t)I$.

11.1. What is the relation between $Q(t)$ and $Q_0(t)$ in the proof of Theorem 11.2?

11.3. The equation $t^2 u'' + (3t - 1)u' + u = 0$ has the formal solution $u = 1 + 1!\,t + 2!\,t^2 + \cdots + k!\,t^k + \ldots$.

12.3. (a) If μ is not an integer, then there are two solutions $J_{\pm\mu}(t)$ of the form (12.12) with $J_\mu = \sum_{k=0}^{\infty} (-1)^k (\tfrac{1}{2}t)^{2k+\mu}/k!\,\Gamma(k + \mu + 1)$. If $\mu \geq 0$ is an integer, there is the solution $J_\mu(t)$ and a linearly independent solution obtained from $u = J_\mu(t) \int^t s^{-1} J_\mu^{-2}(s)\,ds$ which can be taken of the form

$$
u = J_\mu(t) \log t + t^{-\mu} \sum_0^\infty a_{2k} t^{2k},
$$

where the last power series can be determined by a substitution into the Bessel differential equation.

13.2. (a) Apply Lemma 11.2 to $X(t)$. Make the variation of constants $\eta = Z(t)y$ and multiply the resulting system by $P^{-1}(t)$ to obtain a system $t^D \eta' = \ldots$ to which Theorem 13.1 applies.

Chapter V

5.1. Use Lemma 5.1 and its proof.

5.2. Use Stokes' formula (5.3) when S is a rectangle $a_1 \leq y^1 \leq b_1, a_j \leq y^j \leq b_j$ and $y^i = $ const. for $i \neq 1, j$. Differentiate the result with respect to b_j, then b_1.

6.1. Number the components of y so that $(i_1, \ldots, i_d)$ becomes $(0, \ldots, 0, i_{j+1}, \ldots, i_d)$, where $0 \leq j \leq d$ and $i_{j+1}, \ldots, i_d$ are distinct integers on the range $1 \leq i_k \leq d$. Write $y'' = f$ as a system of first order, say $y' = z, z' = f(t, y, z)$ or $dy - z\,dt = 0, dz - f\,dt = 0$. Choose the $2d \times 2d$ matrix $A = A(t, y, z)$ so that

$$
A = \begin{pmatrix} I_{d+j} & 0 \\ Q & P \end{pmatrix},
$$

I_{d+j} is the unit $(d + j) \times (d + j)$ matrix, $P = \text{diag}\,[z^{i_j+1}, \ldots, z^{i_d}]$, and the last $d - j$ components of $A(dy - z\,dt, dz - f\,dt)$ are $z^{i_k}dz^k - f^k\,dy^{i_k}$ for $k = j + 1, \ldots, d$.

6.2. (a) Write the differential equations (6.11) as a system of first order equations for a $2d$-dimensional vector $(x, y) = (x^1, \ldots, x^d, y^1, \ldots, y^d)$ in the form $dx^i - y^i\, dt = 0$, $dy^i + \Sigma\Sigma\Gamma^i_{jk}y^jy^k\, dt = 0$, where $i = 1, \ldots, d$. Choose the $2d \times 2d$ matrix $A = A(x, y)$ to be

$$A = \begin{pmatrix} I & 0 \\ B & I \end{pmatrix},$$

where I is the unit $d \times d$ matrix and $B = (b_{jk})$ is the $d \times d$ matrix with the elements $b_{jk} = \Sigma\Gamma^j_{km}y^m$. Apply Theorem 6.1.

6.2. (c) Write $y^1 = x$, $y^2 = y$, and $ds^2 = h(y)(dx^2 + dy^2)$, where $h = 1 + 9y^{4/3}$. If x is used as the independent variable, the differential equations for geodesics become $d^2y/dx^2 = \frac{1}{2}[1 + (dy/dx)^2]H(y)$, where $H(y) = \partial \log h/\partial y$. The initial value problem $y(0) = 0$, $dy(0)/dx = 0$ has the solutions $y = 0$ and $y = x^3$.

6.3. (a) Let $\omega_j = p_{j1}\, dy^1 + p_{j2}\, dy^2$ and $d\omega_j = q_j\, dy^1 \wedge dy^2$. Determine $\omega_{12} = p_1\, dy^1 + p_2\, dy^2$ by the equations $p_1 p_{22} - p_2 p_{21} = q_1$, $-p_1 p_{12} + p_2 p_{11} = q_2$.

6.3. (b) It can be supposed that $p_{11}p_{22} \neq 0$ at $y = 0$, otherwise renumber ω_1, ω_2. Let $y^1 = \eta^1(y^2, u^1)$ be the solution of the initial value problem $dy^1/dy^2 = -p_{12}/p_{11}$ and $y^1 = u^1$ at $y^2 = 0$. Thus $y^1 = \eta^1(y^2, u^1)$ is of class C^1 and $\partial\eta^1(0, 0)/\partial u^1 \neq 0$. Hence $y^1 = \eta^1(y^2, u^1)$ can be solved for $u^1 = U^1(y^1, y^2)$ for small $\|y\|$. In terms of (y^2, u^1)-coordinates, ω_1 is of the form $\omega_1 = T_1\, du^1$, where $T_1 \neq 0$ is a continuous function of (y^2, u^1), hence of (y^1, y^2). Similarly, let $y^2 = \eta^2(y^1, u^2)$ be the solution of $dy^2/dy^1 = -p_{21}/p_{22}$ and $y^2 = u^2$ when $y^1 = 0$ and let $u^2 = U^2(y^1, y^2)$ be the inverse function. In terms of (y^1, u^2), ω_2 has the form $\omega_2 = T_2\, du^2$ and $T_2 \neq 0$. The transformation $u^j = U^j(y^1, y^2)$ is of class C^1 with a nonvanishing Jacobian at $y = 0$ and has an inverse $y^j = Y^j(u^1, u^2)$. In (u^1, u^2)-coordinates, $\omega_j = T_j\, du^j$ where T_j is a continuous function of (u^1, u^2). Thus $\alpha = 2T\, du^1\, du^2$ and $T = T_1 T_2 \neq 0$. It can now be supposed that $T(u^1, 0) \equiv T(0, u^2) \equiv 1$ for otherwise (u^1, u^2) are replaced by

$$v^1 = \pm |T(0, 0)|^{-1/2} \int_0^{u^1} T(s, 0)\, ds, \qquad v^2 = |T(0, 0)|^{-1/2} \int_0^{u^2} T(0, s)\, ds,$$

where $\pm T(0, 0) > 0$. Under the C^1-transformation $y \to u$, the property of having a continuous exterior derivative is not lost. Hence $\partial T_1/\partial u^2$, $\partial T_2/\partial u^1$ exist and are continuous; Exercise 5.2. Let $h(u^1, u^2) = (T_1/T_2)^{1/2} > 0$. Thus $\omega_1 = hT_1^{1/2}\, du^1$, $\omega_2 = (T^{1/2}/h)\, du^2$, and hence,

$$\omega_{12} = -[\log (T^{1/2}/h)]_{u^1}\, du^1 + [\log (T^{1/2}/h)]_{u^2}\, du^2.$$

Since $d\omega_{12} = K\omega_1 \wedge \omega_2$, Stokes' relation is of the form $\oint \omega_{12} = \iint KT\, du^1\, du^2$.

Apply this to the rectangle with vertices $(0, 0)$, $(u^1, 0)$, (u^1, u^2), $(0, u^2)$ to obtain

$$-\log T(u^1, u^2) = \int_0^{u^1} \int_0^{u^2} KT\, du^1\, du^2.$$

8.1. Cf. the proof of Theorem 8.1 and Exercise III 7.3.

Chapter VI

1.2. Let $x_0 \in E$. Choose the enumeration of the components of x and of $Y_1, \ldots, Y_e$ such that $B_{11} = (b_{ij}(x_0))$, where $i, j = 1, \ldots, e$, is nonsingular and let $B_{21} = (b_{ij}(x))$, where $i = e + 1, \ldots, e + d$ and $j = 1, \ldots, d$. Write

$x = (z^1, \ldots, z^e, y^1, \ldots, y^d)$ and $H(y, z) = B_{21}(x)B_{11}^{-1}(x)$. Then the system (*) can be written as $(\partial u/\partial z)B_{11} + (\partial u/\partial y)B_{21} = 0$ or equivalently as (1.12). Hence (*) is complete on a neighborhood of $x_0 = (z_0, y_0)$ if and only if $\omega = dy - H(y, z)\,dz$ is completely integrable at (y_0, z_0). Define the $(d + e) \times (d + e)$ matrix B_0 by

$$B_0 = \begin{pmatrix} B_{11} & 0 \\ B_{21} & I \end{pmatrix}, \qquad \text{so that} \qquad B_0^{-1} = \begin{pmatrix} B_{11}^{-1} & 0 \\ -H & I \end{pmatrix}.$$

Let $B_0 = (\beta_{jk}(x))$ and $B^{-1} = (\delta_{jk}(x))$, where $j, k = 1, \ldots, d + e$. Thus $Y_k[u] = \sum_j \beta_{jk}(x)\, \partial u/\partial x^j$ for $k = 1, \ldots, d + e$, let $\alpha_j = \sum_k \delta_{jk}(x)\, dx^k$ for $j = 1, \ldots, d + e$. Thus $Y_1 = \cdots = Y_e = 0$ is complete at x_0 if and only if $\alpha_{e+1}, \ldots, \alpha_{d+e}$ is completely integrable at x_0. For any function $u(x)$ of class C^1, the total differential $du = \sum(\partial u/\partial x^i)\, dx^i$ can be written as

(1)
$$du = \sum_{i=1}^{d+e} Y_i[u]\alpha_i \qquad \text{for} \quad u \in C^1,$$

since $(\delta_{jk}) = (\beta_{jk})^{-1}$. If $u(x)$ is of class C^2, a simple calculation of the exterior derivative of du gives

(2)
$$\sum_{k=1}^{d+e} Y_k[u]\, d\alpha_k + \sum_{i=1}^{d+e} \sum_{j=1}^{d+e} Y_i[Y_j(u)]\alpha_i \wedge \alpha_j = 0.$$

There exist (unique) continuous functions $e_{ijk}(x) = -e_{jik}(x)$ such that

(3)
$$d\alpha_k = \sum_{i=1}^{d+e} \sum_{j=1}^{d+e} e_{ijk}(x)\alpha_i \wedge \alpha_j$$

is the expression for $d\alpha_k$ in terms of the base $\alpha_1, \ldots, \alpha_{d+e}$. Consequently, (1) and (2) give

(4)
$$Y_i[Y_j(u)] - Y_j[Y_i(u)] + \sum_{k=1}^{d+e} e_{ijk} Y_k[u] = 0$$

for any function u of class C^2. By Theorems 3.1 and 3.2, the system $\alpha_{e+1}, \ldots, \alpha_{e+d}$ is completely integrable if and only if $\alpha_{e+1} = \cdots = \alpha_{e+d} = 0$ implies that $d\alpha_{e+1} = \cdots = d\alpha_{e+d} = 0$; this is the case if and only if $e_{ijk} = 0$ for $i, j = 1, \ldots, e$ and $k = e + 1, \ldots, e + d$. In view of (4), the desired result follows with $c_{ijk} = -e_{ijk}$ for $i, j, k = 1, \ldots, d$.

3.2. Use Lemma V 5.1.

3.4. See the proof of Lemma 3.1.

3.5. If $d\omega$ exists and $\omega = (\omega_1, \ldots, \omega_d)$, then $d\omega_k$ is of the form $d\omega_k = -\sum\sum(\partial h_{kj}/\partial y^i)\, dy^i \wedge dz^j + \sum\sum p_{kij}\, dz^i \wedge dz^j$, where $p_{kij} = -p_{kji}$. What are the conditions on p_{kij} for (3.1)? Use Exercise V 5.1.

6.1. (b) See Exercise 3.5.

6.2. Show that if such an A exists, the solution of (6.2) for fixed z_0 and $\eta = y_0$ is of the form $y = T(z)\eta + y_1(z)$, where $Y(z)$ is a nonsingular matrix of class C^1 and $y_1(z)$ is a vector of class C^1 for z near z_0. Also, the map $(z, \eta) \to (z, y)$ gives $dy - H\, dz = Y(z)\, d\eta$. Since the form $Y(z)\, d\eta$ in $d\eta$, dz has a continuous exterior derivative, the same is true of the form $dy - H\, dz$ in dy, dz. Apply Lemma 3.1.

8.1. (a) Let $u_0 = u(y_0)$, $p_0 = u_y(y_0)$. By Lemma V 3.1, there are continuous functions a, b^k, c^k of $(u_1, y_1, p_1, u_2, y_2, p_2)$ for small $|u_i - u_0|$, $|y_i - y_0|$, $|p_i - p_0|$ for $i = 1, 2$, such that

(1) $F(u_2, y_2, p_2) - F(u_1, y_1, p_1) = (u_2 - u_1)a$

$$+ \sum_{k=1}^{d} (y_2{}^k - y_1{}^k)b^k + \sum_{k=1}^{d} (p_2{}^k - p_1{}^k)c^k,$$

and that $a = F_u$, $b^k = \partial F/\partial y^k$, $c^k = \partial F/\partial p^k$ if $(u_1, y_1, p_1) = (u_2, y_2, p_2)$. Let $h > 0$ be small. Apply (1) to $(u_1, y_1, p_1) = (u(y), y, u_y(u))$ and $(u_2, y_2, p_2) = (u(y + \Delta y), y + \Delta y, u_y(y + \Delta y))$, where Δy is the vector with its jth component equal to 0 or h according as $j \neq m$ or $j = m$. It follows from (7.1) that

(2) $$a \, \Delta u/h + b^m + \sum_{k=1}^{d} (\Delta p^k/h)c^k = 0.$$

Let $y_h(t)$ be a solution of $y' = c$, $y(0) = y_0$, where $c = (c^1, \ldots, c^d)$ and the argument of c is $(u(y), y, u_y(u), u(y + \Delta y), y + \Delta y, u_y(y + \Delta y))$ for fixed Δy. It is clear that if $\epsilon > 0$ is small and fixed and $h_1 > h_2 > \ldots$ is a suitably chosen sequence, then $y(t) = \lim y_h(t)$, as $h = h_n \to 0$, exists uniformly for $|t| \leq \epsilon$ and is a solution of $y' = F_p(u(y), y, u_y(y))$, $y(0) = y_0$. Note that if $p_h{}^m(t) = [u(y_h(t) + \Delta y) - u(y_h(t))]/h$ and y is replaced by $y_h(t)$ in (2), then (2) can be written as $p_h{}^{m'}(t) = -a\Delta u/h - b^m$ since $(y_h(t) + \Delta y)' = y_h'(t) = c$. Thus, as $h = h_n \to 0$, $p_h{}^m(t) \to \partial u(y(t))/\partial y^m$ and $p_h{}^{m'}(t) \to -F_u \partial u/\partial y^m - \partial F/\partial y_m$ uniformly for $|t| \leq \epsilon$.

8.1. (c) The condition $F_p(u_0, y_0, p_0) \neq 0$ implies that (7.1) can be written in the form (7.17), where y is a real variable if $d = 2$. Application of part (a) gives the desired result; cf. Exercise 7.2 (a) for the corresponding equations of the characteristic strips.

8.2. (b) Let $u(t)$, $y(t)$, $p(t)$ be a characteristic strip for $F = 0$, i.e., a solution of (7.9). Then the relation $G(u(t), y(t), p(t)) = 0$ can be differentiated with respect to t to give $G_u u' + G_y \cdot y' + G_p \cdot p' = 0$. Hence $H(u(t), y(t), p(t)) = 0$ follows from (7.9).

9.1. Introduce the new variables $(x, x + y)$ instead of (x, y).

Chapter VII

3.1. Use (3.1) and a circle J.

3.2. Consider the deformation $(1 - s)f_0(y) + sf(y) = f_0(y) + sf_1(y)$, $0 \leq s \leq 1$ and small $\|y\|$.

7.2. Make a real linear change of variables to bring A into a suitable normal form.

·10.2. Construct C as follows: Let $\alpha > 0$ be large. C consists of the part of the arc C_α: $\frac{1}{2}v^2 + G(u) = \alpha$ for $u \geq a$, the line segments $v = \pm\gamma$ for $|u| \leq a$ with $\gamma > 0$ and $\frac{1}{2}\gamma^2 + G(a) = \alpha$ and the part of the arc C_β: $\frac{1}{2}v^2 + G(u) = \beta$ for $u \leq -a$ with $\beta = \frac{1}{2}\gamma^2 + G(-a)$.

11.1. Put $z(t, y) = g_\alpha(\mu(t, g_\alpha^{-1}(y)))$, where $y = (y^1, y^2)$ and $z = (z^1, z^2)$. Thus $z(t, y)$ is of class C^1 and, for small $|t|$ and $|h|$, $z(t + h, y) = z(t, z(h, y))$. Thus $dz(t, y)/dt = [dz(t + h, y)/dh]_{h=0} = \Sigma (\partial z(t, y)/\partial y^j)[dz^j(h, y)/dh]_{h=0}$.

11.2. (c) By (b), ∂N is empty or $\partial N = N$. Correspondingly, $N^0 = N$ or N^0 is empty. In the first case, N is open and closed, hence $N = M$ since M is connected. In the second case, N^0 is empty implies that N is nowhere dense since N is closed.

13.1. Let $\epsilon > 0$ be arbitrarily fixed. Let i, j be such that $0 < \gamma_j - \gamma_i < \epsilon$.

Then translates of the interval $\gamma_i \leqq y < \gamma_j$ by $m(j - i)\alpha + h$ for $h, m = 0, \pm 1, \ldots$ have endpoints of the form $n\alpha + k$ and cover $-\infty < y < \infty$.

14.1. If the torus is cut along the circle Γ, it becomes a piece of cylindrical surface. If a half-orbit remains in this piece of cylindrical surface for $t \geqq 0$, its set of ω-limit points is a closed orbit; cf. the proof of Theorem 4.1.

14.2. There is an m_0 such that $\Gamma_0 = C^+(m_0)$ is a Jordan curve by Theorem 14.1. Γ_0 cannot be contracted to a point; cf. the proof of Lemma 14.1. If the torus is cut along Γ_0, it becomes a piece of cylindrical surface on which the arguments in the proof of Theorem 4.1 become applicable.

Chapter VIII

2.2. Replace (2.1) by $dy/dx = Y/X$; cf. the proof of Theorem III 6.1.

3.1. Introduce polar coordinates; cf. (3.14) where $\alpha = 0$, $F^1 = xh(r)$, $F^2 = yh(r)$.

3.2. Let $F(z) = (-yh(r), xh(r))$, $r = (x^2 + y^2)^{1/2} \geqq 0$ and $h(r)$ a suitably chosen continuous function for $0 \leqq r \leqq 1$ with $h(0) = 0$.

3.3. Let $F(z) = (-h(r)k(\theta)y, h(r)k(\theta)x)$, where $r = (x^2 + y^2)^{1/2} \geqq 0$, $h(r) = |\log r|^{-1}$ or 0 according as $0 < r \leqq \frac{1}{2}$ or $r = 0$; and for the respective cases (a), (b), or (c), let $k(\theta) = 1$, $|\cos \theta|^{1/2}$, or 0.

3.4. The only characteristic values (mod 2π) are $\theta = 0$, π. Apply Theorem 2.1.

3.6. Cf. Exercise 2.2.

4.2. Consider $x' = -yx^4 - y^5 - \varphi(x, y)y$, $y' = xy^4 - x^3y^2 + \varphi(x, y)x$, where $\varphi(x, y) \geqq 0$ according as $x^2 + y^2 \geqq 0$.

4.3. (b) Note that if $\psi(r) = \alpha(\log 1/r)^{-k/(k-1)}$, then (4.36) has a positive solution $\theta = \epsilon(\log 1/r)^{-1/(k-1)}$ provided that there exists an $\epsilon > 0$ satisfying $\epsilon/(k - 1) - c\epsilon^k \geqq \alpha$. This is the case if $\alpha > 0$ does not exceed the maximum of $\epsilon/(k - 1) - c\epsilon^k$.

4.4. (b) Let $\psi(r) = \alpha(\log 1/r)^{-k/(k-1)}$ and let $\theta = u(\log 1/r)^{-1/(k-1)}$ in (4.38). Then $r(\log 1/r) \, du/dr = \alpha - [u/(k - 1) - cu^k] \geqq$ const. > 0 if α exceeds the maximum of $u/(k - 1) - cu^k$.

4.6. (d) Put $\theta_0 = \displaystyle\int_t^\infty \varphi(s) \, ds + \gamma/t^{\mu-1}$ for a suitable constant γ to be determined

4.6. (e) Put $\theta_0 = t^{-1/(\lambda-1)} \displaystyle\int_t^\infty s^{1/(\lambda-1)} \varphi(s) \, ds$.

Chapter IX

5.3 (b) Suppose that $\eta = E\xi + F(\xi)$ has no solution ξ. Let $0 < r_1 < r_2$ be such that $F(\xi) = 0$ if $\|\xi\| \geqq r_1$, $\|\xi + E^{-1}F(\xi)\| < r_2$ if $\|\xi\| \leqq r_1$, and $\|E^{-1}\eta\| < r_2$. Then $\xi \to E^{-1}(E\xi + F(\xi)) = \xi + E^{-1}F(\xi)$ maps the sphere $\|\xi\| \leqq r_2$ into itself, reduces to the identity on the boundary, and omits the point $E^{-1}\eta$. This is impossible (Brouwer).

5.4. (d) If not, $\|z_n{}^* - z_n\| >$ const. $(c - 2\delta)^n$ for large n [by part (c) applied to $S_n{}^*$ instead of S_n].

5.4. (e) Let $(y_n, z_n) = T_n(y_0, z_0)$, where $z = g(y_0)$; so that $y_n(y_0), z_n(y_0)$ are functions of y_0. Let $y_n{}^h = [y_n(y_0 + he_j) - y_n(y_0)]/h$, $z_n{}^h = [z_n(y_0 + he_j) - z_n(y_0)]/h$. Then $(y_n{}^h, z_n{}^h) = (Ay_{n-1}^h + Y_{1n}^h y_{n-1}^h + Y_{2n}^h z_{n-1}^h, Cz_{n-1}^h + Z_{1n}^h y_{n-1}^h + Z_{2n}^h z_{n-1}^h)$, where the matrices $Y_{1n}^h, Y_{2n}^h, Z_{1n}^h, Z_{2n}^h$ tend to the Jacobian matrices

$\partial_{y,z} Y_n$, Z_n evaluated at (y_{n-1}, z_{n-1}), as $h \to 0$. By (d), the functions $y_n{}^h$, $z_n{}^h$ of y_0 are bounded and equicontinuous (for fixed n) and satisfy $\|z_n{}^h\| \leqq \|y_n{}^h\|$ for $n = 0, 1, \ldots$. Hence, they have limits (u_n, v_n), as h tends to 0 through suitably selected subsequences; $(u_n, v_n) = T_n{}^{**}(u_0, v_0)$, $u_0 = e_j$, $\|v_n\| \leqq \|u_n\|$. By uniqueness, selection is unnecessary.

11.2. If the assertion is false, apply Stokes' formula to $\oint f^2(\xi)\, d\xi^1 - f^1(\xi)\, d\xi^2$.

Chapter X

1.4. (d) In order to obtain the sufficiency of the first criterion, make the change of variables $\xi = [I - G_0(t)]\eta$ for large t and apply part (c). The sufficiency of the second criterion follows from that of the first and from part (b) or can be obtained from (c) by the change of variables $\xi = [I + G_0(t)]^{-1}\eta$.

4.1. (a) For the inequality involving $\sigma(t)$, write $\int_0^t = \int_0^T + \int_T^t$ in (4.18) and apply integration by parts to the second integral.

4.1. (b) Note that $\sigma(t)$ is a solution of the differential equation $\sigma' + (c - \epsilon)\sigma = \psi$ and integrate this equation over the interval $[s, t]$.

4.2. (a) Multiply the differential equation $\sigma' + (c - \epsilon)\sigma = \psi$ for σ by σ^{p-1}, integrate over $[0, t]$, and apply Hölder's inequality to the integral on the right.

4.6. (b) Without loss of generality, it can be supposed that $G(t)$ is continuously differentiable, otherwise $G(t)$ can be approximated arbitrarily closely by such matrices. By Exercise IV 9.1, for large t, there is a continuously differentiable, nonsingular $Q(t)$ such that $\int^\infty \|Q'(t)\| dt < \infty$, $\lim Q(t)$ as $t \to \infty$ is nonsingular; $Q^{-1}[E + G(t)]Q = \text{diag } [\lambda_1(t), \ldots, \lambda_k(t), E_{22}(t)]$, where $E_{22}(t)$ is a $(d - k) \times (d - k)$ matrix and $E_0 = \lim E_{22}(t)$ as $t \to \infty$, is of the form $E_0 = \text{diag } (E_1, E_2)$; E_1 is an $m \times m$ matrix with eigenvalues having positive real parts; E_2 is an $n \times n$ matrix with eigenvalues having negative real parts. The change of variables $\xi = Q(t)\eta$ gives a system

$$\eta' = Q^{-1}(t)[E + G(t)]Q(t)\eta + Q^{-1}(t)Q'(t)\eta.$$

If $\eta = (y, x, z)$, where y is a k-dimensional, x an m-dimensional, and z an n-dimensional vector, then the system for η has the form

$$x' = E_1 x + G_{11}(t)x + G_{12}(t)y + G_{13}(t)z,$$
$$y' = \text{diag } [\lambda_1(t), \ldots, \lambda_k(t)]y + G_{21}(t)x + G_{22}(t)y + G_{23}(t)z,$$
$$z' = E_2 y + G_{31}(t)x + G_{32}(t)y + G_{33}(t)z,$$

where G_{jk} is a rectangular matrix. There exist continuous functions $\psi(t)$, $\psi_0(t)$ such that $\psi(t) \to 0$ as $t \to \infty$ and $\int^\infty \psi_0(t)\, dt < \infty$ and $\|G_{ij}(t)\| \leqq \psi(t) + \psi_0(t)$ for $i, j = 1, 2, 3$ and, in addition, $\|G_{2j}(t)\| \leqq \psi_0(t)$ for $j = 1, 2, 3$. It follows that solutions $\eta(t) = (y(t), x(t), z(t))$ for which $x(t), z(t) = o(\|y(t)\|)$ as $t \to \infty$ are bounded.

4.6. (c) If $\eta(t) \neq 0$ is a solution of the system in (b) such that $x(t), z(t) = o(\|y(t)\|)$, then $\xi = Q(t)\eta(t)$ has the desired form.

4.6. (d) If $y = (y^1, \ldots, y^k)$ in (b), introduce the new variables $u = (u^1, \ldots, u^k)$ in place of y, where $y^j = u^j \exp \int^t \lambda_j(s) \, ds$ for $j = 1, \ldots, k$. Apply Lemma 4.3.

4.7. Use Exercise IX 5.3 (a).

4.8. (a) Let $K_n = \{(x, y, z): \|z\| \leq \frac{1}{2}\theta(\|x\| + \|y\|)\}$, $K_{n0} = \{(x, y, z): \|x\| \leq \theta \|y\|\}$, and $S = \{(x, y, z): x = x_0, \ y = y_0, \ \|z\| \leq \frac{1}{2}\theta(\|x_0\| + \|y_0\|)\}$. Apply Exercise 4.7. To verify that $K_n \cap T_n(S)$ is not empty, use Exercise IX 5.3 (b) and show that there exists a $z_{(n)}$ such that the z-coordinate of $T_n(x_0, y_0, z_{(n)})$ is 0.

4.9. (a) Note that $\|y_{n+1} - y_n\| \leq \psi_0(n)(1 + 2\theta) \|y_n\|$, hence $\|y_{n+1}\| \leq \|y_n\| [1 + \psi_0(n)(1 + 2\theta)]$. Thus $\|y_n\| \leq c_0 \|y_0\|$, where $c_0 = \prod [1 + \psi_0(n)(1 + 2\theta)]$ is a convergent (infinite) product. Hence $\sum \|y_{n+1} - y_n\|$ is majorized by $c_0 \|y_0\| (1 + 2\theta) \sum \psi_0(n)$.

4.9. (b) From the arguments in part (a) and $1 - \psi_0(n)(1 + 2\theta) \geq 1 - 3\psi_0(n) > 0$, note that there exists a constant c_1 such that if the inequality in (4.44) holds for $n = 0, 1, \ldots, j$, then $\|y_k\| \leq c_1 \|y_j\|$ and $\|y_{k+1} - y_j\| \leq c_1 \|y_j\| s_k$ for $k = 0, 1, \ldots, j - 1$, where $s_k = \psi_0(k) + \psi_0(k + 1) + \psi_0(k + 2) + \cdots$. Show that there exists a $(y_{(j)}, z_{(j)})$ such that if $T_n(x_0, y_{(j)}, z_{(j)}) = (x^n, y^n, z^n)$, then $y^j = y_\infty$ and $z^j = 0$. As in Exercise 4.8, $\|z^n\| \leq \frac{1}{2}\theta(\|x^n\| + \|y^n\|)$ for $n = 0, 1, \ldots, j$. For large j, we have $\|x^j\| \leq \theta \|y^j\| = \theta \|y_\infty\|$, hence $\|x^n\| \leq \theta \|y^n\|$ for $n = 0, 1, \ldots, j$. [For suppose, if possible, that $\|x^j\| > \theta \|y^j\|$, then $\delta < (1 - a)\theta/3$ implies that $\|x^k\| > \theta \|y^k\|$ for $k = 0, 1, \ldots, j$. But then $\|x^{k+1}\| \leq \theta_0 \|x^k\|$, where $\theta_0 = a + \delta(2 + \theta)(1 + \theta)/2 < 1$, and so $\|x^j\| \leq \theta_0^j \|x^0\|$. For large j, this contradicts $\|x^j\| > \theta \|y_\infty\|$ and $\|x_0\| = \|x^0\| \leq \theta \|y_\infty\|$.] Thus $\|x^k\| \leq \theta \|y^k\|$, $\|z^k\| \leq \theta \|y^k\|$ for $k = 0, \ldots, j$. It follows that $\|y^k - y_\infty\| \leq c_1 \|y_\infty\| s_{k-1}$ for $k = 1, \ldots, j$. Let (y_0, z_0) be a limit point of the sequence $(y_{(1)}, z_{(1)}), (y_{(2)}, z_{(2)}), \ldots$.

11.2. Use Exercise 4.2.

17.1. Since R is not in a Jordan normal form if $h < d$, let Y be a constant matrix such that $Y^{-1}RY = E = \mathrm{diag} [J(1), \ldots, J(g)]$ is a Jordan normal form. It can be supposed that the diagonal elements of $J(1)$ are $\lambda(1) = 0$, so that $h(1) = h$. Since $a_{d-h+1} = \cdots = a_d = 0$, Y can be chosen of the form

$$Y = \begin{pmatrix} 0 & \cdot & \cdot \\ & \cdot & \cdot \\ I_h & \cdot & \cdot \end{pmatrix},$$

where I_h is the unit $h \times h$ matrix and 0 is the rectangular $(d - h) \times h$ zero matrix; cf. Exercise IV 8.2 where $Y = Y(0)$. The change of variables $\xi = Y\eta$ replaces (17.4) by

$$\eta' = E\eta + Y^{-1}G(t)Y\eta.$$

Note that the only nonzero elements of $G(t)Y$ are in the first row. The first h elements of the first row of GY are $-p_{d-h+1}, \ldots, -p_d$, and the other elements of this row are linear combinations of $p_1, \ldots, p_d$ with constant coefficients. It is then necessary to consider the factor $G(t)YQ(t)$ of $Q^{-1}(t)Y^{-1}G(t)YQ(t)$, where $Q(t)$ is described by (14.1)–(14.2). Actually, $Q(t)$ is of the form

$$Q(t) = \begin{pmatrix} Q_1(t) & 0_2 \\ 0_1 & Q_2(t) \end{pmatrix}$$

where $Q_1(t)$ is an $h \times h$ matrix corresponding to the matrix $Q(t)$ in (14.1) with d replaced by h; $Q_2(t) = t^{1-\beta}D_2$ where $\beta = h - j + \alpha$ and D_2 is a constant diagonal matrix; and $0_1, 0_2$ are rectangular zero matrices.

17.3. (a) Consider the differential equation for the function u'/u and integrate this equation over the interval $[t, s]$.

17.4. (a) and (b) This can be obtained from Exercise 4.6. It is more easily obtained directly by assuming that $q(t)$ has a continuous derivative and replacing the second order equation by a first order system for $y = (y^1, y^2)$ where $y^1 = u' + f^{1/2}u$, $y^2 = u' - f^{1/2}u$, and $f(t) = 1 + q(t)$ in case (a) and $f(t) = 1 - q(t)$ in case (b). Introduce a new independent variable s, $ds = f^{1/2} dt$.

17.5. Write (17.23) as a linear system $y' = H(t)y$ for the binary vector $y = (u' + f^{1/2}u, u' - f^{1/2}u)$. For parts (a) and (b), let $ds = g\, dt$, $z = (y^1/W, y^2 W)$, $f^{1/2} = g + ih$, where $g(t), h(t)$ are real-valued, $g > 0$, $W = \exp \int_0^t ih\, dt$. There are solutions satisfying $z^2 = o(|z^1|)$ and $z^1 = o(|z^2|)$ by Theorem 8.1 and Exercise 11.2. For the end of (b), suppose that $f > 0$ and $f^{1/2} \in L^1$. Since $f^{1/2q} \in L^q$ and $f^{1/2p\,-1/2}f'/f \in L^p$, $f'/f \in L^1$. Hence $\lim f(t) > 0$ exists as $t \to \infty$, but $f^{1/2} \in L^1$. For part (c), suppose $f \in C^2$. Let $y = Q(t)z = (z^1 - z^2 b/(a + c)$, $- z^1 b/(a + c) + z^2)$, $Q^{-1}HQ$ is diagonal where $a = f^{1/2}$, $b = f'/4f$, $c = (a^2 + b^2)^{1/2}$. Introduce the new independent variable $s = \int^t |Re\, f^{1/2}(r)|\, dr$. Note that the assumptions imply that $b/(a + c) \to 0$ as $t \to \infty$ and $\int^\infty |[b/(a + c)]'|\, dt < \infty$.

17.6. Introduce the new dependent variable z defined by $u = ze^{-t^2/4}$; cf. (XI 1.9). Apply Exercise 17.5 (b) to the resulting equation for z.

Chapter XI

4.1. If $u(t)$ is a solution of (4.1λ), (4.2), then $\bar{u}(t)$ is a solution of (4.1λ), (4.2). Apply Green's relation (2.10) to $f = -\lambda u(t)$, $g = -\lambda\bar{u}(t)$.

4.3. (a) Denote by (4.19_n) the equation which results if u is replaced by u_n in (4.19). Multiply (4.19_{n+1}) by u_0, (4.19_0) by u_{n+1}, subtract, and obtain $v_n' = -(\lambda_{n+1} - \lambda_0)r_0u_{n+1}u_0$. Divide this relation by $r_0u_0^2$ and differentiate to get (4.20) with $v = v_n = r_0u_0^2(u_{n+1}/u_0)'$. The last part follows from $v_n'/v_0' = (\lambda_{n+1} - \lambda_0)(\lambda_1 - \lambda_0)^{-1}(u_{u+1}/u_1)$ and L'Hôpital's rule.

4.3. (b) Note that $v_n = p_0u_0^2(u_{n+1}/u_0)'$. Write $u_{0i} = u_i(t)$ for $i = 0, \ldots, k - 1$ and $u_{1i} = v_i(t)$ for $i = 1, \ldots, k - 2$. If $u_{ji}(t)$ for $i = 0, \ldots, k - j - 1$, and $p_j > 0$, $r_j > 0$ have been defined, put $p_{j+1} = 1/r_ju_{j0}^2$. Consider $u = c_0u_0 + \cdots + c_{k-1}u_{k-1}$ or rather $u = c_0u_{00} + \cdots + c_{k-1}u_{0,k-1}$. Then

$$p_0u_{00}^2(u/u_{00})' = c_1u_{10} + \cdots + c_{k-1}u_{1,k-2}.$$

Hence $p_1u_{10}^2[p_0u_{00}^2(u/u_{00})'/u_{10}]' = c_2u_{20} + \cdots + c_{k-1}u_{2,k-3}$. Continuing this process gives the desired result with $a_0 = 1/u_{00}$, $a_1 = p_0u_{00}^2/u_{10}$, $a_2 = p_1u_{10}^2/u_{20}$, ...

4.4. (c) There is a constant c such that $U_n(t) = cD(t, t_0, \ldots, t_{n-1})$, and $D(t, t_0, \ldots, t_{n-1})$ is positive or negative according as the number of inversions in $t, t_0, \ldots, t_{n-1}$ is even or odd.

5.1. Choose $a = 0$, $b = 1$, $0 < s < 1$. It suffices to construct a non-negative continuous $q(t) = q(t; m, s)$ such that (5.1) has a solution $u(t) \not\equiv 0$ satisfying

$u(0) = u(1) = 0$ and $\int_0^1 m(t)q(t)\,dt \geq [m(s) + \epsilon]/s(1 - s)$. To this end, let

$0 < \delta < \min(s, 1 - s)$, $u(t)$ a function having a continuous, nonpositive second derivative on $0 \leq t \leq 1$, such that $u(t) = t$ and $u(t) = (1 - t)(s - \delta)/(1 - s - \delta)$ on $0 \leq t \leq s - \delta$ and $s + \delta \leq t \leq 1$, respectively. Let $q(t) = 0$, $-u''/u$, or 0 according as $0 \leq t \leq s - \delta$, $s - \delta \leq t \leq s + \delta$, or $s + \delta \leq t \leq 1$. Then

$$\int_0^1 m(t)q(t)\,dt = -\int_{s-\delta}^{s+\delta} m(t)u''(t)\,dt/u(t).$$

Use $m(t) \leq m(s) + \epsilon$ if $|t - s| < \delta$ and $\delta > 0$ is small, $u(t) \geq s - \delta$ for $|t - s| < \delta$, and $-u''(t) \geq 0$.

5.3. (a) It can be supposed that $f = f^+ \geq 0$. (Why?) Put $w = u(t)$ and $h = -gu' - fu$, $a = 0$ in (2.18) and differentiate (2.18) to obtain a formula for $u'(t)$. Let $\alpha = \max |u'(t)|$ and note that $|u(t)| \leq \alpha \min(t, b - t)$. The function $G(t) = \int_0^t s|g|\,ds + \int_t^b (b - s)|g|\,ds$ is nonincreasing or nondecreasing according as $0 \leq t \leq b/2$ or $b/2 \leq t \leq b$.

5.3. (b) Multiply the differential equation by u; use $uu'' = (uu')' - u'^2$ and integrate.

5.3. (d) It can be supposed that $u = 0$ at $t = 0, b$. Let $0 = a_1 < \cdots < a_d = b$ be d zeros of u. If $\alpha = \max |u^{(d)}(t)|$, then $|u(t)| \leq \alpha\Pi |t - a_k|/d!$ [This is a consequence of Newton's interpolation formula but can be proved directly: there exists a $\theta = \theta(t_0)$ such that $u(t_0) = u^{(d)}(\theta)\Pi(t_0 - a_k)/d!$ as can be seen by considering $v(t) = u(t) - C\Pi(t - a_k)/d!$, where $C = d!\,u(t_0)/\Pi(t_0 - a_k)$. Hence $v = 0$ at $t = t_0, a_1, \ldots, a_d$, so that its dth derivative vanishes at some $t = \theta$]. Since $a_1 = 0$, $a_d = b$ and $0 \leq t \leq b$, it follows that $\Pi |t - a_j| \leq t^k(b - t)^{d-k}$ when $a_k \leq t \leq a_{k+1}$. For $0 \leq t \leq b$ and $k = 1, \ldots, d - 1$, one has $t^k(b - t)^{d-k} \leq \max[t(b - t)^{d-1}, t^{d-1}(b - t)] \leq b^d(d - 1)^{d-1}/d^d$ by differential calculus. Hence $|u(t)| \leq \alpha(b^d/d!)[(d - 1)^{d-1}/d^d]$. In addition, u' has at least $d - 1$ zeros, say $a_1', \ldots, a_{d-1}'$. Thus $|u'| \leq \alpha\Pi |t - a_j'|/(d - 1)! \leq \alpha b^{d-1}/(d - 1)!$. Similarly, $|u''| \leq \alpha b^{d-2}/(d - 2)!, \ldots, |u^{(d-1)}| \leq \alpha b/1!$. Choose t^* in the differential equation so that $|u^{(d)}(t^*)| = \alpha$.

5.5. (a) Use (2.43) where $p(t) = 1$.

5.5. (b) Use § 2(xiii) and part (a).

5.8. Use $\psi(t) = \arctan u'/v' = \arctan pu'/pv'$ and note that $q \geq 0$ implies that the zeros of u and u' separate each other since $u'' \leq 0$ [or $u'' \geq 0$] if $u \geq 0$ [or $u \leq 0$].

6.1. Assume first that $J: t_1 \leq t \leq t_2$ and let $u_1, u_2 > 0$ in (6.2). Next, if J is not a closed bounded interval, let $u_1(t), u_2(t)$ be linearly indpendent solutions of (6.1), so that by the first step, for any $[t_1, t_2] \in J$, there are constants c_1, c_2 such that $c_1^2 + c_2^2 = 1$ and the solution $u = c_1u_1(t) + c_2u_2(t) > 0$ on $[t_1, t_2]$. Use a limit involving $[t_1, t_2] \to J$.

6.2. Use the device (1.7) to reduce (6.1) to an equation of the form $d^2u/ds^2 + q_0(s)u = 0$ on an interval $0 \leq s < \infty$. From $q_0(s) \geq 0$, it follows that $d^2u/ds^2 \leq 0$ when $u \geq 0$.

6.3. Suppose that the assertion is false. Then (6.1) has a solution $u(t) \not\equiv 0$ with two zeros a, b in J. Suppose that b is not an endpoint of J. Then $u(t)$ changes sign at $t = b$. Let (6.1_ϵ) be the differential equation obtained by replacing $q(t)$ by $q(t) - \epsilon$, $\epsilon > 0$, and $I_\epsilon(\eta, \alpha, \beta)$ the corresponding functional.

Then $I_\epsilon(\eta; \alpha, \beta)$ is positive definite for $\epsilon > 0$ on $A_2(\alpha, \beta)$ for all $[\alpha, \beta] \subset J$. Let $u_\epsilon(t)$ be the solution of (6.1_ϵ) satisfying $u_\epsilon(a) = 0, u_\epsilon'(a) = u'(a)$. Then $u_\epsilon(t) \neq 0$ for $t \neq a$. But $u_\epsilon(t) \to u(t)$ as $\epsilon \to 0$ uniformly on compact intervals of J, and so, for small ϵ, $u_\epsilon(t) = 0$ at a t near b.

6.4. Show that if $u_2(t) \neq 0$ on $t_1 \leqq t \leqq t_2$ [i.e., if (3.1_2) is disconjugate on $t_1 \leqq t \leqq t_2$], then (3.1_1) is disconjugate on $t_1 \leqq t \leqq t_2$. Compare the functionals belonging to (3.1_1) and (3.1_2).

6.5. (a) Let $u(t) \neq 0$ for $T \leqq t < \omega$. It can be supposed that $u(t)$ is a non-principal solution; cf. first part of Corollary 6.3. Then

$$u_{0r}(t) = [u(t)\int_t^r ds/p(s)u^2(s)]/[u(T)\int_T^r ds/p(s)u^2(s)];$$

cf. (6.14).

6.5. (b) Consider $J: 0 \leqq t < \pi$ and the differential equation $u'' + u = 0$. A principal solution is $u = \sin t$. There is no principal solution satisfying $u(0) = 1$.

6.8. This is a consequence of Exercise 6.5(a), but can be obtained directly as follows: Note that $u_T(t) > 0$, $u_T'(t) \leqq 0$ for $a \leqq t < T$. (Why?) Also, if $a < S < T < \omega$, then $u_S(t) \leqq u_T(t)$ for $a \leqq t \leqq S$ since $u_S(t) \neq u_T(t)$ for $t > a$ and $u_T(t) \neq 0$ for $a \leqq t < T$. Hence $u_T'(0) \leqq 0$ is a nondecreasing function of T, so that $\lim u_T'(0)$ exists as $T \to \omega$. This implies the existence of $u_0(t)$ satisfying (6.19). Also $u_0(t)$ is a principal solution by (iii) of Theorem 6.4.

6.10. Make the change of variable $u = vu_1(t)$ in (6.30_2) and apply Corollary 6.4.

6.11. For $0 \leqq t \leqq \frac{1}{2}j\pi$, put $u_{j0} = 1 + j^2 \sin t/j$; $p_j \equiv 1$ and $q_j = -u_{j0}''/u_{j0} = (\sin t/j)/(1 + j^2 \sin t/j)$. For $\frac{1}{2}j\pi < t < \infty$, extend the definitions of q_j, u_{j0} so that u_{j0} becomes the principal solution of (6.33_j), with $p_j \equiv 1$. Thus $q_j(t) \to 0$, $j \to \infty$, uniformly on bounded intervals of $0 \leqq t < \infty$, but $u_{j0}(t) \to \infty$ as $j \to \infty$ for every $t > 0$.

7.2. (b) Introduce the new dependent variable $v = u \exp -c\int^t g(s)\,ds$ and note that, in the resulting differential equation, $v'' + G(t)v' + F(t)v = 0$, we have $F(t) \leqq 0$.

7.3. Let $q(t) = \mu/t^2$ and $Q(t) = \mu/t$.

7.4. Let $u(t)$ be the solution of (7.1) satisfying $u(0) = 1$, $u'(0) = 0$. Then $r = u'/u$ satisfies the Riccati equation (7.4) and $r(0) = 0$. Suppose $u(t) > 0$ for $0 \leqq t \leqq b$. A quadrature of (7.4) gives $r = R - Q$, where $R(t) = -\int_0^t r^2(s)\,ds$ and $R' = -r^2$. Since $Q \geqq 0$ for $a \leqq t \leqq b$, $R' = -r^2 = -(R - Q)^2 \leqq -(R^2 + Q^2)$ and $R \leqq 0$ for $a \leqq t \leqq b$. If $z \neq 0$, its logarithmic derivative $r_1 = z'/z$ satisfies $r_1' = -(r_1^2 + Q^2)$ and $r_1(a) = 0$. Also, $r_1(t) \to -\infty$ as $t \to t_0$ if $t = t_0$ is the first zero of z greater than a. By Theorem III 4.1, $R \leqq r_1$ on the interval $[a, t_0]$.

7.9. Make the change of variables $u = w_j(t)z$ in (7.35_j), where $w_j(t) = \exp \int_a^t c_j' Q_j(s)\,ds$, use Theorem 7.4 and Exercise 7.2(a).

8.1. Introduce the new dependent variable $u = t^{1/2}v$ in Bessel's equation.

8.2. Use (2.22), (2.28), and Exercise X 1.4(d).

8.3. (a) Use § 2(xiii).

8.4. (a) Write (8.8_j) as a first order system for (u, p_ju'), and apply a linear change of variables. It can be supposed that $p_0(u_0'v_0 - u_0v_0') \exp \int^t (r_0/p_0) \, ds = 1$; so that neither $2|p_0u_0'v_0|$ nor $2|p_0u_0v_0'|$ exceed $\frac{1}{2}|p_0(u_0 + v_0)(u_0 + v_0)'| + \frac{1}{2}|p_0(u_0 - v_0)(u_0 - v_0)'| + |\exp - \int^t (r_0/p_0) \, ds|$.

8.4. (b) Assume that $q(t)$ has a continuous derivative. For a fixed choice of $\pm$, note that $v = \exp \pm i \int_0^t q^{1/2}(s) \, ds$ satisfies the differential equation $u'' + [q \mp \frac{1}{2}iq^{-1/2}q']u = 0$. Apply (a), identifying this equation with (8.8_1) and (8.1) with (8.8_0). (In order to show that the conditions on q, including the parenthetical conditions, imply that a solution u of (8.1) and its derivative u' are bounded, let $q = g + ih$, where $g(t)$, $h(t)$ are real-valued. If $E = g|u|^2 + |u'|^2$, then $E' = g'|u|^2 + ih(\bar{u}u' - \bar{u}'u)$, and so $|E'| \leq$ const. $(|g'| + |h|)E$ for large t. This implies that E, hence u and u', are bounded.) For another proof when q is real-valued, see Exercise X 17.4(a).

8.5. Let $z = uf^{1/4}$, so that (8.12) becomes $(f^{-1/2}z')' + (f^{1/2}(t)g(t))z = 0$, where $g(t) = 1 + f''/4f^{3/2} - 5f'^2/16f^{5/2}$; cf. (2.34)–(2.35). If w denotes either of the functions in (8.11), then $w' = \pm if^{1/2}h^{1/4}w$, where $h(t) = 1 + f'^2/16f^3$ Thus $(f^{-1/4}w')' = -f^{1/4}hw \pm i(h^{1/2})'w$. Using $\pm iw = w'/f^{1/2}h^{1/2}$, it follows that w satisfies $(f^{-1/4}w')' - (h^{1/2})'f^{-1/2}h^{-1/2}w' + f^{1/2}hw = 0$. Identify the equations for w and z with (8.8_0) and (8.8_1). Apply Exercise 8.4(a).

8.6. (a) and (b) Consider the case $u(T)u'(T) \geq 0$, say $u'(T) > 0$, $u(T) \geq 0$. Integrate (8.13) over $[T, t]$ to get

$$u'(t) \geq u'(T) - \int_T^t |f| \, d\tau - \int_T^t q(\tau)u(\tau) \, d\tau.$$

Note that

$$\int_T^t q(\tau)u(\tau) \, d\tau = u(T)\int_T^t q(r) \, dr + \int_T^t\left[\int_T^t q(s) \, ds\right] du(r).$$

Hence if $T \leq t \leq U \leq T + \theta/C$ and $u' \geq 0$ on $[T, t]$, $t \leq S$, then

(*) $$u'(t) \geq u'(T) - \int_T^t |f| \, d\tau - Cu(t).$$

Suppose that there is a (first) $t = T^*$, $T < T^* \leq U$, where $u' = 0$, so that

(**) $$u'(t) \leq \int_T^U |f| \, d\tau + Cu(T^*)$$

at $t = T$. This is valid for $T \leq t \leq T^*$. Integrate over $[T, T^*]$ to get an estimate for $u(T^*)$, and hence by (**) with $t = T$, for $u'(T) > 0$. If no T^* exists, then (*) is valid for $T \leq t \leq U$, and an integration over $[T, U]$ gives the desired result.

8.6. (d) The proof of part (a) shows that if $b - a = \Delta < \theta/c$ and $a \leq t \leq b$, then $|u'(t)|$ is majorized by either

$$\int_a^b |f| \, d\tau + C|u(T)|/(1 - \theta)$$

or by one of the two quantities

$$\int_a^b |f|\, d\tau + r(a) \quad \text{or} \quad \int_a^b |f|\, d\tau + r(b).$$

Thus, in any case,

$$(1 - \theta)\,|u'(t)| \le \int_a^b |f|\, d\tau + C\,|u(T^*)| + r(a) + r(b),$$

where $|u(T^*)| = \max |u(\tau)|$ for $a \le \tau \le b$. Integrating this inequality over $[a, T^*]$ gives an estimate for $|u(T^*)|$ which, together with the last inequality, gives the desired result.

8.7. *Case* 1: $g(b) - g(a) \ge 0$. Let $a \le c \le b$. Put $h(t) = P(t) - N(t) + h(c)$ where $P(c) = N(c) = 0$ and $P(t)$, $N(t)$ are the positive and negative variation of h [so that $P(t)$, $N(t)$ are nondecreasing and var $h = P(b) + N(b) - P(a) - N(a)$] Write

$$\int_a^b [h(t) - h(c)]\, dg\,(t) = \int_a^b P(t)d[g(t) - g(b)] - \int_a^b N(t)d[g(t) - g(a)].$$

Integrating by parts and using $P(a) \le 0 \le N(b)$ gives (8.21) with inf h replaced by $h(c)$. *Case* 2: $g(b) - g(a) = 0$. The arguments just used show that, in this case, (8.21) can be improved to

$$\int_a^b h(t)\, dg\,(t) \le (\text{var } h) \sup \int_\alpha^\beta dg\,(t) \quad \text{for} \quad a \le \alpha < \beta \le b.$$

Case 3: $g(b) - g(a) < 0$. It is sufficient to consider the case that $g(t)$ is a polynomial, for otherwise g can be approximated by polynomials and a limit process applied to (8.21). For a polynomial $g(t)$ with $g(a) > g(b)$, the interval $[a, b]$ can be divided in a finite number of subintervals, $a = t_0 < t_1 < \cdots < t_n = b$, so that $g(t)$ is either decreasing on $[t_j, t_{j+1}]$ or $g(t_{j+1}) = g(t_j)$ for each j. If $[\gamma, \delta] = [t_j, t_{j+1}]$, then, by Case 2,

$$\int_\gamma^\delta h\, dg \le 0 \quad \text{or} \quad \int_\gamma^\delta h\, dg \le (\text{var } h) \sup_{[\gamma,\delta]} \int_\alpha^\beta dg \quad \text{for} \quad \gamma \le \alpha < \beta \le \delta.$$

8.8. Reduce (8.20) to the form (8.13) by introducing the new independent variable $s = s(t) = \int_T^t \exp\left(-\int_{T'}^r p(\tau)\, d\tau\right) dr$.

9.6. Apply § 2 (xiii) and Corollary 9.2 with $\lambda = 1$.

9.9. (a) Use Lemma 9.5 and Corollary 9.3.

9.9. (b) Use Lemma 9.5 and Theorem X 17.5 [cf. (X 17.22)].

10.1. Suppose that $I(\eta; a, b) \ge 0$ for all $[a, b] \subset J$ and $\eta \in A_{3/2}(a, b)$ but that (10.2) is not disconjugate on J. Let $(x(t), y(t)) \ne 0$ be a solution of (10.2) and $x(t) = 0$ at points $t = a, b$ of J. Suppose b is not an endpoint of J, say $t = \beta \in J$ and $\beta > b$. Put $\eta_0(t) = x(t)$ for $a \le t \le b$, $\eta_0(t) = 0$ for $b \le t \le \beta$; thus $\eta_0 \in A_{3/2}(a, \beta)$. For an arbitrary $\eta \in A_2(a, b)$, put $\eta_\epsilon(t) = \eta_0(t) + \epsilon \eta(t)$ and $j(\epsilon) = I(\eta_\epsilon; a, \beta)$. Thus $j(\epsilon) \ge 0$, $j(0) = I(\eta_0; a, b) = 0$. Hence $dj(\epsilon)/d\epsilon = 0$ at $\epsilon = 0$. Using (10.22) and integrations by parts, it is easy to see that $(dj/d\epsilon)_{\epsilon=0} = 2P(b)x'(b) \cdot \eta(b)$. If $\eta(b) = x'(b)$, it follows that $x'(b) = 0$, hence $x(t) \equiv 0$. Contradiction.

10.2. If $(P_1 x')' + Q_1 x = 0$ has a solution $x(t) \not\equiv 0$ vanishing at two points $t = \alpha, \beta$, where $a \leq \alpha < \beta \leq b$ then $I(\eta, \alpha, \beta) = \int_\alpha^\beta (P_2 \eta' \cdot \eta' - Q_2 \eta \cdot \eta)\, dt$ is not positive definite on $A_2[\alpha, \beta]$. Let $\eta(t) = x(t)$.

10.4. Note that $B = P^{-1}$. Let $B^{\frac{1}{2}}$ denote the self-adjoint, positive definite square root of B; cf. Exercise XIV 1.2. Since M is nonsingular and $M^{-1} = \int_s^\omega U^{-1} B U^{*-1}\, dt$, it is seen that

$$\int_s^\omega (U^{-1} B U^{*-1} c \cdot c)\, dt = \int_s^\omega \| B^{\frac{1}{2}} U^{*-1} c \|^2\, dt < \infty$$

From the relation $\|c\|^2 = (U^{-1} B^{\frac{1}{2}}) B^{-\frac{1}{2}} Uc \cdot c = B^{-\frac{1}{2}} Uc \cdot B^{\frac{1}{2}} U^{*-1} c$ and Schwarz's inequality $\|c\|^2 \leq \| B^{-\frac{1}{2}} Uc \| \cdot \| B^{\frac{1}{2}} U^{*-1} c \|$, it follows that

$$\| B^{-\frac{1}{2}} Uc \|^{-2} \leq \| B^{\frac{1}{2}} U^{*-1} c \|^2 / \|c\|^4.$$

Finally, $\| B^{-\frac{1}{2}} Uc \|^2 = B^{-1} Uc \cdot Uc = Px \cdot x$.

10.5. Cf. The proof of Theorem 10.4.

10.7. Without loss of generality, it can be supposed that $U_0(a) = U(a) = I$. Since $V(a) U^{-1}(a) = U^{*-1}(a) V^*(a)$, it follows that $\Delta = K_0$ in (10.12) Hence (10.21) holds with $s = a$ and $K_1 = I$. If $K_0 = \Delta \leq 0$, it follows that $\{\dots\}$ in (10.21) is positive definite, hence nonsingular, and so det $U(t) \neq 0$ for $t \geq a$. Conversely, suppose that $K_0 c \cdot c > 0$ for some vector $c \neq 0$ and suppose, if possible, that det $U(t) \neq 0$ for $t \geq a$. Then $S_a(t)$, defined by (10.34) with $s = a$ and $T(r) \equiv I$, satisfies (10.48) with $s = a$. But $M = -K_0 \geq 0$. This is a contradiction.

Chapter XII

3.2. Write (3.12) as a first order system for the $2d$-dimensional vector $(-z' - F^* z, z)$.

3.3. (a) Use the Sturm comparison theorems, e.g., Corollary XI 3.1.

3.3. (b) Repeat the proof of Theorem 3.3 with $r = \|x(t)\|^2 + \int_0^t x(s) \cdot K(s) x(s)\, ds$.

3.3. (c) Use $r = \|x(t)\|$.

4.1. (b) Introduce $x - t x_0 / p$ as a new dependent variable; see Theorem 4.1 and Remark 1 following it.

4.2. Introduce $x - t x_0 / p$ as a new dependent variable in (4.1).

4.3. Let $R_1 > 0$ be arbitrary. Then $\|f\| \leq M$, where $M = aR_1^2 + b$ if $\|x'\| \leq R_1$. Show that if $R_1 = r^*$, then the two inequalities in (4.20) hold.

4.4. If there exist two solutions, let $x(t)$ denote their difference and $r(t) = \|x(t)\|^2$. Show that $r''(t) \geq -(2\theta_0 + \frac{1}{2}\theta_1^2) r(t)$; cf. the proof of Theorem 3.3.

4.6. (a) If $x_1(t), x_2(t)$ are two solutions, the difference $x_1 - x_2$ cannot have a positive maximum at a point t_0, $0 < t_0 < p$. Otherwise $x_1' = x_2'$, $(x_1 - x_2)'' > 0$ at $t = t_0$.

4.6. (b) The boundary value problem $x'' = -3(x')^{\frac{2}{3}}$, $x(0) = x(2) = 0$ has the solutions $x = 0$ and $x = [1 - (1 - t)^4]/4$.

4.6. (c) Suppose that there are two solutions $x_1(t), x_2(t)$ and that $x(t) = x_1(t) - x_2(t)$ has a positive maximum at a point t_0, $0 < t_0 < p$. It can be supposed that $x(t) > 0$ for $0 < t < p$, otherwise $(0, p)$ is replaced by the largest interval containing $t = t_0$ on which $x(t) > 0$. Put

$$\alpha(t) = [f(t, x_1(t), x_1'(t)) - f(t, x_1(t), x_2'(t))]/(x_1'(t) - x_2'(t))$$

or $\alpha(t) = 0$ according as $x_1'(t) - x_2'(t) \neq 0$ or $= 0$ and $\beta(t) = f(t, x_1(t), x_2'(t)) - f(t, x_2(t), x_2'(t))$. Then $x(t)$ satisfies a differential equation of the form $x'' = \alpha(t)x' + \beta(t)$, where $\alpha(t)$ is bounded and measurable and $\beta(t) \geq 0$. Hence, if

$$\gamma(t) = \exp -\int_0^t \alpha(s)\, ds, \quad \text{then} \quad (\gamma(t)x')' = \gamma(t)\beta(t) \geq 0. \quad \text{Introduce the new}$$

independent variable s, where $ds = \gamma(t)\, dt$, so that $dx/ds = \gamma x'$ and $d^2x/ds^2 \geq 0$.

4.7. (a) If there are two, say $x_1(t)$ and $x_2(t)$, let $x = x_1 - x_2$. Then $x''(t) > 0$ [<0] at any t where x has a positive maximum [negative minimum].

4.7. (b) At a point where $x(t)$ has a maximum, $x'(t) = 0$ and $x''(t) \leq 0$. At a point where $x'(t)$ has a maximum, $x''(t) = 0$.

4.7. (c) If $\lambda = 0$, there is the solution $x(t) \equiv 0$. Use Theorem 2.3 to show that the set of λ is open on $0 \leq \lambda \leq 1$ and use part (b) to show that the set of λ is closed.

4.8. Let $h(t)$ be periodic of period p and of class C^1. Then

$$x'' = x\alpha(t, h(t), x') + \beta(t, h(t), x')$$

has a unique solution $x(t) = T_0[h]$ of period p by Exercise 4.7. Show the applicability of Schauder's fixed point theorem.

5.2. On $0 \leq t \leq 1$, consider the family of real-valued functions $x = 1 + \epsilon n^4(t - 1/n)^4$ or $x = 1$ according as $0 \leq t \leq 1/n$ or $1/n \leq t \leq 1$.

5.3. If (5.21) is assumed instead of (5.29), then $r(t)$ has no maximum on $0 < t < p$. Hence if $0 \leq t_0 \leq t \leq p$, then $r(t) \leq \max(r(t_0), r(p)) = r(t_0)$; i.e., $r(t)$ is nonincreasing, so that (5.30) holds. When (5.29) holds, the proof of Theorem 5.1 shows that $r = \|x_\epsilon(t)\|^2$ satisfies (5.30). But the inequalities (5.30) are not lost in the limit process $\epsilon = \epsilon(n) \to 0$.

5.4. Introduce the new dependent variable $x = u - u_0(t)$, then the boundary value problem becomes $x'' = f(t, x, x')$ and $x(0) = u_0 - u_0(0)$, $x(p) = u_p - u_0(p)$, where $f(t, x, x') = h(t, x + u_0(t), x' + u_0'(t)) - h(t, u_0(t), u_0'(t))$. Note that $\pm f(t, \pm R, 0) \geq 0$ for $R > 0$ since h is nondecreasing in u. Also, $|f| \leq \varphi(|x'| + c)$ for a constant $c \geq |u_0'(t)|$. Apply Corollary 5.2.

5.5. (a) Let $m = 1, 2, \ldots$. By Theorem 5.1, (5.26) has a solution on $0 \leq t \leq m$ satisfying $x(0) = x_0$, $x(m) = 0$. For any $p > 0$, the sequences of functions $x_m(t)$, $x_m'(t)$ for $m \geq p$ are uniformly bounded and equicontinuous on $0 \leq t \leq p$.

5.6. Note that $r = \|x(t)\|^2$ satisfies $r'' \geq 0$ for large t. Thus if the assertion is false, there is an $m > 0$ such that $0 < m^2 \leq r(t) \leq R^2$ for large t. Define $q(t) = 2[x \cdot f(t, x, x') + \|x'\|^2]/\|x\|^2$, where $x = x(t)$. Hence $r'' - q(t)r = 0$ and $q(t) \geq 2h(t)/m^2$. Apply the last part of Corollary XI 9.1.

5.8. (b) Define a function $x_1(x_0)$ of x_0 for $\|x_0\| \leq R$ as follows: $x_1 = x(p, x_0)$. This gives a map of the ball $\|x_0\| \leq R$ into itself. Apply Brouwer's fixed point theorem.

8.2. Let $g(t)$ be a bounded, measurable function for $t \geq 0$. Then

$$y(t) = \int_0^\infty G(t, s)g(s)\, ds + U(t)y_0$$

is a bounded solution of (6.1) satisfying $P_0y(0) = y_0$. For $x(t)$ a measurable function on $t \geq 0$ such that $\|x(t)\| \leq \rho$, let $g(t) = f(t, x(t))$ and let $y(t) = T_0[x(t)]$ be defined by the last formula line. Let $\mathfrak{B} = \mathfrak{D} = L^\infty(Y)$ and consider T_0 as a map of Σ_ρ into $\mathfrak{D}$.

9.1. It can be supposed that $y(t) \equiv 0$, otherwise introduce $y - y_1(t)$ as a new dependent variable in (9.5). The problem is reduced to showing existence of solutions [for the transformed equation (9.5)] which exist for large t and tend to 0 as $t \to \infty$.

Chapter XIII

2.1. Suppose first that (2.26), hence (2.28), holds. This is equivalent to

$$\max \left(\|z(r)\|, \|y(t)\|\right) \leq K\|z(s) - y(s)\| \qquad \text{for} \quad 0 \leq r \leq s \leq t$$

if $y(0) \in W_0$, $z(0) \in W_1$. If $z(0) \notin W_0$, let $z(0) = z^0(0) + y(0)$, where $y(0) \in W_0$, $z^0(0) \in W_1$. Follow the proof of Lemma 2.2 to obtain analogue of (2.7). This gives the "if" part under the assumption (2.26). The proof under the assumption (2.27) follows by a modification of the proof of Lemma 2.5.

6.3. Let $0 \leq \varphi(t) \in B$ and

$$\psi(t) = \int_t^{t+1} \varphi \, ds, \qquad \psi_1(t) = \int_0^t e^{-\nu(t-s)}\varphi(s) \, ds, \qquad \psi_2(t) = \int_t^\infty e^{\nu(t-s)}\varphi(s) \, ds.$$

It suffices to show that $\psi_1, \psi_2 \in D$; cf. the proof of Theorem 6.4. Let the integer $n \geq 0$ be such that $n \leq t < n + 1$. Then

$$\psi_1(t) = \sum_{j=1}^n \int_{t-j}^{t-j+1} e^{-\lambda(t-s)}\varphi(s) \, ds + \int_0^{t-n} e^{-\lambda(t-s)}\varphi(s) \, ds$$

$$\leq \sum_{j=1}^n e^{-\lambda(j-1)} \int_{t-j}^{t-j+1} \varphi \, ds + e^{-\lambda n} \int_0^{t-n} \varphi \, ds$$

$$\leq \sum_{j=1}^\infty e^{-\lambda(j-1)} \psi^j(t),$$

where $\psi^j(t) = 0$ or $\psi^j(t) = \psi(t - j)$ according as $t < j$ or $t \geq j$. Thus $\psi_1 \in D$; in fact, $|\psi_1|_D \leq \Sigma \, e^{-j\lambda} |\psi|_D$ by condition (d) for $D \in \mathscr{T}$. Similarly, a use of the first part of (6.9) shows that $\psi_2 \in D$.

7.1. Use Exercise XI 8.8 and Theorems 5.3, 5.4.

13.1. Use Theorem 12.1 and Exercise 7.1.

13.2. Use Theorem 12.3.

Chapter XIV

1.1. If the assertion is false, then (1.3_0) holds for all solutions. Hence $|\det Y(t)|^2$ has a finite limit as $t \to \omega$. This contradicts (1.5_∞) and (1.6).

1.2. (a) Let U be a unitary matrix such that $U^*AU = D$ is diagonal, say $D = \text{diag} \, [\lambda_1^2, \ldots, \lambda_d^2]$, where $\lambda_j > 0$ for $j = 1, \ldots, d$. Put $D^{1/2} = \text{diag} \, (\lambda_1, \ldots, \lambda_d)$. Then $D = D^{1/2}D^{1/2}$ and $A = UDU^* = (UD^{1/2}U^*)(UD^{1/2}U^*)$. Let $A^{1/2} = UD^{1/2}U^*$.

1.2 (b) It is sufficient to suppose that $A = D$ is diagonal, say $D = \text{diag} \, [\lambda_1^2, \ldots, \lambda_d^2]$. (Why?) Let $D^{1/2}$ be any Hermitian, positive definite square root of D. Let the vector y and eigenvalue λ of $D^{1/2}$ satisfy $D^{1/2}y = \lambda y$. Then

$Dy = \lambda^2 y$, so that y is an eigenvector of D belonging to the eigenvalue λ^2. It follows that $D^{1/2} = \text{diag} [\lambda_1, \ldots, \lambda_d]$. (Why?)

1.2. (c) Consider only t near a fixed t_0. Then there exists an $\epsilon > 0$ such that $A(t)y \cdot y \geq \epsilon \|y\|^2$ for all vectors y. If $A(t)$ is multiplied by a positive constant $c > 0$, its square root is multiplied by $c^{1/2}$. Hence it can be supposed that $\epsilon \|y\|^2 \leq A(t)y \cdot y \leq \theta \|y\|^2$, where $0 < \epsilon < \theta < 1$. Then

$$(1 - \epsilon)\|y\|^2 \geq [I - A(t)]y \cdot y \geq (1 - \theta)\|y\|^2 \quad \text{and} \quad \|I - A(t)\| \leq 1 - \epsilon < 1.$$

Show that $A^{1/2}(t) = B(t)$, where

$$B(t) = I - \sum_{n=1}^{\infty} a_n(I - A(t))^n \quad \text{if} \quad (1 - r)^{1/2} \equiv 1 - \sum_{n=1}^{\infty} a_n r^n.$$

Note that $(B(t))^2 = A(t)$ follows from $[(1 - r)^{1/2}]^2 = 1 - r$ since the powers of $I - A(t)$ commute. Also $B(t)$ is Hermitian and positive definite since $a_n > 0$, $a_1 + a_2 + \cdots = 1$. (This gives another proof of the existence of $A^{1/2}$.) Finally, the series for $B(t) = A^{1/2}(t)$ can be differentiated term-by-term.

2.1. (c) Consider the binary system $y^{1\prime} = -y^2, y^{2\prime} = -y_1/t^2$ for $t \geq 1$ and Exercise XI 1.1 (c).

2.1. (d) For necessity, cf. (b); for sufficiency, cf. the proof of Theorem X 1.1.

2.2. Cf. §IV 5.

2.3. (a) Let $y(t)$ be solution given by Theorem 2.1. Verify $(-1)^n y^{(n)}(t) \geq 0$ by induction on n.

2.4. Write the sum of the first two terms $p_0(t)u^{(d)} + p_1(t)u^{(d-1)}$ of (2.7) as $p_0(t)\alpha^{-1}(t)(\alpha(t)u^{(d-1)})'$, where $\alpha(t) = \exp \int_0^t p_1(s) \, ds/p_0(s) > 0$. Divide the resulting equation by $p_0(t)/\alpha(t)$ and consider the result as a first order system for $y = (u, -u', u'', \ldots, (-1)^{d-2}u^{(d-2)}, (-1)^{d-1}\alpha(t)u^{(d-1)})$, i.e., for $y = (y^1, \ldots, y^d)$, where $y^j = (-1)^{j-1}u^{(j-1)}$ for $j = 1, \ldots, d - 1$ and $y^d = (-1)^{d-1}\alpha(t)u^{(d-1)}$. Theorem 2.1 implies the existence of a solution satisfying (2.9) for $n = 0$, $1, \ldots, d - 1$. (Why $u > 0$?) From (2.7), it follows that $(-1)^d(p_0 u^{(d)} + p_1 u^{(d-1)}) \geq 0$.

2.5. Prove (2.9) by induction on n. To this end, first show that if $d > 2$ and (2.9) holds for $n = 1, \ldots, d - 2$, then it holds for $n = d - 1$. Note that $(-1)^d[p_0 u^{(d)} + p_1 u^{(d-1)}] \geq 0$, hence $[(-1)^{d-1}u^{(d-1)} \exp \int_0^t (p_1/p_0) \, ds]' \leq 0$. Show that if $(-1)^{d-1}u^{(d-1)}(a) < 0$ for some a on $0 \leq a < \infty$, then

$$(-1)^{d-1}u^{(d-1)}(t) < 0$$

and $(-1)^{d-2}u^{(d-2)}(t)$ is increasing for $t \geq a$. If $d \geq 3$, this is incompatible with $(-1)^{d-2}u^{(d-2)} \geq 0$ and the fact that $\lim u^{(d-3)}(t)$ exists as $t \to \infty$.

2.6. (a) Use analogue of Corollary 2.3 for $0 < t < \infty$ and the Remark following Theorem 2.1; for uniqueness, use Exercise XI 6.7.

2.7. There exist positive continuous functions $\alpha_0(t), \ldots, \alpha_m(t)$ such that the expression on the left of (2.12) can be written as

$$p_0\alpha_m(\alpha_{m-1}\{\alpha_{m-2}[\ldots(\alpha_0 u)' \ldots]'\}')';$$

§IV 8(ix). Thus (2.7) is

$$p_0\alpha_m(\alpha_{m-1}\{\alpha_{m-2}[\ldots(\alpha_0 u^{(d-m)})' \ldots]'\}')' + \sum_{k=m+1}^{d} (-1)^{k-1}p_k u^{(d-k)} = 0.$$

Write this as a first order system for $y = (y^1, \ldots, y^d)$, where

$$y^k = (-1)^{k-1} u^{(k-1)} \text{ for } k = 1, \ldots, d - m$$

and $y^{d-m+k} = (-1)^{d-m+k-1} \alpha_{k-1} \{\alpha_{k-2}[\ldots(\alpha_0 u^{(d-m)})' \ldots]'\}'$ for $k = 1, \ldots, m$. Apply Theorem 2.1.

2.8. Suppose first that (*) if $T > 0$, then there exists a solution $y_T(t)$ of $y' = -f$ on $0 \leq t \leq T$ satisfying $\|y_T(0)\| = c$ and $y_T(t) \geq 0$, $y_T'(t) \leq 0$ for $0 \leq t \leq T$. Choose a sequence of T-values $T_1 < T_2 < \ldots$ such that $T_n \to \infty$ and $y(0) = \lim y_T(0)$ exists as $n \to \infty$, where $T = T(n)$. Then $y(t) = \lim y_T(t)$, $T = T(n)$ and $n \to \infty$, exists uniformly on bounded t-intervals (why?) and is a solution with the desired properties. Thus it only remains to prove (*) for every fixed T. To this end, suppose first that solutions of $y' = -f$ are uniquely determined by initial conditions and let $y(t, y_0)$ be the solution satisfying $y = y_0$ at $t = T$. Since $y \equiv 0$ is a solution, it follows that $y(t, y_0)$ exists on $0 \leq t \leq T$ if $\|y_0\|$ is sufficiently small; Theorem V 2.1. Since $f \geq 0$ and $y_0 \geq 0$, it follows that $y(t, y_0) \geq y_0$ on $0 \leq t \leq T$ and so $\|y(0, y_0)\| \geq \|y_0\|$. Thus if $c > 0$ is sufficiently small, $\|y(0, y_0)\| = c$ holds for some small $\|y_0\|$. Let S be the set of numbers c_0 such that, for every c on $0 < c < c_0$, there is a y_0 with the property that $\|y(0, y_0)\| = c$. It is clear that S is closed relative to the half-line $0 < c_0 < \infty$ (why?) and also open relative to $0 < c_0 < \infty$ (why?). Hence S contains all $c_0 > 0$. This proves (*) in the case that the solutions of $y' = -f$ are uniquely determined by initial conditions. In other cases, approximate f by smooth functions.

2.9. (a) Let $y_0(t, \alpha)$ be the solution of $y'' = f$ satisfying $y_0 = 0$, $y_0' = \alpha \leq 0$ at $t = 1$. This solution exists for $0 \leq t \leq 1$ if $-\alpha \geq 0$ is sufficiently small. Also $y_0 > 0$, $y_0' \leq 0$, $y_0'' \geq 0$ on $0 \leq t \leq 1$, since $y'' = f \geq 0$. Let $c = y_0(0, \alpha) > 0$ for some small $-\alpha > 0$ and $y(t, \beta)$ be the solution of $y'' = f$ satisfying $y(0) = c$, $y'(0) = \beta \leq 0$. For c fixed, let S be the set of $\beta < 0$ for which there exists a $t_0 = t_0(\beta) > 0$ such that $y(t, \beta)$ exists on $0 \leq t \leq t_0$, $y(t, \beta) > 0$ for $0 \leq \beta \leq t_0$, $y(t_0, \beta) = 0$. Then $y'(t, \beta) \leq 0$, $y''(t, \beta) \geq 0$ for $0 \leq t \leq t_0$. Show that S is not empty and is open. Let $\beta_0 = \sup \beta$ for $\beta \in S$ and show that $y(t, \beta_0)$ is the desired solution.

2.9. (b) Consider $y'' = 3(1 + y'^2)^{1+\lambda} y^2 / 2\lambda$, where $0 < \lambda < 1$. This has no solution satisfying $y(0) > 1$ and $y(t) \geq 0$, $y'(t) \leq 0$ for all $t \geq 0$; cf. Hartman and Wintner [8, pp. 396–397].

3.1. Use Theorem XI 3.2; cf. diagram for Exercise 3.5.

3.2. Introduce the new independent variable s where $ds = dt/p(t)$ and $s(0) = 0$.

3.5. Use Sturm's second comparison Theorem XI 3.2 to show that if the arch of the graph of $u = |u_0(t)|$ on $t_n \leq t \leq t_{n+1}$ is reflected across the line $t = t_{n+1}$, then it lies "over" the arch of $u = |u_0(t)|$ on $t_{n+1} \leq t \leq t_{n+2}$; see diagram. Use "alternating series" argument to prove (3.14).

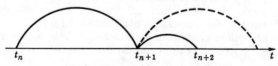

3.8. Let $0 \leq s < \infty$ and let $G(s, t)$ be the solution of $u'' + q(t)u = 0$ satisfying the initial conditions $u = 0$, $u' = 1$ at $t = s$ [e.g., if $q(t) \equiv 1$, then $G(s, t) = \sin(t-s)$]. Show that $v(s) = \displaystyle\int_s^\infty G(s, t)f(t)\,dt$ is uniformly convergent on bounded

s-intervals and is a solution of $v'' + qv = f$. Note that when $q(t) > 0$, then $|G(s, t)|^2 \leq |G_t(s, s)|^2/q(t) \leq 1/q(t)$ and $|G_s(s, t)|^2 \leq |G_s(s, s)|^2 = 1$ for $t \geq s$ by Theorem 3.1_∞. The proof of $v \geq 0$ in the case $n = 0$ depends on arguments similar to those in the proof of Corollary 3.1. When $n = 1$, write $dv/ds =$

$$\int_s^\infty G_s(s, t) f(t) \, dt \text{ and } G_s(s, t) f(t) \text{ as } q(t) G_s(s, t)[f(t)/q(t)] = -G_{stt}(s, t)[f(t)/q(t)];$$

apply integration by parts twice. For $n = 2$, use the fact that $w = v''/q$ is a solution of $w'' + qw = (f/q)''$, for $w + v = f/q$.

3.9. (a) Cf. Exercises XI 7.5 and XI 7.6. Use $r' + r^2 + q = 0$, hence $r' + q \leq 0$, $r' + r^2 \leq 0$.

3.9. (b) Cf. Corollary XI 9.1 for "only if." For the "if" part, use $dE = u^2 \, dq \leq 0$ and that $\int^\infty tq(t) \, dt = \infty$ implies that $u'(\infty) = 0$ for all solutions. Suppose that $E(\infty) > 0$ for some nonprincipal u, so that $r' + E/u^2 = 0$ gives $-r' \geq c/u^2$, $c > 0$. Integrate over $[t, \infty)$, multiply by u to obtain $u'(t) \geq cu_0(t) \geq$ const. > 0.

3.9. (c) First prove the assertions concerning $r_0(\infty)$, $r(\infty)$ by using Corollaries XI 6.4 and 6.5.

3.9. (d) For the first part, cf. the hint for part (c). For the second part, use Exercise X 17.4 (b) and $dE = u^2 \, dq$.

4.1. Let u be as in (4.6), then $u^2 q + u'^2$ is a quadratic form in $(\cos \varphi, \sin \varphi)$ for fixed t. In the case $(4.3 +)$, (4.7) holds and shows that the eigenvalues λ of this form satisfy $\lambda \leq 1$. Using (4.8), it is seen that the equation for the eigenvalues is $\lambda^2 - \lambda(|z|^2 q + |z'|^2) + q = 0$. The desired inequality merely expresses the fact that the largest root λ of this equation satisfies $\lambda \leq 1$. The case $(4.3 -)$ follows from $(4.3 +)$; cf. Lemma 3.1.

4.2. (d) Make the change of independent variables $t = e^s$ in Bessel's equation (4.21); then the change of dependent and independent variables $V = (t^2 - \mu^2)^{1/2} v$, $d\sigma = (t^2 - \mu^2)^{1/2} \, ds$ for $t > \mu$.

4.3. (b) Take a number α such that, up to a constant factor, $u = u_0(t) \cos \alpha + u_1(t) \sin \alpha$: It can be supposed that $u = u_0$, for otherwise replace u_0 and u_1 by $u_0 \cos \alpha + u_1 \sin \alpha$ and $-u_0 \sin \alpha + u_1 \cos \alpha$, respectively. Define a continuous $\theta(t)$ by $\theta(t) = \arctan u_0(t)/u_1(t)$ and $\theta(t_0) = 0$, so that $\theta(t_n) = n\pi$. Then $\theta'(t) = 1/(u_0^2 + u_1^2) > 0$. Let $t = t(\theta)$ be the inverse function of $\theta = \theta(t)$. Then $(-1)^{j+1} d^j t(\theta)/d\theta^j \geq 0$ for $j = 1, \ldots, n + 2$ and $t_n = t(n\pi)$ for $n = 0, 1, \ldots$. Note that a mean value theorem of Hölder implies that if a function $t(\theta)$ has $j \geq 1$ continuous derivatives, then $\Delta^j t(n\pi) = \pi^j (d^j t/d\theta^j)_{\theta=\gamma}$ where $\theta = \gamma$ is some number satisfying $n\pi \leq \gamma \leq (n + j)\pi$. This fact is easily verified

$$\text{from } \Delta^j t(n\pi) = \int_0^\pi [\Delta^{j-1} dt(\theta + n\pi)/d\theta] \, d\theta = \int_0^\pi \cdots \int_0^\pi [d^j t(n\pi + \theta_1 + \cdots + \theta_j)/$$

$d\theta^j] \, d\theta_1 \, d\theta_2 \ldots d\theta_j$.

6.3. The existence proof is similar to that of Theorem 6.1 where Ω^0 is chosen to be $\{(t, y): t, y^1 \text{ arbitrary}, y^2 > 1, y^3 < 0\}$. For uniqueness, use a variant of Exercise III 4.1(d).

12.1 (a) Cf. the Remark following (12.4).

12.1 (b) Introducing polar coordinates with $r = \|y\|$, it is seen that $ds \geq p(y) \, dr \geq p_0(r) \, dr$ if $p_0(r) = \min p(y)$ for $\|y\| = r$. Use Lemma I 4.1.

12.4. (a) The ds is complete by Exercise 12.1(b). If $y(t)$, $0 \leq t \leq 1$, is any C^1

arc joining $y = 0$ and y, then its Riemann length is not less than $\displaystyle\int_0^{\|y\|} p_0(u)\, du$ if $p_0(u) = \min p(y)$ for $\|y\| = u$; cf. the proof of Exercise 12.1(b). Hence $r(0, y)$ is not less than this integral.

14.3. If $\alpha(y) = \lambda_1 + \lambda_2$, verify that, for fixed y and arbitrary vectors x, z,

$$(Jx \cdot x)\|z\|^2 + \|x\|^2 (Jz \cdot z) - (x \cdot z)[Jx \cdot z + x \cdot Jz] \leq \alpha(\|x\|^2 \|z\|^2 - (x \cdot z)^2).$$

To this end, note that the inequality is not affected if J is replaced by J^H and x, z are subject to an orthogonal transformation; thus it can be supposed that $J^H = \operatorname{diag}[\lambda_1, \lambda_2, \ldots, \lambda_d]$. The left side of the desired inequality is

$$\sum_{i=1}^d \sum_{j=1}^d \lambda_i(x^i x^i z^j z^j + x^j x^j z^i z^i - 2x^i x^j z^i z^j) = \sum_{i=1}^d \sum_{j=1}^d \lambda_i(x^i z^j - x^j z^i)^2;$$

or, equivalently,

$$\tfrac{1}{2}\sum_{i \neq j}\sum (\lambda_i + \lambda_j)(x^i z^j - x^j z^i)^2 \leq \tfrac{1}{2}\alpha \sum_{i=1}^d \sum_{j=1}^d (x^i z^j - x^j z^i)^2.$$

This gives the stated inequality. Let $z = f(y)$ and $x \cdot f(y) = 0$.

References

A symbol (V 2) at the end of an entry below indicates a reference in § V 2 or in the notes on § V 2.

N. H. ABEL
[1] *Oeuvres complètes*, I and II, Oslo (1881) (IV 1, IX 8).

L. AMERIO
[1] Soluzioni quasi-periodiche, o limitate, di sistemi differenziale non-lineari quasi-periodici, o limitati, *Ann. Mat. Pura Appl.* (4) **39** (1955) 97–119 (XII 5).

A. ANDRONOW AND C. B. CHAIKIN
[1] *Theory of oscillations* (English ed.), Princeton University Press (1949) (Russian ed., Moscow, 1937) (VII 10).

H. A. ANTOSIEWICZ
[1] A survey of Lyapunov's second method, *Contributions to the theory of nonlinear oscillations* (Ann. Math. Studies) **4** (1958) 141–166 (III 8).

C. ARZELA
[1] Funzioni di linee, *Atti R. Accad. Lincei Rend.* (4) **5** (1889) 342–348 (I 2).
[2] Sulle funzioni di linee, *Mem. R. Accad. Bologna* (5) **5** (1895) 225–244 (I 2).

G. ASCOLI
[1] Le curve limiti di una varietà data di curve, *Mem. R. Accad. Lincei* (3) **18** (1883/4) 521–586 (I 2).

F. V. ATKINSON
[1] Asymptotic formulae for linear oscillations, *Proc. Glasgow Math. Assoc.* **3** (1957) 105–111 (XI 8).
[2] On stability and asymptotic equilibrium, *Ann. Math.* **68** (1958) 690–708 (X 2).

S. BANACH
[1] *Théorie des operations linéaires*, Warsaw (1932) (XII 0).

J. H. BARRETT
[1] A Prüfer transformation for matrix differential equations, *Proc. Amer. Math. Soc.* **8** (1957) 510–518 (XI 10).

R. G. BARTLE
[1] On the openness and inversion of differentiable mappings, *Ann. Acad. Sci. Fenn. Ser. A* no. 257 (1958) (II 2).

R. W. BASS
[1] On the regular solutions at a point of singularity of a system of nonlinear differential equations, *Amer. J. Math.* **77** (1955) 734–742 (Ex. IV 3.3).

R. BELLMAN
[1] On the boundedness of solutions of nonlinear differential and difference equations, *Trans. Amer. Math. Soc.* **62** (1947) 357–386 (X 8).

[2] On an application of a Banach-Steinhaus theorem to the study of the boundedness of solutions of nonlinear differential and difference equations, *Ann. Math.* **49** (1948) 515–522 (XII 6).

[3] On the asymptotic behavior of solutions of $u'' - (1 + f(t))u = 0$, *Ann. Mat. Pura Appl.* (4) **31** (1950) 83–91 (X 17).

[4] *Stability Theory of Differential Equations*, McGraw-Hill, New York (1953) (X).

I. BENDIXSON

[1] Dèmonstration de l'existence de l'integrale d'une équation aux derivées partielles linéaire, *Bull. Soc. Math. France* **24** (1896) 220–225 (V 3).

[2] Sur les courbes définiés par des équations différentielles, *Acta Math.* **24** (1901) 1–88 (VII 0, 4, 6–7, 8–9; VIII 2).

I. BIHARI

[1] A generalization of a lemma of Bellman and its applications to uniqueness problems of differential equations, *Acta Math. Sci. Hungar.* **7** (1956) 71–94 (III 4).

G. D. BIRKHOFF

[1] A simplified treatment of the regular singular point, *Trans. Amer. Math. Soc.* **11** (1910) 199–202 (IV 11).

[2] On a simple type of irregular singular point, *ibid.* **14** (1913) 462–476 (IV 11).

[3] *Dynamical systems*, Amer. Math. Soc. Colloquium Publications, New York (1927) (IV 9, VII 1).

G. D. BIRKHOFF AND O. D. KELLOGG

[1] Invariant points in function space, *Trans. Amer. Math. Soc.* **23** (1922) 96–115 (X 11).

G. A. BLISS

[1] A boundary value problem for a system of ordinary linear differential equations of first order, *Trans. Amer. Math. Soc.* **28** (1926), 561–584 (XII 1).

M. BÔCHER

[1] The theorems of oscillation of Sturm and Klein, I and II, *Bull. Amer. Math. Soc.* **4** (1897/8) 295–313 and 365–376 (XI 3).

[2] On regular singular points of linear differential equations of the second order whose coefficients are not necessarily analytic, *Trans. Amer. Math. Soc.* **1** (1900) 40–53 (X 11, 13, 17; XI 8, 9).

[3] *Leçons sur les méthodes de Sturm*, Gauthier-Villars, Paris (1917) (XI 3).

N. N. BOGOLYUBOV AND N. KRYLOV

[1] *An introduction to nonlinear mechanics*, Ann. Math. Studies 2 (1947) (VII 10).

N. N. BOGOLYUBOV AND YU. A. MITROPOL'SKI

[1] *Asymptotic Methods in the Theory of Nonlinear Oscillations*, State Publishing House, Moscow (1955) (ATIC translation) (VII 10).

P. BOHL

[1] Ueber die hinsichtlich der unabhängigen und abhängigen Variablen periodische Differentialgleichung erster Ordnung, *Acta Math.* **40** (1916) 321–336 (VII 14).

O. BOLZA

[1] *Vorlesungen über Variationsrechnung*, B. G. Teubner, Leipzig and Berlin (1909) (XI 6).

E. BOMPIANI
[1] Un teorema di confronto ed un teorema di unicità per l'equazione differenziale $y' = f(x, y)$, *Atti Accad. Naz. Lincei Rend. Cl. Sci. Fis. Mat. Nat.* (6) **1** (1925) 298–302 (III 6).

G. BORG
[1] Eine Umkehrung der Sturm-Liouvilleschen Eigenwertaufgabe, *Acta Math.* **78** (1946) 1–96 (XI 4).

[2] On a Liapounoff criterion of stability, *Amer. J. Math.* **71** (1949) 67–70 (XI 5).

[3] A condition for the existence of orbitally stable solutions of dynamical systems, *Kungl. Tekn. Högsk. Handl. Stockholm* **153** (1960) (XIV 14).

M. E. BOUNITZKY
[1] Sur la fonction de Green des équations différentielles linéaires ordinaires, *J. Math. Pures Appl.* (6) **5** (1909) 65–125 (XII 1).

J. C. BOUQUET AND C. A. A. BRIOT
[1] Recherches sur les fonctions définies par les équations différentielles, *J. École Polytech.* (*Paris*) **21** cah. 36 (1856) 133–198 (VIII 3).

F. BRAUER
[1] Some results on uniqueness and successive approximations, *Canad. J. Math.* **11** (1959) 527–533 (III 6, 9).

[2] Spectral theory for linear systems of differential equations, *Pacific J. Math.* **10** (1960) 17–34 (XII 1).

F. BRAUER AND S. STERNBERG
[1] Local uniqueness, existence in the large, and the convergence of successive approximations, *Amer. J. Math.* **80** (1958) 421–430; **81** (1959) 797 (III 6, 9).

I. BRINCK
[1] Self-adjointness and spectra of Sturm-Liouville operators, *Math. Scand.* **1** (1929) 219–239 (XI 8, 11).

C. A. A. BRIOT (see J. C. BOUQUET)

L. E. J. BROUWER
[1] On continuous vector distributions. I, II, and III, *Verh. Nederl. Akad. Wetersch. Afd. Natuurk. Sec. I.* **11** (1909) 850–858; **12** (1910) 716–734; and **13**[1] (1910) 171–186 (VII 0, 8–9).

F. E. BROWDER
[1] The solvability of nonlinear functional equations, *Duke Math. J.* **30** (1963) 557–566 (XIV·12).

C. CARATHÉODORY
[1] *Variationsrechnung und partielle Differentialgleichungen erster Ordnung*, B. G. Teubner, Leipzig and Berlin (1935) (VI 6, 7–9).

E. CARTAN
[1] *Leçons sur les invariants intégraux*, Hermann, Paris (1922) (V 5; VI 2, 3).

H. CARTAN
[1] *Algebraic topology*, mimeographed notes, Harvard (1949) (V 5).

A. L. CAUCHY
[1] *Oeuvres complètes* (1) **6**, Gauthier-Villars, Paris (1888) (VI 7–9).

L. CESARI

[1] Un nuovo criterio di stabilità per le soluzioni delle equazioni differenziali lineari, *Ann. Scuola Norm. Sup. Pisa* (2) **9** (1940) 163–186 (IV 9, X 4, XI 8).

[2] *Asymptotic behavior and stability problems in ordinary differential equations,* Springer, Berlin (1959) (X).

C. B. CHAIKIN (SEE A. ANDRONOW)

K. T. CHEN

[1] Equivalence and decomposition of vector fields about an elementary critical point, *Amer. J. Math.* **85** (1963) 693–722 (IX 7, 8–9, 12–14).

A. CLEBSCH

[1] Ueber die Reduktion der zweiten Variation auf ihre einfachste Form, *J. Reine Angew. Math.* **55** (1858) 254–270 (XI 6, 10).

[2] Ueber die simultane Integration linearer partielle Differentialgleichungen, *ibid.* **65** (1866) 257–268 (VI 1).

E. A. CODDINGTON AND N. LEVINSON

[1] Uniqueness and convergence of successive approximations, *J. Indian Math. Soc.* **16** (1952) 75–81 (III 9).

[2] *Theory of Ordinary Differential Equations,* McGraw-Hill, New York (1955) (IV 10, IX 6).

C. V. COFFMAN

[1] Linear differential equations on cones in Banach spaces, *Pacific J. Math.* **12** (1962) 69–75 (XIV 2).

[2] Asymptotic behavior of solutions of ordinary difference equations, *Trans. Amer. Math. Soc.* **110** (1964) 22–51 (IX 5, X 13).

[3] Nonlinear differential equations on cones in Banach spaces, *Pacific J. Math.* **14** (1964) 9–16 (XIV 2).

R. CONTI AND G. SANSONE

[1] *Equazioni differenziali non lineari,* Cremonese, Rome (1956) (VII 10, XII 2).

W. A. COPPEL

[1] On a differential equation of boundary layer theory, *Phil. Trans. Roy. Soc. London Ser. A* **253** (1960) 101–136 (XIV 6, 9).

C. CORDUNEANU

[1] Sur certains systèmes différentielles non-linéaires, *An. Şti. Univ. "Al. I. Cuza" Iaşi. Sec. I* **6** (1960) 257–260 (XII 8).

E. COTTON

[1] Sur les solutions asymptotiques des équation différentielles, *Ann. Sci. École Norm. Sup.* (3) **28** (1911) 473–521 (X 8).

G. DARBOUX

[1] *Leçons sur la théorie générale des surfaces,* IV, Gauthier-Villars, Paris (1896) (V 3).

A. DENJOY

[1] Sur les courbes définies par les équations différentielles à la surface du tore, *J. Math. Pures Appl.* (9) **11** (1932) 333–375 (VII 12, 13–14).

J. DIEUDONNÉ

[1] Sur la convergence des approximations successives, *Bull. Sci. Math.* (2) **69** (1945) 62–72 (III 9).

E. Digel
[1] Zu einem Beispiel von Nagumo und Fukuhara, *Math. Zeit.* **39** (1935) 157–160 (II 4).
[2] Ueber die Bedingungen der Existenz der Integrale partieller Differentialgleichungen erster Ordnung, *ibid.* **44** (1938) 445–451 (VI 7–9).

S. P. Diliberto
[1] On systems of ordinary differential equations, *Contributions to the theory of nonlinear oscillations* (Ann. Math. Studies) **1** (1950) 1–48 (IV 2).
[2] A note on linear ordinary differential equations, *Proc. Amer. Math. Soc.* **8** (1957) 462–464 (IV 2).

G. L. Dirichlet
[1] Ueber die stabilität des gleichgewichts, *J. Reine Angew. Math.* **32** (1846) 85–88 (III 8).

H. Dulac
[1] *Curves definidas por una ecuación diferencial de primer orden y de primer grade*, Madrid (1933) (VIII 3).
[2] Points singuliers des équations différentielles, *Mémor. Sci. Math.* fasc. 61, Gauthier-Villars, Paris (1934) (VIII 3).

O. Dunkel
[1] Regular singular points of a system of homogeneous linear differential equations of the first order, *Proc. Amer. Acad. Arts Sci.* **38** (1902–3) 341–370 (X 1, 11, 13, 17).

G. von Escherich
[1] Die zweite Variation der einfachen Integrale, *S.-B. K. Akad. Wiss. Wien Kl. Math. Natur.* (8) **107** (1898) 1191–1250 (XI 10).
[2] Ueber Systeme von Differentialgleichungen der I. Ordnung, *Abh. Deutsch Akad. Wiss. Berlin Kl. Math. Phys. Tech.* **108** (Abt IIa) (1899) 621–676 (II 1, V 3).

S. Faedo
[1] Il teorema di Fuchs per le equazione differenziale lineari a coefficienti non analitici e proprietà asintotiche delle soluzioni, *Ann. Mat. Pura Appl.* (4) **25** (1946) 111–133 (X 17).
[2] Sulle stabilità delle soluzioni delle equazioni differenziali lineari, I, II, and III, *Atti Accad. Naz. Lincei Rend. Cl. Sci. Fis. Mat. Nat.* (8) **2** (1947) 564–570, 757–764, and (8) **3** (1947) 37–43 (X 17).

D. Flanders and J. J. Stoker
[1] The limit case of relaxation oscillations, *Studies in Nonlinear Vibration Theory*, New York University (1946) (VII 10).

G. Floquet
[1] Sur les équations différentielles linéaires à coefficients périodiques, *Ann. Sci. École Norm. Sup.* **12** (1883) 47–82 (IV 6).

H. Forster
[1] Ueber das Verhalten der Integralkurven einer gewöhnlichen Differentialgleichung erster Ordnung in der Umgebung eines singulären Punktes, *Math. Zeit.* **13** (1938) 271–320 (VIII 4).

A. R. Forsyth
[1] *Theory of Differential Equations*, IV, Cambridge University Press (1902) (IV Appendix).

586   Ordinary Differential Equations

K. O. FRIEDRICHS
[1] Fundamentals of Poincaré's theory, *Proceedings of the Symposium on Nonlinear circuit analysis*, Polytechnic Institute of Brooklyn, **2** (1953) 56–67 (XII 2).

G. FROBENIUS
[1] Ueber die Determinante mehrerer Functionen einer Variablen, *J. Reine Angew. Math.* **77** (1874) 245–257 (IV 3).

[2] Ueber das Pfaffsche Problem, *ibid.* **82** (1877) 230–315 (VI 3).

M. FROMMER
[1] Die Integralkurven einer gewöhnlichen Differentialgleichung erster Ordnung in der umgebung rationaler unbestimmtheitsstellen, *Math. Ann.* **99** (1928) 222–272 (VIII 4).

L. FUCHS
[1] *Gesammelte mathematische Werke*, 1, Berlin (1904) (IV 1, 7, 10, 12).

M. FUKUHARA
[1] Sur les systèmes des équations différentielles ordinaires, *Proc. Imperial Acad. Japan* **4** (1928) 448–449 (II 4).

[2] Sur l'ensemble des courbes intégrales d'un système d'équations différentielles ordinaires, *ibid.* **6** (1930) 360–362 (II 4).

M. FUKUHARA AND M. NAGUMO
[1] Un théorème relatif à l'ensemble des courbes intégrales d'un système d'équations différentielles ordinaires, *Proc. Phys.-Math. Soc. Japan* (3) **12** (1930) 233–239 (II 4).

T. GANELIUS
[1] Un theorème Taubérien pour la transformation de Laplace, *C. R. Acad. Sci. (Paris)* **242** (1956) 719–721 (XI 8).

A. GHIZZETTI
[1] Sul comportamento asintotico degli integrali delle equazioni differenziali ordinarie, lineari ed omogenee, *Giorn. Mat. Battaglini* (4) **1** (77) (1947) 5–27 (X 17).

[2] Un theorema sul comportamento asintotico degli integrali delle equazioni differenziali lineari omogenee, *Rend. Mat. Univ. Roma* (5) **8** (1949) 28–42 (X 17).

P. GILLIS
[1] Sur les équations linéaires aux différentielle totales, *Bull. Soc. Roy. Sci. Liège* **9** (1940) 197–212 (VI 3, 6).

[2] Sur les formes différentielles et la formula de Stokes, *Acad. Roy. Belg. Cl. Sci. Mém.* **20** (1943) (V 5).

L. M. GRAVES
[1] Some mapping theorems, *Duke Math. J.* **17** (1950) 111–114 (XII 0).

M. GREGUŠ
[1] Ueber die asymptotischen Eigenschaften der Lösungen der linearen Differentialgléichung dritter Ordnung, *Ann. Mat. Pura. Appl.* 14, **63** (1963) 1–10 (XIV 2).

D. M. GROBMAN
[1] Homeomorphisms of systems of differential equations, *Dokl. Akad. Nauk SSSR* **128** (1959) 880–881 (IX 7).

[2] Topological classification of the neighborhood of a singular point in *n*-dimensional space, *Mat. Sb.* (N.S.) **56** (98) (1962) 77–94 (IX 7).

D. M. GROBMAN AND R. É. VINOGRAD

[1] On problems of Frommer differentiation, *Uspehi Mat. Nauk (N.S.)* **12** (1957) no. 5 (77) 191–195 (VIII 4).

D. GROHNE AND R. IGLISCH

[1] Die laminare Grenzschicht an der längs angeströmten ebenen Platte mit schrägen Absaugen und Ausblasen, *Veröffentlichung Math. Inst. Tech. Hochschule, Braunschweig* (1945) (XIV 8).

T. H. GRONWALL

[1] Note on the derivatives with respect to a parameter of the solutions of a system of differential equations, *Ann. Math.* (2) **20** (1919) 292–296 (III 1).

W. GROSS

[1] Bemerkung zum Existenzbeweise bei den partiellen Differentialgleichungen erster Ordnung, *S.-B. K. Akad. Wiss. Wien Kl. Math. Nat.* **123** (Abt. IIa) (1914) 2233–2251 (VI 7–9).

A. HAAR

[1] Zur Characteristikentheorie, *Acta Sci. Math. Szeged* **4** (1928) 103–114 (VI 10).

J. HADAMARD

[1] Sur les intégrales d'un system d'équations différentielles ordinaires, considérées comme fonctions des données initiales, *Bull. Soc. Math. France* **28** (1900) 64–66 (V 3).

[2] Sur l'itération et les solutions asymptotiques des équations différentielles, *ibid.* **29** (1901) 224–228 (IX 5).

[3] *Leçons sur la propagation des ondes*, Hermann, Paris (1903) (V 3).

[4] Sur les transformations ponctuelles, *Bull. Soc. Math. France* **34** (1900) 71–84 (XIV 12).

W. HAHN

[1] *Theorie und Anwendung der direkten Methoden von Lyapunov*, Springer, Berlin (1959) (III 8, XIV 11).

J. HALE

[1] *Oscillations in Nonlinear Systems*, McGraw-Hill, New York (1963) (XII 2).

J. HALE AND N. ONUCHIC

[1] On the asymptotic behavior of solutions of a class of differential equations, *Contributions to Differential Equations* **1** (1963) 61–75 (X 1, XII 9).

M. HAMBURGER

[1] Bemerkungen über die Form der Integrale der linearen Differentialgleichungen mit veränderlichen Koeffizienten, *J. Reine Angew. Math.* **76** (1873) 113–120 (IV 10).

P. HARTMAN

[1] On the solutions of an ordinary differential equation near a singular point, *Amer. J. Math.* **68** (1946) 495–504 (VIII 2).

[2] On a theorem of Milloux, *ibid.* **70** (1948) 395–399 (XIV 1).

[3] Differential equations with nonoscillatory eigenfunctions, *Duke Math. J.* **15** (1948) 697–709 (XI 6).

[4] On the linear logarithmico-exponential differential equation of the second order, *Amer. J. Math.* **70** (1948) 764–779 (XI 7).

[5] Unrestricted solution fields of almost separable differential equations, *Trans. Amer. Math. Soc.* **63** (1948) 560–580 (X 4, 8, 11, 17).

[6] A characterization of the spectra of one-dimensional wave equations, *Amer. J. Math.* **71** (1949) 915–920 (XI 4).

[7] The number of L^2-solutions of $x'' + q(t)x = 0$, *ibid.* **73** (1951) 635–645 (XI 5).

[8] On bounded Green's kernels for second order linear ordinary differential equations, *ibid.* **73** (1951) 646–656 (XIII Part I).

[9] On linear second order differential equations with small coefficients, *ibid.* **73** (1951) 955–962 (XI 7).

[10] On nonoscillatory linear differential equations of second order, *ibid.* **74** (1952) 389–400 (XI 7).

[11] On the zeros of solutions of second order linear differential equations, *J. London Math. Soc.* **27** (1952) 492–496 (XI 5).

[12] Self-adjoint, nonoscillatory systems of ordinary, second order, linear differential equations, *Duke Math. J.* **24** (1957) 25–36 (XI 10).

[13] On Jacobi brackets, *Amer. J. Math.* **79** (1957) 187–189 (VI 1).

[14] On integrating factors and on conformal mappings, *Trans. Amer. Math. Soc.* **87** (1958) 387–406 (V 6).

[15] Unrestricted n-parameter families, *Rend. Circ. Mat. Palermo* (2) **7** (1958) 123–142 (Ex. IV 8.4).

[16] On isometries and on a theorem of Liouville, *Math. Zeit.* **69** (1958) 202–210 (Ex. V 6.3).

[17] On exterior derivatives and solutions of ordinary differential equations, *Trans. Amer. Math. Soc.* **91** (1959) 277–292 (V 6, VI 3).

[18] On the ratio $f(t + cf^{-\alpha}(t))/f(t)$, *Boll. Un. Mat. Ital.* **14** (1959) 59–61 (XI 5).

[19] On boundary value problems for systems of ordinary nonlinear, second order differential equations, *Trans. Amer. Math. Soc.* **96** (1960) 493–509 (XII 4, 5).

[20] On local homeomorphisms of Euclidean spaces, *Bol. Soc. Mat. Mexicana* **5** (1960) 220–241 (IX 5, 7, 8–9).

[21] A lemma in the theory of structural stability of differential equations, *Proc. Amer. Math. Soc.* **11** (1960) 610-620 (Ex. IX 7.1, 8.3).

[22] On differential equations and the function $J_\mu{}^2 + Y_\mu{}^2$, *Amer. J. Math.* **83** (1961) 154–188 (XIV 3, 4).

[23] The existence of large or small solutions of linear differential equations, *Duke Math. J.* **28** (1961) 421–430 (XIV 1, 3).

[24] On stability in the large for systems of ordinary differential equations, *Canad. J. Math.* **13** (1961) 480–492 (XIV 11, 12, 13).

[25] On dichotomies for solutions of nth order linear differential equations, *Math. Ann.* **147** (1962) 378–421 (XI 7, 8; XIII Parts I, II).

[26] On uniqueness and differentiability of solutions of ordinary differential equations, *Proceedings of a Symposium on Nonlinear Problems*, Madison (Wis.) (1963) 219–232 (V 6, 7–8, 9).

[27] A differential equation with nonunique solutions, *Amer. Math. Monthly* **70** (1963) 255–259 (II 5).

[28] On the local linearization of differential equations, *Proc. Amer. Math. Soc.* **14** (1963) 568–573 (IX 5, 7, 8, 9).

[29] On the asymptotic behavior of solutions of a differential equation in boundary layer theory, *Z. Angew. Math. Mech.* **44** (1964) 123–128 (XIV 9).

P. HARTMAN AND C. OLECH
[1] On global asymptotic stability of solutions of ordinary differential equations, *Trans. Amer. Math. Soc.* **104** (1962) 154–178 (XIV 14).

P. HARTMAN AND N. ONUCHIC
[1] On the asymptotic integration of ordinary differential equations, *Pacific J. Math.* **13** (1963) 1193–1207 (XII 8, 9).

P. HARTMAN AND C. R. WILCOX
[1] On solutions of the Helmholtz equation in exterior domains, *Math. Zeit.* **75** (1961) 228–255 (Ex. XIV 3.6).

P. HARTMAN AND A. WINTNER
[1] On the asymptotic behavior of the solutions of a nonlinear differential equation, *Amer. J. Math.* **68** (1946) 301–308 (VIII 1, 2; X 2).

[2] The asymptotic arcus variation of solutions of real linear differential equations of second order, *ibid.* **70** (1948) 1–10 (XI 5).

[3] On the asymptotic problems of the zeros in wave mechanics, *ibid.* **70** (1948) 461–480 (XI 6).

[4] On nonconservative linear oscillators of low frequency, *ibid.* **70** (1948) 529–539 (XIV 3).

[5] A criterion for the nondegeneracy of the wave equation, *ibid.* **71** (1949) 206–213 (XI 5).

[6] Oscillatory and nonoscillatory linear differential equations, *ibid.* **71** (1949) 627–249 (XI 6).

[7] On the classical transcendents of mathematical physics, *ibid.* **73** (1951) 381–389 (XIV 2).

[8] On the nonincreasing solutions of $y'' = f(x, y, y')$, *ibid.* **73** (1951) 390–404 (XII 4, 5; XIV 2).

[9] On an oscillation criterion of Liapounoff, *ibid.* **73** (1951) 885–890 (XI 5).

[10] An inequality for the amplitudes and areas in vibration diagrams of time dependent frequency, *Quart. Appl. Math.* **10** (1952) 175–176 (XI 5).

[11] On the behavior of the solutions of real binary differential systems at singular points, *Amer. J. Math.* **75** (1953) 117–126 (VIII 2, 3).

[12] On nonoscillatory linear differential equations, *ibid.* **75** (1953) 717–730 (XI 9).

[13] Linear differential and difference equations with monotone solutions, *ibid.* **75** (1953) 731–743 (XIV 2).

[14] Linear differential equations with completely monotone solutions, *ibid.* **76** (1954) 199–206 (XIV 2).

[15] On nonoscillatory linear differential equations with monotone coefficients, *ibid.* **76** (1954) 207–219 (XIV 3).

[16] On monotone solutions of nonlinear differential equations, *ibid.* **76** (1954) 860–866 (III 9, XIV 2).

[17] Asymptotic integrations of linear differential equations, *ibid.* **77** (1955) 45–87 (X 4, 11, 14, 16, 17; XI 8).

[18] On the assignment of asymptotic values for the solutions of linear differential equations of second order, *ibid.* **77** (1955) 475–483 (XI 6).

590   Ordinary Differential Equations

[19] Asymptotic integrations of ordinary nonlinear differential equations, *ibid.* **77** (1955) 692–724 (VIII 3; X 4, 8, 11, 13).

[20] An inequality for the first eigenvalue of an ordinary boundary value problem, *Quart. Appl. Math.* **13** (1955) 324–326 (XI 7).

[21] On an oscillation criterion of de la Vallée Poussin, *ibid.* **13** (1955) 330–332 (XI 5).

[22] On disconjugate differential systems, *Canad. J. Math.* **8** (1956) 72–81 (XI 10, XII 3).

E. K. HAVILAND

[1] A note on the convergence of the successive approximations to the solution of an ordinary differential equation, *Amer. J. Math.* **54** (1932) 632–634 (III 9).

E. HEINZ

[1] On certain nonlinear elliptic differential equations and univalent mappings, *J. Analyse Math.* **5** (1956/7) 197–272 (XII 5).

[2] Halbbeschränktheit gewöhnlicher Differentialoperatoren höherer Ordnung, *Math. Ann.* **135** (1958) 1–49 (XI 6, 10, 11).

[3] On Weyl's embedding problem, *J. Math. Mech.* **11** (1962) 421–454 (II 2).

E. HILB

[1] Lineare Differentialgleichungen im komplexen Gebiet, *Encyklopädie der mathematischen Wissenschaften*, **II B5** (1913) (IV Appendix, 11, 12).

[2] Ueber diejenigen Integrale linearer Differentialgleichungen, welche sich an einer Unbestimmtheitsstelle bestimmt verhalten, *Math. Ann.* **82** (1921) 40–41 (IV 13).

E. HILLE

[1] Nonoscillation theorems, *Trans. Amer. Math. Soc.* **64** (1948), 234–252 (XI 7).

G. HOHEISEL

[1] Eindeutigkeitskriterien und Knoteninvarianz bei Differentialgleichungen, *Jber. Deutsch. Math. Verein.* **42** (1933) 33–42 (VIII 2, 3).

E. HÖLDER

[1] Mathematische Untersuchungen zur Himmelsmechanik, *Math. Zeit.* **31** (1930) 197–257 (XII 2).

E. HOPF

[1] Closed surfaces without conjugate points, *Proc. Nat, Acad. Sci. U.S.A.* **34** (1948), 47–51 (XI 6).

H. HOPF

[1] Ueber die Drehung der Tangenten und Sehnen ebener Kurven, *Compositio Math.* **2** (1935) 50–62 (VII 2).

H. HOPF AND W. RINOW

[1] Ueber der Begriff der vollstandigen differentialgeometrischen Fläche, *Comment. Math. Helvetici* **3** (1931) 209–225 (XIV 12).

J. HORN

[1] Zur Theorie der Systeme linearer Differentialgleichungen mit einer unäbhängigen Veranderlichen, II, *Math. Ann.* **40** (1892) 527–550 (IV 11).

W. HUREIWICZ AND H. WALLMAN

[1] *Dimension theory*, Princeton University Press (1941) (IX 8, X 2).

R. IGLISCH

[1] Elementarer Existenzbeweis für die Strömung in der laminaren Grenzschicht zur

Potentialströmung $U = u_1 x^m$ mit $m > 0$ bei Absaugen und Ausblasen, *Z. Angew. Math. Mech.* **33** (1953) 143–147 (XIV 6).

[2] Elementarer Beweis für die Eindeutigkeit der Strömung in der laminaren Grenzschicht zur Potentialströmung $U = u_1 x^m$ mit $m \geqq 0$ bei Absaugen und Ausblasen, *ibid.* **34** (1954) 441–443 (XIV 6).

R. IGLISCH (SEE D. GROHNE)

R. IGLISCH AND F. KEMNITZ

[1] Ueber die in der Grenzschichttheorie auftretende Differentialgleichung $f''' + ff'' + \beta(1 - f'^2) = 0$ für $\beta < 0$ bei gewissen Absauge- und Ausblasegezetzen, *50 Jahre Grenzschichtforschung*, Braunschweig (1955) (XIV 7, 8).

K. ISÉ AND M. NAGUMO

[1] On the normal forms of differential equations in the neighborhood of an equilibrium point, *Osaka Math. J.* **9** (1957) 221–234 (IX 7, 12).

S. IYANAGA

[1] Ueber die Unitätsbedingungen der Lösungen der Differentialgleichung: $dy/dx = f(x, y)$, *Jap. J. Math.* **5** (1928) 253–257 (III 6).

C. G. J. JACOBI

[1] *Gesammelte Werke*, IV (1886) and V (1890), Berlin (IV 1, 7; VI 1).

E. KAMKE

[1] *Differentialgleichungen reeller Funktionen*, Akademische Verlagsgesesellschaft, Leipzig (1930) [or Chelsea, New York (1947)] (III 4, 6).

[2] Zur Theorie der Systeme gewöhnlicher Differentialgleichungen, II, *Acta Math.* **58** (1932) 57–85 (II 4, III 4).

[3] A new proof of Sturm's comparison theorems, *Amer. Math. Monthly* **46** (1939) 417–421 (XI 3).

[4] *Differentialgleichungen. Lösungsmethoden und Lösungen, I (Gewöhnliche Differentialgleichungen)* (7th ed.), Akademische Verlagsgesellschaft, Leipzig (1961) (II 0, III 6).

[5] *Differentialgleichungen. Lösungsmethoden und Lösungen, II (Partielle Differentialgleichungen erster Ordnung für eine gesuchte Funktion)* (4th ed.), Akademische Verlagsgesellschaft, Leipzig (1959) (VI 7–9).

E. R. VAN KAMPEN

[1] The topological transformations of a simple closed curve into itself, *Amer. J. Math.* **57** (1935) 142–152 (VII 13–14).

[2] Remarks on systems of ordinary differential equations, *ibid.* **59** (1937) 144–152 (III 7).

[3] Notes on systems of ordinary differential equations, *ibid.* **63** (1941) 371–376 (III 9).

[4] On the argument functions of simple closed curves and simple arcs, *Compositio Math.* **4** (1937) 271–275 (VII 2).

E. R. VAN KAMPEN AND A. WINTNER

[1] On an absolute constant in the theory of variational stability, *Amer. J. Math.* **59** (1937) 270–274 (XI 5).

H. KAUFMAN AND R. L. STERNBERG

[1] Applications of the theory of systems of differential equations to multiple nonuniform transmission lines, *J. Math. Phys.* **31** (1952/3) 244–252 (XI 10).

K. A. KEIL

[1] Das qualitative Verhalten der Integralkurven einer gewöhnlichen Differential-gleichung erster Ordnung in der Umgebung eines singulären Punktes, *Jber. Deutsch. Math. Verein.* **57** (1955) 111–132 (VIII 2).

J. L. KELLEY

[1] *General Topology*, Van Nostrand, New York (1955) (VII 12).

O. D. KELLOGG (SEE G. D. BIRKHOFF)

F. KEMNITZ (SEE R. IGLISCH)

A. KNESER

[1] Untersuchung über die reellen Nullstellen der Integrale linearer Differential-gleichungen, *Math. Ann.* **42** (1893) 409–435 (XI 7).

[2] Untersuchung und asymptotische Darstellung der Integrale gewisser Differential-gleichungen bei grossen reellen Werthen des Arguments, *J. Reine Angew. Math.* **116** (1896) 178–212 (XI 6, XII 5).

H. KNESER

[1] Ueber die Lösungen eines Systems gewöhnlicher Differentialgleichungen das der Lipschitzschen Bedingung nicht genügt, *S. -B. Preuss. Akad. Wiss. Phys. -Math. Kl.* (1923) 171–174 (II 4).

[2] Reguläre Kurvenscharen auf den Ringflächen, *Math. Ann.* **91** (1924) 135–154 (VII 12).

H. W. KNOBLOCH

[1] An existence theorem for periodic solutions of nonlinear ordinary differential equations, *Michigan Math. J.* **9** (1962) 303–309 (XII 2).

G. KOENIGS

[1] Recherches sur les intégrales de certaines équations fonctionelles, *Ann. Sci. École Norm. Sup.* (3) **1** (1884) Suppl. 3–41 (IX 8).

[2] Nouvelles recherches sur les équations fonctionelles, *ibid.* (3) **2** (1885) 385–404 (IX 8).

Z. KOWALSKI

[1] Generalized characteristic directions for a system of differential equations, *Ann. Polonici Math.* **6** (1959) 269–280 (VIII 2).

M. A. KRASNOSEL'SKIĬ AND S. G. KREIN

[1] On a class of uniqueness theorems for the equation $y' = f(x, y)$, *Uspehi Mat. Nauk (N.S.)* **11** (1956) no. 1 (67) 209–213 (III 6).

N. N. KRASOVSKIĬ

[1] On global stability of solutions of a nonlinear system of differential equations, *Prikl. Mat. Meh.* **18** (1954) 735–737 (XIV 11).

[2] Sufficient conditions for the stability of solutions of a system of nonlinear differ-ential equations, *Dokl. Akad. Nauk SSSR* **98** (1954) 901–904 (XIV 11).

[3] On stability for large initial perturbations, *Prikl. Mat. Meh.* **21** (1957) 309–319 (XIV 11).

[4] *Stability of Motion* (English ed.), Stanford University Press, Stanford (1963) (III 8, XIV 11).

M. G. KREIN

[1] On some questions related to the ideas of Lyapunov in the theory of stability, *Uspehi Mat. Nauk (N.S.)* **3** (1948) no. 3 (25) 166–169 (XII 6).

S. G. KREIN (SEE M. A. KRASNOSEL'SKIĬ)

N. KRYLOV (SEE N. N. BOGOLYUBOV)

D. L. KUČER
[1] On some criteria for the boundedness of the solutions of a system of differential equations, *Dokl. Akad. Nauk SSSR* **69** (1949) 603–606.

J. L. LAGRANGE
[1] *Mecanique analytique*, Desaint, Paris (1788) (III 8).
[2] *Oeuvres*, I (1867) and IV (1869), Gauthier-Villars, Paris (IV 1, 2, 7; V 12).

C. E. LANGENHOP
[1] Note on Levinson's existence theorem for forced periodic solutions of a second order differential equation, *J. Math. Phys.* **30** (1951) 36–39 (VII 10).

J. LASALLE
[1] Uniqueness theorems and successive approximations, *Ann. Math.* **50** (1949) 722–730 (III 9).
[2] Relaxation oscillations, *Quart. Appl. Math.* **7** (1949) 1–19 (VII 10).
[3] Some extensions of Lyapunov's second method, *IRE Trans. Circuit Theory*, **CT-7** (1960) 520–527 (XIV 11).

J. LASALLE AND S. LEFSCHETZ
[1] *Stability by Lyapunov's Direct Method with Applications*, Academic Press, New York (1961) (III 8).

S. LATTÉS
[1] Sur les formes réduites des transformations ponctuelles à deux variables, *C.R. Acad. Sci.* (Paris) **152** (1911) 1566–1569 (IX 8–9).
[2] Sur les formes réduites des transformations ponctuelles dans le domaine d'un point double, *Bull. Soc. Math. France* (8) **39** (1911) 309–345 (IX 8–9).

M. LAVRENTIEFF
[1] Sur une équation différentielle du premier ordre, *Math. Zeit.* **23** (1925) 197–209 (II 5).

L. LEAU
[1] Étude sur les equations fonctionelles à une ou à plusieurs variables, *Ann. Fac. Sci. Toulouse* **11E** (1897) 1–110 (IX 8).

M. LEES
[1] A boundary value problem for nonlinear ordinary differential equations, *J. Math. Mech.* **10** (1961) 423–430 (XII 4,5).

S. LEFSCHETZ (SEE ALSO J. LASALLE)
[1] *Differential Equations: Geometric Theory* (2nd ed.), Interscience, New York (1963) (VII 10).

W. LEIGHTON
[1] Principal quadratic functionals, *Trans. Amer. Math. Soc.* **67** (1949) 253–274 (XI 6).
[2] The detection of the oscillation of solutions of a second order linear differential equation, *Duke Math. J.* **17** (1950) 57–62 (XI 7).

W. LEIGHTON AND M. MORSE
[1] Singular quadratic functionals, *Trans. Amer. Math. Soc.* **40** (1936) 252–286 (XI 6).

W. Leighton and Z. Nehari

[1] On the oscillation of solutions of self-adjoint linear differential equations of fourth order, *Trans. Amer. Math. Soc.* **89** (1958) 325–377 (XI 7).

F. Lettenmeyer

[1] Ueber die an einer Unbestimmtheitsstelle regulären Lösungen eines Systemes homogener linearen Differentialgleichungen, *S.-B. Bayer. Akad. Wiss. München Math.-nat. Abt.* (1926) 287–307 (IV 11, 13).

[2] Ueber das asymptotische Verhalten der Lösungen von Differentialgleichungen und Differentialgleichungssystemen, *ibid.* (1929) 201–252 (X 11).

N. Levinson (see also E. A. Coddington)

[1] On the existence of periodic solutions for second order differential equations with a forcing term, *J. Math. Phys.* **22** (1943) 41–48 (VII 10).

[2] Transformation theory of nonlinear differential equations of the second order, *Ann. Math.* **45** (1944) 723–737 (XII 2).

[3] The asymptotic nature of the solutions of linear systems of differential equations, *Duke Math. J.* **15** (1948) 111–126 (IV 9, X 4).

N. Levinson and O. K. Smith

[1] A general equation for relaxation oscillations, *Duke Math. J.* **9** (1942) 382–403 (VII 10).

P. Lévy

[1] *Processus stochastiques et mouvement Brownien*, Gauthier-Villars, Paris (1948) (III 6).

D. C. Lewis

[1] Invariant manifolds near an invariant point of unstable type, *Amer. J. Math.* **60** (1938) 577–587 (IX 5).

[2] Metric properties of differential equations, *ibid.* **71** (1949) 249–312 (V 9, XIV 13).

[3] Differential equations referred to a variable metric, *ibid.* **73** (1951) 48–58 (V 9, XIV 13).

[4] Autosynartetic solutions of differential equations, *ibid.* **83** (1961) 1–32 (XII 2).

A. Libri

[1] Mémoire sur la résolution des équations algébriques dont les racines ont entre elles un rapport donné, et sur l'intégration des équations différentielles linéaires dont les integrales particulières peuvent s'exprimer les unes par les autres, *J. Reine Angew. Math.* **10** (1833) 167–194 (IV 3).

H. Liebmann

[1] Geometrische Theorie die Differentialgleichungen, *Encyklopädie der mathematischen Wissenschaften*, **III D8** (1914) (VII 6, VIII 8).

A. Liénard

[1] Étude des oscillations entretenues, *Revue Générale de l'Électricité* **23** (1928) 901–912, 946–954 (VII 10).

J. C. Lillo

[1] Linear differential equations with almost periodic coefficients, *Amer. J. Math.* **81** (1959) 37–45 (X 8).

E. Lindelöf

[1] Sur l'application des méthodes d'approximations successives à l'étude des

intégrales réeles des équations différentielles ordinaire, *J. Math. Pures Appl.* (4) **10** (1894) 117–128 (II 1).

[2] Démonstration de quelques théorèmes sur les équations différentielles, *ibid.* (5) **6** (1900) 423–441 (V 3).

J. LIOUVILLE

[1] Sur le développement des fonctions ou parties de fonctions en séries dont les divers termes sont assujettis à satisfaire à une même équation différentielles du second ordre contenant un paramètre variable, I and II, *J. Math. Pures Appl.* (1) **1** (1836) 253–265; (1) **2** (1837) 16–35 (XI 1, 4).

[2] Sur la théorie de la variations des constants arbitraires, *ibid.* (1) **3** (1838), 342–349 (IV 1).

R. LIPSCHITZ

[1] Sur la possibilité d'intégrer complétement un système donné d'équations différentielles, *Bull. Sci. Math. Astro.* **10** (1876) 149–159 (II 1).

S. ŁOJASIEWICZ

[1] Sur l'allure asymptotique des intégrales du système d'équations différentielles au voisinage de point singulier, *Ann. Polonici Math.* **1** (1954) 34–72 (X 4–7).

E. R. LONN

[1] Knoteninvarianz bei Differentialgleichungen, *Jber. Deutsch. Math. Verein.* **43** (1934) 232–237 (VIII 4).

[2] Ueber singuläre Punkte gewöhnlicher Differentialgleichungen, *Math. Zeit.* **44** (1939) 507–530 (VIII 4).

L. LORCH AND P. SZEGO

[1] Higher monotonicity properties of certain Sturm-Liouville functions, *Acta Math.* **109** (1963) 55–73 (XIV 4).

W. J. A. LUXEMBURG

[1] On the convergence of successive approximations in the theory of ordinary differential equations, *Canad. Math. Bull.* **1** (1958) 9–20 (III 8).

A. LYAPUNOV

[1] Sur une série relative à la théorie des équations différentielles linéaires à coefficient périodiques, *C. R. Acad. Sci. (Paris)* **123** (1896) 1248–1252 (XI 5).

[2] Problème général de la stabilité du mouvement, *Ann. Fac. Sci. Univ. Toulouse* **9** (1907) 203–475 [reproduced in Ann. Math. Study (17) Princeton (1947)] (III 8, IX 6, X 8).

A. D. MAIZEL'

[1] On the stability of solutions of systems of differential equations, *Trudy Ural'skogo Politehn. Inst.* **5** (1954) 20–50 (XII 6, XIII Part I).

I. G. MALKIN

[1] On stability in the first approximation, *Sbornik Naučnyh Trudov Kazan. Aviacion. Inst.* **3** (1935) 7–17 (XII 6).

A. MAMBRIANI

[1] Su un teorema relativo alle equazioni differenziali ordinarie del 2° ordine, *Atti R. Accad. Naz. Lincei Rend. Cl. Sci. Fis. Mat. Nat.* (6) **9** (1929) 620–622 (XII 5).

L. MARKUS AND H. YAMABE

[1] Global stability criteria for differential systems, *Osaka Math. J.* **12** (1960) 305–317 (XIV 12, 13).

J. L. Massera

[1] The existence of periodic solutions of systems of differential equations, *Duke Math. J.* **17** (1950) 457–475 (XII 1, 2).

[2] Converse theorems of Lyapunov's second method, *Bol. Soc. Mat. Mexicana* (2) **5** (1960) 158–163 (III 8).

[3] Sur l'existence de solutions bornées et périodiques des systèmes quasilinéaires d'équations différentielles, *Ann. Mat. Pura Appl.* (4) **51** (1960) 95–106 (XII 8).

J. L. Massera and J. J. Schäffer

[1] Linear differential equations and functional analysis. I, II, III, and IV, *Ann. Math.* **67** (1958) 517–573; **69** (1959) 88–104; **69** (1959) 535–574; and *Math. Ann.* **139** (1960) 287–342 (XII 6, 7, 8; XIII Parts I, II).

M. Mason

[1] Zur Theorie der Randwertaufgaben, *Math. Ann.* **58** (1904) 528–544 (XII 1).

A. Mayer

[1] Ueber unbeschränkt integrable Systeme von linearen totalen Differentialgleichungen, *Math. Ann.* **5** (1872) 448–470 (VI 1, 6).

H. Milloux

[1] Sur l équation différentielle $x'' + A(t)x = 0$, *Prace Mat.* **41** (1934) 39–53 (XIV 1, 3).

W. E. Milne

[1] On the degree of convergence of expansions in an infinite interval, *Trans. Amer. Math. Soc.* **31** (1929) 907–918 (XI 5).

N. Minorsky

[1] *Introduction to Nonlinear Mechanics*, Edwards Bros., Ann Arbor (Mich.) (1947) (VII 10).

C. Miranda

[1] Un osservazione su un teorema di Brouwer, *Boll. Un. Mat. Ital.* (2) **3** (1940) 5–7 (XII 2).

Yu. A. Mitropol'ski (see N. N. Bogolyubov).

F. Moigno

[1] *Leçons sur le calcul differential et integral (d'apres Cauchy)*, Bachelier, Paris (1840) (II 1).

M. Morse (see also W. Leighton).

[1] A generalization of the Sturm separation and comparison theorems in n-space, *Math. Ann.* **103** (1930) 52–59 (XI 10, XII 3).

[2] *The calculus of variations in the large*, Amer. Math. Soc. Colloquium Publications, New York (1934) (XI 6).

J. Moser

[1] The analytic invariants of an area preserving mapping near a hyperbolic point, *Comm. Pure Appl. Math.* **19** (1956) 673–692 (IX 8–9).

[2] The order of a singularity in Fuchs' theory, *Math. Zeit.* **72** (1959/60) 379–398 (IV 11).

[3] On invariant curves of area preserving mappings of an annulus, *Nachr. Akad. Wiss. Göttingen Math. Phys. Kl.* IIa no. 1 (1962) (IX 8–9).

M. Müller

[1] Ueber das Fundamentaltheorem in der Theorie der gewöhnlichen Differentialgleichungen, *Math. Zeit.* **26** (1927) 619–645 (III 9).

[2] Beweis eines Satzes des Herrn H. Kneser über die Gesamtheit der Lösungen, die ein System gewöhnlicher Differentialgleichungen durch einen Punkt schickt, *ibid.* **28** (1928) 349–355 (II 4).

[3] Neuere Untersuchung über den Fundamentalsatz in der Theorie der gewöhnlichen Differentialgleichungen, *Jber. Deutsch. Math. Verein.* **37** (1928) 33–48 (II 0, III 6).

M. NAGUMO (SEE ALSO M. FUKUHARA, K. ISÉ)

[1] Eine hinreichende Bedingung für die Unität der Lösung von Differentialgleichungen erster Ordnung, *Jap. J. Math.* **3** (1926) 107–112 (III 6).

[2] Ueber die Differentialgleichung $y'' = f(x, y, y')$, *Proc. Phys.-Math. Soc. Japan* (3) **19** (1937) 861–866 (XII 4, 5).

[3] Ueber die Ungleichung $\partial u/\partial y > f(x, y, u, \partial u/\partial x)$, *Japan. J. Math.* **15** (1939) 51–56 (VI 10).

[4] Ueber das Randwertproblem der nicht linearen gewöhnlichen Differentialgleichungen zweiter Ordnung, *Proc. Phys.-Math. Soc. Japan* **24** (1942) 845–851 (XII 4).

B. SZ.-NAGY AND F. RIESZ

[1] *Leçons d'analyse fonctionelle*, Akadémei Kiadó, Budapest (1952) (XI 4, 10).

Z. NEHARI (SEE ALSO W. LEIGHTON)

[1] On the zeros of solutions of second order linear differential equations, *Amer. J. Math.* **76** (1954) 689–697 (XI 5).

[2] On an inequality of Lyapunov, *Studies in Mathematical Analysis and Related Topics*, Stanford University Press, Stanford (1962) 256–261 (XI 5).

V. V. NEMYTSKIĬ AND V. V. STEPANOV

[1] *Qualitative Theory of Differential Equations* (English ed.), Princeton University Press, Princeton (1960) (VIII 2, X 2).

R. NEVANLINNA

[1] Ueber die Methode der sukzessiven Approximationen, *Ann. Acad. Sci. Fennicae Ser. A* no. 291 (1960) (II 2).

M. H. A. NEWMAN

[1] *Elements of the Topology of Plane Sets of Points*, Cambridge University Press, Cambridge (1954) (VII 4).

O. NICCOLETTI

[1] Sugli integrali delle equazioni differenziali considerati come funzioni dei loro valori iniziali, *Atti R. Accad. Lincei. Rend. Cl. Sci. Fis. Mat. Nat.* (5) **4** (1895) 316–324 (V 3).

W. NIKLIBORC

[1] Sur les équations linéaires aux différentielles totales, *Studia Math.* **1** (1929) 41–49 (VI 6).

L. NIRENBERG

[1] *Functional analysis*, mimeographed notes, New York University (1960/61) (XII 4).

C. OLECH (SEE ALSO P. HARTMAN)

[1] On the asymptotic behavior of the solutions of a system of ordinary nonlinear differential equations, *Bull. Acad. Polon. Sci. Cl. III* **4** (1956) 555–561 (X 13).

[2] Remarks concerning criteria for uniqueness of solutions of ordinary differential equations, *ibid.* **8** (1960) 661–666 (III 6, 9).

[3] On the global stability of autonomous systems in the plane, *Contributions to Differential Equations* **1** (1963) 389–400 (XIV 14).

C. OLECH, Z. OPIAL, AND T. WAŻEWSKI

[1] Sur le problème d'oscillation des intégrales de l'équation $y'' + g(t)y = 0$, *Bull. Acad. Polon. Sci. Cl. III* **5** (1957) 621–626 (XI 7).

Z. OPIAL (SEE ALSO C. OLECH)

[1] Sur un système d'inégalités intégrales, *Ann. Polon. Math.* **3** (1957) 200-209 (III 4).

[2] Sur l'allure asymptotique des intégrales de l'équation différentielle $u'' + a(t)u' + b(t)u = 0$, *Bull. Acad. Polon. Sci. Cl. III* **5** (1957) 847–853 (XI 9).

[3] Sur une inégalité de C. de la Vallée Poussin dans le théorie de l'équation différentielle linéaire du second ordre, *Ann. Polon. Math.* **6** (1959) 87–91 (XI 5).

[4] Sur un critère d'oscillation des intégrales de l'équation différentielles $(Q(t)x')' + f(t)x = 0$, *ibid.* **6** (1959) 99–104 (XI 7).

[5] Sur l'allure asymptotique des solutions de l'équation différentielle $u'' + a(t)u' + b(t)u = 0$, *ibid.* **6** (1959) 181–200 (XI 9).

[6] Sur les valeurs asymptotiques des intégrales des équations différentielles linéaires du second ordre, *ibid.* **6** (1959) 201–210 (XI 9).

[7] Démonstration d'un théorème de N. Levinson et C. Langenhop, *ibid.* **7** (1960) 241–246 (VII 10).

[8] Sur la stabilité asymptotique des solutions d'un système d'équations différentielles, *ibid.* **7** (1960) 259–267 (V 9, XIV 13).

N. ONUCHIC(SEE J. HALE, P. HARTMAN)

W. OSGOOD

[1] Beweis der Existenz einer Lösung der Differentialgleichung $dy/dx = f(x, y)$ ohne Hinzunahme der Cauchy-Lipschitzschen Bedingung, *Monatsh. Math. Phys.* **9** (1898) 331–345 (III 6).

A. OSTROWSKI

[1] Sur les conditions de validité d'une classe de relations entre les expressions différentielles linéaires, *Comment. Math. Helv.* **15** (1942–3) 265–286 (VI 1).

P. PAINLEVÉ

[1] Gewöhnliche Differentialgleichungen: Existenz der Lösungen, *Encyklopädie der mathematischen Wissenschaften*, **IIA4a** (II 0, VIII 3).

G. PEANO

[1] Sull' integrabilità delle equazione differenziali di primo ordine, *Atti. R. Accad. Torino* **21** (1885/1886) 677–685 (III 2, 4).

[2] Démonstration de l'integrabilité des équations différentielles ordinaires, *Math. Ann.* **37** (1890) 182–228 (II 2).

[3] Generalità sulle equazioni differenziali ordinarie, *Atti R. Accad. Sci. Torino* **33** (1897) 9–18 (V 3).

O. PERRON

[1] Ueber diejenigen Integrale linearer Differentialgleichungen, welche sich an einer Unbestimmtheitsstelle bestimmt verhalten, *Math. Ann.* **70** (1911) 1–32 (IV 13).

[2] Ueber lineare Differentialgleichungen, bei denen die unabhangig Variable reell ist, I and II, *J. Reine Angew. Math.* **142** (1913) 254–270 and **143** (1913) 25–50 (X 8, 17).

[3] Beweis für die Existenz von Integralen einer gewöhnlichen Differentialgleichung in der Umgebung einer Unstetigkeitsstelle, *Math. Ann.* **75** (1914) 256–273 (VIII 1, 3).

[4] Ein neuer Existenzbeweis für die Integrale der Differentialgleichung $y' = f(x, y)$, *ibid.* **76** (1915) 471–484 (III 2, 4).

[5] Ueber die Gestalt der Integralkurven einer Differentialgleichung erster Ordnung in der Umgebung eines singulären Punkten, I and II, *Math. Zeit.* **15** (1922) 121–146; **16** (1923) 273–295 (VII 6, VIII 3).

[6] Ueber Ein- und Mehrdeutigkeit des Integrales eines Systems von Differentialgleichungen, *Math. Ann.* **95** (1926) 98–101 (III 6).

[7] Ueber Existenz und Nichtexistenz von Integralen partieller Differentialgleichungssysteme im reellen Gebiet, *Math. Zeit.* **27** (1928) 549–564 (VI 9).

[8] Eine hinreichende Bedingung für die Unität der Lösung von Differentialgleichungen erster Ordnung, *ibid.* **28** (1928) 216–219 (III 6).

[9] Ueber Stabilität und asymptotisches Verhalten der Integrale von Differentialgleichungssystemen, *ibid.* **29** (1929) 129–160 (X 8).

[10] Die Ordnungzahlen linearen Differentialgleichungssysteme, *ibid.* **31** (1929) 748–766 (X 8).

[11] Ueber eine Matrixtransformation, *ibid.* **32** (1930) 465–473 (IV 2).

[12] Die Stabilitätsfrage bei Differentialgleichungen, *ibid.* **32** (1930) 703–728 (X 8, XII 6).

[13] Ueber stabilität und asymptotisches Verhalten der Lösungen eines Systemes endlicher Differenzengleichungen, *J. Reine Angew. Math.* **161** (1929) 41–64 (IX 5).

K. P. PERSIDSKIĬ

[1] On stability of motion in the first approximation, *Mat. Sb. (Recueil Math.)* **40** (1933) 284–293 (XII 6).

I. PETROVSKIĬ

[1] Ueber das Verhalten der Integralkurven eines Systems gewöhnlicher Differentialgleichungen in der Nähe eines singulären Punktes, *Mat. Sb. (Receuil Math.)* **41** (1934) 107–155 (IX 6, X 8).

C. M. PETTY

[1] Undirectional boundedness for two dimensional linear systems, Lockheed Missiles and Space Co. Report (1963) (XI 5).

J. F. PFAFF

[1] Methodus generalis, aequationes differentiarum partialum, nec non aequationes differentiales vulgares, utrasque primi ordinis inter quotcunque variabiles, complete integrandi, *Abh. Deutsch. Akad. Wiss. Berlin* (1814/1815) 76–136 (VI 7–9).

É. PICARD

[1] Mémoire sur la théorie des équations aux dérivées partielles et la méthode des approximations successives, *J. Math. Pures Appl.* (5) **6** (1890) 423–441 (II 1).

[2] *Traité d'analyse*, III, Gauthier-Villars, Paris (1896) (XII 2).

[3] *Leçons sur quelques équations fonctionelles*, Gauthier-Villars, Paris (1928) (IX 8).

[4] *Leçons sur quelques problèmes aux limites de la théorie des équations différentielles*, Gauthier-Villars, Paris (1930) (XI 7; XII 3, 4).

M. PICONE

[1] Su un problema al contorno nelle equazioni differenziali lineari ordinarie del secondo ordine, *Ann. R. Scuola Norm. Sup. Pisa* **10** (1908) no. 4 (XI 3).

600　Ordinary Differential Equations

A. PLIŚ

[1] On a topological method for studying the behavior of the integrals of ordinary differential equations, *Bull. Acad. Polon. Sci. Cl. III* **2** (1954) 415–418 (X 2).

[2] Characteristics of nonlinear partial differential equations, *ibid.* **2** (1954) 419–422 (VI 1, 8).

H. POINCARÉ

[1] Sur les propriétés des fonctions définies par les équations aux différences partielles, *Oeuvres*, 1, Gauthier-Villars, Paris (1929) (IX 7).

[2] Sur les courbes définie par les équations différentielles, *C. R. Acad. Sci.* (Paris) **90** (1880) 673–675 (VII Appendix).

[3] Mémoire sur les courbes définie par une équation différentielle, I, II, III, and IV, *J. Math. Pures Appl.* (3) **7** (1881) 375–422; (3) **8** (1882) 251–286; (4) **1** (1885) 167–244; (4) **2** (1886) 151–217 (VII 0, 1, 3, 4, 5, 6–7, 13–14; VIII 3; IX 0).

[4] Sur les équations linéaires aux différentielles ordinaires et aux différences finies, *Amer. J. Math.* **7** (1885) 203–258 (X 8, 17).

[5] *Les méthodes nouvelles de la mécaniques céleste*, I (1892) and III (1899), Gauthier-Villars, Paris (IX 0, 5; XII 2).

B. VAN DER POL

[1] On oscillation hysteresis in a triode generator with two degrees of freedom, *Philos. Mag.* (6) **43** (1922) 700–719 (VII 10).

G. PÓLYA

[1] On the mean value theorem corresponding to a given linear homogeneous differential equation, *Trans. Amer. Math. Soc.* **24** (1922) 312–324 (IV 8).

G. PRODI

[1] Nuovi criteri di stabilita per l'equazione $y'' + A(x)y = 0$, *Atti Accad. Naz. Lincei Rend. Cl. Sci. Fis. Mat. Nat.* **10** (1951) 447–451 (XI 8).

H. PRÜFER

[1] Neue Herleitung der Sturm-Liouvilleschen Reihenentwicklung stetiger Funktionen, *Math. Ann.* **95** (1926) 499–518 (XI 2, 4).

C. PUGH

[1] Cross-sections of solution funnels, *Bull. Amer. Math. Soc.* **70** (1964) (II 4).

C. R. PUTNAM

[1] On isolated eigenfunctions associated with bounded potentials, *Amer. J. Math.* **72** (1950) 135–147 (XIII Part I).

M. RÁB

[1] Asmptotische Formeln für dieLösungen der Differentialglechung $y'' + q(x)y = 0$, *Czech. Math. J.* (14) **89** (1964) 203–221 (XI 9).

G. RASCH

[1] Zur Theorie und Anwendung des Produktintegrales, *J. Reine Angew. Math.* **171** (1934) 65–119 (IV 11).

W. T. REID

[1] Properties of solutions of an infinite system of ordinary linear differential equations of the first order with auxiliary boundary conditions, *Trans. Amer. Math. Soc.* **32** (1930) 284–318 (III 1).

[2] Oscillation criteria for linear differential systems with complex coefficients, *Pacific J. Math.* **6** (1956) 733–751 (XI 10).

[3] Principal solutions of nonoscillatory self-adjoint linear differential systems, *ibid.* 8 (1958) 147–169 (XI 10, 11).

[4] Remarks on a matrix transformation for linear differential equations, *Proc. Amer. Math. Soc.* 8 (1959) 708–712 (IV 2).

[5] Oscillation criteria for self-adjoint differential systems, *Trans. Amer. Math. Soc.* 101 (1961) 91–106 (XI 10).

[6] Riccati matrix differential equations and nonoscillation criteria for associated linear differential systems, *Pacific J. Math.* 13 (1963) 655–686 (XI 10, 11).

B. RIEMANN

[1] *Collected Works* (2nd ed.), Teubner, Leipzig (1892) [or Dover, New York (1953)] (IV 10, 12; VII 2).

F. RIESZ (SEE B. SZ.-NAGY)

W. RINOW (SEE H. HOPF)

A. ROSENBLATT

[1] Ueber die Existenz von Integralen gewöhnlicher Differentialgleichungen, *Ark. Mat. Astro. Fys.* 5 (1909) no. 2 (III 6, 9).

[2] Sur les théorèmes de M. Picard dans la théorie des problèmes aux limites des équations différentielles ordinaires non-linéaires, *Bull. Sci. Math.* 57 (1933) 100–106 (XII 3, 4).

R. SACKSTEDER

[1] Foliations and pseudogroups, *Amer. J. Math.* 86 (1964) (VII 12).

S. SANDOR

[1] Sur l'équation différentielle matricielle de type Riccati, *Bull. Math. Soc. Sci. Math. Phys. R. P. Roumaine* (N.S.) 3 (51) (1959) 229–249 (XI 11).

G. SANSONE (SEE ALSO R. CONTI)

[1] *Equazioni differenziali nel campo reale*, Zanichelli, Bologna (1948) (XI 5, XIV 3).

L. SAUVAGE

[1] Sur les solutions régulières d'un system, *Ann. Sci. École Norm. Sup.* (3) 3 (1886) 391–404 (IV 11).

J. J. SCHÄFFER (SEE ALSO J. L. MASSERA)

[1] Functions spaces with translations; Addendum, *Math. Ann.* 137 (1959) 209–262; 138 (1959) 141–144 (XIII Parts I, II).

[2] Linear differential equations and functional analysis, V and VI, *Math. Ann.* 140 (1960) 308–321; 145 (1962) 354–400 (XII 7, XIII Parts I, II).

P. SCHAFHEITLIN

[1] Die Lage der Nullstellen der Besselschen Funktionen zweiter Art, *S.-B. Berlin Math. Ges.* 5 (1906) 82–93 (XIV 4).

J. SCHAUDER

[1] Der Fixpunktsatz in Funktionalräumen, *Studia Math.* 2 (1930) 171–180 (XII 10).

H. SCHILT

[1] Ueber die isolierten Nullstellen der Flächenkrummung und einige Verbiegbarkeitssätze, *Compositio Math.* 5 (1938) 239–283 (VII 9).

L. SCHLESINGER

[1] *Vorlesungen über lineare Differentialgleichungen*, Teubner, Leipzig and Berlin (1908) (IV 11).

602   Ordinary Differential Equations

[2] Bericht über die Entwicklung der Theorie der linearen Differentialgleichungen seit 1865, *Jber. Deutsch. Math. Verein.* **18** (1909) 133–266 (IV Appendix).

A. SCHMIDT
[1] Neuer Beweis eines Hauptsatzes über Bestimmtheitsstellen linearer Differentialgleichungsysteme, *J. Reine Angew. Math.* **179** (1938) 1–4 (IV Appendix).

E. SCHRÖDER
[1] Ueber unendlich viele Algorithmen zur Auflösung der Gleichungen, *Math. Ann.* **2** (1870) 317–385 (IX 8).

[2] Ueber iterirte Functionen, *ibid.* **3** (1871) 296–322 (IX 8).

A. J. SCHWARTZ
[1] A generalization of a Poincaré-Bendixson theorem to closed two-dimensional manifolds, *Amer. J. Math.* **85** (1963) 453–458 (VII 12).

G. SCORZA-DRAGONI
[1] Sul problema dei valori ai limiti per i systemi di equazioni differenziali del secondo ordine, *Boll. Un. Mat. Ital.* **14** (1935) 225–230 (XII 4).

C. L. SIEGEL
[1] Note on differential equations on the torus, *Ann. Math.* (2) **46** (1945) 423–428 (VII 13–14).

[2] Ueber die Normalform analytischer Differentialgleichungen in der Nähe einer Gleichgewichtslösung, *Nachr. Akad. Wiss. Göttingen. Math.-Phys. Kl.* IIa (1952) 21–30 (IX 7).

[3] Vereinfachter Beweis eines Satzes von J. Moser, *Comm. Pure Appl. Math.* **10,** (1957) 305–309 (IX 8–9).

O. K. SMITH (SEE N. LEVINSON)

V. V. STEPANOV (SEE V. V. NEMITSKIĬ)

R. L. STERNBERG (SEE H. KAUFMAN)

S. STERNBERG (SEE ALSO F. BRAUER)
[1] On the behavior of invariant curves near a hyperbolic point of a surface transformation, *Amer. J. Math.* **77** (1955) 526–534 (IX 5).

[2] On local C^n contractions of the real line, *Duke Math. J.* **24** (1957) 97–102 (IX 8).

[3] Local contractions and a theorem of Poincaré, *Amer. J. Math.* **79** (1957) 809–824 (IX 5–6, 7, 8–9, 12–14).

[4] On the structure of local homeomorphisms of Euclidean *n*-space, II, *ibid.* **80** (1958) 623–631 (IX 7, 8–9, 12–14).

[5] The structure of local homeomorphisms, III, *ibid.* **81** (1959) 578–604 (IX 8–9).

S. STERNBERG AND A. WINTNER
[1] On a class of analogies between differential equations and implicit equations, *J. Analyse Math.* **5** (1956/7) 34–46 (II 2).

J. J. STOKER (SEE ALSO D. FLANDERS)
[1] *Nonlinear Vibrations in Mechanical and Electrical Systems,* Interscience, New York (1950) (VII 10).

C. STURM
[1] Sur les équations différentielles linéaires du second ordre, *J. Math. Pures Appl.* (1) **1** (1836) 106–186 (XI 2, 3, 4).

J. SZARSKI
[1] Remarque sur un critère d'unicité des intégrales d'une équation différentielles ordinaire, *Ann. Polon. Math.* **12** (1962) 203–205 (III 6).

G. SZEGÖ
[1] *Orthogonal polynomials*, Amer. Math. Soc. Colloquium Publication, New York (1950) (XIV 1).

P. SZEGO (SEE L. LORCH)

Z. SZMYDTÓWNA
[1] Sur l'allure asymptotique des intégrales des équations différentielles ordinaires, *Ann. Soc. Polon. Math.* **24** (1951) 17–34 (X 4–7).

L. W. THOMÉ
[1] Zur Theorie der linearen Differentialgleichungen, *J. Reine Angew. Math.* **74** (1872) 193–217 (IV 12).

E. C. TITCHMARSH
[1] *Eigenfunction Expansións*, Clarendon Press, Oxford (1946) (III 1).

L. TONELLI
[1] Sulle equazioni funzionali del tipo di Volterra, *Bull. Calcutta Math. Soc.* **20** (1928) 31–48 (II 2).

S. TURSKI
[1] Sur l'unicité et la limitation des intégrales des équations aux dérivées partielles du premier ordre, *Ann. Soc. Polon. Math.* **120** (1933) 81–86 (VI 10).

A. TYCHONOV
[1] Ein Fixpunktsatz, *Math. Ann.* **111** (1935) 767–776 (XII 0).

C. DE LA VALLÉE POUSSIN
[1] Sur l'équation différentielle linéaire du second ordre, *J. Math. Pures Appl.* (9) **8** (1929) 125–144 (XI 5, 7).

E. VESSIOT
[1] Gewöhnliche Differentialgleichungen; elementare Integrationsmethoden, *Encyklopädie der mathematischen Wissenschaften*, **IIA4b** (II 0).

R. E. VINOGRAD (SEE D. M. GROBMAN)

B. VISWANATHAM
[1] The general uniqueness theorem and successive approximations, *J. Indian Math. Soc.* **16** (1952) 69–74 (III 9).

S. WALLACH
[1] The differential equation $y' = f(y)$, *Amer. J. Math.* **70** (1940) 345–350 (III 6).

H. WALLMAN (SEE H. HUREIWICZ)

G. N. WATSON
[1] A problem of analysis situs, *Proc. London Math. Soc.* (2) **15** (1916) 227–242 (VII 2).
[2] Bessel functions and Kapteyn series, *ibid.* (2) **16** (1917) 150–174 (XIV 1).
[3] *A treatise on the theory of Bessel functions* (2nd ed.), Cambridge University Press, Cambridge (1958) (XIV 1).

T. WAŻEWSKI (SEE ALSO C. OLECH)
[1] Sur l'unicité et la limitation des intégrales des équations aux dérivées partielles du premier ordre, *Atti R. Accad. Naz. Lincei Rend. Cl. Sci. Fis. Mat. Nat.* (6) **18** (1933) 372–376 (VI 10).

[2] Sur l'appréciation du domain d'existence des intégrales de l'équation aux dérivées partielles du premier ordre, *Ann. Soc. Polon. Math.* **14** (1935) 149–177 (VI 9).

[3] Ueber die Bedingungen der Existenz der Integrale partieller Differentialgleichungen erster Ordnung, *Math. Zeit.* **43** (1938) 522–532 (VI 7–9).

[4] Sur l'evaluation du domaine d'existence des fonctions implicites réelles ou complexes, *Ann. Soc. Polon. Math.* **20** (1947) 81–125 (II 2).

[5] Sur un principe topologique de l'examen de l'allure asymptotique des intégrales des équations différentielles ordinaires, *ibid.* **20** (1947) 279–313 (III 8; X 2, 3).

[6] Sur les intégrales d'un système d'équations différentielles ordinaires, *ibid.* **21** (1948) 277–297 (XI 4–7).

[7] Systèmes des équations et des ineqalités différentielles ordinaires aux deuxièmes membres monotones et leurs applications, *ibid.* **23** (1950) 112–166 (III 4).

[8] Sur une extension du procedé de I. Jungermann pour établir la convergence des approximations successive au cas des équations différentielle ordinaires, *Bull. Acad. Polon. Sci. Ser. Math. Astro. Phys.* **8** (1960) 213–216 (III 9).

E. A. WEBER

[1] Partielle Differentialgleichungen, *Encyklopädie der mathematische Wissenschaften*, IIA5 (VI 1).

H. WEYL

[1] Ueber gewöhnliche lineare Differentialgleichungen mit singulären Stellen und ihre Eigenfunktionen, *Nachr. Akad. Wiss. Göttingen. Math.-Phys. Kl.* IIa (1909) 37–63 (XI 9).

[2] *Mathematische Analyse des Raumsproblems*, Springer, Berlin (1923) (VI 6).

[3] On the differential equations of the simplest boundary-layer problems, *Ann. Math.* **43** (1942) 381–407 (XIV 5).

[4] Concerning a classical problem in the theory of singular points of ordinary differential equations, *Revista de Ciencias (Lima)* **46** (1944) 73–112 (VIII 3).

D. V. WIDDER

[1] *The Laplace Transform*, Princeton University Press, Princeton (1941) (XIV 2).

C. R. WILCOX (SEE P. HARTMAN)

A. WIMAN

[1] Ueber die reellen Lösungen der linearen Differentialgleichungen zweiter Ordnung, *Ark. Mat. Astro. Fys.* **12** (1917) no. 14 (X 17, XI 5).

[2] Ueber eine Stabilitätsfrage in der Theorie der linearen Differentialgleichungen, *Acta Math.* **66** (1936) 121–145 (X 17).

A. WINTNER (SEE ALSO P. HARTMAN, E. R. VAN KAMPEN, S. STERNBERG)

[1] The nonlocal existence problem of ordinary differential equations, *Amer. J. Math.* **67** (1945) 277–284 (III 5).

[2] On the convergence of successive approximations, *ibid.* **68** (1946) 13–19 (III 9).

[3] Asymptotic equilibria, *ibid.* **68** (1946) 125–132 (VIII 3, X 1).

[4] The infinities in the nonlocal existence problem of ordinary differential equations, *ibid.* **68** (1946) 173–178 (III 5).

[5] Linear variation of constants, *ibid.* **68** (1946) 185–213 (IV 10).

[6] Asymptotic integration constants in the singularity of Briot-Bouquet, *ibid.* **68** (1946) 293–300 (VIII 3).

[7] An Abelian lemma concerning asymptotic equilibria, *ibid.* **68** (1946) 451–454 (X 1).

[8] Asymptotic integration constants, *ibid.* **68** (1946) 553–559 (VIII 3, X 1).

[9] On the Laplace-Fourier transcendents occurring in mathematical physics, *ibid.* **69** (1947) 87–97 (XI 7).

[10] Asymptotic integrations of the adiabatic oscillator, *ibid.* **69** (1947) 251–272 (X 4, 17; XI 8).

[11] Vortices and nodes, *ibid.* **69** (1947) 815–824 (VIII 3).

[12] On the normalization of characteristic differentials in continuous spectra, *Phys. Rev.* **72** (1947) 516–517 (XI 8, 9).

[13] Asymptotic integrations of the adiabatic oscillator in its hyperbolic range, *Duke Math. J.* **15** (1948) 55–67 (X 17; XI 9).

[14] A norm criterion for nonoscillatory differential equations, *Quart. Appl. Math.* **6** (1948) 183–185 (XI 7).

[15] A criterion of oscillatory stability, *ibid.* **7** (1949) 115–117 (XI 7).

[16] Linear differential equations and the oscillatory property of Maclaurin's cosine series, *Math. Gaz.* **33** (1949) 26–28 (III 9).

[17] On almost free linear motions, *Amer. J. Math.* **71** (1949) 595–602 (XI 9).

[18] On the smallness of isolated eigenfunctions, *ibid.* **71** (1949) 603–611 (XIII Part I).

[19] On the Whittaker functions $W_{km}(x)$, *J. London Math. Soc.* **25** (1950) 351–353 (XIV 2).

[20] On the nonexistence of conjugate points, *Amer. J. Math.* **73** (1951) 368–380 (XI 6, 7).

[21] On a theorem of Bôcher in the theory of ordinary linear differential equations, *ibid.* **76** (1954) 183–190 (X 1).

[22] On the local uniqueness of the initial value problem of the differential equation $d^n x/dt^n = f(t, x)$, *Boll. Un. Mat. Ital.* (3) **11** (1956) 496–498 (III 6).

[23] On a principle of reciprocity between high- and low-frequency problems concerning linear differential equations of second order, *Quart. Appl. Math.* **15** (1957) 314–317 (XIV 3).

[24] On the comparison theorem of Kneser-Hille, *Math. Scand.* **5** (1957) 255–260 (XI 7).

K. G. WOLFSON

[1] On the spectrum of a boundary value problem with two singular endpoints, *Amer. J. Math.* **72** (1950) 713–719 (XI 4).

H. YAMABE (SEE ALSO L. MARKUS)

[1] A proof of a theorem on Jacobians, *Amer. Math. Monthly* **64** (1957) 725–726 (II 2).

M. ZLÁMAL

[1] Oscillation criteria, *Časopis Pěst. Mat. a Fis.* **75** (1950) 213–217 (XI 7).

[2] Ueber asymptotische Eigenschaften der Lösungen der linearen Differentialgleichungen zweiter Ordnung, *Czech. Math. J.* (6) **81** (1956) 75–91 (XI 9).

V. I. ZUBOV

[1] A sufficient condition for the stability of nonlinear systems of differential equations, *Prikl. Mat. Meh.* **17** (1953) 506–508 (XIV 11).

SUPPLEMENT

P. B. Bailey, L.F. Shampine and P.E. Waltman
[S1] Nonlinear two point boundary value problems, Academic Press, New York (1968).

W.A. Coppel
[S1] Stability and asymptotic behavior of differential equations, Heath & Co., Boston (1965) (X17).
[S2] Disconjugacy, Lecture Notes in Math. No.220, Springer, Berlin (1971) (XI 5 and Notes, end).

J.W. Evans and J.A. Feroe
[S1] Successive approximations and the general uniqueness theorem, Amer. J. Math., 96 (1974) 505-510 (III 9).

P. Hartman
[S1] Principal solutions of disconjugate n-th order linear differential equations, Amer. J. Math. 91(1969) 306-362 and 93(1971) 439-451 (XI Notes, end).
[S2] On disconjugacy criteria, Proc. Amer. Math. Soc. 24(1970) 374-381 (XI 5).
[S3] The stable manifold of a point of a hyperbolic map of a Banach space, J. Differential Equations 9(1971) 360-379 (IX, Appendix).
[S4] On the existence of similar solutions of some boundary value problems, SIAM J. Math. Anal. 3(1972) 120-147 (XIV 7).

S.P. Hastings
[S1] An existence theorem for a class of nonlinear boundary value problems including that of Falkner and Skan, J. Differential Equations 9(1971) 580-590 (XIV 7).

S.P. Hastings and S. Siegel
[S1] On some solutions of the Falkner-Skan equation, Mathematika 19(1972) 76-83 (XIV 7).

M.W. Hirsch, C.C. Pugh and M.Shub
[S1] Invariant manifolds, Bull. Amer. Math. Soc. 76(1970) 1015-1019 (IX Appendix).

A.Yu. Levin
[S1] Nonoscillation of solutions of the equation $x^{(n)} + p_1(t)x^{(n-1)} + ... + p_n(t)x = 0$, Uspechi Mat. Nauk 24(1969) No.2(146) 43-96; Russian Math. Surveys 24(1969) 43-99 (XI Notes, end).

L. Lorch and P. Szego

[S1] A singular integral whose kernel involves a Bessel function, Duke Math. J. 22(1955) 407-418 and 24(1957) 683 (XIV 3).

G. Mammana

[S1] Decomposizione della espressioni differenziali lineari omogenee in prodotti di fattori simbolici e applicazione relativo allo studio delle equazioni differenziali lineari, Math. Zeit. 33(1961) 186-231 (IV 8).

C.C. Pugh

[S1] On a theorem of P. Hartman, Amer. J. Math. 91(1969) 363-367 (IX 8).

W.T. Reid

[S1] Ordinary differential equations, Wiley, New York (1968).

G. Stampacchia

[S1] Le trasformazioni funzionali che presentano il fenomeno di Peano, Rend. Accad. Naz. Lincei Cl. Sci. Fis. Mat. Nat. (8) 7(1949) 80-84 (II 4).

C.A. Swanson

[S1] Comparison and oscillation theory of linear differential equations, Academic Press, New York (1968).

C.T. Taam

[S1] Nonoscillatory differential equations, Duke Math. J. 19(1952) 493-497 (XI 7).

A. Wintner

[S1] On linear instability, Quart. Appl. Math. 13(1955) 192-195 (XI 5).

R. Wüst

[S1] Beweis eines Lemmas von Ganelius, Jber. Deutsch Math. Verein 71(1969) 229-230 (XI 8).

Index